Electronic Design
Automation for
IC System Design,
Verification, and Testing

Electronic Design Automation for IC System Design, Verification, and Testing

Edited by

Luciano Lavagno
Politecnico di Torino
Torino, Italy

Grant Martin
Cadence Design Systems, Inc.
San Jose, California, USA

Igor L. Markov
University of Michigan
Ann Arbor, Michigan, USA

Louis K. Scheffer
Howard Hughes Medical Institute
Ashburn, Virginia, USA

CRC Press
Taylor & Francis Group
Boca Raton London New York

CRC Press is an imprint of the
Taylor & Francis Group, an **informa** business

CRC Press
Taylor & Francis Group
6000 Broken Sound Parkway NW, Suite 300
Boca Raton, FL 33487-2742

First issued in paperback 2018

© 2016 by Taylor & Francis Group, LLC
CRC Press is an imprint of Taylor & Francis Group, an Informa business

No claim to original U.S. Government works

ISBN-13: 978-1-4822-5462-4 (hbk)
ISBN-13: 978-1-138-58600-0 (pbk)

Library of Congress Cataloging-in-Publication Data

EDA for IC system design, verification, and testing.
   Electronic design automation for IC system design, verification, and testing / edited by Luciano Lavagno, Igor L. Markov, Grant E. Martin, and Louis K. Scheffer. -- Second edition.
      pages cm
   Revised edition of: EDA for IC system design, verification, and testing. 2006.
   Includes bibliographical references and index.
   ISBN 978-1-4822-5462-4
   1. Integrated circuits--Design and construction. 2. Integrated circuits--Testing 3. Integrated circuits--Computer-aided design. I. Lavagno, Luciano, 1959- editor. II. Title.

TK7874.E26 2016
621.3815--dc23   2015034494

Visit the Taylor & Francis Web site at
http://www.taylorandfrancis.com

and the CRC Press Web site at
http://www.crcpress.com

The editors would like to acknowledge the unsung heroes of EDA, who work to advance the field in addition to their own personal, corporate, or academic agendas. These men and women serve in a variety of ways—they run the smaller conferences, they edit technical journals, and they serve on standards committees, just to name a few. These largely volunteer jobs won't make anyone rich or famous despite the time and effort that goes into them, but they do contribute mightily to the remarkable and sustained advancement of EDA. Our kudos to these folks, who don't get the credit they deserve.

Contents

SECTION I — Introduction

SECTION II — System-Level Design

SECTION III — Microarchitecture Design

SECTION IV — Logic Verification

SECTION V — Test

Preface to the Second Edition

When Taylor & Francis Group (CRC Press) first suggested an update to the 2006 first edition of the *Electronic Design Automation for Integrated Circuits Handbook*, we realized that almost a decade had passed in the electronics industry, almost a complete era in the geological sense. We agreed that the changes in electronic design automation (EDA) and design methods warranted a once-in-a-decade update, and asked our original group of authors to update their chapters. We also solicited some new authors for new topics that seemed of particular relevance for a second edition. In addition, we added a new coeditor, Igor L. Markov, especially since Louis K. Scheffer has moved out of the EDA industry.

Finding all the original authors was a challenge. Some had retired or moved out of the industry, some had moved to completely new roles, and some were just too busy to contemplate a revision. However, many were still available and happy to revise their chapters. Where appropriate, we recruited new coauthors to revise, update, or replace a chapter, highlighting the major changes that occurred during the last decade.

It seems appropriate to quote from our original 2006 preface: "As we look at the state of electronics and IC design in 2005–2006, we see that we may soon enter a major period of change in the discipline." And "Upon further consideration, it is clear that the current EDA approaches have a lot of life left in them." This has been our finding in doing the revision. Some rather revolutionary changes are still coming; but to a great extent, most of the EDA algorithms, tools, and methods have gone through evolutionary, rather than revolutionary, changes over the last decade. Most of the major updates have occurred both in the initial phases of the design flow, where the level of abstraction keeps rising in order to support more functionality with lower NRE costs, and in the final phases, where the complexity due to smaller and smaller geometries is compounded by the slow progress of shorter wavelength lithography.

Major challenges faced by the EDA industry and researchers do not require revising previously accumulated knowledge so much but rather stimulate applying it in new ways and, in some cases, developing new approaches. This is illustrated, for example, by a new chapter on 3D circuit integration—an exciting and promising development that is starting to gain traction in the industry. Another new chapter covers on-chip networks—an area that had been nascent at the time of the first edition but experienced strong growth and solid industry adoption in recent years.

We hope that the readers enjoy the improved depth and the broader topic coverage offered in the second edition.

Luciano Lavagno

Igor L. Markov

Grant E. Martin

Louis K. Scheffer

MATLAB® is a registered trademark of The MathWorks, Inc. For product information, please contact:

The MathWorks, Inc.
3 Apple Hill Drive
Natick, MA 01760-2098 USA
Tel: 508-647-7000
Fax: 508-647-7001
E-mail: info@mathworks.com
Web: www.mathworks.com

Preface to the First Edition

Electronic design automation (EDA) is a spectacular success in the art of engineering. Over the last quarter of a century, improved tools have raised designers' productivity by a factor of more than a thousand. Without EDA, Moore's law would remain a useless curiosity. Not a single billion-transistor chip could be designed or debugged without these sophisticated tools—without EDA, we would have no laptops, cell phones, video games, or any of the other electronic devices we take for granted.

Spurred by the ability to build bigger chips, EDA developers have largely kept pace, and these enormous chips can still be designed, debugged, and tested, even with decreasing time to market.

The story of EDA is much more complex than the progression of integrated circuit (IC) manufacturing, which is based on simple physical scaling of critical dimensions. EDA, on the other hand, evolves by a series of paradigm shifts. Every chapter in this book, all 49 of them, was just a gleam in some expert's eye just a few decades ago. Then it became a research topic, then an academic tool, and then the focus of a start-up or two. Within a few years, it was supported by large commercial EDA vendors, and is now part of the conventional wisdom. Although users always complain that today's tools are not quite adequate for today's designs, the overall improvements in productivity have been remarkable. After all, in which other field do people complain of *only* a 21% compound annual growth in productivity, sustained over three decades, as did the *International Technology Roadmap for Semiconductors* in 1999?

And what is the future of EDA tools? As we look at the state of electronics and IC design in 2005–2006, we see that we may soon enter a major period of change in the discipline. The classical scaling approach to ICs, spanning multiple orders of magnitude in the size of devices over the last 40+ years, looks set to last only a few more generations or process nodes (though this has been argued many times in the past and has invariably been proved to be too pessimistic a projection). Conventional transistors and wiring may well be replaced by new nano- and biologically based technologies that we are currently only beginning to experiment with. This profound change will surely have a considerable impact on the tools and methodologies used to design ICs. Should we be spending our efforts looking at Computer Aided Design (CAD) for these future technologies, or continue to improve the tools we currently use?

Upon further consideration, it is clear that the current EDA approaches have a lot of life left in them. With at least a decade remaining in the evolution of current design approaches, and hundreds of thousands or millions of designs left that must either craft new ICs or use programmable versions of them, it is far too soon to forget about today's EDA approaches. And even if the technology changes to radically new forms and structures, many of today's EDA concepts will be reused and built upon for design of technologies well beyond the current scope and thinking.

The field of EDA for ICs has grown well beyond the point where any single individual can master it all, or even be aware of the progress on all fronts. Therefore, there is a pressing need to create a snapshot of this extremely broad and diverse subject. Students need a way of learning about the many disciplines and topics involved in the design tools in widespread use today. As design grows multidisciplinary, electronics designers and EDA tool developers need to broaden their scope. The methods used in one subtopic may well have applicability to new topics as they arise. All of electronics design can utilize a comprehensive reference work in this field.

With this in mind, we invited many experts from across all the disciplines involved in EDA to contribute chapters summarizing and giving a comprehensive overview of their particular topic or field. As might be appreciated, such chapters represent a snapshot of the state of the art in 2004–2005. However, as surveys and overviews, they retain a lasting educational and reference value that will be useful to students and practitioners for many years to come.

With a large number of topics to cover, we decided to split the handbook into two books. Volume 1, *Electronic Design Automation for IC System Design, Verification, and Testing*, covers system-level design, micro-architectural design, and verification and test. *Electronic Design Automation for IC Implementation, Circuit Design, and Process Technology* is Volume 2 and covers the classical "RTL to GDSII" design flow, incorporating synthesis, placement, and routing, along with analog and mixed-signal design, physical verification, analysis and extraction, and technology CAD topics for IC design. These roughly correspond to the classical "front-end/back-end" split in IC design, where the front end (or logical design) focuses on ensuring that the design does the right thing, assuming it can be implemented, and the back end (or physical design) concentrates on generating the detailed tooling required, while taking the logical function as given. Despite limitations, this split has persisted through the years—a complete and correct logical design, independent of implementation, remains an excellent handoff point between the two major portions of an IC design flow. Since IC designers and EDA developers often concentrate on one side of this logical/physical split, this seemed to be a good place to divide the book as well.

In particular this volume, *Electronic Design Automation for IC System Design, Verification, and Testing*, starts with a general introduction to the topic, and an overview of IC design and EDA. System-level design incorporates many aspects—application-specific tools and methods, special specification and modeling languages, integration concepts including the use of intellectual property (IP), and performance evaluation methods; the modeling and choice of embedded processors and ways to model software running on those processors; and high-level synthesis approaches. ICs that start at the system level need to be refined into micro-architectural specifications, incorporating cycle-accurate modeling, power estimation methods, and design planning. As designs are specified and refined, verification plays a key role—and the handbook covers languages, simulation essentials, and special verification topics such as transaction-level modeling, assertion-based verification, and the use of hardware acceleration and emulation, as well as emerging formal methods. Finally, making IC designs testable and thus cost-effective to manufacture and package relies on a host of test methods and tools, both for digital and analog and mixed-signal designs.

This handbook with its two constituent books is a valuable learning and reference work for everyone involved and interested in learning about electronic design and its associated tools and methods. We hope that all readers will find it of interest and that it will become a well-thumbed resource.

Louis K. Scheffer

Luciano Lavagno

Grant E. Martin

Acknowledgments

Louis K. Scheffer acknowledges the love, support, encouragement, and help of his wife, Lynde, his daughter, Lucynda, and his son, Loukos. Without them, this project would not have been possible.

Luciano Lavagno thanks his wife, Paola, and his daughter, Alessandra Chiara, for making his life so wonderful.

Grant E. Martin acknowledges, as always, the love and support of his wife, Margaret Steele, and his two daughters, Jennifer and Fiona.

Igor L. Markov thanks his parents, Leonid and Nataly, for encouragement and support.

Editors

Luciano Lavagno received his PhD in EECS from the University of California at Berkeley, California, in 1992 and from Politecnico di Torino, Torino, Italy, in 1993. He is a coauthor of two books on asynchronous circuit design, a book on hardware/software codesign of embedded systems, and more than 200 scientific papers. Between 1993 and 2000, he was the architect of the POLIS project, a cooperation between the University of California at Berkeley, Cadence Design Systems, Magneti Marelli, and Politecnico di Torino, which developed a complete hardware/software codesign environment for control-dominated embedded systems. Between 2003 and 2014, he was one of the creators and architects of the Cadence CtoSilicon high-level synthesis system.

Since 2011, he is a full professor with Politecnico di Torino, Italy. He has been serving on the technical committees of several international conferences in his field (e.g., DAC, DATE, ICCAD, ICCD, ASYNC, CODES) and of various workshops and symposia. He has been the technical program chair of DAC, and the TPC and general chair of CODES. He has been an associate editor of IEEE TCAS and ACM TECS. He is a senior member of the IEEE.

His research interests include the synthesis of asynchronous low-power circuits, the concurrent design of mixed hardware and software embedded systems, the high-level synthesis of digital circuits, the design and optimization of hardware components and protocols for wireless sensor networks, and design tools for WSNs.

Igor L. Markov is currently on leave from the University of Michigan, Ann Arbor, Michigan, where he taught for many years. He joined Google in 2014. He also teaches VLSI design at Stanford University. He researches computers that make computers, including algorithms and optimization techniques for electronic design automation, secure hardware, and emerging technologies. He is an IEEE fellow and an ACM distinguished scientist. He has coauthored five books and has four U.S. patents and more than 200 refereed publications, some of which were honored by best-paper awards. Professor Markov is a recipient of the DAC Fellowship, the ACM SIGDA Outstanding New Faculty award, the NSF CAREER award, the IBM Partnership Award, the Microsoft A. Richard Newton Breakthrough Research Award, and the inaugural IEEE CEDA Early Career Award. During the 2011 redesign of the ACM Computing Classification System, Professor Markov led the effort on the Hardware tree. Twelve doctoral dissertations were defended under his supervision; three of them received outstanding dissertation awards.

Grant E. Martin is a distinguished engineer at Cadence Design Systems, Inc. in San Jose, California. Before that, Grant worked for Burroughs in Scotland for 6 years; Nortel/BNR in Canada for 10 years; Cadence Design Systems for 9 years, eventually becoming a Cadence fellow in their Labs; and Tensilica for 9 years. He rejoined Cadence in 2013 when it acquired Tensilica, and has been there since, working in the Tensilica part of the Cadence IP group. He received his bachelor's and master's degrees in mathematics (combinatorics and optimization) from the University of Waterloo, Canada, in 1977 and 1978.

Grant is a coauthor of *Surviving the SOC Revolution: A Guide to Platform-Based Design*, 1999, and *System Design with SystemC*, 2002, and a coeditor of the books *Winning the SoC Revolution: Experiences in Real Design* and *UML for Real: Design of Embedded Real-Time Systems*, June 2003, all published by Springer (originally by Kluwer). In 2004, he cowrote, with Vladimir Nemudrov, the first book on SoC design published in Russian by Technosphera, Moscow. In the middle of the last decade, he coedited *Taxonomies for the Development and Verification of Digital Systems* (Springer, 2005) and *UML for SoC Design* (Springer, 2005), and toward the end of the decade cowrote *ESL Design and Verification: A Prescription for Electronic System-Level Methodology* (Elsevier Morgan Kaufmann, 2007) and *ESL Models and their Application: Electronic System Level Design in Practice* (Springer, 2010).

He has also presented many papers, talks, and tutorials and participated in panels at a number of major conferences. He cochaired the VSI Alliance Embedded Systems study group in the summer of 2001 and was cochair of the DAC Technical Programme Committee for Methods for 2005 and 2006. He is also a coeditor of the Springer Embedded System Series. His particular areas of interest include system-level design, IP-based design of system-on-chip, platform-based design, DSP, baseband and image processing, and embedded software. He is a senior member of the IEEE.

Louis K. Scheffer received his BS and MS from the California Institute of Technology, Pasadena, California, in 1974 and 1975, and a PhD from Stanford University, Stanford, California, in 1984. He worked at Hewlett Packard from 1975 to 1981 as a chip designer and CAD tool developer. In 1981, he joined Valid Logic Systems, where he did hardware design, developed a schematic editor, and built an IC layout, routing, and verification system. In 1991, Valid merged with

Cadence Design Systems, after which Dr. Scheffer worked on place and route, floorplanning systems, and signal integrity issues until 2008.

In 2008, Dr. Scheffer switched fields to neurobiology, studying the structure and function of the brain by using electron microscope images to reconstruct its circuits. As EDA is no longer his daily staple (though his research uses a number of algorithms derived from EDA), he is extremely grateful to Igor Markov for taking on this portion of these books. Lou is also interested in the search for extraterrestrial intelligence (SETI), serves on the technical advisory board for the Allen Telescope Array at the SETI Institute, and is a coauthor of the book *SETI-2020*, in addition to several technical articles in the field.

Contributors

Iuliana Bacivarov
Aviloq
Zurich, Switzerland

Felice Balarin
Cadence Design Systems
San Jose, California

Mike Bershteyn
Cadence Design Systems
San Jose, California

Shuvra Bhattacharyya
Department of Electrical and Computer Engineering
University of Maryland
College Park, Maryland

and

Department of Pervasive Computing
Tampere University of Technology
Tampere, Finland

Joseph T. Buck
Synopsys, Inc.
Mountain View, California

Raul Camposano
Nimbic
Mountain View, California

Naehyuck Chang
Department of Electrical Engineering
KAIST
Daejeon, South Korea

Anupam Chattopadhyay
School of Computer Engineering
Nanyang Technological University
Singapore, Singapore

Kwang-Ting (Tim) Cheng
Department of Electrical and Computer Engineering
University of California, Santa Barbara
Santa Barbara, California

Alain Clouard
STMicroelectronics
Grenoble, France

Marcello Coppola
STMicroelectronics
Grenoble, France

Jérôme Cornet
STMicroelectronics
Grenoble, France

Robert Damiano
Consultant
Hillsboro, Oregon

Robert P. Dick
Electrical Engineering and Computer Science
University of Michigan
Ann Arbor, Michigan

Marco Di Natale
Department of Computer Engineering
Scuola Superiore S. Anna
Pisa, Italy

Nikil Dutt
Donald Bren School of Information and Computer
    Sciences
University of California, Irvine
Irvine, California

Stephen A. Edwards
Computer Science
Columbia University
New York, New York

Limor Fix (retired)
Ithaca, New York

Harry Foster
Mentor Graphics Corporation
Wilsonville, Oregon

Miltos D. Grammatikakis
Department of Applied Informatics & Multimedia
Technological Educational Institute of Crete
Heraklion, Greece

Leopold Haller
Cadence Design Systems
San Jose, California

Norris Ip
Google
Mountain View, California

Ahmed Jerraya
Laboratoire d'électronique des technologies de
    l'information
Commissariat à l'Energie Atomique et aux Energies
    Alternatives
Grenoble, France

Bozena Kaminska
School of Engineering Science
Simon Fraser University
Burnaby, British Colombia, Canada

Yaron Kashai
Cadence Design Systems
San Jose, California

Brion Keller
Cadence Design Systems
San Jose, California

Bernd Koenemann
Independent
San Francisco Bay Area, California

Alex Kondratyev
Xilinx
San Jose, California

Luciano Lavagno
Department of Electronics and Telecommunications
Politecnico di Torino
Torino, Italy

Steve Leibson
Xilinx, Inc.
San Jose, California

Rainer Leupers
Institute for Communication Technologies and Embedded
    Systems
RWTH Aachen University
Aachen, Germany

Huawei Li
Institute of Computing Technology
Chinese Academy of Sciences
Beijing, People's Republic of China

James Chien-Mo Li
Department of Electrical Engineering
National Taiwan University
Taipei, Taiwan, Republic of China

Enrico Macii
Department of Control and Computer Engineering
Politecnico di Torino
Torino, Italy

Laurent Maillet-Contoz
STMicroelectronics
Grenoble, France

Igor L. Markov
Electrical Engineering and Computer Science
University of Michigan
Ann Arbor, Michigan

Erich Marschner
Mentor Graphics Corporation
Wilsonville, Oregon

Grant E. Martin
Cadence Design Systems
San Jose, California

Ken McMillan
Microsoft
Bellevue, Washington

Renu Mehra
Synopsys, Inc.
Mountain View, California

Prabhat Mishra
Department of Computer and Information Science
    and Engineering
University of Florida
Gainesville, Florida

Ralph H.J.M. Otten
Department of Electrical Engineering
Eindhoven University of Technology
Eindhoven, the Netherlands

Eric Paire
STMicroelectronics
Grenoble, France

Antonis Papagrigoriou
Department of Informatics Engineering
Technological Educational Institute of Crete
Heraklion, Greece

Antoine Perrin
STMicroelectronics
Grenoble, France

Polydoros Petrakis
Department of Applied Informatics and Multimedia
Technological Educational Institute of Crete
Heraklion, Greece

Massimo Poncino
Dipartimento di Automatica e Informatica
Politecnico di Torino
Torino, Italy

John Sanguinetti
Forte Design Systems, Inc.
San Jose, California

Louis K. Scheffer
Janelia Research Campus
Howard Hughes Medical Institute
Washington, DC

Frank Schirrmeister
Cadence Design Systems
San Jose, California

Jean-Philippe Strassen
STMicroelectronics
Grenoble, France

Haralampos-G. Stratigopoulos
Laboratoire d'Informatique de Paris 6
Centre National de la Recherche Scientique
Université Pierre et Marie Curie
Sorbonne Universités
Paris, France

Vivek Tiwari
Intel Corporation
Santa Clara, California

Ray Turner
Independent
San Jose, California

Li-C. Wang
Department of Electrical and Computer Engineering
University of California, Santa Barbara
Santa Barbara, California

Yosinori Watanabe
Cadence Design Systems
San Jose, California

Marilyn Wolf
School of Electrical and Computer Engineering
Georgia Institute of Technology
Atlanta, Georgia

I

Introduction

Overview

Luciano Lavagno, Grant E. Martin, Louis K. Scheffer, and Igor L. Markov

CONTENTS

1.1 INTRODUCTION TO *ELECTRONIC DESIGN AUTOMATION FOR INTEGRATED CIRCUITS HANDBOOK,* SECOND EDITION

Modern integrated circuits (ICs) are enormously complicated, sometimes containing billions of devices. The design of these ICs would not be humanly possible without software (SW) assistance at every stage of the process. The tools and methodologies used for this task are collectively called electronic design automation (EDA).

EDA tools span a very wide range, from logic-centric tools that implement and verify functionality to physically-aware tools that create blueprints for manufacturing and verify their feasibility. EDA methodologies combine multiple tools into EDA design flows, invoking the most appropriate software packages based on how the design progresses through optimizations. Modern EDA methodologies can reuse existing design blocks, develop new ones, and integrate entire systems. They not only automate the work of circuit engineers, but also process large amounts of heterogeneous design data, invoke more accurate analyses and more powerful optimizations than what human designers are capable of.

1.1.1 BRIEF HISTORY OF ELECTRONIC DESIGN AUTOMATION

The need for design tools became clear soon after ICs were invented. Unlike a breadboard, an IC cannot be modified easily after fabrication; therefore, testing even a simple change takes weeks (for new masks and a new fabrication run) and requires considerable expense. The internal nodes of an IC are difficult to probe because they are physically small and may be covered by other layers of the IC. Internal nodes with high impedances are difficult to measure without dramatically changing the performance. Therefore, circuit simulators became crucial to IC design almost as soon as ICs came into existence. These programs appeared in the 1960s and are covered in Chapter 17 of *Electronic Design Automation for IC Implementation, Circuit Design, and Process Technology* (hereafter referred to as Volume 2 of this Handbook).

As the circuits grew larger, clerical help was required in producing the masks. At first, the designer drew shapes with colored pencils but the coordinates were transferred to the computer by digitizing programs, written to magnetic tape (hence the handoff from design to fabrication is still called "tapeout"), and then transferred to the mask-making machines. In the 1960s and 1970s, these early programs were enhanced to full-fledged layout editors. Analog designs in the modern era are still largely laid out manually, with some tool assistance, as Chapter 19 of Volume 2 will attest.

As the circuits scaled up further, ensuring the correctness of logic designs became difficult, and logic simulation (Chapter 16) was introduced into the IC design flow. Testing completed chips proved difficult too, since unlike circuit boards, internal nodes could not be observed or controlled through a "bed of nails" fixture. Therefore, automatic test pattern generation (ATPG) programs were developed to generate test vectors that can be entered through accessible pins. Other techniques that modified designs to make them more controllable, observable, and testable were not far behind. These techniques, covered in Chapters 21 and 22, were first available in the mid-1970s. Specialized needs were met by special testers and tools, discussed in Chapter 23.

As the number of design rules, number of layers, and chip size continued to increase, it became increasingly difficult to verify by hand that a layout met all the manufacturing rules and to estimate the parasitics of the circuit. Therefore, as demonstrated in Chapters 20 and 25 of Volume 2 new software was developed, starting in the mid-1970s, to address this need. Increasing numbers of interconnect layers made the process more complex, and the original analytic approximations to R, C, and L values became inadequate, and new techniques for parasitic extraction were required to determine more accurate values, or at least calibrate the parameter extractors.

The next bottleneck was in determining the precise location of each polygon and drawing its detailed geometry. Placement and routing programs for standard-cell designs allowed the user to specify only the gate-level netlist—the computer would then decide on the location of the gates and route the wires connecting them. This greatly improved productivity (with a moderate loss of silicon efficiency), making IC design accessible to a wider group of electronics engineers. Chapters 5 and 8 of Volume 2 cover these programs, which became popular in the mid-1980s.

Even just the gate-level netlist soon proved unwieldy, and synthesis tools were developed to create such a netlist from a higher-level specification, usually expressed in a hardware description language (HDL). This step is called *Logic Synthesis*. It became available in the mid-1980s. In the late 2000s, the issues described in Chapter 2 of Volume 2 have become a major area of concern and the main optimization criterion, respectively, for many designs, especially in the portable and battery-powered categories. Around this time, the large collections of disparate tools required to complete a single design became a serious problem. Electronic design automation *Design Databases* were introduced as common infrastructure for developing interoperable tools. In addition, the techniques described in Chapter 1 of Volume 2 grew more elaborate in how tools were linked together to support design methodologies, as well as use models for specific design groups, companies, and application areas.

In the late 1990s, as transistors continued to shrink, electromagnetic noise became a serious problem. Programs that analyzed power and ground networks, cross-talk, and substrate noise in systematic ways became commercially available. Chapters 23, 24, and 26 of Volume 2 cover these topics.

Gradually through the 1990s and early 2000s, chips and design processes became so complex that yield optimization developed into a separate field called *Design for Manufacturability in the Nanometer Era*, otherwise known as "Design for Yield." In this time frame, the smallest on-chip features dropped below the wavelength of the light used to manufacture them with optical lithography. Due to the diffraction limit, the masks could no longer faithfully copy what the designer intended. The creation of these more complex masks is covered under Chapter 21 of Volume 2.

Developing the manufacturing process itself was also challenging. *Process Simulation* (Volume 2, Chapter 27) tools were developed to explore sensitivities to process parameters. The output of these programs, such as doping profiles, was useful to process engineers but too detailed for electrical analysis. A newly developed suite of tools (see Volume 2, Chapter 28) predicted device performance from a physical description of devices. These models were particularly useful when developing a new process.

System-level design became useful very early in the history of design automation. However, due to the diversity of application-dependent issues that it must address, it is also the least standardized level of abstraction. As Chapter 10 points out, one of the first instruction set simulators appeared soon after the first digital computers did. Yet, until the present day, system-level design has consisted mainly of a varying collection of tricks, techniques, and *ad hoc* modeling tools.

The logic simulation and synthesis processes introduced in the 1970s and 1980s, respectively, are, as was discussed earlier, much more standardized than system-level design. The front-end IC design flow would have been much more difficult without standard HDLs. Out of a huge variety of HDLs introduced from the 1960s to the 1980s, Verilog and VHDL have become the major Design and Verification Languages (Chapter 15). Until the late 1990s, verification of digital designs seemed stuck at standard digital simulation—although at least since the 1980s, a variety of Hardware-Assisted Verification and Software Development (Chapter 19) solutions have been available to designers. However, advances in verification languages and growing design complexity have motivated more advanced verification methods, and the last decade has seen considerable interest in Leveraging Transactional-Level Models in a SoC Design Flow (Chapter 17), Assertion-Based Verification (Chapter 18), and Formal Property Verification (Chapter 20). Equivalence Checking (Volume 2, Chapter 4) has been the formal technique most tightly integrated into design flows, since it allows designs to be compared before and after various optimizations and back-end-related modifications, such as scan insertion.

For many years, specific system-design domains have fostered their own application-specific Tools and Methodologies for System-Level Design (Chapter 3)—especially in the areas of algorithm design from the late 1980s to this day. The late 1990s saw the emergence of and competition between a number of C/C++-based System-Level Specification and Modeling Languages (Chapter 4). With the newly available possibility to incorporate all major functional units of a design (processors, memories, digital and mixed-signal HW blocks, peripheral interfaces, and complex hierarchical buses) onto a single silicon substrate, the last 20 years have seen the rise of the system on chip (SoC). Thus, the area of SoC Block-Based Design and IP Assembly (Chapter 5) has grown, enabling greater complexity with advanced semiconductor processes through the reuse of design blocks. Along with the SoC approach, the last decade saw the emergence of Performance Evaluation Methods for MPSoC Designs (Chapter 6), development of embedded processors through specialized Processor Modeling and Design Tools (Chapter 8), and gradual and still-forming links to Models and Tools for Complex Embedded Software and Systems (Chapter 9*)*. The desire to improve HW design productivity has spawned considerable interest in High-Level Synthesis (Chapter 11) over the years. It is now experiencing a resurgence driven by C/C++/SystemC as opposed to the first-generation high-level synthesis (HLS) tools driven by HDLs in the mid-1990s.

After the system level of design, architects need to descend by one level of abstraction to the microarchitectural level. Here, a variety of tools allow one to look at the three main criteria: timing or delay (Microarchitectural and System-Level Power Estimation and Optimization), power (Chapter 13), and area and cost (Chapter 14). Microarchitects need to make trade-offs between the timing, power, and cost/area attributes of complex ICs at this level.

The last several years have seen a variety of complementary tools and methods added to conventional design flows. Formal verification of design function is only possible if correct timing is guaranteed, and by limiting the amount of dynamic simulation required, especially at the postsynthesis and postlayout gate levels, Static Timing Analysis (Volume 2, Chapter 6) tools provide the assurance that timing constraints are met. Timing analysis also underlies timing optimization of circuits and the design of newer mechanisms for manufacturing and yield. Standard cell–based placement and routing are not appropriate for Structured Digital Design (Volume 2, Chapter 7) of elements such as memories and register files, and this observation motivates specialized tools. As design groups began to rely on foundries and application-specific (ASIC) vendors and as the IC design and manufacturing industry began to "deverticalize," design libraries, covered in Chapter 12 of Volume 2, became a domain for special design flows and tools. Library vendors offered a variety of high-performance and low-power libraries for optimal design choices and allowed some portability of design across processes and foundries. Tools for Chip-Package Co-Design (Volume 2, Chapter 14) began to link more closely the design of IOs on chip, the packages they fit into, and the boards on which they would be placed. For implementation "fabrics," such as field-programmable gate arrays (FPGAs), specialized FPGA

Synthesis and Physical Design Tools (Volume 2, Chapter 16) tools are necessary to ensure good results. A renewed emphasis on Design Closure (Volume 2, Chapter 13) allows a more holistic focus on the simultaneous optimization of design timing, power, cost, reliability, and yield in the design process. Another area of growing but specialized interest in the analog design domain is the use of new and higher-level modeling methods and languages, which are covered in Chapter 18 of Volume 2.

Since the first edition of this handbook appeared, several new areas of design have reached a significant level of maturity. Gate Sizing (Volume 2, Chapter 10) techniques choose the best widths for transistors in order to optimize performance, both in a continuous setting (full-custom-like) and in a discrete setting (library based and FinFET based). Clock Design and Synthesis (Volume 2, Chapter 11) techniques enable the distribution of reliable synchronization to huge numbers of sequential elements. Finally, three-dimensional (3D) integrated circuits are attempting to extend the duration of Moore's law, especially in the elusive domain of improving performance, by allowing multiple ICs to be stacked on top of each other.

A much more detailed overview of the history of EDA can be found in Reference 1. A historical survey of many of the important papers from the International Conference on Computer-Aided Design (ICCAD) can be found in Reference 2.

1.1.2 MAJOR INDUSTRY CONFERENCES AND PUBLICATIONS

The EDA community formed in the early 1960s from tool developers working for major electronics design companies such as IBM, AT&T Bell Labs, Burroughs, and Honeywell. It has long valued workshops, conferences, and symposia, in which practitioners, designers, and later academic researchers could exchange ideas and practically demonstrate the techniques. The Design Automation Conference (DAC) grew out of workshops, which started in the early 1960s and, although held in a number of US locations, has in recent years tended to stay on the west coast of the United States or a bit inland. It is the largest combined EDA trade show and technical conference held annually anywhere in the world. In Europe, a number of country-specific conferences held sporadically through the 1980s, and two competing ones, held in the early 1990s, led to the creation of the consolidated Design Automation and Test in Europe conference, which started in the mid-1990s and has grown consistently in strength ever since. Finally, the Asia-South Pacific DAC started in the mid-1990s to late 1990s and completes the trio of major EDA conferences spanning the most important electronics design communities in the world.

Large trade shows and technical conferences have been complemented by ICCAD, held in San Jose for over 20 years. It has provided a more technical conference setting for the latest algorithmic advances, attracting several hundred attendees. Various domain areas of EDA knowledge have sparked a number of other workshops, symposia, and smaller conferences over the last 20 years, including the International Symposium on Physical Design, International Symposium on Quality in Electronic Design (ISQED), Forum on Design Languages in Europe (FDL), HDL and Design and Verification conferences (HDLCon, DVCon), High-level Design, Verification and Test (HLDVT), International Conference on Hardware–Software Codesign and System Synthesis (CODES+ISSS), and many other gatherings. Of course, the area of Test has its own long-standing International Test Conference (ITC); similarly, there are specialized conferences for FPGA design (e.g., Forum on Programmable Logic [FPL]) and a variety of conferences focusing on the most advanced IC designs such as the International Solid-State Circuits Conference and its European counterpart the European Solid-State Circuits Conference.

There are several technical societies with strong representation of design automation: one is the Institute of Electrical and Electronics Engineers (IEEE, pronounced as "eye-triple-ee") and the other is the Association for Computing Machinery (ACM). The Electronic Design Automation Consortium (EDAC) is an industry group that cosponsors major conferences such as DAC with professional societies.

Various IEEE and ACM transactions publish research on algorithms and design techniques—a more archival-oriented format than conference proceedings. Among these, the IEEE Transactions on computer-aided design (CAD), the IEEE Transactions on VLSI systems, and the ACM Transactions on Design Automation of Electronic Systems are notable. A less-technical, broader-interest magazine is *IEEE Design and Test*.

As might be expected, the EDA community has a strong online presence. All the conferences have Web pages describing locations, dates, manuscript submission and registration procedures, and often detailed descriptions of previous conferences. The journals offer online submission, refereeing, and publication. Online, the IEEE (http://ieee.org), ACM (http://acm.org), and *CiteSeer* (http://citeseer.ist.psu.edu) offer extensive digital libraries, which allow searches through titles, abstracts, and full texts. Both conference proceedings and journals are available. Most of the references found in this volume, at least those published after 1988, can be found in at least one of these libraries.

1.1.3 STRUCTURE OF THE BOOK

In the next chapter, "Integrated Circuit Design Process and Electronic Design Automation," Damiano, Camposano, and Martin discuss the IC design process, its major stages and design flow, and how EDA tools fit into these processes and flows. It particularly covers interfaces between the major IC design stages based on higher-level abstractions, as well as detailed design and verification information. Chapter 2 concludes the introductory section to the Handbook. Beyond that, *Electronic Design Automation for Integrated Circuits Handbook*, Second Edition, comprises several sections and two volumes. Volume 1 (in your hands) is entitled *Electronic Design Automation for IC System Design, Verification, and Testing*). Volume 2 is *Electronic Design Automation for IC Implementation, Circuit Design, and Process Technology*. We will now discuss the division of these two books into sections.

EDA for digital design can be divided into system-level design, microarchitecture design, logic verification, test, synthesis place and route, and physical verification. System-level design is the task of determining which components (bought and built, HW and SW) should comprise a system that can perform required functions. Microarchitecture design fills out the descriptions of each of the blocks and sets the main parameters for their implementation. Logic verification checks that the design does what is intended. Postfabrication test ensures that functional and nonfunctional chips can be distinguished reliably. It is common to insert dedicated circuitry to make test efficient. Synthesis, placement, and routing take the logical design description and map it into increasingly-detailed physical descriptions, until the design is in a form that can be built with a given process. Physical verification checks that such a design is manufacturable and will be reliable. This makes the design flow, or sequence of steps that the users follow to finish their design, a crucial part of any EDA methodology.

In addition to fully digital chips, analog and mixed-signal chips require their own specialized tool sets.

All these tools must scale to large designs and do so in a reasonable amount of time. In general, such scaling cannot be accomplished without behavioral models, that is, simplified descriptions of the behavior of various chip elements. Creating these models is the province of Technology CAD (TCAD), which in general treats relatively small problem instances in great physical detail, starting from very basic physics and building the more efficient models needed by the tools that must handle higher data volumes.

The division of EDA into these sections is somewhat arbitrary. In the following, we give a brief description of each book chapter.

1.2 SYSTEM-LEVEL DESIGN

1.2.1 TOOLS AND METHODOLOGIES FOR SYSTEM-LEVEL DESIGN

Chapter 3 by Bhattacharyya and Wolf covers system-level design approaches and associated tools such as Ptolemy and the MathWorks tools, and illustrates them for video applications.

1.2.2 SYSTEM-LEVEL SPECIFICATION AND MODELING LANGUAGES

Chapter 4 by Edwards and Buck discusses major approaches to specifying and modeling systems, as well as the languages and tools in this domain. It covers heterogeneous specifications, models of computation and linking multidomain models, requirements on languages, and specialized tools and flows in this area.

1.2.3 SoC BLOCK-BASED DESIGN AND IP ASSEMBLY

Chapter 5 by Kashai approaches system design with particular emphasis on SoCs via IP-based reuse and block-based design. Methods of assembly and compositional design of systems are covered. Issues of IP reuse as they are reflected in system-level design tools are also discussed.

1.2.4 PERFORMANCE EVALUATION METHODS FOR MULTIPROCESSOR SYSTEMS-ON-CHIP DESIGN

Chapter 6 by Jerraya and Bacivarov surveys the broad field of performance evaluation and sets it in the context of multiprocessor system on chip (MPSoC). Techniques for various types of blocks—HW, CPU, SW, and interconnect—are included. A taxonomy of performance evaluation approaches is used to assess various tools and methodologies.

1.2.5 SYSTEM-LEVEL POWER MANAGEMENT

Chapter 7 by Chang, Macii, Poncino, and Tiwari discusses dynamic power management approaches, aimed at selectively stopping or slowing down resources, whenever possible while providing required levels of system performance. The techniques can be applied to reduce both power consumption and energy consumption, which improves battery life. They are generally driven by the SW layer, since it has the most precise picture about both the required quality of service and the global state of the system.

1.2.6 PROCESSOR MODELING AND DESIGN TOOLS

Chapter 8 by Chattopadhyay, Dutt, Leupers, and Mishra covers state-of-the-art specification languages, tools, and methodologies for processor development used in academia and industry. It includes specialized architecture description languages and the tools that use them, with a number of examples.

1.2.7 MODELS AND TOOLS FOR COMPLEX EMBEDDED SOFTWARE AND SYSTEMS

Chapter 9 by Di Natale covers models and tools for embedded SW, including the relevant models of computation. Practical approaches with languages such as Simulink® and the Unified Modeling Language are introduced. Embeddings into design flows are discussed.

1.2.8 USING PERFORMANCE METRICS TO SELECT MICROPROCESSOR CORES FOR IC DESIGNS

Chapter 10 by Leibson discusses the use of standard benchmarks and instruction set simulators to evaluate processor cores. These might be useful in nonembedded applications, but are especially relevant to the design of embedded SoC devices where the processor cores may not yet be available in HW, or be based on user-specified processor configurations and extensions.

Benchmarks drawn from relevant application domains have become essential to core evaluation, and their advantages greatly exceed those of the general-purpose benchmarks used in the past.

1.2.9 HIGH-LEVEL SYNTHESIS

Chapter 11 by Balarin, Kondratyev, and Watanabe describes the main steps taken by a HLS tool to synthesize a C/C++/SystemC model into register transfer level (RTL). Both algorithmic techniques and user-level decisions are surveyed.

1.3 MICROARCHITECTURE DESIGN

1.3.1 BACK-ANNOTATING SYSTEM-LEVEL MODELS

Chapter 12 by Grammatikakis, Papagrigoriou, Petrakis, and Coppola discusses how to use system-level modeling approaches at the cycle-accurate microarchitectural level to perform final design architecture iterations and ensure conformance to timing and performance specifications.

1.3.2 MICROARCHITECTURAL POWER ESTIMATION AND OPTIMIZATION

Chapter 13 by Macii, Mehra, Poncino, and Dick discusses power estimation at the microarchitectural level in terms of data paths, memories, and interconnect. *Ad hoc* solutions for optimizing both specific components and entire designs are surveyed, with a particular emphasis on SoCs for mobile applications.

1.3.3 DESIGN PLANNING AND CLOSURE

Chapter 14 by Otten discusses the topics of physical floor planning and its evolution over the years, from dealing with rectangular blocks in slicing structures to more general mathematical techniques for optimizing physical layout while meeting a variety of criteria, especially timing and other constraints.

1.4 LOGIC VERIFICATION

1.4.1 DESIGN AND VERIFICATION LANGUAGES

Chapter 15 by Edwards discusses the two main HDLs in use—VHDL and Verilog—and how they meet the requirements for design and verification flows. More recent evolutions in languages, such as SystemC, SystemVerilog, and verification languages (i.e., OpenVera, e, and PSL), are also described.

1.4.2 DIGITAL SIMULATION

Chapter16 by Sanguinetti discusses logic simulation algorithms and tools, as these are still the primary tools used to verify the logical or functional correctness of a design.

1.4.3 LEVERAGING TRANSACTIONAL-LEVEL MODELS IN A SoC DESIGN FLOW

In Chapter 17, Maillet-Contoz, Cornet, Clouard, Paire, Perrin, and Strassen focus on an industry design flow at a major IC design company to illustrate the construction, deployment, and use of transactional-level models to simulate systems at a higher level of abstraction, with much greater performance than at RTL, and to verify functional correctness and validate system performance characteristics.

1.4.4 ASSERTION-BASED VERIFICATION

Chapter 18 by Foster and Marschner introduces the topic of assertion-based verification, which is useful for capturing design intent and reusing it in both dynamic and static verification methods. Assertion libraries such as OVL and languages including PSL and SystemVerilog assertions are used for illustrating the concepts.

1.4.5 HARDWARE-ASSISTED VERIFICATION AND SOFTWARE DEVELOPMENT

Chapter 19 by Schirrmeister, Bershteyn, and Turner discusses HW-based systems including FPGA, processor-based accelerators/emulators, and FPGA prototypes for accelerated verification. It compares the characteristics of each type of system and typical use models.

1.4.6 FORMAL PROPERTY VERIFICATION

In Chapter 20, Fix, McMillan, Ip, and Haller discuss the concepts and theory behind formal property checking, including an overview of property specification and a discussion of formal verification technologies and engines.

1.5 TEST

1.5.1 DESIGN-FOR-TEST

Chapter 21 by Koenemann and Keller discusses the wide variety of methods, techniques, and tools available to solve design-for-test (DFT) problems. This is a sizable area with an enormous variety of techniques, many of which are implemented in tools that dovetail with the capabilities of the physical test equipment. This chapter surveys the specialized techniques required for effective DFT with special blocks such as memories, as well as general logic cores.

1.5.2 AUTOMATIC TEST PATTERN GENERATION

Chapter 22 by Cheng, Wang, Li, and Li starts with the fundamentals of fault modeling and combinational ATPG concepts. It moves on to gate-level sequential ATPG and discusses satisfiability (SAT) methods for circuits. Moving on beyond traditional fault modeling, it covers ATPG for cross-talk faults, power supply noise, and applications beyond manufacturing test.

1.5.3 ANALOG AND MIXED-SIGNAL TEST

In Chapter 23, Stratigopoulos and Kaminska first overview the concepts behind analog testing, which include many characteristics of circuits that must be examined. The nature of analog faults is discussed and a variety of analog test equipment and measurement techniques surveyed. The concepts behind analog built-in self-test are reviewed and compared with the digital test. This chapter concludes Volume 1 of *Electronic Design Automation for Integrated Circuits Handbook*, Second Edition.

1.6 RTL TO GDSII OR SYNTHESIS, PLACE, AND ROUTE

1.6.1 DESIGN FLOWS

The second volume, *Electronic Design Automation for IC Implementation, Circuit Design, and Process Technology* begins with Chapter 1 by Chinnery, Stok, Hathaway, and Keutzer. The RTL to GDSII flow has evolved considerably over the years, from point tools bundled loosely together to a more integrated set of tools for design closure. This chapter addresses the design-flow challenges based on the rising interconnect delays and new challenges to achieve closure.

1.6.2 LOGIC SYNTHESIS

Chapter 2 by Khatri, Shenoy, Giomi, and Khouja provides an overview and survey of logic synthesis, which has, since the early 1980s, grown to be the vital center of the RTL to GDSII design flow for digital design.

1.6.3 POWER ANALYSIS AND OPTIMIZATION FROM CIRCUIT TO REGISTER-TRANSFER LEVELS

Power has become one of the major challenges in modern IC design. Chapter 3 by Monteiro, Patel, and Tiwari provides an overview of the most significant CAD techniques for low power, at several levels of abstraction.

1.6.4 EQUIVALENCE CHECKING

Equivalence checking can formally verify whether two design specifications are functionally equivalent. Chapter 4 by Kuehlmann, Somenzi, Hsu, and Bustan defines the equivalence-checking problem, discusses the foundation for the technology, and then discusses the algorithms for combinational and sequential equivalence checking.

1.6.5 DIGITAL LAYOUT: PLACEMENT

Placement is one of the fundamental problems in automating digital IC layout. Chapter 5 by Kahng and Reda reviews the history of placement algorithms, the criteria used to evaluate quality of results, many of the detailed algorithms and approaches, and recent advances in the field.

1.6.6 STATIC TIMING ANALYSIS

Chapter 6 by Cortadella and Sapatnekar overviews the most prominent techniques for static timing analysis. It then outlines issues relating to statistical timing analysis, which is becoming increasingly important to handle process variations in advanced IC technologies.

1.6.7 STRUCTURED DIGITAL DESIGN

In Chapter 7, Cho, Choudhury, Puri, Ren, Xiang, Nam, Mo, and Brayton cover the techniques for designing regular structures, including data paths, programmable logic arrays, and memories. It extends the discussion to include regular chip architectures such as gate arrays and structured ASICs.

1.6.8 ROUTING

Routing continues from automatic placement as a key step in IC design. Routing creates the wire traces necessary to connect all the placed components while obeying the process design rules. Chapter 8 by Téllez, Hu, and Wei discusses various types of routers and the key algorithms.

1.6.9 PHYSICAL DESIGN FOR 3D ICs

Chapter 9 by Lim illustrates, with concrete examples, how partitioning the blocks of an IC into multiple chips, connected by through-silicon vias (TSVs), can significantly improve wire length and thus both performance and power. This chapter explores trade-offs between different design options for TSV-based 3D IC integration. It also summarizes several research results in this emerging area.

1.6.10 GATE SIZING

Determining the best width for the transistors is essential to optimize the performance of an IC. This can be done both in a continuous setting, oriented toward full-custom or liquid library approaches, and in a discrete setting, for library-based layout and FinFET circuits. In Chapter 10, Held and Hu emphasize that sizing individual transistors is not very relevant today; the entire gates must be sized.

1.6.11 CLOCK DESIGN AND SYNTHESIS

Chapter 11 by Guthaus discusses the task of distributing one or more clock signals throughout an entire chip, while minimizing power, variation, skew, jitter, and resource usage.

1.6.12 EXPLORING CHALLENGES OF LIBRARIES FOR ELECTRONIC DESIGN

Chapter 12 by Hogan, Becker, and Carney discusses the factors that are most important and relevant for the design of libraries and IP, including standard cell libraries; cores, both hard and soft; and the design and user requirements for the same. It also places these factors in the overall design chain context.

1.6.13 DESIGN CLOSURE

Chapter 13 by Osler, Cohn, and Chinnery describes the common constraints in VLSI design and how they are enforced through the steps of a design flow that emphasizes design closure. A reference flow for ASICs is used and illustrated. Finally, issues such as power-limited design and variability are discussed.

1.6.14 TOOLS FOR CHIP-PACKAGE CO-DESIGN

Chip-package co-design refers to design scenarios, in which the design of the chip impacts the package design or vice versa. In Chapter 14, Franzon and Swaminathan discuss the drivers for new tools; the major issues, including mixed-signal needs; and the major design and modeling approaches.

1.6.15 DESIGN DATABASES

The design database is at the core of any EDA system. While it is possible to build a mediocre EDA tool or flow on *any* database, efficient and versatile EDA tools require more than a primitive database. Chapter 15 by Bales describes the place of a design database in an integrated design system. It discusses databases used in the past, those currently in use as well as emerging future databases.

1.6.16 FPGA SYNTHESIS AND PHYSICAL DESIGN

Programmable logic devices, and FPGAs, have evolved from implementing small glue-logic designs to large complete systems. The increased use of such devices—they now are the majority of design starts—has resulted in significant research in CAD algorithms and tools targeting programmable logic. Chapter 16 by Hutton, Betz, and Anderson gives an overview of relevant architectures, CAD flows, and research.

1.7 ANALOG AND MIXED-SIGNAL DESIGN

1.7.1 SIMULATION OF ANALOG AND RF CIRCUITS AND SYSTEMS

Circuit simulation has always been a crucial component of analog system design and is becoming even more so today. In Chapter 17, Roychowdhury and Mantooth provide a quick tour of modern circuit simulation. This includes circuit equations, device models, circuit analysis, more advanced analysis techniques motivated by RF circuits, new advances in circuit simulation using multitime techniques, and statistical noise analysis.

1.7.2 SIMULATION AND MODELING FOR ANALOG AND MIXED-SIGNAL INTEGRATED CIRCUITS

Chapter 18 by Gielen and Phillips provides an overview of the modeling and simulation methods that are needed to design and embed analog and RF blocks in mixed-signal integrated systems (ASICs, SoCs, and SiPs). The role of behavioral models and mixed-signal methods involving models at multiple hierarchical levels is covered. The generation of performance models for analog circuit synthesis is also discussed.

1.7.3 LAYOUT TOOLS FOR ANALOG ICs AND MIXED-SIGNAL SoCs: A SURVEY

Layout for analog circuits has historically been a time-consuming, manual, trial-and-error task. In Chapter 19, Rutenbar, Cohn, Lin, and Baskaya cover the basic problems faced by those who need to create analog and mixed-signal layout and survey the evolution of design tools and geometric/electrical optimization algorithms that have been directed at these problems.

1.8 PHYSICAL VERIFICATION

1.8.1 DESIGN RULE CHECKING

After the physical mask layout is created for a circuit for a specific design process, the layout is measured by a set of geometric constraints or rules for that process. The main objective of design rule checking (DRC) is to achieve high overall yield and reliability. Chapter 20 by Todd, Grodd, Tomblin, Fetty, and Liddell gives an overview of DRC concepts and then discusses the basic verification algorithms and approaches.

1.8.2 RESOLUTION ENHANCEMENT TECHNIQUES AND MASK DATA PREPARATION

With more advanced IC fabrication processes, new physical effects, negligible in the past, are being found to have a strong impact on the formation of features on the actual silicon wafer. It is now essential to transform the final layout via new tools in order to allow the manufacturing equipment to deliver the new devices with sufficient yield and reliability to be cost-effective. In Chapter 21, Schellenberg discusses the compensation schemes and mask data conversion technologies now available to accomplish the new design for manufacturability (DFM) goals.

1.8.3 DESIGN FOR MANUFACTURABILITY IN THE NANOMETER ERA

Achieving high-yielding designs in state-of-the-art IC process technology has become an extremely challenging task. DFM includes many techniques to modify the design of ICs in order to improve functional and parametric yield and reliability. Chapter 22 by Dragone, Guardiani, and Strojwas discusses yield loss mechanisms and fundamental yield modeling approaches. It then discusses techniques for functional yield maximization and parametric yield optimization. Finally, DFM-aware design flows and the outlook for future DFM techniques are discussed.

1.8.4 DESIGN AND ANALYSIS OF POWER SUPPLY NETWORKS

Chapter 23 by Panda, Pant, Blaauw, and Chaudhry covers design methods, algorithms, tools for designing on-chip power grids, and networks. It includes the analysis and optimization of effects such as voltage drop and electromigration.

1.8.5 NOISE IN DIGITAL ICs

On-chip noise issues and their impact on signal integrity and reliability are becoming a major source of problems for deep submicron ICs. Thus, the methods and tools for analyzing and coping with them, which are discussed by Keller and Kariat in Chapter 24, gained in recent years.

1.8.6 LAYOUT EXTRACTION

Layout extraction is the translation of the topological layout back into the electrical circuit it is intended to represent. In Chapter 25, Kao, Lo, Basel, Singh, Spink, and Scheffer discuss the distinction between designed and parasitic devices and also the three major parts of extraction: designed device extraction, interconnect extraction, and parasitic device extraction.

1.8.7 MIXED-SIGNAL NOISE COUPLING IN SYSTEM-ON-CHIP DESIGN: MODELING, ANALYSIS, AND VALIDATION

Chapter 26 by Verghese and Nagata describes the impact of noise coupling in mixed-signal ICs and reviews techniques to model, analyze, and validate it. Different modeling approaches and computer simulation methods are presented, along with measurement techniques. Finally, the chapter reviews the application of substrate noise analysis to placement and power distribution synthesis.

1.9 TECHNOLOGY COMPUTER-AIDED DESIGN

1.9.1 PROCESS SIMULATION

Process simulation is the modeling of the fabrication of semiconductor devices such as transistors. The ultimate goal is an accurate prediction of the active dopant distribution, the stress distribution, and the device geometry. In Chapter 27, Johnson discusses the history, requirements, and development of process simulators.

1.9.2 DEVICE MODELING: FROM PHYSICS TO ELECTRICAL PARAMETER EXTRACTION

Technology files and design rules are essential building blocks of the IC design process. Development of these files and rules involves an iterative process that crosses the boundaries of technology and device development, product design, and quality assurance. Chapter 28 by Dutton, Choi, and Kan starts with the physical description of IC devices and describes the evolution of TCAD tools.

1.9.3 HIGH-ACCURACY PARASITIC EXTRACTION

Chapter 29 by Kamon and Iverson describes high-accuracy parasitic extraction methods using fast integral equation and random walk-based approaches.

REFERENCES

1. A. Sangiovanni-Vincentelli, The tides of EDA, *IEEE Des. Test Comput.*, 20, 59–75, 2003.
2. A. Kuelhmann, Ed., *20 Years of ICCAD*, Kluwer Academic Publishers (Springer), Dordrecht, the Netherlands, 2002.

Integrated Circuit Design Process and Electronic Design Automation

Robert Damiano, Raul Camposano, and Grant E. Martin

2

CONTENTS

2.1 INTRODUCTION

In this chapter, we describe the design process, its major stages, and how electronic design automation (EDA) tools fit into these processes. We also examine the interfaces between the major integrated circuit (IC) design stages as well as the kind of information—both abstractions upward and detailed design and verification information downward—that must flow between stages. We assume complementary metal oxide semiconductor (CMOS) is the basis for all technologies.

We will illustrate with a continuing example. A company wishes to create a new system on chip (SoC). The company assembles a product team, consisting of a project director, system architects, system verification engineers, circuit designers (both digital and analog), circuit verification engineers, layout engineers, and manufacturing process engineers. The product team determines the target technology and geometry as well as the fabrication facility or foundry. The system architects initially describe the system-level design (SLD) through a transaction-level specification in a language such as C++, SystemC, or Esterel. The system verification engineers determine the functional correctness of the SLD through simulation. The engineers validate the transaction processing through simulation vectors. They monitor the results for errors. Eventually, these same engineers would simulate the process with an identical set of vectors through the system implementation to see if the specification and the implementation match. There is some ongoing research to check this equivalence formally.

The product team partitions the SLD into functional units and hands these units to the circuit design teams. The circuit designers describe the functional intent through a high-level design language (HDL). The most popular HDLs are Verilog and VHDL. SystemVerilog is a recent language, adopted by the IEEE, which contains design, testbench, and assertion syntax. These languages allow the circuit designers to express the behavior of their design using high-level functions such as addition and multiplication. These languages allow expression of the logic at the register transfer level (RTL), in the sense that an assignment of registers expresses functionality. For the analog and analog mixed signal (AMS) parts of the design, there are also HDLs such as Verilog-AMS and VHDL-AMS. Most commonly, circuit designers use Simulation Program with IC Emphasis (SPICE) transistor models and netlists to describe analog components. However, high-level languages provide an easier interface between analog and digital segments of the design, and they allow writing higher-level behavior of the analog parts. Although the high-level approaches are useful as simulation model interfaces, there remains no clear method of synthesizing transistors from them. Therefore, transistor circuit designers usually depend on schematic capture tools to enter their data.

The design team must consider functional correctness, implementation closure (reaching the prioritized goals of the design), design cost, and manufacturability of a design. The product team takes into account risks and time to market as well as choosing the methodology. Anticipated sales volume can reflect directly on methodology; for instance, whether it is better to create a full-custom design, semicustom design or use standard cells, gate arrays, or a field programmable gate array (FPGA). Higher volume mitigates the higher cost of full-custom or semicustom design, while time to market might suggest using an FPGA methodology. If implementation closure for power and speed is tantamount, then an FPGA methodology might be a poor choice. Semicustom designs, depending on the required volume, can range from microprocessor central processor units (CPUs), digital signal processors (DSPs), application-specific standard parts (ASSP), or application-specific ICs (ASIC). In addition, for semicustom designs, the company needs to decide whether to allow the foundry to implement the layout or whether the design team should use customer-owned tooling (COT). We will assume that our product team chooses semicustom COT designs. We will mention FPGA and full-custom methodologies only in comparison.

In order to reduce cost, the product team may decide that the design warrants reuse of intellectual property (IP). IP reuse directly addresses the increasing complexity of design as opposed to feature geometry size. Reuse also focuses on attaining the goals of functional correctness. One analysis estimates that it takes 2000 engineering years and 1 trillion simulation vectors to verify 25 million lines of RTL code. Therefore, verified IP reuse reduces cost and time to market. Moreover, IP blocks themselves have become larger and more complex. For example, the 1176JZ-S ARM core is 24 times larger than the older 7TDI-S ARM core. The USB 2.0 Host is 23 times larger than the Universal Serial Bus (USB) 1.1 Device. PCI Express is 7.5 times larger than PCI v 1.1.

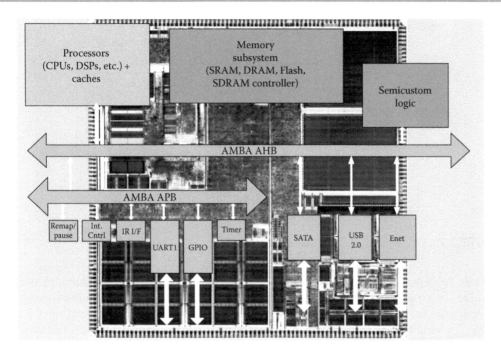

FIGURE 2.1   SoC with IP.

Another important trend is that SoC-embedded memories are an increasingly large part of the SoC real estate. While in 1999 20% of a 180 nm SoC was embedded memory, roadmaps project that by 2005 embedded memory will consume 71% of a 90 nm SoC. These same roadmaps indicate that by 2014 embedded memory will grow to 94% of a 35 nm SoC.

SoCs typically contain one or more CPUs or DSPs (or both), a cache, a large amount of embedded memory, and many off-the-shelf components such as USB, universal asynchronous receiver–transmitter (UART), Serial Advanced Technology Attachment (SATA), and Ethernet (cf. Figure 2.1). The differentiating part of the SoC contains the new designed circuits in the product.

The traditional semicustom IC design flow typically comprises up to 50 steps. On the digital side of design, the main steps are functional verification, logical synthesis, design planning, physical implementation that includes clock-tree synthesis, placement and routing, extraction, design rule checking (DRC) and layout versus schematic (LVS) checking, static timing analysis, insertion of test structures, and test pattern generation. For analog designs, the major steps are as follows: schematic entry, SPICE simulation, layout, layout extraction, DRC, and LVS checking. SPICE simulations can include DC, AC, and transient analysis, as well as noise, sensitivity, and distortion analysis. Analysis and implementation of corrective procedures for the manufacturing process, such as mask synthesis and yield analysis, are critical at smaller geometries. In order to verify an SoC system where many components reuse IP, the IP provider may supply verification IPs, monitors, and checkers needed by system verification.

There are three basic areas where EDA tools assist the design team. Given a design, the first is verification of functional correctness. The second deals with implementation of the design. The last area deals with analysis and corrective procedures so that the design meets all manufacturability specifications. Verification, layout, and process engineers on the circuit design team essentially own these three steps.

2.2 VERIFICATION

The design team attempts to verify that the design under test (DUT) functions correctly. For RTL designs, verification engineers rely heavily on simulation at the cycle level. After layout, EDA tools, such as equivalence checking, can determine whether the implementation matches the RTL functionality. After layout, the design team must check that there are no

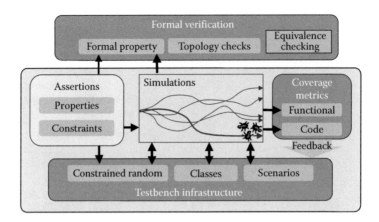

FIGURE 2.2   Digital simulation/formal verification.

problem delay paths. A static timing analysis tool can facilitate this. The team also needs to examine the circuit for noise and delay due to parasitics. In addition, the design must obey physical rules for wire spacing, width, and enclosure as well as various electrical rules. Finally, the design team needs to simulate and check the average and transient power. For transistor circuits, the design team uses SPICE circuit simulation or fast SPICE to determine correct functionality, noise, and power.

We first look at digital verification (cf. Figure 2.2). RTL simulation verifies that the DUT behavior meets the design intent. The verification engineers apply a set of vectors, called a testbench, to the design through an event-driven simulator, and compare the results to a set of expected outputs. The quality of the verification depends on the quality of the testbench. Many design teams create their testbench by supplying a list of the vectors, a technique called directed test. For a directed test to be effective, the design team must know beforehand what vectors might uncover bugs. This is extremely difficult since complex sequences of vectors are necessary to find some corner case errors. Therefore, many verification engineers create testbenches that supply stimulus through random vectors with biased inputs, such as the clock or reset signal. The biasing increases or decreases the probability of a signal going high or low. While a purely random testbench is easy to create, it suffers from the fact that vectors may be illegal as stimulus. For better precision and wider coverage, the verification engineer may choose to write a constrained random testbench. Here, the design team supplies random input vectors that obey a set of constraints.

The verification engineer checks that the simulated behavior does not have any discrepancies from the expected behavior. If the engineer discovers a discrepancy, then the circuit designer modifies the HDL and the verification engineer resimulates the DUT. Since exhaustive simulation is usually impossible, the design team needs a metric to determine quality. One such metric is coverage. Coverage analysis considers how well the test cases stimulate the design. The design team might measure coverage in terms of number of lines of RTL code exercised, whether the test cases take each leg of each decision, or how many "reachable" states are encountered.

Another important technique is for the circuit designer to add assertions within the HDL. These assertions monitor whether the internal behavior of the circuit is acting properly. Some designers embed tens of thousands of assertions into their HDL. Languages like SystemVerilog have extensive assertion syntax based on linear temporal logic. Even for languages without the benefit of assertion syntax, tool providers supply an application program interface (API), which allows the design team to build and attach its own monitors.

The verification engineer needs to run a large amount of simulation, which would be impractical if not for compute farms. Here, the company may deploy thousands of machines, 24/7, to enable the designer to get billions of cycles a day; sometimes the machines may run as many as 200 billion cycles a day. Best design practices typically create a highly productive computing environment. One way to increase throughput is to run a cycle simulation by taking a subset of the chosen verification language that both is synchronous and has a set of registers with clear clock cycles. This type of simulation assumes a uniformity of events and typically uses a time wheel with gates scheduled in a breadth-first manner.

Another way to tackle the large number of simulation vectors during system verification is through emulation or hardware acceleration. These techniques use specially configured hardware to run the simulation. In the case of hardware acceleration, the company can purchase special-purpose hardware, while in the case of emulation the verification engineer uses specially configured FPGA technology. In both cases, the system verification engineer must synthesize the design and testbench down to a gate-level model. Tools are available to synthesize and schedule gates for the hardware accelerator. In the case of an FPGA emulation system, tools can map and partition the gates for the hardware.

Of course, since simulation uses vectors, it is usually a less than exhaustive approach. The verification engineer can make the process complete by using assertions and formal property checking. Here, the engineer tries to prove that an assertion is true or to produce a counterexample. The trade-off is simple. Simulation is fast but by definition incomplete, while formal property checking is complete but may be very slow. Usually, the verification engineer runs constrained random simulation to unearth errors early in the verification process. The engineer applies property checking to corner case situations that can be extremely hard for the testbench to find. The combination of simulation and formal property checking is very powerful. The two can even be intermixed, by allowing simulation to proceed for a set number of cycles and then exhaustively looking for an error for a different number of cycles. In a recent design, by using this hybrid approach, a verification engineer found an error 21,000 clock cycles from an initial state. Typically, formal verification works well on specific functional units of the design. Between the units, the system engineers use an "assume/guarantee" methodology to establish block pre- and postconditions for system correctness.

During the implementation flow, the verification engineer applies equivalence checking to determine whether the DUT preserves functional behavior. Note that functional behavior is different from functional intent. The verification engineer needs RTL verification to compare functional behavior with functional intent. Equivalence checking is usually very fast and is a formal verification technology, which is exhaustive in its analysis. Formal methods do not use vectors.

For transistor-level circuits, such as analog, memory, and radio frequency, the event-driven verification techniques suggested earlier do not suffice (cf. Figure 2.3). The design team needs to compute signals accurately through SPICE circuit simulation. SPICE simulation is very time consuming because the algorithm solves a system of differential equations. One way to get around this cost is to select only a subset of transistors, perform an extraction of the parasitics, and then simulate the subset with SPICE. This reduction gives very accurate results for the subset, but even so, the throughput is still rather low. Another approach is to perform a fast SPICE simulation.

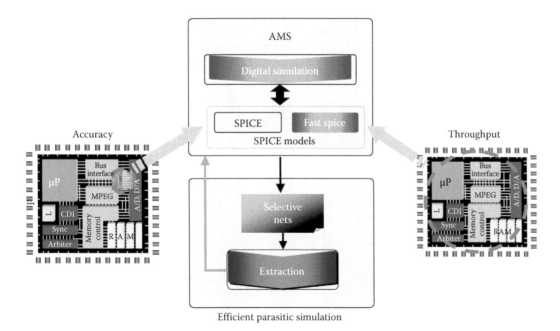

FIGURE 2.3 Transistor simulation with parasitics.

This last SPICE approach trades some accuracy for a significant increase in throughput. The design team can also perform design space exploration by simulating various constraint values on key goals such as gain or phase margin to find relatively optimal design parameters. The team analyzes the multiple-circuit solutions and considers the cost trade-offs. A new generation of tools performs this "design exploration" in an automatic manner. Mixed-level simulation typically combines RTL, gate, and transistor parts of the design and uses a communication backplane to run the various simulations and share input and output values.

Finally, for many SoCs, both hardware and software comprise the real system. System verification engineers may run a hardware–software cosimulation before handing the design to a foundry. All simulation system components mentioned can be part of this cosimulation. In early design stages, when the hardware is not ready, the software can simulate (*execute*) an instruction set model, a virtual prototype (model), or an early hardware prototype typically implemented in FPGAs.

2.3 IMPLEMENTATION

This brings us to the next stage of the design process, the implementation and layout of the digital design. Circuit designers implement analog designs by hand. FPGA technologies usually have a single basic combinational cell, which can form a variety of functions by constraining inputs. Layout and process tools are usually proprietary to the FPGA family and manufacturer. For semicustom design, the manufacturer supplies a precharacterized cell library, either standard cell or gate array. In fact, for a given technology, the foundry may supply several libraries, differing in power, timing, or yield. The company decides on one or more of these as the target technology. One twist on the semicustom methodology is structured ASIC. Here, a foundry supplies preplaced memories, pad rings, and power grids, as well as sometimes preplaced gate array logic, which is similar to the methodology employed by FPGA families. The company can use semicustom techniques for the remaining combinational and sequential logic. The goal is to reduce nonrecurring expenses by limiting the number of mask sets needed and by simplifying physical design.

By way of contrast, in a full-custom methodology, one tries to gain performance and limit power consumption by designing much of the circuit as transistors. The circuit designers keep a corresponding RTL design. The verification engineer simulates the RTL and extracts a netlist from the transistor description. Equivalence checking compares the extracted netlist to the RTL. The circuit designer manually places and routes the transistor-level designs. Complex high-speed designs, such as microprocessors, sometimes use full custom methodology, but the design costs are very high. The company assumes that the high volume will amortize the increased cost. Full-custom designs consider implementation closure for power and speed as most important. At the other end of the spectrum, FPGA designs focus on design cost and time to market. Semicustom methodology tries to balance the goals of timing and power closure with design cost (cf. Figure 2.4).

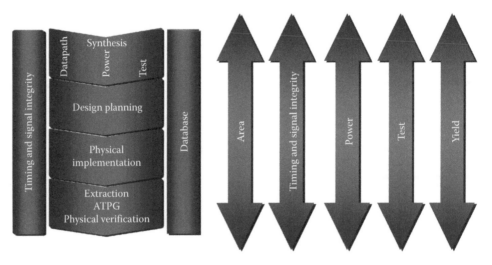

FIGURE 2.4 Multi-objective implementation convergence.

In the semicustom implementation flow, one first attempts to synthesize the RTL design into a mapped netlist. The circuit designers supply their RTL circuit along with timing constraints. The timing constraints consist of signal arrival and slew (transition) times at the inputs and required times and loads (capacitances) at the outputs. The circuit designer identifies clocks as well as any false or multiple-cycle paths. The technology library is usually a file that contains a description of the function of each cell along with delay, power, and area information. The cell description contains the pin-to-pin delay represented either as lookup table functions of input slew, output load, and other physical parameters, such as voltage and temperature, or as polynomial functions that best fit the parameter data. For example, foundries provide cell libraries in Liberty or OLA (Open Library Application Programming Interface) formats. The foundry also provides a wire delay model, derived statistically from previous designs. The wire delay model correlates the number of sinks of a net to capacitance and delay.

Several substages comprise the operation of a synthesis tool. First, the synthesis tool compiles the RTL into technology-independent cells and then optimizes the netlist for area, power, and delay. The tool maps the netlist into a technology. Sometimes, synthesis finds complex functions such as multipliers and adders in parameterized (area/timing) reuse libraries. For example, the tool might select a Booth multiplier from the reuse library to improve timing. For semicustom designs, the foundry provides a standard cell or gate array library, which describes each functional member. In contrast, the FPGA supplier describes a basic combinational cell from which the technology mapping matches functional behavior of subsections of the design. To provide correct functionality, the tool may set several pins on the complex gates to constants. A postprocess might combine these functions for timing, power, or area.

A final substage tries to analyze the circuit and performs local optimizations that help the design meet its timing, area, and power goals. Note that due to finite number of power levels of any one cell, there are limits to the amount of capacitance that functional cell types can drive without the use of buffers. Similar restrictions apply to input slew (transition delay). The layout engineer can direct the synthesis tool by enhancing or omitting any of these stages through scripted commands. Of course, the output must be a mapped netlist.

To get better timing results, foundries continue to increase the number of power variations for some cell types. One limitation to timing analysis early in the flow is that the wire delay models are statistical estimates of the real design. Frequently, these wire delays can differ significantly from those found after routing. One interesting approach to synthesis is to extend each cell of the technology library so that it has an infinite or continuous variation of power. This approach, called gain-based synthesis, attempts to minimize the issue of inaccurate wire delay by assuming cells can drive any wire capacitance through appropriate power level selection. In theory, there is minimal perturbation to the natural delay (or gain) of the cell. This technique makes assumptions such as that the delay of a signal is a function of capacitance. This is not true for long wires where resistance of the signal becomes a factor. In addition, the basic approach needs to include modifications for slew (transition delay).

To allow detection of manufacturing faults, the design team may add extra test generation circuitry. Design for test (DFT) is the name given to the process of adding this extra logic (cf. Figure 2.5). Sometimes, the foundry supplies special registers, called logic-sensitive scan devices. At other times, the test tool adds extra logic called Joint Test Action Group (JTAG) boundary scan logic that feeds the registers. Later in the implementation process, the design team will generate data called scan vectors that test equipment uses to detect manufacturing faults. Subsequently, tools will transfer these data to automatic test equipment (ATE), which perform the chip tests.

As designs have become larger, so has the amount of test data. The economics of the scan vector production with minimal cost and design impact leads to data compression techniques. One of the most widely used techniques is deterministic logic built-in self-test (BIST). Here, a test tool adds extra logic on top of the DFT to generate scan vectors dynamically.

Before continuing the layout, the engineer needs new sets of rules, dealing with the legal placement and routing of the netlist. These libraries, in various exchange formats, for example, LEF for logic, DEF for design, and PDEF for physical design, provide the layout engineer physical directions and constraints. Unlike the technology rules for synthesis, these rules are typically model dependent. For example, there may be information supplied by the circuit designer

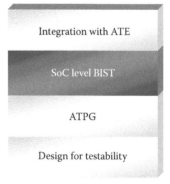

FIGURE 2.5 Design for test.

about the placement of macros such as memories. The routing tool views these macros as blockages. The rules also contain information from the foundry.

Even if the synthesis tool preserved the original hierarchy of the design, the next stages of implementation need to view the design as flat. The design-planning step first flattens the logic and then partitions the flat netlist as to assist placement and routing; in fact, in the past, design planning was sometimes known as floor planning. A commonly used technique is for the design team to provide a utilization ratio to the design planner. The utilization ratio is the percentage of chip area used by the cells as opposed to the nets. If the estimate is too high, then routing congestion may become a problem. If the estimate is too low, then the layout could waste area. The design-planning tool takes the locations of hard macros into account. These macros are hard in the sense that they are rectangular with a fixed length, fixed width, and sometimes a fixed location on the chip. The design-planning tool also tries to use the logical hierarchy of the design as a guide to the partitioning. The tool creates, places, and routes a set of macros that have fixed lengths, widths, and locations. The tool calculates timing constraints for each macro and routes the power and ground grids. The power and ground grids are usually on the chip's top levels of metal and then distributed to the lower levels. The design team can override these defaults and indicate which metal layers should contain these grids. Sometimes design planning precedes synthesis. In these cases, the tool partitions the RTL design and automatically characterizes each of the macros with timing constraints.

After design planning, the layout engineer runs the physical implementation tools on each macro. First, the placer assigns physical locations to each gate of the macro. The placer typically moves gates while minimizing some cost, for example, wire length or timing. Legalization follows the coarse placement to make sure the placed objects fit physical design rules. At the end of placement, the layout engineer may run some more synthesis, like resizing of gates. One of the major improvements to placement over the last decade is the emergence of physical synthesis. In physical synthesis, the tool interleaves synthesis and placement. Recall that previously, logic synthesis used statistical wire capacitance. Once the tool places the gates, it can perform a global route and get capacitances that are more accurate for the wires, based on actual placed locations. The physical synthesis tool iterates this step and provides better timing and power estimates.

Next, the layout engineer runs a tool that buffers and routes the clock tree. Clock-tree synthesis attempts to minimize the delay while assuring that skew, that is the variation in signal transport time from the clock to its corresponding registers, is close to zero.

Routing the remaining nets comes after clock-tree synthesis. Routing starts with a global analysis called global route. Global route creates coarse routes for each signal and its outputs. Using the global routes as a guide, a detailed routing scheme, such as a maze channel or switchbox, performs the actual routing. As with the placement, the tool performs a final legalization to assure that the design obeys physical rules. One of the major obstacles to routing is signal congestion. Congestion occurs when there are too many wires competing for a limited amount of chip wire resource. Remember that the design team gave the design planner a utilization ratio in the hope of avoiding this problem.

Both global routing and detailed routing take the multilayers of the chip into consideration. For example, the router assumes that the gates are on the polysilicon layer, while the wires connect the gates through vias on 3–8 layers of metal. Horizontal or vertical line segments comprise the routes, but some recent work allows 45° lines for some foundries. As with placement, there may be some resynthesis, such as gate resizing, at the end of the detailed routing stage.

Once the router finishes, an extraction tool derives the capacitances, resistances, and inductances. In a 2D parasitic extraction, the extraction tool ignores 3D details and assumes that each chip level is uniform in one direction. This produces only approximate results. In the case of the much slower 3D parasitic extraction, the tool uses 3D field solvers to derive very accurate results. A 2 1/2-D extraction tool compromises between speed and accuracy. By using multiple passes, it can access some of the 3D features. The extraction tool places its results in a standard parasitic exchange format file.

During the implementation process, the verification engineer continues to monitor behavioral consistency through equivalence checking and using LVS comparison. The layout engineer analyzes timing and signal integrity issues through timing analysis tools and uses their results to drive implementation decisions. At the end of the layout, the design team has accurate resistances, capacitances, and inductances for the layout. The system engineer uses a sign-off timing analysis tool to determine if the layout meets timing goals. The layout engineer needs to run a DRC on the layout to check for violations.

Both the Graphic Data System II (GDSII) and the Open Artwork System Interchange Standard (OASIS) are databases for shape information to store a layout. While the older GDSII was the database of choice for shape information, there is a clear movement to replace it by the newer, more efficient OASIS database. The LVS tool checks for any inconsistencies in this translation.

What makes the implementation process so difficult is that multiple objectives need consideration. For example, area, timing, power, reliability, test, and yield goals might and usually cause conflict with each other. The product team must prioritize these objectives and check for implementation closure.

Timing closure—that is meeting all timing requirements—by itself is becoming increasingly difficult and offers some profound challenges. As process geometry decreases, the significant delay shifts from the cells to the wires. Since a synthesis tool needs timing analysis as a guide and routing of the wires does not occur until after synthesis, we have a chicken and egg problem. In addition, the thresholds for noise sensitivity also shrink with smaller geometries. This along with increased coupling capacitances, increased current densities and sensitivity to inductance, make problems like crosstalk and voltage (IR) drop increasingly familiar.

Since most timing analysis deals with worst-case behavior, statistical variation and its effect on yield add to the puzzle. Typically timing analysis computes its cell delay as function of input slew (transition delay) and output load (output capacitance or RC). If we add the effects of voltage and temperature variations as well as circuit metal densities, timing analysis gets to be very complex. Moreover, worst-case behavior may not correlate well with what occurs empirically when the foundry produces the chips. To get a better predictor of parametric yield, some layout engineers use statistical timing analysis. Here, rather than using single numbers (worst case, best case, corner case, nominal) for the delay-equation inputs, the timing analysis tool selects probability distributions representing input slew, output load, temperature, and voltage among others. The delay itself becomes a probability distribution. The goal is to compute the timing more accurately in order to create circuits with smaller area and lower power but with similar timing yield.

Reliability is also an important issue with smaller geometries. Signal integrity deals with analyzing what were secondary effects in larger geometries. These effects can produce erratic behavior for chips manufactured in smaller geometries. Issues such as crosstalk, IR drop, and electromigration are factors that the design team must consider in order to produce circuits that perform correctly.

Crosstalk noise can occur when two wires are close to each other (cf. Figure 2.6). One wire, the aggressor, switches while the victim signal is in a quiet state or making an opposite transition. In this case, the aggressor can force the victim to glitch. This can cause a functional failure or can simply consume additional power. Gate switching draws current from the power and

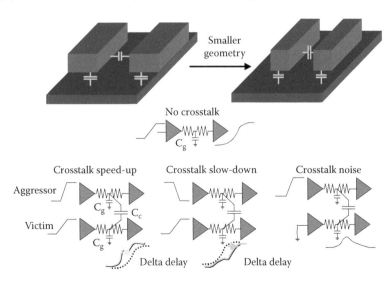

FIGURE 2.6 Crosstalk.

ground grids. That current, together with the wire resistance in the grids, can cause significant fluctuations in the power and ground voltages supplied to gates. This problem, called IR drop, can lead to unpredictable functional errors. Very high frequencies can produce high current densities in signals and power lines, which can lead to the migration of metal ions. This power electromigration can lead to open or shorted circuits and subsequent signal failure.

Power considerations are equally complex. As the size of designs grows and geometries shrink, power increases. This can cause problems for batteries in wireless and handheld devices and thermal management in microprocessor, graphic, and networking applications. Power consumption falls into two areas: (1) dynamic power (cf. Figure 2.7), the power consumed when devices switch value, and (2) leakage power (cf. Figure 2.8), the power leaked through the transistor. Dynamic power consumption grows directly with increased capacitance and voltage. Therefore, as designs become larger, dynamic power increases. One easy way to reduce dynamic power is to decrease voltage. However, decreased voltage leads to smaller noise margins and less speed.

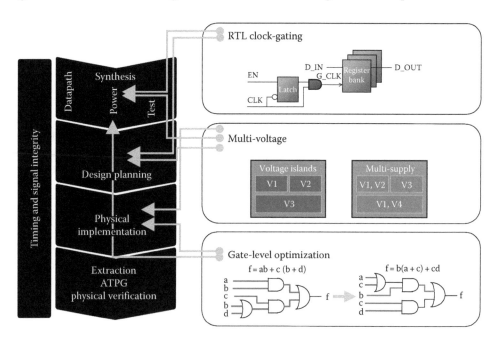

FIGURE 2.7   Dynamic power management.

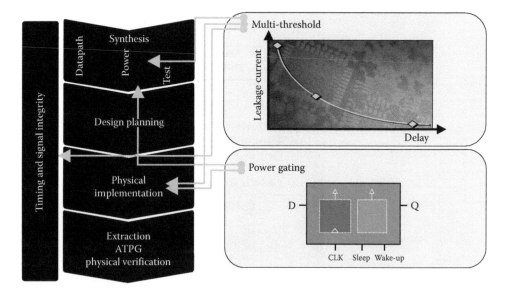

FIGURE 2.8   Static power management (leakage).

A series of novel design and transistor innovations can reduce the power consumption. These include operand isolation, clock gating, and voltage islands. Timing and power considerations are very often in conflict with each other, so the design team must employ these remedies carefully.

A design can have part of its logic clock gated by using logic to enable the bank of registers. The logic driven by the registers is quiescent until the clock-gated logic enables the registers. Latches at the input can isolate parts of a design that implement operations (e.g., an arithmetic logic unit), when results are unnecessary for correct functionality, thus preventing unnecessary switching. Voltage islands help resolve the timing versus power conflicts. If part of a design is timing critical, a higher voltage can reduce the delay. By partitioning the design into voltage islands, one can use lower voltage in all but the most timing-critical parts of the design. An interesting further development is dynamic voltage/frequency scaling, which consists of scaling the supply voltage and the speed during operation to save power or increase performance temporarily.

The automatic generation of manufacturing fault detection tests was one of the first EDA tools. When a chip fails, the foundry wants to know why. Test tools produce scan vectors that can identify various manufacturing faults within the hardware. The design team translates the test vectors to standard test data format and the foundry can inject these inputs into the failed chip through automated test equipment (ATE). Remember that the design team added extra logic to the netlist before design planning, so that test equipment could quickly insert the scan vectors, including set values for registers, into the chip. The most common check is for stuck at 0 or stuck at 1 faults where the circuit has an open or short at a particular cell. It is not surprising that smaller geometries call for more fault detection tests. An integration of static timing analysis with transition/path delay fault automatic test pattern generation (ATPG) can help, for example, to detect contact defects; while extraction information and bridging fault ATPG can detect metal defects.

Finally, the design team should consider yield goals. Manufacturing becomes more difficult as geometries shrink. For example, thermal stress may create voids in vias. One technique to get around this problem is to minimize the vias inserted during routing, and for those inserted, to create redundant vias. Via doubling, which converts a single via into multiple vias, can reduce resistance and produce better yield. Yield analysis can also suggest wire spreading during routing to reduce crosstalk and increase yield. Manufacturers also add a variety of manufacturing process rules needed to guarantee good yield. These rules involve antenna checking and repair through diode insertion as well as metal fill needed to produce uniform metal densities necessary for copper wiring chemical–mechanical polishing (CMP). Antenna repair has little to do with what we typically view as antennas. During the ion-etching process, charge collects on the wires connected to the polysilicon gates. These charges can damage the gates. The layout tool can connect small diodes to the interconnect wires as a discharge path.

Even with all the available commercial tools, there are times when layout engineers want to create their own tool for analysis or small implementation changes. This is analogous to the need for an API in verification. Scripting language and C-language-based APIs for design databases such as MilkyWay and OpenAccess are available. These databases supply the user with an avenue to both the design and rules. The engineer can directly change and analyze the layout.

2.4 DESIGN FOR MANUFACTURING

One of the newest areas for EDA tools is design for manufacturing. As in other areas, the driving force of the complexity is the shrinking of geometries. After the design team translates their design to shapes, the foundry must transfer those shapes to a set of masks. Electron beam (laser) equipment then creates the physical masks for each layer of the chip from the mask information. For each layer of the chip, the foundry applies photoresistive material and then transfers the mask structures by the stepper optical equipment onto the chip. Finally, the foundry etches the correct shapes by removing the excess photoresist material.

Since the stepper uses light for printing, it is important that the wavelength is small enough to transcribe the features accurately. When the chip's feature size was 250 nm, we could use lithography equipment that produced light at a wavelength of 248 nm. New lithography equipment that

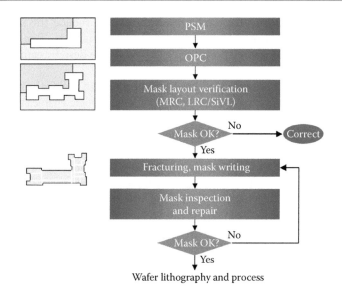

FIGURE 2.9   Subwavelength: from layout to masks.

produces light of lower wavelength needs significant innovation and can be very expensive. When the feature geometry gets significantly smaller than the wavelength, the detail of the reticles (fine lines and wires) transferred to the chip from the mask can be lost. EDA tools can analyze and correct this transfer operation without new equipment, by modifying the shapes of data—a process known as mask synthesis (cf. Figure 2.9). This process uses resolution enhancement techniques and methods to provide dimensional accuracy.

One mask synthesis technique is optimal proximity correction (OPC). This process takes the reticles in the GDSII or OASIS databases and modifies them by adding new lines and wires, so that even if the geometry is smaller than the wavelength, optical equipment adequately preserves the details. This technique successfully transfers geometric features of down to one-half of the wavelength of the light used. Of course given a fixed wavelength, there are limits beyond which the geometric feature size is too small for even these tricks.

For geometries of 90 nm and below, the lithography EDA tools combine OPC with other mask synthesis approaches such as phase shift mask (PSM), off-axis illumination, and assist features (AF). For example, PSM is a technique where the optical equipment images dark features at critical dimensions with 0° illumination on one side and 180° illumination on the other side. There are additional manufacturing process rules needed, such as minimal spacing and cyclic conflict avoidance, to avoid situations where the tool cannot map the phase.

In summary, lithography tools proceed through PSM, OPC, and AF to enhance resolution and make the mask more resistive to process variations. The process engineer can perform a verification of silicon versus layout and a check of lithography rule compliance. If either fails, the engineer must investigate and correct, sometimes manually. If both succeed, another EDA tool "fractures" the design, subdividing the shapes into rectangles (trapezoids), which can be fed to the mask writing equipment. The engineer can then transfer the final shapes file to a database, such as the manufacturing-electron-beam-exposure system (MEBES). Foundry equipment uses the MEBES database (or other proprietary formats) to create the physical masks. The process engineer can also run a *virtual* stepper tool to preanalyze the various stages of the stepper operation. After the foundry manufactures the masks, a mask inspection and repair step ensures that they conform to manufacturing standards.

Another area of design for manufacturing analysis is the prediction of yield (cf. Figure 2.10). The design team would like to correlate some of the activities during route with actual yield. Problems with CMP via voids and crosstalk can cause chips to unexpectedly fail. EDA routing tools offer some solutions in the form of metal fill, via doubling and wire spacing. Library providers are starting to develop libraries for higher yields that take into account several yield failure mechanisms. There are tools that attempt to correlate these solutions with yield. Statistical timing analysis can correlate timing constraints to parametric circuit yield.

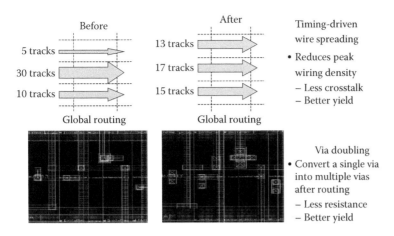

FIGURE 2.10   Yield enhancement features in routing.

Finally, the process engineer can use tools to predict the behavior of transistor devices or processes. Technology computer-aided design (TCAD) deals with the modeling and simulation of physical manufacturing process and devices. Engineers can model and simulate individual steps in the fabrication process. Likewise, the engineer can model and simulate devices, parasitics, or electrical/thermal properties, therefore providing insights into their electrical, magnetic, or optical properties.

For example, because of packing density, foundries may switch isolation technology for an IC from the local oxidation of silicon model toward the shallow trench isolation (STI) model. Under this model, the process engineer can analyze breakdown stress, electrical behavior such as leakage, or material versus process dependencies. TCAD tools can simulate STI effects; extract interconnect parasitics, such as diffusion distance, and determine SPICE parameters.

2.5 UPDATE: A DECADE OF EVOLUTION IN DESIGN PROCESS AND EDA

Reviewing the design process described in this chapter, which was written for the first edition of the handbook published in 2006, it is remarkable how similar today's IC design process is in terms of major tasks and steps, despite the decade that has elapsed, and the huge increase in SoC complexity and process complexity that has occurred. In addition, despite many changes in the EDA industry and its tools, there are remarkable similarities in many ways.

Nevertheless, the past decade has seen some important changes that are worthy of note for their impact on design process and EDA tools. Some of these arguably have more impact in the economics of design than on the technical design processes. In the following update, we try to briefly note some of these major changes and summarize their impact on design process and EDA:

- *SoC design complexity*: From a few tens of blocks and fewer than 10 subsystems in a complex SoC of a decade ago to today's most complex SoCs that may have hundreds of blocks and many complicated subsystems, we have seen the design tasks grow in magnitude. This has many side effects: a large increase in design teams, greater collaboration by design teams around the world, many mergers, acquisitions and restructurings of teams in design companies (especially to acquire design IP and design talent in subsystem areas that are being merged into the more complex SoCs), and much more reliance on third-party IP blocks and subsystems. In addition, less and less can be done by last-minute manual tweaks to the design, hence resulting in more reliance on EDA tools, quality IP, and very well-controlled design processes. The latest *International Technology Roadmap for Semiconductors* (2012 update edition) [27] has substantial discussions of SoC design challenges in its "System Drivers" and "Design" chapters.
- *Reduction in the number of semiconductor fabrication facilities*: The trend, which started in the 1990s and accelerated in the last decade, of system companies shedding their

in-house semiconductor fabrication facilities, and even some semiconductor companies reducing their dependence on in-house fabrication, has led to significant changes in design practices. Fabrication processes have become more standardized with a shrink in the number of potential suppliers; the issue of *second sourcing* has changed its form. It no longer relies on second source fabrication of a single design; rather, these days, systems companies change chip sets for standard functions such as modems or application processors, either by accepting complete systems from providers (e.g., modems) or by using software layers to differentiate their products and recompile it on new application processors. Thus, differentiation is more by software than hardware design. As processes moved down in scale, fewer and fewer companies could afford to stay on the fabrication process treadmill. Several consortia were formed to develop common processes, but this eventually became unaffordable for all but a few companies.

■ *Growing importance of fabless semiconductor industry*: As system companies switched away from in-house process development and fabrication facilities, we have seen the rise of the fabless semiconductor industry and the pure-play foundries, the most notable being TSMC [28]. This has kept alive the dream of ASIC design in system companies, fostered an intermediate fabless semiconductor network of companies (such as Qualcomm, Broadcom, and many more), allowed some semiconductor companies to run a mixed fab/fabless business model, and standardized design processes, which helps design teams deal with the complexity of design in advanced processes at 28 nm and below. It has also fostered the growth of companies, such as eSilicon, that act as intermediaries between fabless IC companies that focus on RTL and above and fabrication. This involves essentially driving the back-end tools, developing a special relationship with fabrication that ensures faster access to technology information, and shielding fabless IC companies from the more intimate knowledge of technology. Keeping this separate from the IP business has also helped reduce the fabless industry concerns about design services.

■ *Consolidation in the EDA industry*: It has become increasingly difficult to fund EDA startups, and the "Big 3" of EDA (Synopsis, Cadence, and Mentor Graphics) have continued buying many of the small startups left in the industry over the last decade, including Forte, Jasper, Springsoft, EVE, and Nimbic [29]. For Cadence and Synopsis, substantial recent mergers and acquisitions include considerable amounts of design IP and IP design teams. This consolidation has good points and bad points in terms of the design process and tools and flows. On the bad side, it may limit some of the innovation that small EDA startups bring to the industry. On the good side, the larger EDA companies have more to invest in new tools and capabilities and flows, are able to work more closely with large IP providers and the various fabless and integrated (with fabrication facilities) semiconductor companies, and can make strategic investments to match tool capabilities to future process generations. In addition, by branching out to include IP as part of the EDA business model, the tools can be improved with in-house designs and the IP offerings themselves have the backing of large EDA companies and their strategic investments. This tends to increase the variety and quality of the IP available to design teams and offers them more choices.

■ *Relative geographic shifts in design activity*: Over the years there has been a tremendous reorganization of design activity. In a geographic sense, design activity in Europe and North America has declined (Europe more than North America), Japan has declined to a lesser extent, and the rest of Asia—Taiwan, Korea, and China—has increased. In India, design activity has also increased. This has had some effect on the adoption and deployment of advanced design methodologies and tools. In particular, electronic system-level (ESL) tools and methods have been of greater interest in Europe and to some extent in North America. This is fostered in Europe by EU programs that fund advanced work in microelectronics and design. The geographic shift has meant that ESL methodologies have had slower adoption than might have been the case if European design activity had remained as strong. However, we have seen more interest in the last few years in the use of system-level methods and models in Asian design companies, ameliorating some of this delay. Japan may be an exception to this, with early and continued interest in ESL.

- *Advanced processes*: The development in process technologies—from 90 nm through 65 nm, 40/45 nm, 28/30/32 nm, 20/22 nm, to 14/16 nm, including new circuit designs such as FinFETs—has continued apace over the last decade, albeit at a somewhat slower pace than traditional Moore's law would indicate (perhaps taking 3 years rather than 2 years to double transistor count). Inherent speeds of transistors no longer increase at the old pace. Some reports claim that even the cost per transistor has reached its minimum value around the 32–22 nm nodes. If true, this would mean that functionality no longer becomes cheaper with every generation and that a completely different strategy would be required in the whole electronic industry. More advanced design techniques are required. Design teams need to grow larger and integrate more diverse subsystems into complex SoCs. Design closure becomes harder and the number of steps required at each phase of design grows.

- *Changes in SoC architectures and more platforms*: The prediction made in 1999 of the move to a platform-based design style for SoCs [30] has come true; almost all complex SoCs follow some kind of platform-based design style in which most new products are controlled derivatives of the platform. This has shifted the hardware/software divide toward relatively fewer custom hardware blocks (only where their performance, power, or area characteristics are vital to the product function) and relatively more processors (whether general purpose or application specific) and memory (for all that software and associated data).

- *Greater use of FPGAs versus ASICs*: Although FPGA vendors have long proclaimed the near-complete takeover of FPGAs from ASICs in electronics product design, it has not unfolded according to their optimistic predictions [31]. However, FPGAs are used more in end-product delivery these days than a decade ago, at least in some classes of products. The new FPGAs with hardened embedded processors and other hard elements such as multiply-accumulate DSP blocks and SERDES look more attractive as product development and delivery vehicles than pure FPGAs. A notable example of this is Xilinx's Zynq platform FPGAs [32]. Use of such design platforms changes aspects of the design process rather significantly.

- *Greater use of IP blocks*: The use of design IP blocks and subsystems, whether sourced from internal design groups or external IP providers, has grown significantly. While reducing new design effort on complex SoCs, the use of IP blocks also changes much of the focus of SoC from new original design to architecting the use of IP and the integration and verification process itself.

- *Greater use of processors and application-specific instruction set processors* (*ASIPs*): The kind of IP block has a big influence on SoC design. Increasingly, platforms are using more processors, both general-purpose control plane processors and dedicated DSPs, accelerators, and ASIPs in the dataplane. In some domains, such as audio decoding, ASIPs are so close in performance, energy consumption, and cost (area) to dedicated hardware blocks, in which their flexibility in supporting multiple standards and new standards makes their use almost obligatory. The shifting boundary between dedicated hardware blocks and processors moves the design boundary between hardware and software in general, thus changing in degree (but not in kind) the design effort devoted to the software side.

- *Growing reliance on software*: This is a natural outgrowth of the increased use of processors discussed earlier. And the shift of the hardware/software divide changes the emphasis in design teams. This handbook, particularly the design process in this chapter, emphasizes hardware design of SoCs, of course.

- *High-level synthesis*: When digital hardware blocks do need to be designed for an SoC, the traditional design methodology has been to design them at RTL level and then use logic synthesis, verification, and normal digital implementation flows to integrate them into SoCs. High-level synthesis has gone through interesting historical evolutions since the late 1980s at least [33], including considerable interest in the mid-1990s using hardware description languages as the input mechanism, a collapse, and growth in the more recent decade using C/C++/SystemC as inputs. It is now used as a serious part of the design flow for various complex hardware blocks, especially those in the dataplane (e.g., various kinds of signal processing blocks). The chapter in the handbook on high-level synthesis is a good reference to the current state of the art here.

■ *Verification growth and change*: Of course, as the SoC design grows in complexity, the verification challenge grows superlinearly. Many methods, some quite old, have seen wider adoption in the last decade. These include formal verification (both equivalence checking and property checking), use of hardware verification engines and emulators, hardware–software coverification, and virtual platforms. In addition, regular RTL and gate-level simulation has exploded for large SoCs. Many chapters in the EDA Handbook touch on these topics.

■ *Back-end design flow*: Several important changes have occurred in back-end design tools and flow. In particular, advances in high-performance place and route, gate sizing, power-grid optimization, and clock gating and multiobjective optimizations are of major impact. For the most important optimizations, industry released large modern benchmark suites and organized multimonth research contests for graduate students. As a result, we now have much better understanding as to which methods work best and why. For participating teams, contests encourage sharper focus on empirical performance and allow researchers to identify simple yet impactful additions to mainstream techniques that would have been difficult to pursue otherwise. In most cases, the technique winning at the contest has been adopted in the industry within 2–3 years and often developed further by other academic researchers. Multiobjective optimization as a domain requires a keen understanding of available components and their empirical performance, as well as rigorous experimentation. It holds the potential for significant improvement in the back-end flow.

■ *More Moore and more than Moore*: Despite the imminent predictions of its demise, Moore's law (or observation) [34] continues apace with each new technology generation. The time required for the number of gates on ICs to double may have stretched beyond the 1 year in the original observation, to 2 years in the 1970s and 3 years in the 2000s. Overall, process speed and thus the maximum MHz achievable with each process generation may have slowed down or stopped (or be limited by power concerns). SoCs are more complex in achievable gate count at 28 nm than at 40 nm, and a greater number of gates are possible at 16 nm. In addition, we have seen new process technologies—such as 3D FinFETs—emerge, which begin to fulfill the "more than Moore" predictions. Whether carbon nanotubes and transistors and gates built on them will become the workhorse technology at 7 or 5 nm or beyond is not clear, but something more than Moore is certainly emerging. These changes clearly influence the design process.

Although many changes in design trends have an impact on the design process, it retains many of the key characteristics and steps of a decade ago. The cumulative effect changes the emphasis on particular design disciplines and moves the dividing lines between them. Fundamentally, it has not changed the nature of the IC design process. This is likely to remain so in the future.

REFERENCES

1. M. Smith, *Application Specific Integrated Circuits*, Addison-Wesley, Reading, MA, 1997.
2. A. Kuehlmann, *The Best of ICCAD*, Kluwer Academic Publishers, Dordrecht, the Netherlands, 2003.
3. D. Thomas and P. Moorby, *The Verilog Hardware Description Language*, Kluwer Academic Publishers, Dordrecht, the Netherlands, 1996.
4. D. Pellerin and D. Taylor, *VHDL Made Easy*, Pearson Education, Upper Saddle River, NJ, 1996.
5. S. Sutherland, S. Davidson, and P. Flake, *SystemVerilog for Design: A Guide to Using SystemVerilog for Hardware Design and Modeling*, Kluwer Academic Publishers, Dordrecht, the Netherlands, 2004.
6. T. Groetker, S. Liao, G. Martin, and S. Swan, *System Design with SystemC*, Kluwer Academic Publishers, Dordrecht, the Netherlands, 2002.
7. G. Peterson, P. Ashenden, and D. Teegarden, *The System Designer's Guide to VHDL-AMS*, Morgan Kaufman Publishers, San Francisco, CA, 2002.
8. K. Kundert and O. Zinke, *The Designer's Guide to Verilog-AMS*, Kluwer Academic Publishers, Dordrecht, the Netherlands, 2004.

9. M. Keating and P. Bricaud, *Reuse Methodology Manual for System-on-a-Chip Designs*, Kluwer Academic Publishers, Dordrecht, the Netherlands, 1998.
10. J. Bergeron, *Writing Testbenches*, Kluwer Academic Publishers, Dordrecht, the Netherlands, 2003.
11. E. Clarke, O. Grumberg, and D. Peled, *Model Checking*, The MIT Press, Cambridge, MA, 1999.
12. S. Huang and K. Cheng, *Formal Equivalence Checking and Design Debugging*, Kluwer Academic Publishers, Dordrecht, the Netherlands, 1998.
13. R. Baker, H. Li, and D. Boyce, *CMOS Circuit Design, Layout, and Simulation*, Series on Microelectronic Systems, IEEE Press, New York, 1998.
14. L. Pillage, R. Rohrer, and C. Visweswariah, *Electronic Circuit and System Simulation Methods*, McGraw-Hill, New York, 1995.
15. J. Elliott, *Understanding Behavioral Synthesis: A Practical Guide to High-Level Design*, Kluwer Academic Publishers, Dordrecht, the Netherlands, 2000.
16. S. Devadas, A. Ghosh, and K. Keutzer, *Logic Synthesis*, McGraw-Hill, New York, 1994.
17. G. DeMicheli, *Synthesis and Optimization of Digital Circuits*, McGraw-Hill, New York, 1994.
18. I. Sutherland, R. Sproull, and D. Harris, *Logical Effort: Defining Fast CMOS Circuits*, Academic Press, New York, 1999.
19. N. Sherwani, *Algorithms for VLSI Physical Design Automation*, Kluwer Academic Publishers, Dordrecht, the Netherlands, 1999.
20. F. Nekoogar, *Timing Verification of Application-Specific Integrated Circuits (ASICs)*, Prentice-Hall PTR, Englewood Cliffs, NJ, 1999.
21. K. Roy and S. Prasad, *Low Power CMOS VLSI: Circuit Design*, Wiley, New York, 2000.
22. C.-K.Cheng, J. Lillis, S. Lin, and N. Chang, *Interconnect Analysis and Synthesis*, Wiley, New York, 2000.
23. W. Dally and J. Poulton, *Digital Systems Engineering*, Cambridge University Press, Cambridge, U.K., 1998.
24. M. Abramovici, M. Breuer, and A. Friedman, *Digital Systems Testing and Testable Design*, Wiley, New York, 1995.
25. A. Wong, *Resolution Enhancement Techniques in Optical Lithography*, SPIE Press, Bellingham, WA, 2001.
26. International Technology Roadmap for Semiconductors (ITRS), 2004, http://public.itrs.net/.
27. International Technology Roadmap for Semiconductors (ITRS), 2012 update, http://www.itrs.net/Links/2012ITRS/Home2012.htm.
28. D. Nenni and P. McLennan, *Fabless: The Transformation of the Semiconductor Industry*, semiwiki.com, 2014.
29. A good source of information on EDA mergers and acquisitions, https://www.semiwiki.com/forum/showwiki.php?title=Semi+Wiki:EDA+Mergers+and+Acquisitions+Wiki.
30. H. Chang, L. Cooke, M. Hunt, G. Martin, A. McNelly, and L. Todd, *Surviving the SOC Revolution: A Guide to Platform-Based Design*, Kluwer Academic Publishers (now Springer), Dordrecht, the Netherlands, 1999.
31. I. Bolsens, FPGA, a future proof programmable system fabric, Talk given at Georgia Tech, Atlanta, GA, March 2005, slides available at: http://users.ece.gatech.edu/limsk/crest/talks/georgiafinal.pdf. (Accessed on October, 2014).
32. I. Bolsens, The all programmable SoC—At the heart of next generation embedded systems, *Electronic Design Process Symposium*, Monterey, CA, April 2013, slides available at: http://www.eda.org/edps/edp2013/Papers/Keynote%200%20FINAL%20for%20Ivo%20Bolsens.pdf. (Accessed on October, 2014).
33. G. Martin and G. Smith, High-level synthesis: Past, present, and future, *IEEE Design and Test*, 26(4), 18–25, July 2009.
34. A good writeup is at http://en.wikipedia.org/wiki/Moore%27s_law, including a reference to the original article.

System-Level Design

Tools and Methodologies for System-Level Design

3

Shuvra Bhattacharyya and Marilyn Wolf

CONTENTS

3.1 INTRODUCTION

System-level design, once the province of board designers, has now become a central concern for chip designers. Because chip design is a less forgiving design medium—design cycles are longer and mistakes are harder to correct—system-on-chip (SoC) designers need a more extensive tool suite than may be used by board designers, and a variety of tools and methodologies have been developed for system-level design of SoCs.

System-level design is less amenable to synthesis than are logic or physical design. As a result, system-level tools concentrate on *modeling, simulation, design space exploration,* and *design verification.* The goal of modeling is to correctly capture the system's operational semantics, which helps with both implementation and verification. The study of models of computation provides a framework for the description of digital systems. Not only do we need to understand a particular style of computation, such as dataflow, but we also need to understand how different models of communication can reliably communicate with each other. Design space exploration tools, such as hardware/software codesign, develop candidate designs to understand trade-offs. Simulation can be used not only to verify functional correctness but also to supply performance and power/energy information for design analysis.

We will use video applications as examples in this chapter. Video is a leading-edge application that illustrates many important aspects of system-level design. Although some of this information is clearly specific to video, many of the lessons translate to other domains.

The next two sections briefly introduce video applications and some SoC architectures that may be the targets of system-level design tools. We will then study models of computation and languages for system-level modeling. We will then survey simulation techniques. We will close with a discussion of hardware/software codesign.

3.2 CHARACTERISTICS OF VIDEO APPLICATIONS

One of the primary uses of SoCs for multimedia today is for video encoding—both compression and decompression. In this section, we review the basic characteristics of video compression algorithms and the implications for video SoC design.

Video compression standards enable video devices to interoperate. The two major lines of video compression standards are MPEG and H.26x. The MPEG standards concentrate on broadcast applications, which allow for a more expensive compressor on the transmitter side in exchange for a simpler receiver. The H.26x standards were developed with videoconferencing in mind, in which both sides must encode and decode. The Advanced Video Codec standard, also known as H.264, was formed by the confluence of the H.26x and MPEG efforts. H.264 is widely used in consumer video systems.

Modern video compression systems combine lossy and lossless encoding methods to reduce the size of a video stream. Lossy methods throw away information such that the uncompressed video stream is not a perfect reconstruction of the original; lossless methods do allow the information provided to them to be perfectly reconstructed. Most modern standards use three major mechanisms:

1. The discrete cosine transform (DCT) together with quantization
2. Motion estimation and compensation
3. Huffman-style encoding

Quantized DCT and motion estimation are lossy, while Huffman encoding is lossless. These three methods leverage different aspects of the video stream's characteristics to more efficiently encode it.

The combination of DCT and quantization was originally developed for still images and is used in video to compress single frames. The DCT is a frequency transform that turns a set of pixels into a set of coefficients for the spatial frequencies that form the components of the image represented by the pixels. The DCT is used over other transforms because a 2D DCT can be computed using two 1D DCTs, making it more efficient. In most standards, the DCT is performed on an 8 × 8 block of pixels. The DCT does not itself lossily compress the image; rather, the quantization phase can more easily pick out information to throw away thanks to the structure of the DCT. Quantization throws out fine detail in the block of pixels, which correspond to the high-frequency coefficients in the DCT. The number of coefficients set to zero is determined on the level of compression desired.

Motion estimation and compensation exploit the relationships between frames provided by moving objects. A reference frame is used to encode later frames through a motion vector, which describes the motion of a macroblock of pixels (16 × 16 in many standards). The block is copied from the reference frame into the new position described by the motion vector. The motion vector is much smaller than the block it represents. Two-dimensional correlation is used to determine the position of the macroblock's position in the new frame; several positions in a search area are tested using 2D correlation. An error signal encodes the difference between the predicted and the actual frames; the receiver uses that signal to improve the predicted picture.

MPEG distinguishes several types of frames: I (inter) frames are not motion compensated, P (predicted) frames have been predicted from earlier frames, and B (bidirectional) frames have been predicted from both earlier and later frames.

The results of these lossy compression phases are assembled into a bit stream and compressed using lossless compression such as Huffman encoding. This process reduces the size of the representation without further compromising image quality.

It should be clear that video compression systems are actually heterogeneous collections of algorithms; this is also true of many other applications, including communications and security. A video computing platform must run several algorithms; those algorithms may perform very different types of operations, imposing very different requirements on the architecture.

This has two implications for tools: first, we need a wide variety of tools to support the design of these applications; second, the various models of computation and algorithmic styles used in different parts of an application must at some point be made to communicate to create the complete system. For example, DCT can be formulated as a dataflow problem, thanks to its butterfly computational structure, while Huffman encoding is often described in a control-oriented style.

Several studies of multimedia performance on programmable processors have remarked on the significant number of branches in multimedia code. These observations contradict the popular notion of video as regular operations on streaming data. Fritts and Wolf [1] measured the characteristics of the MediaBench benchmarks.

They used path ratio to measure the percentage of instructions in a loop body that were actually executed. They found that the average path ratio of the MediaBench suite was 78%, which indicates that a significant number of loops exercise data-dependent behavior. Talla et al. [2] found that most of the available parallelism in multimedia benchmarks came from interiteration parallelism. Exploiting the complex parallelism found in modern multimedia algorithms requires that synthesis algorithms be able to handle more complex computations than simple ideal nested loops.

3.3 PLATFORM CHARACTERISTICS

Many SoCs are heterogeneous multiprocessors and the architectures designed for multimedia applications are no exception. In this section, we review several SoCs, including some general-purpose SoC architectures as well as several designed specifically for multimedia applications.

Two very different types of hardware platforms have emerged for large-scale applications. On the one hand, many custom SoCs have been designed for various applications. These custom SoCs are customized by loading software onto them for execution. On the other hand, platform field-programmable gate arrays (FPGAs) provide FPGA fabrics along with CPUs and other components; the design can be customized by programming the FPGA as well as the processor(s). These two styles of architecture represent different approaches for SoC architecture and they require very different sorts of tools: custom SoCs require large-scale software support, while platform FPGAs are well suited to hardware/software codesign.

3.3.1 CUSTOM SYSTEM-ON-CHIP ARCHITECTURES

The TI OMAP family of processors [3] is designed for audio, industrial automation, and portable medical equipment. All members of the family include a C674x DSP; some members also include an ARM9 CPU.

The Freescale MPC574xP [4] includes 2 e200z4 CPUs operating in delayed lock step for safety checking as well as an embedded floating point unit.

3.3.2 GRAPHICS PROCESSING UNITS

GPUs are widely used in desktop and laptop systems as well as smartphones. GPUs are optimized for graphics rendering but have been applied to many other algorithms as well. GPUs provide SIMD-oriented architectures with floating-point support.

Figure 3.1 shows the organization of the NVIDIA Fermi [5]. Three types of processing elements are provided: cores, each of which has a floating point and an integer unit, load/store units, and special function units. A hierarchy of register files, caches, and shared memory provides very high memory bandwidth. A pair of warp processors controls operation. Each warp scheduler can control a set of 32 parallel threads.

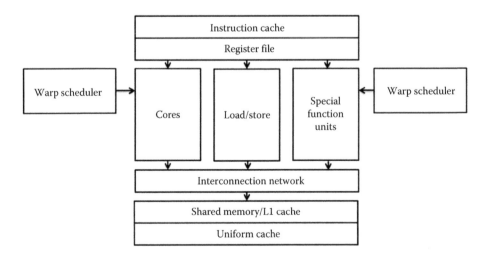

FIGURE 3.1   Organization of the Fermi graphics processing unit.

3.3.3 PLATFORM FPGAs

FPGAs [6] have been used for many years to implement logic designs. The FPGA provides a more general structure than programmable logic devices, allowing denser designs. They are less energy efficient than custom ASICs but do not require the long application specific integrated circuit (ASIC) design cycle.

Many FPGA design environments provide small, customizable *soft processors* that can be embedded as part of the logic in an FPGA. Examples include the Xilinx MicroBlaze and Altera Nios II. Nios II supports memory management and protection units, separate instruction and data caches, pipelined execution with dynamic branch prediction (up to 256 custom instructions), and hardware accelerators. MicroBlaze supports memory management units, instruction and data caches, pipelined operation, floating-point operations, and instruction set extensions.

The term "programmable SoC" refers to an FPGA that provides one or more hard logic CPUs in addition to a programmable FPGA fabric. Platform FPGAs provide a very different sort of heterogeneous platform than custom SoCs. The FPGA fabric allows the system designer to implement new hardware functions. While they generally do not allow the CPU itself to be modified, the FPGA logic is a closely coupled device with high throughput to the CPU and to memory. The CPU can also be programmed using standard tools to provide functions that are not well suited to FPGA implementation. For example, the Xilinx Zynq UltraScale+ family of multiprocessor systems-on-chips [7] includes an FPGA logic array, a quad-core ARM MPCore, dual-core ARM Cortex-R5, graphics processing unit, and DRAM interface.

3.4 ABSTRACT DESIGN METHODOLOGIES

Several groups have developed abstract models for system-level design methodologies. These models help us to compare concrete design methodologies.

An early influential model for design methodologies was the *Gajski–Kuhn Y-chart* [8]. As shown in Figure 3.2, the model covers three types of refinement (structural, behavioral, and physical) at four levels of design abstraction (transistor, gate, register transfer, and system). A design methodology could be viewed as a trajectory through the Y-chart as various refinements are performed at different levels of abstraction.

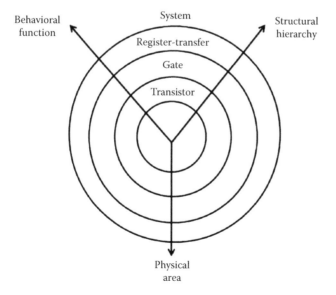

FIGURE 3.2 The Y-chart model for design methodologies. (From Gajski, D.D. and Kuhn, R.H., *Computer*, 16(12), 11, 1983.)

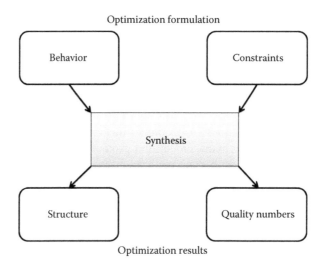

FIGURE 3.3   The X-chart model for design methodologies. (From Gerstlauer, A. et al., *IEEE Trans. Comput. Aided Design Integr. Circuits Syst.*, 28(10), 1517, 2009.)

The X-chart model [9] has been proposed as a model for SoC design methodologies. As shown in Figure 3.3, a system specification is given by the combination of a behavioral description that describe the system function and a set of constraints that describes the nonfunctional requirements on the design. A synthesis procedure generates a structural implementation and a set of quality numbers by which the structure can be judged.

3.5 MODEL-BASED DESIGN METHODOLOGIES

Increasingly, developers of hardware and software for embedded computer systems are viewing aspects of the design and implementation processes in terms of domain-specific models of computation. Models of computation provide formal principles that govern how functional components in a computational specification operate and interact (e.g., see Reference 10). A domain-specific model of computation is designed to represent applications in a particular functional domain such as DSP and image and video processing; control system design; communication protocols or more general classes of discrete, control flow intensive decision-making processes; graphics; and device drivers. For discussions of some representative languages and tools that are specialized for these application domains, see, for example, [11–22]. For an integrated review of domain-specific programming languages for embedded systems, see Reference 23.

Processors expose a low-level Turing model at the instruction set. Traditional *high-level* programming languages like C, C++, and Java maintain the essential elements of that Turing model, including imperative semantics and memory-oriented operation. Mapping the semantics of modern, complex applications onto these low-level models is both time consuming and error prone. As a result, new programming languages and their associated design methodologies have been developed to support applications such as signal/image processing and communications. Compilers for these languages provide correct-by-construct translation of application-level operations to the Turing model, which both improves designer productivity and provides a stronger, more tool-oriented verification path [24].

3.5.1 DATAFLOW MODELS

For most DSP applications, a significant part of the computational structure is well suited to modeling in a dataflow model of computation. In the context of programming models, dataflow refers to a modeling methodology where computations are represented as directed graphs in which vertices (actors) represent functional components and edges between actors represent first-in-first-out (FIFO) channels that buffer data values (tokens) as they pass from an output of one actor to an input of another. Dataflow actors can represent computations of arbitrary complexity; typically in DSP design environments, they are specified using conventional languages such as C or assembly language, and their associated tasks range from simple, *fine-grained* functions such as addition and multiplication to *coarse-grain* DSP kernels or subsystems such as FFT units and adaptive filters.

The development of application modeling and analysis techniques based on dataflow graphs was inspired significantly by the computation graphs of Karp and Miller [25] and the process networks of Kahn [26]. A unified formulation of dataflow modeling principles as they apply to DSP design environment is provided by the dataflow process networks model of computation of Lee and Parks [27].

A dataflow actor is enabled for execution any time it has sufficient data on its incoming edges (i.e., in the associated FIFO channels) to perform its specified computation. An actor can execute at any time when it is enabled (data-driven execution). In general, the execution of an actor results in some number of tokens being removed (consumed) from each incoming edge and some number being placed (produced) on each outgoing edge. This production activity in general leads to the enabling of other actors.

The order in which actors execute, called the "schedule," is not part of a dataflow specification and is constrained only by the simple principle of data-driven execution defined earlier. This is in contrast to many alternative computational models, such as those that underlie procedural languages, in which execution order is *overspecified* by the programmer [28]. The schedule for a dataflow specification may be determined at compile time (if sufficient static information is available), at run time, or when using a mixture of compile-time and run-time techniques. A particularly powerful class of scheduling techniques, referred to as "quasi-static scheduling" (e.g., see Reference 29), involves most, but not all, of the scheduling decisions being made at compile time.

Figure 3.4 shows an illustration of a video processing subsystem that is modeled using dataflow semantics. This is a design, developed using the Ptolemy II tool for model-based embedded system design [30], of an MPEG-2 subsystem for encoding the P frames that are processed

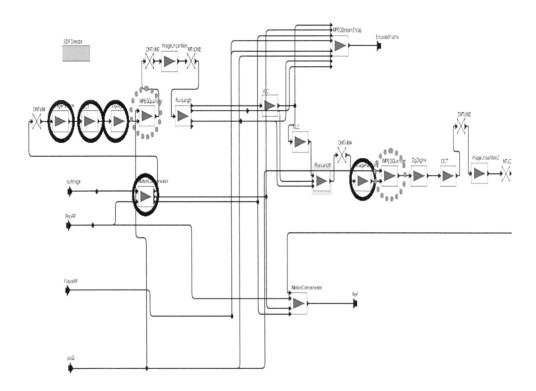

FIGURE 3.4   A video processing subsystem modeled in dataflow.

by an enclosing MPEG-2 encoder system. A thorough discussion of this MPEG-2 system and its comparison to a variety of other modeling representations is presented in Reference 31. The components in the design of Figure 3.4 include actors for the DCT, zigzag scanning, quantization, motion compensation, and run length coding. The arrows in the illustration correspond to the edges in the underlying dataflow graph.

The actors and their interactions all conform to the semantics of *synchronous dataflow* (SDF), which is a restricted form of dataflow that is efficient for describing a broad class of DSP applications and has particularly strong formal properties and optimization advantages [32,33]. Specifically, SDF imposes the restriction that the number of tokens produced and consumed by each actor on each incident edge is constant. Many commercial DSP design tools have been developed that employ semantics that are equivalent to or closely related to SDF. Examples of such tools include Agilent's SystemVue, Kalray's MPPA Software Development Kit, National Instrument's LabVIEW, and Synopsys's SPW. Simulink®, another widely used commercial tool, also exhibits some relationships to the SDF model (e.g., see Reference 34).

3.5.2 DATAFLOW MODELING FOR VIDEO PROCESSING

In the context of video processing, SDF permits accurate representation of many useful subsystems, such as the P-frame encoder shown in Figure 3.4. However, such modeling is often restricted to a highly coarse level of granularity, where actors process individual frames or groups of successive frames on each execution. Modeling at such a coarse granularity can provide compact, top-level design representations, but greatly limits the benefits offered by the dataflow representation since most of the computation is subsumed by the general-purpose, intra-actor program representation. For example, the degree of parallel processing and memory management optimizations exposed to a dataflow-based synthesis tool becomes highly limited at such coarse levels of actor granularity. An example of a coarse-grain dataflow actor that "hides" significant amounts of parallelism is the DCT actor as shown in Figure 3.4.

3.5.3 MULTIDIMENSIONAL DATAFLOW MODELS

A number of alternative dataflow modeling methods have been introduced to address this limitation of SDF modeling for video processing and, more generally, multidimensional signal processing applications. For example, the multidimensional SDF (MD-SDF) model extends SDF semantics to allow constant-sized, *n*-dimensional vectors of data to be transferred across graph edges and provides support for arbitrary sampling lattices and lattice-changing operations [35]. Intuitively, a sampling lattice can be viewed as a generalization to multiple dimensions of a uniformly spaced configuration of sampling points for a 1D signal [36]; hexagonal and rectangular lattices are examples of commonly used sampling lattices for 2D signals. The computer vision (CV) SDF model is designed specifically for CV applications and provides a notion of *structured buffers* for decomposing video frames along graph edges; accessing neighborhoods of image data from within actors, in addition to the conventional production and consumption semantics of dataflow; and allowing actors to efficiently access previous frames of image data [37,38]. Blocked dataflow is a metamodeling technique for efficiently incorporating hierarchical, block-based processing of multidimensional data into a variety of dataflow modeling styles, including SDF and MD-SDF [31]. Windowed SDF is a model of computation that deeply integrates support for sliding window algorithms into the framework of static dataflow modeling [39]. Such support is important in the processing of images and video streams, where sliding window operations play a fundamental role.

3.5.4 CONTROL FLOW

As described previously, modern video processing applications are characterized by some degree of control flow processing for carrying out data-dependent configuration of application tasks and changes across multiple application modes. For example, in MPEG-2 video encoding, significantly different processing is required for I frames, P frames, and B frames. Although the processing for each particular type of frame (I, P, or B) conforms to the SDF model, as illustrated for P frame processing in Figure 3.4, a layer of control flow processing is needed to efficiently integrate these three types of processing methods into a complete MPEG-2 encoder design. The SDF model is not well suited for performing this type of control flow processing and more generally for any functionality that requires dynamic communication patterns or activation/deactivation across actors.

A variety of alternative models of computation have been developed to address this limitation and integrate flexible control flow capability with the advantages of dataflow modeling. In Buck's Boolean dataflow model [40] and the subsequent generalization as integer-controlled dataflow [41], provisions for such flexible processing were incorporated without departing from the framework of dataflow, and in a manner that facilitates the construction of efficient quasi-static schedules. In Boolean dataflow, the number of tokens produced or consumed on an edge is either fixed or is a two-valued function of a control token present on a control terminal of the same actor. It is possible to extend important SDF analysis techniques to Boolean dataflow graphs by employing symbolic variables. In particular, in constructing a schedule for Boolean dataflow actors, Buck's techniques attempt to derive a quasi-static schedule, where each conditional actor execution is annotated with the run-time condition under which the execution should occur. Boolean dataflow is a powerful modeling technique that can express arbitrary control flow structures; however, as a result, key formal verification properties of SDF, such as bounded memory and deadlock detection, are lost (the associated analysis problems are not decidable) in the context of general Boolean dataflow specifications.

3.5.5 INTEGRATION WITH FINITE-STATE MACHINE AND MODE-BASED MODELING METHODS

In recent years, several modeling techniques have also been proposed that enhance expressive power by providing precise semantics for integrating dataflow or dataflow-like representations with finite-state machine (FSM) models and related methods for specifying and transitioning between different modes of actor behavior. These include El Greco [42], which evolved into the Synopsys System Studio and provides facilities for *control models* to dynamically configure specification parameters, *charts (pronounced "starcharts") with heterochronous dataflow as the concurrency model [43], the FunState intermediate representation [44], the DF* framework developed at K. U. Leuven [45], the control flow provisions in bounded dynamic dataflow [46], enable-invoke dataflow [47], and scenario-aware dataflow (SADF) [48].

3.5.6 VIDEO PROCESSING EXAMPLES

Figure 3.5 shows an illustration of a model of a complete MPEG-2 video encoder system that is constructed using Ptolemy, builds on the P-frame-processing subsystem of Figure 3.4, and employs multiple dataflow graphs nested within a FSM representation. Details on this application model can be found in Reference 31.

Figure 3.6 shows a block diagram, adapted from Reference 48, of an MPEG-4 decoder that is specified in terms of *SADF*. Descriptive names for the actors in this example are listed in Table 3.1, along with their SADF component types, which are either shown as "K" for kernel or "D" for detector. Intuitively, kernels correspond to data processing components of the enclosing dataflow graph, while detectors are used for control among different modes of operation (*scenarios*) for the kernels. The FD actor in the example of Figure 3.6 determines the frame type (I or P frame) and is designed as a detector. The other actors in the specification are kernels. For more details on this example and the underlying SADF model of computation, we refer the reader to Reference 48. The "D" symbols that appear next to some of the edges in Figure 3.6 represent dataflow *delays*, which correspond to initial tokens on the edges.

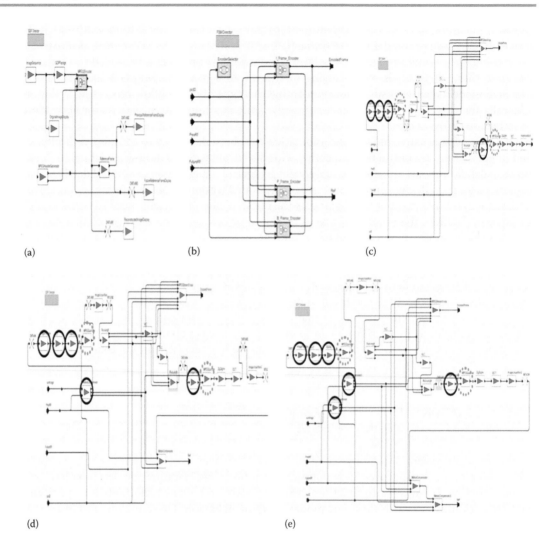

(a)   (b)   (c)

(d)   (e)

FIGURE 3.5   An MPEG-2 video encoder specification. (a) MPEG2 encoder (top); (b) inside the FSM; (c) I frame encoder; (d) P frame encoder; (e) B frame encoder.

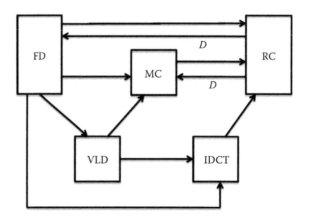

FIGURE 3.6   A block diagram of an MPEG-4 decoder that is specified in terms of scenario-aware dataflow. (Adapted from Theelen, B.D. et al., A scenario-aware data flow model for combined long-run average and worst-case performance analysis, in *Proceedings of the International Conference on Formal Methods and Models for Codesign*, Washington DC, 2006.)

TABLE 3.1   Modeling Components Used in the MPEG-4 Decoder Sample of Figure 3.6

Abbreviation	Descriptive Name	Type
FD	Frame detector	D
IDCT	Inverse discrete cosine transformation	K
MC	Motion compensation	K
RC	Reconstruction	K
VLD	Variable-length decoding	K

3.6 LANGUAGES AND TOOLS FOR MODEL-BASED DESIGN

In this section, we survey research on tools for model-based design of embedded systems, with an emphasis on tool capabilities that are relevant to video processing systems. We discuss several representative tools that employ established and experimental models of computation and provide features for simulation, rapid prototyping, synthesis, and optimization. For more extensive coverage of model-based design tools for video processing systems and related application areas, we refer the reader to Reference 19.

3.6.1 CAL

CAL is a language for dataflow programming that has been applied to hardware and software synthesis and a wide variety of applications, with a particular emphasis on applications in video processing [20]. One of the most important applications of CAL to date is the incorporation of a subset of CAL, called RVC-CAL, as part of the MPEG reconfigurable video coding (RVC) standard [49]. In CAL, dataflow actors are specified in terms of entities that include *actions*, *guards*, *port patterns*, and *priorities*. An actor can contain any number of actions, where each action describes a specific computation that is to be performed by the actor, including the associated consumption and production of tokens at the actor ports, when the action is executed. Whether or not an action can be executed at any given time depends in general on the number of available input tokens, the token values, and the actor state. These *fireability* conditions are specified by input patterns and guards of the action definition. The relatively high flexibility allowed for constructing firing conditions makes CAL a very general model, where fundamental scheduling-related problems become undecidable, as with Boolean dataflow and other highly expressive, "dynamic dataflow" models.

Input patterns also declare variables that correspond to input tokens that are consumed when the action executes and that can be referenced in specifying the computation to be performed by the action. Such deep integration of dataflow-based, actor interface specification with specification of the detailed internal computations performed by an actor is one of the novel aspects of CAL.

Priorities in CAL actor specifications provide a way to select subsets of actions to execute when there are multiple actions that match the fireability conditions defined by the input patterns and guards. For more details on the CAL language, we refer the reader to the CAL language report [50].

A wide variety of tools has been developed to support design of hardware and software systems using CAL. For example, OpenDF was introduced in Reference 22 as an open-source simulation and compilation framework for CAL; the open RVC-CAL compiler (Orcc) is an open-source compiler infrastructure that enables code generation for a variety of target languages and platforms [51]; Boutellier et al. present a plug-in to OpenDF for multiprocessor scheduling of RVC systems that are constructed using CAL actors [52]; and Gu et al. present a tool that automatically extracts and exploits statically schedulable regions from CAL specifications [53].

3.6.2 COMPAAN

MATLAB® is one of the most popular programming languages for algorithm development, and high-level functional simulation for DSP applications. In the Compaan project, developed at Leiden University, systematic techniques have been developed for synthesizing embedded software and FPGA-based hardware implementations from a restricted class of MATLAB programs known as parameterized, static nested loop programs [54]. In Compaan, an input MATLAB specification is first translated into an intermediate representation based on the Kahn process network model of computation [26]. The Kahn process network model is a general model of data-driven computation that subsumes as a special case the dataflow process networks mentioned earlier in this chapter. Like dataflow process networks, Kahn process networks consist of concurrent functional modules that are connected by FIFO buffers with nonblocking writes and blocking reads; however, unlike the dataflow process network model, modules in Kahn process networks do not necessarily have their execution decomposed a priori into well-defined, discrete units of execution [27].

Through its aggressive dependence analysis capabilities, Compaan combines the widespread appeal of MATLAB at the algorithm development level with the guaranteed determinacy, compact representation, simple synchronization, and distributed control features of Kahn process networks for efficient hardware/software implementation.

Technically, the Kahn process networks derived in Compaan can be described as equivalent cyclo-static dataflow graphs [55,56] and therefore fall under the category of dataflow process networks. However, these equivalent cyclo-static dataflow graphs can be very large and unwieldy to work with, and therefore, analysis in terms of the Kahn process network model is often more efficient and intuitive.

The development of the capability for translation from MATLAB to Kahn process networks was originally developed by Kienhuis, Rijpkema, and Deprettere [57], and this capability has since evolved into an elaborate suite of tools for mapping Kahn process networks into optimized implementations on heterogeneous hardware/software platforms consisting of embedded processors and FPGAs [54]. Among the most interesting optimizations in the Compaan tool suite are dependence analysis mechanisms that determine the most specialized form of buffer implementation, with respect to reordering and multiplicity of buffered values, for implementing interprocess communication in Kahn process networks [58].

Commercialization of the Compaan Technology is presently being explored as part of a Big Data effort in the field of astronomy. The Compaan tool set is used to program multi-FPGA boards from C-code for real-time analysis of astronomy data.

At Leiden University, Compaan has been succeeded by the Daedalus project, which provides a design flow for mapping embedded multimedia applications onto multiprocessor SoC devices [59].

3.6.3 PREESM

Parallel and Real-time Embedded Executives Scheduling Method (PREESM) is an extensible, Eclipse-based framework for rapid programming of signal processing systems [21,60]. Special emphasis is placed in PREESM for integrating and experimenting with different kinds of multiprocessor scheduling techniques and associated target architecture models. Such modeling and experimentation is useful in the design and implementation of real-time video processing systems, which must often satisfy stringent constraints on latency, throughput, and buffer memory requirements.

Various types of tools for compilation, analysis, scheduling, and architecture modeling can be integrated into PREESM as Eclipse [61] plug-ins. Existing capabilities of PREESM emphasize the use of architecture models and scheduling techniques that are targeted to mapping applications onto Texas Instruments programmable digital signal processors, including the TMS320C64x+ series of processors. Applications are modeled in PREESM using SDF graphs, while target architectures are modeled as interconnections of abstracted processor cores, hardware coprocessors, and communication media. Both homogeneous and heterogeneous architectures can be modeled, and emphasis also is placed on careful modeling of DMA-based operation associated with the communication media.

Multiprocessor scheduling of actors in PREESM is performed using a form of list scheduling. A randomized version of the list scheduling algorithm is provided based on probabilistic generation of refinements to the schedule derived by the basic list scheduling technique. This randomized version can be executed for an arbitrary amount of time, as determined by the designer, after which the best solution observed during the entire execution is returned. Additionally, the randomized scheduling algorithm can be used to initialize the population of a genetic algorithm, which provides a third alternative for multiprocessor scheduling in PREESM. A plug-in for *edge scheduling* is provided within the scheduling framework of PREESM to enable the application of alternative methods for mapping interprocessor communication operations across the targeted interconnection of communication media.

Pelcat et al. present a study involving the application of PREESM to rapid prototyping of a stereo vision system [62].

3.6.4 PTOLEMY

The Ptolemy project at U.C. Berkeley has had considerable influence on the general trend toward viewing embedded systems design in terms of models of computation [30,41]. The design of Ptolemy emphasizes efficient modeling and simulation of embedded systems based on the interaction of heterogeneous models of computation. A key motivation is to allow designers to represent each subsystem of a design in the most natural model of computation associated with that subsystem, and allow subsystems expressed in different models of computation to be integrated seamlessly into an overall system design.

A key constraint imposed by the Ptolemy approach to heterogeneous modeling is the concept of *hierarchical heterogeneity* [63]. It is widely understood that in hierarchical modeling, a system specification is decomposed into a set C of subsystems in which each subsystem can contain one or more hierarchical components, each of which represents another subsystem in C. Under hierarchical heterogeneity, each subsystem in C must be described using a uniform model of computation, but the nested subsystem associated with a hierarchical component H can be expressed (refined) in a model of computation that is different from the model of computation that expresses the subsystem containing H.

Thus, under hierarchical heterogeneity, the integration of different models of computation must be achieved entirely through the hierarchical embedding of heterogeneous models. A key consequence is that whenever a subsystem S_1 is embedded in a subsystem S_2 that is expressed in a different model of computation, the subsystem S_1 must be abstracted by a hierarchical component in S_2 that conforms to the model of computation associated with S_2. This provides precise constraints for interfacing different models of computation. Although these constraints may not always be easy to conform to, they provide a general and unambiguous convention for heterogeneous integration, and perhaps even more importantly, the associated interfacing methodology allows each subsystem to be analyzed using the techniques and tools available for the associated model of computation.

Ptolemy has been developed through a highly flexible, extensible, and robust software design, and this has facilitated experimentation with the underlying modeling capabilities by many research groups in various aspects of embedded systems design. Major areas of contribution associated with the development of Ptolemy that are especially relevant for video processing systems include hardware/software codesign, as well as contributions in dataflow-based modeling, analysis, and synthesis (e.g., see References 35, 64–66).

The current incarnation of the Ptolemy project, called Ptolemy II, is a Java-based tool that furthers the application of model-based design and hierarchical heterogeneity [30] and provides an even more malleable software infrastructure for experimentation with new techniques involving models of computation.

An important theme in Ptolemy II is the reuse of actors across multiple computational models. Through an emphasis in Ptolemy II on support for *domain polymorphism*, the same actor definition can in general be applicable across a variety of models of computation. In practice, domain polymorphism greatly increases the reuse of actor code. Techniques based on interface automata [67] have been developed to systematically characterize the interactions between actors and models of computation and reason about their compatibility (i.e., whether or not it makes sense to instantiate an actor in specifications that are based on a given model) [68].

3.6.5 SysteMoc

SysteMoc is a SystemC-based library for dataflow-based, hardware/software codesign and synthesis of signal processing systems. (See Section 3.7 for more details about SystemC, the simulation language on which SysteMoc was developed.) SysteMoc is based on a form of dataflow in which the model for each actor A includes a set F of functions and a FSM called the firing FSM of A. Each function $f \in F$ is classified as either an *action* function or a *guard* function. The action functions provide the core data processing capability of the actor, while the guard functions determine the activation of transitions in the firing FSM. Guard functions can access values of tokens present at the input edges of an actor (without consuming them), thereby enabling data-dependent sequencing of actions through the firing FSM. Furthermore, each transition t of the firing FSM has an associated action function $x(t) \in F$, which is executed when the transition t is activated. Thus, SysteMoc provides an integrated method for specifying, analyzing, and synthesizing actors in terms of FSMs that control sets of alternative dataflow behaviors.

SysteMoc has been demonstrated using a design space exploration case study for FPGA-based implementation of a 2D inverse DCT, as part of a larger case study involving an MPEG-4 decoder [69]. This case study considered a 5D design evaluation space encompassing throughput, latency, number of lookup tables (LUTs), number of flip-flops, and a composite resource utilization metric involving the sum of block RAM and multiplier resources. Among the most impressive results of the case study was the accuracy with which the design space exploration framework developed for SysteMoc was able to estimate FPGA hardware resources. For more details on SysteMoc and the MPEG-4 case study using SysteMoc, we refer the reader to Reference 69.

3.7 SIMULATION

Simulation is very important in SoC design. Simulation is not limited to functional verification, as with logic design. SoC designers use simulation to measure the performance and power consumption of their SoC designs. This is due in part to the fact that much of the functionality is implemented in software, which must be measured relative to the processors on which it runs. It is also due to the fact that the complex input patterns inherent in many SoC applications do not lend themselves to closed-form analysis.

SystemC is a simulation language that is widely used to model SoCs [70]. SystemC leverages the C++ programming language to build a simulation environment. SystemC classes allow designers to describe a digital system using a combination of structural and functional techniques. SystemC supports simulation at several levels of abstraction. Register-transfer-level simulations, for example, can be performed with the appropriate SystemC model. SystemC is most often used for more abstract models. A common type of model built in SystemC is a transaction-level model. This style of modeling describes the SoC as a network of communicating machines, with explicit connections between the models and functional descriptions for each model. The transaction-level model describes how data are moved between the models.

Hardware/software cosimulators are multimode simulators that simultaneously simulate different parts of the system at different levels of detail. For example, some modules may be simulated in register-transfer mode, while software running on a CPU is simulated functionally. Cosimulation is particularly useful for debugging the hardware/software interface, such as debugging driver software.

Functional validation, performance analysis, and power analysis of SoCs require simulating large numbers of vectors. Video and other SoC applications allow complex input sequences. Even relatively compact tests can take up tens of millions of bytes. These long input sequences are necessary to run the SoC through a reasonable number of the states implemented in the system. The large amounts of memory that can be integrated into today's systems, whether they be on-chip or off-chip, allow the creation of SoCs with huge numbers of states that require long simulation runs.

Simulators for software running on processors have been developed over the past several decades. The Synopsys Virtualizer, for example, provides a transaction-level modeling interface for software prototyping. Both computer architects and SoC designers need fast simulators to run the large benchmarks required to evaluate architectures. As a result, a number of simulation techniques covering a broad range of accuracy and performance have been developed.

A simple method of analyzing a program's execution behavior is to sample the program counter (PC) during program execution. The Unix *prof* command is an example of a PC-sampling analysis tool. PC sampling is subject to the same limitations on sampling rate as any other sampling process, but sampling rate is usually not a major concern in this case. A more serious limitation is that PC sampling gives us relative performance but not absolute performance. A sampled trace of the PC tells us where the program spent its time during execution, which gives us valuable information about the relative execution time of program modules that can be used to optimize the program. But it does not give us the execution time on a particular platform—especially if the target platform is different than the platform on which the trace is taken—and so we must use other methods to determine the real-time performance of programs.

Some simulators concentrate on the behavior of the cache, given the major role of the cache in determining overall system performance. The Dinero simulator (http://pages.cs.wisc.edu/~markhill/DineroIV/) is a well-known example of a cache simulator. These simulators generally work from a trace generated from the execution of a program. The program to be analyzed is augmented with additional code that records the execution behavior of the program. The Dinero simulator then reconstructs the cache behavior from the program trace. The architect can view the cache in various states or calculate cache statistics.

Some simulation systems model the behavior of the processor itself. A functional CPU simulator models instruction execution and maintains the state of the programming model, that is, the set of registers visible to the programmer. The functional simulator does not, however, model the performance or energy consumption of the program's execution.

A cycle-accurate simulator of a CPU is designed to accurately predict the number of clock cycles required to execute every instruction, taking into account pipeline and memory system effects. The CPU model must therefore represent the internal structure of the CPU accurately enough to show how resources in the processor are used. The SimpleScalar simulation tool [71] is a well-known toolkit for building cycle-accurate simulators. SimpleScalar allows a variety of processor models to be built by a combination of parameterization of existing models and linking new simulation modules into the framework.

Power simulators are related to cycle-accurate simulators. Accurate power estimation requires models of the CPU microarchitecture at least as detailed as those used for performance evaluation. A power simulator must model all the important wires in the architecture since capacitance is a major source of power consumption. Wattch [72] and SimplePower [73] are the two best-known CPU power simulators.

3.8 HARDWARE/SOFTWARE COSYNTHESIS

Hardware/software cosynthesis tools allow system designers to explore architectural trade-offs. These tools take a description of a desired behavior that is relatively undifferentiated between hardware and software. They produce a heterogeneous hardware architecture and the architecture for the software to run on that platform. The software architecture includes the allocation of software tasks to the processing elements of the platform and the scheduling of computation and communication.

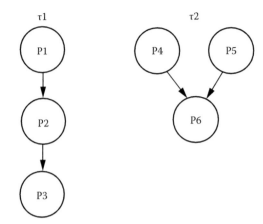

FIGURE 3.7 A task graph.

The functional description of an application may take several forms. The most basic is a task graph, as shown in Figure 3.7. The graph describes data dependencies between a set of processes. Each component of the graph (i.e., each set of connected nodes) forms a task. Each task runs periodically and every task can run at a different rate. The task graph model generally does not concern itself with the details of operations within a process. The process is characterized by its execution time. Several variations of task graphs that include control information have been developed. In these models, the output of a process may enable one of several different processes.

Task graph models are closely related to the dataflow graph models introduced in Section 3.5.1. The difference often lies in how the models are used. A key difference is that in dataflow models, emphasis is placed on precisely characterizing and analyzing how data are produced and consumed by computational components, while in task graph models, emphasis is placed on efficiently abstracting the execution time or resource utilization of the components or on analyzing real-time properties. In many cases, dataflow graph techniques can be applied to the analysis or optimization of task graphs and vice versa. Thus, the terms "task graph" and "dataflow graph" are sometimes used interchangeably.

An alternative representation for behavior is a programming language. Several different codesign languages have been developed and languages like SystemC have been used for cosynthesis as well. These languages may make use of constructs to describe parallelism that were originally developed for parallel programming languages. Such constructs are often used to capture operator-level concurrency. The subroutine structure of the program can be used to describe task-level parallelism.

The most basic form of hardware/software cosynthesis is hardware/software partitioning. As shown in Figure 3.8, this method maps the design into an architectural template. The basic system architecture is bus based, with a CPU and one or more custom hardware processing elements attached to the bus. The type of CPU is determined in advance, which allows the tool to accurately estimate software performance. The tool must decide what functions go into the custom processing elements; it must also schedule all the operations, whether implemented in hardware or software. This approach is known as hardware/software partitioning because the bus divides the architecture into two partitions and partitioning algorithms can be used to explore the design space.

Two important approaches to searching the design space during partitioning were introduced by early tools. The Vulcan system [74] starts with all processes in custom processing elements and iteratively moves selected processes to the CPU to reduce the system cost. The COSYMA system [75] starts with all operations running on the CPU and moves selected operations from loop nests into the custom processing element to increase performance.

Hardware/software partitioning is ideally suited to platform FPGAs, which implement the bus-partitioned structure and use FPGA fabrics for the custom processing elements. However,

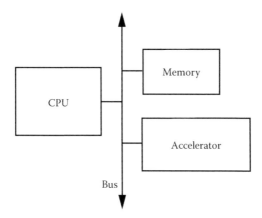

FIGURE 3.8   A template for hardware/software partitioning.

the cost metric is somewhat different than in custom designs. Because the FPGA fabric is of a fixed size, using more or less of the fabric may not be important, so long as the design fits into the amount of logic available.

Other cosynthesis algorithms have been developed that do not rely on an architectural template. Kalavade and Lee [64] alternately optimize for performance and cost to generate a heterogeneous architecture. Wolf [76] alternated cost reduction and load balancing while maintaining a performance-feasible design. Dick and Jha [77] used genetic algorithms to search the design space.

Scheduling is an important task during cosynthesis. A complete system schedule must ultimately be constructed; an important aspect of scheduling is the scheduling of multiple processes on a single CPU. The study of real-time scheduling for uniprocessors was initiated by Liu and Layland [78], who developed rate-monotonic scheduling (RMS) and earliest-deadline-first (EDF) scheduling. RMS and EDF are priority-based schedulers, which use priorities to determine which process to run next. Many cosynthesis systems use custom, state-based schedulers that determine the process to execute based upon the state of the system.

Design estimation is an important aspect of cosynthesis. While some software characteristics may be determined by simulation, hardware characteristics are often estimated using high-level synthesis. Henkel and Ernst [79] used forms of high-level synthesis algorithms to quickly synthesize a hardware accelerator unit and estimate its performance and size. Fornaciari et al. [80] used high-level synthesis and libraries to estimate power consumption.

Software properties may be estimated in a variety of ways, depending on the level of abstraction. For instance, Li and Wolf [81] built a process-level model of multiple processes interacting in the cache to provide an estimate of the performance penalty due to caches in a multitasking system. Tiwari et al. [82] used measurements to build models of the power consumption of instructions executing on processors.

3.9 SUMMARY

System-level design is challenging because it is heterogeneous. The applications that we want to implement are heterogeneous in their computational models. The architectures on which we implement these applications are also heterogeneous combinations of custom hardware, processors, and memory. As a result, system-level tools help designers manage and understand complex, heterogeneous systems. Models of computation help designers cast their problem in a way that can be clearly understood by both humans and tools. Simulation helps designers gather important design characteristics. Hardware/software cosynthesis helps explore design spaces. As applications become more complex, expect to see tools continue to reach into the application space to aid with the transition from algorithm to architecture.

REFERENCES

1. Fritts, J. and W. Wolf. Evaluation of static and dynamic scheduling for media processors. In *Proceedings of the MICRO-33 MP-DSP2 Workshop*, 2000.
2. Talla, D. et al. Evaluating signal processing and multimedia applications on SIMD, VLIW and super-scalar architectures. In *Proceedings of the International Conference on Computer Design*, 2000.
3. Texas Instruments, http://www.ti.com/lsds/ti/processors/dsp/c6000_dsp-arm/omap-l1x/overview.page, accessed November 23, 2015.
4. Freescale Semiconductor, MPC574xP: Ultra-reliable MPC574xP MCU for automotive & industrial safety applications, http://www.freescale.com/products/power-architecture-processors/mpc5xxx-5xxx-32-bit-mcus/mpc57xx-mcus/ultra-reliable-mpc574xp-mcu-for-automotive-industrial-safety-applications:MPC574xP, accessed November 23, 2015.
5. Whitepaper: NVIDIA's next generation CUDA compute architecture: Fermi. NVIDIA Corporation, 2009.
6. Wolf, W. *FPGA-Based System Design*. Prentice Hall, 2004.
7. Xilinx, http://www.xilinx.com/products/silicon-devices/soc/zynq-ultrascale-mpsoc.html, accessed November 23, 2015.
8. Gajski, D.D. and R.H. Kuhn. Guest editors' introduction: New VLSI tools. *Computer*, **16**(12): 11–14, 1983.
9. Gerstlauer, A. et al. Electronic system-level synthesis methodologies. *IEEE Transactions on Computer-Aided Design of Integrated Circuits and Systems*, **28**(10): 1517–1530, 2009.
10. Lee, E.A. and A. Sangiovanni-Vincentelli. A framework for comparing models of computation. *IEEE Transactions on Computer-Aided Design of Integrated Circuits and Systems*, **17**(12): 1217–1229, 1998.
11. Basu, A. et al. A language-based approach to protocol construction. In *ACM SIGPLAN Workshop on Domain-Specific Languages*, 1997.
12. Conway, C.L. and S.A. Edwards. NDL: A domain-specific language for device drivers. In *Proceedings of the Workshop on Languages Compilers and Tools for Embedded Systems*, 2004.
13. Konstantinides, K. and J.R. Rasure. The Khoros software-development environment for image-processing and signal-processing. *IEEE Transactions on Image Processing*, **3**(3): 243–252, 1994.
14. Lauwereins, R. et al. Grape-II: A system-level prototyping environment for DSP applications. *Computer*, **28**(2): 35–43, 1995.
15. Lee, E.A. et al. Gabriel: A design environment for DSP. *IEEE Transactions on Acoustics, Speech, and Signal Processing*, **37**(11): 1751–1762, 1989.
16. Manikonda, V., P.S. Krishnaprasad, and J. Hendler. Languages, behaviors, hybrid architectures and motion control. In *Essays in Mathematical Control Theory (in Honor of the 60th Birthday of Roger Brockett)*, J. Baillieul and J.C. Willems, Editors, Springer-Verlag, pp. 199–226, 1998.
17. Proudfoot, K. et al. A real-time procedural shading system for programmable graphics hardware. In *Proceedings of SIGGRAPH*, 2001.
18. Thibault, S.A., R. Marlet, and C. Consel. Domain-specific languages: From design to implementation application to video device drivers generation. *IEEE Transactions on Software Engineering*, **25**(3): 363–377, 1999.
19. S.S. Bhattacharyya et al., Editors. *Handbook of Signal Processing Systems*, 2nd edn. Springer, 2013.
20. Eker, J. and J.W. Janneck. Dataflow programming in CAL—Balancing expressiveness, analyzability, and implementability. In *Proceedings of the IEEE Asilomar Conference on Signals, Systems, and Computers*, 2012.
21. Pelcat, M. et al. *Physical Layer Multi-Core Prototyping*. Springer, 2013.
22. Bhattacharyya, S.S. et al. OpenDF—A dataflow toolset for reconfigurable hardware and multicore systems. In *Proceedings of the Swedish Workshop on Multi-Core Computing*, Ronneby, Sweden, 2008.
23. Edwards, S.A., *Languages for Digital Embedded Systems*. Kluwer Academic Publishers, 2000.
24. Jantsch, A. *Modeling Embedded Systems and SoC's: Concurrency and Time in Models of Computation*. Morgan Kaufmann Publishers, Inc., 2003.
25. Karp, R.M. and R.E. Miller. Properties of a model for parallel computations: Determinacy, termination, queuing. *SIAM Journal of Applied Mathematics*, **14**(6): 1390–1411, 1966.
26. Kahn, G. The semantics of a simple language for parallel programming. In *Proceedings of the IFIP Congress*, 1974.
27. Lee, E.A. and T.M. Parks. Dataflow process networks. *Proceedings of the IEEE*, **83**(5): 773–799, 1995.
28. Ambler, A.L., M.M. Burnett, and B.A. Zimmerman. Operational versus definitional: A perspective on programming paradigms. *Computer*, **25**(9): 28–43, 1992.
29. Ha, S. and E.A. Lee. Compile-time scheduling and assignment of data-flow program graphs with data-dependent iteration. *IEEE Transactions on Computers*, **40**(11): 1225–1238, 1991.
30. Eker, J. et al. Taming heterogeneity—The Ptolemy approach. *Proceedings of the IEEE*, **91**(1): 127–144, 2003.

31. Ko, D. and S.S. Bhattacharyya. Modeling of block-based DSP systems. *Journal of VLSI Signal Processing Systems for Signal, Image, and Video Technology,* **40**(3): 289–299, 2005.

32. Lee, E.A. and D.G. Messerschmitt. Synchronous dataflow. *Proceedings of the IEEE,* **75**(9): 1235–1245, 1987.

33. Bhattacharyya, S.S., P.K. Murthy, and E.A. Lee. *Software Synthesis from Dataflow Graphs.* Kluwer Academic Publishers, 1996.

34. Lublinerman, R. and S. Tripakis. Translating data flow to synchronous block diagrams. In *Proceedings of the IEEE Workshop on Embedded Systems for Real-Time Multimedia,* 2008.

35. Murthy, P.K. and E.A. Lee. Multidimensional synchronous dataflow. *IEEE Transactions on Signal Processing,* **50**(8): 2064–2079, 2002.

36. Vaidyanathan, P.P. *Multirate Systems and Filter Banks.* Prentice Hall, 1993.

37. Stichling, D. and B. Kleinjohann. CV-SDF—A model for real-time computer vision applications. In *Proceedings of the IEEE Workshop on Applications of Computer Vision,* 2002.

38. Stichling, D. and B. Kleinjohann. CV-SDF—A synchronous data flow model for real-time computer vision applications. In *Proceedings of the International Workshop on Systems, Signals and Image Processing,* 2002.

39. Keinert, J., C. Haubelt, and J. Teich. Modeling and analysis of windowed synchronous algorithms. In *Proceedings of the International Conference on Acoustics, Speech, and Signal Processing,* 2006.

40. Buck, J.T. and E.A. Lee. Scheduling dynamic dataflow graphs using the token flow model. In *Proceedings of the International Conference on Acoustics, Speech, and Signal Processing,* 1993.

41. Buck, J.T. Static scheduling and code generation from dynamic dataflow graphs with integer-valued control systems. In *Proceedings of the IEEE Asilomar Conference on Signals, Systems, and Computers,* 1994.

42. Buck, J. and R. Vaidyanathan. Heterogeneous modeling and simulation of embedded systems in El Greco. In *Proceedings of the International Workshop on Hardware/Software Co-Design,* 2000.

43. Girault, A., B. Lee, and E.A. Lee. Hierarchical finite state machines with multiple concurrency models. *IEEE Transactions on Computer-Aided Design of Integrated Circuits and Systems,* **18**(6): 742–760, 1999.

44. Thiele, L. et al. FunState—An internal representation for codesign. In *Proceedings of the International Conference on Computer-Aided Design,* 1999.

45. Cossement, N., R. Lauwereins, and F. Catthoor. DF*: An extension of synchronous dataflow with data dependency and non-determinism. In *Proceedings of the Forum on Specification and Design Languages,* 2000.

46. Pankert, M. et al. Dynamic data flow and control flow in high level DSP code synthesis. In *Proceedings of the International Conference on Acoustics, Speech, and Signal Processing,* 1994.

47. Plishker, W. et al. Functional DIF for rapid prototyping. In *Proceedings of the International Symposium on Rapid System Prototyping,* Monterey, CA, 2008.

48. Theelen, B.D. et al. A scenario-aware data flow model for combined long-run average and worst-case performance analysis. In *Proceedings of the International Conference on Formal Methods and Models for Codesign,* Washington DC, 2006.

49. Janneck, J.W. et al. Reconfigurable video coding: A stream programming approach to the specification of new video coding standards. In *Proceedings of the ACM SIGMM Conference on Multimedia Systems,* 2010.

50. Eker, J. and J.W. Janneck. CAL language report, Language version 1.0—Document edition 1. Electronics Research Laboratory, University of California at Berkeley, Berkeley, CA, 2003.

51. Yviquel, H. et al. Orcc: Multimedia development made easy. In *Proceedings of the ACM International Conference on Multimedia,* 2013.

52. Boutellier, J. et al. Scheduling of dataflow models within the reconfigurable video coding framework. In *Proceedings of the IEEE Workshop on Signal Processing Systems,* 2008.

53. Gu, R. et al. Exploiting statically schedulable regions in dataflow programs. *Journal of Signal Processing Systems,* **63**(1): 129–142, 2011.

54. Stefanov, T. et al. System design using Kahn process networks: The Compaan/Laura approach. In *Proceedings of the Design, Automation and Test in Europe Conference and Exhibition,* 2004.

55. Bilsen, G. et al. Cyclo-static dataflow. *IEEE Transactions on Signal Processing,* **44**(2): 397–408, 1996.

56. Deprettere, E.F. et al. Affine nested loop programs and their binary cyclo-static dataflow counterparts. In *Proceedings of the International Conference on Application Specific Systems, Architectures, and Processors,* Steamboat Springs, CO, 2006.

57. Kienhuis, B., E. Rijpkema, and E. Deprettere. Compaan: Deriving process networks from MATLAB for embedded signal processing architectures. In *Proceedings of the International Workshop on Hardware/Software Co-Design,* 2000.

58. Turjan, A., B. Kienhuis, and E. Deprettere. Approach to classify inter-process communication in process networks at compile time. In *Proceedings of the International Workshop on Software and Compilers for Embedded Systems,* 2004.

59. Nikolov, H. et al. Daedalus: Toward composable multimedia MP-SoC design. In *Proceedings of the Design Automation Conference*, 2008.
60. Pelcat, M. et al. An open framework for rapid prototyping of signal processing applications. *EURASIP Journal on Embedded Systems*, **2009**: Article ID 11, 2009.
61. Holzner, S. *Eclipse*. O'Reilly & Associates, Inc., 2004.
62. Pelcat, M. et al. *Dataflow-Based Rapid Prototyping for Multicore DSP Systems*. Institut National des Sciences Appliquees de Rennes, 2014.
63. Buck, J.T. et al. Ptolemy: A framework for simulating and prototyping heterogeneous systems. *International Journal of Computer Simulation*, **4**: 155–182, 1994.
64. Kalavade, A. and E.A. Lee. A hardware/software codesign methodology for DSP applications. *IEEE Design and Test of Computers Magazine*, **10**(3): 16–28, 1993.
65. Neuendorffer, S. and E. Lee. Hierarchical reconfiguration of dataflow models. In *Proceedings of the International Conference on Formal Methods and Models for Codesign*, 2004.
66. Bhattacharyya, S.S. et al. Generating compact code from dataflow specifications of multirate signal processing algorithms. *IEEE Transactions on Circuits and Systems—I: Fundamental Theory and Applications*, **42**(3): 138–150, 1995.
67. de Alfaro, L. and T. Henzinger. Interface automata. In *Proceedings of the Joint European Software Engineering Conference and ACM SIGSOFT International Symposium on the Foundations of Software Engineering*, 2001.
68. Lee, E.A. and Y. Xiong. System-level types for component-based design. In *Proceedings of the International Workshop on Embedded Software*, 2001.
69. Haubelt, C. et al. A SystemC-based design methodology for digital signal processing systems. *EURASIP Journal on Embedded Systems*, **2007**: Article ID 47580, 22 pages, 2007.
70. Black, D.C. et al. *SystemC: From the Ground Up*, 2nd edn. Springer, 2010.
71. Burger, D.C. and T.M. Austin. The SimpleScalar tool set, Version 2.0. Department of Computer Sciences, University of Wisconsin at Madison, Madison, WI, 1997.
72. Brooks, D., V. Tiwari, and M. Martonosi. Wattch: A framework for architectural-level power analysis and optimizations. In *International Symposium on Computer Architecture*, 2000.
73. Vijaykrishnan, N. et al. Energy-driven integrated hardware-software optimizations using SimplePower. In *International Symposium on Computer Architecture*, 2000.
74. Gupta, R.K. and G. De Micheli. Hardware-software cosynthesis for digital systems. *IEEE Design and Test of Computers Magazine*, **10**(3): 29–41, 1993.
75. Ernst, R., J. Henkel, and T. Benner. Hardware-software cosynthesis for microcontrollers. *IEEE Design and Test of Computers Magazine*, **10**(4): 64–75, 1993.
76. Wolf, W.H. An architectural co-synthesis algorithm for distributed, embedded computing systems. *IEEE Transactions on Very Large Scale Integration (VLSI) Systems*, **5**(2): 218–229, 1997.
77. Dick, R.P. and N.K. Jha. MOGAC: A multiobjective genetic algorithm for hardware-software cosynthesis of distributed embedded systems. *IEEE Transactions on Computer-Aided Design of Integrated Circuits and Systems*, **17**(10): 920–935, 1998.
78. Liu, C.L. and J.W. Layland. Scheduling algorithms for multiprogramming in a hard-real-time environment. *Journal of the Association for Computing Machinery*, **20**(1): 46–61, 1973.
79. Henkel, J. and R. Ernst. A path-based technique for estimating hardware runtime in HW/SW-cosynthesis. In *Proceedings of the International Symposium on System Synthesis*, 1995.
80. Fornaciari, W. et al. Power estimation of embedded systems: A hardware/software codesign approach. *IEEE Transactions on Very Large Scale Integration (VLSI) Systems*, **6**(2): 266–275, 1998.
81. Li, Y. and W. Wolf. Hardware/software co-synthesis with memory hierarchies. *IEEE Transactions on Computer-Aided Design of Integrated Circuits and Systems*, **18**(10): 1405–1417, 1999.
82. Tiwari, V., S. Malik, and A. Wolfe. Power analysis of embedded software: A first step towards software power minimization. *IEEE Transactions on Very Large Scale Integration (VLSI) Systems*, 1994.

System-Level Specification and Modeling Languages

4

Stephen A. Edwards and Joseph T. Buck

CONTENTS

4.1 INTRODUCTION

While a digital integrated circuit can always be specified and modeled correctly and precisely as a network of Boolean logic gates, the complexity of today's multibillion transistor chips would render such a model unwieldy, to say the least. The objective of system-level specification and modeling languages—the subject of this chapter—is to describe such behemoths in more abstract terms, making them easier to code, evaluate, test, and debug, especially early during the design process when fewer details are known.

Our systems compute—perform tasks built from Boolean decision making and arithmetic—so it is natural to express them using a particular model of computation. Familiar models of computation include networks of Boolean logic gates and imperative software programming languages (e.g., C, assembly, Java). However, other, more specialized models of computation are often better suited for describing particular kinds of systems, such as those focused on digital signal processing.

Specialized models of computation lead to a fundamental conundrum: while a more specialized model of computation encourages more succinct, less buggy specifications that are easier to reason about, it may not be able to model the desired system. For example, while many dataflow models of computation are well suited to describing signal processing applications, using them to describe a word processor application would be either awkward or impossible. We need models of computation that are specialized enough to be helpful, yet are general enough to justify their creation.

4.2 PARALLELISM AND COMMUNICATION: THE SYSTEM MODELING PROBLEMS

Digital systems can be thought of as a set of components that operate in parallel. While it can be easier to conceptualize a system described in a sequential model such as the C software language, using such a model for a digital system is a recipe for inefficiency. All the models described in this chapter are resolutely parallel.

Describing how parallel components communicate provides both the main challenge in system modeling and the largest opportunity for specialization. A system is only really a system if its disparate components exchange information (otherwise, we call them multiple systems), and there are many ways to model this communication. Data can be exchanged synchronously, meaning every component agrees to communicate at a periodic, agreed-upon time, or asynchronously. Communication among components can take place through point-to-point connections or arbitrary networks. Communication channels can be modeled as memoryless or be allowed to store multiple messages at a time.

How communication is performed in each model of computation provides a convenient way to classify the models. In the following, we discuss a variety of models of computation following this approach. Others, including Simulink® [18,19], LabView, and SysML, which are more difficult to classify and less formal, are discussed in Chapter 9.

4.3 SYNCHRONOUS MODELS AND THEIR VARIANTS

Digital hardware designers are most familiar with the *synchronous model*: a network of Boolean logic gates and state-holding elements (e.g., flip-flops) that march to the beat of a global clock. At the beginning of each clock cycle, each state-holding element updates its state, potentially changing its output and causing a cascade of changes in the Boolean gates in its fanout.

The synchronous model imposes sequential behavior on what are otherwise uncoordinated, asynchronously running gates. The clock's frequent, periodic synchronization forces the system's components to operate in lockstep. Provided that the clock's period is long enough to allow the combinational network to stabilize, a synchronous logic network behaves as a finite-state machine that takes one step per clock tick.

While most digital systems are ultimately implemented in a synchronous style, the model has two big disadvantages. Uniformly distributing a multi-GHz clock signal without skew (time difference among components, which limits maximum speed) presents a practical challenge; assigning each computation to a cycle and limiting communication to one message per cycle presents a more conceptual challenge.

A number of system design languages are based on the synchronous model of computation. Each takes a different approach to describing the behavior of processes, and many have subtle differences in semantics. Benveniste et al. [3] discuss the history of many of these languages.

4.3.1 DATAFLOW: LUSTRE AND SIGNAL

The Lustre language [30] is a textual *dataflow language*. A Lustre program consists of arithmetic, decision, and delay operators applied to *streams* of Booleans, integers, and other types. For example, the following (recursive) Lustre equation states that the stream n consists of 0 followed by the stream n delayed by one cycle with 1 added to every element. This equation produces the sequence 0, 1, 2, ...

```
n = 0 -> pre(n) + 1
```

Lustre permits such recursive definitions provided that every loop is broken by at least one delay (pre), a requirement similar to requiring combinational logic to be cycle-free. This rule ensures the behavior is always well defined and easy to compute.

Functions are defined in Lustre as a group of equations with named ports. Here is a general-purpose counter with initial, step, and reset inputs from Halbwachs et al. [30].

```
node COUNTER(val_init, val_incr: int,

            reset: bool) returns (n: int);
let
  n = val_init -> if reset then val_init

                          else pre(n) + val_incr;
tel.
```

Lustre's when construct subsamples a stream under the control of a Boolean-valued stream, similar to the downsampling operation in digital signal processing. Its complementary current construct interpolates a stream (using a sample-and-hold mechanism) to match the current basic clock, but Lustre provides no mechanism for producing a faster clock.

The Signal language [40] is a similar dataflow-oriented language based on synchronous semantics, but departs from Lustre in two important ways. First, it has much richer mechanisms for sub- and oversampling data streams that allow it to create faster clocks whose rates are data dependent. While this is difficult to implement in traditional logic, it provides Signal with expressibility approaching that of a software programming language. For example, Signal can describe a module that applies Euclid's greatest common divisor algorithm to each pair of numbers in an input stream. A designer would have to manually add additional handshaking to model such behavior in Lustre.

The other main difference is that Signal allows arbitrary recursive definitions, including those that have no solution along with those with multiple solutions. Signal's designers chose this route to enable the modeling of more abstract systems, including those with *nondeterminism*.

4.3.2 IMPERATIVE: ESTEREL

The Esterel synchronous language [7] provides an imperative modeling style rather different from the aforementioned dataflow languages. Such an approach simplifies the description of discrete control applications. To illustrate, Berry's [5] ABRO example expresses a resettable synchronizer that waits for the arrival of two signals (A and B) before outputting O:

```
module ABO:
input A, B, R;
output O;
loop
    [ await A || await B ];
    emit O
each R
end module
```

The semantics of Esterel uniquely define the cycle-level timing of such a program, which are actually fairly complex because of all the edge cases. Here, each `await` construct waits one or more cycles for the Boolean signal A or B to be true. In the cycle that the second signal is true, the parallel construct `||` passes control immediately to the `emit` statement, which makes O true in that cycle (it is false by default). Note that A and B can arrive in either order or in the same cycle. The `loop` construct immediately restarts this process once O has been emitted or in a cycle where R is present. Finally, if A, B, and R arrive in the same cycle, the semantics of `loop...each` gives priority to R; O will not be true in that cycle.

Esterel's precise semantics makes it a good target for model checkers. For example, XEVE [11] was able to achieve high state coverage when verifying the control-dominated parts of a commercial digital signal processor (DSP) [2]. The reference model was written in a mixture of C for the datapath and Esterel for control.

The techniques used to compile Esterel have evolved substantially [56]. While synchronous models always imply an underlying finite-state machine, the earliest Esterel compilers enumerated the program's states, which could lead to exponential code sizes. The next compilers avoided the exponential code size problem by translating the program into a circuit and generating code that simulated the circuit. Berry described this approach in 1991 [4], which formed the basis of the successful V5 compiler. Later work by Edwards [22] and Potop-Butucaru [55] better preserved the control structure of the source program in the compiled code, greatly speeding its execution.

An Esterel program is allowed to have, within a single clock cycle, cyclic dependencies, provided that they can be resolved at compile time. This solution falls between that in Lustre, which simply prohibits cyclic dependencies, and Signal, which defers their resolution to runtime. This subtle point in Esterel's semantics was not clearly understood until the circuit translation of Esterel emerged. Shiple et al. [62] describes the techniques employed, which evolved from earlier work by Malik [47].

Seawright's *production-based* specification system, Clairvoyant [59], shares many ideas with Esterel. Its productions are effectively regular expressions that invoke actions (coded in VHDL or Verilog), reminiscent of the Unix tool *lex* [46]. Seawright et al. later extended this work to produce a commercial tool called Protocol Compiler [60], whose circuit synthesis techniques were similar to those used for compiling Esterel. While not a commercial success, the tool was very effective for synthesizing and verifying complex controllers for SDH/Sonet applications [51].

4.3.3 GRAPHICAL: STATECHARTS AND OTHERS

Harel's Statecharts [31] describe hierarchical state machines using a graphical syntax. Ultimately, a synchronous language shares Esterel's focus on discrete control and even some of its semantic constructs, but its graphical approach is fundamentally different. I-Logix, which Harel cofounded, commercialized the technology in a tool called Statemate, which has since been acquired by IBM.

Statecharts assemble hierarchical extended finite-state machines from OR and AND components. An OR component is a familiar bubble-and-arc diagram for a state machine, with guard conditions and actions along the arcs, but the states may be either atomic or Statechart models (i.e., hierarchical). As in a standard FSM, the states in an OR component are mutually exclusive. By contrast, an AND component consists of two or more Statechart models that operate in parallel.

In Harel's original version, transition arrows can cross levels of hierarchy, providing great expressive power (something like exceptions in software) while breaking modularity.

The Statechart below is an OR component (states C and AB); AB is an AND component consisting of two OR components. It starts in state AB at states A1 and B1. The arcs may have conditions and actions (not shown here).

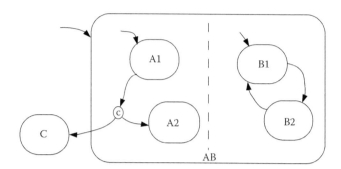

Statecharts can be used for generation of C or HDL (hardware description language) code for implementation purposes, but the representation is most commonly used today for creating executable specifications [32] as opposed to final implementations.

The form of Statecharts has been widely accepted; it has even been adopted as part of the Unified Modeling Language [58], but there has been a great deal of controversy about the semantics. Many variants have been created as a result of differing opinions about the best approach to take: von der Beeck [68] identifies 20 and proposes another.

The following are the main points of controversy:

- *Modularity*: Many researchers, troubled by Harel's level-crossing transitions, eliminated them and came up with alternative approaches to make them unnecessary. In many cases, signals are used to communicate between levels.
- *Microsteps*: Harel's formulation uses delta cycles to handle cascading transitions that occur based on one transition of the primary inputs, much like in VHDL and Verilog. Others, such as Argos [48] and SyncCharts [1], have favored synchronous-reactive semantics and find a fixpoint, so that what requires a series of microsteps in Harel's Statecharts becomes a single atomic transition. These formulations reject a specification if its fixpoint is not unique.
- *Strong preemption vs. weak preemption*: A race of sorts occurs when a system is in a state where both an inner and outer state have a transition triggered by the same event. With strong preemption, the outer transition "wins"; the inner transition is completely preempted. With weak preemption, both transitions take place (meaning that the action associated with the inner transition is performed), with the inner action taking place first (this order is required because the outer transition normally causes the inner state to terminate). Strong preemption can create causality violations, since the action on an inner transition can cause an outer transition that would preempt the inner transition. Many Statechart variants reject specifications with this kind of causality violation as ill formed. Some variants permit strong or weak preemption to be specified separately for each transition.
- *History and suspension*: When a hierarchical state is exited and reentered, does it remember its previous state? If it does, is the current state remembered at all levels of hierarchy (deep history) or only at the top level (shallow history)? In Harel's Statecharts, shallow history or deep history can be specified as an attribute of a hierarchical state. In some other formulations, a suspension mechanism is used instead, providing the equivalent of deep history by freezing a state (analogous to gating a clock in a hardware implementation).

The Statechart below, a traffic light controller with the ability to pause, illustrates the difference between history and suspension. The controller on the left uses a Harel-style history mechanism that commands the controller to resume in a stored state after a pause; on the right, a SyncCharts-like suspension mechanism holds the controller's state while the pause input is asserted.

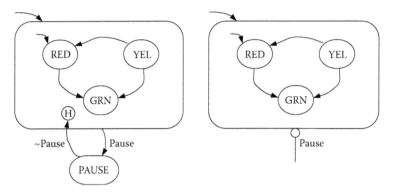

4.4 TRANSACTION-LEVEL MODELS AND DISCRETE-EVENT SIMULATION

The constraints of the synchronous model that guarantee determinism—lockstep operation coupled with fixed, periodic communication—demand a higher level of detail than is often desired; transaction-level modeling (TLM) is a popular approach that relaxes both of these requirements, providing a model of computation comprised of asynchronous processes that communicate (reliably) at will. The result is more abstract than synchronous models, but can be nondeterministic, or at least sensitive to small variations in timing.

Grötker et al. [29] define TLM as "a high-level approach to modeling digital systems where details of communication among modules are separated from the details of the implementation of the functional units or of the communication architecture." The original SystemC development team coined the term "transaction-level modeling"; an alternative, "transaction-based modeling," might have been preferable, as TLM does not correspond to a particular level of abstraction in the same sense that, for example, register-transfer level (RTL) does [28]. However, distinctions between TLM approaches and RTL can clearly be made: while in a RTL model of a digital system the detailed operation of the protocol, address, and data signals on a bus are represented in the model, with TLM, a client of a bus-based interface might simply issue a call to high-level `read()` or `write()` functions. The TLM approach therefore not only reduces model complexity but also reduces the need for simulating parallel threads, both of which improve simulation speed. Grötker et al. [29, Chapter 8] give a simple but detailed example of a TLM approach to the modeling of a bus-based system with multiple masters, slaves, and arbiters.

While the SystemC project coined the TLM term, the idea is much older. For example, the SpecC language [26], through its `channel` feature, permits the details of communication to be abstracted away in much the same manner. Furthermore, the SystemVerilog language [64] now supports TLM through ports declared with its `interface` keyword. Even without direct language support, it is almost always possible to separate functional units from communication as cleaner way of organizing a system design, even in all its detailed RTL glory; Sutherland et al. [64] recommend exactly this.

TLM is commonly used in a discrete-event (DE) setting: a simulation technique for synchronous models whose processes are idle unless one of their inputs change. By demanding such behavior, a DE simulator only needs to do work when something changes. Broadly, a DE simulator advances simulation time (i.e., the clock cycle being simulated, typically a measure of real time) until it reaches the next scheduled event, then it executes that event, which may cause a change and schedule another future event, and so on. The common Verilog and VHDL hardware modeling languages, for example, are based on this model, so it is semantically easy to mix TLM with detailed RTL models of certain components, which is useful at various steps in the design refinement process.

The TLM methodology for system design continues to mature, yet disagreement remains about how many levels of abstraction to consider during a design flow. Cai and Gajski [17] propose a rigorous approach in which the designer starts with an untimed (i.e., purely functional) specification, then separately refines the model's computation and communication. They identify intermediate points such as bus-functional models with accurate communication timing but only approximate computation timing.

4.5 HOARE'S COMMUNICATING SEQUENTIAL PROCESSES

Communicating sequential processes with rendezvous (CSP) is an untimed model of computation first proposed by Hoare [34]. Unlike synchronous models, CSP does not assume a global clock and instead ensures synchronization among parts of a system through rendezvous communication. A pair of processes that wish to communicate must do so through a rendezvous point or channel. Each process must post an I/O request before data are transferred; whichever process posts first waits for its peer.

Such a lightweight communication mechanism is easy to implement in both hardware and software and has formed the basis for high-level languages such as Occam [36], the remote-procedure call style of interprocess communication in operating systems, and even asynchronous hardware design [49].

Nondeterminism is a central issue in CSP: nothing prohibits three or more processes to attempt to rendezvous on the same channel. This situation produces a race that is won by the first two processes to arrive; changing processes' execution rate (but not their function) can dramatically change a system's behavior and I/O. Such races can be dangerous or a natural outcome of a system's structure. Server processes naturally wait for a communication from any of a number of other processes and respond to the first one that arrives. This behavior can be harmless if requests do not update the server's state, or can mask hard-to-reproduce bugs that depend on timing, not just function.

4.6 DATAFLOW MODELS AND THEIR VARIANTS

Dataflow models of computation allow for varying communication rates among components, a more expressive alternative to synchronous models, which usually assume a constant, unit communication rate among components. Most dataflow models also provide some facility for buffering communication among components, which adds more flexibility in execution rates and implementations since the whole system no longer needs to operate in lockstep.

Dataflow's flexibility allows the easier description of multirate and data-dependent communication, often convenient abstractions. While both can be "faked" in synchronous models (e.g., by adding the ability to hold the state of certain components so they perceive fewer clock cycles), expressing this in a synchronous model requires adding exactly the sort of implementation details designers would prefer to omit early in the design process.

Nondeterminism can be an issue with dataflow models. In general, allowing parallel processes to read from and write to arbitrary channels at arbitrary times produces a system whose function depends on relative execution rates, which are often difficult to control. Such nondeterminism can make it very difficult to verify systems since testing can be inconclusive: even if a system is observed to work, it may later behave differently even when presented with identical inputs.

Whether the model of computation guarantees determinacy is thus a key question to ask of any dataflow model. Note that nondeterministic models of computation can express deterministic systems; the distinction is that they do not provide any guarantee. In 1974, Gilles Kahn [37] proposed a very flexible deterministic dataflow model that now underlies most dataflow models; most successful ones are careful restrictions of this.

4.6.1 KAHN PROCESS NETWORKS: DETERMINISM WITH UNBOUNDED QUEUES

A Kahn process network (KPN) [37] is a network of processes that communicate by means of unbounded first-in-first-out queues (FIFOs). Each process has zero or more input FIFOs, and zero or more output FIFOs. Each FIFO is connected to exactly one input process and one output process.

When a process writes data to an output FIFO, the write always happens immediately; the FIFO grows as needed to accommodate the data written. However, a process may only read one of its inputs by means of a blocking read: if there are insufficient data to satisfy a request, the reading process blocks until the writing process for the FIFO provides more. Furthermore, a process can only read from a single FIFO at any time: it cannot, say, wait for the first available data on multiple FIFOs. The data written by a process depend only on the data read by that process and the initial state of the process. Under these conditions, Kahn showed that the trace of data written on each FIFO is independent of process scheduling order, that is, the I/O behavior of a Kahn network is deterministic even if the relative speeds of the processes are uncontrolled.

Because any implementation that preserves the semantics will compute the same data streams, KPN representations, or special cases of them, are a useful starting point for the system architect and are often used as executable specifications.

Dataflow process networks are a special case of Kahn process networks in which the behavior of each process (often called an "actor" in the literature) can be divided into a sequence of execution steps called "firings" by Lee and Parks [44]. A firing consists of zero or more read operations from input FIFOs, followed by a computation, and then by zero or more write operations to output queues (and possibly a modification of the process's internal state). This model is widely used in both commercial and academic software tools, such as Scade [6], SPW, COSSAP [38] and its successor, System Studio [12], and Ptolemy [20]. The subdivision into firings, which are treated as indivisible quanta of computation, can greatly reduce context switching overhead in simulation, and can enable synthesis of software and hardware. In some cases (e.g., Yapi [21], Compaan [63]), the tools permit processes to be written as if they were separate threads, and then split the threads into individual firings by means of analysis; the thread representation allows read and write directives to occur anywhere, while the firing representation can make it easier to understand the data rates involved, which is important for producing consistent designs.

While mathematically elegant, Kahn process networks in their strictest form are unrealizable because of their need for unbounded buffers. Even questions like whether a network can run in bounded memory are generally undecidable. Parks [52] presents a scheduling policy to run a KPN in bounded memory if it can be done, but it is an online algorithm that provides no compile-time guarantees.

4.6.2 STATIC DATAFLOW: DATA-INDEPENDENT FIRING

In an important special case of dataflow process networks, the number of values read and written by each firing of each process is fixed, and does not depend on the data. This model of computation was originally called synchronous dataflow (SDF) [43], an unfortunate choice because it is untimed, yet suggests the synchronous languages discussed earlier. In fact, the term was originally used for the Lustre language described earlier. The term "static dataflow" is now considered preferable; Lee himself uses it [44]. Fortunately, the widely used acronym SDF still applies.

Below is a simple SDF graph that upsamples its input (block A) by a factor of 100, applies a function C to the signal, and then downsamples it down by 100. Each number indicates how many data values are read or written to the attached FIFO each time the actor fires. For example, every time the B actor fires, it consumes 1 value (on the FIFO from A) and produces 10 values (on the FIFO to C). While not shown in this example, each edge may have initial logical delays, which can be thought of as initial values in the queues. Such initial values are compulsory to avoid deadlock if the graph contains cycles.

Compile-time analysis is a central advantage of SDF over more general Kahn networks: scheduling, buffer sizes, and the possibility of deadlock can all be established precisely and easily before the system runs [43]. Analysis begins by determining consistent relative actor firing rates, if they exist. To avoid buffer over- or underflow, the number of values read from a FIFO must asymptotically match the number of values written. Define entry (i,j) of the *topology matrix* Γ to be the net number of values written (reads count negatively; zero represents no connection) per firing of actor j on FIFO i. Note that Γ is not necessarily square. To avoid unbounded data accumulation or systematic underflow on each FIFO, these production and consumption rates must balance, that is, if q is a column vector and each q_j is the number of times actor j fires, we must ultimately have

$$\Gamma q = 0$$

If the graph is connected, q has nonzero solutions only if rank(Γ) is $s - 1$, where s is the number of actors in the graph [42].

For example, the SDF graph below is inconsistent. There is no way both A and B can fire continuously without the data in the top FIFO growing unbounded. Equivalently, the only solution to $\Gamma q = 0$ is $q = 0$.

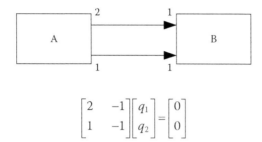

$$\begin{bmatrix} 2 & -1 \\ 1 & -1 \end{bmatrix} \begin{bmatrix} q_1 \\ q_2 \end{bmatrix} = \begin{bmatrix} 0 \\ 0 \end{bmatrix}$$

By contrast, if Γ has rank $s - 1$, the system has a 1D family of solutions that corresponds to running with bounded buffers. In general, s actors demands Γ have rank $s - 1$; rank s implies inconsistent rates; and a rank below $s - 1$ implies disconnected components—usually a design error. For the up-/downsampler example given earlier,

$$\begin{bmatrix} 10 & -1 & 0 & 0 & 0 \\ 0 & 10 & -1 & 0 & 0 \\ 0 & 0 & 1 & -10 & 0 \\ 0 & 0 & 0 & 1 & -10 \end{bmatrix} \begin{bmatrix} q_1 \\ q_2 \\ q_3 \\ q_4 \\ q_5 \end{bmatrix} = 0$$

We are interested in the smallest nonzero integral solution to this system, which gives the number of firings in the shortest repeatable schedule. For this example,

$$q = x[1 \quad 10 \quad 100 \quad 10 \quad 1],$$

where x is an arbitrary real number. This solution means that, in the smallest repeatable schedule, actors A and E fire once, B and D fire 10 times, and C fires 100 times. While a schedule listing all 122 firings is correct, a single-appearance schedule [8], which includes loops and lists each actor exactly once, produces a more efficient implementation. For this example, $A(BC^{10}D)^{10}E$ is such a schedule.

A nontrivial solution to the balance equations $\Gamma q = 0$ is necessary but does not guarantee a feasible schedule; the system may still deadlock. The graph below has a simple family of solutions (A and B fire equally often), yet it immediately deadlocks because there are no initial data in either FIFO. It is always possible to make a system with consistent rates avoid underflow by adding initial tokens to its FIFOs, but doing so usually changes the data passed between the actors.

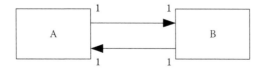

For static dataflow networks, efficient static schedules are easily produced, and bounds can be determined for all of the FIFO buffers, whether for a single programmable processor, multiple processors, or hardware implementations. Bhattacharyya et al. [9] provide an excellent overview of the analysis of SDF designs, as well as the synthesis of software for a single processor from such designs.

For multiple processors, one obvious alternative is to form a task dependence graph from the actor firings that make up one iteration of the SDF system, and apply standard task scheduling techniques such as list scheduling to the result. Even for uniform-rate graphs, where there is no looping, the scheduling problem is NP-hard. However, because it is likely that implicit loops are present, a linear schedule is likely to be too large to handle successfully. Efficient multiprocessor solutions usually require preservation of the hierarchy introduced by looping; Pino et al. [53] propose a way to do this.

Engels et al. [25] propose cyclo-static dataflow: a subtle but powerful extension of static dataflow that allows the input–output pattern of an actor to vary periodically. Complete static schedules can still be obtained, but in most cases the interconnecting FIFO queues can be made much shorter because more details of an actor's behavior can be exposed, reducing unneeded buffering otherwise imposed by the SDF model.

The Gabriel system [10] was one of the earliest examples of a design environment that supported the SDF model of computation for both simulation and code generation for DSPs. Gabriel's successor, Ptolemy, extended and improved Gabriel's dataflow simulation and implementation capabilities [54]. Descartes [57] was another early, successful SDF-based code generation system, which was later commercialized along with COSSAP by Cadis (later acquired by Synopsys). The GRAPE-II system [39] supported implementation using the cyclo-static dataflow model.

While many graphical SDF design systems exist, others have employed textual languages whose semantics are close to SDF. The first of these was Silage [33], a declarative language in which all loops are bounded; it was designed to allow DSP algorithms to be efficiently implemented in software or hardware. The DFL (dataflow) language [69] was derived from Silage, and a set of DFL-based implementation tools was commercialized by Mentor Graphics as DSP station, now defunct.

4.6.3 DYNAMIC DATAFLOW: DATA-DEPENDENT FIRING

The term dynamic dataflow is often used to describe dataflow systems that include data-dependent firing and are therefore not static. COSSAP [38] may have been the first true dynamic dataflow simulator; Messerschmitt's Blosim [50], while older, required the user to specify sizes of all FIFO buffers and writes to full FIFOs blocked (while COSSAP buffers grow as needed), so it was not a true dataflow (or Kahn process network) simulator. However, the method used by COSSAP for dynamic dataflow execution was suitable for simulation only, not for embedded systems implementation, and did not provide the same guarantees as Parks [52].

While there are many algorithmic problems or subproblems that can be modeled as SDF, at least some dynamic behavior is required in most cases, so there has long been interest in providing for at least some data-dependent execution of actors in tools, without paying for the cost of full dynamic dataflow.

The original SPW tool from Comdisco (later Cadence, now Synopsys) [16] used a dataflow-like model of computation that was restricted in a different way: each actor had an optional hold signal. The actor would always read the hold signal. If hold was *true*, the actor did not execute; otherwise, it would read one value from each of its other inputs and write one value to each of its outputs. This model is more cumbersome than SDF for static multirate operation, but can express dynamic behaviors that SDF cannot express, and the one-place buffers simplified the generation of hardware implementations. It is a special case of dynamic dataflow (though limited to one-place buffers). Later versions of SPW added full support for SDF and dynamic dataflow.

Boolean-controlled dataflow (BDF) [13], later extended to integer dataflow (IDF) [14] to allow for integer control streams, was an attempt to extend SDF analysis and scheduling techniques to a subset of dynamic dataflow. While in SDF the number of values read or written by each I/O port is fixed, in BDF the number of values read by any port can depend on the value of a Boolean data value read or written by some other port called a "control port." In the BDF model, as in SDF, each port of each actor is annotated with the number of values transferred (read or written) during one firing. However, in the case of BDF, instead of a compile-time constant the number of values transferred can be an expression containing Boolean-valued variables. These variables are the data values that arrive at, or are written by a control port, a port of the BDF actor that must transfer one value per firing. The SPW model, then, is a special case of BDF where there is one control port that controls all other ports, and the number of values transferred must be one or zero. IDF allows control streams to be integers.

While BDF is still restricted compared to general dataflow, it is sufficiently expressive to be Turing equivalent. Unfortunately, this means that a number of analysis problems, including the important question of whether buffer sizes can be bounded, are undecidable in general [13]. Nevertheless, clustering techniques can be used to convert a BDF graph into a reduced graph consisting of clusters; each individual cluster has a static or quasi-static schedule, and only a subset of the buffers connecting clusters can potentially grow to unbounded size. This approach was taken in the dataflow portion of Synopsys's System Studio [12], for example.

Zepter and Grötker [70] used BDF for hardware synthesis. Their approach can be thought of as a form of interface synthesis: given a set of predesigned components that read and write data periodically, perhaps with different periods and perhaps controlled by enable signals, together with a behavioral description of each component as a BDF model, their Aden tool synthesized the required control logic and registers to correctly interface the components.

Later work based on Zepter's concept relaxed the requirement of periodic component behavior and provided more efficient solutions, but only handled the SDF case [35].

4.6.4 DATAFLOW WITH FIXED QUEUES

Bounding queue sizes between dataflow actors makes a dataflow model finite state (and thus easier to analyze than Kahn's Turing-complete networks), although it can impose deadlocks that would not occur in the more general model. This is often a reasonable trade-off, however, and is a more realistic model of practical implementations.

SHIM (software/hardware integration medium) [23] is one such model: complementary to SDF, it allows data-dependent communication patterns but requires FIFO sizes to be statically bounded. Its data-dependent communication protocols allow it to model such things as variable-bitrate decompression (e.g., as used in JPEG and related standards), which is beyond static dataflow models. The SHIM model guarantees the same I/O determinacy of Kahn but because it is finite state, it enables very efficient serial code generation based on static analysis [24], recursion [65], deterministic exceptions [66], and static deadlock detection [61], none of which would be realistic in a full Kahn network setting.

Perhaps the most commonly used fixed-sized queues are SystemC's `sc_fifo<t>` channels [29]. The task-transfer-level approach [67] illustrates how such queues can be used. It starts from an executable Kahn network (i.e., with unbounded queues) written in Yapi [21]. The model is then refined to use fixed-sized queues.

4.7 HETEROGENEOUS MODELS

Each of the models of computation discussed earlier is a compromise because most advantages arise from restrictions that limit what a model can express, making certain behaviors either awkward or impossible to express.

Heterogeneous models attempt to circumvent this trade-off by offering the best of multiple worlds. Ptolemy [15] was the first systematic approach that accepted the value of domain-specific tools and models of computation, yet sought to allow designers to combine more than one model in the same design. Its approach to heterogeneity was to consistently use block diagrams for designs but to assign different semantics to a design based on its *domain*, or model of computation. Primitive actors were designed to function only in particular domains, and hierarchical designs were simulated based on the rules of the current domain. To achieve heterogeneity,

Ptolemy allowed a hierarchical design belonging to one domain (e.g., SDF) to appear as an atomic actor following the rules of another domain (e.g., DE simulation). For this to work correctly, mechanisms had to be developed to synchronize schedulers operating in different domains.

The original Ptolemy, now called Ptolemy Classic, was written in C++ as an extensible class library; its successor, Ptolemy II [20], was thoroughly redesigned and written in Java. Ptolemy Classic treated domains as either untimed (e.g., dataflow) or timed (e.g., DE). From the perspective of a timed domain, actions in an untimed domain appeared to be instantaneous. In the case of a mixture of SDF and DEs, for example, one might represent a computation with SDF components and the associated delay in the DE domain, thus separating the computation from the delay involved in a particular implementation. When two distinct timed domains are interfaced, a global time is maintained, and the schedulers for the two domains are kept synchronized. Buck et al. [15] describe the details of synchronizing schedulers across domain boundaries.

Ptolemy Classic was successful as a heterogeneous simulation tool, but it possessed a path to implementation (in the form of generated software or HDL code) only for dataflow domains (SDF and BDF). Furthermore, all of its domains shared the characteristic that atomic blocks represented processes and connections represented data signals.

One of the more interesting features added by Ptolemy II was its approach to hierarchical control [27]. The concept, flowing logically out of the Ptolemy idea, was to extend Statecharts to allow for complete nesting of data-oriented domains (e.g., SDF and DE) as well as synchronous-reactive domains, working out the semantic details as required. Ptolemy II calls designs that are represented as state diagrams *modal designs*. If the state symbols represent atomic states, we simply have an extended finite-state machine (extended because, as in Statecharts, the conditions and actions on the transition arcs are not restricted to Boolean signals). However, the states can also represent arbitrary Ptolemy subsystems. When a state is entered, the subsystem contained in the state begins execution. When an outgoing transition occurs, the subsystem halts its execution. The so-called time-based signals that are presented to the modal model propagate downward to the subsystems that are inside the states.

Girault et al. [27] claim that the semantic issues with Statechart variants identified by von der Beek [68] can be solved by orthogonality: nesting FSMs together with domains providing the required semantics, thereby obtaining, for example, either synchronous-reactive or DE behavior. But this only solves the problem in part because there are choices to be made about the semantics of FSMs nested inside of other FSMs.

A similar project to allow for full nesting of hierarchical FSMs and dataflow, with a somewhat different design, was part of Synopsys's System Studio [12]. System Studio's functional modeling combines dataflow models with Statechart-like control models that have semantics very close to those of SyncCharts [1], as both approaches started with the Esterel semantics. Like Ptolemy II, System Studio permits any kind of model to be nested inside of any other, and state transitions cause interior subsystems to start and stop. One unique feature of System Studio is that parameter values, which in Ptolemy are set at the start of a simulation run or are compiled in when code generation is performed, can be reset to different values each time a subsystem inside a state transition diagram is started.

There have been several efforts to make the comparison and the combination of models of computation more theoretically rigorous. Lee and Sangionvanni-Vincentelli [45] introduce a metamodel that represents signals as sets of events. Each event is a pair consisting of a value and a tag, where the tags can come from either a totally ordered or a partially ordered set. Synchronous events share the same tag. The model can represent important features of a wide variety of models of computation, including most of those discussed in this chapter; however, in itself it does not lead to any new results.

4.8 CONCLUSIONS

This chapter has described a wide variety of approaches, all of them deserving of more depth than could be presented here. Some approaches have been more successful than others.

It should not be surprising that there is resistance to learning new languages and models. It has long been argued that system-level languages are most successful when the user does not realize that a new language is being proposed. Accordingly, there has sometimes been less resistance to graphical approaches (especially when the text used in such an approach is from a familiar programming language such as C++ or an HDL), and to class libraries that extend C++ or Java. Lee [41] argues that such approaches are really languages, but acceptance is sometimes improved if users are not told this.

Approaches based on Kahn process networks and dataflow have been highly successful in a variety of application areas that require digital signal processing. These include wireless; audio, image, and video processing; radar, 3D graphics; and many others. Chapter 3 gives a detailed comparison of the tools used in this area. Hierarchical control tools, such as those based on Statechart and Esterel, have also been successful, though their use is not quite as widespread. Most of the remaining tools described here have been found useful in smaller niches, though some of these are important. Nevertheless, it seems higher-level tools are underused. As the complexity of systems to be implemented continues to increase, designers that exploit domain-specific system-level approaches will benefit by doing so.

REFERENCES

1. C. André. Representation and analysis of reactive behaviors: A synchronous approach. In *Proceedings of Computational Engineering in Systems Applications (CESA)*, pp. 19–29, Lille, France, July 1996.
2. L. Arditi, A. Bouali, H. Boufaied, G. Clave, M. Hadj-Chaib, L. Leblanc, and R. de Simone. Using Esterel and formal methods to increase the confidence in the functional validation of a commercial DSP. In *Proceedings of the ERCIM Workshop on Formal Methods for Industrial Critical Systems (FMICS)*, Trento, Italy, June 1999.
3. A. Benveniste, P. Caspi, S. A. Edwards, N. Halbwachs, P. L. Guernic, and R. de Simone. The synchronous languages 12 years later. *Proceedings of the IEEE*, 91(1):64–83, January 2003.
4. G. Berry. A hardware implementation of pure Esterel. In *Proceedings of the International Workshop on Formal Methods in VLSI Design*, Miami, FL, January 1991.
5. G. Berry. The foundations of Esterel. In *Proof, Language and Interaction: Essays in Honour of Robin Milner*. G. Plotkin, C. Stirling, and M. Tofte (Eds.) MIT Press, Cambridge, MA, 2000.
6. G. Berry. SCADE: Synchronous design and validation of embedded control software. In S. Ramesh and P. Sampath, editors, *Next Generation Design and Verification Methodologies for Distributed Embedded Control Systems*, pp. 19–33. Springer, Dordrecht, the Netherlands, 2007.
7. G. Berry and G. Gonthier. The Esterel synchronous programming language: Design, semantics, implementation. *Science of Computer Programming*, 19(2):87–152, November 1992.
8. S. S. Bhattacharyya and E. A. Lee. Looped schedules for dataflow descriptions of multirate signal processing algorithms. *Journal of Formal Methods in System Design*, 5(3):183–205, December 1994.
9. S. S. Bhattacharyya, R. Leupers, and P. Marwedel. Software synthesis and code generation for signal processing systems. *IEEE Transactions on Circuits and Systems—II: Analog and Digital Signal Processing*, 47(9):849–875, September 2000.
10. J. C. Bier, E. E. Goei, W. H. Ho, P. D. Lapsley, M. P. O'Reilly, G. C. Sih, and E. A. Lee. Gabriel: A design environment for DSP. *IEEE Micro*, 10(5):28–45, October 1990.
11. A. Bouali. XEVE, an Esterel verification environment. In *Proceedings of the International Conference on Computer-Aided Verification (CAV)*, volume 1427 of Lecture Notes in Computer Science, pp. 500–504, University of British Columbia, Vancouver, CA, 1998.
12. J. Buck and R. Vaidyanathan. Heterogeneous modeling and simulation of embedded systems in El Greco. In *Proceedings of the Eighth International Workshop on Hardware/Software Codesign (CODES)*, San Diego, CA, May 2000.
13. J. T. Buck. Scheduling dynamic dataflow graphs with bounded memory using the token flow model. PhD thesis, University of California, Berkeley, CA, 1993. Available as UCB/ERL M93/69.
14. J. T. Buck. Static scheduling and code generation from dynamic dataflow graphs with integer-valued control streams. In *Proceedings of the Asilomar Conference on Signals, Systems & Computers*, pp. 508–513, Pacific Grove, CA, October 1994.
15. J. T. Buck, S. Ha, E. A. Lee, and D. G. Messerschmitt. Ptolemy: A framework for simulating and prototyping heterogeneous systems. *International Journal of Computer Simulation*, 4:155–182, April 1994.
16. Cadence Design Systems. *Signal Processing Work System (SPW)*, Cadence Design Systems, San Jose, CA, 1994.
17. L. Cai and D. Gajski. Transaction level modeling: An overview. In *Proceedings of the International Conference on Hardware/Software Codesign and System Synthesis (CODES+ISSS)*, pp. 19–24, Newport Beach, CA, October 2003.
18. R. Colgren. *Basic MATLAB, Simulink, and Stateflow*. AIAA, Reston, VA, 2007.

19. J. B. Dabney and T. L. Harman. *Mastering Simulink*. Prentice Hall, Upper Saddle River, NJ, 2003.

20. J. Davis, R. Galicia, M. Goel, C. Hylands, E. A. Lee, J. Liu, X. Liu et al. Heterogeneous concurrent modeling and design in Java. Technical report UCB/ERL M98/72, EECS, University of California, Berkeley, CA, November 1998.

21. E. A. de Kock, W. J. M. Smits, P. van der Wolf, J.-Y. Brunel, W. M. Kruijtzer, P. Lieverse, K. A. Vissers, and G. Essink. YAPI: Application modeling for signal processing systems. In *Proceedings of the 37th Design Automation Conference*, pp. 402–405, Los Angeles, CA, June 2000.

22. S. A. Edwards. An Esterel compiler for large control-dominated systems. *IEEE Transactions on Computer-Aided Design of Integrated Circuits and Systems*, 21(2):169–183, February 2002.

23. S. A. Edwards. Concurrency and communication: Lessons from the SHIM project. In *Proceedings of the Workshop on Software Technologies for Future Embedded and Ubiquitous Systems (SEUS)*, Vol. 5860 of Lecture Notes in Computer Science, pp. 276–287, Newport Beach, CA, November 2009. Springer, Berlin, Heidelberg, and New York.

24. S. A. Edwards and O. Tardieu. Efficient code generation from SHIM models. In *Proceedings of Languages, Compilers, and Tools for Embedded Systems (LCTES)*, pp. 125–134, Ottawa, Canada, June 2006.

25. M. Engels, G. Bilsen, R. Lauwereins, and J. Peperstraete. Cyclo-static dataflow: Model and implementation. In *Proceedings of the Asilomar Conference on Signals, Systems & Computers*, Vol. 1, pp. 503–507, Pacific Grove, CA, October 1994.

26. D. D. Gajski, J. Zhu, R. Dömer, A. Gerstlauer, and S. Zhao. *SpecC: Specification Language and Methodology*. Kluwer, Boston, MA, 2000.

27. A. Girault, B. Lee, and E. A. Lee. Hierarchical finite state machines with multiple concurrency models. *IEEE Transactions on Computer-Aided Design of Integrated Circuits and Systems*, 18(6):742–760, June 1999.

28. T. Grötker. Private communication, 2004.

29. T. Grötker, S. Liao, G. Martin, and S. Swan. *System Design with SystemC*. Kluwer, Boston, MA, 2002.

30. N. Halbwachs, P. Caspi, P. Raymond, and D. Pilaud. The synchronous data flow programming language LUSTRE. *Proceedings of the IEEE*, 79(9):1305–1320, September 1991.

31. D. Harel. Statecharts: A visual formalism for complex systems. *Science of Computer Programming*, 8(3):231–274, June 1987.

32. D. Harel and E. Gery. Executable object modeling with statecharts. *IEEE Computer*, 30(7):31–42, July 1997.

33. P. N. Hilfinger, J. Rabaey, D. Genin, C. Scheers, and H. D. Man. DSP specification using the Silage language. In *Proceedings of the IEEE International Conference on Acoustics, Speech, & Signal Processing (ICASSP)*, pp. 1057–1060, Albuquerque, NM, April 1990.

34. C. A. R. Hoare. *Communicating Sequential Processes*. Prentice Hall, Upper Saddle River, NJ, 1985.

35. J. Horstmannshoff and H. Meyr. Efficient building block based RTL code generation from synchronous data flow graphs. In *Proceedings of the 37th Design Automation Conference*, pp. 552–555, Los Angeles, CA, 2000.

36. INMOS Limited. *Occam 2 Reference Manual*. Prentice Hall, Upper Saddle River, NJ, 1988.

37. G. Kahn. The semantics of a simple language for parallel programming. In *Information Processing 74: Proceedings of IFIP Congress 74*, pp. 471–475, Stockholm, Sweden, August 1974. North-Holland, Amsterdam.

38. J. Kunkel. COSSAP: A stream driven simulator. In *IEEE International Workshop on Microelectronics in Communications*, Interlaken, Switzerland, March 1991.

39. R. Lauwereins, M. Engels, M. Ade, and J. Perperstraete. GRAPE-II: A tool for the rapid prototyping of multi-rate asynchronous DSP applications on heterogeneous multiprocessors. *IEEE Computer*, 28(2):35–43, February 1995.

40. P. Le Guernic, T. Gautier, M. Le Borgne, and C. Le Maire. Programming real-time applications with SIGNAL. *Proceedings of the IEEE*, 79(9):1321–1336, September 1991.

41. E. A. Lee. Embedded software. In M. Zelowitz, editor, *Advances in Computers*, Vol. 56. Academic Press, Waltham, MA, September 2002.

42. E. A. Lee and D. G. Messerschmitt. Static scheduling of synchronous data flow programs for digital signal processing. *IEEE Transactions on Computers*, C-36(1):24–35, January 1987.

43. E. A. Lee and D. G. Messerschmitt. Synchronous data flow. *Proceedings of the IEEE*, 75(9):1235–1245, September 1987.

44. E. A. Lee and T. M. Parks. Dataflow process networks. *Proceedings of the IEEE*, 83(5):773–801, May 1995.

45. E. A. Lee and A. Sangiovanni-Vincentelli. A framework for comparing models of computation. *IEEE Transactions on Computer-Aided Design of Integrated Circuits and Systems*, 17(12):1217–1229, December 1998.

46. M. E. Lesk and E. Schmidt. LEX: A lexical analyzer generator. Computing Science Technical Report 39, AT&T Bell Laboratories, Murry Hill, NJ, 1975.

47. S. Malik. Analysis of cyclic combinational circuits. *IEEE Transactions on Computer-Aided Design of Integrated Circuits and Systems*, 13(7):950–956, July 1994.

48. F. Maraninchi. The Argos language: Graphical representation of automata and description of reactive systems. In *Proceedings of the IEEE Workshop on Visual Languages*, Kobe, Japan, October 1991.

49. A. J. Martin. Programming in VLSI: From communicating processes to delay-insensitive circuits. In C. A. R. Hoare, editor, *UT Year of Programming Institute on Concurrent Programming*, pp. 1–64. Addison-Wesley, Boston, MA, 1989.

50. D. G. Messerschmitt. A tool for structured functional simulation. *IEEE Journal on Selected Areas in Communications*, SAC-2(1):137–147, January 1984.

51. W. Meyer, A. Seawright, and F. Tada. Design and synthesis of array structured telecommunication processing applications. In *Proceedings of the 34th Design Automation Conference*, pp. 486–491, Anaheim, CA, 1997.

52. T. M. Parks. Bounded scheduling of process networks. PhD thesis, University of California, Berkeley, CA, 1995. Available as UCB/ERL M95/105.

53. J. L. Pino, S. S. Bhattacharyya, and E. A. Lee. A hierarchical multiprocessor scheduling system for DSP applications. In *Proceedings of the Asilomar Conference on Signals, Systems & Computers*, Pacific Grove, CA, November 1995.

54. J. L. Pino, S. Ha, E. A. Lee, and J. T. Buck. Software synthesis for DSP using Ptolemy. *Journal of VLSI Signal Processing*, 9(1):7–21, January 1995.

55. D. Potop-Butucaru. Optimizations for faster execution of Esterel programs. In *Proceedings of the International Conference on Formal Methods and Models for Codesign (MEMOCODE)*, pp. 227–236, Mont St. Michel, France, June 2003.

56. D. Potop-Butucaru, S. A. Edwards, and G. Berry. *Compiling Esterel*. Springer, Berlin, January 2007.

57. S. Ritz, M. Pankert, and H. Meyr. High level software synthesis for signal processing systems. In *International Conference on Application Specific Array Processors*, pp. 679–693, Berkeley, CA, August 1992.

58. J. Rumbaugh, I. Jacobson, and G. Booch. *The Unified Modeling Language Reference Manual*, 2nd edn. Addison-Wesley, Boston, MA, 2004.

59. A. Seawright and F. Brewer. Clairvoyant: A synthesis system for production-based specification. *IEEE Transactions on Very Large Scale Integration (VLSI) Systems*, 2(2):172–185, June 1994.

60. A. Seawright, U. Holtmann, W. Weyer, B. Pangrle, R. Verbrugghe, and J. Buck. A system for compiling and debugging structured data processing controllers. In *Proceedings of the European Design Automation Conference*, Geneva, Switzerland, September 1996.

61. B. Shao, N. Vasudevan, and S. A. Edwards. Compositional deadlock detection for rendezvous communication. In *Proceedings of the International Conference on Embedded Software (Emsoft)*, pp. 59–66, Grenoble, France, October 2009.

62. T. R. Shiple, G. Berry, and H. Touati. Constructive analysis of cyclic circuits. In *Proceedings of the European Design and Test Conference*, pp. 328–333, Paris, France, March 1996.

63. T. Stefanov, C. Zissulescu, A. Turjan, B. Kienhuis, and E. Deprettere. System design using Kahn process networks: The Compaan/Laura approach. In *Proceedings of Design, Automation, and Test in Europe (DATE)*, Paris, France, February 2004.

64. S. Sutherland, S. Davidmann, and P. Flake. *SystemVerilog for Design: A Guide to Using SystemVerilog for Hardware Design and Modeling*, 2nd edn. Springer, Berlin, 2006.

65. O. Tardieu and S. A. Edwards. R-SHIM: Deterministic concurrency with recursion and shared variables. In *Proceedings of the International Conference on Formal Methods and Models for Codesign (MEMOCODE)*, p. 202, Napa, CA, July 2006.

66. O. Tardieu and S. A. Edwards. Scheduling-independent threads and exceptions in SHIM. In *Proceedings of the International Conference on Embedded Software (Emsoft)*, pp. 142–151, Seoul, Korea, October 2006.

67. P. van der Wolf, E. de Kock, T. Henriksson, W. M. Kruijtzer, and G. Essink. Design and programming of embedded multiprocessors: An interface-centric approach. In *Proceedings of the International Conference on Hardware/Software Codesign and System Synthesis (CODES+ISSS)*, pp. 206–216, Stockholm, Sweden, September 2004.

68. M. von der Beeck. A comparison of statecharts variants. In *Formal Techniques in Real-Time and Fault-Tolerant Systems: Third International Symposium Proceedings*, Vol. 863 of Lecture Notes in Computer Science. Springer, Lübeck, Germany, 1994.

69. P. Willekens, D. Devisch, M. V. Canneyt, P. Conflitti, and D. Genin. Algorithm specification in DSP station using data flow language. *DSP Applications*, 3(1):8–16, January 1994.

70. P. Zepter and T. Grötker. Generating synchronous timed descriptions of digital receivers from dynamic data flow system level configurations. In *Proceedings of the European Design Automation Conference (EDAC)*, p. 672, February 1994.

SoC Block-Based Design and IP Assembly

Yaron Kashai

5

CONTENTS

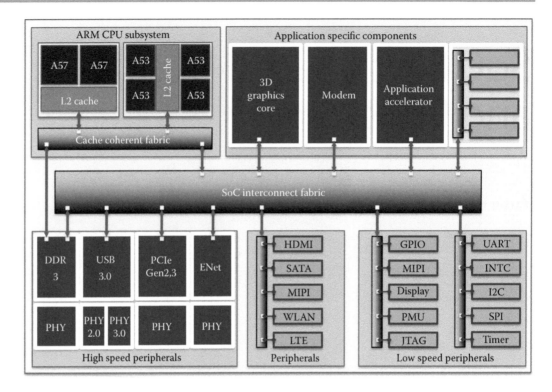

FIGURE 5.1   An ARM-based multicore system-on-chip block diagram.

System-on-chip (SoC) designs are ubiquitous—they can be found at the core of every cell phone and tablet [1]. Most SoCs are *platform based* [2]: built on a foundation architecture with a processing subsystem comprising numerous cores and a large number of support and peripheral *intellectual property* (IP) blocks. Processors, IP blocks, and memories communicate via a complex interconnect, typically a cascade of sophisticated *network-on-chip* (NoC) modules. Figure 5.1 depicts a block diagram of an ARM-based SoC. SoCs require embedded software in order to perform basic functionality such as booting, power management, and clocking control. The embedded software is tightly matched to the hardware and is considered an integral part of the design. The whole design, NoCs, and memory subsystem in particular are highly tuned for the desired SoC functionality and performance.

Complexity grows with every new generation of SoC designs. The newest crop features up to eight general-purpose compute cores, with 64-bit cores rapidly becoming the norm, along with several special-purpose processors, coherent interconnect with high-bandwidth memory access, and hundreds of integrated peripheral modules. While some of the complexities address mobile computing needs, future designs will undoubtedly become even more highly integrated. The Internet of Things,* which is identified as the next driver for computing [3], poses new challenges. Extreme low power and integration of major analog modules such as radio transceivers will continue to push up complexity.

SoC design projects face severe time constraints due to ever-present time-to-market pressure. This conflicts with the need to integrate a growing number of features: a large number of wired and wireless communication protocols, a variety of sensors, more processors, higher-bandwidth interconnect, and more sophisticated system-wide capabilities such as power and security management. This is made possible by relying on a block-based design methodology, where IP blocks—predesigned and verified modules—are composed to create the SoC. While this methodology is not new [4], the extent to which it is used is unprecedented.

---

* The *Internet of Things* is a vision of computing devices embedded in common objects such as household appliances, sensors embedded in buildings, cars, the electric grid, and medical devices. All these computing devices are connected to the Internet and therefore can communicate with each other and the existing World Wide Web.

5.1 MAJOR CHALLENGES WITH BLOCK-BASED DESIGN METHODOLOGY

Knowledge transfer: IP block-based design methodology faced an uphill struggle when it was first introduced in the early 2000s: what seemed initially like an obvious design accelerator turned out in practice to be a problematic and error-prone proposition. While using a ready-made design saves on design time, integration time typically balloons, with rampant integration and configuration errors. The contrast between the optimistic 2006 view expressed in Reference 5 and the more recent experience reported in Reference 6 highlights the difficulty. The reason is simple: the knowledge needed to configure and integrate the IP block resides with the original designer and not the system integrator. Packaging IP blocks properly such that all the information necessary for integration is available remains a major challenge.

Correct configuration: Composing an SoC out of hundreds of configurable components produces an astronomically large configuration space. Many commonly used IP blocks, such as standard communication protocols, have tens of parameters each. There are complex interdependencies between configuration parameters of each IP block, and composing these together into a coherent system is even more challenging. Therefore, configuration requires automated assistance and validation. Many of the hard-to-find bugs at the integration level turn out to be configuration related [7].

Integrating software: Software is an indispensable part of an SoC design. Some of the SoC functionality is unavailable without firmware being present. A case in point is a power management processor that provides software control over switching low power modes for the various SoC modules. Low-level drivers are often required as well, because some of the hardware modules require complex initialization sequences. The ubiquity of software has several major implications:

- Some software must be ready before SoC simulation can commence.
- The software must match the specific SoC configuration.
- Simulation of SW execution is inherently slow, posing a major throughput challenge.

Functional verification and performance validation: Functional verification is arguably the biggest challenge for block-based SoC designs. Verification methodology is adopting modularization with the introduction of *verification IP blocks* (VIPs) [8], which are paired with matching design IP blocks. Such VIPs are most common for communication protocols such as Ethernet, USB, on-chip busses, and memory interfaces. Yet the impact of VIPs on reducing the verification effort is considerably smaller than the impact of IP blocks on the design process. One key reason is the huge functional space presented by modern SoCs, that is, the combination of computation and communication with auxiliary functions such as controlling power, reset, clocking, and security. Many of the auxiliary functions are partially controlled by software, making their verification more challenging still. Some of the system properties, performance first and foremost, are *emergent*—they can only be observed when the system is integrated.

5.2 METADATA-DRIVEN METHODOLOGY

Design composition used to be a manual task in which a design engineer assembles the SoC by editing hardware description language (HDL) files, instantiating IP blocks, and connecting them to each other. Complexity and time constraints make this approach impractical—virtually all SoCs today are constructed using *generator programs*. These programs receive as input some high-level definition of the component blocks, as well as the desired configuration of the SoC. The term "metadata" is often used to describe this information [9]. The generator programs produce the top-level HDL of the composed design as well as other necessary artifacts for verification and implementation. Such generator programs range in sophistication from basic assembly [10] producing standard HDL output to experimental high-level design languages that are compositional and directly synthesizable like Chisel [11] and metagenerators like Genesis-2 [12].

The ability to generate new design configurations with little manual effort and minimal time investment is sometimes used to explore design alternatives, most often focusing on the SoC interconnect. In contrast with high-level architectural models, designs obtained by metadata-driven generation are typically timing accurate, and this improves the accuracy of performance estimation.

Many of the IP blocks are highly configurable, and some of them are created on demand by their own specific generator programs, customized by additional metadata input. Two broadly used examples are Cadence Tensilica configurable processor cores, whose instruction sets can be optimized to meet desired computation, power, and area requirements [13], and NoCs, whose topology and throughput are customized for every instance [14,15].

Besides hardware design, other artifacts are produced by automatic generation. These include generating verification environments, test benches for performance validation, and some low-level software drivers as well. Such generators require additional metadata representing verification components, test scenarios, and HW/SW interfaces. Novel work is attempting to capture aspects of the architecture and leverage that for verification. An example of such a formalism is iPave [16].

A common form of metadata used to represent the design composition and its memory architecture is the IP-XACT standard, which has been under development for over a decade [17]. Early development of this standard was performed by the Spirit initiative; see discussion in the previous version of this chapter [18]. Several automation tools are focused on managing IP-XACT metadata [19]. IP-XACT primary objects represent the following design artifacts:

- Components and component instances
- Bus interfaces/interface connections
- Ports/ad hoc connections
- Design configurations
- Filesets, listing source files, and directories
- Registers

In addition, IP-XACT accepts *vendor extensions* that can be used to add metadata that is interpreted by individual tools (the semantics of such extensions is not defined by the standard). Design teams and EDA companies utilize IP-XACT with vendor extensions to implement various integration flows. An example of such homegrown automation is reported in Reference 20.

While IP-XACT was conceived as a complete standard for IP packaging and integration, it has several shortcomings. It does not provide sufficient detail for registers, memory architectures, interrupts, and resets. It does not address cross-cutting design concerns such as timing, power management, and design for test. Because the standard is focused on design modeling, capturing verification artifacts is only possible in a rudimentary way.

Other standards have been created to supplement metadata where needed. SystemRDL [21] is a standard focused on detailed representation of registers and memories addressing both the hardware and software viewpoints. CPF [22] and UPF [23] are standards for power management.

In addition to standard formats, there are many upcoming and ad hoc formats ranging from XML to spreadsheets capturing various supplementary aspects required for composition, verification, and analysis. The extensive reliance on generators makes metadata creation, validation, and management a central requirement for SoC integration. Means of verifying the compatibility of metadata with respective design data are emerging in commercial products, leveraging formal methods. Examples include the Cadence Jasper connectivity verification application [24] and Atrenta's SpyGlass [25]. Support and automation of metadata remains an active development area.

5.3 IP PACKAGING

The block-based methodology creates a separation between the team developing the IP block and the team integrating it in a context of an SoC. The two teams often belong to different organizations, may be geographically distant, and can have limitations on information sharing due to IP concerns. This separation requires a formal handoff between the teams, such that all necessary information is available at the time of integration. Primarily, the design itself, along with a test bench and software drivers are packaged. Another major aspect of this handoff is metadata describing the IP block in terms of structure, interfaces, configurability, low power behavior, and test plan. The process of assembling the data for the handoff is called "IP packaging." It is part of an automatic release procedure of the IP. Several commercial products such as Magillem IP Packager [26] and ARM Duolog Socrates [27] are aimed at automating this task.

Since an SoC integrates many IP blocks, it is highly desired that all IPs are packaged the same way, providing the same contents. Such uniformity is a strong requirement when considering automation. The uniformity requirement is rarely met, however, because the form and contents of metadata are dependent on the IP vendor, while the automation requirements can be specific to the integrator's design methodology. This state of affairs is owing to the rapid development of integration technology, with standards and tools struggling to catch up with the leading integration projects. As a result of this disparity, SoC design teams must devote effort and resources to repackaging IP blocks, filling in missing metadata aspects [28].

5.4 IP CONFIGURATION AND COMPOSITION

Many IP blocks are configurable—they may be adapted to work with several interface widths at varying frequencies, can enable or disable certain protocol extensions, could optionally be switched to a low-power mode, and so on. Some of these configuration options are static and determined during instantiation. Others are dynamic and sometimes software controlled.

Configuration options are often codependent: a certain mode of operation may only be available for particular interface configurations. Such interdependencies are spelled out in the IP block's functional specification or datasheet. At the same time, the complexity of configuring even a single IP can be significant to the point that automation-guided configuration is often desired [29]. Configurability of VIPs is more complex still: these blocks need to match their design counterparts and in addition their verification-related features could be configured. For instance, a protocol VIP can be active (driving stimulus) or passive (monitoring only). Verification features such as enabling the collection of functional coverage, and the logging of events to a trace file can be configured with finer granularity, the selection of test sequences, the hookup of clock and reset conditions, and the application of additional constraints.

System composition is typically automated by one of the metadata-driven generators, most commonly using IP-XACT as input. This requires capturing all configuration information in metadata to ensure that the IP block instances are configured correctly. The composition process results in a top-level HDL file instantiating and connecting the IPs, as well as a top-level IP-XACT file. Similarly, automatically generated verification environments must instantiate properly configured VIPs. Verification IP configuration is not covered by current standards; instead there are various vendor-provided formats and GUIs for that purpose. An example of a configuration GUI is provided in Figure 5.2.

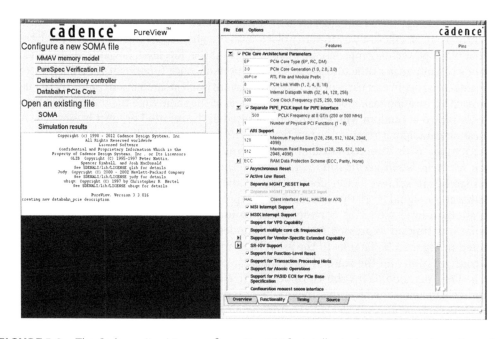

FIGURE 5.2   The Cadence PureView configuration GUI for intellectual property block configuration.

A *composed* system is partially configured:

- Some static configuration aspects such as connectivity are determined.
- Other static configuration options may be parameterized and will be fully determined only when a simulation model is built and elaborated.
- Dynamic configuration options are yet undetermined and will be finalized by the boot code executing during SoC initialization.

SoCs are configurable as a whole, in addition to the configuration of the integrated IPs. An SoC can have multiple modes of operation, some of which are required for functional testing as discussed in Section 5.5 below. Other configurations may be needed to support different market requirements. These configurations are mostly software controlled, but some modes may be hardwired depending on pin or pad ring connectivity.

The many configuration choices of individual IP blocks and the system as a whole span a huge *configuration space*, which turns out to be very sparse. Legal configurations, yielding a functional system, are rare—most configuration combinations are illegal and will result in a failure. Unfortunately, such failures can be obscure and hard to detect, being virtually indistinguishable from functional design errors. Recognizing that the identification of configuration errors through functional verification is hugely inefficient and expensive, there are growing efforts to ensure "correct by construction" configuration. Approaches range from rule-based automated guidance to autogeneration of configurations based on higher-level requirements [30]. An example for autogeneration is the OpenCores SOC_maker tool [31]. Both approaches rely on capturing configuration parameters and their relationships.

5.5 FUNCTIONAL VERIFICATION AND PERFORMANCE VALIDATION

Some design modules, most notably NoCs, are created by automatic generators. While presumed *correct by construction*, there is still a need to verify the functionality of such modules to demonstrate that the metadata driving the generator was correct and the generated result meets the expected functional requirements. This calls for additional automation, generating the test environment and the functional tests to preserve and enhance the time and labor savings. Tools such as the Cadence Interconnect WorkBench [32] pair with NoC generators like ARM CoreLink AMBA Designer [15] to meet this need.

Dynamic functional verification converged over the last two decades to a finely honed metric-driven methodology that is making extensive use of directed-random stimulus generation. This is currently the most efficient way to verify block-level designs, augmented by formal model checking that offers powerful complementary capabilities. At the SoC level, however, these block-level methods run into major difficulties [33]. The SoC presents huge functional variability, extremely large sequential depth and parts of the behavior that are implemented in software. These qualities make each verification task much more difficult, and at the same time the probability that some behavior picked at random will actually be exercised in the released product is very low. A new verification paradigm is required to practically verify SoC functionality.

Difficulties with directed-random test generation led to the development of *software-driven scenario-based verification* [34]. This methodology leverages the SoC software programmability to achieve better controllability and observability. Rather than randomizing stimulus at the signal level, this approach uses a programming model of the SoC to randomly generate high-level scenarios corresponding to interesting use cases. An example of such a scenario definition expressed in a unified modeling language (UML) activity diagram is depicted in Figure 5.3. Thanks to the associated model, the resulting tests are aware of the SoC configuration and the state of its resources, ensuring only legitimate stimulus is being driven. This methodology is shown to be effective at validating complex user scenarios that are unlikely to be reached by other means, blending in both mainline functionality and auxiliary events such as power cycles, resets, and dynamic reconfigurations [35]. An added benefit is the ability to use this methodology in emulation and even postsilicon (i.e., on the actual manufactured chip) as means of failure debug.

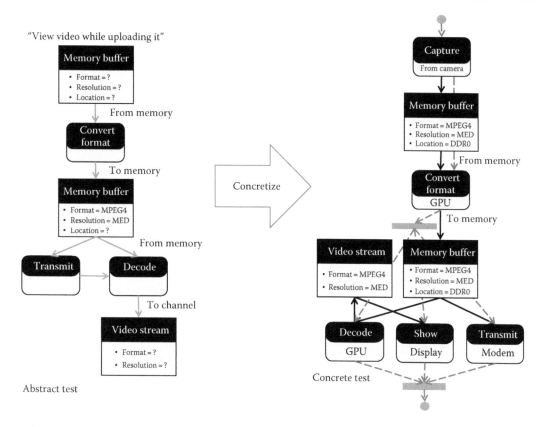

FIGURE 5.3   An abstract scenario captured as a UML activity diagram is automatically concretized to become a system-level test.

Another practical challenge with SoC verification is the throughput of simulation. SoC timing-accurate simulation models are huge, owning to the large scale of integration. The addition of software as firmware and low-level drivers makes the situation much worse: the huge SoC model must spin, simulating numerous CPU cycles processing the embedded software. The throughput constraint is too severe to allow booting an operating system on a cycle-accurate RTL model. At the same time, it is absolutely necessary to exercise the system, including low-level software, to flush out configuration problems and logic errors. The industry's current solution is leveraging hardware acceleration, which is in growing demand and is ubiquitous in SoC verification. However, even with acceleration, the performance of software on the simulated SoC remains poor.

A new hybrid-acceleration methodology is offering a significant improvement over plain acceleration. The key to the speedup is migrating the compute subsystem to a virtual platform while maintaining a coherent memory representation between the accelerated portion and the virtual platform. This can be done economically because only a small part of the memory is actually accessed by both portions. This partition lets the main processors run at virtual platform speed (same or better than real time), while the rest of the design executes in timing-accurate, accelerated simulation. Execution in the virtual platform is interrupted when memory locations mapped to accelerated modules are accessed, triggering a backdoor mechanism that synchronizes both memory representations (the one in the accelerator and the relevant portion of the virtual platform memory). Accelerated code can generate interrupts that will cause synchronization as well. As long as the frequency of events requiring synchronization is sufficiently low, performance gains are substantial. Results of this methodology show a 10-fold speedup over *plain* acceleration on major industrial designs [36].

Besides acceleration speed, the capacity of the hardware accelerator is a concern as well. Accelerator capacity is at a premium and it is often best utilized when a design is thoughtfully trimmed down to exclude modules that do not participate in the test scenario. Creating a whole new design configuration per test scenario is prohibitively expensive if done manually, but thanks to metadata-driven generation such configurations can be automatically created, following a very small change in the metadata. See Chapter 19, for a more in-depth discussion of acceleration techniques.

Much of the SoC architecture design effort is aimed at optimizing performance, which is a major selling feature for most SoCs. High-level models of the architecture are developed and tested against expected usage scenarios. Many design variations are considered and carefully refined. Yet once the design is implemented, validating SoC performance is a major challenge. On the one hand, after the design is released in product form, it is not hard to assess the customer-perceived performance—as evident by the myriad smartphone performance benchmarks. On the other hand, it is really hard to extract meaningful performance predictions from a simulation involving massive concurrent activity, typical of functional verification scenarios. Instead, a dedicated performance suite is needed, targeting specific critical communication and computation paths in a systematic way over a set of configurations chosen to represent the most common use cases [37]. Such performance suites typically focus on the SoC interconnect and memory interface, and may be tuned specifically to target market applications such as streaming video or high-end graphics for gaming. Not only does this approach provide reliable performance predictions that should correlate to the architectural predictions, but it also helps identify and isolate specific bottlenecks [38].

5.6 CONCLUSIONS

A decade ago it seemed as if higher degrees of integration could be achieved by "more of the same" tools and methodologies, relying on ever-increasing computing power. Difficulties with block-based design and related metadata formats slowed initial adoption. While some of the challenges that affected them still remain, the scale and complexity of integration saw a tremendous growth, making metadata-driven block-based design indispensable.

SoC design and verification today are driven by metadata. Much of the actual top-level HDL design, as well as key modules such as NoCs, are automatically generated. The verification environment portions of the embedded and driver software are generated and configured automatically. This makes metadata a *golden source* impacting critical aspects. At the same time, much of the metadata is nonstandard, using ad hoc formats such as spreadsheets. Software tools range from commercially available EDA tools through IP provider generators to homegrown automation developed by the SoC integrator. The EDA landscape will continue to evolve as metadata-related tools and formats mature.

Verification is undergoing a disruptive change as well. The scale of functionality the design is capable of and the addition of software as an integral part of the design require a major shift in our approach to verification. Scenario-based software-driven testing emerges as a leading test generation methodology, augmented by sophisticated acceleration techniques. These changes are just starting to take root at the most advanced design projects and will likely continue to evolve and trickle down to the rest of the design community. SoC integration and verification automation remain an active research and development field with a big potential impact on electronic design.

REFERENCES

1. SEMICO, ASIC design starts for 2014 by key end market applications, SEMICO market research report, 2014, online at: http://semico.com/content/asic-design-starts-2014-key-end-market-applications (Accessed on August, 2015).
2. K. Keutzer et al., System-level design: Orthogonalization of concerns and platform-based design, *IEEE Transactions on Computer-Aided Design of Integrated Circuits and Systems*, 19(12), 1523–1543, December 2000.
3. O. Vermesan and P. Friess, editors, *Internet of Things—From Research and Innovation to Market Deployment*, Chapter 3, River Publishing, Aalberg, Denmark, 2014, online at: http://www.internet-of-things-research.eu/pdf/IERC_Cluster_Book_2014_Ch.3_SRIA_WEB.pdf (Accessed on August, 2015).

4. D. Gajski, IP-based design methodology, in *Proceedings of the 36th Design Automation Conference*, New Orleans, LA, 1999, pp. 43–47.

5. R. Saleh et al., System-on-chip: Reuse and integration, *Proceedings of the IEEE*, San Francisco, CA, 94(6), 1050–1069, June 2006.

6. D. Murry and S. Boylan, Lessons from the field—IP/SoC integration techniques that work, White paper, Kyoto, Japan, 2014, online at: http://www.duolog.com/wp-content/uploads/IP-Integration-of-a-Complex-ARM-IP-Based-System.pdf (Accessed on August, 2015).

7. E. Sperling, SoC integration mistakes, February 2014, online at: http://semiengineering.com/experts-at-the-table-soc-integration-mistakes/ (Accessed on August, 2015).

8. D. Mathaikutty, Metamodeling driven IP reuse for system-on-chip integration and microprocessor design, Doctoral thesis, Virginia Polytechnic, Blacksburg VA, November 2007, online at: http://scholar.lib.vt.edu/theses/available/etd-11152007-105652/unrestricted/dissertation.pdf (Accessed on August, 2015).

9. B. Bailey and G. Martin, IP meta-models for SoC assembly and HW/SW interfaces, in *ESL Models and Their Application, Embedded Systems*, Springer, Cham, Switzerland, 2010, pp. 33–82.

10. ARM, Socrates Weaver data-sheet, June 2014, online at: http://community.arm.com/docs/DOC-4264 (Accessed on August, 2015).

11. J. Bachrach et al., Chisel: Constructing hardware in a Scala embedded language, in *Proceeding of the 49th Design Automation Conference*, San Francisco, CA, 2012.

12. O. Shacham, Creating chip generators for efficient computing at low NRE design costs, in *High Performance Embedded Computing* (*HPEC*), 2011.

13. G. Ezer, Xtensa with user defined DSP coprocessor microarchitectures, in *Proceedings of the 2000 Conference on Computer Design*, Austin, TX, 2000, pp. 335–342.

14. A. Mishra et al., A heterogeneous multiple network-on-chip design: An application-aware approach, in *Proceedings of the 50th Design Automation Conference*, Austin, TX, 2013.

15. ARM, CoreLink AMBA designer, online at: http://www.arm.com/products/system-ip/amba-design-tools/amba-designer.php, consulted December 2014.

16. R. Fraer et al., From visual to logical formalisms for SoC validation, in *The 12th ACM-IEEE International Conference on Formal Methods and Models for System Design* (*MEMOCODE'14*), Lusanne, Switzerland, 2014.

17. IEEE Standard for IP-XACT, Standard structure for packaging, integrating, and reusing IP within tools flows, IEEE Std 1685-2009, February 2010, pp. C1–C360.

18. J. Willson, SoC block-based design and IP assembly, in L. Scheffer et al., editors, *Electronic Design Automation for Integrated Circuits Handbook*, CRC Press, Boca Raton, FL, 2006.

19. W. Kruijtzer et al., Industrial IP integration flows based on IP-XACT standards, in *DATE'08*, Munich, Germany, March 2008, pp. 32–37.

20. A. Dey et al., Churning the most out of IP-XACT for superior design quality, in *2014 Design Automation Conference*, San Francisco, CA, online at: http://www.dac.com/App_Content/files/50/50th%20DT%20Slides/02D_3.ppt.

21. SystemRDL Working Group Site, online at: http://www.accellera.org/activities/committees/systemrdl/, consulted December 2014.

22. SI2 Common Power Format Specification, online at: http://www.si2.org/?page=811, consulted December 2014.

23. IEEE1801 Unified Power Format, http://www.p1801.org/ (Accessed on September, 2015).

24. JasperGold Connectivity Verification App, http://www.cadence.com/products/fv/jaspergold_connectivity/pages/default.aspx (Accessed on September, 2015).

25. SpyGlass Products, http://www.synopsys.com/Tools/Verification/SpyGlass/Pages/default.aspx (Accessed on September, 2015).

26. Magillem IP-Xact Packager, http://www.magillem.com/wp-content/uploads/2015/02/Magillem-IP-XACT-Packager_1-8.pdf (Accessed on September, 2015).

27. Socrates Design Environment, https://www.arm.com/products/system-ip/ip-tooling/socrates.php (Accessed on September, 2015).

28. E. Sperling, IP-XACT becoming more useful, online at: http://semiengineering.com/IP-XACT/, consulted December 2014.

29. D. Murray and S. Rance, Solving next generation IP configurability, White paper, online at: http://www.duolog.com/wp-content/uploads/Solving_Next_Generation_IP_Configuability.pdf, consulted December 2014.

30. J. Lee et al., A framework for automatic generation of configuration files for a custom hardware/software RTOS, in *Proceedings of the International Conference on Engineering of Reconfigurable Systems and Algorithms* (*ERSA'02*), Las Vegas, NV, 2002.

31. OpenCores SOC_Maker home page, September 2014, online at: http://opencores.org/project,soc_maker.

32. Y. Lurie, Verification of interconnects, in *Haifa Verification Conference* (*HVC*), Haifa, Israel, 2013, online at: http://www.research.ibm.com/haifa/conferences/hvc2013/present/Cadence_Verification-of-Interconnects.pdf (Accessed on August, 2015).

33. Y. Yun et al., Beyond UVM for practical SoC verification, in *Proceedings of the SoC Design Conference (ISOCC)*, Jeju, South Korea, 2011, pp. 158–162.
34. L. Piccolboni and G. Pravadelli, Simplified stimuli generation for scenario and assertion based verification, in *Test Workshop—LATW, 2014 15th Latin American*, Fortaleza, Brazil, pp. 1–6.
35. J. Koesters et al., Verification of non-mainline functions in todays processor chips, in *Proceedings of the 51st Design Automation Conference*, San Francisco, CA, 2014.
36. C. Chuang and C. Liu, Hybrid testbench acceleration for reducing communication overhead, *IEEE Design & Test of Computers*, 28(2), 40–50, March 2011.
37. J. Liu et al. A NoC traffic suite based on real applications, in *2011 IEEE Computer Society Annual Symposium on VLSI*, Kyoto, Japan, 2011, pp. 66–71.
38. W. Orme and N. Heaton, How to measure and optimize the system performance of a smartphone RTL design, October 2013, online at: http://community.arm.com/groups/soc-implementation/blog/2013/10/29/how-to-measure-and-optimize-the-system-performance-of-a-smartphone-rtl-design (Accessed on August, 2015).

Performance Evaluation Methods for Multiprocessor System-on-Chip Designs

6

Ahmed Jerraya and Iuliana Bacivarov

CONTENTS

6.1 INTRODUCTION

Multiprocessor systems-on-chip (MPSoCs) require the integration of heterogeneous components (e.g., microprocessors, DSP, ASIC, memories, buses, etc.) on a single chip. The design of MPSoC architectures requires the exploration of a huge space of architectural parameters for each component. The challenge of building high-performance MPSoCs is closely related to the availability of fast and accurate performance evaluation methodologies. This chapter provides an overview of the performance evaluation methods developed for specific subsystems. It then proposes to combine subsystem performance evaluation methods to deal with MPSoC.

Performance evaluation is the process that analyzes the capabilities of a system in a particular context, that is, a given behavior, a specific load, or a specific set of inputs. Generally, performance evaluation is used to validate design choices before implementation or to enable architecture exploration and optimization from very early design phases.

A plethora of performance evaluation tools have been reported in the literature for various subsystems. Research groups have approached various types of subsystems, that is, software (SW), hardware (HW), or interconnect, differently, by employing different description models, abstraction levels, performance metrics, or technology parameters. Consequently, there is currently a broad range of methods and tools for performance evaluation, addressing virtually any kind of design and level of hierarchy, from very specific subsystems to generic, global systems.

Multi-processor system-on-chip (MPSoC) is a concept that aims at integrating multiple subsystems on a single chip. Systems that put together complex HW and SW subsystems are difficult to analyze. Additionally, in this case, the design space exploration and the parameter optimization can quickly become intractable. Therefore, the challenge of building high-performance MPSoCs is closely related to the availability of fast and accurate performance evaluation methodologies.

Existing performance evaluation methods have been developed for specific subsystems. However, MPSoCs require new methods for evaluating their performance. Therefore the purpose of this study is to explore different methodologies used for different subsystems in order to propose a general framework that tackles the problem of performance evaluation for heterogeneous MPSoC. The long-term goal of this work is to build a global MPSoC performance evaluation by composing different tools. This kind of evaluation will be referred to as holistic performance evaluation.

The chapter is structured as follows: Section 6.2 defines the key characteristics of performance evaluation environments. It details the analyzed subsystems, their description models and environments, and the associated performance evaluation tools and methods. Section 6.3 is dedicated to the study of MPSoC performance evaluation. Section 6.4 proposes several trends that could guide future research toward building efficient MPSoC performance evaluation environments.

6.2 OVERVIEW OF PERFORMANCE EVALUATION IN THE CONTEXT OF SYSTEM DESIGN FLOW

This section defines typical terms and concepts used for performance evaluation. First, the performance-evaluation process is positioned within a generic design flow. Three major axes define existing performance evaluation tools: the subsystem under analysis, the performance model, and the performance evaluation methodology. They are detailed in this section. An overview of different types of subsystems is provided, focusing on their abstraction levels, performance metrics, and technology parameters. In the end, the main performance evaluation approaches are introduced.

6.2.1 MAJOR STEPS IN PERFORMANCE EVALUATION

This section analyzes the application domain of the performance evaluation process within the systems design flow. A designed system is evaluated by a suitable performance evaluation tool where it is represented by a performance model. Next, this section presents how evaluation results may influence decisions during system design.

A design flow may include one or more performance evaluation tools. These evaluation tools could be used for different purposes, e.g., to verify if a system meets the constraints imposed or runs properly and to help making design choices.

Figure 6.1 presents a generic design flow. The initial specification is split into a functional and a nonfunctional part of the subsystem to be analyzed. The functional part contains the behavior of the subsystem under analysis, described it as an executable program or as a formal model (e.g., equation). However, the set of evaluation constraints or quality criteria selected from the initial specification constitute the nonfunctional part.

The performance model cannot be separated from the evaluation methodology, because it provides the technology characteristics used to compute performance results prior to the real implementation. Moreover, it selects the critical characteristics to be analyzed, such as the model performance metrics: timing, power, and area. Eventually it chooses the measurement strategy (e.g., an aggregation approach). Both performance metrics and technology parameters may be built into the evaluation tool or given as external libraries.

The design process may be made of several iterations when the system needs to be tuned or partially redesigned. Several design loops may modify the performance model. Each time a new calculation of the metrics (e.g., the chip area can be enlarged in order to meet timing deadlines) or technological parameters (e.g., a new underlying technology or increased clock frequency) is initiated.

6.2.2 KEY CHARACTERISTICS OF PERFORMANCE EVALUATION

Figure 6.2 details on three axes, the three main characteristics of the performance evaluation process: the subsystem under analysis, its abstraction level, and the performance evaluation methodology. Analysis of five kinds of subsystems will be considered: HW, SW, CPUs, interconnect subsystems, and MPSoCs. Each basic subsystem has specific design methods and evaluation tools. They may be designed at different abstraction levels, of which we will consider only three. Also, three performance evaluation methodologies will be considered, with metrics and technology parameters specific to different subsystems.

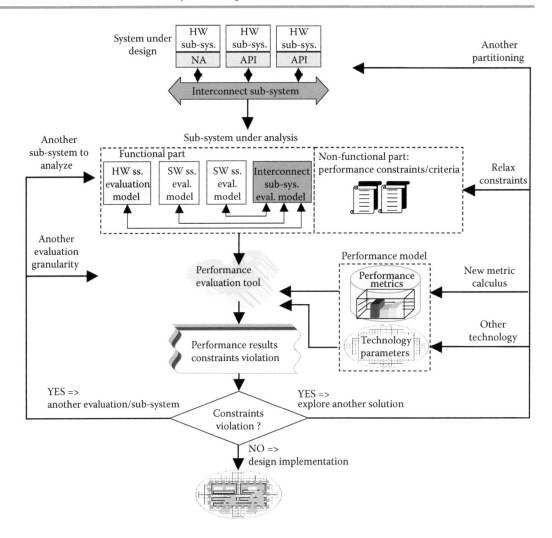

FIGURE 6.1   General performance evaluation and design optimization flow.

A subsystem under analysis is characterized by its type (e.g., HW, SW, etc.), and its abstraction level. The performance evaluation may be applied to different kinds of subsystems varying from simple devices to sophisticated modules. We will consider five main kinds of subsystems: HW subsystems, CPU subsystems, SW subsystems, interconnect subsystems, and constituent parts of MPSoC. Traditionally they are studied by five different research communities.

Classically, the HW community [1] considers HW subsystems as HDL models. They are designed with electronic design automation (EDA) tools that include specific performance analysis [2–5]. The computer architecture community e.g., [6] considers CPU subsystems as complex microarchitectures. Consequently, specialized methodologies for CPU design and performance evaluation have been developed [7]. The SW community [8,9] considers SW subsystems as programs running parallel tasks. They are designed with computer-aided SW engineering (CASE) tools and evaluated with specific methods [10–22]. The networking community [23–30] considers interconnect subsystems as a way to connect diverse HW or SW components. The network performance determines the overall system efficiency, and consequently it is an intensively explored domain.

Each of these communities uses different *abstraction levels* to represent their subsystem. Without any loss of generality, in this study only three levels will be considered: register transfer level (RTL), virtual-architecture level, and task level. These may be adapted for different kinds of subsystems.

Performance evaluation uses a specific methodology and a system performance model. The methodology may be simulation-based, analytic (i.e., using a mathematical description),

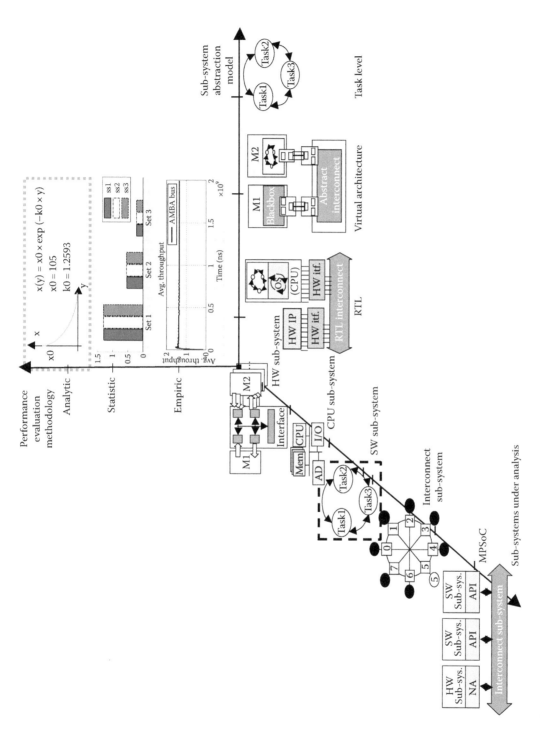

FIGURE 6.2   Performance evaluation environment characteristics.

or statistical. The system performance model takes into account the performance metrics and technology parameters.

The performance metrics are used for assessing the system under analysis. They may be physical metrics related to real system functioning or implementation (e.g., execution timings, occupied area, or consumed power), or quality metrics that are related to nonfunctional properties (e.g., latency, bandwidth, throughput, jitter, or errors).

The technology parameters are required to fit the performance model to an appropriate analysis domain or to customize given design constraints. The technology parameters may include architectural features of higher level (e.g., the implemented parallelism, the network topology), or lower level (e.g., the silicon technology, the voltage supply).

6.2.3 PERFORMANCE EVALUATION APPROACHES

The two main classes of performance evaluation reported in literature are: statistical approaches and deterministic approaches. For statistical approaches, the performance is a random variable characterized by several parameters such as a probability distribution function, average, standard deviation, and other statistical properties. Deterministic approaches are divided into empirical and analytical approaches. In this case, the performance cost function is defined as a deterministic variable, a function of critical parameters. Each of these approaches is defined as follows:

The *statistical approach* [17,19] proceeds in two phases. The first phase finds the most suitable model to express the system performance. Usually parameters are calibrated by running random benchmarks. The second phase makes use of the statistical model previously found to predict the performance of new applications. In most cases, this second phase provides a feedback for updating the initial model.

The *empirical* approach can be accomplished either by measurement or simulation. *Measurement* is based on the real measurement of an already built or prototyped system. It generally provides extremely accurate results. Because this approach can be applied only late in the design cycle when a prototype can be made available, we do not include it in this study. The *simulation approach* [3,16,21,24–28,31] relies on the execution of the complete system using input scenarios or representative benchmarks. It may provide very good accuracy. Its accuracy and speed depend on the abstraction level used to describe the simulated system.

The *analytical approach* [2,10–12,14,15,32] formally investigates system capabilities. The subsystem under analysis is generally described at a high level of abstraction by means of algebraic equations. Mathematical theories applied to performance evaluation make possible a complete analysis of the full system performance at an early design stage. Moreover, such approaches provide fast evaluation because they replace time-consuming system compilation and execution. Building an analytical model could be very complex. The dynamic behavior (e.g., program context switch and wait times due to contentions or collisions) and refinement steps (e.g., compiler optimizations) are hard to model. However, this approach may be useful for worst-case analysis or to find corner cases that are hard to cover with simulation.

6.2.4 HARDWARE SUBSYSTEMS

6.2.4.1 DEFINITION

A HW subsystem is a cluster of functional units with a low programmability level like FPGA or ASIC devices. It can be specified by finite state machines (FSMs) or logic functions. In this chapter, the HW concept excludes any modules that are either CPUs or interconnection like. We also restrain the study to digital HW.

6.2.4.2 ABSTRACTION LEVELS

HW abstraction is related to system timing, of which we consider three levels: high-level language (HLL), bus cycle-accurate (BCA), and RTL. At HLL, the behavior and communication may hide clock cycles, by using abstract channels and high-level communication of primitives, e.g., a system described by untimed computation and transaction-level communication. At BCA level, only the communication of the subsystem is detailed to the clock cycle level, while the computation may still be untimed. At RTL, both the computation and communication of the system are detailed to clock cycle level. A typical example is a set of registers and some combinatorial functions.

6.2.4.3 PERFORMANCE METRICS

Typical performance metrics are power, execution time, or size, which could accurately be extracted during low-level estimations and used in higher abstraction models.

6.2.4.4 TECHNOLOGY PARAMETERS

Technology parameters abstract implementation details of the real HW platform, depending on the abstraction level. At HLL, as physical signals and behavior are abstracted, the technology parameters denote high-level partitioning of processes with granularity of functions (e.g., C function) and with reference to the amount of exchanged transactions. At BCA level, the technology parameters concern data formats (e.g., size, coding, etc.), or behavioral data processing (e.g., number of bytes transferred, throughputs, occupancies, and latencies). At RTL, the HW subsystem model is complete. It requires parameters denoting structural and timing properties (e.g., for the memory or communication subsystems) and implementation details (e.g., the FPGA mapping or ASIC implementation).

There are different performance evaluation tools for HW subsystems, which make use of different performance evaluation approaches: simulation-based approaches [3], analytical approaches [2], mixed analytical and statistical approaches [18], mixed simulation and statistical approaches [5], and mixed analytical, simulation, and statistical approaches [4].

6.2.5 CPU MODULES

6.2.5.1 DEFINITION

A CPU module is a hardware module executing a specific instruction set. It is defined by an instruction set architecture (ISA) detailing the implementation and interconnection of the various functional units, the set of instructions, register utilization, and memory addressing.

6.2.5.2 ABSTRACTION LEVELS

For CPU modules, three abstraction levels can be considered: RTL, also known as the micro-architecture level, the cycle-accurate ISA level, and the assembler (ASM) ISA level. The RTL (or micro-architecture level) offers the most detailed view of a CPU. It contains the complete detailed description of each module, taking into account the internal data, control, and memory hierarchy. The cycle-accurate ISA level details the execution of instructions with clock accuracy. It exploits the real instruction set model and internal resources, but in an abstract CPU representation. The ASM ISA level increases the abstraction, executing programs on a virtual CPU representation, with abstract interconnections and parameters, e.g., an instruction set simulator.

6.2.5.3 PERFORMANCE METRICS

The main performance metrics for CPU subsystems are related to timing behavior. We can mention among these the throughput that expresses the number of instructions executed by CPU per time unit, the utilization that represents the time ratio spent on executing tasks, and the time dedicated to the execution of a program or to respond to a peripheral. Other performance evaluation metrics are power consumption and memory size.

6.2.5.4 TECHNOLOGY PARAMETERS

Technology parameters abstract the CPU implementation details, depending on the abstraction level. At RTL, only the technology for the CPU physical implementation is abstract. The ISA level abstracts the control and data path implementation, but it still details the execution with clock-cycle accuracy using the real instruction set (load/store, floating point, or memory management instructions), the internal register set, and internal resources. And finally, the ASM level abstracts the micro-architecture (e.g., pipeline and cache memory), providing only the instruction set to program it.

Different performance evaluation tools for CPU subsystems exist, making use of different performance evaluation approaches: simulation-based approaches [31,33] analytical approaches [10,32], statistical approaches [19], mixed analytical and statistical approaches [34], and mixed analytical, simulation, and statistical approaches [7].

Chapter 10 has as its main objective the measurement of CPU performance. For a thorough discussion on CPU performance evaluation utilizing benchmarking techniques, we refer the reader to Chapter 10.

6.2.6 SOFTWARE MODULES

6.2.6.1 DEFINITION

A software module is defined by the set of programs to be executed on a CPU. They may have different representations (procedural, functional, object-oriented, etc.), different execution models (single- or multi-threaded), different degrees of responsiveness (real time, nonreal time), or different abstraction levels (from HLL down to ISA level).

6.2.6.2 ABSTRACTION LEVELS

We will consider three abstraction levels for SW modules. At HLL, parallel programs run independently on the underlying architecture, interacting by means of abstract communication models. At transaction-level modeling (TLM) level, parallel programs are mapped and executed on generic CPU subsystems. They communicate explicitly but their synchronization remains implicit. And finally, at ISA level, the code is targeted at a specific CPU and it targets explicit interconnects.

6.2.6.3 PERFORMANCE METRICS

The metrics most used for SW performance evaluation are run time, power consumption, and occupied memory (footprint). Additionally, SW performance may consider concurrency, heterogeneity, and abstraction at different levels [35].

6.2.6.4 TECHNOLOGY PARAMETERS

For SW performance evaluation, technology parameters abstract the execution platform. At HLL, technology parameters abstract the way different programs communicate using, for example, coarse-grain send()/receive() primitives. At TLM level, technology parameters hide the technique or resources used for synchronization, such as using a specific Operating System (OS), application program interfaces (APIs) and mutex_lock()/unlock()-like primitives. At ISA level, technology parameters abstract the data transfer scheme, the memory mapping, and the addressing mode.

Different performance evaluation tools for SW models exist, making use of different performance evaluation approaches: simulation-based [21], analytical [12], and statistical [17].

6.2.6.5 SOFTWARE SUBSYSTEMS

6.2.6.5.1 Definition

When dealing with system-on-chip design, the SW is executed on a CPU subsystem, made of a CPU and a set of peripherals. In this way, the CPU and the executed SW program are generally combined in an SW subsystem.

6.2.6.5.2 Abstraction Levels

The literature presents several classifications for the abstraction of SW subsystems, among which we will consider three abstraction levels: The HLL, the OS level, and the hardware abstraction layer (HAL) level. At the HLL, the application is composed of a set of tasks communicating through abstract HLL primitives provided by the programming languages (e.g., send()/receive()). The architecture, the interconnections, and the synchronization are abstract. At the OS level, the SW model relies on specific OS APIs, while the interconnections and the architecture still remain abstract. Finally, at the HAL level, the SW is bound to use a specific CPU insturtion set and may run on an RTL model of the CPU. In this case, the interconnections are described as an HW model, and the synchronization is explicit.

6.2.6.5.3 Performance Metrics

The main performance metrics are the timing behavior, the power consumption, and the occupied area. They are computed by varying the parameters related to the SW program and to the underlying CPU architecture.

6.2.6.5.4 Technology Parameters

In SW subsystem performance evaluation, technology parameters abstract the execution platform, and characterize the SW program. At HLL, technology parameters mostly refer to application behavioral features, abstracting the communication details. At the OS level, technology parameters include OS features (e.g., interrupts, scheduling, and context switching delays), but their implementation remains abstract. At HLL, technology parameters abstract only the technology of implementation, while all the other details, such as the data transfer scheme, the memory mapping or the addressing mode are explicitly referred to.

Different performance evaluation tools for SW subsystems exist, making use of different performance evaluation approaches: simulation-based approaches [16,20], analytical approaches [11,13], statistical approaches [19], and mixed analytical, simulation, and statistical approaches [22].

6.2.7 INTERCONNECT SUBSYSTEMS

6.2.7.1 DEFINITION

The interconnect subsystem provides the media and the necessary protocols for communication between different subsystems.

6.2.7.2 ABSTRACTION LEVELS

In this study, we will consider RTL, transactional, and service or HLL models. At the HLL, different modules communicate by requiring services using an abstract protocol, via abstract network topologies. The transactional level still uses abstract communication protocols (e.g., send/receive) but it fixes the communication topology. The RTL communication is achieved by means of explicit interconnects like physical wires or buses, driving explicit data.

6.2.7.3 PERFORMANCE METRICS

The performance evaluation of interconnect subsystem focuses on the traffic, interconnection topology (e.g., network topology, path routing, and packet loss within switches), interconnection technology (e.g., total wire length and the amount of packet switch logic), and application demands (e.g., delay, throughput, and bandwidth).

6.2.7.4 TECHNOLOGY PARAMETERS

A large variety of technology parameters emerge at different levels. Thus, at HLL, parameters are the throughput or latency. At TLM level, the parameters are the transaction times and the arbitration strategy. At the RTL, the wires and pin-level protocols allow system delays to be measured accurately.

Simulation is the performance evaluation strategy most used for interconnect subsystems at different abstraction levels: behavioral [24], cycle-accurate level [25], and TLM level [27]. Interconnect simulation models can be combined with HW/SW co-simulation at different abstraction levels for the evaluation of full MPSoC [26,28].

6.2.8 MULTIPROCESSOR SYSTEMS-ON-CHIP MODELS

6.2.8.1 DEFINITION

An MPSoC is a heterogeneous system built of several different subsystems like HW, SW, and interconnect, and it takes advantage of their synergetic collaboration.

6.2.8.2 ABSTRACTION LEVELS

MPSoCs are made of subsystems that may have different abstraction levels. For example, in the same system, RTL HW components can be coupled with HLL SW components, and they can communicate at the RTL or by using transactions. In this study, we will consider the interconnections, synchronization, and interfaces between the different subsystems. The abstraction levels considered are the functional level, the virtual architecture model level, and the level that combines RTL models of the hardware with instruition set architecture models of the CPU. At the functional level (like message passing interface (MPI)) [36], the HW/SW interfaces, the interconnections, and synchronization are abstracted and the subsystems interact through high-level primitives (send, receive). For the virtual architecture level, the interconnections and synchronization are explicit but the HW/SW interfaces are still abstract. The lowest level considered deals with an RTL architecture for the HW-related sections coupled with the ISA level for the SW. This kind of architecture explicitly presents the interconnections, synchronization, and interfaces. In order to master the complexity, most existing methods used to assess heterogeneous MPSoC systems are applied at a high abstraction level.

6.2.8.3 PERFORMANCE METRICS

MPSoC performance metrics can be viewed as the union of SW, HW and interconnect subsystems performance metrics, for instance, execution time, memory size, and power consumption.

6.2.8.4 TECHNOLOGY PARAMETERS

A large variety of possible technology parameters emerge for each of the subsystems involved, mostly at different levels and describing multiple implementation alternatives. They are considered during system analysis, and exploited for subsystem performance optimization.

Different performance-evaluation tools for MPSoC exist. They are developed for specific subsystems [37–41], or for the complete MPSoC [42–44]. Section 6.3 deals with performance-evaluation environments for MPSoCs.

6.3 MPSoC PERFORMANCE EVALUATION

As has been shown in previous sections, many performance evaluation methods and tools are available for different subsystems: HW, SW, interconnect and even for MPSoCs. They include a large variety of measurement strategies, abstraction levels, evaluated metrics, and techniques. However, there is still a considerable gap between particular evaluation tools that consider only isolated components and performance evaluation for a full MPSoC.

The evaluation of a full MPSoC design containing a mixture of HW, SW and interconnect subsystems, needs to cover the evaluation of all the subsystems, at different abstraction levels.

Few MPSoC evaluation tools are reported in the literature [37–39,41,42,44,45]. The key restriction with existing approaches is the use of a homogeneous model to represent the overall system, or the use of slow evaluation methods that cannot allow the exploration of the architecture by evaluating a massive number of solutions.

For example, the SymTA/S approach [45] makes use of a standard event model to represent communication and computation for complex heterogeneous MPSoC. The model allows taking into account complex behavior such as interrupt control and data dependant computation. The approach allows accurate performance analysis; however, it requires a specific model of the MPSoC to operate.

The ChronoSym approach [41] makes use of a time-annotated native execution model to evaluate SW execution times. The timed simulation model is integrated into an HW/SW co-simulation framework to consider complex behaviors such as interactions with the HW resources and OS performance. The approach allows fast and accurate performances analysis of the SW subsystem. However, for the entire MPSoC evaluation, it needs to be combined with other approaches for the evaluation of interconnect and HW subsystems.

Co-simulation approaches are also well suited for the performance evaluation of heterogeneous -systems. The co-simulation offers flexibility and modularity to couple various subsystem executions at different abstraction levels and even specified in different languages. The accuracy of performance evaluation by co-simulation depends on the chosen subsystem model and on their global synchronization.

A complete approach aiming at analyzing the power consumption for the entire MPSoC by using several performance models is presented in Reference 42. It is based on interconnecting different simulations (e.g., SystemC simulation and ISS execution) and different power models for different components, in a complete system simulation platform named MPARM. A related work is [43], where the focus is on MPSoC communication-performance analysis.

The co-estimation approach in Reference 44 is based on the concurrent and synchronized execution of multiple power estimators for HW/SW system-on-chip-power analysis. Various power estimators can be plugged into the co-estimation framework, possibly operating at different levels of abstraction. The approach is integrated in the POLIS system design environment and PTOLEMY simulation platform. The power co-estimation framework drives system design trade-offs, e.g., HW/SW partitioning, component, or parameter selection. A similar approach is represented in References 46 and 47. The tool named ROSES allows different subsystems that may be described at different abstraction levels and in different languages to be co-simulated.

However, when applied to low abstraction levels, evaluation approaches based on co-simulation [43,44,46] appear to be slow. They cannot explore large solution spaces. An alternative would be to combine co-simulation with analytical methods in order to achieve faster evaluation. This is similar to methods used for CPU architecture exploration [31,33].

Figure 6.3 shows such a scheme for MPSoC. The key idea is to use co-simulation for computing extensive data for one architectural solution. The results of co-simulation will be further used to parameterize an analytical model. A massive number of new design solutions can be evaluated faster using the newly designed analytic model.

The left branch of Figure 6.3 represents the performance evaluation of the full system by co-simulation. This can be made by using any existing co-simulation approach [43,44,46]. The input of the evaluation process is the specification of the MPSoC architecture to be analyzed. The specification defines each subsystem, the communication model, and the interconnection interfaces.

The right branch of Figure 6.3 describes the evaluation of the full system using an analytical model. The figure represents the analytical model construction with dotted lines. This is done through component characterizations and parameter extraction from the base co-simulation model.

After the construction phase, the analytical branch can be decoupled from the co-simulation. The stand-alone analytical performance evaluation provides quick and still accurate evaluations for new designs. The two branches of Figure 6.3, that is, the co-simulation and the analytical approach, lead to similar performance results, but they are different in terms of evaluation speed and accuracy.

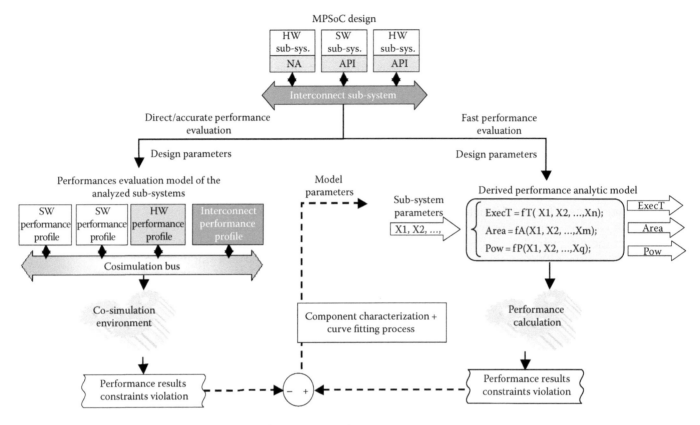

FIGURE 6.3 Global heterogeneous MPSoC evaluation approach.

The proposed strategy is based on the composition of different evaluation models, for different MPSoC subsystems. It combines simulation and analytical models for fast and accurate evaluation of novel MPSoC designs. The further objective is to develop a generic framework for design space exploration and optimization, where different simulation based or analytical evaluation methods could be applied to different subsystems and at different levels of abstraction.

6.4 CONCLUSION

MPSoC is an emerging community trying to integrate multiple subsystems on a single chip and consequently requiring new methods for performance evaluation [48]. Therefore, the aim of this study was to explore different methodologies for the different subsystems that may compose the MPSoC.

We defined a general taxonomy to handle the heterogeneity and diversity of performance-evaluation solutions. This taxonomy introduced the attributes of an evaluation tool: abstraction levels, modeling techniques, measured metrics, and technology parameters. Finally, we proposed an evaluation framework based on the composition of different methods in which simulation and analytical methods could be combined in an efficient manner, to guide the design space exploration and optimization.

REFERENCES

1. G. de Micheli, R. Ernst, and W. Wolf, *Readings in Hardware/Software Co-Design*, 1st edn., Morgan Kaufmann, San Francisco, CA, June 1, 2001.
2. S. Dey and S. Bommu, Performance analysis of a system of communicating processes, *International Conference on Computer-Aided Design (ICCAD 1997)*, ACM and IEEE Computer Society, San Jose, CA, 1997, pp. 590–597.

3. V. Agarwal, M.S. Hrishikesh, S.W. Keckler, and D. Burger, Clock rate versus IPC: The end of the road for conventional microarchitectures, *Proceedings of the 27th International Symposium on Computer Architecture*, Vancouver, British Columbia, Canada, June 2000, pp. 248–259.

4. H. Yoshida, K. De, and V. Boppana, Accurate pre-layout estimation of standard cell characteristics, *Proceedings of ACM/IEEE Design Automation Conference (DAC)*, San Diego, CA, June 2004, pp. 208–211.

5. C. Brandolese, W. Fornaciari, and F. Salice, An area estimation methodology for FPGA based designs at SystemC-level, *Proceedings of ACM/IEEE Design Automation Conference (DAC)*, San Diego, CA, June 2004, pp. 129–132.

6. A.D. Patterson and L.J. Hennessy, *Computer Organization and Design, the Hardware/SW Interface*, 2nd edn., Morgan-Kaufmann, San Francisco, CA, 1998.

7. D. Ofelt and J.L. Hennessy, Efficient performance prediction for modern microprocessors, *SIGMETRICS*, Santa Clara, CA, 2000, pp. 229–239.

8. B. Selic, An efficient object-oriented variation of the statecharts formalism for distributed real-time systems, *CHDL 1993, IFIP Conference on Hardware Description Languages and Their Applications*, Ottawa, Ontario Canada, 1993, pp. 28–28.

9. B. Selic and J. Rumbaugh, *Using UML for Modeling Complex Real-Time Systems*, Whitepaper, Rational Software Corp., 1998, http://www.ibm.com/developerworks/rational/library/139.html.

10. J. Walrath and R. Vemuri, A performance modeling and analysis environment for reconfigurable computers, *IPPS/SPDP Workshops*, Orlando, FL, 1998, pp. 19–24.

11. B. Spitznagel and D. Garlan, Architecture-based performance analysis, *Proceedings of the 1998 Conference on Software Engineering and Knowledge Engineering*, San Francisco, CA, 1998.

12. P. King and R. Pooley, Derivation of Petri net performance models from UML specifications of communications software, *Proceedings of Computer Performance Evaluation Modelling Techniques and Tools: 11th International Conference, TOOLS 2000*, Schaumburg, IL, 2000.

13. F. Balarin, STARS of MPEG decoder: A case study in worst-case analysis of discrete event systems, *Proceedings of the International Workshop on HW/SW Codesign*, Copenhagen, Denmark, April 2001, pp. 104–108.

14. T. Schuele and K. Schneider, Abstraction of assembler programs for symbolic worst case execution time analysis, *Proceedings of ACM/IEEE Design Automation Conference (DAC)*, San Diego, CA, June 2004, pp. 107–112.

15. C. Lu, J.A. Stankovic, T.F. Abdelzaher, G. Tao, S.H. Son, and M. Marley, Performance specifications and metrics for adaptive real-time systems, *IEEE Real-Time Systems Symposium (RTSS 2000)*, Orlando, FL, 2000.

16. M. Lajolo, M. Lazarescu, and A. Sangiovanni-Vincentelli, A compilation-based SW estimation scheme for hardware/software co-simulation, *Proceedings of the 7th IEEE International Workshop on Hardware/Software Codesign*, Rome, Italy, May 3–5, 1999, pp. 85–89.

17. E.J. Weyuker and A. Avritzer, A metric for predicting the performance of an application under a growing workload, *IBM Syst. J.*, 2002, 41, 45–54.

18. V.D. Agrawal and S.T. Chakradhar, Performance estimation in a massively parallel system, *Proceedings of Super Computing 1990*, New York, November 12–16, 1990, pp. 306–313.

19. E.M. Eskenazi, A.V. Fioukov, and D.K. Hammer, Performance prediction for industrial software with the APPEAR method, *Proceedings of STW PROGRESS Workshop*, Utrecht, the Netherlands, October 2003.

20. K. Suzuki and A.L. Sangiovanni-Vincentelli, Efficient software performance estimation methods for hardware/software codesign, *Proceedings ACM/IEEE Design Automation Conference (DAC)*, Los Vegas, NV, 1996, pp. 605–610.

21. Y. Liu, I. Gorton, A. Liu, N. Jiang, and S.P. Chen, Designing a test suite for empirically-based middleware performance prediction, *The Proceedings of TOOLS Pacific 2002*, Sydney, Australia, 2002.

22. A. Muttreja, A. Raghunathan, S. Ravi, and N.K. Jha, Automated energy/performance macromodeling of embedded software, *Proceedings ACM/IEEE Design Automation Conference (DAC)*, San Diego, CA, June 2004, pp. 99–102.

23. K. Lahiri, A. Raghunathan, and S. Dey, Fast performance analysis of bus-based system-on-chip communication architectures, *Proceedings ACM/IEEE Design Automation Conference (DAC)*, San Jose, CA, June 1999, pp. 566–572.

24. M. Lajolo, A. Raghunathan, S. Dey, L. Lavagno, and A. Sangiovanni-Vincentelli, A case study on modeling shared memory access effects during performance analysis of HW/SW systems, *Proceedings of the Sixth IEEE International Workshop on Hardware/Software Codesign*, Seattle, WA, March 15, 1998 to March 18, 1998, pp. 117–121.

25. J.A. Rowson and A.L. Sangiovanni-Vincentelli, Interface-based design, *Proceedings of the 34th Conference on Design Automation*, Anaheim Convention Center, ACM Press, Anaheim, CA, June 9, pp. 566–572 13, 1997, pp. 178–183.

26. K. Hines and G. Borriello, Optimizing communication in embedded system co-simulation, *Proceedings of the Fifth International Workshop on Hardware/Software Co-Design*, Braunschweig, Germany, March 24–March 26, 1997, p. 121.

27. S.G. Pestana, E. Rijpkema, A. Radulescu, K.G.W. Goossens, and O.P. Gangwal, Cost-performance trade-offs in networks on chip: A simulation-based approach, *DATE*, Paris, France, 2004, pp. 764–769.

28. F. Poletti, D. Bertozzi, L. Benini, and A. Bogliolo, Performance analysis of arbitration policies for SoC communication architectures, *Kluwer J. Des. Autom. Embed. Syst.*, 2003, 8, 189–210.

29. L. Benini and G.D. Micheli, Networks on chips: A new SoC paradigm, *IEEE Comp.*, 2002, 35, 70–78.

30. E. Bolotin, I. Cidon, R. Ginosar, and A. Kolodny, QNoC: QoS architecture and design process for network on chip, *J. Syst. Archit.*, 2003, 49. Special issue on Networks on Chip.

31. K. Chen, S. Malik, and D.I. August, Retargetable static timing analysis for embedded software, *International Symposium on System Synthesis ISSS*, Montreal, Canada, 2001, pp. 39–44.

32. A. Hergenhan and W. Rosenstiel, Static timing analysis of embedded software on advanced processor architectures, *Proceedings of Design, Automation and Test in Europe*, Paris, France, 2000, pp. 552–559.

33. V. Tiwari, S. Malik, and A. Wolfe, Power analysis of embedded software: A first step towards software power minimization, *IEEE Trans. Very Large Scale Integr. Syst.*, 1994, 2, 437–445.

34. P.E. McKenney, Practical performance estimation on shared-memory multiprocessors, *Proceedings of Parallel and Distributed Computing and Systems Conference*, Fort Lauderdale, FL, 1999, 125–134.

35. M.K. Nethi and J.H. Aylor, Mixed level modelling and simulation of large scale HW/SW systems, *High Performance Scientific and Engineering Computing: Hardware/Software Support*, Tianruo Y., Laurence, Y.P. (eds.) Kluwer Academic Publishers, Norwell, MA, 2004, pp. 157–166.

36. The MPI Forum, The MPI standard 3.1, MPI-3.1, June 4, 2015, http://www.mpi-forum.org/docs/: 3.1 standard URL is http://www.mpi-forum.org/docs/mpi-3.1/mpi31-report.pdf, (Accessed November 22, 2015).

37. V. Gupta, P. Sharma, M. Balakrishnan, and S. Malik, Processor evaluation in an embedded systems design environment, *13th International Conference on VLSI Design (VLSI-2000)*, Calcutta, India, 2000, pp. 98–103.

38. A. Maxiaguine, S. Künzli, S. Chakraborty, and L. Thiele, Rate analysis for streaming applications with on-chip buffer constraints, *ASP-DAC*, Yokohama, Japan, 2004.

39. R. Marculescu, A. Nandi, L. Lavagno, and A. Sangiovanni-Vincentelli, System-level power/performance analysis of portable multimedia systems communicating over wireless channels, *Proceedings of IEEE/ACM International Conference on Computer-Aided Design*, San Jose, CA, 2001.

40. Y. Li and W. Wolf, A task-level hierarchical memory model for system synthesis of multiprocessors, *Proceedings of the ACM/IEEE Design Automation Conference (DAC)*, Anaheim, CA, June 1997, pp. 153–156.

41. I. Bacivarov, A. Bouchhima, S. Yoo, and A.A. Jerraya, ChronoSym—A new approach for fast and accurate SoC cosimulation, *Int. J. Embed. Syst.*, Interscience Publishers, 1(1/2), 103–111, 2005.

42. M. Loghi, M. Poncino, and L. Benini, Cycle-accurate power analysis for multiprocessor systems-on-a-chip, *ACM Great Lakes Symposium on VLSI*, Boston, MA, 2004, pp. 401–406.

43. M. Loghi, F. Angiolini, D. Bertozzi, L. Benini, and R. Zafalon, Analyzing on-chip communication in a MPSoC environment, *Proceedings of the Design, Automation and Test in Europe (DATE)*, Paris, France, February 2004, Vol. 2, pp. 752–757.

44. M. Lajolo, A. Raghunathan, S. Dey, and L. Lavagno, Efficient power co-estimation techniques for systems-on-chip design, *Proceedings of Design Automation and Test in Europe*, Paris, France, 2000.

45. R. Henia, A. Hamann, M. Jersak, R. Racu, K. Richter, and R. Ernst, System level performance analysis—The SymTA/S approach, *IEE Proc. Comput. Dig. Tech.*, 2005, 152, 148–166.

46. W. Cesario, A. Baghdadi, L. Gauthier, D. Lyonnard, G. Nicolescu, Y. Paviot, S. Yoo, A.A. Jerraya, and M. Diaz-Nava, Component-based design approach for multicore SoCs, *Proceedings of the ACM/IEEE Design Automation Conference (DAC)*, New Orleans, LA, June 2002, pp. 789–794.

47. A. Baghdadi, D. Lyonnard, N.-E. Zergainoh, and A.A. Jerraya, An efficient architecture model for systematic design of application-specific multiprocessor SoC, *Proceedings of the Conference on Design, Automation and Test in Europe (DATE)*, Munich, Germany, March 2001, pp. 55–63.

48. *Fourth International Seminar on Application-Specific Multi-Processor SoC Proceedings* Saint-Maximin la Sainte Baume, France, 2004, http://tima.imag.fr/mpsoc/2004/index.html.

System-Level Power Management

7

Naehyuck Chang, Enrico Macii, Massimo Poncino, and Vivek Tiwari

CONTENTS

Abstract

Early consideration of power as a metric in the very early stages of the design flow allows to drastically reduce it and to simplify later design stages. Power-aware system design has thus become one of the most important areas of research in the recent past, although only a few techniques that have been proposed to address the problem have found their way into automated tools.

One of the approaches that have received a lot of attention, both from the conceptual and the implementation sides, is the so-called dynamic power management (DPM). The fundamental idea behind this technique, which is very broad and thus may come in very many different flavors, is that of selectively stopping or underutilizing the system resources that for some periods are not executing useful computation or not providing result at maximum speed.

The landscape of DPM techniques is very wide; an exhaustive survey of the literature goes beyond the scope of this handbook. For additional technical details, the interested reader may refer to the excellent literature on the subject, for instance [1–4]. This chapter describes the principles, the models, and the implementation issues of DPM.

7.1 INTRODUCTION

In general terms, a system is a collection of components whose combined operations provide a service to a user. In the specific case of electronic systems, the components are processor cores, DSPs, accelerators, memories, buses, and devices. Power efficiency in an electronic system can be achieved by optimizing (1) the architecture and the implementation of the individual components, (2) the communication between the components, and (3) the usage of components.

In this chapter, we will restrict our attention to the last point, that is, to the issues of achieving power efficiency in an electronic system by means of properly managing its resources during the execution of the operations in which the system is designed for.

The underlying assumption for the solutions we will discuss is that the activity of the system components is event driven; for example, the activity of display servers, communication interfaces, and user interface functions is triggered by external events and it is often interleaved with long periods of quiescence. An intuitive way of reducing the average power dissipated by the whole system consists therefore of shutting down (or reducing the performance of) the components during their periods of inactivity (or underutilization). In other words, one can adopt a *DPM* strategy that dictates how and when the various components should be powered (or slowed) down in order to minimize the total system power budget under some performance/throughput constraints.

DPM can essentially take two forms, depending on the available degree of reconfiguration of the resource that is power managed. In its simplest variant, the resource has an *on* and an *off* states, and DPM essentially entails the decision of when to turn the resource off. In a more elaborated variant, the resource can provide its functionality at different power/performance levels, that is, it features multiple *on* states; in this case, DPM becomes a more complicated problem in which the objective is to provide the requested services and performance levels with the least energy consumption. Therefore, in the former scenario, only the idleness of the resource matters and affects the on–off decisions. In the latter one, conversely, also its underutilization can be successfully exploited to reduce power consumption.

For both variants of DPM, appropriate hardware support is required to implement the *off* and *on* states. Typical implementations of the *off* states include clock gating or power/supply gating, whereas multiple *on* states are usually made available by tuning the supply voltage, the frequency, or, more typically, both of them.

The core of this chapter focuses on DPM and the relative policies, for the two earlier discussed variants (Section 7.2). A specific section is devoted to the issue of DPM in the context of battery-operated systems (Section 7.3). Some of the techniques described in Sections 7.2 and 7.3 apply also to the software domain, described in Section 7.4, which also includes a description of available strategies for software-level power estimation.

7.2 DYNAMIC POWER MANAGEMENT

7.2.1 POWER MODELING FOR DPM: POWER AND ENERGY STATE MACHINES

The formalism of finite-state machines can be extended to model power consumption of an electronic system. The power state machine (PSM) is the most widely used model to describe a power-managed resource [5]. At any given point in time, each resource is in a specific power state, which consumes a given amount of power.

Figure 7.1 shows an example of PSM, namely, that of the Intel XScale processor. Each power state denotes the power value and/or performance of the system in that state. Transitions are labeled with the time required to complete them.

In addition to the time cost, the PSM formalism can be enhanced to account for the power cost of the state transition, which depends on the specific implementation of the destination power state, for example, the energy required to change the clock frequency and/or the supply voltage.

Given the PSM model of all the power-managed components in a system, we can easily track the current power values of all resources as the state of the system changes. This approach enables us to have the average or peak power dissipation by looking at the power state transition over time for a given set of environmental conditions. The power consumption in each state can be a designer's estimate, a value obtained from simulation, or a manufacturer's specification.

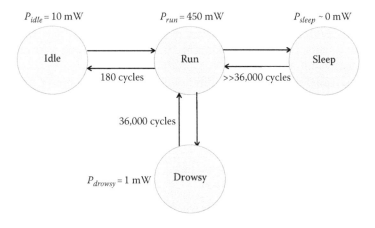

FIGURE 7.1 Power state machine for Intel XScale.

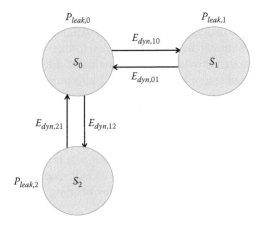

FIGURE 7.2   Energy state machine.

Given the increased relevance of static (i.e., leakage) power in deeply scaled technologies, it can be useful to separately denote the dynamic energy and leakage power contributions in a power-managed component. Furthermore, leakage power in a state in fact depends on frequency, and it increases as the clock frequency of the device decreases.

Energy state machines (ESMs) are an alternative formalism that provides such a separation, which is not possible in PSMs [6]. To eliminate time dependency, the dynamic portion is represented by energy, while the leakage portion is represented by power. Figure 7.2 shows the concept of ESM. Dynamic energy consumption, $E_{dyn,ij}$, is associated with a transition from state i to state j, and leakage power consumption, $P_{leak,i}$, is associated with a state. Each state change clearly requires different amount of dynamic energy. If we slow down the clock frequency, only the storage time of the states becomes longer.

The ESM exactly represents the actual behavior of a system and its energy consumption. However, it is not trivial to acquire leakage power and dynamic energy consumption separately without elaborated power estimation. Special measurement and analysis tools that separately handle dynamic energy and leakage power in system level are often required to annotate the ESM [7,8].

7.2.2 REQUIREMENTS AND IMPLEMENTATION OF DPM

In addition to models for power-managed systems, as those discussed in the previous section, there are several other issues that must be considered when implementing DPM in practice. These include the choice of an implementation style for the power manager that must guarantee accuracy in measuring interarrival times and service times for the system components, flexibility in monitoring multiple types of components, low perturbation, and marginal impact on component usage.

Another essential aspect concerns the choice of the style for monitoring component activity; options here are offline (traces of events are dumped for later analysis) and online (traces of events are analyzed while the system is running, and statistics related to component usage are constantly updated) monitoring.

Finally, of utmost importance is the definition of an appropriate power management policy, which is essential for achieving good power/performance trade-offs. The remainder of this section is devoted to the discussion of various options for DPM policies, including those regarding dynamic voltage scaling (DVS).

7.2.3 DPM POLICIES

The function of a DPM policy is to decide (1) when to perform component state transitions and (2) which transition should be taken, depending on workload, system history, and performance constraints. The fundamental requirement of any DPM policy is that they must be computationally simple, so that their execution requires as little as possible time and power. For this reason, most practical policies are heuristic and do not have any optimality claim.

The rest of this section surveys the most popular DPM policies and highlights their benefits and pitfalls. We classify power management approaches into four major categories: *static*, *predictive*, *stochastic*, and *adaptive* policies. Within each class, we introduce approaches being applied to the system design and described in the literature.

7.2.3.1 STATIC POLICIES

The simplest DPM policy is a static one based on a *timeout T*. Let us assume a simple PSM with an *on* and an *off* state; the component is put in its *off* state T time units after the start of an idle interval. This scheme assumes that if an idle T has elapsed, there will be a very high chance for the component to be idle for an even longer time. The policy is static because the timeout for a given power state is chosen once and for all. The nicest feature of a timeout policy lies in its simplicity of implementation.

However, timeout policies can be inefficient for three reasons: First, the assumption that if the component is idle for more than T time units, it will be so for much longer may not be true in many cases. Second, while waiting for timeout to elapse, the component is actually kept in the *on* state for at least T time units, wasting a considerable amount of power. Third, speed and power degradations due to shutdowns performed at inappropriate times are not taken into account; in fact, it should be kept in mind that the transition from power down to fully functional mode has an overhead. It takes some time to bring the system up to speed, and it may also take more power than the average, steady-state power.

7.2.3.2 PREDICTIVE POLICIES

Improving the performance and energy efficiency of static policies implies somehow predicting the length of the next incoming idle interval. *Predictive* policies exploit the correlation that exists between the past history of the workload and its near future in order to make reliable predictions about future events. Specifically, we are interested in *predicting idle periods* that are long enough to justify a transition into a low-power state.

An essential parameter for the design of efficient policies is the *breakeven time* T_{BE}, that is, the minimum time for which the component should be turned off so that it compensates for the overhead associated with shutting it down and turning it on the component. Each power state has its own T_{BE}, whose value can be calculated from the power values in the involved states and the power and performance costs of the relative transitions and does not depend on the workload. In particular, the value of T_{BE} for a power state is lower bounded by the time needed to complete the transition to and from that state [3]. As an example, consider the simple PSM of Figure 7.3; the breakeven time $T_{BE,Off}$ of the off state is at least as large as $T_{On,Off} + T_{Off,On}$. The accurate calculation of T_{BE} for a power state is discussed in details in Reference 3.

In practice, we should therefore turn the resource in a given state if the next incoming idle period T_{idle} is greater than the T_{BE} for that state. However, since T_{idle} is unknown, we have to predict it in some way. If we call T_{pred} our prediction of T_{idle}, the policy becomes now "go to state S if $T_{pred} > T_{BE,S}$."

It is clear that we need good predictors in order to minimize mispredictions. We define overprediction (underprediction) as a predicted idle period longer (shorter) than the actual

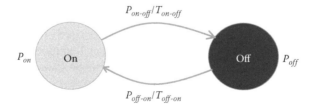

FIGURE 7.3   Example of power state machine for T_{BE} calculation.

one. Overpredictions originate a performance penalty, while underpredictions imply a waste of power without incurring performance penalty. The quality of a predictor is measured by two figures: *safety* and *efficiency*. The safety is the complement of the probability of making overpredictions, and the efficiency is the complement of the probability of making underpredictions. We call a predictor with maximum safety and efficiency an *ideal predictor*. Unfortunately, predictors of practical interest are neither safe nor efficient, thus resulting in suboptimum policies.

Two examples of predictive policy are proposed in Reference 9. The first policy relies on the observation of past history of idle and busy periods. A nonlinear regression equation is built offline by characterization of various workloads, which expresses the next idle time T_{pred} as a function of the sequence of previous idle and active times. The power manager stores then the history of idle and active times, and when an idle period starts, it calculated the prediction T_{pred} based on that equation: if $T_{pred} > T_{BE}$, the resource is immediately put into that power state. This policy has the main disadvantage of requiring offline data collection and analysis for constructing and fitting the regression model.

In the second policy, the idle period is predicted based on a threshold $T_{threshold}$. The duration of the busy period preceding the current idle period is observed. If the previous busy period is longer than the threshold, the current idle period is assumed to be longer than T_{BE}, and the system is shut down. The rationale of this policy is that short busy periods are often followed by long idle periods. The critical design decision parameter is the choice of the threshold value $T_{threshold}$.

7.2.3.3 STOCHASTIC POLICIES

Although all the predictive techniques address workload uncertainty, they assume deterministic response time and transition time of a system. However, the system model for policy optimization is very abstract, and abstraction introduces uncertainty. Hence, it is safer and more general to assume a stochastic model for the system as well as the workload. Moreover, real-life systems support multiple power states, which cannot be handled by simple predictive techniques. Predictive techniques are based on heuristic algorithms, and their optimality can be gauged only through comparative simulations. Finally, predictive techniques are geared toward only the power minimization and cannot finely control performance penalty.

Stochastic control techniques formulate policy selection as an optimization problem in the presence of uncertainty both in the system and in the workload. They assume that both the system and the workload can be modeled as Markov chains and offer significant improvement over previous power management techniques in terms of theoretical foundations and of robustness of the system model. Using stochastic techniques allows one to (1) model the uncertainty in the system power consumption and the state transition time, (2) model complex systems with multiple power states, and (3) compute globally optimum power management policies, which minimize the energy consumption under performance constraints or maximize the performance under power constraints.

A typical Markov model of a system consists of the following entities [14]:

- A service requester (SR) that models the arrival of service requests
- A service provider (SP) that models the operation states of the system
- A power manager that observes the state of the system and the workload, makes a decision, and issues a command to control the next state of the system
- A cost metrics that associate power and performance values with each command

The basic stochastic policies [14] perform policy optimization based on the fixed Markov chains of the SR and of the SP. Finding a globally power-optimal policy that meets given performance constraints can be formulated as a linear program (LP), which can be solved in polynomial time in the number of variables. Thus, policy optimization for Markov processes is exact and computationally efficient. However, several important points must be understood: (1) The performance and power obtained by a policy are expected values, and there is no guarantee of the optimality for a specific workload instance. (2) We cannot assume that we always know the SR model beforehand. (3) The

Markov model for the SR or SP is just an approximation of a much more complex stochastic process, and thus the power-optimal policy is also just an approximate solution.

The Markov model in Reference 14 assumes a finite set of states, a finite set of commands, and discrete time. Hence, this approach has some shortcomings: (1) The discrete-time Markov model limits its applicability since the power-managed system should be modeled in the discrete-time domain. (2) The power manager needs to send control signals to the components in every time slice, which results in heavy signal traffic and heavy load on the system resources (therefore more power dissipation). Continuous-time Markov models [15] overcome these shortcomings by introducing the following characteristics: (1) A system model based on continuous-time Markov decision process is closer to the scenarios encountered in practice. (2) The resulting power management policy is asynchronous, which is more suitable for implementation as a part of the operating system (OS).

7.2.3.4 ADAPTIVE POLICIES

All the DPM policies discussed cannot effectively manage workloads that are unknown a priori or nonstationary, since they strongly rely on the workload statistics. For this reason, several adaptive techniques have been proposed to deal with nonstationary workloads. Notice that adaptivity is a crosscut feature over the previous categories: we can therefore have timeout-based, predictive, or stochastic adaptive policies.

Adaptive Timeout-Based Policies: One option is to maintain a set of timeout values, each associated with an index indicating how successful it has been Reference 11. The policy chooses, at each idle time, the timeout value that would have performed best among the set of available ones. Alternatively, a weighted list of timeout values is kept, where the weight is determined by relative performance of past requests with respect to the optimal strategy [12], and the actual timeout is calculated as a weighted average of all the timeout values in the list [13]. The timeout value can be increased when it results in too many shutdowns, and it can be decreased when more shutdowns are desirable.

Adaptive Predictive Policies: Another adaptive shutdown policy has been proposed in Reference 10. The idle time prediction is calculated as a weighted sum of the last idle period and the last prediction (exponential average):

$$T_{pred}^n = a T_{idle}^{n-1} + (1-a)T_{pred}^{n-1}$$

This prediction formula can effectively predict idle periods in most cases; however, when a single very long idle period occurs in between a sequence of nearly uniform idle periods, the prediction of the upcoming long idle period and the following one will generally be quite inaccurate. This is because the single long idle period represents an outlier in the sequence and alters the estimation process. The authors of Reference 10 introduce appropriate strategy to correct mispredictions for this particular scenario that are able to solve the problem but also taint the simplicity of the basic estimator.

Adaptive Stochastic Policies: An adaptive extension of stochastic control technique is proposed to overcome a limitation of the (static) stochastic techniques. It is not possible to know complete knowledge of the system (SP) and its workload (SR) a priori. Even though it is possible to construct a model for the SP once and for all, the system workload is generally much harder to characterize in advance. Furthermore, workloads are often nonstationary. Adaptation consists of three phases: policy precharacterization, parameter learning, and policy interpolation [16]. Policy precharacterization builds a *n*-dimensional table addressed by *n* parameters for the Markov model of the workload. The table contains the optimal policy for the system under different workloads. Parameter learning is performed online during system operation by short-term averaging techniques. The parameter values are then used for addressing the table and for obtaining the power management policy. If the estimated parameter values do not exactly match the discrete set of values used for addressing the table, interpolation may obtain an appropriate policy as a combination of the policies in the table.

7.2.3.5 LEARNING-BASED POLICIES

More recently, a few works have proposed the use of various learning algorithms to devise policies that naturally adapt to nonstationary workloads [17–19]. These policies represent a sort of hierarchical generalization of adaptive policies, in the sense that instead of using a single adaptive policy, they take a set of well-known policies and rather design a *policy selection* mechanism. The approaches essentially differ in the type of policy selection strategy used: Markov decision processes [17], machine learning [18], or reinforcement Q-learning [19].

7.2.4 DYNAMIC VOLTAGE SCALING

Supply voltage scaling is one of the most effective techniques in energy minimization of CMOS circuits because the dynamic energy consumption of CMOS is quadratically related to the supply voltage V_{dd} as follows:

$$(7.1) \qquad\qquad E_{dynamic} = CV_{dd}^2 R$$

$$(7.2) \qquad\qquad P_{dynamic} = CV_{dd}^2 f$$

where

$E_{dynamic}$ is the dynamic energy consumption for a task execution
C is the average switching load capacitance of target circuit
V_{dd} is the controlled supply voltage
R is the number of cycles required in the execution of a given task
f is the maximum operating frequency corresponding to the V_{dd}

However, the supply voltage has a strong relation with the circuit delay. The circuit delay of CMOS can be approximated to be linearly proportional to the inverse of the supply voltage [20]. That is, a lower supply voltage results in a larger circuit delay and a smaller maximum operating frequency, which may degrade the performance of the target system:

$$(7.3) \qquad\qquad \tau_{delay} \propto \frac{V_{dd}}{(V_{dd} - V_{th})^2} \approx \frac{1}{V_{dd}},$$

where

τ_{delay} is the circuit delay
V_{th} is the threshold voltage of the CMOS transistors

DVS is the power management strategy that controls the supply voltage according to the current workload at runtime to minimize the energy consumption without having an adverse effect on system performance. DVS can be viewed as a variant of DPM in which DPM is applied not just to *idle* components but also to those resources that are *noncritical* in terms of performance, running the resource at different power/speed points. In other words, DVS introduces the concept of *multiple active states*, as opposed to DPM, where there can be multiple *idle* states but only a single active state.

Since in DVS power/speed trade-off points are defined by different supply voltage (and/or frequency) levels, DVS is traditionally applied to digital programmable components rather than generic devices and is usually associated with a trade-off between *computation speed* and power consumption. In order to apply DVS to real systems, hardware support for voltage scaling is required [22], as well as the software support that monitors the task execution and gives the voltage control command to the DVS hardware.

DVS is typically managed at the task granularity, where "task" refers to a generic computation burst with an associated maximum duration (e.g., a deadline). It does not necessarily coincide with a task or thread of an OS. DVS conceptually exploits the slack time (i.e., idleness) of tasks to avoid performance degradation: if the current task has more remaining time than the expected execution time at the maximum frequency, the voltage scheduler lowers the supply voltage and extends the execution time of this task up to the arrival time of the next task.

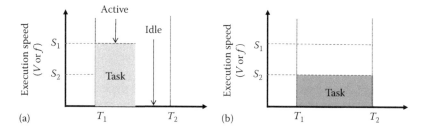

FIGURE 7.4 Dynamic power management vs. dynamic voltage scaling: the concept.

As an example, consider the conceptual scenario of Figure 7.4, where a core with two execution speeds S_1 and S_2 ($S_1 > S_2$) executes a task in a time slot between T_1 and T_2. The two speeds are obtained by powering the core with different supply voltages $V_{dd,1}$ and $V_{dd,2}$ ($V_{dd,1} > V_{dd,2}$). Let us assume also that the computational demand of this task is such that, when executing at S_1, the task finishes earlier than T_2.

In a DPM scenario (Figure 7.4a), we turn the system into an *off* state when it finishes execution (provided that the remaining idle time is long enough to justify the transition). With DVS (Figure 7.4b), conversely, we execute the task at the lower speed S_2, which in this specific case allows to finish exactly at T_2, thus leaving no idleness.

The power and energy benefits of this strategy are evident when considering the quadratic dependency of power and energy on supply voltage in Equations 7.1 and 7.2. Let us do a rough evaluation of the energy for the two case of Figure 7.4. For DPM, assuming that the time slot $T_2 - T_1$ corresponds to N cycles, that the task executed at S_1 terminates after $N/2$ cycles (staying thus idle for the remaining $N/2$ cycles), and that the power in the *off* state is zero, the total energy is therefore $E_{dynamic,a} = C \cdot V_1^2 \cdot \dfrac{N}{2}$. In the DVS case, assuming, for instance, that S_2 is achieved with a supply voltage $V_2 = 0.5V_1$, we get $E_{dynamic,b} = C(0.5V_1)^2 N = 0.5E_{dynamic,a}$. The calculation obviously neglects all the overheads in the state (for DPM) or speed (for DVS) transitions.

7.2.4.1 TASK SCHEDULING SCHEMES FOR DVS

The vast majority of the DVS strategies proposed in the literature focus on systems with real-time requirements, because they can be well characterized in terms of task start and finish times, durations, and deadlines, thus allowing the possibility of devising formal energy-optimal *scheduling algorithms*. The various algorithms suggest different runtime slack estimation and distribution schemes that are trying to achieve the theoretical lower bound of energy consumption that is calculated statically for a given workload [24]. Several DVS scheduling schemes are summarized and evaluated in Reference 23.

The rest of this section analyzes two important issues concerning these DVS scheduling algorithms for hard real-time system, namely, the *granularity* of the voltage adjustments and the slack estimation methods.

1. *Voltage scheduling granularity*: DVS scheduling schemes are classified by voltage scheduling granularity and fall into two categories: intertask DVS algorithm and intratask DVS algorithm. In the intratask DVS algorithms, a task is partitioned into multiple pieces such as time slots [25] or basic blocks [26], and a frequency and consequent voltage assignment is applied during the task execution. The actual execution time variation is estimated at the boundary of each time slot or each basic block and used for the input of adaptive operation frequency and voltage assignment.

   In intertask DVS algorithms, voltage assignment occurs at the task boundaries. After a task is completed, a new frequency and consequent voltage setting is applied by static or dynamic slack time estimation. The slack time estimation method has to be aggressive to effectively achieve system energy reduction. At the same time, it must be conservative so that every task is successfully scheduled within its deadline. These two rules are conflicting with each other, making it difficult to develop an effective slack estimation algorithm.

2. *Slack time estimation*: DVS techniques for hard real-time system enhance the traditional earliest deadline first or rate monotonic scheduling to exploit slack time that is used

to adjust voltage and frequency of voltage-scalable components. Therefore, the primary objective is to estimate the slack time accurately for more energy reduction. Various kind of static and dynamic slack estimation methods have been proposed to exploit the most of slack time without violating the hard real-time constraints.

Most approaches for intertask DVS algorithms deal with static slack estimation methods [27–31]. These methods typically aim at finding the lowest possible operating frequency at which all the tasks meet their deadlines. These methods rely on the worst-case execution time (WCET) to guarantee hard real-time demands. Therefore, the operating frequency can be lowered to the extent that each task's WCET does not exceed the deadline. The decision of each task's frequency can be done statically because it is the function of WCET and deadline, which are not changed during runtime.

In general, the actual execution time is much shorter than WCET. Therefore, WCET-based static slack time estimation cannot fully exploit the actual slacks. To overcome this limitation, various dynamic slack estimation methods have been proposed. Cycle-conserving RT-DVS technique utilizes the extra slack time to run other remaining tasks at a lower clock frequency [29]. Using this approach, the operating frequency is scaled by the CPU utilization factor (a value between 0 [always idle] and 1 [100% active]). The utilization factor is updated when the task is released or completed. When any task is released, the utilization factor is calculated according to the task's WCET. After a task is completed, the utilization factor is updated by using the actual execution time. The operation frequency may be lowered until the next arrival time of that task.

The next release time of a task can be used to calculate the slack budget [28]. This approach maintains two queues: the *run queue* and the *delay queue*. The former holds tasks that are waiting sorted by their priority order, while the latter holds tasks that are waiting for next periods, sorted by their release schedule. When the active queue is empty and the required execution time of an active task is less than its allowable time frame, the operation frequency is lowered using that slack time. As shown in Reference 28, the allowable time frame is defined as the minimum between the deadline of the current active task and the release time of the first element of the delay queue.

Path-based slack estimation for intratask DVS algorithms is also possible [26]. The control flow graph (CFG) of the task is used for slack time estimation; each node of the CFG is a basic block of the task, and each edge indicates control dependency between basic blocks. When the thread of execution control branches to the next basic block, the expected execution time is updated. If the expected execution time is smaller than the task's WCET, the operating frequency can be lowered.

7.2.4.2 PRACTICAL CONSIDERATIONS IN DVS

Most DVS studies focused on task scheduling assume a target system that (1) consists of all voltage-scalable components whose supply voltage can be set to any value within a given range of supply voltage, (2) in which only dynamic energy is considered, and (3) in which the speed setting of the tasks do not affect other components of the system. Although these assumptions simplify calculation of energy consumption and development of a scheduling scheme, the derived scheduling may not perform well because the assumption does not reflect a realistic setting. In fact, in deeply scaled CMOS technology leakage power is become a significant portion of total power, which does not fit exactly in the relations of Equations 7.1 and 7.2. Recent studies have addressed the impact of these nonidealities on DVS.

Discretely Variable Voltage Levels: Most scheduling algorithms assume that the supply voltage can be regulated in a continuous way; however, commercial DVS-enabled microprocessors usually support only a small number of predefined voltages. To get a more practical DVS schedule, some scheduling techniques have been proposed for discretely variable supply voltage levels as opposed to continuously variable one.

An optimal voltage allocation technique for a single task with discrete voltage levels can be obtained by solving an integer LP [32]. In case that only a small number of discretely variable voltages can be used, this work shows that using the two voltages in that are the immediate neighbors to the ideal voltage that minimizes the energy consumption. By iterating this "interpolation" of adjacent voltages, it is possible to yield an energy-optimal schedule for multiple tasks [33].

Leakage-aware DVS: As the supply voltage of CMOS devices becomes lower, the threshold voltage should be also reduced, which results in dramatic increase of the subthreshold leakage current. Therefore, especially for deeply scaled technologies, both static and dynamic energy should be considered.

However, when adding the static energy contribution, total energy is not a monotonically increasing function of the supply voltage anymore. Since the performance degradation due to reduction of the supply voltage may increase the execution time, this may result in the increase of static energy consumption. Consequently, if the supply voltage is reduced below a certain limit, the energy consumption becomes larger again. Inspired by this convex energy curve, which is not monotonically increasing with respect to the supply voltage, it is possible to identify an optimal *critical* voltage value [34] below which it is not convenient to scale supply voltage, even though there is still slack time left over. For this leftover idle time, extra power can be saved by turning the resource off (provided that this is convenient, based on breakeven time analysis).

Memory-aware DVS: As the complexity of modern system increases, the other components except microprocessor, for example, memory devices, contribute more to system power consumption. Thus, their power consumption must be considered when applying a power management technique. Unfortunately, many off-chip components do not allow supply voltage scaling. Since they are controlled by the microprocessor, their active periods may become longer when we slow down the microprocessor using DVS. The delay increase due to lower supply voltage of a microprocessor may increase the power consumption of non-supply-voltage-scalable devices eventually. In these cases, the energy gain achieved from DVS on the processor must be traded off against the energy increase of non-voltage-scalable devices.

There are some related studies about the DVS in the system including non-supply-voltage-scalable devices, especially focusing on memory devices. The works of References 35 and 36 show that aggressive reduction of supply voltage of microprocessor results in the increase of total energy consumption because the static energy consumption of memory devices becomes larger as the execution time gets longer, and thus chances of power down decrease. In Reference 36, an analytical method to obtain an energy-optimal frequency assignment of non-supply-voltage-scalable memory and supply-voltage-scalable CPU is proposed.

7.2.4.3 NONUNIFORM SWITCHING LOAD CAPACITANCES

Even if multiple tasks are scheduled, most studies characterize these tasks only by the timing constraints. This assumes that all tasks with the same voltage assignment and the same period will consume the same amount of energy irrespective of their operation. This means that uniform switching load capacitances C in the energy consumption equation are assumed for all the tasks. However, different tasks may utilize different datapath units that result in different energy consumption even for the tasks with the same period. Scheduling algorithms for multiple tasks that account for nonuniform switching load capacitances can be obtained by adapting the basic algorithm of Reference 24 [33].

7.2.4.4 DVS SCHEDULING FOR NON-REAL-TIME WORKLOADS

Unlike real-time workloads, generic workloads cannot be characterized in terms of a task-based model using start and finish times, deadlines, and WCETs. As a consequence, it is not possible to devise a universal, energy-optimal scheduling algorithm for the real-time case. The traditional approach is to consider execution as split in uniform-length time intervals of relatively large size (in the order of several thousand cycles), and update processor speed (via voltage/frequency setting) at the beginning of each time interval based on the characteristics of the workload in the previous interval(s) [37,38].

An example of policy is the PAST policy of Reference 37, in which the speed of the next time interval is increased by 20% if the utilization factor in the previous interval was greater than 70%, and speed is decreased by 60% minus the utilization factor if the latter was lower than 50%. The algorithm considers also the computation that, due to a too low speed in the previous interval, can carry over the current interval; the utilization factor includes this excess computation.

The heuristic nature of these algorithms is evident from this example—the thresholds are calibrated based on empirical evaluation on typical workloads and no optimality can be done.

These methods are simple implementations of a more general principle that provides minimum overall energy, that is, to *average the workload* over multiple time periods and use the closest speed that allows to execute leaving the smallest idle time. This approach, besides reducing the number of speed changes, can be shown to improve average energy because of the convex relationship between energy and speed [21].

7.2.4.5 APPLICATIONS OF DVS

Besides the approaches listed earlier, several extensions have been made to DVS for various application areas. One of them is an extension of DVS to multiple processing elements. Although a single processor is often assumed, recent studies try to derive the energy-optimal task schedule and allocation on multiple DVS-enabled processing elements such as a multiprocessor system on chip [39]. The dependency between each processing element makes DVS more complex, and thus needs more future study.

Some studies focus DVS specialization for specific applications, for example, a multimedia workload. In Reference 40, the decoding time of an MPEG video frame is predicted using the characteristics of MPEG-2. The task scheduler utilizes this information to determine the most appropriate supply voltage for decoding the current frame.

DVS is often applied to a system that must meet fault tolerance requirements as well as real-time constraints [41,42]. Fault tolerance can be achieved through the checkpointing and rollback recovery, which also require slack time like DVS. Therefore, the slack times should be carefully distributed among the fault tolerance features and the power management techniques, respectively, to minimize the energy consumption maintaining fault tolerance.

7.3 BATTERY-AWARE DYNAMIC POWER MANAGEMENT

The power management techniques described in the previous section do assume an ideal power supply, that is, (1) it can satisfy any power demand by the load and (2) the available battery energy is always available regardless of the load demand. While this simplifying assumption may be considered reasonable for systems connected to a fixed power supply, it is simply wrong in the case of battery-operated systems.

7.3.1 BATTERY PROPERTIES

Batteries are in fact nowhere close to be ideal charge storage units, as pointed out in any battery handbook [47]. From a designer's standpoint, two are the main nonidealities of real-life battery cells that need to be considered:

- *The capacity of a battery depends on the discharge current*. At higher currents, a battery is less efficient in converting its chemically stored energy into available electrical energy. This effect is called the "rated capacity effect": it is shown in the left plot of Figure 7.5, where the capacity of the battery is plotted as a function of the average current load. We observe that, for increasing load currents, the battery capacity progressively decreases: at higher rates, the cell is less efficient at converting its stored energy into available electrical energy.
- A consequence of the rated capacity effect is that battery lifetime will be negatively correlated with the *variance* of the current load; for a given average current value, a constant load (i.e., with no variance) will result in the longest battery lifetime of all load profiles.
- *Batteries have some (limited) recovery capacity when they are discharged at high current loads*. A battery can recover some of its deliverable charge if periods of discharge are interleaved with rest periods, that is, periods in which no current is drawn. The right plot in Figure 7.5 shows how an intermittent current load (dashed line) results in a longer battery lifetime than a constant current load (solid line), for identical discharge rate. The x-axis represents the actual elapsed time of discharge, that is, it does not include the time during which the battery has been disconnected from the load.

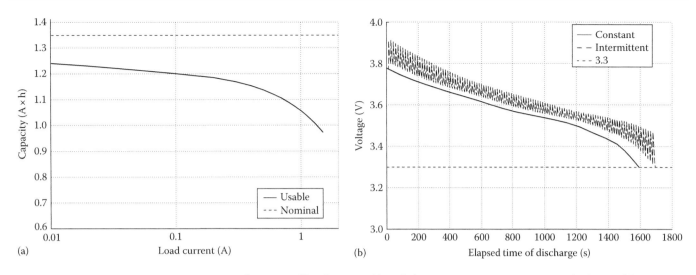

FIGURE 7.5 Battery capacity variation as a function of load current (a) and charge recovery in intermittent discharge (b).

Accounting for these nonidealities is essential, since it can be shown that power management techniques (both with and without DVS) that neglect these issues may actually result in an increase in energy [48,49].

7.3.2 BATTERY-AWARE DPM

The most intuitive solution consists thus of incorporating *battery-driven* policies into the DPM framework, either implicitly (i.e., using a battery-driven metric for a conventional policy) [51] or explicitly (i.e., a truly battery-driven policy) [50,52]. A simple example of the latter type could simply be a policy whose decision rules used to control the system operation state are based on the observation of a battery output voltage, which is (nonlinearly) related with the charge state [50].

More generally, it is possible to directly address the aforementioned nonidealities to shape the load current profile so as to increase as much as possible the effective capacity of the battery.

The issue of load-dependent capacity can be tackled along two dimensions. The dependency on the average current can be tackled by shaping the current profile in such a way that highly current-demanding operations are executed first (i.e., with fully charged battery), and low-current-demanding ones are executed later (i.e., with a reduced-capacity battery).

This principle fits well at the task granularity, where the shaping of the profile corresponds to task scheduling. Intuitively, the solution that maximizes battery efficiency and thus its lifetime is the one in which tasks are scheduled in nondecreasing order of their average current demand (Figure 7.6), compatibly with deadline or response time constraints [55].

The issue of charge recovery can be taken into account by properly arranging the idle times in the current profile. In particular, idle periods can be *inserted* between the executions of tasks. Notice that this is different from typical current profiles, where active and idle intervals typically alternate in relatively long bursts (Figure 7.7a).

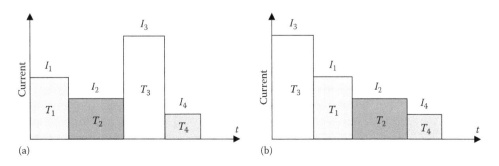

FIGURE 7.6 A set of tasks (a) and its optimal sequencing (b).

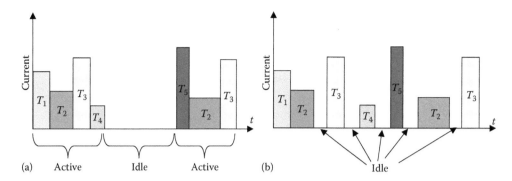

FIGURE 7.7   Active/idle bursts (a) and idle time insertion (b).

Inserting idle slots between task execution will allow the battery to recover some of the charge so that lifetime can be increased (Figure 7.6b). In the example, it may happen that after execution of T_2 the battery is almost exhausted and execution of T_3 will not complete; conversely, the insertion of an idle period will allow the battery to recover part of the charge so that execution of T_3 becomes possible. Idle time insertion can obviously be combined with the ordering of tasks to exploit both properties and achieve longer battery lifetimes [56].

Availaibility of appropriate battery models are therefore essential to assess the actual impact of the system's operations on the available energy stored in a battery. Stated in other terms, how the energy consumed by the hardware actually reflects on the energy drawn from the battery. Many types of battery models have been proposed in the literature, with a large variety of abstraction levels, degrees of accuracy, and semantics ([57]–[63]).

7.3.3 BATTERY-AWARE DVS

The possibility of scaling supply voltages at the task level adds further degrees of freedom in choosing the best shaping of the current loads. Voltage scaling can be in fact viewed as another opportunity to reduce the current demand of a task, at the price of increased execution time. Under a simple, first-order approximation, the drawn current I is proportional to V^3, while delay increases proportionally with V. Therefore, the trade-off is between a decrease (increase) in the discharge current and an increase (decrease) in the duration of the stress [56].

Battery-aware DVS tries to enforce the property of the load-dependent capacity, since it can be proved that scaling voltage is always more efficient than inserting idle periods [55]. Therefore, anytime a slack is available, it should be filled by scaling the voltage (Figure 7.8), compatibly with possible constraints.

This is equivalent to stating that the impact on lifetime of the rate-dependent behavior of batteries dominates that due to the charge recovery effect. Notice that this property is consistent with the results derived in Section 7.2 for non-battery-operated systems: Scaling voltage is always more convenient, energy-wise, than shutting down.

Solutions proposed in the literature typically start from a nondecreasing current profile and reclaim the available slack from the tail of the schedule by slowing down tasks according to the specified constraints [53–55].

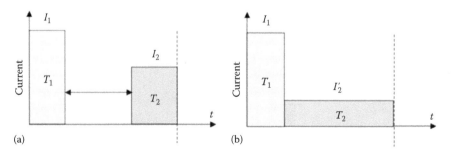

FIGURE 7.8   Idle period insertion (a) vs. voltage scaling (b).

7.3.4 MULTIBATTERY SYSTEMS

In the case of supply systems consisting of multiple batteries, the load-dependent capacity of batteries has deeper implications, which open additional opportunities for optimization. In fact, since at a given point in time the load is connected to only one battery, the other ones are idle. DPM with multiple batteries amounts thus to the problem of assigning of *battery* idle times (as opposed to task idle times). In other words, since this is equivalent to schedule battery usage over time, this problem is called *battery scheduling*.

The default battery scheduling policy in use in most devices is a nonpreemptive one that sequentially discharges one battery after another, in some order. Because of the aforementioned nonidealities, this is clearly a suboptimal policy [64].

Similarly to task scheduling, battery scheduling can be either workload independent or workload dependent. In the former case, batteries are attached to the load in a round-robin fashion for a fixed amount of time. Unlike task scheduling, the choice of this quantity is dictated by the physical property of batteries. It can be shown, in fact, that the smaller this interval, the higher is the equivalent capacity of the battery sets [65]. This is because by rapidly switching between full load and no load, each battery perceives an effective averaged discharge current proportional to the fraction of time it is connected to the load. In other words, if a battery is connected to the load current I for a fraction $\alpha < 1$ of the switching period, it will perceive a load current $\alpha \cdot I$. This is formally equivalent to connecting the two batteries in parallel, without incurring into the problem of mutually discharging the batteries.

In the latter case, we assign a battery to the load depending on its characteristics. More precisely, one should select which battery to connect to the load based on runtime measurement of current draw, in an effort to match a load current to the battery that better responds to it [50,64,65]. A more sophisticated workload-dependent scheme consists of adapting the round-robin approach to heterogeneous (i.e., having different nominal capacities and discharge curves) multibattery supplies. In these cases, the current load should be split nonuniformly over all the cells in the power supply. Therefore, the round-robin policy can be modified in such a way that the time slice has different durations. For instance, in a two-battery system, this is equivalent to connect batteries to the load following a square wave with unbalanced duty cycle [66].

7.4 SOFTWARE-LEVEL DYNAMIC POWER MANAGEMENT

7.4.1 SOFTWARE POWER ANALYSIS

Focusing solely on the hardware components of a design tends to ignore the impact of the software on the overall power consumption of the system. Software constitutes a major component of systems where power is a constraint. Its presence is very visible in a mobile computer, in the form of the system software and application programs running on the main processor. But software also plays an even greater role in general digital applications. An ever-growing fraction of these applications are now being implemented as embedded systems. In these systems, the functionality is partitioned between a hardware and a software component.

The software component usually consists of application-specific software running on a dedicated processor, while the hardware component usually consists of application-specific circuits. The basic formulation of DPM described in the prior sections is general enough for its application to generic hardware components. However, for the software component it is more effective to view the power consumption from the point of view of the software running on the programmable element. Relating the power consumption in the processor to the instructions that execute on it provides a direct way of analyzing the impact of the processor on the system power consumption. Software impacts the system power consumption at various levels. At the highest level, this is determined by the way functionality is partitioned between hardware and software. The choice of the algorithm and other higher-level decisions about the design of the software component can affect system power consumption significantly. The design of the system software, the actual application source code, and the process of translation into machine instructions—all of these determine the power cost of the software component.

In order to systematically analyze and quantify this cost, however, it is important to start at the most fundamental level, that is, at the level of the individual instructions executing on the processor. Just as logic gates are the fundamental units of computation in digital hardware circuits, instructions can be thought of as the fundamental unit of software. Accurate modeling and analysis at this level is therefore essential. Instruction-level models can then be used to quantify the power costs of the higher constructs of software (application programs, system software, algorithm, etc.).

It would be helpful to define the terms "power" and "energy," as they relate to software. The average power consumed by a processor while running a given program is given by

$$P = I \cdot V_{dd}$$

where
  P is the average power
  I is the average current
  V_{dd} is the supply voltage

Power is also defined as the rate at which energy is consumed. Therefore, the energy consumed by a program is given by

$$E = P \cdot T$$

where T is the execution time of the program. This in turn is given by $T = N \cdot \tau$, where N is the number of clock cycles taken by the program and τ is the clock period. Energy is thus given by

$$E = I \cdot V_{dd} \cdot N \cdot \tau$$

Note that if the processor supports dynamic voltage and frequency switching, then V_{dd} and τ can vary over the execution of the program. It is then best to consider the periods of code execution with different (V_{dd}, τ) combinations as separate components of the power/energy cost. As it can be seen from the earlier discussion, the ability to obtain an estimate of the current drawn by the processor during the execution of the program is essential for evaluating the power/energy cost of software. These estimates can either be obtained through simulations or through direct measurements.

7.4.1.1 SOFTWARE POWER ESTIMATION THROUGH SIMULATION

The most commonly used method for power analysis of circuits is through specialized power analysis tools that operate on abstract models of the given circuits. These tools can be used for software power evaluation too. A model of the given processor and a suitable power analysis tool are required. The idea is to simulate the execution of the given program on the model. During simulation, the power analysis tool estimates the power (current) drawn by the circuit using predefined power estimation formulas, macromodels, heuristics, and/or algorithms.

However, this method has some drawbacks. It requires the availability of models that capture the internal implementation details of the processor. This is proprietary information, which most software designers do not have access to. Even if the models are available, there is an accuracy vs. efficiency trade-off. The most accurate power analysis tools work at the lower levels of the design—switch level or circuit level. These tools are slow and impractical for analyzing the total power consumption of a processor as it executes entire programs. More efficient tools work at the higher levels—register transfer or architectural. However, these are limited in the accuracy of their estimates.

7.4.1.2 MEASUREMENT-BASED INSTRUCTION-LEVEL POWER MODELING

The earlier problems can be overcome if the current being drawn by the processor during the execution of a program is physically measured. A practical approach to current measurement as applied to the problem of instruction-level power analysis has been proposed [46].

Using this approach, empirical instruction-level power models were developed for three commercial microprocessors. Other researchers have subsequently applied these concepts to other microprocessors. The basic idea is to assign power costs to individual instructions (or instruction pairs to account for interinstruction switching effects) and to various interinstruction effects like pipeline stalls and cache misses. These power costs are obtained through experiments that involve the creation of specific programs and measurement of the current drawn during their execution. These costs are the basic parameters that define the instruction-level power models. These models form the basis of estimating the energy cost of entire programs. For example, for the processors studied in Reference 46, for a given program, the overall energy cost is given by

$$E = \sum_i (B_i \cdot N_i) + S_{i,j}(O_i, j \cdot N_{i,j}) + S_k E_k$$

The base power cost, B_i, of each instruction, i, weighted by the number of times it will be executed, N_i, is added up to give the base cost of the program. To this, the circuit state switching overhead, $O_{i,j}$, for each pair of consecutive instructions, i, j, weighted by the number of times the pair is executed, $N_{i,j}$, is added. The energy contribution, E_k, of the other interinstruction effects, k (stalls and cache misses) that would occur during the execution of the program, is finally added.

The base costs and overhead values are empirically derived through measurements. The other parameters in the previous formula vary from program to program. The execution counts N_i and $N_{i,j}$ depend on the execution path of the program. This is dynamic, runtime information that has to be obtained using software performance analysis techniques. In certain cases, it can be determined statically, but in general it is best obtained from a program profiler. For estimating E_k, the number of times pipeline stalls and cache misses occur has to be determined. This is again dynamic information that can be statically predicted only in certain cases. In general, this information is obtained from a program profiler and cache simulator.

The processors whose power models have been published using the ideas mentioned earlier have so far been in-order machines with relatively simple pipelines. Out-of-order superscalar machines present a number of challenges to an instruction-oriented modeling approach and provide good opportunities for future research.

7.4.1.3 IDLE TIME EVALUATION

The earlier discussion is for the case when the processor is active and is constantly executing instructions. However, a processor may not always be performing useful work during program execution. For example, during the execution of a word processing program, the processor may simply be waiting for keyboard input from the user and may go into a low-power state during such idle periods. To account for these low-power periods, the average power cost of a program is thus given by

$$P = P_{active} \cdot T_{active} + P_{idle} \cdot T_{idle}$$

where
  P_{active} is the average power consumption when the processor is active
  T_{active} is the fraction of the time the processor is active
  P_{idle} and T_{idle} are the corresponding parameters for when the processor is idle and has been put in a low-power state

T_{active} and T_{idle} need to be determined using dynamic performance analysis techniques. In modern microprocessors, a hierarchy of low-power states is typically available, and the average power and time spent for each state would need to be determined.

7.4.2 SOFTWARE-CONTROLLED POWER MANAGEMENT

For systems in which part of the functionality is implemented in software, it is natural to expect that there is potential for power reduction through modification of software. Software power analysis (whether achieved through physical current measurements or through simulation of models of processors) as described in Section 7.2.4 helps to identify the reasons for variation in power from one program to another. These differences can then be exploited in order to search for low-power alternatives for each program. The information provided by the instruction-level analysis can guide higher-level design decisions like hardware–software partitioning and choice of algorithm. It can also be directly used by automated tools like compilers, code generators, and code schedulers for generating code targeted toward low power. Several ideas in this regard have been published, starting with the work summarized in Reference 46. Some of these ideas are based on specific architectural features of the subject processors and memory systems. The most important conclusion though is that in modern general-purpose CPUs, software energy and performance track each other, that is, for a given task, a faster program implementation will also have lower energy. Specifically, it is observed that the difference in average current for instruction sequences that perform the same function is not large enough to compensate for any difference in the number of execution cycles. Thus, given a function, the least energy implementation for it is the one with the faster running time. The reason for this is that CPU power consumption is dominated by a large common cost factor (power consumption due to clocks, caches, instruction fetch hardware, etc.) that for the most part is independent of instruction functionality and does not vary much from one cycle to the other. This implies that the large body of work devoted to software performance optimization provides direct benefits for energy as well. Power management techniques such as increased use of clock gating and multiple on-chip voltages indicate that future CPUs may show greater variation in power consumption from cycle to cycle. However, CPU design and power consumption trends do suggest that the relationship between software energy and power that was observed before will continue to hold. In any case, it is important to realize that software directly impacts energy/power consumption, and thus it should be designed to be efficient with respect to these metrics. A classic example of inefficient software is "busy wait loops." Consider an application such as a spreadsheet that requires frequent user input. During the times when the spreadsheet is recalculating values, high CPU activity is desired in order to complete the recalculation in a short time. In contrast, when the application is waiting for the user to type in values, the CPU should be inactive and in a low-power state. However, a busy wait loop will prevent this from happening and will keep the CPU in a high-power state. The power wastage is significant. Converting the busy wait loops to an instruction or system call that puts the CPU into a low-power state from which it can be woken up on an I/O interrupt will eliminate this wasted power.

7.4.2.1 OS-DIRECTED DYNAMIC POWER MANAGEMENT

An important trend in modern CPUs is the ability for software to control the operating voltage and frequency of the processor. These different voltage/frequency operating points, which represent varying levels of power consumption, can be switched dynamically under software control. This opens up additional opportunities for energy-efficient software design, since the CPU can be made to run at the lowest power state that still provides enough performance to meet the task at hand. In practice, the DPM and DVS techniques described in Section 7.2.4 from a hardware-centric perspective (because they managed multiple power states of *hardware components*) can be also viewed from the software perspective.

In this section, we describe implementation issues of power management techniques. In general, the OS is the best software layer where the DPM policy can be implemented. OS-directed power management (OSPM) has several advantages: (1) the power/performance dynamic control is performed by the software layer that manages computational, storage, and I/O tasks of the system; (2) power management algorithms are unified in the OS, yielding much better integration between the OS and the hardware; and (iii) moving the power management functionality into the OS makes it available on every machine on which the OS is installed. Implementation of OSPM is a hardware/software codesign problem because the hardware resources need to interface with the OS-directed software power manager, and because both hardware resources and the software applications need to be designed so that they could cooperate with OSPM.

The advanced configuration and power interface (ACPI) specification [43] was developed to establish industry common interfaces enabling a robust OSPM implementation for both devices and entire systems. Currently, it is standardized by Hewlett-Packard, Intel, Microsoft, Phoenix, and Toshiba. It is the key element of an OSPM, since it facilitates and accelerates the codesign of OSPM by providing a standard interface to control system components. From a power management perspective, OSPM/ACPI promotes the concept that systems should conserve energy by transitioning unused devices into lower-power states, including placing the entire system in a low-power state (sleep state) when possible. ACPI-defined interfaces are intended for wide adoption to encourage hardware and software designers to build ACPI-compatible (and thus, OSPM compatible) implementations. Therefore, ACPI promotes the availability of power manageable components that support multiple operational states. It is important to notice that ACPI specifies neither how to implement hardware devices nor how to realize power management in the OS. No constraints are imposed on implementation styles for hardware and on power management policies. The implementation of ACPI-compliant hardware can leverage any technology or architectural optimization as long as the power-managed device is controllable by the standard interface specified by ACPI. The power management module of the OS can be implemented using any kind of power management policies including predictive techniques, and stochastic control techniques. A set of experiments were carried out by Lu et al. to measure the effectiveness of different power management policies on ACPI-compliant computers [44,45].

7.5 CONCLUSIONS

The design of energy-efficient systems goes through the optimization of the architectures of the individual components, the communications between them, and their usage. Of these three dimensions, the optimization of components usage seems to better fit to a *system*-level context, since it may rely on very abstract models of the components and does not usually require detailed information about their implementation.

In this chapter, we discussed and analyzed the problem of optimizing the usage of components of a system, which is usually called dynamic power management. DPM aims at dynamically adapting the multiple states of operation of various components to the required performance level, in an effort to minimize the power wasted by idle or underutilized components.

In its simplest form, DPM entails the decision between keeping the component active or turning it off, where the *off* state may consists of different power/performance trade-off values. When combined with the possibility of dynamically varying the supply voltage, DPM generalizes to the more general problem of DVS, in which multiple active states are possible. DVS is motivated by the fact that it is always more efficient, energy-wise, to slow down a component rather than shutting it down, because of the quadratic dependency of power on voltage. Nonidealities can also be incorporated into DVS, such as the use of a discrete set of voltage levels, the impact of leakage power, and the presence of nonideal supply source (i.e., a battery).

Finally, we revisited the DPM/DVS problem from the software perspective, illustrating how this can be managed entirely by the software, possibly the OS running on the system.

REFERENCES

1. L. Benini, G. De Micheli, *Dynamic Power Management: Design Techniques and CAD Tools*, Kluwer Academic Publishers, 1998.
2. L. Benini, G. De Micheli, System level power optimization: Techniques and tools, *ACM Transactions on Design Automation of Electronic Systems*, 5(2), 115–192, April 2000.
3. L. Benini, A. Bogliolo, G. De Micheli, A survey of design techniques for system-level dynamic power management, *IEEE Transactions on Very Large Scale Integration (VLSI) Systems*, 8(3), 299–316, June 2000.
4. E. Macii (Editor), *IEEE Design & Test of Computers*, Special Issue on Dynamic Power Management, 18(2), April 2001.

5. L. Benini, R. Hodgson, P. Siegel, System-level power estimation and optimization, in *ISLPED-98: ACM/IEEE International Symposium on Low Power Electronics and Design*, pp. 173–178, August 1998.

6. H. Shim, Y. Joo, Y. Choi, H. G. Lee, K. Kim, N. Chang, Low-energy off-chip SDRAM memory systems for embedded applications, *ACM Transactions on Embedded Computing Systems*, Special Issue on Memory Systems, 2(1), 98–130, February 2003.

7. N. Chang, K. Kim, H. G. Lee, Cycle-accurate energy consumption measurement and analysis: Case study of ARM7TDMI, *IEEE Transaction on Very Large Scale Integration (VLSI) Systems*, 10(4), 146–154, April 2002.

8. I. Lee, Y. Choi, Y. Cho, Y. Joo, H. Lim, H. G. Lee, H. Shim, N. Chang, A web-based energy exploration tool for embedded systems, *IEEE Design & Test of Computers*, 21, 572–586, November–December 2004.

9. M. Srivastava, A. Chandrakasan, R. Brodersen, Predictive system shutdown and other architectural techniques for energy efficient programmable computation, *IEEE Transactions on Very Large Scale Integration Systems*, 4(3), 42–55, March 1996.

10. C.-H. Hwang, A. Wu, A predictive system shutdown method for energy saving of event-driven computation, in *ICCAD'97: International Conference on Computer-Aided Design*, pp. 28–32, November 1997.

11. P. Krishnan, P. Long, J. Vitter, Adaptive disk spin-down via optimal rent-to-buy in probabilistic environments, in *International Conference on Machine Learning*, pp. 322–330, July 1995.

12. D. Helmbold, D. Long, E. Sherrod, Dynamic disk spin-down technique for mobile computing, in *International Conference on Mobile Computing*, pp. 130–142, November 1996.

13. F. Douglis, P. Krishnan, B. Bershad, Adaptive disk spin-down policies for mobile computers, in *2nd USENIX Symposium on Mobile and Location-Independent Computing*, pp. 121–137, April 1995.

14. L. Benini, G. Paleologo, A. Bogliolo, G. D. Micheli, Policy optimization for dynamic power management, *IEEE Transactions on Computer-Aided Design*, 18(6), 813–833, June 1999.

15. Q. Qiu, M. Pedram, Dynamic power management based on continuous-time Markov decision processes, in *DAC-36: 36th Design Automation Conference*, pp. 555–561, June 1999.

16. E. Chung, L. Benini, A. Bogliolo, G. D. Micheli, Dynamic power management for non-stationary service requests, in *DATE'99: Design Automation and Test in Europe Conference*, pp. 77–81, March 1999.

17. Z. Ren, B. H. Krogh, R. Marculescuo, Hierarchical adaptive dynamic power management, *DATE'04: Design, Automation and Test in Europe Conference*, pp. 136–141, March 2004.

18. G. Dhiman, T. Simunic Rosing, Dynamic power management using machine learning, in *ICCAD'06: International Conference on Computer-Aided Design*, pp. 747–754, November 2006.

19. Y. Tan, W. Liu, Q. Qiu, Adaptive power management using reinforcement learning, in *ICCAD'09: International Conference on Computer-Aided Design*, pp. 461–467, November 2009.

20. A. P. Chandrakasan, S. Sheng, R. W. Brodersen, Low-power CMOS digital design, *IEEE Journal of Solid-State Circuits*, 27(4), 473–484, April 1992.

21. A. Sinha, A. P. Chandrakasan, Dynamic voltage scheduling using adaptive filtering of workload traces, in *VLSID'01: 14th International Conference on VLSI Design*, pp. 221–226, January 2001.

22. T. Burd, R. Brodersen, Design issues for dynamic voltage scaling, in *ISLPED'00: International Symposium on Low Power Electronics and Design*, pp. 9–14, July 2000.

23. W. Kim, D. Shin, H. S. Yun, J. Kim, S. L. Min, Performance comparison of dynamic voltage scaling algorithms for hard real-time systems, in *Real-Time and Embedded Technology and Applications Symposium*, pp. 219–228, September 2002.

24. F. Yao, A. Demers, A. Shenker, A scheduling model for reduced CPU energy, in *FOCS'95: IEEE Foundations of Computer Science*, pp. 374–382, October 1995.

25. S. Lee, T. Sakurai, Run-time voltage hopping for low-power real-time systems, in *DAC-37: 37th Design Automation Conference*, pp. 806–809, June 2000.

26. D. Shin, J. Kim, S. Lee, Low-energy intra-task voltage scheduling using static timing analysis, in *DAC-38: 38th Design Automation Conference*, pp. 438–443, June 2001.

27. Y.-H. Lee, C. M. Krishna, Voltage-clock scaling for low energy consumption in real-time embedded systems, in *Real-Time Computing Systems and Applications*, pp. 272–279, December 1999.

28. Y. Shin, K. Choi, T. Sakurai, Power optimization of real-time embedded systems on variable speed processors, in *ICCAD'00: International Conference on Computer-Aided Design*, pp. 365–368, November 2000.

29. P. Pillai, K. G. Shin, Real-time dynamic voltage scaling for low-power embedded operating systems, in *18th ACM Symposium on Operating Systems Principles*, pp. 89–102, October 2001.

30. F. Gruian, Hard real-time scheduling using stochastic data and DVS processors, in *ISLPED'01: International Symposium on Low Power Electronics and Design*, pp. 46–51, August 2001.

31. G. Quan, X. S. Hu, Energy efficient fixed-priority scheduling for real-time systems on variable voltage processors, in *DAC-38: 38th Design Automation Conference*, pp. 828–833, June 2001.

32. T. Ishihara, H. Yasuura, Voltage scheduling problem for dynamically variable voltage processors, in *ISLPED'98: International Symposium on Low Power Electronics and Design*, pp. 197–202, August 1998.

33. W. Kwon, T. Kim, Optimal voltage allocation techniques for dynamically variable voltage processors, in *DAC-40: 40th Design Automation Conference*, pp. 125–130, June 2003.

34. R. Jejurikar, C. Pereira, R. Gupta, Leakage aware dynamic voltage scaling for real-time embedded systems, in *DAC-41: 41st Design Automation Conference*, pp. 275–280, June 2004.

35. X. Fan, C. Ellis, A. Lebeck, The synergy between power-aware memory systems and processor voltage, in *Workshop on Power-Aware Computer Systems*, pp. 130–140, December 2003.

36. Y. Cho, N. Chang, Memory-aware energy-optimal frequency assignment for dynamic supply voltage scaling, in *ISLPED'04: International Symposium on Low Power Electronics and Design*, pp. 387–392, August 2004.

37. M. Weiser, B. Welch, A. Demers, S. Shenker, Scheduling for reduced CPU energy, in *USENIX Symposium on Operating Systems Design and Implementation*, pp. 13–23, November 1994.

38. K. Govil, E. Chan, H. Wasserwan, Comparing algorithm for dynamic speed-setting of a low-power CPU, in *MobiCom '95: First International Conference on Mobile Computing and Networking*, pp. 13–25, November 1995.

39. M. Schmitz, B. Al-Hashimi, P. Eles, Energy-efficient mapping and scheduling for DVS enabled distributed embedded systems, in *DATE'02: Design, Automation and Test in Europe Conference*, pp. 514–521, March 2002.

40. K. Choi, K. Dantu, W. Cheng, M. Pedram, Frame-based dynamic voltage and frequency scaling for a MPEG decoder, in *ICCAD'02: International Conference on Computer Aided Design*, pp. 732–737, November 2002.

41. Y. Zhang, K. Chakrabarty, Energy-aware adaptive checkpointing in embedded real-time systems, in *DATE'03: Design, Automation and Test in Europe Conference*, pp. 918–923, March 2003.

42. Y. Zhang, K. Chakrabarty, V. Swaminathan, Energy-aware fault-tolerance in fixed-priority real-time embedded systems, in *ICCAD'03: International Conference on Computer Aided Design*, pp. 209–213, November 2003.

43. *Advanced Configuration and Power Interface Specification Revision 3.0*, Hewlett-Packard Corporation, Intel Corporation, Microsoft Corporation, Phoenix Technologies Ltd., and Toshiba Corporation, September 2004.

44. Y. Lu, T. Simunic, G. D. Micheli, Software controlled power management, in *CODES'99: International Workshop on Hardware-Software Codesign*, pp. 151–161, May 1999.

45. Y. Lu, E. Y. Chung, T. Simunic, L. Benini, G. D. Micheli, Quantitative comparison of power management algorithms, in *DATE'00: Design Automation and Test in Europe Conference*, pp. 20–26, March 2000.

46. V. Tiwari, S. Malik, A. Wolfe, T. C. Lee, Instruction level power analysis and optimization software, *Journal of VLSI Signal Processing*, 13(2), 223–238, August 1996.

47. D. Linden, *Handbook of Batteries*, 2nd edn., McGraw Hill, Hightstown, NJ, 1995.

48. T. Martin, D. Sewiorek, Non-ideal battery and main memory effects on CPU speed-setting for low power, *IEEE Transactions on VLSI Systems*, 9(1), 29–34, February 2001.

49. M. Pedram, Q. Wu, Battery-powered digital CMOS design, in *DATE-99: Design Automation and Test in Europe Conference*, Munich, Germany, pp. 72–76, March 1999.

50. L. Benini, G. Castelli, A. Macii, R. Scarsi, Battery-driven dynamic power management, *IEEE Design & Test of Computers*, 18(2), 53–60, March–April 2001.

51. M. Pedram, Q. Wu, Design considerations for battery-powered electronics, *IEEE Transactions on VLSI Systems*, 10(5), 601–607, October 2002.

52. P. Rong, M. Pedram, Extending the lifetime of a network of battery-powered mobile devices by remote processing: A Markovian decision-based approach, in *DAC-40: 40th Design Automation Conference*, pp. 906–911, June 2003.

53. J. Luo, N. Jha, Battery-aware static scheduling for distributed real-time embedded systems, in *DAC-38: 38th Design Automation Conference*, pp. 444–449, June 2001.

54. D. Rakhmatov, S. Vrudhula, C. Chakrabarti, Battery-conscious task sequencing for portable devices including voltage/clock scaling, in *DAC-39: 39th Design Automation Conference*, pp. 211–217, June 2002.

55. P. Chowdhury, C. Chakrabarti, Static task-scheduling algorithms for battery-powered DVS systems, *IEEE Transactions on VLSI Systems*, 13(2), 226–237, February 2005.

56. D. Rakhmatov, S. Vrudhula, Energy management for battery-powered embedded systems, *ACM Transactions on Embedded Computing Systems*, 2(3), 277–324, August 2003.

57. M. Doyle, T. F. Fuller, J. Newman, Modeling of galvanostatic charge and discharge of the lithium/polymer/insertion cell, *Journal of Electrochemical Society*, 140(6), 1526–1533, 1993.

58. T. F. Fuller, M. Doyle, J. Newman, Simulation and optimization of the dual lithium ion insertion cell, *Journal of Electrochemical Society*, 141(1), 1–10, 1994.

59. M. Glass, Battery electro-chemical nonlinear/dynamic SPICE model, in *Energy Conversion Engineering Conference*, Washington, DC, pp. 292–297, August 1996.

60. L. Benini, G. Castelli, A. Macii, E. Macii, M. Poncino, R. Scarsi, Discrete-time battery models for system-level low-power design, *IEEE Transactions on Very Large Scale Integration (VLSI) Systems*, 9(5), 630–640, October 2001.

61. D. Rakhmatov, S. Vrudhula, An analytical high-level battery model for use in energy management of portable electronic systems, in *International Conference on Computer Aided Design*, pp. 488–493, November 2001.

62. C. F. Chiasserini, R. R. Rao, Energy efficient battery management, *IEEE Journal in Selected Areas in Communications*, 19(7), 1235–1245, July 2001.

63. P. Rong, M. Pedram, An analytical model for predicting the remaining battery capacity of lithium-ion batteries, in *DATE'03: Design, Automation and Test in Europe Conference*, pp. 1148–1149, March 2003.

64. Q. Wu, Q. Qiu, M. Pedram, An interleaved dual-battery power supply for battery-operated electronics, in *ASPDAC'00: Asia and South Pacific Design Automation Conference*, pp. 387–390, January 2000.

65. L. Benini, G. Castelli, A. Macii, E. Macii, M. Poncino, R. Scarsi, Scheduling battery usage in mobile systems, *IEEE Transactions on Very Large Scale Integration (VLSI) Systems*, 11(6), 1136–1143, December 2003.

66. L. Benini, D. Bruni, A. Macii, E. Macii, M. Poncino, Extending lifetime of multi-battery mobile systems by discharge current steering, *IEEE Transactions on Computers*, 53(5), 985–995, May 2003.

Processor Modeling and Design Tools

Anupam Chattopadhyay, Nikil Dutt, Rainer Leupers, and Prabhat Mishra

CONTENTS

8.1 INTRODUCTION

Computing is an integral part of daily life. We encounter two types of computing devices every day: desktop-based computing devices and embedded computer systems. Desktop-based computing systems encompass traditional *computers*, including personal computers, notebook computers, workstations, and servers. Embedded computer systems are ubiquitous—they run the computing devices hidden inside a vast array of everyday products and appliances such as smartphones, toys, handheld PDAs, cameras, and cars. Both types of computing devices use programmable components such as processors, coprocessors, and memories to execute application programs. These programmable components are also referred as "programmable accelerators." Figure 8.1 shows an exemplary embedded system with programmable accelerators. Depending on the application domain, the embedded system can have application-specific accelerators, interfaces, controllers, and peripherals. The complexity of the programmable accelerators is increasing at an exponential rate due to technological advances as well as demand for realization of ever more complex applications in communication, multimedia, networking, and entertainment. Shrinking time to market coupled with short product lifetimes creates a critical need for design automation of increasingly sophisticated and complex programmable accelerators.

Modeling plays a central role in the design automation of processors. It is necessary to develop a specification language that can model complex processors at a higher level of abstraction and also enable automatic analysis and generation of efficient prototypes. The language should be powerful enough to capture high-level description of the programmable architectures. On the other hand, the language should be simple enough to allow correlation of the information between the specification and the architecture manual.

Specifications widely in use today are still written informally in natural languages like English. Since natural language specifications are not amenable to automated analysis, there are possibilities of ambiguity, incompleteness, and contradiction: all problems that can lead to different interpretations of the specification. Clearly, formal specification languages are suitable for analysis and verification. Some have become popular because they are input languages for powerful verification tools such as a model checker. Such specifications are popular among verification engineers with expertise in formal languages. However, these specifications are not acceptable by designers and other tool developers. Therefore, the ideal specification language should have formal (unambiguous) semantics as well as easy correlation with the architecture manual.

Architecture description languages (ADLs) have been successfully used as a specification language for processor development. Development of a processor is associated with multiple steps as embodied in the Mescal design methodology [1]. These steps—*benchmarking, architectural*

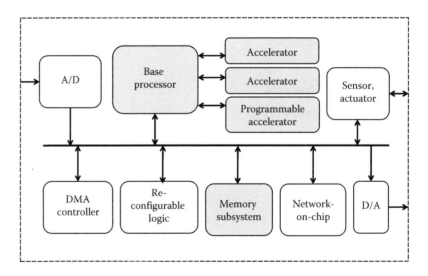

FIGURE 8.1   An exemplary embedded system.

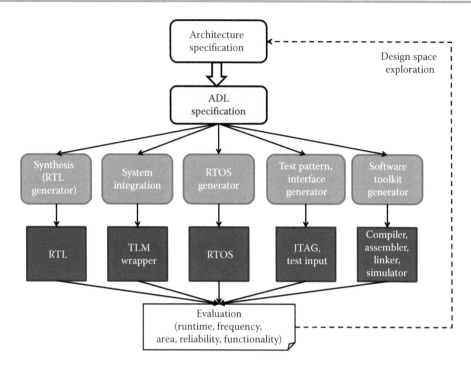

FIGURE 8.2   ADL-driven exploration, synthesis, and validation of programmable architectures.

space identification, *design evaluation*, *design exploration*, and *deployment*—are correlated with corresponding tools and implementations. Mescal methodology, in practice, is represented by the ADL-based processor design flow. The ADL specification is used to perform design representation, design evaluation, design validation, and synthesis to a more detailed abstraction such as register transfer level (RTL). This is shown in Figure 8.2.

More specifically, the ADL specification is used to derive a processor toolsuite such as instruction-set (IS) simulator, compiler, debugger, profiler, assembler, and linker. The specification is used to generate detailed and optimized hardware description in an RTL description [2,3]. The ADL specification is used to validate the design by formal, semiformal, and simulation-based validation flows [4], as well as for the generation of test interfaces [5]. The specification can also be used to generate device drivers for real-time operating systems (OSs) [6].

The rich modeling capability of ADL is used to design various kinds of processor architectures ranging from programmable coarse-grained reconfigurable architectures (CGRAs) to superscalar processors [2,7,8]. The design exploration capability of an ADL-driven toolsuite is extended to cover high-level power and reliability estimations [9]. The prominence of ADL is also palpable in its widespread market acceptance [10,11]. In this chapter, we attempt to cover the entire spectrum of processor design tools, from the perspective of ADL-based design methodology.

The rest of this chapter is organized as follows: Section 8.2 describes processor modeling using ADLs. Section 8.3 presents ADL-driven methodologies for software toolkit generation, hardware synthesis, exploration, and validation of programmable architectures. Finally, Section 8.4 concludes the chapter.

8.2 PROCESSOR MODELING USING ADLs

The phrase "architecture description language" has been used in the context of designing both software and hardware architectures. Software ADLs are used for representing and analyzing software architectures [12]. They capture the behavioral specifications of the components and their interactions that comprise the software architecture. However, hardware ADLs (also known as processor description languages) capture the structure, that is, hardware components and their connectivity, and the behavior (IS) of processor architectures. The concept of using high-level languages (HLLs) for specification of architectures has been around for a long time. Early ADLs such as ISPS [13] were used for simulation, evaluation, and synthesis of computers and other digital systems. This section gives a short overview of prominent ADLs and tries to define a taxonomy of ADLs.

8.2.1 ADLs AND OTHER LANGUAGES

How do ADLs differ from programming languages, hardware description languages (HDLs), modeling languages, and the like? This section attempts to answer this question. However, it is not always possible to answer the following question: given a language for describing an architecture, what are the criteria for deciding whether it is an ADL or not?

In principle, ADLs differ from programming languages because the latter bind all architectural abstractions to specific point solutions whereas ADLs intentionally suppress or vary such binding. In practice, architecture is embodied and recoverable from code by reverse engineering methods. For example, it might be possible to analyze a piece of code written in C and figure out whether it corresponds to a *Fetch* unit or not. Many languages provide architecture-level views of the system. For example, C++ offers the ability to describe the structure of a processor by instantiating objects for the components of the architecture. However, C++ offers little or no architecture-level analytical capabilities. Therefore, it is difficult to describe architecture at a level of abstraction suitable for early analysis and exploration. More importantly, traditional programming languages are not a natural choice for describing architectures due to their unsuitability for capturing hardware features such as parallelism and synchronization.

ADLs differ from modeling languages (such as UML) because the latter are more concerned with the behaviors of the whole rather than the parts, whereas ADLs concentrate on representation of components. In practice, many modeling languages allow the representation of cooperating components and can represent architectures reasonably well. However, the lack of an abstraction would make it harder to describe the IS of the architecture. Traditional HDLs, such as VHDL and Verilog, do not have sufficient abstraction to describe architectures and explore them at the system level. It is possible to perform reverse engineering to extract the structure of the architecture from the HDL description. However, it is hard to extract the IS behavior of the architecture.

8.2.2 PROMINENT ADLs

This section briefly surveys some of the prominent ADLs in the context of processor modeling and design automation. There are many comprehensive ADL surveys available in the literature including ADLs for retargetable compilation [14], programmable embedded systems [15], and system-on-chip (SoC) design [16]. A definitive compilation of the ADLs can be found in Reference 17.

Figure 8.3 shows the classification of ADLs based on two aspects: *content* and *objective*. The content-oriented classification is based on the nature of the information an ADL can capture,

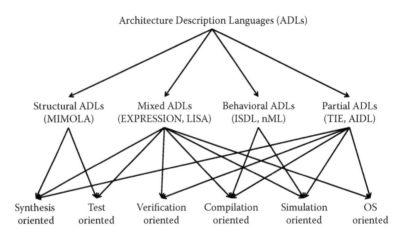

FIGURE 8.3 Taxonomy of ADLs.

whereas the objective-oriented classification is based on the purpose of an ADL. ADLs can be classified into six categories based on the objective: simulation oriented, synthesis oriented, test oriented, compilation oriented, validation oriented, and OS oriented.

ADLs can be classified into four categories based on the nature of the information: structural, behavioral, mixed, and partial. This classification can be also linked with the objective and origin of the different ADLs. The structural ADLs capture the structure in terms of architectural components and their connectivity. The behavioral ADLs capture the IS behavior of the processor architecture. The mixed ADLs capture both structure and behavior of the architecture. These ADLs capture complete description of the structure or behavior or both. However, the partial ADLs capture specific information about the architecture for the intended task. For example, an ADL intended for interface synthesis does not require internal structure or behavior of the processor.

Early ADLs are inspired from RTL abstraction. An example is the ADL MIMOLA [18], which is categorized within structural ADLs. A brief description of MIMOLA is provided here.

8.2.2.1 MIMOLA

MIMOLA [18] was developed at the University of Dortmund, Germany. It was originally proposed for microarchitecture design. One of the major advantages of MIMOLA is that the same description can be used for synthesis, simulation, test generation, and compilation. A tool chain including the MSSH hardware synthesizer, the MSSQ code generator, the MSST self-test program compiler, the MSSB functional simulator, and the MSSU reservation table (RT)-level simulator was developed based on the MIMOLA language [18]. MIMOLA has also been used by the RECORD [18] compiler.

MIMOLA description contains three parts: the algorithm to be compiled, the target processor model, and additional linkage and transformation rules. The software part (algorithm description) describes application programs in a PASCAL-like syntax. The processor model describes microarchitecture in the form of a component netlist. The linkage information is used by the compiler in order to locate important modules such as program counter and instruction memory. The following code segment specifies the program counter and instruction memory locations [18]:

```
LOCATION_FOR_PROGRAMCOUNTER PCReg;
LOCATION_FOR_INSTRUCTIONS IM[0..1023];
```

The algorithmic part of MIMOLA is an extension of PASCAL. Unlike other HLLs, it allows references to physical registers and memories. It also allows use of hardware components using procedure calls. For example, if the processor description contains a component named MAC, programmers can write the following code segment to use the multiply–accumulate operation performed by MAC:

```
res:= MAC(x, y, z);
```

The processor is modeled as a netlist of component modules. MIMOLA permits modeling of arbitrary (programmable or nonprogrammable) hardware structures. Similar to VHDL, a number of predefined, primitive operators exist. The basic entities of MIMOLA hardware models are modules and connections. Each module is specified by its port interface and its behavior. The following example shows the description of a multifunctional ALU module [18]:

```
MODULE ALU
   (IN inp1, inp2: (31:0);
    OUT outp: (31:0);
    IN ctrl;
   )
CONBEGIN
       outp <- CASE ctrl OF
           0: inp1 + inp2;
           1: inp1 - inp2;
           END;
CONEND;
```

The CONBEGIN/CONEND construct includes a set of concurrent assignments. In the example mentioned previously, a conditional assignment to output port *outp* is specified, which depends on the two-bit control input *ctrl*. The netlist structure is formed by connecting ports of module instances. For example, the following MIMOLA description connects two modules: *ALU* and accumulator *ACC*.

```
CONNECTIONS ALU.outp -> ACC.inp
            ACC.outp -> ALU.inp
```

The MSSQ code generator extracts IS information from the module netlist. It uses two internal data structures: connection operation graph (COG) and instruction tree (I-tree). It is a very difficult task to extract the COG and I-trees even in the presence of linkage information due to the flexibility of an RT-level structural description. Extra constraints need to be imposed in order for the MSSQ code generator to work properly. The constraints limit the architecture scope of MSSQ to microprogrammable controllers, in which all control signals originate directly from the instruction word. The lack of explicit description of processor pipelines or resource conflicts may result in poor code quality for some classes of very long instruction word (VLIW) or deeply pipelined processors.

The difficulty of IS extraction can be avoided by abstracting behavioral information from the structural details. Behavioral ADLs, such as nML [19] and instruction set description language (ISDL) [20], explicitly specify the instruction semantics and put less emphasis on the microarchitectural details.

8.2.2.2 nML

nML was designed at Technical University of Berlin, Germany. nML has been used by code generators CBC [21] and CHESS [22] and IS simulators Sigh/Sim [23] and CHECKERS. CHESS/CHECKERS environment is used for automatic and efficient software compilation and IS simulation [23].

nML developers recognized the fact that several instructions share common properties. The final nML description would be compact and simple if the common properties are exploited. Consequently, nML designers used a hierarchical scheme to describe ISs. The instructions are the topmost elements in the hierarchy. The intermediate elements of the hierarchy are partial instructions (PIs). The relationship between elements can be established using two composition rules: AND-rule and OR-rule. The AND-rule groups several PIs into a larger PI and the OR-rule enumerates a set of alternatives for one PI. Therefore, instruction definitions in nML can be in the form of an AND–OR tree. Each possible derivation of the tree corresponds to an actual instruction.

To achieve the goal of sharing instruction descriptions, the IS is enumerated by an attributed grammar. Each element in the hierarchy has few attributes. A nonleaf element's attribute values can be computed based on its children's attribute values. Attribute grammar is also adopted by other ADLs such as ISDL [20]. The following nML description shows an example of instruction specification [19]:

```
op numeric_instruction(a:num_action, src:SRC, dst:DST)
action {
   temp_src = src;
   temp_dst = dst;
   a.action;
   dst = temp_dst;
}
op num_action = add | sub
op add()
action = {
   temp_dst = temp_dst + temp_src
}
```

The definition of *numeric_instruction* combines three PIs with the AND-rule: *num_action*, SRC, and DST. The first PI, *num_action*, uses OR-rule to describe the valid options for actions: *add* or *sub*. The number of all possible derivations of *numeric_instruction* is the product of the size of *num_action*, *SRC*, and *DST*. The common behavior of all these options is defined in the *action* attribute of *numeric_instruction*. Each option for *num_action* should have its own action attribute defined as its specific behavior, which is referred by the *a.action* line. For example, the aforementioned code segment has action description for *add* operation. Object code image and assembly syntax can also be specified in the same hierarchical manner.

nML also captures the structural information used by IS architecture. For example, storage units should be declared since they are visible to the IS. nML supports three types of storages: RAM, register, and transitory storage. Transitory storage refers to machine states that are retained only for a limited number of cycles, for example, values on buses and latches. Computations have no delay in nML timing model—only storage units have delay. Instruction delay slots are modeled by introducing storage units as pipeline registers. The result of the computation is propagated through the registers in the behavior specification.

nML models constraints between operations by enumerating all valid combinations. The enumeration of valid cases can make nML descriptions lengthy. More complicated constraints, which often appear in digital signal processors (DSPs) with irregular instruction-level parallelism (ILP) constraints or VLIW processors with multiple issue slots, are hard to model with nML. For example, nML cannot model the constraint that operation *I1* cannot directly follow operation *I0*. nML explicitly supports several addressing modes. However, it implicitly assumes an architecture model that restricts its generality. As a result, it is hard to model multicycle or pipelined units and multiword instructions explicitly. A good critique of nML is given in Reference 24.

Several ADLs endeavored to capture both structural and behavioral details of the processor architecture. We briefly describe two such *mixed* ADLs: EXPRESSION [25] and Language for Instruction-Set Architecture (LISA) [26].

8.2.2.3 EXPRESSION

The aforementioned mixed ADLs require explicit description of RTs. Processors that contain complex pipelines, large amounts of parallelism, and complex storage subsystems typically contain a large number of operations and resources (and hence RTs). Manual specification of RTs on a per-operation basis thus becomes cumbersome and error prone. The manual specification of RTs (for each configuration) becomes impractical during rapid architectural exploration. The EXPRESSION ADL [25] describes a processor as a netlist of units and storages to automatically generate RTs based on the netlist [27]. Unlike MIMOLA, the netlist representation of EXPRESSION is of coarse granularity. It uses a higher level of abstraction similar to a block-diagram-level description in an architecture manual.

EXPRESSION ADL was developed at the University of California, Irvine. The ADL has been used by the retargetable compiler (EXPRESS [28]) and simulator (SIMPRESS [29]) generation framework. The framework also supports a graphical user interface (GUI) and can be used for design space exploration of programmable architectures consisting of processor cores, coprocessors, and memories [30]. An EXPRESSION description is composed of two main sections: behavior (IS) and structure. The behavior section has three subsections: operations, instruction, and operation mappings. Similarly, the structure section consists of three subsections: components, pipeline/data-transfer paths, and memory subsystem.

The operation subsection describes the IS of the processor. Each operation of the processor is described in terms of its opcode and operands. The types and possible destinations of each operand are also specified. A useful feature of EXPRESSION is operation group that groups similar operations together for the ease of later reference. For example, the following code segment shows an operation group (*alu_ops*) containing two ALU operations: *add* and *sub*.

```
(OP_GROUP alu_ops
  (OPCODE add
    (OPERANDS (SRC1 reg) (SRC2 reg/imm) (DEST reg))
    (BEHAVIOR DEST = SRC1 + SRC2)
  ...)
  (OPCODE sub
    (OPERANDS (SRC1 reg) (SRC2 reg/imm) (DEST reg))
    (BEHAVIOR DEST = SRC1 - SRC2)
  ...)
)
```

The instruction subsection captures the parallelism available in the architecture. Each instruction contains a list of slots (to be filled with operations), with each slot corresponding to a functional unit. The operation mapping subsection is used to specify the information needed by instruction selection and architecture-specific optimizations of the compiler. For example, it contains mapping between generic and target instructions.

The component subsection describes each RT-level component in the architecture. The components can be pipeline units, functional units, storage elements, ports, and connections. For multicycle or pipelined units, the timing behavior is also specified.

The pipeline/data-transfer path subsection describes the netlist of the processor. The *pipeline path description* provides a mechanism to specify the units that comprise the pipeline stages, while the *data-transfer path description* provides a mechanism for specifying the valid data transfers. This information is used to both retarget the simulator and to generate RTs needed by the scheduler [27]. An example path declaration for the DLX architecture [31] (Figure 8.4) is shown as follows. It describes that the processor has five pipeline stages. It also describes that the *Execute* stage has four parallel paths. Finally, it describes each path, for example, it describes that the *FADD* path has four pipeline stages.

```
(PIPELINE Fetch Decode Execute MEM WriteBack)
(Execute (ALTERNATE IALU MULT FADD DIV))
(MULT (PIPELINE MUL1 MUL2... MUL7))
(FADD (PIPELINE FADD1 FADD2 FADD3 FADD4))
```

The memory subsection describes the types and attributes of various storage components (such as register files, SRAMs, DRAMs, and caches). The memory netlist information can be used to generate memory-aware compilers and simulators [32]. Memory-aware compilers can exploit the detailed information to hide the latency of the lengthy memory operations. EXPRESSION captures the datapath information in the processor. The control path is not explicitly modeled. The instruction model requires extension to capture interoperation constraints such as sharing of common fields. Such constraints can be modeled by ISDL through cross-field encoding assignment.

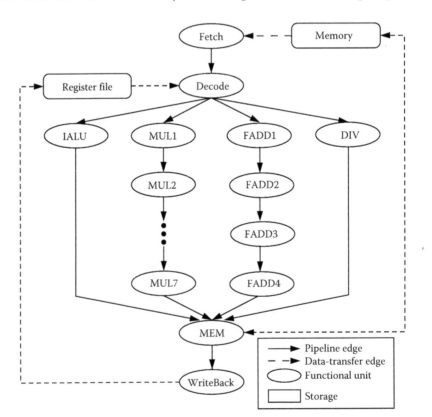

FIGURE 8.4   The DLX architecture.

8.2.2.4 LISA

LISA [26] was developed at RWTH Aachen University, Germany, with the original goal of developing fast simulators. The language has been used to produce production–quality simulators [33]. Depending on speed and accuracy constraints, different modes of IS simulator, that is, compiled, interpretive, just-in-time cache compiled, can be generated. An important aspect of LISA is its ability to stepwise increase the abstraction details. A designer may start from an instruction-accurate LISA description, perform early design exploration, and then move toward a detailed, cycle-accurate model. In this stepwise improvement of architectural details, application profiler, automatic IS encoding [34], and custom instruction identification [35] play an important role. From a cycle-accurate LISA description, optimized, low-power RTL [36,37] generation is permitted. LISA also provides a methodology for automated test-pattern and assertion generation [38]. LISA has been used to generate retargetable C compilers [39,40]. LISA descriptions are composed of two types of declarations: resource and operation. The resource declarations cover hardware resources such as registers, pipelines, and memory hierarchy. An example pipeline description for the architecture shown in Figure 8.4 is as follows:

```
PIPELINE int = {Fetch; Decode; IALU; MEM; WriteBack}
PIPELINE flt = {Fetch; Decode; FADD1; FADD2;
               FADD3; FADD4; MEM; WriteBack}
PIPELINE mul = {Fetch; Decode; MUL1; MUL2; MUL3; MUL4;
               MUL5; MUL6; MUL7; MEM; WriteBack}
PIPELINE div = {Fetch; Decode; DIV; MEM; WriteBack}
```

Operations are the basic objects in LISA. They represent the designer's view of the behavior, the structure, and the IS of the programmable architecture. Operation definitions capture the description of different properties of the system such as operation behavior, IS information, and timing. These operation attributes are defined in several sections. *Coding* and *Syntax* sections cover the instruction encoding and semantics. *Behavior* section contains the datapath of the instruction and *Activation* section dictates the timing behavior of the instruction across the pipeline stages. LISA exploits the commonality of similar operations by grouping them into one. The following code segment describes the decoding behavior of two immediate-type (i_type) operations (ADDI and SUBI) in the DLX *Decode* stage. The complete behavior of an instruction can be obtained by combining the behavior definitions along the operation-activation chain. In this regard, an entire LISA model can be conceptualized as a directed acyclic graph, where the nodes are the LISA operations and the edges are formed by the LISA Activation section.

```
OPERATION i_type IN pipe_int.Decode {
    DECLARE {
    GROUP opcode={ADDI || SUBI}
GROUP rs1, rd = {register};
    }
    CODING {opcode rs1 rd immediate}
    SYNTAX {opcode rd ``,'' rs1 ``,'' immediate}
    BEHAVIOR {rd = rs1; imm = immediate; cond = 0;}
    ACTIVATION {opcode, writeback}
}
```

Recently, the language has been extended to cover a wide range of processor architectures such as weakly programmable ASICs, CGRAs [41], and partially reconfigurable ASIPs [42]. One particular example of the language extension is to efficiently represent VLIW architectures. This is done by template operation declaration as shown in the following. Using this description style, multiple datapaths can be described in a very compact manner. The actual instantiation of the datapath with different values of *tpl* for, say different VLIW slots, take place during simulator/RTL generation.

```
OPERATION alu_op<tpl> IN pipe_int.EX {
    DECLARE {
        GROUP opcode={ADDI<tpl> || SUBI<tpl>}
        GROUP rs1, rd = {register};
    }
    CODING {opcode<tpl> rs1 rd immediate}
    SYNTAX {opcode rd '','' rs1 '','' immediate}
    BEHAVIOR {rd = alu_op<tpl>(rs1, immediate);}
    ACTIVATION {writeback}
}
```

Within the class of partial ADLs, an important entry is Tensilica Instruction Extension (TIE) ADL. TIE [43] captures the details of a processor to the extent it is required for the customization of a base processor.

8.2.2.5 TIE

To manage the complexity of processor design, customizable processor cores are provided by Tensilica [11] and ARC [44]. For Tensilica's XTensa customizable cores, the design space is restricted by discrete choices such as number of pipeline stages, and width and size of the basic address register file. On the other hand, users are able to model, for example, arbitrary custom IS extensions, register files, VLIW formats, and vectorization rules by an ADL, known as TIE. Exemplary TIE description for a vector add instruction is shown in the following. Following the automatic parsing of TIE by XTensa C compiler generation, the instruction *add_v* can be used in the application description as a C intrinsic call. Detailed description of TIE is available at References 43 and 45 (Chapter 6).

```
// large register file
Regfile v 128 16

// custom vector instruction
operation add_v {out v v_out, in v v_a, in v v_b} {}
{
assign v_out = {v_a[127:96] + v_b[127:96],
        v_a[95:64] + v_b[95:64],
        v_a[64:32] + v_b[64:32],
        v_a[31:0] + v_b[31:0];}
}
```

8.3 ADL-DRIVEN METHODOLOGIES

This section describes the ADL-driven methodologies used for processor development. It presents the following three methodologies that are used in academic research as well as industry:

- Software toolsuite generation
- Optimized generation of hardware implementation
- Top-down validation

8.3.1 SOFTWARE TOOLSUITE GENERATION

Embedded systems present a tremendous opportunity to customize designs by exploiting the application behavior. Rapid exploration and evaluation of candidate architectures are necessary due to time-to-market pressure and short product lifetimes. ADLs are used to specify processor and memory architectures and generate software toolkit including compiler, simulator, assembler, profiler, and debugger. Figure 8.5 shows an ADL-based design space exploration flow. The application programs are compiled to machine instructions and simulated, and the feedback is used to modify the ADL specification with the goal of finding

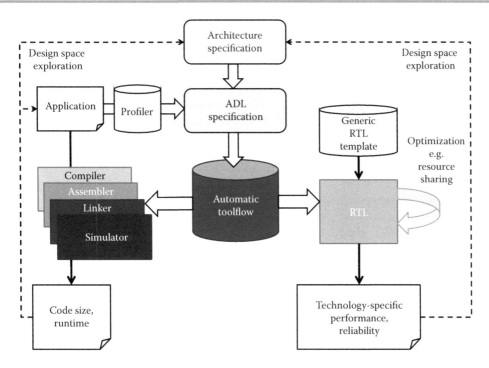

FIGURE 8.5   ADL-driven design space exploration.

the best possible architecture for the given set of application programs under various design constraints such as area, power, performance, and reliability.

An extensive body of recent work addresses ADL-driven software toolkit generation and design space exploration of processor-based embedded systems, in both academia (ISDL [20], Valen-C [46], MIMOLA [18], LISA [26], nML [19], Sim-nML [47], and EXPRESSION [25]) and industry (ARC [44], RADL [48], Target [23], Processor Designer [10], Tensilica [11], and MDES [49]).

One of the main purposes of an ADL is to support automatic generation of a high-quality software toolkit including a cycle-accurate simulator. For supporting fast design space exploration, the simulator needs to balance speed and accuracy. In the same manner, the C/C++ compiler to be generated from the ADL specification needs to support advanced features such as specification of ILP. This section describes some of the challenges in automatic generation of software tools (focusing on compilers and simulators) and surveys some of the approaches adopted by current tools.

8.3.1.1 COMPILERS

Traditionally, software for embedded systems was hand-tuned in assembly. With increasing complexity of embedded systems, it is no longer practical to develop software in assembly language or to optimize it manually except for critical sections of the code. Compilers that produce optimized machine-specific code from a program specified in a HLL such as C/C++ and Java are necessary in order to produce efficient software within the time budget. Compilers for embedded systems have been the focus of several research efforts recently [50,51].

The compilation process can be broadly broken into two steps: analysis and synthesis. During analysis, the program (in HLL) is converted into an intermediate representation (IR) that contains all the desired information such as control and data dependences. During synthesis, the IR is transformed and optimized in order to generate efficient target-specific code. The synthesis step is more complex and typically includes the following phases: instruction selection, scheduling, resource allocation, code optimizations/transformations, and code generation. The effectiveness of each phase depends on the algorithms chosen and the target architecture. A further problem during the synthesis step is that the optimal ordering between these phases is highly dependent on the target architecture and the application program. As a result, traditionally, compilers have been painstakingly hand-tuned to a particular architecture (or architecture class) and application domain(s).

However, stringent time-to-market constraints for SoC designs no longer make it feasible to manually generate compilers tuned to particular architectures. Automatic generation of an efficient compiler from an abstract description of the processor model becomes essential.

A promising approach to automatic compiler generation is the *retargetable compiler* approach. A compiler is classified as retargetable if it can be adapted to generate code for different target processors with significant reuse of the compiler source code. Retargetability is typically achieved by providing target machine information (in an ADL) as input to the compiler along with the program corresponding to the application. The complexity in retargeting the compiler depends on the range of target processors it supports and also on its optimizing capability. Due to the growing amount of ILP features in modern processor architectures, the difference in quality of code generated by a naive code conversion process and an optimizing ILP compiler can be enormous. Recent approaches on retargetable compilation have focused on developing optimizations/ transformations that are *retargetable* and capturing the machine-specific information needed by such optimizations in the ADL. The retargetable compilers can be classified into three broad categories, based on the type of the machine model accepted as input.

1. *Architecture template based*: Such compilers assume a limited architecture template that is parameterizable for customization. The most common parameters include operation latencies, number of functional units, and number of registers. Architecture template-based compilers have the advantage that both optimizations and the phase ordering between them can be manually tuned to produce highly efficient code for the limited architecture space. Examples of such compilers include the Valen-C compiler [46] and the GNU-based C/C++ compiler from Tensilica, Inc. [11]. The Tensilica GNU-based C/C++ compiler is geared toward the Xtensa parameterizable processor architecture. One important feature of this system is the ability to add new instructions (described through an Instruction Extension Language) and automatically generate software tools tuned to the new IS.

2. *Explicit behavioral information based*: Most compilers require a specification of the behavior in order to retarget their transformations (e.g., instruction selection requires a description of the semantics of each operation). Explicit behavioral information–based retargetable compilers require full information about the IS as well as explicit resource conflict information. Examples include the Aviv [52] compiler using ISDL, CHESS [22] using nML, and Elcor [49] using MDES. The Aviv retargetable code generator produces machine code, optimized for minimal size, for target processors with different ISs. It solves the phase ordering problem by performing a heuristic branch-and-bound step that performs resource allocation/assignment, operation grouping, and scheduling concurrently. CHESS is a retargetable code generation environment for fixed-point DSPs. CHESS performs instruction selection, register allocation, and scheduling as separate phases (in that order). Elcor is a retargetable compilation environment for VLIW architectures with speculative execution. It implements a software pipelining algorithm (modulo scheduling) and register allocation for static and rotating register files.

3. *Behavioral information generation based*: Recognizing that the architecture information needed by the compiler is not always in a form that may be well suited for other tools (such as synthesis) or does not permit concise specification, some research has focused on the extraction of such information from a more amenable specification. Examples include the MSSQ and RECORD compiler using MIMOLA [18], retargetable C compiler based on LISA [39], and the EXPRESS compiler using EXPRESSION [25]. MSSQ translates Pascal-like HLL into microcode for microprogrammable controllers, while RECORD translates code written in a DSP-specific programming language, called data flow language, into machine code for the target DSP. The EXPRESS compiler tries to bridge the gap between explicit specification of all information (e.g., Aviv) and implicit specification requiring extraction of IS (e.g., RECORD), by having a mixed behavioral/structural view of the processor. The retargetable C compiler generation using LISA is based on reuse of a powerful C compiler platform with many built-in code optimizations and generation of mapping rules for code selection using the instruction semantics information [39]. The commercial Processor Designer environment based on LISA [10] is extended to support a retargetable compiler generation back end based on LLVM [53] or CoSy [54] (see Figure 8.6).

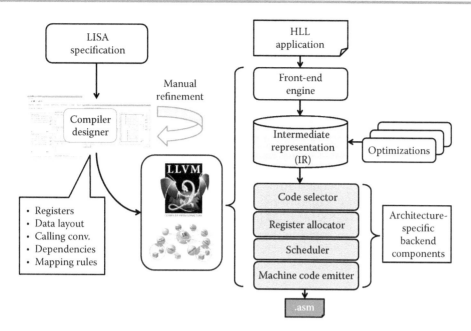

FIGURE 8.6   Retargetable compiler generation from LISA.

Custom instruction synthesis: Although there are embedded processors being designed completely from scratch to meet stringent performance constraints, there is also a trend toward partially predefined, configurable embedded processors [11], which can be quickly tuned to given applications by means of *custom instruction* and/or custom feature synthesis. The custom instruction synthesis tool needs to have a front end, which can identify the custom instructions from an input application under various architectural constraints and a flexible back end, which can retarget the processor tools and generate the hardware implementation of the custom instruction quickly. Custom instruction, as a stand-alone problem, has been studied in depth [55–57]. In the context of processor, custom instruction synthesis with architectural back end is provided in References 11 and 35. Custom instruction syntheses with hardware-oriented optimizations are proposed in Reference 58, while custom instructions are mapped onto a reconfigurable fabric in Reference 59.

8.3.1.2 SIMULATORS

Simulators are critical components of the exploration and software design toolkit for the system designer. They can be used to perform diverse tasks such as verifying the functionality and/or timing behavior of the system (including hardware and software) and generating quantitative measurements (e.g., power consumption [60]) that can aid the design process.

Simulation of the processor system can be performed at various abstraction levels. At the highest level of abstraction, a functional simulation of the processor can be performed by modeling only the IS. Such simulators are termed instruction-accurate (IA) simulators. At lower levels of abstraction are the cycle-accurate and phase-accurate simulation models that yield more detailed timing information. Simulators can be further classified based on whether they provide bit-accurate models, pin-accurate models, exact pipeline models, and structural models of the processor.

Typically, simulators at higher levels of abstraction are faster but gather less information as compared to those at lower levels of abstraction (e.g., cycle accurate, phase accurate). Retargetability (i.e., ability to simulate a wide variety of target processors) is especially important in the arena of embedded SoC design with emphasis on the design space exploration and codevelopment of hardware and software. Simulators with limited retargetability are very fast but may not be useful in all aspects of the design process. Such simulators typically incorporate a fixed architecture template and allow only limited retargetability in the form of parameters such as number of registers and ALUs. Examples of such simulators are numerous in the industry and include the HPL-PD [49] simulator using the MDES ADL. The model of simulation adopted has significant impact on the simulation speed and flexibility of the simulator. Based on the simulation model, simulators can be classified into three types: interpretive, compiled, and mixed.

1. *Interpretation based*: Such simulators are based on an interpretive model of the processors IS. Interpretive simulators store the state of the target processor in host memory. It then follows a fetch, decode, and execute model: instructions are fetched from memory, decoded, and then executed in serial order. Advantages of this model include ease of implementation, flexibility, and the ability to collect varied processor state information. However, it suffers from significant performance degradation as compared to the other approaches primarily due to the tremendous overhead in fetching, decoding, and dispatching instructions. Almost all commercially available simulators are interpretive. Examples of interpretive retargetable simulators include SIMPRESS [29] using EXPRESSION, and GENSIM/XSIM [61] using ISDL.

2. *Compilation based*: Compilation-based approaches reduce the runtime overhead by translating each target instruction into a series of host machine instructions that manipulate the simulated machine state. Such translation can be done either at compile time (static compiled simulation), where the fetch–decode–dispatch overhead is completely eliminated, or at load time (dynamic compiled simulation) that amortizes the overhead over repeated execution of code. Simulators based on the static compilation model are presented by Zhu and Gajski [62] and Pees et al. [63]. Examples of dynamic compiled code simulators include the Shade simulator [64] and the Embra simulator [65].

3. *Interpretive + compiled*: Traditional interpretive simulation is flexible but slow. Instruction decoding is a time-consuming process in software simulation. Compiled simulation performs compile time decoding of application programs to improve the simulation performance. However, all compiled simulators rely on the assumption that the complete program code is known before the simulation starts and is further more runtime static. Due to the restrictiveness of the compiled technique, interpretive simulators are typically used in embedded systems design flow. Several simulation techniques (just-in-time cache-compiled simulation [JIT-CCS] [33] and IS-CS [66]) combine the flexibility of interpretive simulation with the speed of the compiled simulation.

The *JIT-CCS* technique compiles an instruction during runtime, *just-in-time* before the instruction is going to be executed. Subsequently, the extracted information is stored in a simulation cache for direct reuse in a repeated execution of the program address (Figure 8.7). The simulator recognizes if the program code of a previously executed address has changed and initiates a recompilation. The *IS compiled simulation* (IS-CS)

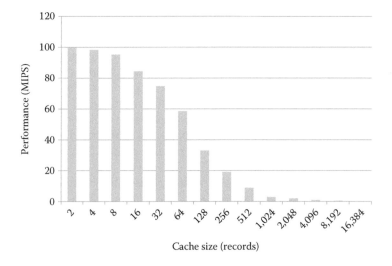

FIGURE 8.7   Performance of the just-in-time cache-compiled simulation. (From Nohl, A. et al., A universal technique for fast and flexible instruction-set architecture simulation, in *Proceedings of the 39th Annual Design Automation Conference, DAC'02*, 2002, pp. 22–27.)

technique performs time-consuming instruction decoding during compile time. In case an instruction is modified at runtime, the instruction is redecoded prior to execution. It also uses an *instruction abstraction* technique to generate aggressively optimized decoded instructions that further improves simulation performance [66].

Hybrid Simulation (HySim): Normally, a software developer does not need high simulation accuracy all the time. Some parts of an application just need to be functionally executed to reach a zone of interest. For those parts a fast but inaccurate simulation is completely sufficient. For the region on interest, however, a detailed and accurate simulation might be required to understand the behavior of the implementation of a particular function. The HySim [67] concept—a simulation technique—addresses this problem. It allows the user to switch between a detailed simulation, using previously referred ISS techniques, and direct execution of the application code on the host processor. This gives the designer the possibility to trade simulation speed against accuracy. The tricky part in hybrid simulation is to keep the application memory synchronous between both simulators. By limiting the switching to function borders, this problem becomes easier to handle.

8.3.2 GENERATION OF HARDWARE IMPLEMENTATION

For a detailed performance evaluation of the processor as well as the final deployment on a working silicon, a synthesizable RTL description is required. There are two major approaches in the literature for synthesizable HDL generation. The first one is based on a parameterized processor core. These cores are bound to a single processor template whose architecture and tools can be modified to a certain degree. The second approach is based on high-level processor specification, that is, ADLs.

1. *Template-based RTL generation*: Examples of template-based RTL generation approaches are Xtensa [11], Jazz [68], and PEAS [69]. Xtensa [11] is a scalable RISC processor core. Configuration options include the width of the register set, caches, and memories. New functional units and instructions can be added using the Tensilica Instruction Language (TIE) [11]. A synthesizable hardware model along with software toolkit can be generated for this class of architectures. Improv's Jazz [68] processor was supported by a flexible design methodology to customize the computational resources and IS of the processor. It allows modifications of data width, number of registers, depth of hardware task queue, and addition of custom functionality in Verilog. PEAS [69,70] is a GUI-based hardware/software codesign framework. It generates HDL code along with software toolkit. It has support for several architecture types and a library of configurable resources.
2. *ADL-based RTL generation*: Figure 8.5 includes the flow for HDL generation from processor description languages. Structure-centric ADLs such as MIMOLA are suitable for hardware generation. Some of the behavioral languages (such as ISDL and nML) are also used for hardware generation. For example, the HDL generator HGEN [61] uses ISDL description, and the synthesis tool GO [23] is based on nML. Itoh et al. [71] have proposed a microoperation description–based synthesizable HDL generation. It can handle simple processor models with no hardware interlock mechanism or multicycle operations.

The synthesizable HDL generation approach based on LISA [3] produces an HDL model of the architecture. The designer has the choice to generate a VHDL, Verilog, or SystemC representation of the target architecture [3]. The commercial offering [10], based on LISA [26], allows the designer to select between a highly optimized code with poor readability or an unoptimized code. Different design options like resource sharing [72], localization of storage, and decision minimization [73] can be enabled. The HDL generation methodology, based on EXPRESSION ADL, is demonstrated to have excellent performance [2].

8.3.3 TOP-DOWN VALIDATION

Validation of microprocessors is one of the most complex and important tasks in the current SoC design methodology. Traditional top-down validation methodology for processor architectures would start from an architectural specification and ensure that the actual implementation is in sync.

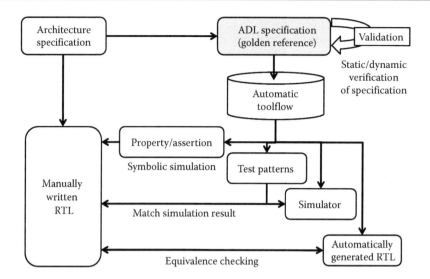

FIGURE 8.8   Top-down validation flow.

with the specification. The advent of ADL provided an option to have an *executable specification*, thereby allowing top-down validation flow [38,74]. This is shown graphically in Figure 8.8.

This methodology is enabled in two phases. First, the ADL specification is validated for completeness. Second, the ADL specification is used for driving simulation-based verification with increasingly detailed abstraction level.

1. *Validation of ADL specification*: It is important to verify the ADL specification to ensure the correctness of the architecture specified and the generated software toolkit. Both static and dynamic behavior need to be verified to ensure that the specified architecture is well formed. The static behavior can be validated by analyzing several static properties such as connectedness, false pipeline and data-transfer paths, and completeness using a graph-based model of the pipelined architecture [75]. The dynamic behavior can be validated by analyzing the instruction flow in the pipeline using a finite-state machine–based model to verify several important architectural properties such as determinism and in-order execution in the presence of hazards and multiple exceptions [76]. In Reference 38, assertions are generated from a LISA description for detecting incorrect dynamic behavior, for example, multiple write access to the same storage element.
2. *Specification-driven validation*: The validated ADL specification can be used as a golden reference model for top-down validation of programmable architectures. The top-down validation approach has been demonstrated in two directions: functional test program generation and design validation using a combination of equivalence checking and symbolic simulation.

Test generation for functional validation of processors has been demonstrated using MIMOLA [18], EXPRESSION [77], LISA [38], and nML [23]. A model checking–based approach is used to automatically generate functional test programs from the processor specification using EXPRESSION [77]. It generates graph model of the pipelined processor from the ADL specification. The functional test programs are generated based on the coverage of the pipeline behavior. Further test generation covering pipeline interactions and full-coverage test-pattern generations have been demonstrated for EXPRESSION [78] and LISA [38], respectively. ADL-driven design validation using equivalence checking is proposed in Reference 74. This approach combines ADL-driven hardware generation and validation. The generated hardware model (RTL) is used as a reference model to verify the hand-written implementation (*RTL design*) of the processor. To verify that the implementation satisfies certain properties, the framework generates the intended properties. These properties are applied using symbolic simulation [74].

Note that the functional validation problem is relatively simpler for configurable processor cores. There, for each possible configuration, which is a finite set, the processor specification needs to be validated. This is presented for Tensilica configurable cores [79].

8.4 CONCLUSIONS

The emergence of heterogeneous multiprocessor SoCs has increased the importance of application-specific processors/accelerators. The complexity and tight time-to-market constraints of such accelerators require the use of automated tools and techniques. Consequently, over the last decade, ADLs have made a successful transition from pure academic research to widespread acceptance in industry [10,11,23,44,80].

Indeed, the academic evolution and stepwise transition of ADLs to industrial usage makes an interesting study in the history of technology. In the academic research, starting from the early effect of nML [19], Target Compiler Technologies was founded, which is eventually acquired by Synopsys [23]. LISA, another prominent ADL, commercially ventured out as LISATek, before being acquired by CoWare, Inc. and finally by Synopsys [10]. The ASIP design environment based on PEAS [69] is commercialized [70], too. Many other notable ADLs such as MIMOLA [18] and ISDL [20] are not pursued for research or commercial usage anymore. EXPRESSION [30], ArchC [80], and MDES [49] are freely available. Among configurable cores, Tensilica was acquired by Cadence [11] and ARC configurable cores are now available via Synopsys [44].

Nowadays, the processor is modeled using an ADL or is chosen from a range of configuration options. The selected configuration/specification is used to generate software tools including compiler and simulator to enable early design space exploration. The ADL specification is also used to perform other design automation tasks including synthesizable hardware generation and functional verification.

Academic research with ADLs has reached a saturation point as far as the aforementioned performance metrics are considered. However, new research directions have emerged with time. The wide range of microarchitectures to be supported in a heterogeneous SoC demands a flexible modeling platform. However, standard ADLs were designed with the focus on processor microarchitectures. Offbeat structures such as CGRA, partially reconfigurable processors, and weakly configurable ASICs are yet to be efficiently and automatically designed. For example, mapping a CDFG on a CGRA is an important research problem at this moment. With decreasing device sizes, physical effects are percolating to the upper design layers. As a result, thermal/reliability-aware processors are being designed. This demand is being reflected on the ADL-based design methodologies as well.

This chapter provided a short overview of the high-level processor architecture design methodologies. Detailed treatment of ADLs can be found in Reference 17. Custom processor architectures and language-based partially reconfigurable processor architectures are discussed in detail in References 81 and 42, respectively. Technically inquisitive readers can fiddle with the open-source, academic ADLs [30,80] or commercial ADLs [10,23], as well as commercial template-based processor design flows [11].

REFERENCES

1. M. Gries and K. Keutzer. *Building ASIPs: The Mescal Methodology.* Springer, New York, 2005.
2. P. Mishra, A. Kejariwal, and N. Dutt. Synthesis-driven exploration of pipelined embedded processors. In *Proceedings of the 17th International Conference on VLSI Design, 2004*, pp. 921–926, 2004.
3. O. Schliebusch, A. Chattopadhyay, R. Leupers, G. Ascheid, H. Meyr, M. Steinert, G. Braun, and A. Nohl. RTL processor synthesis for architecture exploration and implementation. In *Design, Automation and Test in Europe Conference and Exhibition, 2004. Proceedings*, Vol. 3, pp. 156–160, February 2004.
4. P. Mishra and N. Dutt. Functional coverage driven test generation for validation of pipelined processors. In *Proceedings of the Conference on Design, Automation and Test in Europe—Vol. 2, DATE'05*, pp. 678–683, 2005.
5. O. Schliebusch, D. Kammler, A. Chattopadhyay, R. Leupers, G. Ascheid, and H. Meyr. Automatic generation of JTAG interface and debug mechanism for ASIPs. In *GSPx, 2004. Proceedings*, 2004.
6. S. Wang and S. Malik. Synthesizing operating system based device drivers in embedded systems. In *Proceedings of the First IEEE/ACM/IFIP International Conference on Hardware/Software Codesign and System Synthesis, CODES+ISSS'03*, pp. 37–44, 2003.
7. Z.E. Rakossy, T. Naphade, and A. Chattopadhyay. Design and analysis of layered coarse-grained reconfigurable architecture. In *2012 International Conference on Reconfigurable Computing and FPGAs (ReConFig)*, pp. 1–6, December 2012.

8. Z.E. Rakossy, A.A. Aponte, and A. Chattopadhyay. Exploiting architecture description language for diverse ip synthesis in heterogeneous mpsoc. In *2013 International Conference on Reconfigurable Computing and FPGAs (ReConFig)*, pp. 1–6, December 2013.

9. C. Chen, Z. Wang, and A. Chattopadhyay. Fast reliability exploration for embedded processors via high-level fault injection. In *2013 14th International Symposium on Quality Electronic Design (ISQED)*, pp. 265–272, March 2013.

10. Synopsys Processor Designer. http://www.synopsys.com/systems/blockdesign/processordev/pages/default.aspx (formerly CoWare Processor Designer).

11. *Cadence Tensilica Customizable Processor IP*. http://ip.cadence.com/ipportfolio/tensilica-ip.

12. P.C. Clements. A survey of architecture description languages. In *Proceedings of the 8th International Workshop on Software Specification and Design, IWSSD'96*, pp. 16–25, 1996.

13. M.R. Barbacci. Instruction Set Processor Specifications (ISPS): The notation and its applications. *IEEE Trans. Comput.*, 30(1):24–40, January 1981.

14. W. Qin and S. Malik. Architecture description languages for retargetable compilation. In *Compiler Design Handbook: Optimizations & Machine Code Generation*, Y.N. Srikant and P. Shankar (eds.), pp. 535–564. CRC Press, Boca Raton, FL, 2002.

15. P. Mishra and N. Dutt. Architecture description languages for programmable embedded systems. In *IEE Proceedings on Computers and Digital Techniques*, 2005.

16. H. Tomiyama, A. Halambi, P. Grun, N. Dutt, and A. Nicolau. Architecture description languages for systems-on-chip design. In *The Sixth Asia Pacific Conference on Chip Design Language*, pp. 109–116, 1999.

17. P. Mishra and N. Dutt (editors). *Processor Description Languages*. Morgan Kaufmann Publishers, Inc., San Francisco, CA, 2008.

18. R. Leupers and P. Marwedel. Retargetable code generation based on structural processor description. *Des. Autom. Embed. Syst.*, 3(1):75–108, 1998.

19. M. Freericks. The nML machine description formalism. TU Berlin Computer Science Technical Report TR SM-IMP/DIST/08, 1993.

20. G. Hadjiyiannis, S. Hanono, and S. Devadas. ISDL: An instruction set description language for retargetability. In *Proceedings of the 34th Annual Design Automation Conference, DAC'97*, pp. 299–302, 1997.

21. A. Fauth and A. Knoll. Automated generation of DSP program development tools using a machine description formalism. In *1993 IEEE International Conference on Acoustics, Speech, and Signal Processing*, Minneapolis, MN, Vol. 1, pp. 457–460, April 1993.

22. D. Lanneer, J. Praet, A. Kifli, K. Schoofs, W. Geurts, F. Thoen, and G. Goossens. CHESS: Retargetable code generation for embedded DSP processors. *Code Generation for Embedded Processors*, Springer, US, pp. 85–102, 1995.

23. Synopsys IP designer, IP programmer and MP designer (formerly Target Compiler Technologies). http://www.synopsys.com/IP/ProcessorIP/asip/ip-mp-designer/Pages/default.aspx.

24. M.R. Hartoog, J.A. Rowson, P.D. Reddy, S. Desai, D.D. Dunlop, E.A. Harcourt, and N. Khullar. Generation of software tools from processor descriptions for hardware/software codesign. In *Proceedings of the 34th Annual Design Automation Conference, DAC'97*, pp. 303–306, 1997.

25. A. Halambi, P. Grun, V. Ganesh, A. Khare, N. Dutt, and A. Nicolau. EXPRESSION: A language for architecture exploration through compiler/simulator retargetability. In *Design, Automation and Test in Europe Conference and Exhibition 1999. Proceedings*, pp. 485–490, March 1999.

26. H. Meyr, A. Chattopadhyay, and R. Leupers. LISA: A uniform ADL for embedded processor modelling, implementation and software toolsuite generation. *Processor Description Languages*, edited by P. Mishra and N. Dutt, pp. 95–130, 2008.

27. P. Grun, A. Halambi, N. Dutt, and A. Nicolau. RTGEN: An algorithm for automatic generation of reservation tables from architectural descriptions. In *Proceedings. 12th International Symposium on System Synthesis, 1999*, pp. 44–50, November 1999.

28. A. Halambi, A. Shrivastava, N. Dutt, and A. Nicolau. A customizable compiler framework for embedded systems. In *Proceedings of Software and Compilers for Embedded Systems (SCOPES)*, 2001.

29. A. Khare, N. Savoiu, A. Halambi, P. Grun, N. Dutt, and A. Nicolau. V-SAT: A visual specification and analysis tool for system-on-chip exploration. In *Proceedings of the 25th EUROMICRO Conference, 1999.* Vol. 1, pp. 196–203, 1999.

30. Exploration framework using EXPRESSION. http://www.ics.uci.edu/~express.

31. J. Hennessy and D. Patterson. *Computer Architecture: A Quantitative Approach.* Morgan Kaufmann Publishers, Inc., San Mateo, CA, 1990.

32. P. Mishra, M. Mamidipaka, and N. Dutt. Processor-memory coexploration using an architecture description language. *ACM Trans. Embed. Comput. Syst.*, 3(1):140–162, February 2004.

33. A. Nohl, G. Braun, O. Schliebusch, R. Leupers, H. Meyr, and A. Hoffmann. A universal technique for fast and flexible instruction-set architecture simulation. In *Proceedings of the 39th Annual Design Automation Conference, DAC'02*, pp. 22–27, 2002.

34. A. Nohl, V. Greive, G. Braun, A. Hoffman, R. Leupers, O. Schliebusch, and H. Meyr. Instruction encoding synthesis for architecture exploration using hierarchical processor models. In *Design Automation Conference, 2003. Proceedings*, pp. 262–267, June 2003.

35. R. Leupers, K. Karuri, S. Kraemer and M. Pandey. A design flow for configurable embedded processors based on optimized instruction set extension synthesis. In *DATE'06: Proceedings of the Conference on Design, Automation and Test in Europe*, pp. 581–586, European Design and Automation Association, Leuven, Belgium, 2006.

36. A. Chattopadhyay, D. Kammler, E.M. Witte, O. Schliebusch, H. Ishebabi, B. Geukes, R. Leupers, G. Ascheid, and H. Meyr. Automatic low power optimizations during ADL-driven ASIP design. In *2006 International Symposium on VLSI Design, Automation and Test*, pp. 1–4, April 2006.

37. A. Chattopadhyay, B. Geukes, D. Kammler, E.M. Witte, O. Schliebusch, H. Ishebabi, R. Leupers, G. Ascheid, and H. Meyr. Automatic ADL-based operand isolation for embedded processors. In *Proceedings of the Conference on Design, Automation and Test in Europe: Proceedings, DATE'06*, pp. 600–605, 2006.

38. A. Chattopadhyay, A. Sinha, D. Zhang, R. Leupers, G. Ascheid, and H. Meyr. Integrated verification approach during ADL-driven processor design. In *Seventeenth IEEE International Workshop on Rapid System Prototyping, 2006.* pp. 110–118, June 2006.

39. M. Hohenauer, H. Scharwaechter, K. Karuri, O. Wahlen, T. Kogel, R. Leupers, G. Ascheid, H. Meyr, G. Braun, and H. van Someren. A methodology and tool suite for C compiler generation from ADL processor models. In *Proceedings of the Conference on Design, Automation and Test in Europe—Vol. 2, DATE'04*, 2004.

40. O. Wahlen, M. Hohenauer, R. Leupers, and H. Meyr. Instruction scheduler generation for retargetable compilation. *IEEE Des. Test Comput.*, 20(1):34–41, January 2003.

41. A. Chattopadhyay, X. Chen, H. Ishebabi, R. Leupers, G. Ascheid, and H. Meyr. High-level modelling and exploration of coarse-grained re-configurable architectures. In *Proceedings of the Conference on Design, Automation and Test in Europe, DATE'08*, pp. 1334–1339, 2008.

42. A. Chattopadhyay, R. Leupers, H. Meyr, and G. Ascheid. *Language-Driven Exploration and Implementation of Partially Re-configurable ASIPs*. Springer, Netherlands, 2009.

43. H. Sanghavi and N. Andrews. TIE: An ADL for designing application-specific instruction-set extensions. *Processor Description Languages*, edited by P. Mishra and N. Dutt, pp. 183–216, 2008.

44. Synopsys. DesignWare ARC processor cores. http://www.synopsys.com/IP/ProcessorIP/ARCProcessors/Pages/default.aspx.

45. B. Bailey and G. Martin. *ESL Models and Their Application*. Springer US, 2010.

46. A. Inoue, H. Tomiyama, E. Fajar, N.H. Yasuura, and H. Kanbara. A programming language for processor based embedded systems. In *Proceedings of the APCHDL*, pp. 89–94, 1998.

47. V. Rajesh and R. Moona. Processor modeling for hardware software codesign. In *Proceedings. Twelfth International Conference on VLSI Design, 1999*, pp. 132–137, January 1999.

48. C. Siska. A processor description language supporting retargetable multi-pipeline DSP program development tools. In *Proceedings of the 11th International Symposium on System Synthesis, ISSS'98*, pp. 31–36, 1998.

49. The MDES User Manual. http://www.trimaran.org.

50. M. Hohenauer, F. Engel, R. Leupers, G. Ascheid, H. Meyr, G. Bette, and B. Singh. Retargetable code optimization for predicated execution. In *Proceedings of the Conference on Design, Automation and Test in Europe, DATE'08*, pp. 1492–1497, 2008.

51. M. Hohenauer, C. Schumacher, R. Leupers, G. Ascheid, H. Meyr, and H. van Someren. Retargetable code optimization with SIMD instructions. In *CODES+ISSS'06: Proceedings of the Fourth International Conference on Hardware/Software Codesign and System Synthesis*, ACM, New York, pp. 148–153, 2006.

52. S. Hanono and S. Devadas. Instruction selection, resource allocation, and scheduling in the aviv retargetable code generator. In *Proceedings of the 35th Annual Design Automation Conference, DAC'98*, pp. 510–515, 1998.

53. The LLVM compiler infrastructure. llvm.org.

54. Associated compiler experts. http://www.ace.nl.

55. K. Atasu, W. Luk, O. Mencer, C. Ozturan, and G. Dundar. FISH: Fast instruction SyntHesis for custom processors. *IEEE Trans. Very Large Scale Integr. Syst.*, 20(1):52–65, January 2012.

56. N. Pothineni, A. Kumar, and K. Paul. Exhaustive enumeration of legal custom instructions for extensible processors. In *VLSID 2008. 21st International Conference on VLSI Design, 2008*, pp. 261–266, January 2008.

57. P. Biswas, N.D. Dutt, L. Pozzi, and P. Ienne. Introduction of Architecturally visible storage in instruction set extensions. *IEEE Trans. Comput. Aided Design Integr. Circuits Systems*, 26(3):435–446, March 2007.

58. K. Karuri, A. Chattopadhyay, M. Hohenauer, R. Leupers, G. Ascheid, and H. Meyr. Increasing data-bandwidth to instruction-set extensions through register clustering. In *IEEE/ACM International Conference on Computer-Aided Design (ICCAD)*, 2007.

59. K. Karuri, A. Chattopadhyay, X. Chen, D. Kammler, L. Hao, R. Leupers, H. Meyr, and G. Ascheid. A design flow for architecture exploration and implementation of partially reconfigurable processors. *IEEE Trans. Very Large Scale Integr. Syst.*, 16(10):1281–1294, October 2008.

60. H. Xie Z. Wang, L. Wang, and A. Chattopadhyay. Power modeling and estimation during ADL-driven embedded processor design. In *2013 Fourth Annual International Conference on Energy Aware Computing Systems and Applications (ICEAC)*, pp. 97–102, December 2013.

61. G. Hadjiyiannis, P. Russo, and S. Devadas. A methodology for accurate performance evaluation in architecture exploration. In *Proceedings of the 36th Annual ACM/IEEE Design Automation Conference, DAC'99*, pp. 927–932, 1999.

62. J. Zhu and D.D. Gajski. A retargetable, ultra-fast instruction set simulator. In *Proceedings of the Conference on Design, Automation and Test in Europe, DATE'99*, 1999.

63. S. Pees, A. Hoffmann, and H. Meyr. Retargetable compiled simulation of embedded processors using a machine description language. *ACM Trans. Des. Autom. Electron. Syst.*, 5(4):815–834, October 2000.

64. B. Cmelik and D. Keppel. Shade: A fast instruction-set simulator for execution profiling. In *Proceedings of the 1994 ACM SIGMETRICS Conference on Measurement and Modeling of Computer Systems, SIGMETRICS'94*, pp. 128–137, 1994.

65. E. Witchel and M. Rosenblum. Embra: Fast and flexible machine simulation. In *Proceedings of the 1996 ACM SIGMETRICS International Conference on Measurement and Modeling of Computer Systems, SIGMETRICS'96*, pp. 68–79, 1996.

66. M. Reshadi, P. Mishra, and N. Dutt. Instruction set compiled simulation: A technique for fast and flexible instruction set simulation. In *Proceedings of the 40th Annual Design Automation Conference, DAC'03*, pp. 758–763, 2003.

67. S. Kraemer, L. Gao, J. Weinstock, R. Leupers, G. Ascheid, and H. Meyr. HySim: A fast simulation framework for embedded software development. In *CODES+ISSS'07: Proceedings of the Fifth IEEE/ACM International Conference on Hardware/Software Codesign and System Synthesis*, pp. 75–80, 2007.

68. Improv, Inc. http://www.improvsys.com (now defunct).

69. M. Itoh, S. Higaki, J. Sato, Y. Takeuchi, A. Kitajima, and M. Imai. PEAS-III: An ASIP design environment. In *Proceedings. 2000 International Conference on Computer Design, 2000*, pp. 430–436, 2000.

70. ASIP Solutions. http://www.asip-solutions.com/.

71. M. Itoh, Y. Takeuchi, M. Imai, and A. Shiomi. Synthesizable HDL generation for pipelined processors from a micro-operation description. *IEICE Trans. Fundam.*, E83-A(3), 394–400, March 2000.

72. E.M. Witte, A. Chattopadhyay, O. Schliebusch, D. Kammler, R. Leupers, G. Ascheid, and H. Meyr. Applying resource sharing algorithms to adl-driven automatic asip implementation. In *Proceedings. 2005 IEEE International Conference on Computer Design: VLSI in Computers and Processors, 2005. ICCD 2005*, pp. 193–199, October 2005.

73. O. Schliebusch, A. Chattopadhyay, E.M. Witte, D. Kammler, G. Ascheid, R. Leupers, and H. Meyr. Optimization techniques for adl-driven rtl processor synthesis. In *Proceedings of the 16th IEEE International Workshop on Rapid System Prototyping, RSP'05*, pp. 165–171, 2005.

74. P. Mishra. Processor validation: A top-down approach. *IEEE Potentials*, 24(1):29–33, February 2005.

75. P. Mishra and N. Dutt. Modeling and validation of pipeline specifications. *ACM Trans. Embed. Comput. Syst.*, 3(1):114–139, February 2004.

76. P. Mishra, N. Dutt, and H. Tomiyama. Towards automatic validation of dynamic behavior in pipelined processor specifications. *Design Autom. Embed. Syst.*, 8(2–3):249–265, 2003.

77. P. Mishra and N. Dutt. Graph-based functional test program generation for pipelined processors. In *Design, Automation and Test in Europe Conference and Exhibition, 2004. Proceedings, Vol. 1*, pp. 182–187, February 2004.

78. T.N. Dang, A. Roychoudhury, T. Mitra, and P. Mishra. Generating test programs to cover pipeline interactions. In *Proceedings of the 46th Annual Design Automation Conference, DAC'09*, pp. 142–147, 2009.

79. M. Puig-Medina, G. Ezer, and P. Konas. Verification of configurable processor cores. In *Design Automation Conference, 2000. Proceedings 2000*, pp. 426–431, 2000.

80. ArchC (Architecture Description Language). http://archc.sourceforge.net/.

81. P. Ienne and R. Leupers. *Customizable Embedded Processors*. Morgan Kaufmann Publishers, Inc., 2006.

Models and Tools for Complex Embedded Software and Systems

Marco Di Natale

9

CONTENTS

Abstract

The development of correct complex software for reactive embedded systems requires the verification of properties by formal analysis or by simulation and testing. Reduced time to market, increased complexity, and the push for better quality also demand the reuse of composable software components. Currently, no language (or design methodology) provides all the desirable features for distributed cyber-physical systems. Languages and modeling standards are catching up with the need for a separation of concerns between the model of the functionality, the model of the platform, and the description of the functional implementation or mapping. Also, verification and automatic generation of implementations require a strong semantics characterization of the modeling language. This chapter provides an overview of existing models and tools for embedded software, libraries, the starting from an introduction to the fundamental concepts and the basic theory of existing models of computation, both synchronous and asynchronous. The chapter also features an introduction to the Unified Modeling Language, the Systems Modeling Language, Simulink® and Esterel languages, and commercial (and open source when applicable) tools. Finally, the chapter provides a quick peek at research work in software models and tools to give a firm understanding of future trends and currently unsolved issues.

9.1 INTRODUCTION

The increasing cost necessary for the design and fabrication of ASICs, together with the need for the reuse of functionality, adaptability, and flexibility, is among the causes for an increasing share of software-implemented functions in embedded projects. Figure 9.1 represents a typical architectural framework for embedded systems, where application software runs on top of a real-time operating system (RTOS) (and possibly a middleware layer) that abstracts from the hardware and provides a common application programming interface (API) for reuse of functionality (such as the AUTOSAR standard [1] in the automotive domain).

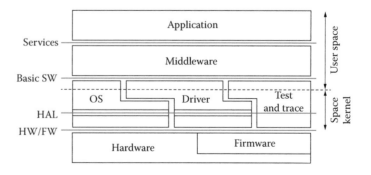

FIGURE 9.1   Common architecture for embedded software.

Unfortunately, mechanisms for improving the software reuse at the level of programming code, such as libraries, the RTOS- or middleware-level APIs, currently fall short of achieving the desired productivity [2], and the error rate of software programs is exceedingly high. Today, model-based design of software carries the promise of a much needed step-up in productivity and reuse.

The use of abstract software models may significantly increase the chances that the design and its implementation are correct, when used at the highest possible level in the development process. Correctness can be mathematically proved by formal reasoning upon the model of the system and its desired properties (provided the model is built on solid mathematical foundations and its properties are expressed by some logic functions). Unfortunately, in many cases, formal model checking is simply impractical, because of the lack of models with a suitable semantics or because of the excessive number of possible states of the system. In this case, the modeling language should at least provide abstractions for the specification of reusable components, so that software modules can be clearly identified with the provided functionality or services.

When exhaustive proof of correctness cannot be achieved, a secondary goal of the modeling language should be to provide support for simulation and testing. In this view, formal methods can also be used to guide the generation of the test suite and to guarantee coverage. Finally, modeling languages and tools should help ensure that the model of the software checked formally or by simulation is correctly implemented in a programming language executed on the target hardware (this requirement is usually satisfied by automatic code generation tools.)

Industrial and research groups have been working for decades in the software engineering area looking for models, methodologies, and tools to increase the reusability of software components and reduce design errors. Traditionally, software models and formal specifications focused on behavioral properties and have been increasingly successful in the verification of functional correctness. However, modern embedded software is characterized by concurrency, resource constraints, and nonfunctional properties, such as deadlines or other timing constraints, which ultimately depend upon the computation platform.

This chapter attempts to provide an overview of (visual and textual) languages and tools for embedded software modeling and design. The subject is so wide and rapidly evolving that only a short survey is possible in the limited space allocated to this chapter. Despite all efforts, existing methodologies and languages fall short in achieving most of the desirable goals, and they are continuously being extended in order to allow for the verification of at least some properties of interest.

We outline the principles of functional and nonfunctional modeling and verification, the languages and tools available on the market, and the realistic milestones with respect to practical designs. Our description of (some) commercial languages, models, and tools is supplemented with a survey of the main research trends and far-reaching results. The bibliography covers advanced discussions of the key issues involved.

The organization of this chapter is as follows: the introduction section defines a reference framework for the discussion of the software modeling problem and reviews abstract models for functional and temporal (schedulability) analysis. The second section provides a quick glance at the two main categories of available languages and models: purely synchronous and general asynchronous. An introduction to the commercial modeling (synchronous) languages Esterel and Simulink and to the Unified Modeling Language (UML) and Systems Modeling Language (SysML) standards is provided. Discussion of what can be achieved with both, with respect to formal analysis of functional properties, schedulability analysis, simulation, and testing, is provided later. The chapter also discusses the recent extensions of existing methodologies to achieve the desirable goal of component-based design. Finally, a quick glance at the research work in the area of embedded software design, methods, and tools closes the chapter.

9.1.1 CHALLENGES IN THE DEVELOPMENT OF EMBEDDED SOFTWARE

With a typical development process (Figure 9.2), an embedded system is the result of refinement stages encompassing several levels of abstraction, from user requirements to (code) implementation, followed by the testing and validation stages. At each stage, the system is

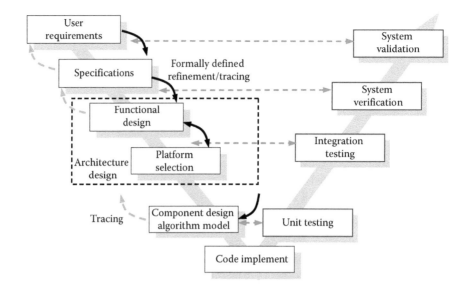

FIGURE 9.2 Typical embedded SW development process.

described using an adequate formalism, starting from abstract models for the user domain entities at the requirements level. Lower-level models, typically developed in later stages, provide an implementation of the abstract model by means of design entities, representing hardware and software components. The implementation process is a sequence of steps that constrain the generic specification by exploiting the possible options (such as the possible nondeterminism, in the form of optional or generic behavior that is allowed by specifications) available from higher levels.

The designer must ensure that the models of the system developed at the different stages satisfy the required properties and that low-level descriptions of the system are correct implementations of higher-level specifications. This task can be considerably easier when the models of the system at the different abstraction levels are homogeneous, that is, if the computational models on which they are based share common semantics and notation.

The problem of correct mapping from a high-level specification, employing an abstract model of the system, to a particular software and hardware architecture or *platform* is one of the key aspects in embedded system design.

The separation of the two main concerns of functional and architectural specification and the subsequent mapping of functions to architecture elements are among the founding principles of many design methodologies such as platform-based design [3] and frameworks like Ptolemy and Metropolis [4,5], as well as emerging standards and recommendations, such as the AUTOSAR automotive initiative [1] and the UML MARTE (Modeling and Analysis of Real-Time and Embedded) profile [6] from the Object Management Group (OMG) [7], and industry best practices, such as the V-cycle of software development [8]. A keyhole view of the corresponding stages is represented in Figure 9.3.

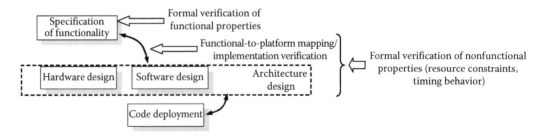

FIGURE 9.3 Mapping formal specifications to an HW/SW platform.

The main design activities taking place at this stage and the corresponding challenges can be summarized as follows:

- *Specification of functionality* is concerned with the development of logically correct system abstractions. If the specification is defined using a formal model, *formal verification* allows checking that the functional behavior satisfies a given set of properties.
- The *system software and hardware* platform components are defined in the *architecture design* level.
- The definition of logical and physical resources available for the execution of the functional model allows the definition of the *mapping of functional model elements onto the platform (architecture elements)* executing them. When this mapping step is complete, *formal verification of nonfunctional properties*, such as timing properties and schedulability analysis, may be performed.

Complementing these two steps, *implementation verification* checks that the behavior of the logical model, when mapped onto the architecture model, correctly implements the high-level formal specifications.

9.1.2 FORMAL MODELS AND LANGUAGES AND SCHEDULABILITY ANALYSIS

A short review of the most common MOCs (formal languages) proposed with the objective of formal or simulation-based verification is fundamental for understanding commercial models and languages, and it is also important for understanding today's and future challenges [9,10].

9.1.2.1 MODELS OF COMPUTATION

Formal models are precisely defined languages that specify the semantics of computation and communication (also defined as "model of computation" [MOC] [9]). MOCs may be expressed, for example, by means of a language or automaton formalism.

A system-level MOCs is used to describe the system as a (possibly hierarchical) collection of design entities (blocks, actors, tasks, processes) performing units of computation represented as transitions or actions, characterized by a state and communicating by means of events (tokens) carried by signals. Composition and communication rules, concurrency models, and time representation are among the most important characteristics of an MOC. MOCs are discussed at length in several books, such as Reference 10.

A first classification may divide models by their representation of time and whether the behavior is discrete or continuous:

- *Continuous time (CT)*: In these models, computational events can occur at any point in time. CT models are often used (as ordinary differential equations or differential algebraic equations) for modeling physical systems interacting with or controlled by the electronic/software system.
- *Discrete time*: Events occur at points in time on a periodic time lattice (integer multiples of some base period). The behavior of the controller is typically assumed as unspecified between any two periodic events or assumed compliant with a sample-and-hold assumption.
- *Discrete event (DE)*: The system reacts to events that are placed at arbitrary points in time. However, in contrast to CT systems, the time at which events occur is typically not important for specifying the behavior of the system, and the evolution of the system does not depend on the time that elapses between events.

Hybrid systems feature a combination of the aforementioned models. In reality, practically all systems of interest are hybrid systems. However, for modeling convenience, discrete-time or DT abstractions are often used.

Once the system specifications are given according to a formal MOC, *formal methods* can be used to achieve design-time verification of *properties* and *implementation* as in Figure 9.3. In general, properties of interest fall under the two general categories of *ordered execution* and *timed execution*:

1. Ordered execution relates to the verification of event and state ordering. Properties such as *safety, liveness, absence of deadlock, fairness,* and *reachability* belong to this category.
2. Timed execution relates to event enumeration, such as checking that no more that *n* events (including time events) occur between any two events in the system. *Timeliness* and some notions of *fairness* are examples.

Verification of desirable system properties may be quite hard or even impossible to achieve by logical reasoning on formal models. Formal models are usually classified according to the *decidability* of properties. Decidability in timed and untimed models depends on many factors, such as the type of logic (propositional or first order) for conditions on transitions and states; the real-time semantics, including the definition of the time domain (discrete or dense); and the linear or branching time logic that is used for expressing properties (the interested reader may refer to Reference 11 for a survey on the subject). In practice, decidability should be carefully evaluated [9]. In some cases, even if it is decidable, the problem cannot be practically solved since the required runtime may be prohibitive and, in other instances, even if undecidability applies to the general case, it may happen that the problem at hand admits a solution.

Verification of model properties can take many forms. In the *deductive* approach, the system and the property are represented by statements (clauses) written in some logic (e.g., expressed in the linear temporal logic [LTL] [12] or in the branching temporal CTL [13]), and a theorem proving tool (usually under the direction of a designer or some expert) applies deduction rules until (hopefully) the desired property reduces to a set of axioms or a counterexample is found. In model checking, the system and possibly the desirable properties are expressed by using an automaton or some other kind of executable formalism. The verification tool ensures that no executable transition nor any system state violates the property. To do so, it can generate all the potential (finite) states of the system (exhaustive analysis). When the property is violated, the tool usually produces the (set of) counterexample(s).

The first model checkers worked by explicitly computing the entire structure of all the reachable system states prior to property checking, but modern tools are able to perform verification as the states are produced (on-the-fly model). This means that the method does not necessarily require the construction of the whole state graph and can be much more efficient in terms of the time and memory that is needed to perform the analysis. On-the-fly model checking and the SPIN [14] toolset provide, respectively, an instance and an implementation of this approach.

To give some examples (Figure 9.4), checking a system implementation **I** against a specification of a property **P** in case both are expressed in terms of automata (*homogeneous verification*) requires the following steps. The implementation automaton A_I is composed with the complementary automaton $\neg A_P$ expressing the negation of the desired property. The implementation **I** violates the specification property if the product automaton $A_I \parallel \neg A_P$ has some possible run and it is verified if the composition has no runs. *Checking by observers* can be considered as a particular instance of this method, very popular for synchronous models.

In the very common case, the property specification consists of a logical formula and the implementation of the system is given by an automaton. Then the verification problem can be solved algorithmically or deductively by transforming it into an instance of the previous cases, for example, by transforming the negation of a specification formula f_S into the corresponding automaton and by using the same techniques as in homogeneous verification.

Verification of implementation correctness is usually obtained by exploiting simulation and bisimulation properties.

A very short survey of formal system models is provided, starting with finite-state machines (FSMs), probably the most popular and the basis for many extensions.

In FSM, the behavior is specified by enumerating the (finite) set of possible system states and the transitions among them. Each transition connects two states and it is labeled with the subset of input variables (and possibly the guard condition upon their values) that triggers its execution. Furthermore, each transition can produce output variables. In Mealy FSMs, outputs depend on

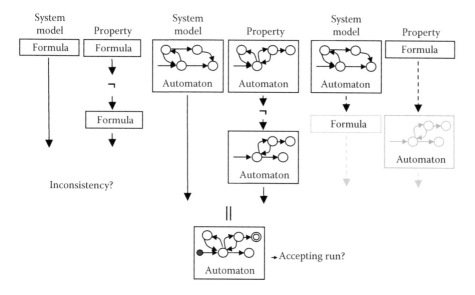

FIGURE 9.4   Checking the system model against some property.

both state and input variables, while in the Moore model, outputs only depend on the process state. Guard conditions can be expressed according to different logics, such as propositional logic, first-order logic, or even (Turing-complete) programming code.

In the *synchronous* FSM model, signal propagation is assumed to be instantaneous. Transitions and the evaluation of the next state occur for *all the system components* at the same time. Synchronous languages, such as Esterel and Lustre, are based on this model. In the *asynchronous* model, two asynchronous FSMs never execute a transition at the same time except when explicit rendezvous is explicitly specified (a pair of transitions of the communicating FSMs occur simultaneously). The Specification and Description Language (SDL) process behavior is an instance of this general model.

The composition of FSMs is obtained by construction of a product transition system, that is, a single FSM machine where the set of states is the product of the sets of the states of the component machines. The difference between synchronous and asynchronous execution semantics is quite clear when compositional behaviors are compared. Figure 9.5 illustrates the differences between the synchronous and asynchronous composition of two FSMs.

When there is a cyclic dependency among variables in interconnected synchronous FSMs, the Mealy model, like any other model where outputs are instantaneously produced based on the input values, may result in a fixed-point problem and possibly inconsistency (Figure 9.6 shows a simple functional dependency). The existence of a unique fixed-point solution (and its evaluation) is a problem that must be solved in all models in which the composition of synchronous (Mealy-type) FSMs results in a loop.

In large, complex systems, composition may easily result in a huge number of states, the problem is often referred to as *state explosion.*

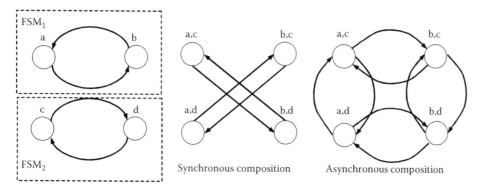

FIGURE 9.5   Composition of synchronous and asynchronous finite-state machines.

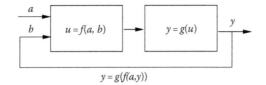

$$y = g(f(a,y))$$

FIGURE 9.6   A fixed-point problem arising from composition and cyclic dependencies.

In its *statechart* extension [15], Harel proposed three mechanisms to reduce the size of an FSM for modeling practical systems: state hierarchy, simultaneous activity, and nondeterminism. In *statecharts*, a state can possibly represent an enclosed state machine. In this case, the machine is in one of the states enclosed by the superstate (or-states) and concurrency is achieved by enabling two or more state machines to be active simultaneously (and-states, such as lap and off in Figure 9.7).

In *Petri net* (PN) models, the system is represented by a graph of places connected by transitions. Places represent unbounded channels that carry *tokens* and the state of the system is represented at any given time by the number of tokens existing in a given subset of places. Transitions represent the elementary reactions of the system. A transition can be executed (fired) when it has a fixed, prespecified number of tokens in its input places. When fired, it consumes the input tokens and produces a fixed number of tokens on its output places. Since more than one transition may originate from the same place, one transition can execute while disabling another one by removing the tokens from shared input places. Hence, the model allows for nondeterminism and provides a natural representation of concurrency by allowing simultaneous and independent execution of multiple transitions (Figure 9.8a).

The FSM and PN models have been originally developed with no reference to time or time constraints, but the capability of expressing and verifying timing requirements is key in many design domains (including embedded systems). Hence, both have been extended to support time-related specifications. Time extensions differ according to the time model assumed (discrete or CT). Furthermore, proposed extensions differ in how time references should be used in the system, whether a global clock or local clocks are assumed, and how time should be used in guard conditions on transitions or states.

When computing the set of reachable states, time adds another dimension, further contributing to the state explosion problem. In general, discrete-time models are easier to analyze compared to CT models, but synchronization of signals and transitions results in fixed-point evaluation problems whenever the system model contains cycles without delays.

Discrete-time systems naturally lead to an implementation based on the *time-triggered* paradigm, where all actions are bound to happen at multiples of a time reference (usually implemented by means of a response to a timer interrupt) and CT (asynchronous systems) conventionally corresponds to implementations based on the event-based design paradigm, where system actions can happen at any time instant. This does not imply a correspondence between time-triggered systems and

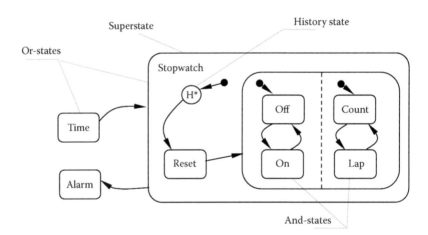

FIGURE 9.7   An example of statechart. (From the example in Harel, D., *Sci. Comput. Program.*, 8, 231, 1987.)

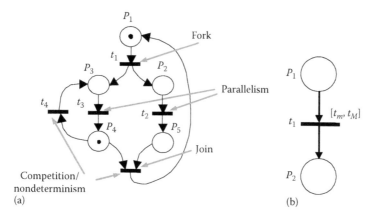

FIGURE 9.8 A sample Petri net showing examples of concurrent behavior and nondeterminism (a) and notations for the time Petri net model (b).

synchronous systems. The latter are characterized by the additional constraint that all system components must perform an action synchronously (at the same time) at each tick in a periodic time base.

Many models have been proposed in the research literature for time-related extensions. Among those, time PNs (TPNs) [16,17] and timed automata (TAs) [18] are probably the best known.

TAs (see, e.g., Figure 9.9) operate with a finite set of locations (states) and a finite set of real-valued clocks. All clocks proceed at the same rate and measure the amount of time that passed since they were started (reset). Each transition may reset some of the clocks and each defines a restriction on the value of the symbols as well as on the clock values required for it to happen. A state may be reached only if the values of the clocks satisfy the constraints and the proposition clause defined on the symbols evaluates to true.

Timed PNs [19] and TPNs are extensions of the PN formalism allowing for the expression of time-related constraints. The two differ in the way time advances: in timed PNs time advances in transitions, thus violating the instantaneous nature of transitions (which makes the model much less prone to analysis). In the timed PN model, time advances while token(s) are in places. Enabling and deadline times can be associated with transitions, the enabling time being the time a transition must be enabled before firing and the deadline being the time instant by which the transition must be taken (Figure 9.8b).

The additional notion of stochastic time allows the definition of the (generalized) stochastic PNs [20,21] used for the purpose of performance evaluation (Figure 9.10).

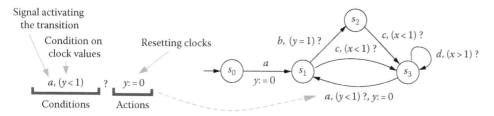

FIGURE 9.9 An example of timed automata. (From Alur, R. and Dill, D.L., *Theor. Comput. Sci.*, 126, 183, 1994.)

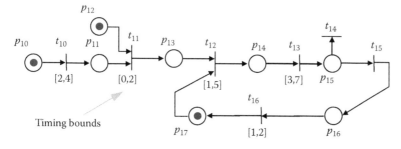

FIGURE 9.10 A sample time Petri net.

Many further extensions have been proposed for both TAs and TPNs. The task of comparing the two models for expressiveness should take into account all the possible variants and is probably not particularly interesting in itself. For most problems of practical interest, however, both models are essentially equivalent when it comes to expressive power and analysis capability [22].

A few tools based on the TA paradigm have been developed and are very popular. Among those, we cite Kronos [23] and Uppaal [24]. The Uppaal tool allows modeling, simulation, and verification of real-time systems modeled as a collection of nondeterministic processes with finite control structure and real-valued clocks, communicating through channels or shared variables [24,25]. The tool is free for nonprofit and academic institutions.

TAs and TPNs allow the formal expression of requirements for logical-level resources, timing constraints, and timing assumptions, but timing analysis only deals with abstract specification entities, typically assuming infinite availability of physical resources (such as memory or CPU speed). If the system includes an RTOS, with the associated scheduler, the model needs to account for preemption, resource sharing, and the nondeterminism resulting from them. Dealing with these issues requires further evolution of the models.

For example, in TAs, clock variables can be used for representing the execution time of each action. In this case, however, only the clock associated with the action scheduled on the CPU should advance, with all the others being stopped.

The hybrid automata model [26] combines discrete transition graphs with continuous dynamic systems. The value of system variables may change according to a discrete transition or it may change continuously in system states according to a trajectory defined by a system of differential equations. Hybrid automata have been developed for the purpose of modeling digital systems interacting with (physical) analog environments, but the capability of stopping the evolution of clock variables in states (first derivative equal to 0) makes the formalism suitable for the modeling of systems with preemption.

TPNs and TAs can also be extended to cope with the problem of modeling finite computing resources and preemption. In the case of TAs, the extension consists of the stopwatch automata model, which handles suspension of the computation due to the release of the CPU (because of real-time scheduling), implemented in the HyTech tool [27] (for linear hybrid automata). Alternatively, the scheduler is modeled with an extension to the TA model, allowing for clock updates by subtraction inside transitions (besides normal clock resetting). This extension, available in the Uppaal tool, avoids the undecidability of the model where clocks associated with the actions not scheduled on the CPU are stopped.

Likewise, TPNs can be extended to the preemptive TPN model [28], as supported by the ORIS tool [29]. A tentative correspondence between the two models is traced in Reference 30. Unfortunately, in all these cases, the complexity of the verification procedure caused by the state explosion poses severe limitations upon the size of the analyzable systems.

Before moving on to the discussion of formal techniques for the analysis of time-related properties at the architecture level (schedulability), the interested reader may refer to Reference 31 for a survey on formal methods, including references to industrial examples.

9.1.2.2 MAPPING FUNCTIONAL MODEL ONTO IMPLEMENTATIONS

Formal MOCs are typically used to represent the desired system (or subsystem component) functionality, abstracted from implementation considerations.

Functionality is implemented on an execution (computation and communication) platform, realized as software code, firmware, or dedicated hardware components. This process involves (at least) three layers of the system representation with the corresponding models: functional, platform, and implementation.

Ideally, these three (sets of) models should be constructed and relationships among their elements should be defined in such a way that the following applies:

1. The platform model abstracts the platform properties of interest for evaluating the mapping solution with respect to quantitative properties that relate to performance (time), reliability, cost, extensibility, and power (and possibly others).
2. The platform model permits verifying the existence of a mapping solution that provably preserves the semantics properties of interest of the functional model. This property is

of particular interest when formal safety properties are demonstrated on the functional model. In order for these properties to carry over to the implementation, a formal proof of correctness must be provided.

3. The implementation model complies with the programming and implementation standards in use and allows the analysis of the properties of interest. An example of such analysis is schedulability analysis that refers to the capability of a multiprogrammed system to meet timing constraints or deadlines.

9.1.2.3 SCHEDULABILITY ANALYSIS

If specification of functionality aims at producing a logically correct representation of system behavior, the (architecture-level) implementation representation is where physical concurrency and schedulability requirements are expressed and evaluated. At this level, the units of computation are the *processes* or *threads* (the distinction between these two operating systems concepts is not relevant for the purpose of this chapter, and in the following, the generic term "task" will be optionally used for both), executing concurrently in response to environmental stimuli or prompted by an internal clock. Threads cooperate by exchanging data and synchronization or activation signals and contend for use of the execution resource(s) (the processor) as well as for the other resources in the system. The physical architecture level allows the definition of the mapping of the concurrent entities onto the target hardware. This activity entails the selection of an appropriate scheduling policy (e.g., offered by a RTOS) and possibly supports by timing or schedulability analysis tools.

Formal models, exhaustive analysis techniques, and model checking are now evolving toward the representation and verification of time and resource constraints together with the functional behavior. However, applicability of these models is strongly limited by state explosion. In this case, when exhaustive analysis and joint verification of functional and nonfunctional behavior are not practical, the designer could seek the lesser goal of analyzing only the *worst-case timing* behavior of coarse-grain design entities representing concurrently executing threads.

Software models for time and schedulability analysis deal with preemption, physical and logical resource requirements, and resource management policies and are typically limited to a quite simplified view of functional (logical) behavior, mainly limited to synchronization and activation signals.

To give an example: if, for the sake of simplicity, we limit the discussion to single processor systems, the scheduler assigns the execution engine (the CPU) to threads (tasks) and the main objective of real-time scheduling policies is to formally guarantee the timing constraints (deadlines) on the thread response to external events.

In this case, the software architecture can be represented as a set of concurrent tasks (threads). Each task τ_i executes periodically or according to a sporadic pattern and it is typically represented by a simple set of attributes, such as the tuple $(C_i, \theta_i, p_i, D_i)$, representing the worst-case computation time, the period (for periodic threads) or minimum interarrival time (for sporadic threads), the priority, and the relative (to the release time r_i) deadline of each thread instance.

Fixed-priority scheduling and rate monotonic analysis (RMA) [32,33] are by far the most common real-time scheduling and analysis methodologies. RMA provides a very simple procedure for assigning static priorities to a set of *independent periodic tasks* together with a formula for checking schedulability against deadlines.

The highest priority is assigned to the task having the highest rate and schedulability is guaranteed by checking the worst-case scenario that can possibly happen. If the set of tasks is schedulable in that condition, then it is schedulable under all circumstances. For RMA the critical condition happens when all tasks are released at the same time instant initiating the largest busy period (CT interval when the processor is busy executing tasks of a given priority level).

By analyzing the busy period (from $t = 0$), it is possible to derive the worst-case response (sometime also called completion) time W_i for each task τ_i. If the task can be proven to complete before or at the deadline ($W_i \leq D_i$), then it is defined as *guaranteed* with respect to its timing constraints.

The worst-case response time can be computed by considering the critical condition for each task. Consider, for example, task τ_i of Figure 9.11. Its response time is the sum of the time

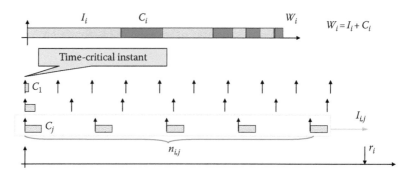

FIGURE 9.11   Computing the worst-case response time from the critical instant.

needed for its own execution C_i and the processor time used by higher-priority tasks, denoted as interference I_i.

The interference I_i can be obtained as the sum of the interferences $I_{i,j}$ from all the higher-or equal-priority tasks (indicated by the notation $j \in he(i)$) as in

$$W_i = C_i + \sum_{\forall j \in he(i)} I_{i,j}$$

Each term $I_{i,j}$ can be further refined by considering that each time a higher-or equal-priority task is activated, it will be executed for its entire worst-case time C_j. If $n_{i,j}$ denotes the number of times the higher-or equal-priority task τ_j is activated before τ_i completes, then

$$W_i = C_i + \sum_{\forall j \in he(i)} n_{i,j} C_j$$

The iterative formula for computing W_i (in case $\tau_i, \leq D_i$) is obtained with the final consideration that in the critical instant, the number of activations of τ_j before τ_i completes is

$$n_{i,j} = \left\lceil \frac{W_i}{\theta_j} \right\rceil$$

The worst-case response time can be computed as the lowest value solution (if it exists) of

(9.1)
$$W_i = C_i + \sum_{\forall j \in he(i)} \left\lceil \frac{W_i}{\theta_j} \right\rceil C_j$$

Rate monotonic scheduling was developed starting from a very simple model where all tasks are periodic and independent. In reality, tasks require access to shared resources (apart from the processor) that can only be used in an exclusive way, such as communication buffers shared among asynchronous threads.

In this case, one task may be blocked because another task holds a lock on shared resources. When the blocked task enjoys a priority higher than the blocking task, blocking and *priority inversion* may occur; finding the optimal-priority assignment becomes an NP-hard problem and the previous formula for computing W_i is not valid anymore, since each task can now be delayed not only by interference but also when trying to use a critical section that is used by a lower-priority task. In addition, with traditional semaphores, this blocking can occur multiple times and the time spent each time a task needs to wait is also unbounded (think of the classical deadlock problem).

Real-time scheduling theory settles at finding resource assignment policies that provide at least a worst-case bound upon the blocking time. The priority inheritance (PI) and the (immediate) priority ceiling (PC) protocols [34] belong to this category.

The essence of the PC protocol (which has been included in the real-time OS OSEK standard issued by the automotive industry) consists of raising the priority of a thread entering a critical section to the highest among the priorities of all threads that may possibly request access to the same critical section. The thread returns to its nominal priority as soon as it leaves the critical section. The PC protocol ensures that each thread can be blocked at most once and bounds the duration of the blocking time to the largest critical section shared between itself or higher-priority threads and lower-priority threads.

When the blocking time due to priority inversion is bounded for each task and its worst-case value is B_i, the evaluation of the worst-case completion time in the schedulability test becomes

$$(9.2) \qquad W_i = C_i + \sum_{\forall j \in he(i)} \left\lceil \frac{W_i}{\theta_j} \right\rceil C_j + B_i$$

9.1.2.4 MAPPING FUNCTIONAL REACTIONS ONTO THREADS

The mapping of the actions defined in the functional model onto architectural model entities is the critical design activity where the two views are reconciled. In practice, the actions or transitions defined in the functional part must be executed in the context of one or more system threads. The definition of the architecture model (number and attributes of threads) and the selection of resource management policies, the mapping of the functional model into the corresponding architecture model, and the validation of the mapped model against functional and nonfunctional constraints is probably one of the major challenges in software engineering.

Single-threaded implementations are quite common and a simple choice for several tools that can provide (practical) verification and a semantics-preserving implementation for a given MOC. Schedulability analysis degenerates to the simple condition that the execution time of the implementation thread is less than its execution period. The entire functional specification is executed in the context of a single thread performing a never-ending cycle where it serves events in a noninterruptable fashion according to the run-to-completion paradigm. The thread waits for an event (either external, like an interrupt from an I/O interface, or internal, like a call or signal from one object or FSM to another), fetches the event and the associated parameters, and, finally, executes the corresponding code.

All the actions defined in the functional part need to be scheduled (statically or dynamically) for execution inside the thread. The schedule is usually driven by the partial order of the execution of the actions, as defined by the MOC semantics. Commercial implementations of this model range from code produced by the Esterel compiler [35] to single-threaded implementations of Simulink models produced by the Embedded Coder toolset from the MathWorks [36], or to the single-threaded code generated by Rational Rhapsody Architect for Software [37] for the execution of UML models.

The scheduling problem is much simpler than in the multithreaded case, since there is no need to account for thread scheduling and preemption and resource sharing usually are addressed trivially.

On the other extreme, one could associate *one thread with every functional block* or every possible action. Each thread can be assigned its own priority, depending on the criticality and on the deadline of the corresponding action. At runtime, the operating system scheduler properly synchronizes and sequentializes the tasks so that the order of execution respects the functional specification.

Both approaches may easily prove inefficient. The single-threaded implementation requires to complete the processing of each event before the next event arrives. The one-to-one mapping of functions or actions to threads suffers from excessive scheduler overhead caused by the need for a context switch at each action. Considering that the action specified in a functional block can be very short and that the number of functional blocks is usually quite high (in many applications it is in the order of hundreds), the overhead of the operating system could easily prove unbearable.

The designer essentially tries to achieve a compromise between these two extremes, balancing responsiveness with schedulability, flexibility of the implementation, and performance overhead.

9.1.3 PARADIGMS FOR REUSE: COMPONENT-BASED DESIGN

One more dimension can be added to the complexity of the software design problem if the need for maintenance and reuse is considered. To this purpose, component-based and object-oriented (OO) techniques have been developed for constructing and maintaining large and complex systems.

A component is a product of the analysis, design, or implementation phases of the lifecycle and represents a prefabricated solution that can be reused to meet (sub)system requirement(s). A component is commonly used as a vehicle for the reuse of two basic design aspects:

1. *Functionality* (or behavior): The functional syntax and semantics of the solution the component represents.
2. *Structure*: The structural abstraction the component represents. These can range from "small grain" to architectural features, at the subsystem or system level. Common examples of structural abstractions are not only OO classes and packages but also Simulink subsystems.

The generic requirement for *reusability* maps into a number of issues. Probably the most relevant property that components should exhibit is *abstraction*, meaning the capability of hiding implementation details and describing relevant properties only. Components should also be easily adaptable to meet changing processing requirements and environmental constraints through controlled modification techniques (like *inheritance* and *genericity*). *Composition* rules must be used to build higher-level components from existing ones. Hence, an ideal component-based modeling language should ensure that properties of components (functional properties, such as liveness, reachability, deadlock avoidance, or nonfunctional properties such as timeliness and schedulability) are preserved or at least decidable after composition. Additional (practical) issues include support for implementation, separate compilations, and imports.

Unfortunately, reconciling the standard issues of software components, such as context independence, understandability, adaptability, and composability, with the possibly conflicting requirements of timeliness, concurrency, and distribution, typical of hard real-time system development, is not an easy task and is still an open problem.

OO design of systems has traditionally embodied the (far from perfect) solution to some of these problems. While most (if not all) OO methodologies, including the UML, offer support for inheritance and genericity, adequate abstraction mechanisms and especially composability of properties are still subject of research.

Starting with its 2.0 release, the UML language has reconciled the abstract interface abstraction mechanism with the common box–port–wire design paradigm. Lack of an explicit declaration of required interfaces and absence of a language feature for structured classes were among the main deficiencies of classes and objects, if seen as components. In UML 2.0, ports allow for a formal definition of a required as well as a provided interface. Association of protocol declaration with ports further improves and clarifies the semantics of interaction with the component. In addition, the concept of a structured class allows for a much better definition of a component.

Of course, port interfaces and the associated protocol declarations are not sufficient for specifying the semantics of the component. In UML, the Object Constraint Language (OCL) can also be used to define behavioral specifications in the form of invariants, preconditions, and postconditions, in the style of the contract-based design methodology (implemented in Eiffel [38]). Other languages have emerged for supporting the specification of behaviors in UML, such as the Action Language for Foundational UML (ALF) [39] and the Executable UML Foundation (fUML) [40]. However, the UML action languages are still not in widespread use and are not further discussed here.

9.2 SYNCHRONOUS VS. ASYNCHRONOUS MODELS

The verification of functional and nonfunctional properties of software demands a formal semantics and a strong mathematical foundation of the models. Among the possible choices, the *synchronous-reactive (SR)* model enforces determinism and provides a sound methodology for checking functional and nonfunctional properties at the price of expensive implementation and performance limitations.

In the *SR model*, time advances at discrete instants and the program progresses according to successive *atomic reactions* (sets of synchronously executed actions), which are performed instantaneously (zero computation time), meaning that the reaction is fast enough with respect to the environment. The resulting discrete-time model is quite natural to many domains, such as control engineering and (hardware) synchronous digital logic design (Verilog or VHDL).

The synchronous assumption (each computation or reaction completes before the next event of interest for the system, or, informally, computation times are negligible with respect to the environment dynamics and synchronous execution) does not always apply to the controlled environment and to the architecture of the system.

Asynchronous or general models typically allow for (controlled) nondeterminism and more expressiveness, at the price of strong limitations on the extent of the functional and nonfunctional verification that can be performed.

Some modeling languages, such as UML, are deliberately general enough to specify systems according to a generic asynchronous or synchronous paradigm using suitable sets of extensions (semantics restrictions).

By the end of this chapter, it will hopefully be clear how neither of the two design paradigms (synchronous or asynchronous) is currently capable of facing all the implementation challenges of complex systems. The requirements of the synchronous assumption (on the environment and the execution platform) are difficult to meet in large, distributed systems, and true component-based design, where each property of a composite can be simply inferred by the black-box properties of its components (without breaking encapsulation), making it very difficult (if not impossible). The asynchronous paradigm, on the other hand, results in implementations that are very difficult to analyze for logical and time behavior.

9.3 SYNCHRONOUS MODELS

In a synchronous system, the composition of system blocks implies the product combination of the states and the conjunction of the reactions for each component. In general, this results in a fixed-point problem and the composition of the function blocks is a relation, not a function, as outlined in Section 9.2.

The synchronous languages *Signal*, *Esterel*, and *Lustre* are probably the best representatives of the synchronous modeling paradigm.

Lustre [41,42] is a declarative language based on the dataflow model where nodes are the main building block. In Lustre, each flow or stream of values is represented by a variable, with a distinct value for each tick in the discrete-time base. A node is a function of flows: it takes a number of typed input flows and defines a number of output flows using a system of equations.

A Lustre node (an example in Figure 9.12) is a pure functional unit except for the preinitialization and initialization (->) expressions, which allow referencing the previous element of a given

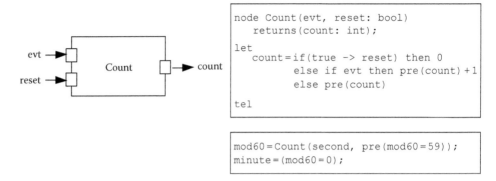

```
node Count(evt, reset: bool)
    returns(count: int);

let
    count=if(true -> reset) then 0
          else if evt then pre(count)+1
          else pre(count)

tel
```

```
mod60=Count(second, pre(mod60=59));
minute=(mod60=0);
```

FIGURE 9.12   An example of Lustre node and its program. (From Caspi, P., LUSTRE: A declarative language for programming synchronous systems, in: *ACM SIGACT-SIGPLAN Symposium on Principles of Programming Languages (POPL)*, Munich, Germany, 1987 pp. 178–188.)

stream or forcing an initial value for a stream. Lustre allows streams at different rates, but in order to avoid nondeterminism, it forbids syntactically cyclic definitions.

Esterel [43] is an imperative language, more suited for the description of control. An Esterel program consists of a collection of nested, concurrently running threads. Execution is synchronized to a single, global clock. At the beginning of each reaction, each thread resumes its execution from where it paused (e.g., at a pause statement) in the last reaction, executes imperative code (e.g., assigning the value of expressions to variables and making control decisions), and finally either terminates or pauses waiting for the next reaction.

Esterel threads communicate exclusively through signals representing globally broadcast events. A signal does not persist across reactions and it is present in a reaction if and only if it is emitted by the program or by the environment.

Esterel allows cyclic dependencies and treats each reaction as a fixpoint equation, but the only legal programs are those that behave functionally in every possible reaction. The solution of this problem is provided by *constructive causality* [44], which amounts to checking if, regardless of the existence of cycles, the output of the program (the binary circuit implementing it) can be formally proven to be causally and deterministically dependent from the inputs for all possible sequences of inputs.

The language allows for conceptually sequential (operator ;) or concurrent (operator ||) execution of reactions, defined by language expressions handling signal identifiers (as in the example of Figure 9.13). All constructs take zero time except await and loop… each…, which explicitly produce a program pause. Esterel includes the concept of preemption, embodied by the loop… each R statement in the example of Figure 9.13 or the abort *action* when *signal* statement. The reaction contained in the body of the loop is preempted (and restarted) when the signal R is set. In case of an abort statement, the reaction is preempted and the statement terminates.

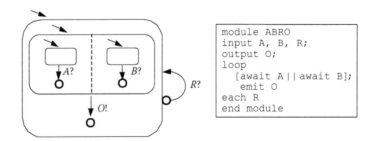

FIGURE 9.13   An example showing features of the Esterel language and an equivalent statechart-like visual formalization. (From Boussinot, F. and de Simone, R., *Proc. IEEE*, 79, 1293, September 1991.)

Formal verification was among the original objectives of Esterel. In synchronous languages, verification of properties typically requires the definition of special programs called "observers" that observe the variables or signals of interest and at each step decide if the property is fulfilled. A program satisfies the property if and only no observer ever complains during any execution.

The verification tool takes the program implementing the system, an observer of the desired property (or assertion observer, since it checks that the system complies with its asserted properties), and another observer program modeling the assumptions on the environment. The three programs are combined in a synchronous product, as in Figure 9.14, and the tool explores the set of reachable states. If the assertion observer never reaches a state where the system property is invalid before reaching a state where the assumption observer declares violation of the environment assumptions, then the system is correct. The process is described in detail in Reference 45.

The commercial package *Simulink/Stateflow* by the MathWorks [46] allows modeling and simulation of control systems according to an SR MOC. Although its semantics is not formally defined (in mathematical terms) in any single document, it consolidated over time into a set of execution rules that are available throughout the user manuals.

The system is a network of functional blocks b_j. Blocks can be of two types: regular (dataflow) or Stateflow (state machine) blocks. Dataflow blocks can be of type continuous, discrete, or triggered. Continuous-type blocks process CT signals and produce as output other continuous signal functions according to the provided block function description (typically a set of differential equations). This includes the rules to update the state of the block (if there is one). Discrete blocks are activated at periodic time instants and process input signals, sampled at periodic time instants producing a set of periodic output signals and state updates. Finally, triggered blocks are only executed on the occurrence of a given event (a signal transition or a *function call*). When the event arrives, the current values on the input signals are evaluated, producing values on the output signals and updating (possibly) the block state.

When a system model must be simulated, the update of the outputs and of the state of continuous-type blocks must be computed by an appropriate solver, which could use a variable or a fixed step. When the system model is used for the automatic generation of an implementation, only fixed step solvers can be used. This means that even continuous blocks are evaluated on a discrete-time base (the step of the solver), which must be selected as an integer divider of any other period in the system and takes the name of *base period*. Taken aside triggered blocks (for now), this means that all blocks in the system are activated periodically (or at events belonging to a periodic time base) in case the automatic generation of an implementation is required.

Regular blocks process a set of input signals at times that are multiples of a period T_j, which is in turn an integer multiple of a system-wide *base period* T_b (in the case of triggered blocks, the model could be extended to sporadic activations).

We denote the inputs of block b_j by $i_{j,p}$ ($\boldsymbol{i}_j$ as vector) and its outputs by $o_{j,q}$ ($\boldsymbol{o}_j$). At all times kT_j the block reads the signal values on its inputs and computes two functions: an *output update*

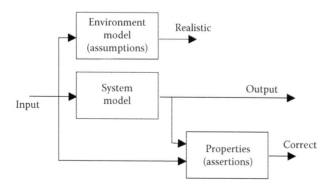

FIGURE 9.14   Verification by observers.

function $\boldsymbol{o}_j = f_o(\boldsymbol{i}_j, S_j)$ and a *state update function* $S_{\text{New}j} = f_s(i_j, S_j)$, where S_j ($S_{\text{New}j}$) is the current (next) state of b_j. Often, the two update functions can be considered as one:

$$(o_j, S_{\text{New}j}) = f_u(i_j, S_j)$$

Signal values are persistent until updated (each signal can be imagined as stored in a shared buffer). Therefore, each input and output is formally a right-continuous function, possibly sampled at periodic time instants by a reading block.

Figure 9.15 shows an example model with regular and Stateflow (labeled as supervisor in the figure) blocks.

The execution periods of discrete blocks are not shown in the model views, but block rates can be defined in the block properties as fixed or inherited. Inherited rates are resolved at model compilation time based on rules that compose (according to the block type) the rates of its predecessors or successors. Inheritance can proceed forward or backward in the input–output model flow. For example, a constant block providing (with an unspecified rate) a constant value to a block will inherit the rate of the block that reads the constant value (if unspecified, this block will in turn inherit from its successor). Conversely, if two blocks with assigned periods are feeding a simple adder block, with inherited rate, this block will be executed at the greatest common divisor of its input block periods.

Stateflow (or FSM) blocks can have multiple activation events $e_{j,v}$ (as shown in Figure 9.16, with two events of periods 2 and 5 entering the top of the Stateflow block *chart*; if no input event is defined, as for the *supervisor* block of Figure 9.16, the block implicitly reacts to periodic events at every base period). At any integer multiple of one of the events' periods $kT_{j,v}$, an update function is computed depending on the current state, the subset of input events that are active, and the set of input values. Update functions are extended by allowing the execution of generic functions whenever a given event is active on a given state. When multiple events are active, an ordering is provided to

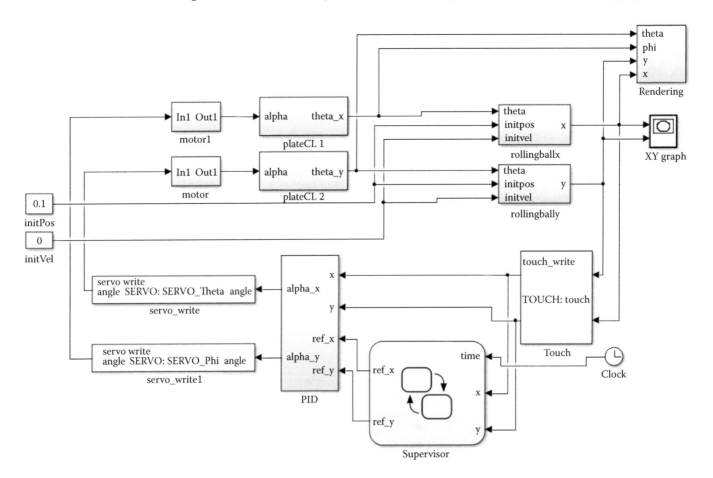

FIGURE 9.15   An example of Simulink model.

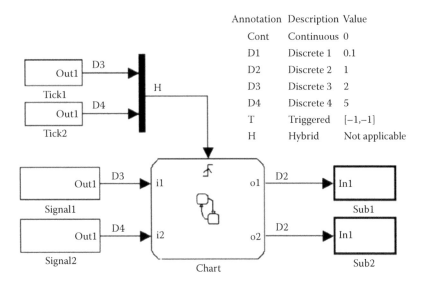

Annotation	Description	Value
Cont	Continuous	0
D1	Discrete 1	0.1
D2	Discrete 2	1
D3	Discrete 3	2
D4	Discrete 4	5
T	Triggered	[−1,−1]
H	Hybrid	Not applicable

FIGURE 9.16 Execution periods for Stateflow blocks are computed from their activation events.

give precedence to some events (for the given state) over others (thereby guaranteeing determinism). Figure 9.16 represents an FSM with two input signals, two output signals, and two incoming events with periods 2 and 5, obtained from the rising edge of two signals. The figure also shows the result of the model compilation with the computed signal periods. The outputs of the Stateflow block are labeled as periodic with a rate of 1 (the greatest common divisor of the activation events' periods).

A fundamental part of the model semantics is the rules dictating the evaluation order of the blocks. Any block for which the output is directly dependent on its input (i.e., any block with direct *feedthrough*) cannot execute until the block driving its input has executed. Some blocks set their outputs based on values acquired in a previous time step or from initial conditions specified as a block parameter. The set of topological dependencies implied by the direct feedthrough defines a partial order of execution among blocks. The partial order must be accounted for in the simulation and in the runtime execution of the model.

If two blocks b_i and b_j are in an input–output relationship (the output of b_j depends on its input coming from one of the outputs of b_i, and b_j is of type feedthrough, Figure 9.17), then there is a communication link between them, denoted by $b_i \rightarrow b_j$. In case b_j is not of type feedthrough, then the link has a delay, as indicated by $b_i \xrightarrow{-1} b_j$. Let $b_i(k)$ represent the kth occurrence of block b_i (belonging to the set of time instants $\cup_v kT_{i,v}$ for a state machine block, or kT_i for a dataflow block), then a sequence of activation times $a_i(k)$ is associated with b_i.

Given $t \geq 0$, we define $n_i(t)$ to be the number of times that b_i has been activated before or at t. In case of a link $b_i \rightarrow b_j$, if $i_j(k)$ denotes the input of the kth occurrence of b_j, then the SR semantics specifies that this input is equal to the output of the last occurrence of b_i that is no later than the kth occurrence of b_j, that is,

(9.3) $$i_j(k) = o_i(m); \text{ where } m = n_i(a_j(k))$$

This implies a partial order in the execution of the block functions. If $b_i \xrightarrow{-1} b_j$, then the previous output value is read, that is,

(9.4) $$i_j(k) = o_i(m-1)$$

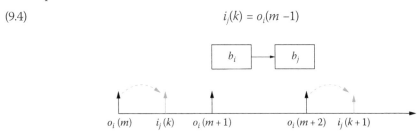

FIGURE 9.17 Input–output dependencies between communicating blocks.

Any implementation that verifies properties (9.3) or (9.4) will preserve the correspondences between the values produced or written by a predecessor block and the subset of them that is read by a successor, and it is therefore called "flow preserving."

The timeline on the bottom of Figure 9.17 illustrates the execution of a pair of blocks with SR semantics. The horizontal axis represents time. The vertical arrows capture the time instants when the blocks are activated and compute their outputs from the input values. In the figure, it is $i_j(k) = o_i(m)$.

A cyclic dependency among blocks where output values are instantaneously produced based on the inputs (all blocks in the cycle of type feedthrough) results in a fixed-point problem and possibly inconsistency (in general, a violation of the SR functional behavior).

This problem (called "algebraic loops") may be treated in different ways. Previous Esterel compilers tried to detect the cases in which loops still resulted in a deterministic behavior (the fixed-point equation allows a single solution). Today, most tools and languages (Simulink, among others) simply disallow algebraic loops in models for which an implementation must be automatically produced. The problem can be solved by the designer by adding a block that is not of type feedthrough (a Moore-type FSM or simply a delay block) in the loop.

The behavior of a Simulink/Stateflow block follows the semantics and notation of extended (hierarchical and concurrent) state machines. The Stateflow semantics have evolved with time and are described (unfortunately not in mathematical terms) in the user manual of the product, currently of more than 800 pages. The resulting language, a mix of a graphical and textual notation, tends to give the user full control of the programmability of the reactions and is de facto a Turing-complete language, with the potential for infinite recursion and nonterminating reactions.

The formal definition of a subset of the Stateflow semantics (now partly outdated) is provided in References 47 and 48, and compositionality is explored in Reference 49. Analogies between Stateflow and the general model of hierarchical automata are discussed in Reference 50.

Given the industrial recognition of the tool, the synchronous language community initially studied the conditions and the rules for the translation of Simulink [51,52] and then Stateflow [53] models into Lustre. In all papers (but especially the last one), the authors describe all the semantics issues they discovered in the mapping, which may hamper the predictability and determinism of Simulink/Stateflow models, outlining recommendations for the restricted use of a "safe" subset of the languages. Previously cited research on flow-preserving communication mechanisms also applies to Simulink systems. In particular, in Reference 54, the extension of the Simulink rate transition (RT) block (a mechanism for ensuring flow preservation under restricted conditions) to multicore platforms is discussed and possible solutions are presented. The formal description of input/output dependencies among components as reusable systems is provided in Reference 55.

9.3.1 SEMANTICS-PRESERVING IMPLEMENTATIONS

In References 56 and 57, the very important problem of how to map a zero-execution time Simulink semantics into a software implementation of concurrent threads where each computation necessarily requires a finite execution time is discussed.

The update functions of blocks and their action extensions are executed by program functions (or lines of code), executed by a task.

For regular blocks, the implementation may consist of two functions or sequences of statements, one for the state update and another for the output update (the output update parts must be executed before the state update). The two functions are often merged into a single update function, typically called "step."

Similarly, the implementation of a Stateflow (FSM) block consists of two functions or statement sets, evaluating the state and output update for the state machine (possibly merged). The update function is executed every time there is a possible incoming event computed from a transition of a signal. In the commercial code generator, the code is embedded in a periodic function, executing at the greatest common divisor of the periods of the activation events. This function first checks if there is any active event resulting from a value transition of one of the incoming signals. Then, it passes the set of active events to a monolithic function that computes the actions and the new state based on the current active state, the active events, and the value of the variables (and the guard conditions). Contrary to the Esterel and Lustre compilers, the code generator from Stateflow does not flatten the state machine or precomputes all the states of the chart (or the system). Flattening the model representation before code generation has advantages and disadvantages. The main disadvantages are that opportunities for reusing code (of multiple instances of the same subsystem) are lost and the code typically computes values (and reactions) that are not needed for a specific event/point in time. The main advantage, however, is that the code generation rules that are applied to the flattened model are very simple and it is much easier to generate code that corresponds to the model (i.e., it does not introduce any unintended behavior).

The code generation framework follows the general rules of the simulation engine and must produce an implementation with the same behavior (preserving the semantics). In many cases, what may be required from a system implementation is not the preservation of the synchronous assumption, that is, *the reaction (the outputs and the next state) of the system must be computed before the next event in the system,* but the looser, *flow preservation* property. Formally, the implementation must guarantee that every source block is executed without losing any execution instance that Equation 9.3 or 9.4 holds for any signal exchanged between two blocks and the correct untimed behavior of all (regular and Stateflow) blocks. The example implementation of Figure 9.19 does not satisfy the synchronous assumption (the output of block E is produced after real-time 1), but is flow preserving.

The Simulink Coder (formerly Real-Time Workshop)/Embedded Coder code generator of MathWorks allows two different code generation options: single task and fixed-priority multitask. Single-task implementations are guaranteed to preserve the simulation-time execution semantics but impose stricter conditions on system schedulability. A single periodic task replicates the same schedule of the block functions performed by the simulator, only in real time instead of logical time. It is triggered by a clock running at the base period of the system. Each time the task is activated, it checks which block reactions need to execute by scanning the global order list of the block functions. After its execution completes, the task goes back to sleep. In this case, a task is unnecessary. What is required is a mechanism for triggering a periodic execution, the global list of the system reactions (with their periods), and (possibly) a mechanism for checking task overruns and timing errors.

The possible problem with this implementation is described at the bottom of Figure 9.18. If the time that is required to compute the longest reaction in the system is larger than the base period (for the implementation on the selected execution platform), then the single-task implementation is not feasible since it *violates the synchronous assumption.*

By reasoning on the example system in the figure, a possible solution could be outlined for a multitask implementation on a single-core platform: blocks C and E (with period 4) could execute with a deadline equal to their periods. Figure 9.19 shows the multitask solution produced by the MathWorks code generator. One task is defined for each period in the system. Each task executes the blocks with the same period, in the order in which blocks are executed in the global order list. The task priorities are assigned using the rate monotonic rule to maximize the chances of schedulability.

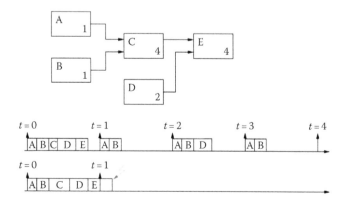

FIGURE 9.18   A single-task implementation of a network of blocks.

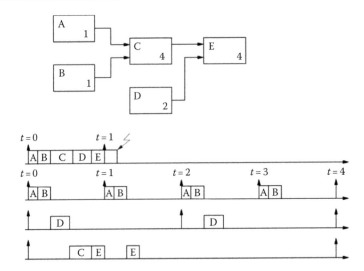

FIGURE 9.19   An example of multitask implementation of a network of blocks.

In general, considering not only single-core platforms, but arbitrary execution architectures and a generic scheduling policy on each node, the stage of the design process in which the functional model is mapped into a task (thread) model is the starting point of several optimization problems. Examples of design decisions that may be subject to constraints or determine the quality of the implementation are

- Mapping functions into tasks
- Assigning the execution order of functions inside tasks
- Assigning the task parameters (priority, deadline, and offset) to guarantee semantics preservation and schedulability
- Assigning scheduling attributes to functions (including preemptability and preemption threshold)
- Designing communication mechanisms that ensure flow preservation while minimizing the amount of memory used (an example of possible issues with flow preservation is shown in Figure 9.20).

In multitask implementations, the runtime execution of the model is performed by running the code in the context of a set of threads under the control of a priority-based RTOS. The function-to-task mapping consists of a relationship between a block update function (or each one of them in the case of an FSM block) and a task and a static scheduling (execution order) of the function code inside the task. The ith task is denoted as τ_i. $M(f_j, k, i)$ indicates that the step function f_j of block b_j is executed as the kth segment of code in the context of τ_i.

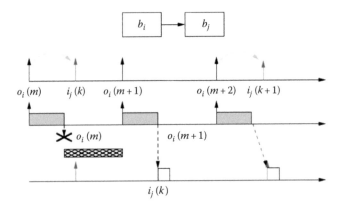

FIGURE 9.20   An implementation with a violation of signal flows.

The function order is simply the order in which the implementation code is invoked (typically by a call to its Step() function) from inside the task code.

Because of preemption and scheduling, in a multirate system, the signal flows of the implementation can differ from the model flows. The bottom part of Figure 9.20 shows the possible problems with flow preservation in multitask implementations because of preemption and scheduling.

The timeline on top represents the execution of the block reactions in logical time. The schedule at the bottom shows a possible execution when the computation of the update function is performed in real time by software code, and the code (task) implementing the kth instance of the reader is delayed because of preemption.

In this case, the writer finishes its execution producing the output $o_i(m)$. If the reader performs its read operation before the preemption by the next writer instance, then (correctly) $i_j(k) = o_i(m)$. Otherwise, it is preempted and a new writer instance produces $o_i(m + 1)$. In the latter case, the reader reads $o_i(m + 1)$, in general different from $o_i(m)$.

Problems can possibly arise when two communicating blocks are mapped into different tasks. In this case, we expect that the two blocks execute with different rates (there may be exceptions to this rule and will be handled in similar ways).

In case a high-priority block/task τ_1 drives a low-priority block/task τ_2 (left side of Figure 9.21), *there is uncertainty* about which instance of τ_1 produces the data consumed by τ_2 (τ_2 should read the values produced by the first instance of τ_1, not the second). In the oversampling case, there is uncertainty on the reader instance that is consuming the data. In addition, if the reader has higher priority than the writer, the communication must have an associated delay.

Clearly, the shared variables implementing the communication channel must be protected for data consistency. However, lock-based mechanisms are in general not suited for the implementation of flow-preserving communication, since they are based on the assumption that the execution order of the writer and reader is unknown and there is the possibility of one preempting the other while operating on the shared resource and flow preservation in the case of a multirate/multitask generation mode for single-core platforms. The design element is called a "rate transition" block [46] and is in essence a special case of a wait-free communication method. In a multitask implementation, the MathWorks code generator creates one task (identifier) for each block rate in the model and assigns its priority according to the rate monotonic rule (as in the example of Figure 9.20).

RT blocks are placed between any two blocks with different rates. In the case of high-to-low-rate/priority transitions, the RT block output update function executes at the rate of the receiver block (left side of Figure 9.22), but within the task and at the priority of the sender block. In low-to-high-priority transitions (right side of Figure 9.22), the RT block state update function executes in the context of the low-rate task. The RT block output update function runs in the context of the high-rate task, but at the rate of the sender task, feeding the high-rate receiver. The output function uses the state of the RT that was updated in the previous instance of the low-rate task. RT blocks can only be applied in a more restricted way than generic wait-free methods [58]: the periods of the writer and reader need to be harmonic, meaning one is an integer multiple of the other; also, they only apply to one-to-one communication links. One-to-n communication

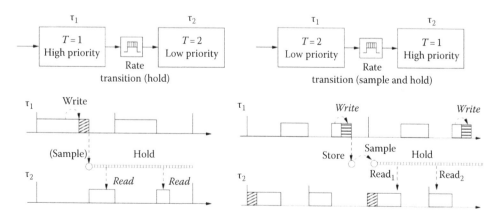

FIGURE 9.21   The rate transition blocks.

is regarded as n separate links and each of them is buffered independently. The RT mechanism has the clear advantage of being completely implemented in user space, but cannot be used when the reader and the writer do not have harmonic periods (or are not periodic, but triggered), or are not executed with statically assigned priorities, or even when they execute on different cores. In the following, we will analyze the mechanisms for guaranteeing data consistency in a general single-core single-writer–multiple-reader case. The same mechanisms can then be extended to deal with intercore communication. In this last case, several implementation options are possible.

9.3.2 ARCHITECTURE DEPLOYMENT AND TIMING ANALYSIS

Synchronous models are typically implemented as a single task that executes according to an event server model. Reactions are decomposed into atomic actions that are partially ordered by the causality analysis of the program. The scheduling is generated at compile time trying to exploit the partial causality order among the functions that implement the block reactions, in order to make the best possible use of hardware and shared resources. The main concern is checking that the synchrony assumption holds, that is, ensuring that the longest chain of reactions ensuing from any internal or external event is completed within the step duration. Static scheduling means that critical applications are deployed without the need for any operating system (and the corresponding overhead). This reduces system complexity and increases predictability avoiding preemption, dynamic contention over resources, and other nondeterministic operating system functions.

The timing analysis of task sets implementing Simulink/Stateflow models can be quite challenging. At the highest level of abstraction, the multitasking implementation of a Simulink model simply consists of a set of periodic independent tasks with implicit (equal to their periods) deadlines, exchanging information using wait-free communication mechanisms (such as the RT blocks).

A system like this can easily be analyzed using the basic response time formulation outlined in the previous sections.

However, a deeper analysis reveals a number of additional issues. When a task is implementing Stateflow reactions, an abstraction as a simple periodic task with a worst-case execution time can be quite pessimistic. In Reference 59, the authors discuss the problem and show an analogy with the analysis of the task digraph model [60] (a very general model for the analysis of branching jobs). A more efficient analysis for this type of systems is also presented in Reference 61. Alternative methods for implementing synchronous state machines as a set of concurrent tasks are discussed in References 62 and 63.

Also, RT blocks are only applicable in single-core architectures and under the condition that the rates of the communicating blocks are harmonic. A discussion on the extension of the RT block mechanism to multicore platforms can be found in Reference 64.

In References 65–67, the authors show how existing wait-free communication mechanisms can be extended to provide guaranteed flow preservation and how to optimize the sizing of the buffers and/or optimize the design with respect to timing performance [68]. The use of a priority assignment different from rate monotonic to selectively avoid the use of RT blocks is discussed in References 69 and 70.

9.3.3 TOOLS AND COMMERCIAL IMPLEMENTATIONS

Lustre is implemented by the commercial toolset SCADE, offering an editor that manipulates both graphical and textual descriptions; two code generators, one of which is accepted by certification authorities for qualified software production; a simulator; and an interface to verification tools such as the plug-in from Prover [71].

The early *Esterel* compilers had been developed by Gerard Berry's group at INRIA/CMA and freely distributed in binary form. The commercial version of Esterel was first marketed in 1998 and then distributed by Esterel Technologies, which later acquired the SCADE environment. SCADE has been used in many industrial projects, including integrated nuclear protection systems (Schneider Electric), flight control software (Airbus A340–600), and track control systems (CS Transport). Dassault Aviation was one of the earliest supporters of the Esterel project and has long been one of its major users.

Several verification tools use the synchronous observer technique for checking Esterel programs [72]. It is also possible to verify implementations of Esterel programs with tools exploiting explicit state space reduction and bisimulation minimization (FC2Tools), and finally, tools can be used to automatically generate test sequences with guaranteed state/transition coverage.

The *Simulink* tool by MathWorks was developed with the purpose of simulating control algorithms and has been since its inception extended with a set of additional tools and plug-ins, such as the Stateflow plug-in for the definition of the FSM behavior of a control block, allowing modeling of hybrid systems, and a number of automatic code generation tools, such as the Simulink Coder and Embedded Coder.

9.3.4 CHALLENGES

The main challenges and limitations that synchronous languages must face when applied to complex systems are the following:

- Despite improvements, the space and time efficiency of the compilers is still not satisfactory.
- Embedded applications can be deployed on architectures or control environments that do not comply with the SR model.
- Designers are familiar with other dominant methods and notations. Porting the development process to the synchronous paradigm and languages is not easy.

Efficiency limitations are mainly due to the formal compilation process and the need to check for constructive causality. The first three Esterel compilers used automata-based techniques and produced efficient code for small programs, but they did not scale to large-scale systems because of state explosion. Versions 4 and 5 were based on translation into digital logic and generated smaller executables at the price of slow execution (The program generated by these compilers wastes time evaluating each gate in every clock cycle.) This inefficiency can produce code 100 times slower than that from previous compilers [72].

Version 5 of the compiler allowed cyclic dependencies by exploiting Esterel's constructive semantics. Unfortunately, this requires evaluating all the reachable states by symbolic state space traversal [73], which makes it extremely slow.

As for the difficulty in matching the basic paradigm of synchrony with system architectures, the main reasons of concern are

- The bus and communication lines, if not specified according to a synchronous (time-triggered) protocol and the interfaces with the analog world of sensors and actuators
- The dynamics of the environment, which can possibly invalidate the instantaneous execution semantics

The former has been discussed at length in a number of papers (such as [74,75]), giving conditions for providing a synchronous implementation on top of distributed platforms.

A possible solution consists in the adoption of a desynchronization approach based on the loosely time-triggered architecture (LTTA), which allows to provide a flow-preserving implementation by means of constraints on the rates of the processing actors or backpressure mechanisms for controlling the flow of data (when the transmitter can be stalled). The timing analysis of LTTA implementations on distributed platforms is discussed in References 76 and 77.

Finally, in order to integrate synchronous languages with the mainstream commercial methodologies and languages, translation and import tools are required. For example, it is possible from SCADE to import discrete-time Simulink diagrams and Sildex allows importing Simulink/Stateflow discrete-time diagrams. Another example is the attempt to integrate Esterel and UML based on a proposal for coupling Esterel Studio and Rational Rose drafted by Dassault [78] and adopted by commercial Esterel tools.

The commercial Esterel compilers were first acquired by Synfora and then by Synopsys, but finally, despite early interest in a standardization by IEEE, the product appears as discontinued.

The SCADE Suite is still commercially available from Esterel Technologies (now acquired by Ansys) and quite popular for applications that require certification (or a certified code generator), such as under the DO-178B and DO-178C or the upcoming standard for automotive systems ISO 26262 (ASIL D and C).

9.4 ASYNCHRONOUS MODELS

UML (an OMG standard) and SDL (an ISO-ITU standard) are languages developed, respectively, in the context of general-purpose computing and in the context of (large) telecommunication systems. UML is the merger of many OO design methodologies aimed at the definition of generic software systems. Its semantics is not completely specified and intentionally retains many variation points in order to adapt to different application domains. In order to be practically applicable to the design of embedded systems, further characterization (a specialized *profile* in UML terminology) is required. In the 2.0 revision of the language, the system is represented by a (transitional) model where active and passive components, communicating by means of connections through port interfaces, cooperate in the implementation of the system behavior. Each reaction to an internal or external event results in the transition of a *statechart* automaton describing the object behavior.

SDL has a more formal background since it was developed in the context of software for telecommunication systems for the purpose of easing the implementation of verifiable communication protocols. An SDL design consists of blocks cooperating by means of asynchronous signals. The behavior of each block is represented by one or more (conceptually concurrent) processes. Each process, in turn, implements an extended FSM.

Until the development of the UML profile for schedulability, performance, and time (SPT) and its follower MARTE systems profile, UML did not provide any formal means for specifying time or time-related constraints nor for specifying resources and resource management policies. The deployment diagrams were the only (inadequate) means for describing the mapping of software unto the hardware platform and tool vendors had tried to fill the gap by proposing nonstandard extensions.

The situation with SDL is not much different, although SDL offers at least the notion of global and external time. Global time is made available by means of a special expression and can be stored in variables or sent in messages.

Implementation of asynchronous languages typically (but not necessarily) relies on an operating system. The latter is responsible for scheduling, which is necessarily based on static (design-time) priorities if a commercial product is used to this purpose. Unfortunately, as it will be clear in the following, real-time schedulability techniques are only applicable to very simple models and are extremely difficult to generalize to most models of practical interest or even to the implementation model assumed by most (if not all) commercial tools.

9.4.1 UML

UML represents a collection of engineering practices that have proven successful in the modeling of large and complex systems and has emerged as the software industry's dominant OO modeling language.

Born at Rational in 1994, UML was taken over in 1997 at version 1.1 by the OMG Revision Task Force (RTF), which became responsible for its maintenance. The RTF released UML version 1.4 in September 2001 and a major revision, UML 2.0, appeared in 2003. UML is now at version 2.5 (beta) [7].

UML has been designed as a wide-ranging, general-purpose modeling language for specifying, visualizing, constructing, and documenting the artifacts of software systems. It has been successfully applied to a wide range of domains, ranging from health and finance to aerospace and e-commerce, and its application goes even beyond software, given recent initiatives in areas such as systems engineering, testing, and hardware design. A joint initiative between the OMG and the International Council on Systems Engineering defined a profile for systems engineering, which later produced the standard specification for the SysML. The SysML 1.4 specification was adopted in March 2014 [79]. At the time of this writing, over 60 UML CASE tools can be listed from the OMG resource page (http://www.omg.org).

After its major revision 2.0, the UML language specification consists of four parts:

1. UML Infrastructure, defining the foundational language constructs and the language semantics in a more formal way than it was in the past
2. UML Superstructure, which defines the user-level constructs
3. OCL Object Constraint Language, which is used to describe expressions (constraints) on UML models
4. UML Diagram Interchange, including the definition of the XML-based XMI format, for model interchange among tools

UML consists of a metamodel definition and a graphical representation of the formal language, but it intentionally refrains from including any design process. The UML language in its general form is deliberately semiformal and even its state diagrams (a variant of statecharts) retain sufficient semantics variation points in order to ease adaptability and customization.

The designers of UML realized that complex systems cannot be represented by a single design artifact. According to UML, a system model is seen under different views, representing different aspects. Each view corresponds to one or more of diagrams, which, taken together, represent a unique model. Consistency of this multiview representation is ensured by the UML metamodel definition. The diagram types included in the UML 2.0 specification are represented in Figure 9.22, as organized in the two main categories that relate to *structure* and *behavior*.

When domain-specific requirements arise, more specific (more semantically characterized) concepts and notations can be provided as a set of stereotypes and constraints and packaged in the context of a *profile*.

Structure diagrams show the static structure of the system, that is, specifications that are valid irrespective of time. Behavior diagrams show the dynamic behavior of the system. The main diagrams are

- *Use case diagram*, indicating a high-level (user requirements–level) description of the interaction of the system with external agents
- *Class diagram*, representing the static structure of the software system, including the OO description of the entities composing the system and of their static properties and relationships
- *Behavior diagrams* including *interaction or sequence diagrams* and *state diagrams* as variants of message sequence charts [MSCs] and statecharts), providing a description of the dynamic properties of the entities composing the system, using various notations
- *Architecture diagrams* (including *composite* and *component diagrams*, showing a description of reusable components), showing a description of the internal structure of classes and objects and a better characterization of the communication superstructure, including communication paths and interfaces.
- *Implementation diagrams*, containing a description of the physical structure of the software and hardware components

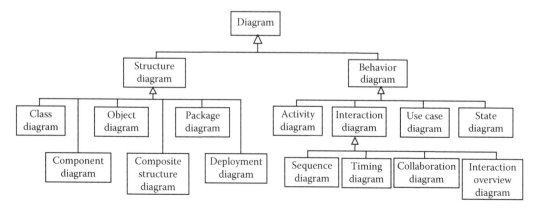

FIGURE 9.22 A taxonomy of UML 2.0 diagrams.

The class diagram is typically the core of a UML specification, as it shows the logical structure of the system. The concept of classifier (class) is central to the OO design methodology. Classes can be defined as user-defined types consisting of a set of attributes defining the internal state and a set of operations (signature) that can be possibly invoked on the class objects resulting in an internal transition. As units of reuse, classes embody the concepts of encapsulation (or information hiding) and abstraction. The signature of the class abstracts the internal state and behavior and restricts possible interactions with the environment. Relationships exist among classes, with special names and notations in relevant cases, such as aggregation and composition, use, and dependency. The generalization (or refinement) relationship allows controlled extensions of the model by letting a derived class specification inherit all the characteristics of the parent class (attributes and operations, and also, selectively, relationships) while providing new ones (or redefining existing ones).

Objects are instances of the type defined by the corresponding class (or classifier). As such, they embody all of the classifier attributes, operations, and relationships. Several books [80–82] have been dedicated to the explanation of the full set of concepts in OO design. The interested reader is invited to refer to the literature on the subject for a more detailed discussion.

All diagram elements can be annotated with constraints, expressed in OCL or in any other formalism that the designer sees as appropriate. A typical class diagram showing dependency, aggregation, and generalization associations is shown in Figure 9.23.

UML finally acknowledged the need for a more formal characterization of the language semantics and for better support for component specifications. In particular, it became clear that simple classes provide a poor match for the definition of a reusable component (as outlined in previous sections).

As a result, necessary concepts, such as the means to clearly identify provided and (especially) required *interfaces*, have been added by means of the port construct. An interface is an abstract class declaring a set of functions with their associated signature. Furthermore, structured classes and objects allow the designer to formally specify the internal communication structure of a component configuration.

UML classes, structured classes, and components are now encapsulated units that model active system components and can be decomposed into contained classes communicating by signals exchanged over a set of *ports*, which model communication terminals. A port carries both structural information on the connection between classes or components and *protocol* information that specifies what messages can be exchanged across the connection (an example in Figure 9.24). A state machine and/or a UML sequence diagram (similar to MSCs) may be associated with a protocol to express the allowable message exchanges. Two components can interact if there is a connection between any two ports that they own and that support the same protocol in complementary (or *conjugated*) roles. The behavior or reaction of a component to an incoming message or signal is typically specified by means of one or more statechart diagrams.

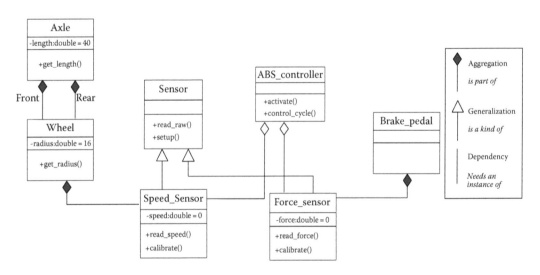

FIGURE 9.23 A sample class diagram.

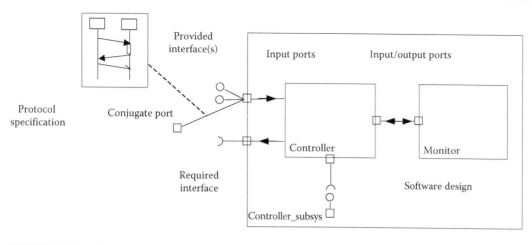

FIGURE 9.24   Ports and components in UML 2.0.

Behavior diagrams comprise *statechart diagrams, sequence diagram,* and *collaboration diagrams.*

Statecharts [15] describe the evolution in time of an object or an interaction between objects by means of a hierarchical state machine. UML statecharts are an extension of Harel's statecharts, with the possibility of defining actions upon entry into or exit from a state as well as actions to be executed when a transition is taken. Actions can be simple expressions or calls to methods of the attached object (class) or entire programs. Unfortunately, not only does the Turing completeness of actions prevent decidability of properties in the general model, but UML does not even clarify most of the semantics variations left open by the standard statechart formalism.

Furthermore, the UML specification explicitly gives actions a run-to-completion execution semantics, which makes them nonpreemptable and makes the specification (and analysis) of typical RTOS mechanisms such as interrupts and preemption impossible.

To give an example of UML statecharts, Figure 9.25 shows a sample diagram where, upon entry of the composite state (the outermost rectangle), the subsystem enters into three concurrent (and-type) states, named Idle, WaitForUpdate, and Display_all, respectively. Upon entry in the WaitForUpdate state, the variable count is also incremented. In the same portion of the diagram, reception of message msg1 triggers the exit action setting the variable `flag` and the (unconditioned) transition with the associated *call action* `update()`. The `count` variable is finally incremented upon reentry in the state WaitForUpdate.

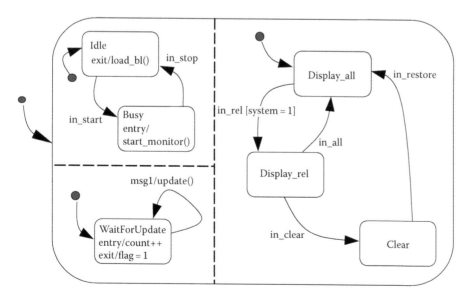

FIGURE 9.25   An example of UML statechart.

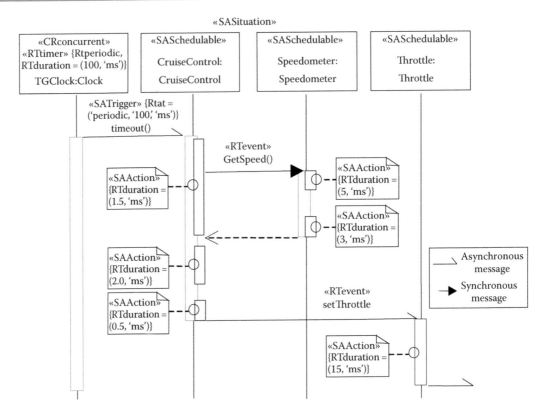

FIGURE 9.26 A sample sequence diagram with annotations showing timing constraints.

Statechart diagrams provide the description of the state evolution of a single object or class, but neither they are meant to represent the emergent behavior deriving from the cooperation of more objects nor are they appropriate for the representation of timing constraints. *Sequence diagrams* partly fill this gap. *Sequence diagrams* show the possible message exchanges among objects, ordered along a time axis. The timepoints corresponding to message-related events can be labeled and referred to in constraint annotations. Each sequence diagram focuses on one particular scenario of execution and provides an alternative to temporal logic for expressing timing constraints in a visual form (Figure 9.26).

Collaboration diagrams also show message exchanges among objects, but they emphasize structural relationships among objects (i.e., "who talks with whom") rather than time sequences of messages.

Collaboration diagrams (Figure 9.27) are also the most appropriate way for representing logical resource sharing among objects. Labeling of messages exchanged across links defines the sequencing of actions in a similar (but less effective) way to what can be specified with sequence diagrams.

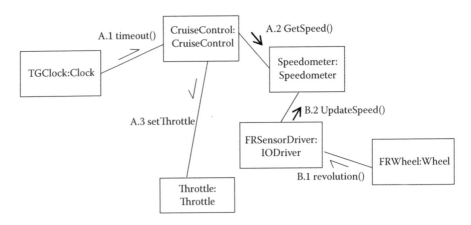

FIGURE 9.27 Collaboration diagram.

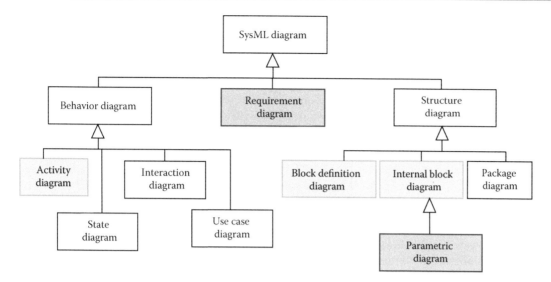

FIGURE 9.28 SysML diagrams.

Despite availability of multiple diagram types (or maybe because of it), the UML metamodel is quite weak when it comes to the specification of dynamic behavior. The UML metamodel concentrates on providing structural consistency among the different diagrams and provides sufficient definition for the static semantics, but the dynamic semantics are never adequately addressed, up to the point that a major revision of the UML action semantics became necessary. In 2008, this originated the first adopted specification of the Foundational UML (fUML) language [40]. The intention is for UML to eventually become an executable modeling language, which would, for example, allow early verification of system functionality. Within the OMG, it was clear that a graphical language was not the best candidate for the UML executable semantics. The Concrete Syntax for a UML Action Language RFP, issued by OMG in 2008, asked for a textual language solution and resulted in the Action Language for fUML (Alf) [39]. Alf is a formal textual description for UML behaviors that can be attached to a UML model element. Alf code can be used to specify the behaviors of state machine actions or the method of an operation.

Semantically, Alf maps to the fUML subset. Syntactically, Alf looks a lot like typical C++ or Java. This is because the language proposers realized that it was only reasonable to define Alf in a way that would be familiar to practitioners, to ease the adoption of the new language.

As of today, Alf is still not widely used to model behavior and tool support for its execution (for simulation and verification) is missing. An early attempt at the definition and implementation of transformation mechanisms toward the translational execution of ALF can be found in Reference 83. An extended discussion on how the previous approach is extended to full C++ code generation from UML profiles and Alf is discussed in Reference 84.

However, simulation or verification of (at least) some behavioral properties and (especially) automatic production of code are features that tool vendors cannot ignore if UML is not to be relegated to the role of simply documenting software artifacts. Hence, CASE tools provide an interpretation of the variation points. This means that validation, code generation, and automatic generation of test cases are tool specific and depend upon the semantics choices of each vendor.

Concerning formal verification of properties, it is important to point out that UML neither provide any clear means for specifying the properties that the system (or components) is expected to satisfy nor give any means for specifying assumptions on the environment. The proposed use of OCL in an explicit contract section to specify assertions and assumptions acting upon the component and its environment (its users) can hopefully fill this gap in the future.

Research groups are working on the definition of a formal semantics restriction of UML behavior (especially by means of the statechart formalism), in order to allow for formal verification of system properties [85,86]. After the definition of such restrictions, UML models can be translated into the formats used by existing validation tools for timed MSCs or TAs.

Finally, the last type of UML diagrams is the *implementation diagrams*, which can be either component diagrams or deployment diagrams. Component diagrams describe the physical structure of the software in terms of software components (modules) related with each other by dependency and containment relationships. Deployment diagrams describe the hardware architecture in terms of processing or data storage nodes connected by communication associations and show the placement of software components onto the hardware nodes.

The need to express in UML timeliness-related properties and constraints and the pattern of hardware and software resource utilization as well as resource allocation policies and scheduling algorithms found a (partial) response with the OMG MARTE profile. The specification of timing attributes and constraints in UML designs will be discussed in Section 9.4.5.

9.4.2 SysML

To meet the modeling needs of (large-scale embedded) systems, the OMG issued the SysML modeling standard [79]. SysML derives several of its metamodeling features and diagrams from UML. However, SysML is formally not a UML refinement even though it reuses a large share of it.

SysML provides a number of additional modeling elements in its metamodel (among others, see Figure 9.29):

> *Blocks*: They are the basic structure elements. Based on the UML composite classes, they provide a unifying concept to describe the structure of an element or system and support SysML new features (e.g., flow ports, value properties). In system-level modeling, blocks are used to represent any type of system or element, including hardware, software, data, facilities, people, and possibly signals and other physical quantities.
> *Flow ports*: These ports are used to represent interfaces for physical quantities or signals, or even for data-oriented communication, providing a much closer correspondence to the concept of a port in SR modeling languages.
> *Flow specifications*: These are extended specifications of the type of signals, data, or physical items exchanged over flow ports. In very general terms, flow specifications allow the specification of not only the type of the flow items but also their direction.
> *Item flows*: These flows allow the identification of the specific item in a flow specification that is communicated in a connection over two ports.
> *Allocations*: They allow the representation of general relationships that map one model element to another, such as behavioral mapping (i.e., function to component), structural mapping (i.e., logical to physical), or software to hardware.
> *Requirements*: The *requirement* stereotype represents a text-based requirement, which includes the definition of a requirement identifier and of its text properties.
> *Parametric* constraints: They are associated with flow port values and expressed as equations or any other suitable language.
> *Continuous flows*. A refinement of data flows used for CT signals or physical quantities.

SysML reuses a number of diagrams from UML, even if some of them have a different name. The UML class diagram becomes the block definition diagram (BDD), and the composite structure diagram becomes the internal block diagram (IBD). The package, state machine, activity,

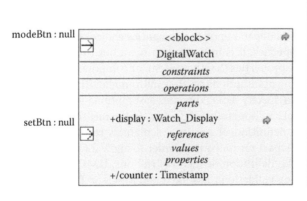
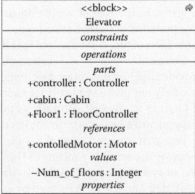

FIGURE 9.29 Block compartments as shown in a block definition diagram.

use case, and sequence diagrams (the last with the name interaction diagram) are also reused from UML.

In addition, SysML provides some new diagrams:

Parametric diagrams are used to express constraints among (signal) values of ports. They provide support for engineering analysis (e.g., performance, reliability) on design models. A parametric diagram represents the usage of the constraints in an analysis context and may be used for identification of critical performance properties. Parametric diagrams are still structural and declarative. They are not meant for the equations to be simulated or integrated.

Requirement diagrams are provided to let the system developer enter textual requirements and the relationship among them. Also, requirements may be traced to their refinements and linked to the test descriptions.

Blocks are used in the BDDs and IBDs. BDDs describe the relationship among blocks (e.g., composition, association, specialization); IBDs describe the internal structure of blocks in terms of properties and connectors.

In the BDD, blocks appear with a set of compartments (in the classification of Figure 9.29, BDDs are shown—as all new diagrams—in shaded rectangles; an example of BDD is in Figure 9.30) that are used to describe the block characteristics. Compartments allow the definition of

- Properties
- Parts
- References
- Values
- Ports
- Operations
- Constraints
- Allocations from/to other model elements (e.g., activities)
- Requirements that the block satisfies

An example of a BDD is shown in Figure 9.30 with composition and generalization dependencies (similar to their UML counterparts).

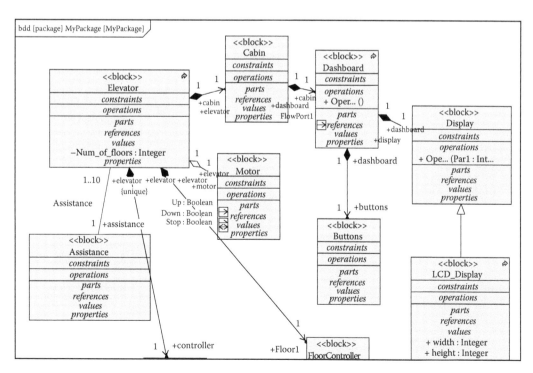

FIGURE 9.30   An example of block definition diagram.

BDDs cannot define completely the communication dependencies and the composition structure (the topology of the system). Given that blocks are used to specify hierarchical architecture structures, there is a need to represent the internal structure of a composite and the interconnections among its parts (the usage of a block in the context of a composite block, characterized by a role). The internal structure is made explicit by the IBDs, where the communication and signaling topology become explicit (an example in Figure 9.31).

Parametric diagrams are used to express constraints (possibly through equations) between value properties (Figure 9.32). Constraints can be defined by other blocks. The expression language can be formal (e.g., MathML [87], OCL) or informal and binds the constraint parameters to value properties of blocks or their ports (e.g., vehicle mass bound to parameter "m" in F = m × a). The computational or analysis engine that verifies the constraints should be provided by a specific analysis tool and not by SysML.

Requirement diagrams are used to define a classification of requirements categories (e.g., functional, interface, performance) or a hierarchy of requirements, describing the refinement relationships among requirements in a specification. Requirement relationships include DeriveReqt, Satisfy, Verify, Refine, Trace, and Copy. An example of requirement diagram is shown in Figure 9.33.

Finally, different ways are provided in SysML for specifying allocations. Graphical and tabular representations can be used for «allocation» on IBDs (an example of IBD that describes an execution architecture is shown in Figure 9.34). BDDs can be used to define the deployment of software and data to hardware.

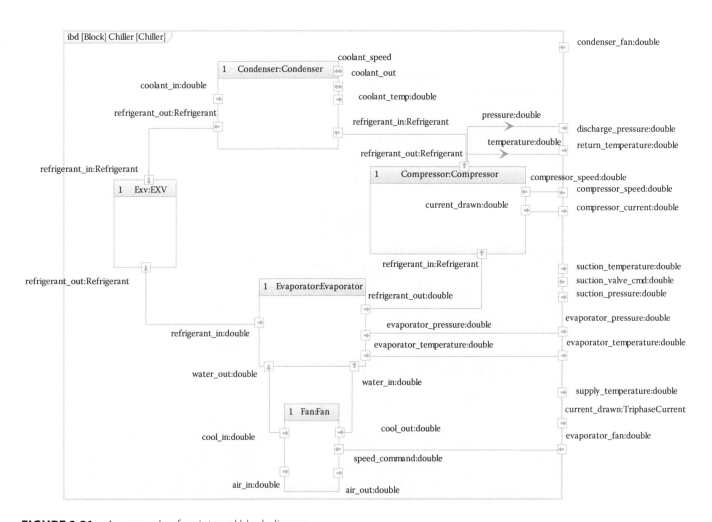

FIGURE 9.31 An example of an internal block diagram.

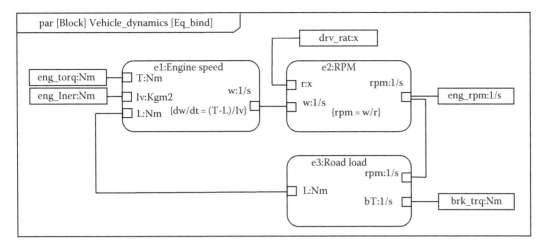

FIGURE 9.32   An example of a parametric diagram.

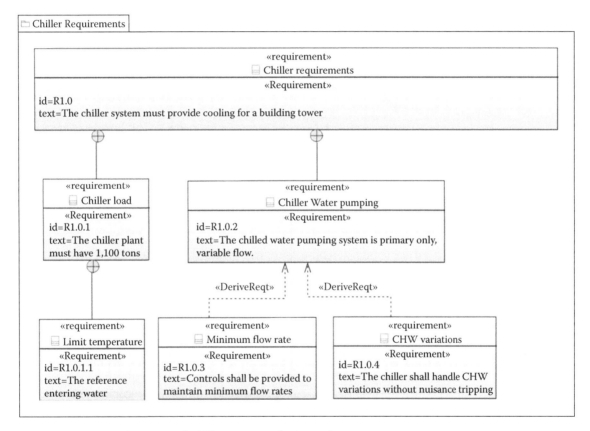

FIGURE 9.33   A requirement diagram with different types of relationships.

9.4.3 OCL

The object constraint language (OCL) [88] is a formal language used to describe constraint expressions on UML models. An OCL expression is typically used to specify invariants or other type of constraint conditions that must hold for the system. OCL expressions refer to the *contextual instance,* that is, the model element to which the expression applies, such as classifiers, for example, types, classes, interfaces, associations (acting as types), and datatypes. Also all attributes, association ends, methods, and operations without side effects that are defined on these types can be used.

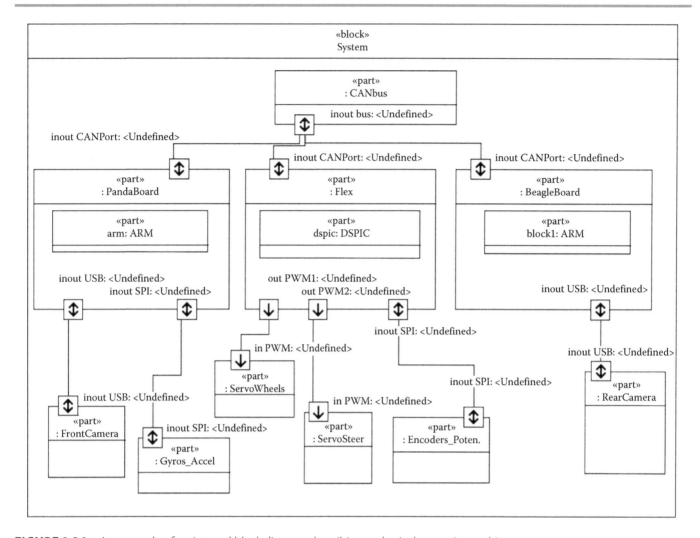

FIGURE 9.34 An example of an internal block diagram describing a physical execution architecture.

OCL can be used to specify *invariants* associated with a classifier. In this case, it returns a Boolean type and its evaluation must be true for each instance of the classifier at any moment in time (except when an instance is executing an operation).

Preconditions and *postconditions* are other types of OCL constraints that can be possibly linked to an operation of a classifier, and their purpose is to specify the conditions or contract under which the operation executes. If the caller fulfills the precondition before the operation is called, then the called object ensures the postcondition to hold after execution of the operation, but of course, only for the instance that executes the operation (Figure 9.35).

9.4.4 SDL

The *SDL* is an International Telecommunications Union (ITU-T) [89] standard promoted by the SDL Forum Society for the specification and description of systems [55].

Since its inception, a formal semantics has been part of the SDL standard (Z.100), including visual and textual constructs for the specification of both the architecture and the behavior of a system. The behavior of (active) SDL objects is described in terms of concurrently operating and asynchronously communicating abstract state machines. SDL provides the formal behavior semantics that enables tool support for simulation, verification, validation, testing, and code generation.

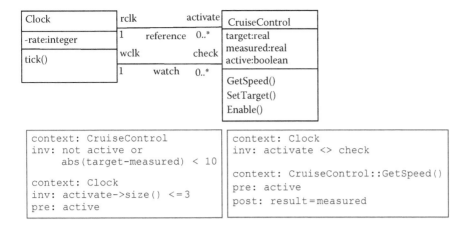

```
context: CruiseControl
inv: not active or
     abs(target-measured) < 10

context: Clock
inv: activate->size() <=3
pre: active
```

```
context: Clock
inv: activate <> check

context: CruiseControl::GetSpeed()
pre: active
post: result=measured
```

FIGURE 9.35 OCL examples adding constraints or defining behavior of operations in a class diagram.

In SDL, systems are decomposed into a hierarchy of *block agents* communicating via (unbounded) channels that carry typed signals. Agents may be used for structuring the design and can in turn be decomposed into subagents until leaf blocks are decomposed into *process agents*. Block and process agents differ since blocks allow internal concurrency (subagents), while process agents only have an interleaving behavior.

The behavior of process agents is specified by means of extended finite and communicating state machines (SDL services) represented by a connected graph consisting of states and transitions. Transitions are triggered by external stimuli (signals, remote procedure calls) or conditions on variables. During a transition, a sequence of actions may be performed, including the use and manipulation of data stored in local variables or asynchronous interaction with other agents or the system environment via signals that are placed into and consumed from channel queues.

Figure 9.36 shows a process behavior and a matching representation by means of an extended FSM (right side).

Channels are asynchronous (as opposed to synchronous or rendezvous) FIFO queues (one for each process) and provide a reliable, zero- or finite-delay transmission of communication elements from a sender to a receiver agent.

Signals sent to an agent will be delivered to the input port of the agent. Signals are consumed in the order of their arrival either as a trigger of a transition or by being discarded in case there is no transition defined for the signal in the current state. Actions executed in response to the reception of input signals include expressions involving local variables or calls to procedures.

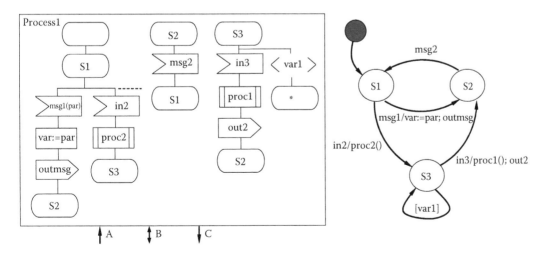

FIGURE 9.36 A Specification and Description Language process behavior and the corresponding extended finite-state machine.

In summary, an agent definition consists of

- A *behavior specification* given by the agent extended FSM
- A *structure and interaction diagram* detailing the hierarchy of system agents with their internal communication infrastructure
- The variables (*attributes*) under the control of each agent
- The black-box or *component view* of the agent defining the interaction points (ports) with the provided and required interfaces

SDL 2000 extends the language by including typical features of OO and component-based modeling techniques, including

- Encapsulation of services by ports (gates) and interfaces
- Classifiers and associations
- Specialization (refinement) of virtual class structure and behavior

SDL offers native mechanisms for representing external (global) time. Time is available by means of the predefined variable **now**, the now() primitive and timer constructs. Process actions can set timers, that is, the specification of a signal at a predefined point in time, wait, and eventually receive a timer expiry signal. SDL timer timeouts are always received in the form of asynchronous messages and timer values are only meant to be minimal bounds, meaning that any timer signal may remain enqueued for an unbounded amount of time.

In SDL, processes inside a block are meant to be executed concurrently, and no specification for a sequential implementation by means of a scheduler, necessary when truly concurrent hardware is not available, is given. Activities implementing the processes behavior (transitions between states) are executed in a run-to-completion fashion. From an implementation point of view, this raises the same concerns that hold for implementation of UML models.

Other language characteristics make verification of time-related properties impossible: the Z.100 SDL semantics says that external time progresses in both states and actions. However, each action may take an unbounded amount of time to execute, and each process can remain in a state for an indeterminate amount of time before taking the first available transition.

Furthermore, for timing constraint specification, SDL does not include the explicit notion of event; therefore, it is impossible to define a time tagging of most events of interest such as transmission and reception of signals, although MSCs (similar in nature to UML sequence diagrams) are typically used to fill this gap since they allow expression of constraints on time elapsing between events.

Incomplete specification of time events and constraints prevents timing analysis of SDL diagrams, but the situation is not much better for untimed models. Properties of interest are in general undecidable because of infinite data domains, unbounded message buffers, and the semiformal modeling style, where SDL is mixed with code fragments inside conditions and actions (the formal semantics of SDL is aimed at code generation rather than at simulation and verification).

9.4.5 ARCHITECTURE DEPLOYMENT, TIMING SPECIFICATION, AND ANALYSIS

UML has been developed outside the context of embedded systems design and it is clear from the previous sections that it neither cope with the modeling of resource allocation and sharing nor deal with the minimum requirements for timing analysis. In fact, almost nothing exists in the standard UML (the same could be said for SDL) for modeling or analyzing nonfunctional aspects, and neither scheduling nor placement of software components on hardware resources can be specified or analyzed.

The MARTE profile for the Modeling and Analysis of Real-Time Embedded Systems for UML [6] enhances the standard language by defining timed models of systems, including time assumptions on the environment and platform-dependent aspects like resource availability and scheduling. Such model extensions should allow formal or simulation-based validation of the timing behavior of the software.

SDL contains primitives for dealing with global time, but nevertheless, when the specification of time behavior and resource usage is an issue, it makes capturing most time specifications and constraints practically impossible.

Another problem of SDL is that it does not provide deployment and resource usage information, and it does not support the notion of preemptable or nonpreemptable actions that is necessary for schedulability analysis. Deficiencies in the specification of time also clearly affect simulation of SDL models with time constraints. Any rigorous attempt to construct the simulation graph of an SDL system (the starting point for simulation and verification) must account for all possible combinations of execution times, timer expirations, and resource consumptions. Since no control over time progress is possible, many undesirable executions might be obtained during this exhaustive simulation.

9.4.6 MARTE UML PROFILE

The OMG Real-Time Embedded systems MARTE profile aims at substituting a number of proposals for time-related extensions that appeared in recent years (such as the OMG SPT profile [90]). In order to better support the mapping of active objects into concurrent threads, many research and commercial systems introduced additional nonstandard diagrams.

An UML profile is a collection of language extensions or semantics restrictions of generic UML concepts. These extensions are called "stereotypes" and indicated with their names in between guillemets, as in «TimedEvent». The profile concept is itself a stereotype of the standard UML package.

The MARTE profile defines a comprehensive conceptual framework that uses stereotypes built on the UML metamodel providing a much broader scope than any other real-time extension and that applies to all diagrams. MARTE consists mostly of a notation framework or vocabulary, with the purpose of providing the necessary concepts for schedulability and performance analysis of (timed) behavioral diagrams or scenarios. However, MARTE inherits from UML the deficiencies related to its incomplete semantics and, at least as of today (2015), it lacks a sufficiently established practice. The current version of the profile is based on extensions (stereotyped model elements, tagged values, and constraints) belonging to four main framework packages, further divided into subpackages (as in Figure 9.37).

Of the four frameworks, the *foundation* package contains the fundamental definitions for modeling *time*, clocks, and timed events. The GRM package contains the Generic Resource Model

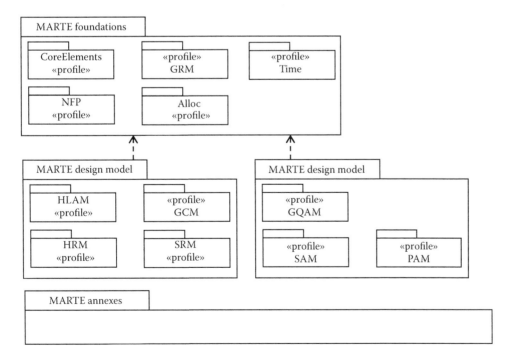

FIGURE 9.37 The framework packages and the subpackages in the OMG MARTE profile.

specification and usage patterns. The NFP package contains the stereotypes for nonfunctional properties. The *design* model contains the extensions for modeling *concurrency* and resources in the generic concurrency GCM, software resource SRM, and hardware resource HRM packages. The *analysis model* package contains specialized concepts for modeling *schedulability* (SAM) and *performance* (PAM) analysis.

In MARTE, the *time model* provides for both continuous and discrete-time models, as well as global and local clocks, including drift and offset specifications. The profile allows referencing to time instances (associated with events), of *time* type, and to the *duration* time interval between any two instances of time in attributes or constraints inside any UML diagram.

The *time* package (some of its stereotypes are shown in Figure 9.38) contains not only definitions for a formal model of time but also stereotyped definitions for the two basic mechanisms of timer and clock. Timers can be periodic; they can be set or reset, paused or restarted, and, when expired, they send timeout signals. Clocks are specialized periodic timers capable of generating tick events.

Time values are typically associated with events, defined in UML as a "specification of a type of observable occurrence" (change of state). A pairing of an event with the associated time instance (time tag) is defined in the MARTE profile as a TimedEvent.

The GRM *resource model* package defines the main resource types as well as generic resource managers and schedulers (its main packages with their relationships in Figure 9.39 and a detail of some stereotypes in Figure 9.40).

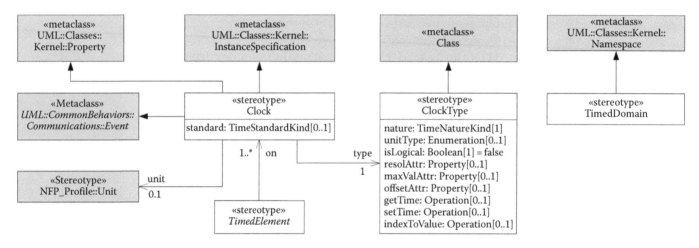

FIGURE 9.38   Stereotypes for clocks in the time package of MARTE.

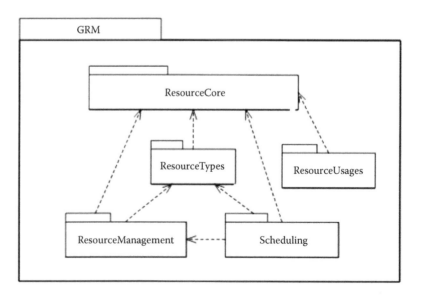

FIGURE 9.39   The packages for resource modeling in the GRM package.

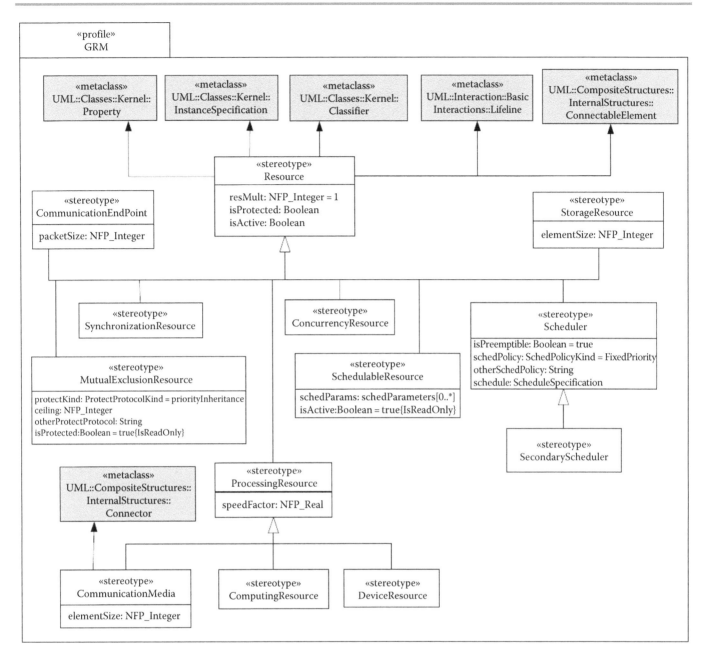

FIGURE 9.40   Stereotypes for resources in the GRM package.

In the MARTE profile, the mapping between the logical entities and the physical architecture supporting their execution is a form of *realization layering* (synonymous of *deployment*). The semantics of the mapping provides a further distinction between the "deploys" mapping, indicating that instances of the supplier are located on the client and the "requires" mapping, which is a specialization indicating that the client provides a minimum deployment environment as required by the supplier.

Although a visual representation of both is possible, it is not clear in which of the existing diagrams it should appear. Hence, the MARTE recommendation for expressing the mappings consists of a table of relationships expressing the deploy and require associations among logical and engineering components.

The GRM profile package is used for the definition of the stereotypes for software and hardware resources. The software resource modeling is much more detailed and comprehensive than the hardware modeling package, which only contains stereotypes for the basic concepts (Figure 9.41).

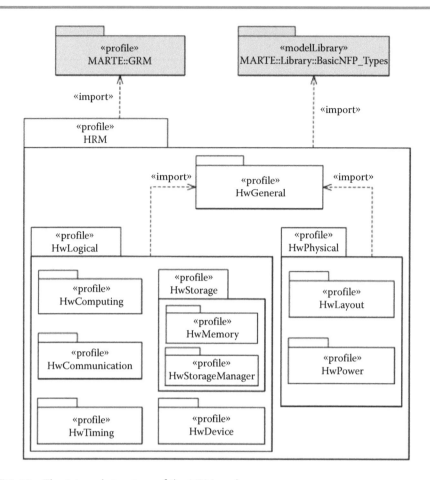

FIGURE 9.41 The internal structure of the HRM package.

The *schedulability analysis* model is based on stereotyped scenarios. Each scheduling situation is in practice a sequence, collaboration, or activity diagram, where one or more trigger events result in actions to be scheduled within the deadline associated with the trigger.

RMA is the method of choice for analyzing simple models with a restricted semantics, conforming to the so-called "task-centric" design paradigm. This requires updating the current definition of UML actions, in order to allow for preemption (which is a necessary prerequisite of RMA, see also Section 9.1.2.3).

In this task-centric approach, the behavior of each active object or task consists of a combination of reading input signals, performing computation, and producing output signals. Each active object can request the execution of actions of other passive objects in a synchronous or asynchronous fashion (Figure 9.42).

Figure 9.42 shows an example with two activities that are logically concurrent, activated periodically, and handle a single event. The MARTE stereotypes provide for the specification of the execution times and deadlines. As long as the active objects cooperate only by means of pure asynchronous messages, simple schedulability analysis formulas, such as (9.1) or (9.2), can be used. These messages can be implemented by means of memory mailboxes, which are a kind of protected (shared resource) object.

Unfortunately, the task-centric model, even if simple and effective in some contexts, only allows analysis of simple models where active objects do not interact with each other. See also other design methodologies [91–93], where the analysis of UML models is made possible by restricting the semantics of the analyzable models. In general UML models, each action is part of multiple end-to-end computations (with the associated timing constraints) and the system consists of a complex network of cooperating active objects, implementing state machines and exchanging asynchronous messages.

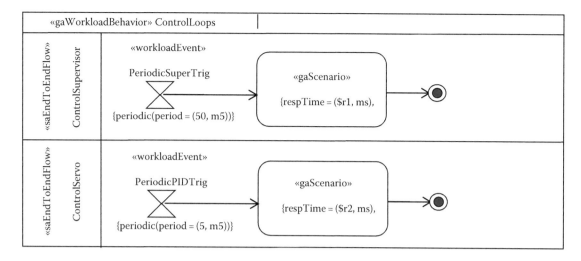

FIGURE 9.42 A sample activity diagram representing a situation suitable for rate monotonic analysis (from the MARTE profile).

In their work [94–96], Saksena, Karvelas, and Wang present an integrated methodology that allows dealing with more general OO models where multiple events can be provided as inputs to a single thread. According to their model, each thread has an incoming queue of events, possibly representing real-time transactions and the associated timing constraints (Figure 9.43).

Consequently, scheduling priority (usually a measure of time urgency or criticality) is attached to events rather than threads. This design and analysis paradigm is called "event-centric design." Each event has an associated priority. For example, a deadline monotonic priority assignment can be used, where the priority associated to an event is inversely proportional to the time by which the event processing must be completed (its deadline). Event queues are ordered by priority (i.e., threads process events based on their priority) and threads inherit the priority of the events they are currently processing. This model entails a two-level scheduling: the events enqueued as input to a thread need to be scheduled to find their processing order. At system level, the threads are scheduled by the underlying RTOS (a preemptive priority-based scheduler is assumed).

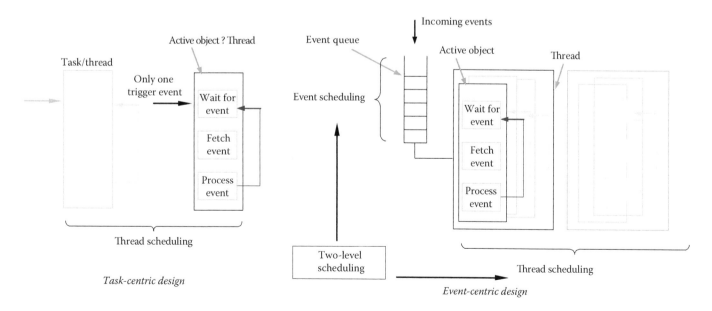

FIGURE 9.43 The dual scheduling problem in the event-centric analysis.

Schedulability analysis of the general case is derived from the analysis methodology for generic deadlines (possibly greater than the activation rates of the triggering events) by computing response times of actions relative to the arrival of the external event that triggers the end-to-end reaction (transaction) T containing the action. The analysis is based on the standard concepts of critical instant and busy period for task instances with generic deadlines (adapted to the transaction model).

In Reference 96, many implementation options for the event-centric model are discussed, namely, single-threaded implementation and multithread implementation with fixed-priority or dynamic-priority multithreaded implementation. A formula or procedure for schedulability analysis is also provided or at least discussed for each of these models. The interested reader should refer to Reference 96 for details.

From a time-analysis point of view, it makes sense to define the set of nested actions that are invoked synchronously in response to an event. This set is called the "synchronous set."

In a *single-threaded implementation*, the only application thread processes pending events in priority order. Since there is only one thread, there is only one level of scheduling. Actions inherit their priority from the priority of the triggering event. Also, a synchronously triggered action inherits its priority from its caller. In single-threaded implementations, any synchronous set that starts executing is completed without interruption. Hence, the worst-case blocking time of an action starting at $t = 0$ is bounded by the longest synchronous set of any lower-priority action that started prior to $t = 0$ and interference can be computed as the sum of the interference terms from other transactions and the interference from actions in the same transaction.

The single-thread model can be analyzed for schedulability (details in Reference 96) and it is also conformant to the nonpreemptable semantics of UML actions. Most of the existing CASE tools support a single-threaded implementation and some of them (such as Rational Rose RT) support the priority-based enqueuing of activation events (messages).

In *multithreaded implementations*, each event represents a request for an end-to-end sequence of actions, executed in response to its arrival. Each action is executed in the context of a thread. Conceptually, we can reverse the usual meaning of the thread concept, considering events (and the end-to-end chain of actions) as the main scheduling item and the threads required for the execution of actions as special mutual exclusion (mutex) resources, because of the impossibility of having a thread preempt itself. This insight allows using threads and threading priorities in a way that facilitates response time analysis.

If threads behave as mutexes, then it makes sense to associate with each thread a ceiling priority as the priority at which the highest-priority event is served. As prescribed by PI or PCP, *threads inherit the priority of the highest-priority event* in their waiting queue and this allows bounding priority inversion. In this case, the worst-case blocking time is restricted to the processing of a lower-priority event. Furthermore, before processing an event, a thread locks the active object within which the event is to be processed (this is necessary when multiple threads may handle events that are forwarded to the same object), and a ceiling priority and a PI rule must be defined for active objects as well as for threads.

For the multithreaded implementation, a schedulability analysis formula is currently available for the aforementioned case in which threads inherit the priority of the events. The schedulability analysis formula generally results in a reduced priority inversion with respect to the single-threaded implementation, but its usefulness is hindered by the fact that (to the author's knowledge) there is no CASE tool supporting generation of code with the possibility of assigning priorities to events and runtime support for executing threads inheriting the priority of the event they are currently handling.

Assigning *static priorities to threads* is tempting (static priority scheduling is supported by most RTOSs) and it is a common choice for multithreaded implementations (Rational Rose RT and other tools use this method). Unfortunately, when threads are required to handle multiple events, it is not clear what priority should be assigned to them and the computation of the worst-case response time is in general very hard, because of the two-level scheduling and the difficulty of constructing an appropriate critical instant. The analysis needs to account for multiple priority inversions that arise from all the lower-priority events handled by a higher-priority thread. Still, however, it may be possible to use static thread priorities in those cases when it is easy to estimate the amount of priority inversion.

The discussion of the real-time-analysis problem bypasses some of the truly fundamental problems in designing a schedulable software implementation of a functional (UML) model. The three main degrees of freedom in a real-time UML design subject to schedulability analysis are

- Assigning priorities to events
- Defining the optimal number of tasks
- Especially defining a mapping of methods (or entire objects) to tasks for their execution

Timing analysis (as discussed here) is only the last stage, after the mapping decisions have been taken.

What is truly required is a set of design rules, or even better an automatic synthesis procedure that helps in the definition of the set of threads and especially in the mapping of the logical entities to threads and to the physical platform.

The problem of synthesizing a schedulable implementation is a complex combinatorial optimization problem. In Reference 96, the authors propose an automatic procedure for synthesizing the three main degrees of freedom in a real-time UML design. The synthesis procedure uses a heuristic strategy based on a decomposition approach, where priorities are assigned to events/actions in a first stage and mapping is performed in a separate stage. This problem, probably the most important in the definition of a software implementation of real-time UML models, is still an open problem.

9.4.7 TOOLS AND COMMERCIAL IMPLEMENTATIONS

There is a large number of specialized CASE tools for UML modeling of embedded systems. Among those, Rhapsody from IBM [97], Real-Time Studio from ATEGO [98], and Enterprise Architect from Sparxx [99] are probably the most common commercial solutions. However, in recent years, several open-source implementations started to appear. Among those, Topcased [100] and Papyrus [101] are probably the most popular. Both are based on the Eclipse Modeling Framework (EMF [102]) and leverage its capabilities, including the options for model transformations. In contrast with general-purpose UML tools, all these try to answer the designer's need for automated support for model simulation, verification, and testing. In order to do so, they all provide an interpretation of the dynamic semantics of UML. Common features include interactive simulation of models and automatic generation of code. Nondeterminism is forbidden or eliminated by means of semantics restrictions. Some timing features and design of architecture-level models are usually provided, although in a nonstandard form if compared to the specifications of the MARTE profile. Furthermore, support for OCL and user-defined profiles is often quite weak.

Formal validation of untimed and timed models is typically not provided, since commercial vendors focus on providing support for simulation, automatic code generation, and (partly) automated testing. Third-party tools interoperating by means of the standard XMI format provide schedulability analysis. Research work on restricted UML semantics models demonstrates that formal verification techniques can be applied to UML behavioral models. This is usually done by the transformation of UML models into a formal MOC and subsequent formal verification by existing tools (for both untimed and timed systems).

9.5 METAMODELING AND EMF

A very interesting trend in recent years has been the availability of tools that support the creation of metamodels and the generation of a toolset infrastructure based on the defined metamodel, including editors, serialization/deserialization, diff/merge, constraint checkers, and transformations from model to model and from model to text (including programming code).

EMF [102] is a very popular open-source solution for the creation of domain-specific modeling languages based on metamodels. The Eclipse core facility (Ecore) allows one to create a metamodel specification compliant with the metaobject facility (MOF) [103] OMG standard metamodeling language.

EMF provides tools and runtime support to produce a set of Java classes for the model, along with a set of adapter classes that enable viewing and command-based editing of the model, serialization, and a basic editor. The EMF project (now Eclipse modeling tool [102]) includes a number of additional open-source tools and plug-ins to create and manage models in a production environment.

Among the metamodel and model *graphical editors, Sirius* enables the specification of a modeling workbench with graphical, table, or tree editors that include support for the definition of validation rules and actions. *EcoreTools* is another plug-in providing a diagram editor for Ecore and support for documenting models, the definition of dependencies, and the specification and verification of constraints on models and metamodels.

For managing model versions, *EMF Compare* provides generic support to compare and merge models constructed according to any type of EMF-supported (MOF) metamodel. *EMF Diff/Merge* is another diff/merge tool for models. The plug-in can be used as a utility for tools that need to merge models based on consistency rules. Typical usage includes model refactoring, iterative model transformations, bridges between models or modeling tools, collaborative modeling environments, or versioning systems.

EMF also includes a validation component that provides an API for defining constraints for any EMF metamodel and support for customizable model traversal algorithm, as well as parsing of constraints specified in Java or OCL.

The metamodeling capability of EMF has been leveraged by several tools that provide open-source modeling solutions for standard languages, such as real-time OO modeling (ROOM), UML, and SysML. Among those, *eTrice* provides an implementation of the ROOM [104] modeling language together with editors, code generators for Java, C++ and C code, and exemplary target middleware. *Topcased* is an open-source SysML modeler that includes an editor, a constraint checker, and a plug-in for the automatic generation of documentation. Finally, *Papyrus* is today probably the best known open-source UML/SysML modeling tool.

9.6 TRANSFORMATIONS

The Eclipse EMF also supports a number of model-to-model transformation (MMT) languages that allow the parsing of a model constructed according to some language and, based on the detection of patterns or features in it, constructing another model, possibly according to a different metamodel, or even updating the source model.

Among the model transformation languages supported by EMF, Query/View/Transformation (QVT) is a standard set of languages defined by the OMG [105]. The QVT standard defines three model transformation languages that operate on models that conform to MOF 2.0 metamodels. The QVT standard integrates and extends OCL 2.0. There are three QVT languages:

1. QVT-Operational: An imperative language designed for writing unidirectional transformations. This language is supported by the Eclipse EMF, which offers an implementation in its MMT project as a QVT-Operational Mappings Language.
2. QVT-Relations (QVTr) is a declarative language designed to permit both unidirectional and bidirectional model transformations. A transformation embodies a consistency relation on the sets of models. Consistency can be checked by executing the transformation in check-only mode or in enforce mode. In the last mode, the engine will modify one of the models so that the set of relations between models are consistent.
3. QVT-Core (QVTc) is a declarative language designed to act as the target of translation from QVTr. Eclipse also provides an implementation [105] for the QVT Declarative Languages, including QVTc and QVTr.

In addition to QVT, the Eclipse EMF also supports the ATL transformation language [94].

A different set of transformation languages can be used for model-to-text transformations. The Eclipse model-to-text project includes a number of such plug-ins [106]. Special uses of model-to-text transformations include the automatic generation of code from models according to user-selected rules and the automatic generation of documentation, including pdf or word formats.

Among the model-to-text projects supported by Eclipse are

■ *Epsilon*, a family of languages and tools for code generation, MMT, model validation, comparison, migration, and refactoring
■ *Acceleo*, an implementation of the OMG MOF Model to Text Language standard

9.7 RESEARCH ON MODELS FOR EMBEDDED SOFTWARE

The models and tools described in the previous sections are representatives of a larger number of methods, models, languages, and tools (at different levels of maturity) that are currently being developed to face the challenges posed by embedded software design.

This section provides an insight on other (often quite recent) proposals for solving advanced problems related to the modeling, simulation, and/or verification of functional and nonfunctional properties. They include support for compositionality and possibly for integration of heterogeneous models, where heterogeneity applies to both the execution model and the semantics of the component communications or, in general, interactions. Heterogeneity is required because there is no clear winner among the different models and tools—different parts of complex embedded systems may be better suited to different modeling and analysis paradigms (such as dataflow models for data-handling blocks and FSMs for control blocks). The first objective, in this case, is to reconcile by unification the synchronous and asynchronous execution paradigms.

The need to handle timing and resource constraints (hence preemption) together with the modeling of system functionality is another major requirement for modern methodologies [120].

Performance analysis by simulation of timed (control) systems with scheduling and resource constraints is, for example, the main goal of the TrueTime Toolset from the Technical University in Lund [107].

The TrueTime Toolset is one example of a modeling paradigm where scheduling and resource handling policies can be handled as separate modules to be plugged in together with the blocks expressing the functional behavior. TrueTime consists of a set of Simulink blocks that simulate real-time kernel policies as found in commercial RTOSs. The system model is obtained by connecting kernel blocks, network blocks, and ordinary Simulink blocks representing the functional behavior of the control application. The toolbox is capable of simulating the system behavior of the real-time system with interrupts, preemption, and scheduling of resources. The T-Res framework [108] is a successor of TrueTime, providing essentially similar analysis capability, but a better integration with models that are natively developed as a network of Simulink blocks.

The TimesTool [109] is another research framework (built on top of Uppaal and available at http://www.timestool.org) attempting an integration of the TA formalism with methods and algorithms for task and resource scheduling. In TimesTool, a timed automaton can be used to define an arbitrarily complex activation model for a set of deadline-constrained tasks, to be scheduled by fixed or dynamic (earliest deadline first [32]) priorities. Model checking techniques are used to verify the schedulability of the task set, with the additional advantage (if compared to standard worst-case analysis) of providing the possible runs that fail the schedulability test.

Other design methodologies bring further distinction among at least three different design dimensions representing

1. The *functional behavior* of the system and the timing constraints imposed by the environment and/or resulting from design choices
2. The *communication network* and the *semantics of communication* defining interactions upon each link
3. The *platform* onto which the system is mapped and the timing properties of the system blocks resulting from the binding

The Ptolemy and the Metropolis environments are probably the best known examples of design methodologies founded on the previous principles.

Ptolemy (http://www.ptolemy.eecs.berkeley.edu/) is a simulation and rapid prototyping framework for heterogeneous systems developed at the Center for Hybrid and Embedded Software Systems in the Department of Electrical Engineering and Computer Sciences of the University of California at Berkeley [4]. Ptolemy II targets complex heterogeneous systems encompassing different domains and functional needs, such as signal processing, feedback control, sequential decision making, and possibly even user interfaces.

Prior to UML 2.0, Ptolemy II had already introduced the concept of actor-oriented design, by complementing traditional object orientation with concurrency and abstractions of communication between components.

In Ptolemy II, the model of the application is built as a hierarchy of interconnected actors (most of which are domain polymorphic). Actors are units of encapsulation (components) that can be composed hierarchically producing the design tree of the system. Actors have an interface abstracting their internals and providing bidirectional access to functions. The interface includes ports that represent points of communication and parameters that are used to configure the actor operations.

Communication channels pass data from one port to another according to some communication semantics. The abstract syntax of actor-oriented design can be represented concretely in several ways, one example being the graphical representation provided by the Ptolemy II Vergil GUI (Figure 9.44).

Ptolemy is not built on a single, uniform, MOC, but it rather provides a finite library of *directors* implementing different MOCs. A director, when placed inside a Ptolemy II composite actor, defines its abstract execution semantics (MOC) and the execution semantics of the interactions among its component actors (its *domain*).

A component that has a well-defined behavior under different MOCs is called a domain polymorphic component. This means that its behavior is polymorphic with respect to the domain or MOC that is specified at each node of the design tree.

The available MOCs in Ptolemy II include

- CSP with synchronous rendezvous
- CT, where behavior is specified by a set of algebraic or differential equations
- DEs, where an event consists of a value and a time stamp
- FSMs

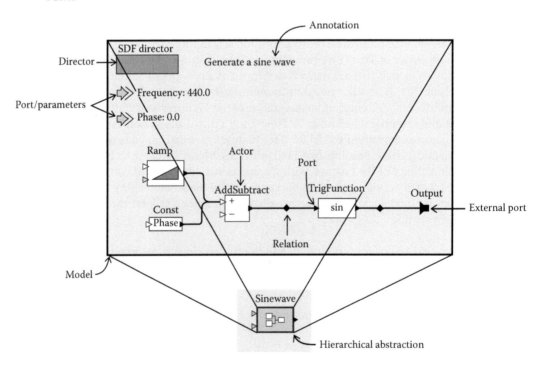

FIGURE 9.44 A Ptolemy actor and its black-box (higher-level) representation.

- (Kahn) Process networks (PN)
- Synchronous dataflow
- Synchronous/reactive (SR)
- Time-triggered synchronous execution (the Giotto framework [110])

The CT and SR domains have fixed-point semantics, meaning that in each iteration, the domain may repeatedly fire the components (execute the available active transitions inside them) until a fixed point is found.

Of course, key to the implementation of polymorphism is proving that an aggregation of components under the control of a domain defines in turn a polymorphic component. This is possible for a large number of combinations of MOCs [111].

In Ptolemy II, the implementation language is Java, and an experimental module for automatic Java code generation from a design is now available at http://ptolemy.eecs.berkeley.edu/. The source code of the Ptolemy framework is available for free from the same website. Currently, the software has hundreds of active users at various sites worldwide in industry and academia.

The Metropolis environment embodies the platform-based design methodology [3] for design representation, analysis, and synthesis under development at the University of California at Berkeley.

In the Metropolis and its follow-up Metro II toolsets [112,113], system design is seen as the result of a progressive refinement of high-level specifications into lower-level models, possibly heterogeneous in nature. The environment deals with all the phases of design from conception to final implementation.

Metropolis and Metro II are designed as a very general infrastructure based on a metamodel with precise semantics that is general enough to support existing computation models and to accommodate new ones. The metamodel supports not only functionality capture and analysis through simulation and formal verification but also architecture description and the mapping of functionality to architectural elements (Figure 9.45).

The Metropolis metamodel (MMM) language provides basic building blocks that are used to describe communication and computation semantics in a uniform way. These components represent *computation*, *communication*, and *coordination* or *synchronization*:

- *Processes* for describing computation
- *Media* for describing communication between processes
- *Quantity managers* for enforcing a scheduling policy for processes
- *Netlists* for describing interconnections of objects and for instantiating and interconnecting quantity managers and media

A process is an active object (it possesses its own thread of control) that executes a sequence of actions (instructions, subinstructions, function calls, and awaits [114]) and generates a

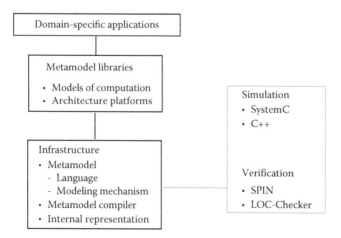

FIGURE 9.45   The Metropolis and Metro II frameworks.

sequence of events. Each process in a system evolves by executing one action after the other. At each step (which is formally described in [114] in terms of a global execution index), each process in the system executes one action and generates the corresponding event. This synchronous execution semantics is relaxed to asynchrony by means of a special event called NOP that can be freely interleaved in the execution of a process (providing nondeterminism).

Each *process* defines its *ports* as the possible interaction points. *Ports* are typed objects associated with an interface, which declares the services that can be called by the process or called from external processes (bidirectional communication) in a way much similar to what it is done in UML 2.0.

Processes cannot connect directly to other processes but the interconnection has to go through a *medium*, which has to define (i.e., implement) the services declared by the interface associated with the ports. The separation of the three orthogonal concepts of process, port, and medium provides maximum flexibility and reusability of behaviors, that is, the meaning of the communication can be changed and refined without changing the computation description that resides in the processes (Figure 9.46).

A model can be viewed as a variant of a system-encompassing timed automaton where the transition between states is labeled by event vectors (one event per process). At each step, there is a set of event vectors that could be executed to make a transition from the current state to the next state. Unless a suitably defined quantity manager restricts the set of possible executions by means of scheduling constraints, the choice among all possible transitions is performed nondeterministically.

Quantity managers define the scheduling of actions by assigning tags to events. A tag is an abstract quantity from a partially order set (for instance, time). Multiple requests for action execution can be issued to a quantity manager that has to resolve them and schedule the processes in order to satisfy the ordering relation on the set of tags. The tagged-signal model [115] is the formalism that defines the unified semantics framework of signals and processes that stand at the foundation of the MMM.

By defining different communication primitives and different ways of resolving concurrency, the user can, in effect, specify different MOCs. For instance, in a synchronous MOC, all events in an event vector have the same tag.

The mapping of a functional model into a platform is performed by enforcing a synchronization of function and architecture. Each action on the function side is correlated with an action on the architecture side using synchronization constraints.

The precise semantics of Metropolis allows for system simulation but also for synthesis and formal analysis. The metamodel includes constraints that represent in abstract form requirements not yet implemented or assumed to be satisfied by the rest of the system and its environment.

Metropolis uses a logic language to capture nonfunctional constraints (e.g., time or energy constraints). Constraints are declarative formulas of two kinds: LTL and logic of constraints.

Although two verification tools are currently provided to check that constraints are satisfied at each level of abstraction, the choice of the analysis and synthesis methods or tools depends on the application domain and the design phase. Metropolis clearly cannot possibly provide algorithms and tools for all possible design configurations and phases, but it provides mechanisms to compactly store and communicate all design information, including a parser that reads metamodel

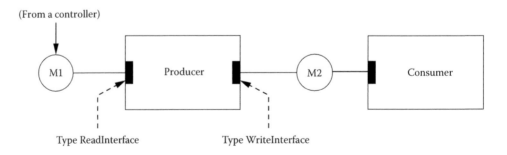

FIGURE 9.46   Processes, ports, and media in Metropolis and Metro II.

designs and a standard API that lets developers browse and modify design information so that they can plug in the required algorithms for a given application domain. This mechanism has been exploited to integrate the Spin software verification tool [14].

Separation of behavior, communication, or interaction and execution models are among the founding principles of the component-based design and verification procedure outlined in References 116 and 117. The methodology is based on the assumption that the system model is a transition system with dynamic priorities restricting nondeterminism in the execution of system actions. Priorities define a strict partial order among actions ($a_1 \prec a_2$). A priority rule is a set of pairs $\{(C^j, \prec^j)\}_{j \in J}$ such that $\prec^j$ is a priority order and C^j is a state constraint specifying when the rule applies.

In Reference 117, a component is defined as the superposition of three models defining its behavior, its interactions, and the execution model. The transition-based formalism is used to define the behavior. Interaction specification consists of a set of connectors, defining maximal sets of interacting actions (i.e., sets of actions that must occur simultaneously in the context of an interaction). The set of actions that are allowed to occur in the system consists of the *complete actions* (triangles in Figure 9.47). Incomplete actions (circles in Figure 9.47) can only occur in the context of an interaction defining a higher-level complete action (as in the definition of $IC[K_1]^+$ in Figure 9.47, which defines the complete action $a_5|a_9$, meaning that incomplete action a_9 can only occur synchronously with a_5). This definition of interaction rules allows for a general model of asynchronous components, which can be synchronized upon a subset of their actions if and when needed. Finally, the execution model consists of a set of dynamic-priority rules.

Composing a set of components means applying a set of rules at each of the three levels. The rules defined in Reference 118 define an associative and commutative composition operator. Further, the composition model allows for *composability* (properties of components are preserved after composition) and *compositionality* (properties of the compound can be inferred from properties of the components) with respect to deadlock freedom (liveness).

In Reference 118, the authors propose a methodology for analysis of timed systems (and a framework for composition rules for component-based real-time models) based on the framework represented in Figure 9.48. According to the proposed design methodology, the system architecture consists of a number of layers, each capturing a different set of properties.

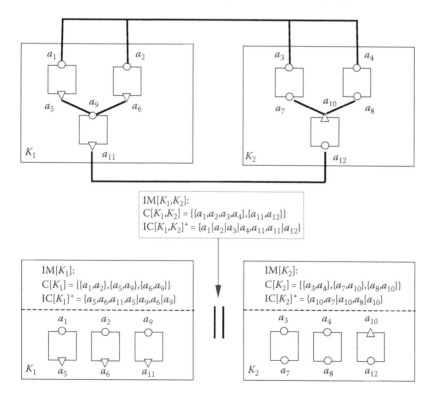

FIGURE 9.47   Interaction model among system components in Reference 105.

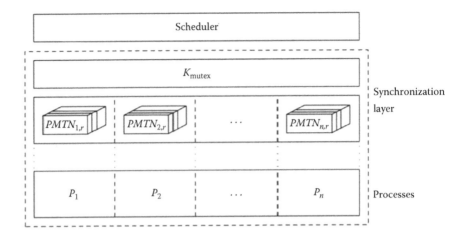

FIGURE 9.48 The architecture of the system model as defined in Reference 118.

At the lowest level, threads (processes) are modeled according to the TA formalism, capturing the functional behavior and the timing properties and constraints.

At the preemption level, system resources and preemptability are accounted for by adding one or more preemption transitions (Figure 9.49a), one for each preemptable resource and mutual exclusion rules are explicitly formulated as a set of constraints acting upon the transitions of the processes.

Finally, the resource management policies and the scheduling policies are represented as additional constraints $K_{pol} = K_{adm} \wedge K_{res}$, where K_{adm} are the admission control constraints and K_{res} are the constraints specifying how resource conflicts are to be resolved.

Once scheduling policies and schedulability requirements are in place, getting a correctly scheduled system amounts to finding a nonempty control invariant K such that $K \Rightarrow K_{sched} \wedge K_{pol}$.

The last example of research framework for embedded software modeling is the Generic Modeling Environment (GME), developed at Vanderbilt University [119], which is a configurable toolkit offering a metamodeling facility for creating domain-specific models and program synthesis environments. The configuration of a domain-specific metamodel can be achieved by defining the syntactic, semantics, and presentation information regarding the domain. This implies defining the main domain concepts, the possible relationships, and the constraints restricting the possible system configurations as well as the visibility rules of object properties.

The vocabulary of the domain-specific languages implemented by different GME configurations is based on a set of generic concepts built into GME itself. These concepts include hierarchies, multiple aspects, sets, references, and constraints. Models, atoms, references, connections, and sets are first-class objects.

Models are compound objects that can have parts and inner structure. Each part in a container is characterized by a role. The modeling instance determines what parts are allowed and in which roles. Models can be organized in a hierarchy, starting with the root object. Aspects provide

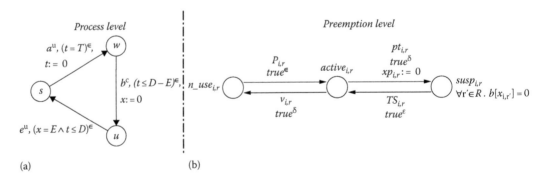

(a)   (b)

FIGURE 9.49 Process and preemption modeling. (From Gossler, G. and Sifakis, J., Composition for component-based modeling.)

visibility control. Relationships can be directed or undirected connections, further characterized by attributes. The model specification can define several kinds of connections, which objects can participate in a connection and further explicit constraints. Connections only appear between two objects in the same model entity. References to model-external objects help establish connections to external objects as well.

9.8 CONCLUSIONS

This chapter discusses the use of models for the design and verification of embedded software systems. It attempts at a classification and a survey of existing formal MOCs, following the classical divide between synchronous and asynchronous models and between models for functionality as opposed to models for software architecture specification. Problems like formal verification of system properties, both timed and untimed, and schedulability analysis are discussed. The chapter also provides an overview of the commercially relevant modeling languages such as Simulink, UML, SysML, Lustre, and SDL.

The discussion of each topic is supplemented with an indication of the available tools that implement the methodologies and analysis algorithms.

The situation of software modeling languages appears to be consolidating. Formal verification is making its way into commercial practice (especially for safety-critical systems), but rather than being extended to more general MOCs, it is successfully applied to systems that adhere to a constrained and controlled MOC (usually synchronous and time triggered).

Similarly, availability of executable specifications is making simulation and code generation a reality. However, these technologies are mostly applied to single-core targets for the generated code, given that the modeling support for distributed platforms is limited.

Several open issues remain: most notably, how to successfully model execution architectures for complex, distributed systems and how to deploy functional models onto execution platform models while preserving the semantics properties of interest.

Even though several academic proposals have been put forward in these years, none of them has yet achieved the level of maturity for a successful use in industrial practice.

REFERENCES

1. The AUTOSAR Consortium, AUTOSAR 4.2 specification, available at http://www.autosar.org/specifications/.
2. Charette, R. N., Why software fails: We waste billions of dollars each year on entirely preventable mistakes, *IEEE Spectrum*, September 2, 2005 (online edition http://spectrum.ieee.org/computing/software/why-software-fails).
3. Sangiovanni-Vincentelli, A., Defining platform-based design, *EE Design of EE Times*, February 2002.
4. Lee, E. A., Overview of the Ptolemy project, Technical memorandum UCB/ERL M03/25, University of California, Berkeley, CA, July 2, 2003.
5. Balarin, F., Hsieh, H., Lavagno, L., Passerone, C., Sangiovanni-Vincentelli, A., and Watanabe, Y., Metropolis: An integrated environment for electronic system design, *IEEE Computer*, 36(2), 42–52. April 2003.
6. The OMG Consortium MARTE Profile, available at http://www.omg.org/spec/MARTE/2015.
7. The OMG Consortium, *Unified Modeling Language™ (UML®) Resource Page.* OMG Adopted Specification, September, 9, 2013, available at http://www.uml.org. 2015.
8. Beck, T., Current trends in the design of automotive electronic systems, in *Proceedings of the Design Automation and Test in Europe Conference*, Munich, Germany, 2001.
9. Edwards, S., Lavagno, L., Lee, E. A., and Sangiovanni-Vincentelli, A., Design of embedded systems: Formal models, validation and synthesis, *Proceedings of the IEEE*, March 1997.
10. Lee, E. A. and Varaiya, P., *Structure and Interpretation of Signals and Systems*, 2nd edn., LeeVaraiya.org, 2011.
11. Alur, R. and Henzinger, T. A., Logics and models of real time: A survey, *Real-Time: Theory in Practice, REX Workshop*, LNCS 600, pp. 74–106, 1991.
12. Pnueli, A., The temporal logic of programs, in *Proceedings of the 18th Annual Symposium on the Foundations of Computer Science*, pp. 46–57. IEEE Computer Society, Washington, DC, November 1977.

13. Emerson, E. A., Temporal and modal logics, in van Leeuwen, J. editor, *Handbook of Theoretical Computer Science*, Vol. B, pp. 995–1072. Elsevier, 1990.

14. Holzmann, G. J., *Design and Validation of Computer Protocols*. Prentice-Hall, 1991.

15. Harel, D., Statecharts: A visual approach to complex systems. *Science of Computer Programming*, 8:231–275, 1987.

16. Merlin P. M. and Farber D. J., Recoverability of communication protocols. *IEEE Transactions of Communications*, 24(9):36–103, September 1976.

17. Sathaye, A. S. and Krogh, B. H., Synthesis of real-time supervisors for controlled time Petri nets. *Proceedings of the 32nd IEEE Conference on Decision and Control*, Vol. 1, San Antonio, TX, pp. 235–238, 1993.

18. Alur, R. and Dill, D. L., A theory of timed automata. *Theoretical Computer Science*, 126:183–235, 1994.

19. Ramchandani, C., Analysis of asynchronous concurrent systems by timed Petri nets. PhD thesis, Department of Electrical Engineering, MIT, Cambridge, MA, 1974.

20. Molloy, M. K., Performance analysis using stochastic Petri nets. *IEEE Transactions on Computers*, 31(9):913–917, 1982.

21. Ajmone Marsan, M., Conte, G., and Balbo, G. A class of generalized stochastic Petri nets for the performance evaluation of multiprocessor systems. *ACM Transactions on Computer Systems*, 2(2):93–122, 1984.

22. Haar, S., Kaiser, L., Simonot-Lion, F., and Toussaint, J., On equivalence between timed state machines and time Petri nets. Research report, *Rapport de recherche de l'INRIA*, Lorraine, France, November 2000.

23. Yovine, S. Kronos: A verification tool for real-time systems. *International Journal of Software Tools for Technology Transfer*, 1(1):123–133, 1997.

24. Larsen, K. G., Pettersson, P., and Yi, W., Uppaal in a nutshell. *International Journal of Software Tools for Technology Transfer*, 1(1):134–152, 1997.

25. Yi, W., Pettersson, P., and Daniels, M., Automatic verification of real-time communicating systems by constraint solving, *Proceedings of the 7th International Conference on Formal Description Techniques*, Berne, Switzerland, October, 4–7, 1994.

26. Henzinger, T. A., The theory of hybrid automata, *Proceedings of the 11th Annual Symposium on Logic in Computer Science (LICS)*, IEEE Computer Society Press, 1996, pp. 278–292.

27. Henzinger, T. A., Ho, P.-H., and Wong-Toi, H., HyTech: A model checker for hybrid systems, *Software Tools for Technology Transfer*, 1:110–122, 1997.

28. Vicario, E., Static analysis and dynamic steering of time-dependent systems using time Petri nets, *IEEE Transactions on Software Engineering*, 27(8):728–748, July 2001.

29. Vicario, E., et al. The ORIS tool web page, available at http://stlab.dinfo.unifi.it/oris1.0/ (2015).

30. Lime, D. and Roux, O. H., A translation based method for the timed analysis of scheduling extended time Petri nets, *The 25th IEEE International Real-Time Systems Symposium*, Lisbon, Portugal, December 5–8, 2004.

31. Clarke, E. M. and Wing, J. M., Formal methods: State of the art and future directions, Technical report, CMU-CS-96-178, Carnegie Mellon University (CMU), Pittsburg, PA, September 1996.

32. Liu, C. and Layland, J., Scheduling algorithm for multiprogramming in a hard real-time environment, *Journal of the ACM*, 20(1):46–61, January 1973.

33. Klein, M. H. et al., *A Practitioner's Handbook for Real-Time Analysis: Guide to Rate Monotonic Analysis for Real-Time Systems*, Kluwer Academic Publishers, Hingham, MA, 1993.

34. Rajkumar, R., Synchronization in multiple processor systems, *Synchronization in Real-Time Systems: A Priority Inheritance Approach*, Kluwer Academic Publishers, Norwell, MA, 1991.

35. Benveniste, A., Caspi, P., Edwards, S. A., Halbwachs, N., Le Guernic, P., and de Simone, R., The synchronous languages 12 years later, *Proceedings of the IEEE*, 91(1):64–83, January 2003.

36. The Mathworks Co., Embedded Coder web page, available at http://www.mathworks.com/products/embedded-coder/ (2015).

37. IBM Corp., Rational Rhapsody Architect for Software, available at http://www-03.ibm.com/software/products/it/ratirhaparchforsoft (2015).

38. Meyer, B., An overview of Eiffel, *The Handbook of Programming Languages*, Vol. 1, Object-Oriented Languages, Salus, P. H., editor, Macmillan Technical Publishing, London, UK 1998.

39. The OMG Consortium, Concrete syntax for a UML action language: Action language for foundational UML (ALF), available at www.omg.org/spec/ALF 2015.

40. The OMG Consortium, Semantics of a foundational subset for executable UML models (FUML), available at www.omg.org/spec/FUML.

41. Caspi, P., Pilaud, D., Halbwachs, N., and Plaice, J. A., LUSTRE: A declarative language for programming synchronous systems, *ACM Symposium on Principles of Programming Languages (POPL)*, Munich, Germany, pp. 178–188, 1987.

42. Halbwachs, N., Caspi, P., Raymond, P., and Pilaud, D., The synchronous data flow programming language LUSTRE, *Proceedings of the IEEE*, 79:1305–1320, September 1991.

43. Boussinot, F. and de Simone, R., The Esterel language, *Proceedings of the IEEE*, 79:1293–1304, September 1991.
44. Berry, G., The constructive semantics of pure Esterel, *Algebraic Methodology and Software Technology*, Conference, Munich, Germany, pp. 225–232, 1996.
45. Westhead, M. and Nadjm-Tehrani, S., Verification of embedded systems using synchronous observers, *Formal Techniques in Real-Time and Fault-Tolerant Systems*, Lecture Notes in Computer Science, Vol. 1135, Springer-Verlag, Heidelberg, Germany, 1996.
46. The Mathworks Simulink and StateFlow, available at http://www.mathworks.com. (2015).
47. Hamon, G. and Rushby, J., An operational semantics for stateflow. *Proceedings of Fundamental Approaches to Software Engineering* (*FASE*), Barcelona, Spain, March 2004.
48. Tiwari, A., Formal semantics and analysis methods for Simulink StateFlow models, Technical report, SRI International, Pittsburgh, PA, 2002.
49. Luttgen, G., von der Beeck, M., and Cleaveland, R., A compositional approach to statecharts semantics, Rosenblum, D. editor, *Proceedings of the Eighth ACM SIGSOFT International Symposium on Foundations of Software Engineering*, ACM Press, pp. 120–129, 2000.
50. Mikk, E., Lakhnech, Y., and Siegel, M., Hierarchical automata as a model for statecharts, *Asian Computing Science Conference (ASIAN97)*, number 1345 in Lecture Notes in Computer Science, Springer, December 1997.
51. Caspi, P., Curic, A., Maignan, A., Sofronis, C., and Tripakis, S., Translating discrete-time Simulink to Lustre, in Alur, R. and Lee, I., editors, *EMSOFT03*, Lecture Notes in Computer Science, Springer Verlag, 2003.
52. Caspi, P., Curic, A., Maignan, A., Sofronis, C., Tripakis, S., and Niebert, P., From Simulink to SCADE/Lustre to TTA: A layered approach for distributed embedded applications, in *ACMSIGPLAN Languages, Compilers, and Tools for Embedded Systems (LCTES03)*, San Diego, CA, 2003.
53. Scaife, N., Sofronis, C., Caspi, P., Tripakis, S., and Maraninchi, F., Defining and translating a "safe" subset of Simulink/Stateflow into Lustre, in *Proceedings of 2004 Conference on Embedded Software*, *EMSOFT'04*, Pisa, Italy, September 2004.
54. Zeng, H. and Di Natale, M., Mechanisms for guaranteeing data consistency and time determinism in AUTOSAR software on multi-core platforms, in *Proceedings of the Sixth IEEE Symposium on Industrial Embedded Systems* (SIES), Vasteras, Sweden, June 2011.
55. Lublinerman, R. and Tripakis, S., Modularity vs. reusability: Code generation from synchronous block diagrams, *Proceedings of the Date Conference*, Munich, Germany, 2008.
56. Scaife, N. and Caspi, P., Integrating model-based design and preemptive scheduling in mixed time- and event-triggered systems, in *16th Euromicro Conference on Real-Time Systems (ECRTS'04)*, Catania, Italy, pp. 119–126, June–July 2004.
57. Caspi, P., Scaife, N., Sofronis, C., and Tripakis, S., Semantics preserving multitask implementation of synchronous programs, *ACM Transactions on Embedded Computing Systems*, 7(2):1–40, January 2008.
58. Chen, J. and Burns, A., Loop-free asynchronous data sharing in multiprocessor real-time systems based on timing properties, in *Proceedings of the Sixth International Conference on Real-Time Computing Systems and Applications (RTCSA'99)*, Hong Kong, China, pp. 236–246, 1999.
59. Zeng, H. and Di Natale, M., Schedulability analysis of periodic tasks implementing synchronous finite state machines, in *Proceedings of the 23rd Euromicro Conference on Real-Time Systems*, Pisa, Italy, 2012.
60. Stigge, M., Ekberg, P., Guan, N., and Yi, W., The digraph real-time task model, in *Proceedings of the 16th IEEE Real-Time and Embedded Technology and Applications Symposium*, Chicago, IL, 2011.
61. Zeng, H. and Di Natale, M., Using max-plus algebra to improve the analysis of non-cyclic task models, in *Proceedings of the 24th Euromicro Conference on Real-Time Systems*, Paris, France, July 2013.
62. Di Natale, M. and Zeng, H., Task implementation of synchronous finite state machines, in *Proceedings of the Conference on Design, Automation, and Test in Europe*, Dresden, Germany, 2012.
63. Zhu, Q., Deng, P., Di Natale, M., and Zeng, H., Robust and extensible task implementations of synchronous finite state machines, in *Proceedings of the DATE Conference*, 2013.
64. Zeng, H. and Di Natale, M., Mechanisms for guaranteeing data consistency and flow preservation in AUTOSAR software on multi-core platforms, in *Proceedings of the SIES Conference*, Vasteras, Sweden, 2011.
65. Tripakis, S., Sofronis, C., Scaife, N., and Caspi, P., Semantics-preserving and memory-efficient implementation of inter-task communication on static-priority or EDF schedulers, in *Proceedings of the Fifth ACM EMSOFT Conference*, Jersey City, NJ, 2005.
66. Sofronis, C., Tripakis, S., and Caspi, P., A memory-optimal buffering protocol for preservation of synchronous semantics under preemptive scheduling, in *Proceedings of the Sixth ACM International Conference on Embedded Software*, Seoul, Korea, 2006.
67. Di Natale, M., Wang, G., and Sangiovanni-Vincentelli, A., Improving the size of communication buffers in synchronous models with time constraints, *IEEE Transactions on Industrial Informatics*, 5(3):229–240, 2009.

68. Wang, G., Di Natale, M., and Sangiovanni-Vincentelli, A. L., Optimal synthesis of communication procedures in real-time synchronous reactive models. *IEEE Transactions on Industrial Informatics*, 6(4):729–743, 2010.

69. Di Natale, M. and Pappalardo, V., Buffer optimization in multitask implementations of Simulink models. *ACM Transactions on Embedded Computing Systems*, 7(3):1–32, 2008.

70. Di Natale, M., Guo, L., Zeng, H., and Sangiovanni-Vincentelli, A., Synthesis of multi-task implementations of Simulink models with minimum delays, *IEEE Transactions on Industrial Informatics*, 6(4):637–651, November 2010.

71. Prover Technology, available at http://www.prover.com/2015.

72. Edwards, S. A., An Esterel compiler for large control-dominated systems, *IEEE Transactions on Computer-Aided Design*, 21, 169–183, February 2002.

73. Shiple, T. R., Berry, G., and Touati, H., Constructive analysis of cyclic circuits, *European Design and Test Conference*, Paris, France, 1996.

74. Benveniste, A., Caspi, P., Le Guernic, P., Marchand, H., Talpin, J.-P., and Tripakis, S., A protocol for loosely time-triggered architectures, in Sifakis, J. and Sangiovanni-Vincentelli, A., editors, *Proceedings of 2002 Conference on Embedded Software, EMSOFT'02*, Grenoble, France, Vol. 2491 of LNCS, pp. 252–265, Springer Verlag.

75. Benveniste, A., Caillaud, B., Carloni, L., Caspi, P., and Sangiovanni-Vincentelli, A., Heterogeneous reactive systems modeling: Capturing causality and the correctness of loosely time-triggered architectures (LTTA), in Buttazzo, G. and Edwards, S., editors, *Proceedings of 2004 Conference on Embedded Software, EMSOFT'04*, Pisa, Italy, September 27–29, 2004.

76. Tripakis, S., Pinello, C., Benveniste, A., Sangiovanni-Vincentelli, A., Caspi, P., and Di Natale, M., Implementing synchronous models on loosely time-triggered architectures, in *IEEE Transactions on Computers*, 57(10):1300–1314, October 2008.

77. Lin, C. W., Di Natale, M., Zeng, H., Phan, L., and Sangiovanni-Vincentelli, A., Timing analysis of process graphs with finite communication buffers, *RTAS Conference* 2013, Philadelphia, PA.

78. Biannic, Y. L., Nassor, E., Ledinot, E., and Dissoubray, S., UML object specification for real-time software. *RTS Show 2000*.

79. SysML, SysML version 1.3 specification, available at http://www.omg.org/spec/SysML/1.3/2015.

80. Fowler, M., *UML Distilled: A Brief Guide to the Standard Object Modeling Language* Addison Wesley, 2015.

81. Gamma, E., Helm, R., Johnson, R., and Vissides, J., *Design Patterns: Elements of Reusable Object-Oriented Software*, Addison Wesley.

82. Douglass, B. P., *Real Time UML: Advances in the UML for Real-Time Systems*, Addison-Wesley, 2004.

83. Ciccozzi, F., Cicchetti, A., and Sjodin, M., Towards translational execution of action language for foundational UML, in *39th EUROMICRO Conference on Software Engineering and Advanced Applications (SEAA)* Santander, Spain, 2013.

84. Ciccozzi, F., Cicchetti, A., and Sjodin, M., On the generation of full-fledged code from UML profiles and ALF for complex systems, in *12th International Conference on Information Technology: New Generations*, Las Vegas, NV, February 2015.

85. Latella, D., Majzik, I., and Massink, M., Automatic verification of a behavioural subset of UML statechart diagrams using the SPIN modelchecker, *Formal Aspects of Computing*, 11(6):637–664, 1999.

86. del Mar Gallardo, M., Merino, P., and Pimentel, E., Debugging UML designs with model checking, *Journal of Object Technology*, 1(2), 101–117, July–August 2002.

87. The W3C Consortium, Mathematical Markup Language (MathML) version 3.0, April 2014, available at http://www.w3.org/TR/MathML/2015.

88. The OMG Consortium, OCL 2.4 Final adopted specification, available at http://www.omg.org/spec/OCL/2.4/.

89. ITU-T, Recommendation Z.100, Specification and Description Language (SDL), Z-100, International Telecommunication Union Standardization Sector, 2000.

90. UML profile for schedulability, performance and time specification, OMG adopted specification, July 1, 2002, available at http://www.omg.org. 2015.

91. Gomaa, H., *Software Design Methods for Concurrent and Real-Time Systems*, Addison-Wesley Publishing Company, 1993.

92. Burns, A. and Wellings, A. J., HRT-HOOD: A design method for hard real-time, *Journal of Real-Time Systems*, 6(1):73–114, 1994.

93. Awad, M., Kuusela, J., and Ziegler, J., *Object-Oriented Technology for Real-Time Systems: A Practical Approach Using OMT and Fusion.* Prentice Hall, 1996.

94. Saksena, M., Freedman, P., and Rodziewicz, P., Guidelines for automated implementation of executable object oriented models for real-time embedded control systems, in *Proceedings, IEEE Real-Time Systems Symposium* 1997, San Francisco, CA, pp. 240–251, December 1997.

95. Saksena, M. and Karvelas, P., Designing for schedulability: Integrating schedulability analysis with object-oriented design, in *Proceedings of the Euromicro Conference on Real-Time Systems*, Stockholm, Sweden, June 2000.

96. Saksena, M., Karvelas, P., and Wang, Y., Automatic synthesis of multi-tasking implementations from real-time object-oriented models, in *Proceedings, IEEE International Symposium on Object-Oriented Real-Time Distributed Computing*, Newport Beach, CA, March 2000.

97. IBM, Rational Rhapsody product family, available at http://www-03.ibm.com/software/products/en/ratirhapfami (2015).

98. ATEGO Modeler (implemented as Real-Time Studio) product, available at http://www.atego.com/products/atego-modeler/(2015).

99. Sparxx Systems, Enterprise architect, available at http://www.sparxsystems.com.au/(2015).

100. Topcased, Open source toolkit for critical systems, available at http://www.topcased.org/ (2015).

101. Papyrus UML, available at http://www.papyrusuml.org (2015).

102. The Eclipse Modeling Project, available at http://eclipse.org/modeling/(2015).

103. OMG, Meta Object Facility (MOF) 2.0 query/view/transformation, v1.1 specification, available at http://www.omg.org/spec/QVT/1.1/(2015).

104. Selic, B., Gullekson, G., and Ward, P. T., *Real-Time Object-Oriented Modeling*, John Wiley & Sons, 1994.

105. The Eclipse, MMT, available at http://wiki.eclipse.org/Model_to_Model_Transformation_-_MMT. (2015).

106. The Eclipse, Model-to-text project, available at https://www.eclipse.org/modeling/m2t/(2015).

107. Henriksson, D., Cervin, A., and Årzén, K.-E., True-Time: Simulation of control loops under shared computer resources, in *Proceedings of the 15th IFAC World Congress on Automatic Control*, Barcelona, Spain, July 2002.

108. Cremona, F., Morelli, M., and Di Natale, M., TRES: A modular representation of schedulers, tasks, and messages to control simulations in Simulink, in *Proceedings of the SAC Conference*, Salamanca, Spain, April 2015.

109. Amnell, T. et al., Times—A tool for modelling and implementation of embedded systems, in *Proceedings of Eighth International Conference, TACAS 2002*, Grenoble, France, April 8–12, 2002.

110. Henzinger, T. A., Giotto: A time-triggered language for embedded programming, in *Proceedings of the 1st International Workshop on Embedded Software (EMSOFT'2001)*, Tahoe City, CA, October 2001, Vol. 2211 of Lecture Notes in Computer Science, pp. 166–184, Springer Verlag, 2001.

111. Lee, E. A. and Xiong, Y., System-level types for component-based design, in *Proceedings of the 1st International Workshop on Embedded Software (EMSOFT'2001)*, Tahoe City, CA, October 2001, Vol. 2211 of Lecture Notes in Computer Science, pp. 237–253, Springer Verlag, 2001.

112. Balarin, F., Lavagno, L., Passerone, C., and Watanabe, Y., Processes, interfaces and platforms. Embedded software modeling in metropolis, in *Proceedings of the EMSOFT Conference 2002*, Grenoble, France, pp. 407–416.

113. Densmore, D., Meyerowitz, T., Davare, A., Zhu, Q., and Yang, G., Metro II execution semantics for mapping, Technical report, University of California, Berkeley, CA, UCB/EECS-2008-16, February 2008.

114. Balarin, F., Lavagno, L., Passerone, C., Sangiovanni-Vincentelli, A., Sgroi, M., and Watanabe, Y., Modeling and design of heterogeneous systems, Vol. 2549 of *Lecture Notes in Computer Science*, pp. 228–273, Springer Verlag, 2002.

115. Lee, E.A. and Sangiovanni-Vincentelli, A., Comparing models of computation, in *Proceedings of the 1996 IEEE/ACM International Conference on Computer-Aided Design*, San Jose, CA, 1996.

116. Gossler, G. and Sifakis, J., Composition for component-based modeling, in *Proceedings of FMCO'02*, Leiden, the Netherlands, LNCS 2852, pp. 443–466, November 2002.

117. Altisen, K., Goessler, G., and Sifakis, J., Scheduler modeling based on the controller synthesis paradigm, *Journal of Real-Time Systems*, special issue on "Control Approaches to Real-Time Computing", 23:55–84, 2002.

118. Gossler, G. and Sifakis, J., Composition for component-based modeling, *Proceedings of FMCO'02*, Leiden, The Netherlands, LNCS 2852, pp. 443–466, November 2002.

119. Ledeczi, A., Maroti, M., Bakay, A., Karsai, G., Garrett, J., Thomason IV, C., Nordstrom, G., Sprinkle, J., and Volgyesi, P., The generic modeling environment, in *Workshop on Intelligent Signal Processing*, Budapest, Hungary, May 17, 2001.

120. Jantsch, A., *Modeling Embedded Systems and SoCs—Concurrency and Time in Models of Computation*, Morgan Kaufman, June 2003

Using Performance Metrics to Select Microprocessor Cores for IC Designs

Steve Leibson

10.1 INTRODUCTION

Ten years after their introduction in 1971, microprocessors became essential for board-level electronic system design. In the same manner, microprocessors are now absolutely essential components for designers using field-programmable gate arrays (FPGAs) and for integrated circuit (IC) design. The reason is simple. Processor cores are the most reusable on-chip components because of their easy software programmability and because of the extensive development tool environment and ecosystem that surround any good processor architecture. Although it is easy to envision using processor cores to implement many tasks on an IC, it is often difficult to select an appropriate processor from the many cores now available because contemporary processor cores are complex, multifunction elements. Consequently, there is no single, simple, meaningful way to evaluate the suitability of a processor core for specific embedded tasks on a board, in an FPGA, or in an IC. Many factors must be considered including processor performance, gate count, power dissipation, availability of suitable interfaces, and the processor's software development support tools and ecosystem. This chapter deals with the objective measurement of processor core performance with respect to the selection of processor cores for use in IC design and on FPGAs.

Designers often measure microprocessor and processor core performance through comprehensive benchmarking. Processor chips already realized in silicon are considerably easier to benchmark than processor cores. Processors used in FPGAs fall somewhere in the middle. All major microcontroller and microprocessor vendors offer their processors in evaluation kits, which include evaluation boards and software development tool suites (which include compilers, assemblers, linkers, loaders, instruction-set simulators [ISSs], and debuggers). Some of these kits are quite inexpensive, having dropped to the $10 level and below in the last few years. Consequently, benchmarking chip-level microcontroller or microprocessor performance consists of porting and compiling selected benchmarks for the target processor, downloading and running those benchmarks on the evaluation board, and recording the results.

Processor cores for use in the design of ICs are incorporeal and are not usually available as stand-alone chips. In fact, if these cores were available as chips, they would be far less useful to IC designers because on-chip processors can have vastly greater I/O capabilities than processors realized as individual chips. Numerous wide and fast buses are the norm for processor cores—to accommodate instruction and data caches, local memories, and private I/O ports—which is not true for microprocessor chips because these additional buses would greatly increase pin count to an impractical level. However, extra I/O pins are not costly for processor cores, so they usually have many buses and ports. A case in point is the Xilinx Zynq All Programmable SoC. This device fuses a dual-core ARM Cortex-A9 MPCore processor complex called the PS (processor system) with an FPGA programmable logic fabric (called the PL). The PS and PL intercommunicate using multiple interfaces and other signals using a combined total of more than 3000 connections. That sort of connectivity is simply impossible using a separately packaged microprocessor.

Benchmarking processor cores has become increasingly important in the IC design flow because any significant twenty-first century IC design incorporates more than one processor core—at least two, often several, and occasionally hundreds. Figure 10.1 shows a Xilinx Zynq-7000 All Programmable SoC. This device combines an entire dual-processor subsystem based on the ARM Cortex-A9 MPCore processor and a large chunk of programmable logic (FPGA). There are several members of the Zynq SoC device family with varying amounts of programmable logic depending on the family member.

The Xilinx Zynq SoC is interesting in the context of this chapter because it contains two ARM Cortex-A9 processor cores running at clock rates of 800 MHz to 1 GHz and because you can instantiate other processor architectures in the FPGA portion of the device. Xilinx design tools include the configurable MicroBlaze processor core with a configuration tool. Configurable

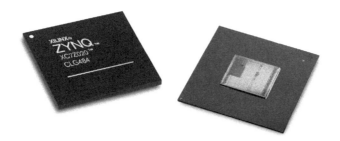

FIGURE 10.1 The Xilinx Zynq-7000 All Programmable SoC fuses a dual-core ARM Cortex-A9 MPCore processor complex called the PS (processor system) with an FPGA programmable logic fabric (called the PL) on one piece of silicon. (Copyright Xilinx Inc. Reproduced by permission.)

processors make interesting benchmarking targets, which is discussed in much more detail later in this chapter. Figure 10.2 shows a block diagram of a Zynq SoC discussed earlier.

As the number of processor cores used on a chip increases and as these processor cores perform more on-chip tasks, measuring processor core performance becomes an increasingly important and challenging task in the overall design of the IC. Benchmarking the hardened

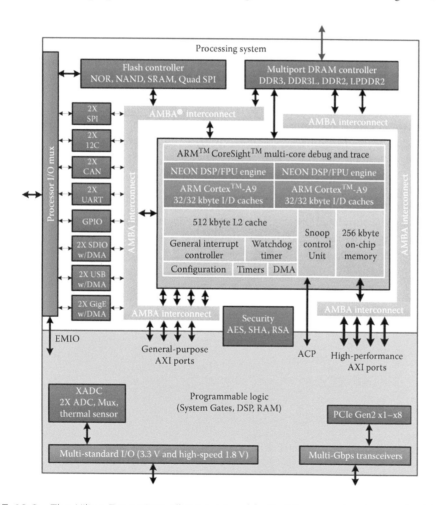

FIGURE 10.2 The Xilinx Zynq-7000 All Programmable SoC incorporates two embedded ARM Cortex-A9 MPCore processor cores in a processing block called the PS (processor system) with an FPGA programmable logic fabric (called the PL). Additional processors can be instantiated in the Zynq SoC's PL. The PS and PL intercommunicate using multiple interfaces and other signals using a combined total of more than 3000 connections. (Copyright Xilinx Inc. Reproduced by permission.)

processor cores in an FPGA is not so difficult if the device initially powers up as though it is a processor chip. That happens to be the case for the Xilinx Zynq SoC. However, measuring the performance of processor Internet Protocol (IP) cores intended for use in ASIC and SoC designs is a bit more complex than benchmarking microprocessor chips: another piece of software, an ISS must stand in for a physical processor chip until the processor is realized in silicon. Each processor core and each member of a microprocessor family require its own specific ISS because the ISS must provide a cycle-accurate model of the processor's architecture to produce accurate benchmark results.

Note that it is also possible to benchmark a processor core using gate-level or net-list simulation, but this approach is three orders of magnitude slower (because gate-level simulation is much slower than instruction-set simulation) and is therefore used infrequently [1]. For processor cores, the selected benchmarks are compiled and run on the ISS to produce the benchmark results. All that remains is to determine where the ISS and the benchmarks are to be obtained, how they are to be used, and how the results will be interpreted. These are the subjects of this chapter.

10.2 ISS AS BENCHMARKING PLATFORM

ISSs serve as benchmarking platforms because processor cores as realized on an IC rarely exist as a chip. ISSs run quickly—often 1000 times faster than gate-level processor simulations running on HDL simulators—because they simulate the processor's software-visible state without employing a gate-level processor model. The earliest ISS was created for the Electronic Delay Storage Automatic Calculator (EDSAC), which was developed by a team led by Maurice V. Wilkes at the University of Cambridge's Computer Laboratory (shown in Figure 10.3). The room-sized EDSAC I was the world's first fully operational, stored-program computer (the first von Neumann machine) and went online in 1949. EDSAC II became operational in 1958.

The EDSAC ISS was first described in a paper on debugging EDSAC programs, which was written and published in 1951 by S. Gill, one of Wilkes' research students [2]. The paper describes the operation of a *tracing simulator* that operates by

- Fetching the simulated instruction
- Decoding the instruction to save trace information
- Updating the simulated program counter if the instruction is a branch or else placing the nonbranch instruction in the middle of the simulator loop and executing it directly
- Returning to the top of the simulator loop

FIGURE 10.3   The EDSAC I, developed at the Computer Laboratory, University of Cambridge, U.K., became the first fully functional stored-program computer in 1948. This computer also had the first instruction-set simulator, which was described in an article published in 1951. (Copyright 2015, Computer Laboratory, University of Cambridge, Cambridge, U.K., Reproduced with permission.)

Thus, the first operational processor ISS predates the introduction of the first commercial microprocessor (Intel's 4004) by some 20 years.

Cycle-accurate ISSs, the most useful simulator category for processor benchmarking, compute exact instruction timing by accurately modeling the processor's pipeline. All commercial processor cores have at least one corresponding ISS. They are obtained in different ways. Some core vendors rely on third parties to provide an ISS for their processors. Other vendors offer an ISS as part of their tool suite. Some ISSs must be purchased and some are available in evaluation packages from the processor core vendor.

To serve as an effective benchmarking tool, an ISS accurately simulates the operation of the processor core and it must provide the instrumentation needed to provide critical benchmarking statistics. These statistics include cycle counts for the various functions and routines executed and main-memory and cache-memory usage. The better the ISS instrumentation, the more comparative information the benchmark will produce.

10.3 IDEAL VERSUS PRACTICAL PROCESSOR BENCHMARKS

IC designers use benchmarks to help them pick the *best* processor core for a given task to be performed on the IC. The original definition of a benchmark was literally a mark on a craftsperson's wooden workbench that provided some measurement standard. Eventually, the first benchmarks (carved into the workbench) were replaced with standard measuring tools such as yardsticks. Processor benchmarks provide yardsticks for measuring processor performance. The ideal yardstick would be one that could measure any task running on any processor. The ideal processor benchmark produces results that are relevant, reliable, repeatable, objective, comparable, and applicable. Unfortunately, no such processor benchmark exists so we must make do with approximations of the ideal.

The ideal processor benchmark would be the actual application code that the processor must run. No other piece of code can possibly be as representative of the processor's ability to execute the actual task to be performed. No other piece of code can possibly replicate the instruction-use distribution; register, memory, and bus use; or the data-movement patterns of the actual application code. However, the actual application code is less than ideal as a benchmark in many ways.

First and foremost, the actual application code may not even exist when candidate processor cores are benchmarked because processor benchmarking and selection must occur early in the project. A benchmark that does not exist is truly worthless.

Next, the actual application code serves as an extremely specific benchmark. It will indeed give a very accurate prediction of processor performance for a specific task *and for no other task*. In other words, the downside of a highly specific benchmark is that the benchmark will give a less-than-ideal indication of processor performance for other tasks. Because on-chip processor cores are often used for a variety of tasks, the ideal benchmark may well be a suite of application programs and not just one program.

Yet another problem with application-code benchmarks is their lack of instrumentation. The actual application code has almost always been written to execute the task, not to measure a processor core's performance. Appropriate measurements may require instrumentation of the application code. This modification consumes time and resources, which may not be readily available, and it could well break the application code. Technically, that is known as a "bad thing." Even with all of these issues, the application code (if it exists) provides invaluable information on processor core performance and should be used whenever possible to help make a processor core selection.

10.4 STANDARD BENCHMARK TYPES

Given that the *ideal* processor benchmark proves less than ideal, the industry has sought *standard* benchmarks for use when the target application code is either not available or not appropriate. There are four types of standard benchmarks: full-application or *real-world* benchmarks and benchmark suites, synthetic or small-kernel benchmarks, hybrid or derived benchmarks that mix and match aspects of the full-application and synthetic benchmarks, and microbenchmarks (not to be confused with microprocessor benchmarks).

Full-application benchmarks and benchmark suites employ existing system- or application-level code drawn from real applications, although probably not precisely the application of interest to any given ASIC or SoC design team. To be sufficiently informative, these benchmarks may incorporate many thousands of lines of code, have large instruction-memory footprints, and consume large amounts of data memory.

Synthetic benchmarks tend to be smaller than full, application-based benchmarks. They consist of smaller code sections representing commonly used algorithms. Code sections may be extracted from working code or they may be written specifically for use as a benchmark. Writers of synthetic benchmarks try to approximate instruction mixes of real-world applications without replicating the entire application. It is important to keep in mind that these benchmarks are necessarily approximations of real-world applications.

Hybrid benchmarks mix and match large application programs and smaller blocks of synthetic code to create a sort of microprocessor torture track (with long straight sections and tight curves to use a racetrack analogy). Hybrid benchmark code is augmented with test data sets taken from real-world applications. A processor core's performance around this torture track can give a good indication of the processor's abilities over a wide range of situations, although probably not the specific use to which the processor will be put.

Microbenchmarks are very small code snippets designed to exercise some particular processor feature or to characterize a particular machine characteristic in isolation from all other processor features and characteristics. Microbenchmark results can delineate a processor's peak capabilities and reveal potential architectural bottlenecks, but peak performance is not a very good indicator of overall application performance. Nevertheless, a suite of microbenchmarks may come close to approximating an ideal benchmark for certain applications, if the application is a common one with many standardized, well-defined, well-understood functions.

10.5 PREHISTORIC PERFORMANCE RATINGS: MIPS, MOPS, AND MFLOPS

Lord Kelvin might have been predicting processor performance measurements when he said, "When you can measure what you are speaking about, and express it in numbers, you know something about it; but when you cannot measure it, when you cannot express it in numbers, your knowledge is of a meager and unsatisfactory kind. It may be the beginning of knowledge, but you have scarcely, in your thoughts, advanced to the stage of science." [3]

The need to rate processor performance is so great that, at first, microprocessor vendors grabbed any and all numbers at hand to rate performance. The first figures of merit used were likely clock rate and memory bandwidth. These prehistoric ratings measured processor performance divorced of any connection to running code. Consequently, these performance ratings are not benchmarks. Just as an engine's revolutions per minute (RPM) reading is not sufficient to measure vehicle performance (you need engine torque plus transmission gearing, differential gearing, and tire diameter to compute speed), the prehistoric, clock-related processor ratings of MIPS, MOPS, MFLOPS, and VAX MIPS (defined below) tell you almost nothing about a processor's true performance potential.

Before it was the name of a microprocessor architecture, the term "MIPS" was an acronym for *millions of instructions per second*. If all processors had the same instruction-set architecture (ISA) and used the same compiler to generate code, then a MIPS rating might possibly be used as a performance measure. However, all processors do not have the same ISA and they most definitely do not use the same compiler, so they are not equally efficient when it comes to task execution speed versus clock rate. In fact, microprocessors and processor cores have very different ISAs and compilers for these processors are differently abled when generating code. Consequently, some processors can do more with one instruction than other processors, just as large automobile engines can do more than smaller ones running at the same RPM.

This problem was already bad in the days when only complex-instruction-set computer (CISC) processors roamed the earth. The problem went from bad to worse when reduced-instruction-set computer (RISC) processors arrived on the scene. One CISC instruction would often do the work of several RISC instructions (by design) so that a CISC microprocessor's MIPS rating did not correlate at all with a RISC microprocessor's MIPS rating because of the work differential between RISC's simple instructions and CISC's more complex instructions.

The next prehistoric step in creating a usable processor performance rating was to switch from MIPS to VAX MIPS, which was accomplished by setting the extremely successful VAX 11/780 minicomputer—introduced in 1977 by the now defunct Digital Equipment Corporation (DEC)—as the benchmark against which all other processors are measured. So, if a microprocessor executed a set of programs twice as fast as a VAX 11/780, it was said to be rated at 2 VAX MIPS. The original term "MIPS" then became "native MIPS," so as not to confuse the original ratings with VAX MIPS. DEC referred to VAX MIPS as VAX units of performance (VUPs) just to keep things interesting or confusing depending on your point of view.

Both native and VAX MIPS are woefully inadequate measures of processor performance because they were usually provided without specifying the software (or even the programming language) used to create the rating. Because different programs have different instruction mixes, different memory usage patterns, and different data-movement patterns, the same processor could easily earn one MIPS rating on one set of programs and quite a different rating on another set. Because MIPS ratings are not linked to a specific benchmark program suite, the *MIPS* acronym now stands for *meaningless indication of performance* for those in the know.

A further problem with the VAX MIPS measure of processor performance is that the concept of using a VAX 11/780 minicomputer as the gold performance standard is an idea that is more than a bit long in the tooth in the twenty-first century. There are no longer many (or any) VAX 11/780s available for running benchmark code and DEC effectively disappeared when Compaq Computer Corp. purchased what was left of the company in January 1998, following the decline of the minicomputer market. Hewlett-Packard absorbed Compaq in May 2002, submerging DEC's identity even further. VAX MIPS is now the processor equivalent of a furlongs-per-fortnight speed rating—woefully outdated.

Even more tenuous than the MIPS performance rating is the concept of MOPS, an acronym that stands for "millions of operations per second." Every algorithmic task requires the completion of a certain number of fundamental operations, which may or may not have a one-to-one correspondence with machine instructions. Count these fundamental operations in the millions and they become MOPS. If they are floating-point operations, you get MFLOPS. One thousand MFLOPS equals one GFLOPS. The MOPS, MFLOPS, and GFLOPS ratings suffer from the same drawback as the MIPS rating: there is no standard software to serve as the benchmark that produces the ratings. In addition, the *conversion factor* for computing how many operations a processor performs per clock (or how many processor instructions constitute one operation) is somewhat fluid as well, which means that the processor vendor is free to develop a conversion factor on its own. Consequently, MOPS and MFLOPS performance ratings exist for various processor cores but they really do not help an IC design team pick a processor core because they are not true benchmarks.

10.6 CLASSIC PROCESSOR BENCHMARKS (THE STONE AGE)

Like ISSs, standardized processor performance benchmarks predate the 1971 introduction of Intel's 4004 microprocessor, but just barely. The first benchmark suite to attain de facto *standard* status was a set of programs known as the Livermore Kernels (also popularly called the Livermore Loops).

10.6.1 LIVERMORE FORTRAN KERNELS/LIVERMORE LOOPS BENCHMARK

The Livermore Kernels were first developed in 1970 and consist of 14 numerically intensive application kernels written in FORTRAN. Ten more kernels were added during the early 1980s and the final suite of benchmarks was discussed in a paper published in 1986 by F. H. McMahon of the Lawrence Livermore National Laboratory (LLNL), located in Livermore, CA [4]. The Livermore

Kernels actually constitute a supercomputer benchmark, measuring a processor's floating-point computational performance in terms of MFLOPS (millions of floating-point operations per second). Because of the somewhat frequent occurrence of floating-point errors in many computers, the Livermore Kernels test both the processor's speed and the system's computational accuracy. Today, the Livermore Kernels are called the Livermore FORTRAN Kernels (LFK) or the Livermore Loops.

The Livermore Loops are real samples of floating-point computation taken from a diverse workload of scientific applications extracted from operational program code used at LLNL. The kernels were extracted from programs in use at LLNL because those programs were generally far too large to serve as useful benchmark programs; they included hardware-specific subroutines for performing functions such as I/O, memory management, and graphics that were not appropriate for benchmark testing of floating-point performance; and they were largely classified, due to the nature of the work done at LLNL. Some kernels represent widely used, generic computations such as dot and matrix (SAXPY) products, polynomials, first sum and differences, first-order recurrences, matrix solvers, and array searches. Some kernels typify often-used FORTRAN constructs while others contain constructs that are difficult to compile into efficient machine code. These kernels were selected to represent both the best and worst cases of common FORTRAN programming practice to produce results that measure a realistic floating-point performance range by challenging the FORTRAN compiler's ability to produce optimized machine code. Table 10.1 lists the 24 Livermore Loops.

A complete LFK run produces 72 timed results, produced by timing the execution of the 24 LFK kernels three times using different DO-loop lengths.

The LFK kernels are a mixture of vectorizable and nonvectorizable loops and test the computational capabilities of the processor hardware and the software tools' ability to generate efficient machine code. The Livermore Loops also tests a processor's vector abilities and the associated software tools' abilities to vectorize code.

TABLE 10.1   Twenty-Four Kernels in the Livermore Loops

LFK Kernel Number	Kernel Description
Kernel 1	An excerpt from a hydrodynamic application
Kernel 2	An excerpt from an Incomplete Cholesky-Conjugate Gradient program
Kernel 3	The standard inner-product function from linear algebra
Kernel 4	An excerpt from a banded linear equation routine
Kernel 5	An excerpt from a tridiagonal elimination routine
Kernel 6	An example of a general linear recurrence equation
Kernel 7	Equation of state code fragment (as used in nuclear weapons research)
Kernel 8	An excerpt of an alternating direction, implicit integration program
Kernel 9	An integrate predictor program
Kernel 10	A difference predictor program
Kernel 11	A first sum
Kernel 12	A first difference
Kernel 13	An excerpt from a 2D particle-in-cell program
Kernel 14	An excerpt from a 1D particle-in-cell program
Kernel 15	A sample of casually written FORTRAN to produce suboptimal machine code
Kernel 16	A search loop from a Monte Carlo program
Kernel 17	An example of an implicit conditional computation
Kernel 18	An excerpt from a 2D explicit hydrodynamic program
Kernel 19	A general linear recurrence equation
Kernel 20	An excerpt from a discrete ordinate transport program
Kernel 21	A matrix product calculation
Kernel 22	A Planckian distribution procedure
Kernel 23	An excerpt from a 2D implicit hydrodynamic program
Kernel 24	A kernel that finds the location of the first minimum in X

At first glance, the Livermore Loops benchmark appears to be nearly useless for the benchmarking of embedded processor cores in ICs. It is a floating-point benchmark written in FORTRAN that looks for good vector abilities. FORTRAN compilers for embedded processor cores are quite rare—essentially nonexistent. Today, very few real-world applications run tasks like those appearing in the 24 Livermore Loops, which are far more suited to research on the effects of very high-speed nuclear reactions than the development of commercial, industrial, or consumer products. It is unlikely that the processor cores in a mobile phone handset, a digital camera, or a tablet will ever need to perform 2D hydrodynamic calculations. Embedded-software developers working in FORTRAN are also quite rare. Consequently, processor core vendors are quite unlikely to tout Livermore Loops benchmark results for their processor cores. The Livermore Loops benchmarks are far more suited to testing supercomputers. However, as on-chip gate counts grow, as processor cores gain single-instruction, multiple-data (SIMD) and floating-point execution units, and as IC designs increasingly tackle tougher number-crunching applications including the implementation of audio and video codecs, the Livermore Loops benchmark could become valuable for testing more than just supercomputers.

10.6.2 LINPACK

LINPACK is a collection of FORTRAN subroutines that analyze and solve linear equations and linear least-squares problems. Jack Dongarra assembled the LINPACK collection of linear algebra routines at the Argonne National Laboratory in Argonne, IL. The first versions of LINPACK existed in 1976 but the first users' guide was published in 1977 [5,6]. The package solves linear systems whose matrices are general, banded, symmetric indefinite, symmetric positive definite, triangular, and tridiagonal square. In addition, the package computes the QR and singular value decompositions of rectangular matrices and applies them to least-squares problems. The LINPACK routines are not, strictly speaking, a benchmark but they exercise a computer's floating-point capabilities. As of 2014, LINPACK benchmark has been largely supplanted by the Linear Algebra Package (LAPACK) benchmark, which is designed to run efficiently on shared-memory, vector supercomputers. The University of Tennesee maintains these programs at www.netlib.org.

Originally, LINPACK benchmarks performed computations on a 100×100 matrix, but with the rapid increase in computing performance that has occurred, the size of the arrays grew to 1000×1000. The original versions of LINPACK were written in FORTRAN but there are now versions written in C and Java as well. Performance is reported in single- and double-precision MFLOPS, which reflects the benchmark's (and the national labs') focus on "big iron" machines (supercomputers). LINPACK and LAPACK benchmarks are not commonly used to measure microprocessor or processor core performance.

10.6.3 WHETSTONE BENCHMARK

The Whetstone benchmark was written by Harold Curnow of the now defunct Central Computer and Telecommunications Agency (CCTA), which was tasked with computer procurement in the British government. The CCTA used the Whetstone benchmark to test the performance of computers being considered for purchase. The Whetstone benchmark is the first program to appear in print that was designed as a synthetic benchmark for testing processor performance. It was specifically developed to test the performance of only one computer: the hypothetical Whetstone machine.

The Whetstone benchmark is based on application-program statistics gathered by Brian A. Wichmann at the National Physical Laboratory in England. Wichmann was using an Algol-60 compilation system that compiled Algol statements into instructions for the hypothetical Whetstone computer system, which was named after the small town of Whetstone located just outside the city of Leicester, England, where the compilation system was developed. Wichmann compiled statistics on instruction usage for a wide range of numerical computation programs then in use.

An Algol-60 version of the Whetstone benchmark was released in November 1972 and single- and double-precision FORTRAN versions appeared in April 1973. The FORTRAN versions became the first widely used, general-purpose, synthetic computer performance benchmarks. Information about the Whetstone benchmark was first published in 1976 [7].

The Whetstone benchmark suite consists of several small loops that perform integer and floating-point arithmetic, array indexing, procedure calls, conditional jumps, and elementary function evaluations. These loops reside in three subprograms (called p3, p0, and pa) that are called from a main program. The subprograms call trigonometric (sine, cosine, and arctangent) functions and other math-library functions (exponentiation, log, and square root). The benchmark authors empirically managed the Whetstone's instruction mix by manipulating loop counters within the modules to match Wichmann's instruction-mix statistics. Empirical Whetstone loop weights range from 12 to 899.

The Whetstone benchmark produces speed ratings in terms of thousands of Whetstone instructions per second (KWIPS), thus using the hypothetical Whetstone computer as the gold standard for this benchmark in much the same way that the actual VAX 11/780 was used as the golden reference for VAX MIPS and VUPs. In 1978, self-timed versions of the Whetstone benchmark written by Roy Longbottom (also of CCTA) produced speed ratings in MOPS and MFLOPS and an overall rating in MWIPS.

As with the LFK and LINPACK benchmarks, the Whetstone benchmark focuses on floating-point performance. Consequently, it is not commonly applied to embedded processor cores that are destined to execute integer-oriented and control tasks on an IC because such a benchmark really would not tell you much about the performance of the processor(s) in question. However, the Whetstone benchmark's appearance spurred the creation of a plethora of "stone" benchmarks, and one of those benchmarks, known as the "Dhrystone," became the first widely used benchmark to rate processor performance.

10.6.4 DHRYSTONE BENCHMARK

Reinhold P. Weicker, working at Siemens-Nixdorf Information Systems, wrote and published the first version of the Dhrystone benchmark in 1984 [8]. The benchmark's name, *Dhrystone*, is a pun derived from the Whetstone benchmark. The Dhrystone benchmark is a synthetic benchmark that consists of 12 procedures in one measurement loop. It produces performance ratings in Dhrystones per second. Originally, the Dhrystone benchmark was written in a "Pascal subset of Ada." Subsequently, versions in Pascal and C have appeared and the C version is the one most used today.

The Dhrystone benchmark differs significantly from the Whetstone benchmark and these differences made the Dhrystone far more suitable as a processor benchmark. First, the Dhrystone is strictly an integer program. It does not test a processor's floating-point abilities because most microprocessors in the early 1980s (and even those in the early twenty-first century) had no native floating-point computational abilities. Floating-point capabilities are now far more common in microprocessor cores, however. The Dhrystone devotes a lot of time executing string functions (copy and compare), which microprocessors often execute in real applications.

The original version of the Dhrystone benchmark quickly became successful. One indication of the Dhrystone's success was the attempts by microprocessor vendors to unfairly inflate their Dhrystone ratings by "gaming" the benchmark (cheating). Weicker made no attempt to thwart compiler optimizations when he wrote the Dhrystone benchmark because he reasonably viewed those optimizations as typical of real-world programming. However, version 1 of the Dhrystone benchmark did not print or use the results of its computations allowing *properly* written optimizing compilers using dead-code-removal algorithms to optimize away almost all of the benchmark by removing the benchmark program's computations. The results were not used, which the optimizers detected. Clearly, such optimizations go far beyond the spirit of benchmarking. Weicker corrected this problem by publishing version 2.1 of the Dhrystone benchmark in 1988 [9]. At the same time, he acknowledged some of the benchmark's limitations and published his criteria for using the Dhrystone benchmarks to compare processors [10].

Curiously, DEC's VAX also plays a role in the saga of the Dhrystone benchmarks. The VAX 11/780 minicomputer could run 1757 version 2.1 Dhrystones/sec. Because the VAX 11/780 was (erroneously) considered a 1 MIPS computer, it became the Dhrystone standard machine, which resulted in the emergence of the DMIPS or D-MIPS (Dhrystone MIPS) rating. By dividing a processor's Dhrystone 2.1 performance rating by 1757, processor vendors could produce an official-looking DMIPS rating for their products. Thus, DMIPS manages to fuse two questionable rating systems (Dhrystones and VAX MIPS) to create a third, derivative, equally questionable microprocessor-rating system.

The Dhrystone benchmark's early success as a marketing tool further encouraged abuse, which became rampant. For example, vendors quickly (though unevenly) added Dhrystone-optimized string routines written in machine code to some of their compiler libraries because accelerating these heavily used library routines boosted Dhrystone ratings. Actual processor performance did not change at all but the Dhrystone ratings suddenly improved. Some compilers started in-lining these machine-coded string routines for even better performance ratings.

As the technical marketing teams at the microprocessor vendors continued to study the benchmark, they found increasingly better ways of improving their products' ratings by introducing compiler optimizations that only applied to the benchmark. Some compilers even had pattern recognizers that could recognize the Dhrystone benchmark source code, or there was a special Dhrystone command-line switch that caused the compiler to *generate* the entire program using a previously prewritten, precompiled, hand-optimized version of the Dhrystone benchmark program.

It is not even necessary to alter a compiler to produce wildly varying Dhrystone results using the same microprocessor. Using different compiler optimization settings can drastically alter the outcome of a benchmark test, even if the compiler has not been *Dhrystone optimized*. For example, Bryan Fletcher of Memec taught a class titled "FPGA Embedded Processors: Revealing True System Performance" at the 2005 Embedded Systems Conference in San Francisco where he showed that the Xilinx MicroBlaze soft core processor produced Dhrystone benchmark results that differ by almost 9:1 in terms of DMIPS/MHz depending on configuration and compiler settings as shown in Table 10.2 [11].

As Fletcher's results show, compiler optimizations and processor configuration can alter a processor's Dhrystone performance results significantly (9:1 variation in this example alone). These results clearly delineate a problem with using Dhrystone as a benchmark: there's no organization to enforce Weicker's published rules for using the Dhrystone benchmark, so there is no way to ensure that processor vendors benchmark fairly and there is no way to force vendors to fully disclose the conditions that produced their benchmark results. If benchmarks are not conducted by objective third parties under controlled conditions that are disclosed along with the benchmark results, the results must always be at least somewhat suspect even if there are published rules for the benchmark's use because no one enforces the rules. Without enforcement, history teaches us that some vendors will not follow rules when the sales of a new microprocessor are at stake as discussed in the following.

TABLE 10.2   Performance Differences in Xilinx MicroBlaze Dhrystone Benchmark Performance Based on Processor Configuration and Compiler Settings

Conditions	DMIPS	Clock Frequency (MHz)	DMIPS(MHz)
Initializing optimized system, running from SRAM	6.85	75	0.091
Increasing compiler optimization level to Level 3	7.14	75	0.095
Adding instruction and data caches	29.13	60	0.486
Removing data cache and moving stack and small data sections to local memory	33.17	60	0.553
Adding remaining data sections to local memory	47.81	60	0.797
Removing instruction cache, moving large data sections back to SRAM, and moving instructions to local memory	41.48	75	0.553
Moving entire program to local memory	59.79	75	0.797

Over time, other technical weaknesses have appeared in the Dhrystone benchmark. For example, the actual Dhrystone code footprint is quite small. As their instruction caches grew, some processors were able to run the Dhrystone benchmark entirely from their caches, which boosted their performance on the Dhrystone benchmark but did not represent real performance gains because most application programs would not fit entirely into a processor's cache. Table 10.2 demonstrates the sensitivity of the Dhrystone benchmark to cache and memory size. Processors with wider data buses and wide-load/store instructions that move many string bytes per bus transaction also earn better Dhrystone performance ratings. Although this instruction behavior also benefits application code, it can mislead the unwary who think that the benchmark is comparing the relative computational performance of processors.

One of the worst abuses of the Dhrystone benchmark occurred when processor vendors ran benchmarks on their own evaluation boards and on competitors' evaluation boards, and then published the results. The vendor's own board would have fast memory and the competitors' boards would have slow memory, but this difference would not be revealed when the scores were published. Differences in memory speed do affect Dhrystone results, so publishing the Dhrystone results of competing processors without also disclosing the use of dissimilar memory systems to compare the processors is clearly dishonest, or at least mean spirited. Other sorts of shenanigans were used to distort Dhrystone comparisons as well. One of the Dhrystone benchmark's biggest weaknesses is therefore the absence of a governing body or an objective, third party that can review, approve, and publish Dhrystone benchmark results.

Despite all of these weaknesses, microprocessor and processor core vendors continue to publish Dhrystone benchmark results—or at least they have Dhrystone results handily available—because Dhrystone ratings are relatively easy to generate and the benchmark code is very well understood after three decades of use. Even with its flaws, microprocessor and processor core vendors keep Dhrystone ratings handy because the lack of such ratings might raise a red flag for some prospective customers. (*"Why won't they tell me? The numbers must be bad."*)

Here is a stated opinion from ARM Application Note 273 [12], "Dhrystone Benchmarking for ARM Cortex Processors":

> Today, Dhrystone is no longer seen as a representative benchmark and ARM does not recommend using it. In general, it has been replaced by more complex processor benchmarks such as SPEC and CoreMark.

Dhrystone's explosive popularity as a tool for making processor comparisons coupled with its technical weaknesses and the abuses heaped upon the benchmark program helped to spur the further evolution of processor benchmarks in the 1990s, producing new benchmark programs including SPEC and CoreMark mentioned in the ARM application note quoted earlier. Section 10.7 of this chapter discusses several newer processor benchmarks including SPEC and Embedded Microprocessor Benchmark Consortium (EEMBC) CoreMark.

10.6.5 EDN MICROPROCESSOR BENCHMARKS

EDN magazine, a trade publication for the electronics industry published from 1956 through 2015, was a very early advocate of microprocessor use for general electronic system design. Editors at EDN wrote extensively about microprocessors starting with their introduction in the early 1970s. EDN published the first comprehensive microprocessor benchmarking article in 1981, just after the first wave of commercial 16-bit microprocessor chips had been introduced [13]. This article compared the performance of the four leading 16-bit microprocessors of the day: DEC's LSI-11/23, Intel's 8086, Motorola's 68000, and Zilog's Z8000. (The Motorola 68000 microprocessor actually employed a 16/32-bit architecture.) The Intel, Motorola, and Zilog processors heavily competed for system design wins during the early 1980s and this article provided design engineers with some of the very first microprocessor benchmark ratings to be published by an objective third party.

This article was written by Jack E. Hemenway, a consulting editor for EDN, and Robert D. Grappel of MIT's Lincoln Laboratory. Hemenway and Grappel summed up the reason that the industry needs objective benchmark results succinctly in their article:

> Why the need for a benchmark study at all? One sure way to start an argument among computer users is to compare each one's favorite machine with the others. Each machine has strong points and drawbacks, advantages and liabilities, but programmers can get used to one machine and see all the rest as inferior. Manufacturers sometimes don't help: Advertising and press releases often imply that each new machine is the ultimate in computer technology. Therefore, only a careful, complete and unbiased comparison brings order out of the chaos.

Those sentiments are still true; nothing has changed after more than three decades.

The EDN article continues with an excellent description of the difficulties associated with microprocessor benchmarking:

> Benchmarking anything as complex as a 16-bit processor is a very difficult task to perform fairly. The choice of benchmark programs can strongly affect the comparisons' outcome so the benchmarker must choose the test cases with care.

These words also remain accurate after more than 30 years.

Hemenway and Grappel used a set of benchmark programs created in 1976 by a Carnegie–Mellon University (CMU) research group. These benchmarks were first published by the CMU group in 1977 in a paper at the National Computer Conference [14]. The benchmark tests—interrupt handlers, string searches, bit manipulation, and sorting—are very representative of the tasks microprocessors and processor cores must perform in most embedded applications. The EDN authors excluded the CMU benchmarks that tested floating-point computational performance because none of the processors in the EDN study had native floating-point resources. The seven programs in the 1981 EDN benchmark study appear in Table 10.3.

Significantly, the authors of this article allowed programmers from each microprocessor vendor to code each benchmark for their company's processor. Programmers were required to use assembly language to code the benchmarks, which eliminated compiler inefficiencies from the performance results and allowed the processor architectures to compete directly. Considering the relatively immature state of microprocessor compilers in 1981, this was probably an excellent decision.

The authors refereed the benchmark tests by reviewing each program to ensure that no corners were cut and that the programs faithfully executed each benchmark algorithm. The authors did not force the programmers to code in a specific way and allowed the use of special instructions (such as instructions designed to perform string manipulation) because using such instructions fairly represented actual processor use.

The study's authors ran the programs and recorded the benchmark results to ensure fairness. They published the results including the program size and execution speed. Publishing the program size recognized that memory was costly and limited, a situation that still holds true today in IC design. The authors published the scores of each of the seven benchmark tests separately, acknowledging that a combined score would tend to mask important information about a processor's specific abilities.

TABLE 10.3   Programs in the 1981 EDN Benchmark Article by Hemenway and Grappel

Component	Benchmark Description
Benchmark A	Simple priority interrupt handler
Benchmark B	FIFO interrupt handler
Benchmark E	Text string search
Benchmark F	Primitive bit manipulation
Benchmark H	Linked-list insertion (test addressing modes and 32-bit operations)
Benchmark I	Quicksort algorithm (tests stack manipulation and addressing modes)
Benchmark K	Matrix transposition (tests bit manipulation and looping)

TABLE 10.4   Twelve Programs in EDN's 1988 DSP Benchmark Article

Program	Benchmark Description
Benchmark 1	20-tap FIR filter
Benchmark 2	64-tap FIR filter
Benchmark 3	67-tap FIR filter
Benchmark 4	8-pole canonic IIR filter (4×) (direct form II)
Benchmark 5	8-pole canonic IIR filter (5×) (direct form II)
Benchmark 6	8-pole transpose IIR filter (direct form I)
Benchmark 7	Dot product
Benchmark 8	Matrix multiplication, 2 × 2 times 2 × 2
Benchmark 9	Matrix multiplication, 3 × 3 times 3 × 1
Benchmark 10	Complex 64-point FFT (radix-2)
Benchmark 11	Complex 256-point FFT (radix-2)
Benchmark 12	Complex 1024-point FFT (radix-2)

EDN published a similar benchmarking article in 1988 covering DSPs [15]. The article was written by David Shear, one of EDN's regional editors. Shear benchmarked 18 DSPs using 12 benchmark programs selected from 15 candidate programs. The DSP vendors participated in the selection of the 12 DSP benchmark programs used from the 15 candidates. The final dozen DSP benchmark programs included six DSP filters, three math benchmarks (a simple dot product and two matrix multiplications), and three fast Fourier transforms (FFTs). The benchmark programs that Shear used to compare DSPs differed substantially from the programs used by Grappel and Hemenway to compare general-purpose microprocessors because, as a class, DSPs are applied quite differently than general-purpose processors. Shear's benchmark set of 12 DSP programs appear in Table 10.4.

Notably, Shear elected to present the benchmark performance results (both execution speed and performance) as bar charts instead of in tabular form, noting that the bar charts emphasized large performance differences while masking differences of a few percent, which he dismissed as "not important."

Shear's EDN article also recognized the potential for cheating by repeating the "Three Not-So-golden Rules of Benchmarking" that had appeared in 1981. EDN Editor Walt Patstone wrote and published these rules as a follow-up to the Hemenway and Grappel article [16]:

- *Rule 1*: All's fair in love, war, and benchmarking.
- *Rule 2*: Good code is the fastest possible code.
- *Rule 3*: Conditions, cautions, relevant discussion, and even actual code never make it to the bottom line when results are summarized.

These two EDN microprocessor benchmark articles established and demonstrated all of the characteristics of an effective, modern microprocessor benchmark:

- Conduct a series of benchmark tests that exercise relevant processor features for a class of tasks.
- Use benchmark programs that are appropriate to the processor class being studied.
- Allow experts to code the benchmark programs.
- Have an objective third-party check and run the benchmark code to ensure that the vendors and their experts do not cheat.
- Publish benchmark results that maximize the amount of available information about the tested processor.
- Publish both execution speed and memory use for each processor on each benchmark program because there is always a trade-off between a processor's execution speed and its memory consumption.

These characteristics shaped the benchmarking organizations and their benchmark programs during the 1990s.

10.7 MODERN PROCESSOR PERFORMANCE BENCHMARKS

As the number of available microprocessors mushroomed in the 1980s, the need for good benchmarking standards became increasingly apparent. The Dhrystone benchmarking experiences showed how useful a standard benchmark could be and these experiences also demonstrated the lengths (both fair and unfair) to which processor vendors would go to earn top benchmark scores. The EDN articles demonstrated that an objective third party could bring order out of benchmarking chaos. As a result, both private companies and industry consortia stepped forward with the goal of producing *fair and balanced* processor benchmarking standards.

10.7.1 SPEC: THE STANDARD PERFORMANCE EVALUATION CORPORATION

One of the first organizations to seriously tackle the need for good microprocessor benchmarking standards was SPEC (www.spec.org). (Originally, the abbreviation SPEC stood for the "System Performance Evaluation Cooperative." It now stands for the "Standard Performance Evaluation Corporation"). SPEC was founded in 1988 by a group of workstation vendors including Apollo, Hewlett-Packard, MIPS Computer Systems, and Sun Microsystems working in conjunction with the trade publication *Electronic Engineering Times*.

One year later, the cooperative produced its first processor benchmark, SPEC89, which was a standardized measure of compute-intensive processor performance with the express purpose of replacing the existing, vague MIPS, and MFLOPS rating systems. Because high-performance microprocessors were primarily used in high-end workstations at the time and because SPEC was formed by workstation vendors as a cooperative, the SPEC89 benchmark consisted of source code that was to be compiled for the UNIX operating system (OS). At the time, UNIX was strictly a workstation OS. The open-source version of UNIX named Linux would not appear for another 2 years and it would not become a popular embedded OS until the next millennium.

The Dhrystone benchmark had already demonstrated that benchmark code quickly rots over time due to rapid advances in processor architecture and compiler technology. (It is no small coincidence that Reinhold Weicker, Dhrystone's creator and one of the people most familiar with processor benchmarking, became a key member of SPEC.) The likelihood of *benchmark rot* was no different for SPEC89, so the SPEC organization has regularly improved and expanded its benchmarks, producing SPEC92, SPEC95 (with separate integer and floating-point components called CINT95 and CFP95), SPEC CPU2000 (consisting of CINT2000 and CFP2000), SPEC CPUv6, and finally SPEC CPU2006 (consisting of CINT2006 and CFP2006). Here is what the SPEC website says about SPEC CPU2006 as of mid-2014:

> CPU2006 is SPEC's next-generation, industry-standardized, CPU-intensive benchmark suite, stressing a system's processor, memory subsystem, and compiler. SPEC designed CPU2006 to provide a comparative measure of compute-intensive performance across the widest practical range of hardware using workloads developed from real user applications. These benchmarks are provided as source code and require the user to be comfortable using compiler commands as well as other commands via a command interpreter using a console or command prompt window in order to generate executable binaries. The current version of the benchmark suite is V1.2, released in September 2011; submissions using V1.1 are only accepted through December 19, 2011.

Tables 10.5 and 10.6 list the SPEC CINT2006 and CFP2006 benchmark component programs, respectively.

As Tables 10.5 and 10.6 show, SPEC benchmarks are application-based benchmarks, not synthetic benchmarks. The SPEC benchmark components are excellent as workstation/server benchmarks because they use actual applications that are likely to be assigned to these machines. SPEC publishes benchmark performance results for various computer systems on its website and sells its benchmark code. As of 2014, the CPU2006 benchmark suite costs $800 ($200 for nonprofit use). A price list for the CPU2006 benchmarks as well as a wide range of other computer benchmarks appears on the SPEC website.

Because the high-performance microprocessors used in workstations and servers are sometimes used as embedded processors and some of them are available as processor cores for use

TABLE 10.5   SPEC CINT2006 Benchmark Component Programs

Component	Language	Category
400.perlbench	C	A set of work tools written in the Perl programming language
401.bzip2	C	Compression
403.gcc	C	C programming language compiler
429.mcf	C	Combinatorial optimization, vehicle scheduling
445.gobmk	C	Game playing: go
456.hmmr	C	Protein sequence analysis using hidden Markov models
458.sjeng	C	Artificial intelligence: chess
462.libquantum	C	Quantum computer simulation
464.h264ref	C	Video compression using h.264 encoder
471.omnetpp	C++	Discrete event simulator that emulates an Ethernet network
473.astar	C++	Pathfinding library for 2D maps
483.xalancbmk	C++	XML document processing

TABLE 10.6   SPEC CFP2006 Benchmark Component Programs

Component	Language	Category
410.bwaves	Fortran	Computes 3D transonic transient laminar viscous flow
416.gamess	Fortran	Quantum chemical computations
433.milc	C	Gauge field generator for lattice gauge theory
434.zeusmp	Fortran	Computational fluid dynamics for astrophysical simulation
435.gromacs	C, Fortran	Molecular dynamics simulation
436.cactusADM	C, Fortran	Einstein evolution equation solver using numerical methods
437.leslie3d	Fortran	Computational fluid dynamics using large Eddy simulations
444.namd	C++	Large biomolecular simulator
447.dealll	C++	Helmholtz-type equation solver with nonconstant coefficients
450.soplex	C++	Linear program solver using simplex algorithm
453.povray	C++	Image rendering
454.calculix	C, Fortran	Finite element solver for linear and nonlinear 3D structures
459.GemsFDTD	Fortran	3D solver for Maxwell's equations
465.tonto	Fortran	Quantum chemistry
470.lbm	C	Incompressible fluid simulation
481.wrf	C, Fortran	Weather modeling
482.sphinx3	C	Speech recognition

on SoCs, microprocessor and processor core vendors sometimes quote SPEC benchmark scores for their products. You should use these performance ratings with caution because the SPEC benchmarks are not necessarily measuring performance that is meaningful to embedded applications. In addition, memory and storage subsystem performance will significantly affect the results of these benchmarks. No mobile phone is likely to be required to simulate seismic-wave propagation, except for the possible exception of handsets sold in California, and the memory and storage subsystems in mobile phones do not resemble the ones used in workstations.

A paper written by Jakob Engblom at Uppsala University compared the static properties of the SPECint95 benchmark programs with code from 13 embedded applications consisting of 334,600 lines of C source code [17]. Engblom's static analysis discovered several significant differences between the static properties of the SPECint95 benchmark and the 13 embedded programs. Noted differences include the following:

- *Variable sizes*: Embedded programs carefully control variable size to minimize memory usage. Workstation-oriented software like SPECint95 does not limit variable size nearly as much because workstation memory is roomy by comparison.
- Unsigned data are more common in embedded code.
- Logical (as opposed to arithmetic) operations occur more frequently in embedded code.

- Many embedded functions only perform side effects (such as flipping an I/O bit). They do not return values.
- Embedded code employs global data more frequently for variables and for large constant data.
- Embedded programs rarely use dynamic memory allocation.

Although Engblom's intent in writing this paper was to show that SPECint95 code was not suitable for benchmarking embedded development tools such as compilers, his observations about the characteristic differences between SPECint95 code and typical embedded application code are also significant for embedded processors and processor cores. Engblom's observations underscore the maxim that the best benchmark code is always the actual target application code.

10.7.2 BERKELEY DESIGN TECHNOLOGY, INC.

Berkeley Design Technology, Inc. (BDTI) is a technical services company that has focused exclusively on the applications of DSP processors since 1991. BDTI helps companies select, develop, and use DSP technology by providing expert advice and consulting, technology analysis, and highly optimized software development services. As part of those services, BDTI has spent more than a decade developing and applying DSP benchmarking tools. As such, BDTI serves as a private third party that develops and administers DSP benchmarks.

BDTI introduced its core suite of DSP benchmarks, formally called the "BDTI Benchmarks," in 1994. The BDTI Benchmarks consist of a suite of 12 algorithm kernels that represent key DSP operations used in common DSP applications. BDTI revised, expanded, and published information about the 12 DSP algorithm kernels in its BDTI Benchmark in 1999 [18]. The 12 DSP kernels in the BDTI Benchmark appear in Table 10.7.

BDTI's DSP benchmark suite is an example of a specialized processor benchmark. DSPs are not general-purpose processors and therefore require specialized benchmarks. BDTI Benchmarks are not full applications as are SPEC benchmark components. The BDTI Benchmark is a hybrid benchmark that uses code (kernels) extracted from actual DSP applications.

TABLE 10.7   BDTI Benchmark Kernels

Function	Function Description	Example Applications
Real block FIR	FIR filter that operates on a block of real (not complex) data	Speech processing (e.g., G.728 speech coding)
Complex block FIR	FIR filter that operates on a block of complex data	Modem channel equalization
Real single-sample FIR	FIR filter that operates on a single sample of real data	Speech processing and general filtering
LMS adaptive FIR	Least-mean-square adaptive filter that operates on a single sample of real data	Channel equalization, servo control, and linear predictive coding
IIR	IIR filter that operates on a single sample of real data	Audio processing and general filtering
Vector dot product	Sum of the point-wise multiplication of two vectors	Convolution, correlation, matrix multiplication, and multidimensional signal processing
Vector add	Point-wise addition of two vectors, producing a third vector	Graphics, combining audio signals or images
Vector maximum	Find the value and location of the maximum value in a vector	Error control coding and algorithms using block floating point
Viterbi decoder	Decode a block of bits that have been convolutionally encoded	Error control coding
Control	A sequence of control operations (test, branch, push, pop, and bit manipulation)	Virtually all DSP applications including some control code
256-point, in-place FFT	FFT that converts a time domain signal to the frequency domain	Radar, sonar, MPEG audio compression, and spectral analysis
Bit unpack	Unpacks variable-length data from a bit stream	Audio decompression and protocol handling

BDTI calls its benchmarking methodological approach "algorithm kernel benchmarking and application profiling." The algorithm kernels used in the BDTI Benchmarks are functions that constitute the building blocks used by most signal processing applications. BDTI extracted these kernels from DSP application code. The kernels are the most computationally intensive portions of the donor DSP applications. They are written in assembly code and therefore must be hand ported to each new DSP architecture. BDTI believes that the extracted kernel code is more rigorously defined than the large DSP applications, such as a V.90 modem or a Dolby AC-3 audio codec, which perform many functions in addition to the core DSP algorithms.

The *application-profiling* portion of the BDTI Benchmark methodology refers to a set of techniques used by BDTI to either measure or estimate the amount of time, memory, and other resources that an application spends executing various code subsections, including subsections that correspond to the DSP kernels in the BDTI Benchmark. BDTI develops kernel-usage estimates from a variety of information sources including application-code inspections, instrumented code-run simulations, and flow-diagram analysis.

BDTI originally targeted its benchmark methodology at actual DSP processor chips, not processor cores. The key portion of the benchmark methodology that limited its use to processors and not cores was the need to conduct the benchmark tests at actual clock speeds because BDTI did not want processor manufacturers to test this year's processors and then extrapolate the results to next year's clock speeds. However, ASIC and SoC designers must evaluate and compare processor core performance long before a chip has been fabricated. Processor core clock speeds will depend on the IC fabrication technology and cell libraries used to create the IC. Consequently, BDTI adapted its benchmarking methodology to evaluate processor cores. BDTI uses results from a cycle-accurate simulator and worst-case projected clock speeds (based on gate-level processor core simulation and BDTI's evaluation of the realism of that projection) to obtain benchmark results from processor cores.

BDTI does not release the BDTI Benchmark code. Instead, it works with processor vendors (for a fee) to port the benchmark code to new processor architectures. BDTI acts as the objective third party; conducts the benchmark tests; and measures execution time, memory usage, and energy consumption for each benchmark kernel.

BDTI rolls a processor's execution times on all 12 DSP kernels into one composite number that it has dubbed the BDTImark2000 to satisfy people who prefer using one number to rank candidate processors. To prevent the mixing of processor benchmark scores verified with hardware and simulated processor core scores, BDTI publishes the results of simulated benchmark scores under a different name: the BDTIsimMark2000. Both the BDTImark2000 and the BDTIsimMark2000 scores are available without charge on BDTI's website.

BDTI publishes many of the results on its website (www.BDTI.com) and the company sold some of the results of its benchmark tests in reports on specific DSP processors and in a book called the "Buyers Guide to DSP Processors." As of mid 2014, the company is no longer actively selling industry reports. However, the company does still offer the BDTI Benchmark Information Service (www.bdti.com/Services/Benchmarks/BIS).

10.7.3 EMBEDDED MICROPROCESSOR BENCHMARK CONSORTIUM

EDN magazine's legacy of microprocessor benchmark articles grew into full-blown realization when EDN editor Markus Levy founded the nonprofit, embedded-benchmarking organization called EEMBC (www.eembc.org). EEMBC stands for the EDN Embedded Benchmark Consortium, which dropped the "EDN" from its name but not the corresponding "E" from its abbreviation in 1997. EEMBC's stated goal was to produce accurate and reliable metrics based on real-world embedded applications for evaluating embedded processor performance. Levy drew remarkably broad industry support from microprocessor and DSP vendors for his concept of a benchmarking consortium, which had picked up 21 founding corporate members by the end of 1998: Advanced Micro Devices, Analog Devices, ARC, ARM, Hitachi, IBM, IDT, Lucent Technologies, Matsushita, MIPS, Mitsubishi Electric, Motorola, National Semiconductor, NEC, Philips, QED, Siemens, STMicroelectronics, Sun Microelectronics, Texas Instruments, and Toshiba [19]. EEMBC (pronounced "embassy") spent nearly 3 years working on a suite of benchmarks for testing embedded microprocessors and introduced its first benchmark suite at the Embedded Processor Forum in 1999. EEMBC released its first certified scores in 2000 and, during the same year, announced that it would start to certify simulation-based benchmark scores so that processor cores could be benchmarked. As of 2014, EEMBC has nearly 50 corporate members, 100 commercial licensees, and more than 120 university licensees.

The current EEMBC processor benchmarks are contained in 10 suites loosely grouped according to application:

- *CoreMark*: EEMBC CoreMark [20] is a simple, yet sophisticated, benchmark that is designed specifically to test the functionality of a processor core. CoreMark consists of one integer workload with four small functions written in easy-to-understand ANSI C code with a realistic mixture of read/write operations, integer operations, and control operations. CoreMark has a total binary size of no more than 16 kB when compiled with the gcc compiler on an x86 machine. This small size makes CoreMark more convenient to run using simulation tools. Unlike EEMBC's primary benchmark suites, CoreMark is not based on any real application, but the workload consists of several commonly used algorithms that include matrix manipulation to exercise hardware multiplier/accumulators (MACs) and common math operations, linked-list manipulation to exercise the common use of pointers, state machine operation to exercise data-dependent branch instructions, and CRC (cyclic redundancy check). CoreMark is not system dependent. It exhibits the same behavior regardless of the platform (big/little endian, high-end or low-end processor). Running CoreMark produces a single-number score for making quick comparisons between processors. The EEMBC CoreMark was specifically designed to prevent the kind of cheating that took place with the Dhrystone benchmark. The CoreMark work cannot be optimized away or there will be no results. Furthermore, CoreMark does not use special libraries that can be artificially manipulated and it makes no library calls from within the timed portion of the benchmark, so *special* benchmark-optimized compiler libraries will not artificially boost a processor's CoreMark score.
- *CoreMark-HPC*: Released by EEMBC in 2014, the CoreMark-HPC is a more advanced, wider scope benchmark that includes five integer and four floating-point workloads, with data set sizes that range from 42 kB to more than 3 MB (per context when used in a multicore implementation).
- *Ultralow-Power (ULP) Microcontrollers*: The EEMBC ULPBench-CoreProfile performs a variety of functions commonly found in ULP applications; among them are memory and math operations, sorting, GPIO interaction, etc. Unlike the other EEMBC performance benchmarks, the goal of ULPBench is to measure microcontroller energy efficiency. To go along with benchmark software of ULPBench, EEMBC developed its EnergyMonitor, an inexpensive, yet highly accurate energy measuring device. Using ULPBench, the microcontroller wakes up once per second and performs the given workload of approximately 20,000 cycles, then goes back to sleep. Hence, ULPBench is a measure of the energy consumed over a 1 s duty cycle, allowing microcontrollers to take advantage of their low power modes.
- *Floating Point*: Many embedded applications including audio, DSP/math, graphics, automotive, and motor control employ floating-point arithmetic. In the same way that the EEMBC CoreMark benchmark is a "better Dhrystone," the EEMBC FPMark provides a *better* embedded floating-point benchmark than Whetstone and LINPACK, which are not really embedded benchmarks. Uniquely, the FPMark provides a combination of single- and double-precision workloads, as well as a mixture of small, medium, and large data sets; this makes this benchmark useful for testing the range of low-end microcontrollers to high-end processors.
- *Multicore*: EEMBC MultiBench is a suite of embedded benchmarks that allows processor and system designers to analyze, test, and improve multicore architectures and platforms. It leverages EEMBC's proven library of application-focused benchmarks in hundreds of workload combinations using a thread-based API to establish a common programming model. The suite's individual benchmarks target three forms of concurrency:

- *Data decomposition*: Allows multiple threads to cooperate on achieving a unified goal and demonstrates a processor's support for fine-grained parallelism
- *Processing of multiple-data streams*: Uses common code running on multiple threads to demonstrate how well a multicore solution scales with multiple-data inputs
- *Multiple workload processing*: Shows the scalability of a solution for general-purpose processing and demonstrates concurrency over both code and data

- *Automotive Industrial*: EEMBC AutoBench is a suite of benchmarks that predict the performance of microprocessors and microcontrollers in automotive and industrial applications, but because of the diverse algorithms contained in this suite, it is more commonly used as a general-purpose benchmark. Its 16 benchmark kernels include the following:
  - *Generic Workload Tests*: These tests include bit manipulation, matrix mapping, a specific floating-point tester, a cache buster, pointer chasing, pulse-width modulation, multiplication, and shift operations (typical of encryption algorithms).
  - *Basic Automotive Algorithms*: These tests include controller area network (CAN), tooth to spark (locating the engine's cog when the spark is ignited), angle-to-time conversion, road speed calculation, and table lookup and interpolation.
  - *Signal Processing Algorithms*: These tests include sensor algorithms used for engine-knock detection, vehicle stability control, and occupant safety systems. The algorithms in this group include FFTs and inverse FFTs (iFFT), finite impulse response (FIR) filter, inverse discrete cosine transform (iDCT), and infinite impulse response (IIR) filter.

- *Digital Entertainment*: The EEMBC DENBench is a suite of benchmarks that approximates the performance of processor subsystems in multimedia tasks such as image, video, and audio file compression and decompression. Other benchmarks in the suite focus on encryption and decryption algorithms commonly used in digital rights management and eCommerce applications. The DENBench components include the following algorithms and minisuites:
  - *MPEG*: Includes MP3 decode, MPEG-2 encode and decode, and MPEG-4 encode and decode, each of which are applied to five different data sets for a total of 25 results
  - *Cryptography*: A collection of four benchmark tests for common cryptographic standards and algorithms including Advanced Encryption Standard (AES), Data Encryption Standard, Rivest–Shamir–Adleman algorithm for public-key cryptography, and Huffman decoding for data decompression
  - *Digital Image Processing*: JPEG compression and color-space-conversion tests including JPEG compress, JPEG decompress, RGB to YIQ, RGB to CMYK, and RGB to HPG. Seven different data sets are applied to each of these tests producing a total of 35 results
  - *MPEG Encode Floating Point*: A floating-point version of the MPEG-2 Encode benchmark, using single-precision floating-point arithmetic instead of integer functions

- *Networking*: A suite of benchmarks that allow users to approximate the performance of processors tasked with moving packets in networking applications. The suite's seven benchmark kernels include the following:
  - *IP Packet Check*: Tests the correct processing of IP packets. The first step is to validate the IP header information of all packets. The RFC1812 standard defines the requirements for packet checks carried out by IP routers and EEMBC created its Packet Check benchmark to model a subset of the IP header validation work specified in that standard.
  - *IP Reassembly*: Packets are often fragmented before being transmitted over the Internet from one part of the network to another and then reassembled upon arrival. The IP Reassembly benchmark measures processor performance when reconstructing these disjointed packets.
  - *IP Network Address Translator (NAT)*: Follows NAT rules to rewrite the IP addresses and port numbers of packets. Because rewriting each packet modifies its source IP address and port chosen by the algorithm, the NAT benchmark simulates an important part of network processing for many router designs.
  - *Route Lookup*: Receives and forwards IP packets using a mechanism commonly applied to commercial network routers employing a Patricia Tree data structure,

which is a compact binary tree that allows fast and efficient searches with long or unbounded length strings. The benchmark monitors the processor's ability to check the tree for the presence of a valid route and measures the time needed to walk through the tree to find the destination node for packet forwarding.

- *Open Shortest Path First (OSPF)*: Implements the Dijkstra shortest path first algorithm, which is widely used in routers and other networking equipment.
- *Quality of Service (QoS)*: Transporting voice, video, and multimedia packets present a greater challenge than transporting simple text and files because packet timing and order are critical for smooth multimedia playback. QoS processing tests measure data transfer and error rates to ensure that they are suitable to support such applications.
- *TCP*: EEMBC's TCP benchmark is designed to reflect performance in three different network scenarios. The first component, Gigabit Ethernet (*TCP Jumbo*), represents the likely workload of Internet backbone equipment using large packet transfers. The second (*TCP Bulk*) concentrates on large transfers of packets using protocols such as FTP. The last component (*TCP Mixed*) focuses on the relay of mixed traffic types, including Telnet, FTP, and HTTP. The benchmark processes all of the packet queues through a server task, network channel, and client task. Simulating the data transfers through the connections reveals how the processor will realistically cope with various forms of TCP-based traffic.

- *Text and Image Processing*: EEMBC OABench approximates the performance of processors in printers, plotters, and other office automation systems that handle text and image processing tasks. Its five benchmark kernels include the following:
  - *Bezier*: Benchmarks the classic Bezier curve algorithm by interpolating a set of points defined by the four points of a Bezier curve.
  - *Dithering*: Uses the Floyd–Steinberg error diffusion dithering algorithm.
  - *Ghostscript*: Provides an indication of the potential performance of an embedded processor running a PostScript printer engine. OABench can be run with or without this benchmark.
  - *Image Rotation*: Uses a bitmap rotation algorithm to perform a clockwise 90° rotation on a binary image.
  - *Text Parsing*: Parses Boolean expressions made up of text strings and tests bit manipulation, comparison, and indirect reference capabilities.
- *Telecomm (DSP)*: A suite of benchmarks that allows users to approximate the performance of processors in modem and related fixed-telecom applications. Its benchmark kernels include representative DSP algorithms:
  - *Autocorrelation*: A mathematical tool used frequently in signal processing for analyzing functions or series of values, such as time domain signals. Produces scores from three different data sets: pulse, sine, and speech.
  - *Convolutional Encoder*: Supports a type of error-correcting code based on an algorithm often used to improve the performance of digital radio, mobile phones, satellite links, and Bluetooth implementations.
  - *Bit Allocation*: Tests the target processor's ability to spread a stream of data over a series of buffers (or *frequency bins*) and then modulates and transmits these buffered streams on a telephone line in a simulated ADSL application.
  - *iFFT*: Tests the target processor's ability to convert frequency domain data into time domain data.
  - *FFT*: Tests the target processor's ability to convert time domain data into frequency domain data.
  - *Viterbi Decoder*: Tests the processor's ability to recover an output data packet from an encoded input data packet in embedded IS-136 channel coding applications.
- *Consumer*: The EEMBC ConsumerBench is a suite of benchmarks that approximates the performance of processors in digital still cameras, printers, and other embedded systems that handle digital imaging tasks. Its four benchmark kernels include the following:
  - *Image Compression and Decompression*: Consists of industry-standard JPEG compression and decompression algorithms. The EEMBC benchmarks provide

standardization of the JPEG options as well as the input data set to ensure level comparisons.

- *Color Filtering and Conversion*: Tests include a high-pass grayscale filter, as well as RGB-to-CMYK and RGB-to-YIQ color conversions.

EEMBC's processor benchmark suites with their individual benchmark programs allow designers to select the benchmarks that are relevant to a specific design, rather than lumping all of the benchmark results into one number [21,22]. However, many companies use all of the EEMBC processor benchmark suites combined to provide the most comprehensive level of testing for balancing instruction-type distribution, memory footprints, cache missing, branch predictions, and instruction-level parallelism (ILP). EEMBC's benchmark suites are developed by separate subcommittees, each working on one application segment. Each subcommittee selects candidate applications that represent the application segment and dissects each application for the key kernel code that performs the important work. This kernel code coupled with a test harness becomes the benchmark. Each benchmark has published guidelines in an attempt to force the processor vendors to play fair.

However, the industry's Dhrystone experience proved conclusively that some processor vendors would not play fair without a fair and impartial referee, so EEMBC created one: the EEMBC Technology Center (ETC). Just as major sports leagues hire referee organizations to conduct games and enforce the rules of play, EEMBC's ETC conducts benchmark tests and enforces the rules of EEMBC benchmarking to produce certified EEMBC scores. The ETC is also responsible for verifying the exact environment under which each test was run; this includes processor frequency, wait states, and compiler version and switches.

The EEMBC website lists certified and uncertified scores but EEMBC only guarantees the reliability of scores that have been officially certified by the ETC. During its certification process, the ETC reestablishes the manufacturer's benchmark environment, verifies all settings, rebuilds the executable, and runs CoreMark according to the specific run rules. EEMBC certification ensures that scores are repeatable, accurate, obtained fairly, and derived according to EEMBC's rules. Scores for devices that have been tested and certified can be searched from EEMBC's CoreMark Benchmark search page.

EEMBC publishes benchmark scores on its website (www.EEMBC.org) and the organization's work is ongoing. To prevent benchmark rot, EEMBC's subcommittees constantly evaluate revisions to the benchmark suites and have added benchmarks including EEMBC's system-level benchmarks, which include the following:

- *AndEBench and AndEBench-Pro*: An industry-accepted method of evaluating Android hardware and platform performance. These benchmarks are a free download in the Google Play market. AndEBench is designed to be easy to run with just the push of a button.
- *BrowsingBench*: It provides a standardized, industry-accepted method of evaluating web browser performance. This benchmark
  - Targets smartphones, netbooks, portable gaming devices, navigation devices, and IP set-top boxes
  - Measures the complete user experience—from the click/touch on a URL to the final page rendered and scrolled on the screen
  - Factors in Internet content diversity as well as various network profiles used to access the Internet
- *DPIBench (in development as of the mid-2014)*: This provides a standardized, industry-accepted method of evaluating systems and processors performing deep-packet inspection:
  - Targets network security appliances
  - Measures throughput and latency and quantifies number of flows
  - Enables consumers of firewall technologies with an objective means of selecting a solution from the myriad of vendor offerings

Except for CoreMark and CoreMark-HPC, the EEMBC benchmark code can be licensed for nonmember evaluation, analysis, and criticism by industry analysts, journalists, and independent third parties. Such independent evaluations would be unnecessary if all of the EEMBC corporate members tested their processors and published their results. However, fewer than half of the

EEMBC corporate members have published any benchmark results for their processors and few have tested all of their processors [23].

10.7.4 MODERN PROCESSOR BENCHMARKS FROM ACADEMIC SOURCES

The EEMBC CoreMark benchmark has become the de facto industry standard for measuring embedded microprocessor performance but its use is limited to EEMBC licensees, industry analysts, journalists, and independent third parties. Consequently, two university sources produced freely available benchmarks to give everyone access to nontrivial processor benchmarking code.

10.7.4.1 UCLA'S MEDIABENCH 1.0

The first academic processor benchmark was MediaBench, originally developed by the Computer Science and Electrical Engineering Departments at the University of California at Los Angeles (UCLA) [24]. The MediaBench 1.0 benchmark suite was created to explore compilers' abilities to exploit ILP in processors with very long instruction word and SIMD structures. The MediaBench benchmark suite is aimed at processors and compilers targeting new-media applications and consists of several applications culled from image processing, communications, and DSP applications and includes a set of input test data files to be used with the benchmark code.

Since then, a MediaBench II has appeared, supported by the MediaBench Consortium on a website at Saint Louis University (http://euler.slu.edu/~fritts/mediabench/); however, there does not seem to have been much activity on the site for a while. Applications used in MediaBench II include the following:

- *Composite*: The *composite* benchmark, *MBcomp*, suite will contain 1 or 2 of the most advanced applications from each of the individual media benchmarks (except kernels).
- *Video and Image*: This image- and video-centric media benchmark suite is composed of the encoders and decoders from six image/video compression standards: JPEG, JPEG-2000, H.263, H.264, MPEG-2, and MPEG-4.
- *Audio and Speech*: Listed as under development.
- *Security*: Listed as under development.
- *Graphics*: Listed as under development.
- *Analysis*: Listed as under development.
- *Kernels*: Listed as under development.

Benchmarks for the composite and video and image suites appear in Tables 10.8 and 10.9.

TABLE 10.8   MediaBench II Composite Benchmark Programs

Benchmark	Application Description	Example Applications
H.264	Video Codec	Video
JPEG-2000	Image Codec	Still images

TABLE 10.9   MediaBench II Video and Image Benchmark Programs

MediaBench 1.0 Application	Application Description	Example Applications
H.264/MPEG-4 AVC	Video codec	Digital video cameras and HD and 4K2K displays
MPEG-4 ASP	MPEG-4 video codec	Digital television and DVD
JPEG-2000	Image codec	Printers
MPEG-2	Video codec	Digital cameras and displays
H.263	Video codec	Videoconferencing
JPEG	Image codec	Digital cameras, displays, and printers

10.7.4.2 MIBENCH FROM THE UNIVERSITY OF MICHIGAN

A comprehensive benchmark suite called MiBench was presented by its developers from the University of Michigan at the IEEE's 4th annual Workshop on Workload Characterization in December 2001 [25]. MiBench intentionally mimics EEMBC's benchmark suite. It includes a set of 35 embedded applications in six application-specific categories: automotive and industrial, consumer devices, office automation, networking, security, and telecommunications. Table 10.10 lists the 35 benchmark programs (plus two repeated benchmark tests) in the six categories. All of the benchmark programs in the MiBench suite are available in standard C source code at http://www.eecs.umich.edu/mibench.

The MiBench automotive and industrial benchmark group tests a processor's control abilities. These routines include

- *basicmath*: Cubic function solving, integer square root, and angle conversions
- *bitcount*: Counts the number of bits in an integer array
- *qsort*: Quicksort algorithm applied to a large string array
- *susan*: Image recognition program developed to analyze MRI brain images

MiBench applications in the consumer device category include

- *JPEG encode/decode*: Still-image codec
- *tiff2bw*: Converts a color tiff image into a black-and-white image
- *tiff2rgba*: Converts a color image into one formatted into red, green, and blue
- *tiffdither*: Dithers a black-and-white image to reduce its size and resolution
- *tiffmedian*: Reduces an image's color palette by taking several color-palette medians
- *lame*: The Lame MP3 music encoder
- *mad*: A high-quality audio decoder for the MPEG1 and MPEG2 video formats
- *typeset*: A front-end processor for typesetting HTML files

Applications in the MiBench office automation category include

- *Ghostscript*: a PostScript language interpreter minus a graphical user interface
- *stringsearch*: Searches a string for a specific word or phrase
- *ispell*: A fast spelling checker
- *rsynth*: A text-to-speech synthesis program
- *sphinx*: A speech decoder

The set of MiBench networking applications includes

- *Dijkstra*: Computes the shortest path between nodes
- *Patricia*: A routing-table algorithm based on Patricia tries

The networking group also reuses the CRC32, sha, and Blowfish applications from the MiBench suite's security application group.

The security group of MiBench applications includes

TABLE 10.10   MiBench Benchmark Programs

Auto/Industrial	Consumer Devices	Office Automation	Networking	Security	Telecom
basicmath	JPEG	Ghostscript	Dijkstra	Blowfish encoder	CRC32
bitcount	lame	ispell	patricia	Blowfish decoder	FFT
qsort	mad	rsynth	(CRC32)	PGP sign	IFFT
susan (edges)	tiff2bw	sphinx	(sha)	PGP verify	ADPCM encode
susan (corners)	tiff2rgba	stringsearch	(Blowfish)	Rijndael encoder	ADPCM decoder
susan (smoothing)	tiffdither			Rijndael decoder	GSM encoder
	tiffmedian			sha	GSM decoder
	typeset				

- *Blowfish encryption/decryption*: A symmetric block cipher with a variable-length key
- *sha*: A secure hashing algorithm
- *Rijndale encryption/decryption*: The cryptographic algorithm used by the AES encryption standard
- *PGP sign/verify*: A public-key cryptographic system called "Pretty Good Privacy"

MiBench's telecommunications application suite includes the following applications:

- *FFT/IFFT*: FFT and the iFFT
- *GSM encode/decode*: European voice codec for mobile telephony
- *ADPCM encode/decode*: Adaptive differential pulse code modulation speech-compression algorithm
- *CRC32*: A 32-bit cyclic redundancy check

With its 35-component benchmark applications, MiBench certainly provides a thorough workout for any processor/compiler combination. Both MediaBench and MiBench solve the problem of proprietary code for anyone who wants to conduct a private set of benchmark tests by providing a standardized set of benchmark tests at no cost and with no use restrictions. Industry analysts, technical journalists, and researchers can publish results of independent tests conducted with these benchmarks, although none seem to have done so, to date.

The trade-off made with these processor benchmarks from academia is that there is no official body to enforce benchmarking rules, to provide result certification, or even to drive future development. Nothing seems to have happened on the MiBench website for more than a decade, yet the state of the art in algorithm development has certainly advanced over that time, as has processor capability. For someone conducting their own set of benchmark tests, self-certification may be sufficient. For anyone else, the entity publishing the results must be scrutinized for fairness in the conducting of tests, for bias in test comparisons among competing processors, and for bias in any conclusions.

10.8 CONFIGURABLE PROCESSORS AND THE FUTURE OF PROCESSOR CORE BENCHMARKS

For the past 30 years, processor benchmarks have attempted to show how well specific processor architectures work. Myriad benchmark programs developed over that period have had mixed success in achieving this goal. One thing that has been constant over the years is the use of benchmarks to compare microprocessors and processor cores with fixed ISAs. Nearly all processors realized in silicon have fixed architectures and the processor cores available for use in FPGAs, ASICs, and SoCs generally have had fixed architectures as well. However, the transmutable silicon of FPGAs, ASICs, and SoCs makes it feasible to employ configurable processor cores instead of fixed ones, which greatly complicates the use of comparison benchmark programs because configurable processors can be tuned to specific applications by adding instructions and registers, which sort of invalidates the whole idea of benchmarking. Nevertheless, you can use benchmarking tools to guide you to the best processor configuration for a specific application or set of applications.

Processor vendors take two fundamental approaches to making configurable ISA processors available to FPGA users and ASIC and SoC designers. The first approach provides tools that allow the designer to modify a base processor architecture to boost the processor's performance on a target application or set of applications. Xilinx and Altera both offer configurable processor cores for their FPGAs (MicroBlaze and NIOS, respectively). Companies taking this approach for ASIC and SoC designers include the portion of Imagination Technologies formerly known as MIPS, the portion of Synopsys formerly known as ARC/Virage, and the portion of Cadence formerly known as Tensilica. (Imagination acquired MIPS; Synopsys acquired Virage, which had formerly acquired ARC; and Cadence acquired Tensilica.) The second approach employs tools that compile entirely new processor architectures for each application. Two companies subsequently acquired by Synopsys took this approach: CoWare and Target Compilers.

The ability to tailor a processor architecture to a target application (which includes benchmarks) allows you to trade off performance against on-chip resources. Tailoring can drastically

TABLE 10.11 **Xilinx Configurable MicroBlaze Processor Relative Performance on EEMBC Benchmarks**

EEMBC Benchmark	Xilinx MicroBlaze Processor Minimum Resources (Normalized)	Xilinx MicroBlaze Processor Maximum Performance (Normalized)	Xilinx MicroBlaze Processor Intermediate Performance (Normalized)
Angle-to-time conversion	1.00	11.48	4.65
Basic floating point	1.00	5.42	3.63
Bit Manipulation	1.00	11.38	6.29
Cache buster	1.00	11.44	7.45
Response to remote request (CAN)	1.00	10.23	8.51
FFT	1.00	18.52	7.94
FIR filter	1.00	9.21	8.55
iDCT	1.00	21.97	11.94
Low-pass filter (IIR) and DSP functions	1.00	35.55	10.88
iFFT	1.00	20.27	8.47
Matrix math	1.00	5.99	3.20
Pointer chasing	1.00	2.47	1.99
Pulse-width modulation	1.00	11.24	10.12
Road speed calculation	1.00	16.04	14.71
Table lookup	1.00	15.61	4.51
Tooth to spark	1.00	8.25	4.61
Automark™	1.00	11.35	6.49
JPEG compression benchmark	1.00	1.67	0.92
JPEG decompression benchmark	1.00	1.67	1.07
Grayscale image filter	1.00	1.67	1.68
RGBCMY01 (consumer RGB to CMYK)	1.00	1.07	1.00
RGBYIQ01 (consumer RGB to YIQ)	1.00	47.88	44.71
Consume mark™	1.00	2.99	2.37
TCP-BM jumbo	1.00	2.92	1.84
TCP-BM bulk	1.00	4.60	2.90
TCP-BM mixed	1.00	6.16	3.62
TCPmark™	1.00	4.36	2.68
Networking: IP Packet Check Benchmark	1.00	8.55	6.21
Networking: IP Packet Check Benchmark	1.00	8.05	6.19
Networking: IP Packet Check Benchmark	1.00	7.92	6.19
Networking: IP Packet Check Benchmark	1.00	7.57	5.96
Networking: QoS	1.00	6.69	5.95
Networking: Route Lookup Benchmark	1.00	10.07	7.77
Networking: OSPF Benchmark	1.00	4.61	3.24
Networking: IP Reassembly Benchmark	1.00	3.45	3.49
Network Address Translation	1.00	6.85	3.87
IPmark™	1.00	6.78	5.22

increase a program's execution speed and it can drastically increase or decrease the amount of on-chip resources needed to implement the processor. For example, Table 10.11 shows the relative performance for three different Xilinx MicroBlaze processor configurations (minimal, intermediate, and maximum performance) over four EEMBC benchmark suites. These MicroBlaze processor configurations (see Tables 10.12 through 10.14 for configuration details) were created specifically to illustrate benchmark performance.

Note that the benchmark performance varies among the processor configurations by more than an order of magnitude depending on the benchmark. So which configuration is the best? That is a trick question. Benchmark performance alone would not tell you which processor configuration to select. You must also consider the resources needed to achieve these performance scores.

TABLE 10.12   Xilinx Configurable MicroBlaze Processor: Minimum Resources Configuration

MicroBlaze v9.3, Vivado 2014.1—247 MHz

Minimum Resources: Three-Stage Pipeline, No Barrel Shifter, No Hardware Multiplier, No FPU

Certification report	None
Type of platform	FPGA xc7k325tffg900-3, KC705 board
Type of certification	Optimized
Certification date	N/A
Benchmark notes	C optimized
Hardware type	Production silicon
Native data type	32 bits
Architecture type	RISC
Processor cooling method	Fan
Maximum power dissipation	Not measured
Names and nominal voltage levels	VDD 1.5 V, VDD DDR 1.5 V, VDD I/O 2.5 V
Processor process technology	28 nm
L1 instruction cache size (kB)	0.25
L1 data cache size (kB)	0.25
External data bus width (bits)	32
Memory clock (MHz)	400
Memory configuration	4-3-3-3
L2 cache size (kB)	0
L2 cache clock	N/A

TABLE 10.13   Xilinx Configurable MicroBlaze Processor: Intermediate Performance and Intermediate Resources Configuration

MicroBlaze v9.3, Vivado 2014.1—247 MHz

Intermediate Resources: Five-Stage Pipeline, Barrel Shifter, Hardware Multiplier, No FPU

Certification report	None
Type of platform	FPGA xc7k325tffg900-3, KC705 board
Type of certification	Optimized
Certification date	N/A
Benchmark notes	C optimized
Hardware type	Production silicon
Native data type	32 bits
Architecture type	RISC
Processor cooling method	Fan
Maximum power dissipation	Not measured
Names and nominal voltage levels	VDD 1.5 V, VDD DDR 1.5 V, VDD I/O 2.5 V
Processor process technology	28 nm
L1 instruction cache size (kB)	4
L1 data cache size (kB)	4
External data bus width (bits)	32
Memory clock (MHz)	400
Memory configuration	4-3-3-3
L2 cache size (kB)	0
L2 cache clock	N/A

TABLE 10.14 Xilinx Configurable MicroBlaze Processor: Maximum Performance Configuration

MicroBlaze v9.3, Vivado 2014.1—247 MHz

Maximum Performance: Five-Stage Pipeline, Barrel Shifter, Hardware Multiplier, FPU

Certification report	None
Type of platform	FPGA xc7k325tffg900-3, KC705 board
Type of certification	Optimized
Certification date	N/A
Benchmark notes	C optimized
Hardware type	Production silicon
Native data type	32 bits
Architecture type	RISC
Processor cooling method	Fan
Maximum power dissipation	Not measured
Names and nominal voltage levels	VDD 1.5 V, VDD DDR 1.5 V, VDD I/O 2.5 V
Processor process technology	28 nm
L1 instruction cache size (kB)	16
L1 data cache size (kB)	16
External data bus width (bits)	32
Memory clock (MHz)	400
Memory configuration	4-3-3-3
L2 cache size (kB)	0
L2 cache clock	N/A

In the case of the three Xilinx MicroBlaze processor configurations shown in Table 10.11, the maximum performance configuration consumes approximately four times as many on-chip FPGA resources as the minimum resources configuration and the intermediate performance configuration consumes roughly half as much FPGA resource as the maximum performance configuration and twice as much resource as the minimum resources configuration.

Design teams should always pick the configuration that achieves the project's performance goals while consuming as little resource as possible. Benchmarks give design teams the guidance they need to make these configuration choices when the actual application code is not available.

The Xilinx MicroBlaze soft processor core is a 32-bit RISC Harvard architecture processor core included with Xilinx FPGA design tools (Vivado Design Edition, Vivado Webpack Edition, and ISE). The MicroBlaze core has more than 70 configurable options, allowing designers using Xilinx FPGAs to create many, many processor configurations for applications ranging from small-footprint state machines and microcontroller-like processors to high-performance, compute-intensive processors that can run the Linux OS. MicroBlaze processors can have three- or five-stage pipelines to optimize speed and have many other configuration options including ARM AXI or Xilinx processor local bus interfaces, memory management units, instruction and data-side caches, and an floating-point unit (FPU)—yet all of these configurations are still based on the same basic MicroBlaze processor architecture. Tables 10.12 through 10.14 show the salient configuration parameters for the three Xilinx MicroBlaze processor configurations used to generate the relative EEMBC benchmark performance numbers in Table 10.11. Of these configuration parameters, the key differences among the three processor configurations include

- Pipeline stages
  - *Minimum resources configuration*: three-stage pipeline
  - *Intermediate resources configuration*: five-stage pipeline
  - *Maximum performance configuration*: five-stage pipeline
- Hardware execution units
  - *Minimum resources configuration*: without barrel shifter, hardware multiplier, or FPU
  - *Intermediate resources configuration*: with barrel shifter and hardware multiplier, but without FPU
  - *Maximum performance configuration*: with barrel shifter, hardware multiplier, and FPU

- Cache sizes
  - *Minimum resources configuration*: 256 byte instruction and data caches
  - *Intermediate resources configuration*: 4 kB instruction and data caches
  - *Maximum performance configuration*: 16 kB instruction and data caches

Note that GCC compiler settings were the same for all Xilinx MicroBlaze processor Benchmark testing:

$$-O3 \ -mcpu = 9.3 \ -mlittle\text{-}endian \ -fno\text{-}asm \ -fsigned\text{-}char$$

Also note that the performance numbers shown in Table 10.11 are all normalized to an operating clock rate of 247 MHz. In general, you can expect processor configurations with more pipeline stages to operate at faster maximum clock rates than processors with fewer pipeline stages because there is less logic in each pipeline stage. However, carrying that differential into the benchmark results adds yet another dimension to the complexity of evaluating the various configurations. The clock rates for the three MicroBlaze processor configurations are normalized to simplify comparisons in these benchmark results. In many designs, maximum achievable clock rate will be a further consideration, not directly tied to but certainly related to the benchmark results.

10.9 CONCLUSION

As a growing number of processor cores are employed to implement a variety of on-chip tasks, development teams creating designs with FPGAs, ASICs, and SoCs increasingly rely on benchmarking tools to evaluate processor core performance. The tool of first recourse is always the actual application code for the target task. However, this particular evaluation tool is not always conveniently at hand when the processor selection is needed to start hardware design. Standard processor benchmark programs must often stand in as replacement tools for processor evaluation. These tools have undergone more than two decades of evolution and their use (and abuse) is now well understood.

Benchmarks have evolved over time. Initially, they were used to compare mainframes and supercomputers. When microprocessors first appeared, benchmarks quickly became a way of comparing processor offerings from different semiconductor vendors. Expansion into the embedded markets placed new emphasis on the ability to benchmark processors as the number of microprocessors and microcontrollers exploded. Benchmark evolution has not ended. More processor benchmarks are always in the wings. EEMBC is continually expanding its reach with new benchmarks, for example. The transitions to multicore processors, nanometer lithography, mobile systems, and the low-power portion of the Internet of Things have placed a new emphasis on the amount of power and energy required to execute on-chip tasks, so power-oriented benchmarks are growing in importance and benchmarking organizations including EEMBC and SPEC are developing new processor benchmarks to measure power dissipation on standardized tasks. EEMBC's ULPBench (ULP benchmark) is one example of such a benchmark.

As long as there are microprocessors and processor cores, we will need benchmarks to compare them.

REFERENCES

1. J. Rowson, Hardware/software co-simulation, in *Proceedings of the 31st Design Automation Conference (DAC'94)*, San Diego, CA, June 1994.
2. S. Gill, The diagnosis of mistakes in programmes on the EDSAC, in *Proceedings of the Royal Society Series A Mathematical and Physical Sciences*, Cambridge University Press, London, U.K., pp. 538–554, May 22, 1951.
3. Lord Kelvin (William Thomson), *Popular Lectures and Addresses*, Vol. 1, Macmillan and Co, 1889, p. 73 (originally: Lecture to the Institution of Civil Engineers, May 3, 1883).
4. F. McMahon, The Livermore FORTRAN Kernels: A computer test of the numerical performance range, Technical report, Lawrence Livermore National Laboratory, Livermore, CA, December 1986.

5. J.J. Dongarra, LINPACK working note #3, FORTRAN BLAS timing, Argonne National Laboratory, Argonne, IL, November 1976.

6. J.J. Dongarra, J.R. Bunch, C.M. Moler, and G.W. Stewart, LINPACK working note #9, Preliminary LINPACK user's guide, ANL TM-313, Argonne National Laboratory, Argonne, IL, August 1977.

7. H.J. Curnow and B.A. Wichmann, A synthetic benchmark, *Computer Journal*, 19(1), 43–49, 1976.

8. R.P. Weicker, Dhrystone: A synthetic systems programming benchmark, *Communications of the ACM*, 27(10), 1013–1030, October 1984.

9. R.P. Weicker, Dhrystone benchmark: Rationale for version 2 and measurement rules, *SIGPLAN Notices*, 23(8), 49–62, August 1988.

10. R.P. Weicker, Understanding variations in dhrystone performance, Microprocessor report, pp. 16–17, May 1989.

11. B.H. Fletcher, FPGA embedded processors: Revealing true system performance, in *Embedded Training Program, Embedded Systems Conference*, San Francisco, CA, 2005, ETP-367, available at http://www.xilinx.com/products/design_resources/proc_central/resource/ETP-367paper.pdf.

12. Dhrystone benchmarking for ARM cortex processors: Application note 273, July 2011, ARM Ltd, available at http://infocenter.arm.com/help/index.jsp?topic=/com.arm.doc.dai0273a, last accessed November 30, 2015.

13. R.D. Grappel and J.E. Hemenway, A tale of four μPs: Benchmarks quantify performance, *EDN*, April 1, 1981, pp. 179–232.

14. S.H. Fuller, P. Shaman, D. Lamb, and W. Burr, Evaluation of computer architectures via test programs, in *AFIPS Conference Proceedings*, Vol. 46, pp. 147–160, June 1977.

15. D. Shear, EDN's DSP benchmarks, *EDN*, September 29, 1988, pp. 126–148.

16. W. Patstone, 16-bit μP benchmarks—An update with explanations, *EDN*, September 16, 1981, p. 169.

17. J. Engblom, Why SpecInt95 should not be used to benchmark embedded systems tools, in *Proceedings of the ACM SIGPLAN 1999 Workshop on Languages, Compilers, and Tools for Embedded Systems (LCTES'99)*, Atlanta, GA, May 5, 1999.

18. *Evaluating DSP Processor Performance*, Berkeley Design Technology, Inc. (BDTI), Berkeley, CA, 1997–2000, http://www.bdti.com/MyBDTI/pubs/benchmk_2000.pdf, last accessed November 30, 2015.

19. M. Levy, At last: Benchmarks you can believe, *EDN*, November 5, 1988, pp. 99–108.

20. S. Gal-On and M. Levy, Exploring CoreMark™—A benchmark maximizing simplicity and efficacy, 2010, available at www.eembc.org/techlit/whitepaper.php, last accessed November 30, 2015.

21. J. Poovey, T. Conte, M. Levy, and S. Gal-On, A benchmark characterization of the EEMBC benchmark suite, *IEEE Micro*, 29(5), 18–29, September–October 2009.

22. S. Gal-On and M. Levy, Creating portable, repeatable, realistic benchmarks for embedded systems and the challenges thereof, in *LCTES '12 Proceedings of the 13th ACM SIGPLAN/SIGBED International Conference on Languages, Compilers, Tools and Theory for Embedded Systems, ACM SIGPLAN Notices—LCTES '12*, 47(5), 149–152, May 2012.

23. T. Halfhill, Benchmarking the benchmarks, Microprocessor Report, August 30, 2004.

24. C. Lee, M. Potkonjak et al., MediaBench: A tool for evaluating and synthesizing multimedia and communications systems, in *MICRO 30, Proceedings of the 30th Annual IEEE/ACM International Symposium on Microarchitecture*, Research Triangle Park, NC, December 1–3, 1997.

25. M. Guthaus, J. Ringenberg et al., MiBench: A free, commercially representative embedded benchmark suite, in *IEEE Fourth Annual Workshop on Workload Characterization*, Austin, TX, December 2001.

High-Level Synthesis

Felice Balarin, Alex Kondratyev, and Yosinori Watanabe

CONTENTS

11.1 INTRODUCTION

The history of high-level synthesis (HLS) is long. HLS established its status as an active research topic in the EDA community by the late 1970s and was often introduced as "the next big thing" by the early 1990s, following the significant and very successful adoption of logic synthesis [1]. However, only recently have commercial design projects started using HLS as the primary vehicle in the hardware design flow. Even then, the commercial use of HLS was limited to design applications that were historically considered the sweet spot of HLS, dominated by data-processing operations with little control logic. This might suggest that HLS has had limited commercial success. On the other hand, industry users who have adopted HLS in their commercial design projects unanimously state that they would never go back to the register-transfer level (RTL)-based design flow. For them, HLS is an indispensable technology that enables them to achieve the quality of designs unattainable with an RTL-based design flow. Designs synthesized with HLS typically outperform manually written RTL designs in commercial design projects. Sometimes the HLS tools can better explore the design space to find more optimal implementations, but the main source of improvements is helping the design engineers identify the best microarchitectures for the target designs. When using HLS, designers can specify the intention by focusing only on a few key aspects of the architectures while leaving the tools to fill out the details automatically. With RTL design, all of those details must be explicitly specified, and while it is theoretically possible to spell out the details manually, the engineers or their management find it too time consuming and error prone to do so, especially when considering the effort needed to revise and maintain the code across multiple projects or to cope with specification changes.

The complexity of designs and stringent project schedules make the high reusability of design and verification assets an imperative requirement. Therefore, the type of designs that HLS is used for has been expanding from the traditional datapath-dominant designs to more control-centric designs. There are several major semiconductor companies today that use HLS in control and communication IPs extensively—those IPs are integrated into a broad range of SoCs that impose very different requirements for the IPs in terms of implementation such as clock frequencies or performance constraints, as well as functionality on specific features or I/O interface configurations. These companies concluded that they would not meet the project schedules if they needed to support all these diverse requirements by manually writing RTL models and verifying them. Instead, they write behavioral descriptions of the IPs in a highly configurable manner, so that the functionality and microarchitecture for individual design targets can be specified through simple reconfiguration of the same models, and HLS provides detailed implementations necessary to realize the specification. As a consequence, HLS is used in a diverse range of production designs of such IPs as memory and DMA controllers, cache controllers, power controllers, interconnect fabrics, bus interfaces, and network traffic switches, in addition to computation-acceleration IPs such as image and graphic processors and DSP algorithms.

In this chapter, we present technical details of a state-of-the-art HLS technology. We start by illustrating the steps that HLS tools execute in the synthesis procedure, and then we go into details of the individual steps using specific examples. Before doing so, let us first provide an overview, in the rest of this section, of the level of abstraction at which the behavioral descriptions for HLS are written, as well as types of design decisions and design transformations that HLS technology typically realizes. This overview will help the readers understand the kinds of automation that HLS technology focuses on and the kinds of design tasks that are left for humans to complete.

11.1.1 HOW "HIGH" IS HLS ANYWAY?

As its name suggests, HLS usually starts from a design description that is at a higher level of abstraction than RTL, the most common design-entry point in use today. We will highlight several specific design aspects that engineers must explicitly deal with in RTL code, but can be abstracted in HLS.

Some of the abstraction in HLS comes from the input language. The input language of most of the HLS tools currently in use is based on C++, either the language directly or via the SystemC extension of C++ [25]. Tools based on proprietary languages have not gained much traction mostly because of the lack of compilers and debuggers and the lack of legacy code. SystemC is an extension of C++ that can be implemented by a C++ library, so it can be developed using standard C++ compilers and debuggers. The C++ language was developed to design complex software systems and has a number of features to facilitate abstraction and encapsulation, for example, templates, polymorphism, class inheritance, and abstract interfaces. C/C++ and SystemC-based HLS tools support most of these mechanisms. They allow the hardware designer to apply abstraction techniques developed for software that go far beyond the capabilities of typical hardware design languages (HDLs) such as Verilog.

The examples used in this chapter will generally be in SystemC, but the issues discussed are generic to HLS, regardless of the input languages to be used. When language-specific differences arise, we point them out and examine them both from the point of view of C/C++ and SystemC-based tools.

In addition to C++ abstraction mechanisms, there are hardware-specific design issues that HLS tools can optimize, allowing additional abstraction of the design description. These capabilities can improve designers' productivity by automating tedious and error-prone tasks, such as specifying low-level details. They also make the input description easier to reuse by removing implementation-specific details and allowing tools to identify them instead, under specific implementation requirements provided to the tools.

HLS tools help RTL designers by automatically optimizing resource sharing. This makes it possible to better encapsulate the functionality, making it easier to understand and extend. They also do the *scheduling*, that is, the assignment of operations to cycles. This transformation is somewhat similar to retiming in logic synthesis, but HLS scheduling often changes the design much more significantly than logic synthesis retiming.

For example, consider the following code segments:

```
res = p1*c1+p3*c3-p0*c0-p2*c2;
```

and

```
t1 = (p0+p2-2*p1)*c4 +p1;
t2 = (p1+p3-2*p2)*c4 +p2;
res = t1*c5 + t2*(256-c5);
```

They represent alternative interpolation formulas for points **p0,...,p3**. A typical design may support different modes, and different formulas may be used in each mode. The first formula contains four multiplications. The second formula contains 6, but two of them are by a constant 2, which does not require a multiplier for implementation. Thus, to implement either formula over two cycles, two multipliers are sufficient. Furthermore, since the two formulas are never evaluated at the same time, these multipliers can be shared. To achieve this in RTL, all the multiplexing of intermediate results would have to be done explicitly and it would be very hard to separate and recognize the original formulas. In HLS, the formulas can be encapsulated in separate functions, and all the details about scheduling the multiplications in two cycles and multiplexing the right operands to each one based on the state and active mode are handled by HLS tools. Furthermore, if in the future it becomes necessary to implement yet another formula using the same multipliers, all one needs to do is to add another function, and all the multiplexing details will be automatically regenerated.

Starting with design descriptions that do not describe pipelining, HLS tools can create hardware that is pipelined, with computations that are properly balanced between the stages and a latency that is adjusted to optimize the design. If some of the design parameters, such as clock cycles or technology library, change, it is easy to rerun the tool and come up with a different design with a different number of stages and different computations in each stage. Generating such an alternative from an RTL implementation may take weeks.

```
void DUT::proc1() {                    void DUT::f1() {
  ...                                    rsp2.put(a * b * c * d);
  while (true)                         }
    req_code_t sel = req1.get();
    rsp1.put(sel);                     int DUT::f2() {
                                         return a + b + c + d;
    if (sel==0) {                      }
      status = 1;
    } else if (sel==1) {               void DUT::f3() {
      f1();                              rsp2.put(a * b - c * d);
    } else if (sel==2) {                 wait();
      int v = f2();                      rsp2.put(a * c - b * d);
      rsp2.put(v);                     }
    } else if (sel==3) {
      f3();
    } else {
      flag=1;
      status = 2;
      wait();
    }
    wait();
  }
}
```

FIGURE 11.1   Example of procedural code.

It is common in HLS to separate out the part of the design description that describes the computation from the parts that describe the communication. The communication parts can be developed separately, encapsulated in a library, and then reused in many different designs. These communication primitives can range from single point-to-point handshaking channels, to complex bus interfaces such as AXI4, relieving the designer from learning the details of the bus interface design, or even just from understanding all the details of the bus protocols. This leads to major savings in the overall design effort. We must note that it is possible to predesign bus interfaces in RTL as well. However, having these separate predefined components usually carries a performance penalty. In contrast, using HLS, even though communication primitives are specified separately, the tool will synthesize and optimize them together with computational parts, often leading to better designs.

Last but not least, HLS allows the designer to abstract the notion of state. In RTL, the designer needs to model design states explicitly by defining and manipulating state variables. HLS tools allow the designer to write in the more natural procedural way commonly used in software. HLS tools can automatically construct and implement control finite-state machine (FSMs) from such a description. Typically, the tools offer several FSM implementation options, which make it very easy to explore design alternatives that would require significant code changes if one needs to do in RTL.

Consider, for example, the code in Figure 11.1. HLS tools can process such code and extract the FSM(s) for it. The FSM for the main thread **proc1** will have a state for each of the two **wait()** statements in its body, but there will also be additional states associated with functions called by **proc1**. In particular, if the **put** and **get** functions have one state each, which is true in a typical implementation, then function **f1** adds one state and function **f3** adds 3: two from calls to **put** and one for **wait()**. The FSM implementation options may include a single one-hot encoded FSM or separate coordinated FSMs for **proc1**, **f1,** and **f3**, possibly with different encodings. In addition, an HLS tool may choose to create additional states if that helps it satisfy design objectives. Note also that in this code, the details of pin-level communication have been completely hidden in a library that implements the **put** and **get** functions. This allows the designer to focus on the intended functionality without spending time on details of interface protocols.

11.1.2 DESIGN DECISIONS MADE BY HLS

The output of HLS is RTL. It is therefore necessary that HLS tools make design decisions on those aspects described in the previous section, which must exist in the RTL but are abstracted away in the input of HLS.

Specifically, there are four primary types of design decisions made by HLS. The first type of decision is the structure of FSMs and the interactions among them. RTL specifies a set of interacting FSMs, but states are not specified explicitly in the HLS input. The nature of interactions between the FSMs determines the concurrency of hardware components where each component is associated with a single FSM. The HLS tools decide which parts of the behavior of the input description should be captured as single components and how these components should interact. Then for each component, the tools decide the number of states and their transitions to form an FSM.

In practice, today's HLS tools often use structures given in the input description to define individual components. For example, functions in the input description could become single components, or a language-like SystemC can explicitly specify concurrent threads in the input description, for which the tools define separate components. The designers can take advantage of this to express the intended concurrency in the design. When they have a clear idea about how the component-level concurrency should be created in the resulting hardware, they can write the input description in such a way that the underlying structure of the description leads the HLS tools to make this decision according to the designers' intention. The designers can also specify this intention separately from the input description. For example, some HLS tools allow the designers to specify which functions should be treated as single components and which ones should be absorbed into the components that call the functions.

The decisions about the interaction of these components determine how they communicate data between them. Let us consider a case where a function f() is called in another function, say main_control(), defined in the input behavioral description. Suppose that f() was selected to become a dedicated component, separate from main_control(). The simplest way to define the communication between them is that the component for main_control() assigns data used as the input of f() to some signals and asserts a control signal, indicating that the new data are available. The component for f() then retrieves the data and executes the function. Once completed, it assigns the results of the function to some signals and asserts a control signal to indicate that the result is available. The main_control() identifies that the control signal is asserted, and at that point it retrieves the data and acknowledges the retrieval. The component for f() then resets the control signal. There are some variations in this interaction. For example, the component for main_control() could wait for f() to complete its computation, whenever it assigns the input data for f(). Alternatively, it could proceed to other parts of the behavior of main_control() that do not depend on the result of f(), so that main_control() and f() could be executed in parallel in the resulting hardware. Further, it might be decided that main_control() sends new data to f(), before receiving the result from f() for the previous input data, which in effect implements a pipelined execution between the two components.

Similar to the structure of components, today's HLS tools often provide these implementation options to the users of the tools, so that the designers can express their intentions. The tools usually also have some heuristics to determine which implementation options should be used based on their analysis of the input behavior.

Now the structure of the hardware components is defined, as well as the manner in which they interact. At this point, HLS tools decide the structure of the FSM associated with each component. There are multiple factors to consider in defining the FSM structure, and we will introduce them gradually in the subsequent sections, starting from the simplest case.

The second type of decision made by HLS tools is to define hardware resources used for implementing the behavior. We often refer to this type of design decision as resource allocation. Resources are of two kinds: computational resources and data storage resources. Computational resources implement operations given in the input behavior, such as additions, multiplications, comparisons, or compositions of such. Data storage resources determine how data should be retained. Computational resources access these resources for retrieving inputs of the computation or writing the outputs. Data storage resources could be simply wires, or ports if the data are at the inputs or outputs of the components, or registers if the data need to remain intact across multiple states of the FSMs, or some sort of memories that keep data and provide a particular way to access them. We will describe how the decisions are made to choose particular kinds of data storage resources in HLS tools in the succeeding sections, again starting from the simplest case.

Computational resources also have variations. Even if they are all for the same type of computation, such as the addition of two numbers, there are different kinds of adders in terms of how the addition is done, each of which results in different implementation characteristics such as area or timing. HLS tools determine what kind of resources should be used, based on the context in which the computations are performed.

One of the contexts in which the resource allocation is determined is called scheduling, and this is the third type of design decision made in HLS. Scheduling decides which computations should be done at which states of the FSMs. Sometimes, there is no flexibility in making such a decision. For example, the communication between the hardware components might require that the result of a computation needs to be produced in the same state where the input data for the computation become available. In this case, this computation needs to be scheduled at this specific state. On the other hand, there are cases where the result of the computation is used in a state that can be reached only after many state transitions. In such a case, there are multiple possible states at which this computation might be done.

As you can see, these first three types of design decisions are all interdependent. This makes HLS a complex problem to solve, because decisions made for one type will impose certain consequences on choices one can make for the other types. In the rest of this chapter, we will provide more formal descriptions of the problems for which these types of design decision are made.

We stated in the beginning of this section that there are four types of decision made in HLS. What is the fourth one? It is the decision about interpreting the behavioral description given as input to HLS and defining the "objects" for which one can make the three types of design decisions described earlier. In this sense, this type of design decision has a slightly different nature from the other three types. That is, it is a decision about how HLS tools transform the behavioral descriptions to some data structures that represent the semantics of the behavior to be processed, in a manner suitable for making the aforementioned design decisions.

This process usually involves not only identifying the underlying behavior of the design but also representing it using some internal data structures so that good design decisions can be made in the subsequent steps in HLS. For example, many behavioral descriptions used for today's HLS employ procedural programming paradigms such as those found in C++. In these descriptions, many computations that could be executed in parallel are written in a sequential manner. During the transformation of the description to the internal data structures, HLS tools analyze the data dependencies among the individual computations, so that such potential parallel execution could be explored as a part of the subsequent synthesis process. Similarly, some simplifications of the representation are also made, if those are good for the rest of the synthesis process regardless of specific design decisions. For example, if some constant numbers are used as a part of the behavior, then the tools propagate this information throughout the design, which could reduce the number of bits required for data storage resources. Some simplifications are also possible on the control structure of the behavior. For example, a complex branch structure could be simplified if the condition used in one branch can be subsumed by another branch.

11.1.3 DESIGN DECISIONS NOT MADE BY HLS

State-of-the-art HLS tools implement advanced techniques for effectively exploring the design space to make decisions on the aspects described in the previous section. At the same time, there are other aspects that need to be addressed, in order to produce high-quality hardware. Today's HLS tools mostly leave design decisions on these aspects to humans, in such a way that designers can explicitly specify how the design should be implemented, while the tools perform the actual implementation accordingly.

These human design decisions are generally macroarchitectural decisions, corresponding to the types of decisions that go into creating a high-level block diagram of the hardware. There are four aspects to the architectural decisions—block-level parallelism and hierarchy, communication mechanisms between the blocks, usage of memories for storage, and timing budgets.

The first aspect, the parallelism, is about how the target behavior is decomposed so that their components can be executed together. In hardware design, this parallelism can be customized for each specific target behavior, and this provides significant advantages for timing and power, compared to software implementation. HLS tools are good at exploring possible parallel implementations for certain cases. For example, if there are two addition operators in the design scope that is currently synthesized, the tools can often recognize that these operators could be executed in parallel and decide whether this should be indeed the option to use in the final implementation. However, the tools are not effective when the target objects are defined at a coarser level. If there is an iterative loop in the behavioral description, for example, today's HLS tools are not always good at evaluating whether the individual iterations could be implemented in parallel or if it is better to implement each of the iterations sequentially. Also, they are not good at extracting hierarchy out of a given behavioral description and then deciding whether those hierarchical components should be implemented in parallel or not. In these cases, humans often specify how the iterations of a loop should be implemented or create design hierarchy explicitly in the behavioral description and specify what sort of parallel implementation should be used for the resulting design components.

The second aspect is about how those design components should interact. Whether they are implemented for parallel execution or not, the protocol and interfaces between two components need to be determined for implementation. In the previous section, we looked at the interaction between two functions, main_control() and f(). They need to exchange data to realize the target behavior, but there are multiple ways to organize this exchange, depending upon the amount of parallelism that must be achieved. HLS tools usually present multiple options for possible implementations, so that designers can choose one or explicitly write implementation mechanisms so that the tools implement them exactly as specified.

These two aspects are about implementation options on flows of operations, but a similar consideration exists also for data to be used for operations. In hardware design, it is necessary to decide at design time which data should be stored using what kind of storage resources. Individual data could be stored in dedicated storage resources, or resources could be shared for storing multiple data. HLS tools can make effective decisions in general, but there is a case where designers know better the ideal implementation choices. This specific case is when data are stored in array constructs in the behavioral description. Multiple data can be stored in the elements of an array, and the index in the array is used to access them. Hardware can implement this exact mechanism by using a memory, where the index is used to define the addresses of the memory. However, operations that use these data might be implemented in parallel, and thus one might want to implement this array so that multiple data can be accessed in parallel. There are different ways to implement the parallel accesses. One way is to decompose the array elements so that they can be implemented with independent registers. Alternatively, one might want to keep the array as a memory, but multiple access ports are created. The adequate number of ports and access mechanisms depend upon how the operations that use the data are implemented, and one might want the HLS tool to decide the port structures based on the parallelism used in the implementation. Further, such memories might be already implemented efficiently and available as libraries, and one might want to use such a memory for the array rather than implementing it from scratch. Typically, HLS tools present the designer with possible implementation options for arrays, so that they can choose the adequate options and provide the information necessary to realize the implementation.

The fourth aspect that designers sometimes specify to the HLS tool explicitly is the timing budget. In design applications in which latency is important, the overall latency of the design is provided as a design constraint. Sometimes, it is difficult for tools to decompose this constraint for individual design components that constitute the whole design, and in such a case the designer specifies timing constraints for those components explicitly.

In all these aspects, it is not necessarily the case that the designer knows the best implementation options in the beginning. It is rather an interactive process between the human and the tools, where the designer can evaluate the synthesis results after some given implementation choices and then change some of the choices to iterate the synthesis. Therefore, it is very important for HLS tools to provide an effective analysis capability, with which the designer can quickly focus on critical regions of the design and steer the synthesis process.

11.1.4 DEFINING BEHAVIOR OF HIGH-LEVEL MODELS

In most general terms, HLS transforms a high-level model (say a SystemC model) into a lower-level model (typically RTL). Clearly, this transformation must preserve the behavior of the high-level model, but depending on the precise definition of what exactly is preserved, very different synthesis problems with different issues and benefits may arise. In this section, we describe several commonly used alternatives.

All of these alternatives aim at preserving only the relation between inputs and outputs, leaving to the synthesis tool complete freedom to rearrange the internal computation. Thus, the first step is to create a mapping between the inputs and the outputs of the two models. Doing so in SystemC is straightforward because the notion of sc_in and sc_out is very similar to the notion of port in HDLs (like input and output regs in Verilog). Still, there are some differences between ports in SystemC and HDLs that need to be carefully considered. First of all, C and C++ (and thus also SystemC) support only the binary representation where each bit can be 0 or 1. HDLs extend this by additional values, most notably X indicating undefined values. Also, Verilog supports only primitive data types like bit vectors and integers, while C/C++ support composite data types like structs.

For pure C/C++ high-level models, a mapping is defined from arguments of a function to ports of an RTL module. This mapping may be fully automatic or it may require some additional user input.

RTL assigns a cycle-based semantics to ports, where a port has a single definite value in every cycle. Thus, to be able to compare inputs and outputs of the two models, similar semantics need to be defined for high-level models. SystemC already defines such semantics, as long as the clock edges are the only events that explicitly depend on time. In practice, all HLS tools consider only such SystemC models. There is no universally accepted cycle-based semantics of pure C models of hardware, so one must be defined by each tool.

In addition to ports, hardware designs often use memories for communication between modules. In high-level models, memories are typically modeled as C/C++ arrays, although some tools also offer alternative modeling styles with special objects.

The simplest and the most restrictive notion of preserving behavior is a cycle-accurate interpretation of high-level models. In this modeling approach, we assume that inputs of the two models have the same values at each cycle boundary, and then we require that the outputs also have cycle-by-cycle equivalent values. This approach is the most restrictive because it does not allow the HLS tool to make any changes to the timing of I/O operations. It is also the least abstract approach because the user needs to explicitly ensure that the timing of I/O operations is such that there is sufficient time to complete all the computations that need to be done. Nevertheless, this approach is still significantly more abstract than RTL. Most of the main abstraction mechanisms are available at this level as well: sharing is abstracted, operations can be scheduled at different states, FSMs can be specified implicitly by procedural code, and there could be several options for their implementation. In addition, one can still use all the C++ abstraction techniques. However, compared to other approaches, the scheduling freedom is limited and automatic pipelining by HLS is not an option because exact I/O timing must be preserved.

In the strict cycle-accurate interpretation, the operations of reading and writing from a memory that is shared with other design components are treated as I/O operations. This means that in RTL they must appear exactly in the same cycle as in the high-level model.

While modeling at cycle-accurate level seems to limit the power of HLS technology, there are many design applications in practice for which this modeling style is natural. In those designs, exact timing requirements for the I/O operations are defined as a part of the design specifications. Hence, the designers' primary value of using HLS is not about changing the I/O timing, but about the optimization of internal behavior and the coding style of the input models that allows more configurations and reuse compared to RTL, due to the abstraction made for the behavior. Another key advantage of the cycle-accurate approach is on the verification side. Since high-level and RTL models have equivalent I/O behavior, the verification environment can use either one of them without much adjustment, and so most of the verification tasks can be done even before RTL is available. Also, in this model, it is straightforward to formalize the notion of equivalence between the high-level and RTL models.

In certain application domains, designs have strong throughput requirements (the number of inputs processed in a period of time), but the latency (the time spent processing a single input) is often not as relevant. This naturally leads to pipelined designs where processing of multiple inputs is overlapped. To abstract pipelining details from the high-level model, one needs to relax the cycle-accuracy requirement on the I/O behavior. We call this relaxation a *cycle-delayed* model. In this model, an output in RTL does not have to be cycle-by-cycle equivalent to the matching output in the high-level model, but it has to be equivalent to a delayed version of that output. More precisely, if h(n) and r(n) denote matching outputs in the high-level and RTL models, respectively (n = 0, 1, 2, represents the cycles), then we require that

$$h(n) = r(n + d)$$

holds for all n. The parameter d corresponds to the pipeline depth. Usually, it is chosen by the HLS tool, but the user has the ability to limit it. It can be different for different outputs in the design. Clearly, the aforementioned relation says nothing about the first d values of the RTL output. These cycles correspond to loading of the pipeline. Usually, the outputs are required to hold their reset value, but in some designs styles these values do not matter and an HLS tool may be able to take advantage of this to optimize the design.

While cycle-delayed models may accurately represent some simple pipelined designs, many more complex pipelining techniques may not preserve this simple relation. In particular, if pipelines are allowed to stall or the pipelined implementation has a throughput that is different from the high-level model, then the cycle-delayed model is no longer applicable.

Cycle-delayed models require a more complex verification environment, because it must be aware of the pipeline depth. Further complexities arise if the pipeline depth is different for each output. This could be dealt with by making the verification environment parameterizable by pipeline depth(s), or additional signals can be added to the interface to indicate when the pipelines are loaded.

Another common class of designs are *latency-insensitive* designs [19,22] that can be seen as an implementation of *dataflow process networks* [21]. In this approach, communication between design components is not interpreted as waveforms carrying a value in every cycle, but rather as a stream of data tokens that is obtained by sampling the waveforms at certain points. Auxiliary handshaking signals controlled by both the sender and the receiver are used to determine at which point to sample the waveform, that is, at which point the token is transferred. Each design component processes input tokens and generates output tokens. When there are no tokens to process, the component idles.

The concept of streaming can be extended to components communicating through a shared memory. Typically, the sender and the receiver never access the memory at the same time. The handshaking defines how to transfer the access rights back and forth between the sender and the receiver. The content of the memory at the moment of access right transfer represents the token that has been exchanged.

To specify a high-level model of a latency-insensitive design, one needs to specify the protocol to exchange data tokens between design components that define the computation processing input tokens and generating output tokens. In C/C++-based HLS tools, the protocol is specified by selecting one of the predefined options built into the tool and the processing is specified by C/C++ code. In SystemC-based tools, both the protocols and the processing are specified by SystemC code. The part of the code describing the protocols needs to be specially designated because it is interpreted in a cycle-accurate way. In contrast, timing is abstracted from the code describing the data processing. Every tool provides some way to specify timing constraints for such code, but the tool is free to choose any timing consistent with these constraints. The parts of code describing the protocols are often referred to as protocol regions or simply protocols. The name for this design style comes from the fact that the functionality of the system is not changed if the latency of the individual streams or computations is changed. This is the major advantage of this style. It allows for optimizing design components without affecting the overall functionality. The disadvantage is that unless special care is taken to balance the system, a lot of components may spend a lot of time idling, which is not the optimal use of resources. Also, there is the overhead of the handshaking logic. In general, this design style is well suited for applications where timing is not critical or when combined with pipelining, for designs with strict throughput but loose latency constraints.

RTL implementations of high-level latency-insensitive models need to preserve only the streams of data tokens at each output. More precisely, HLS needs to ensure that if the input streams in the high-level and RTL models are the same, then the streams at the outputs must also be the same. The timing of individual tokens and the ordering between tokens on different outputs do not have to be preserved. The verification environments for latency-insensitive designs also have to be latency insensitive. They also need to take into account that RTL and high-level models may operate with very different latencies. In fact, for some designs under this class of modeling, it is not possible to determine an upper bound on the difference of latencies between the input and output models of HLS, and in that case it is known that the notion of equivalence between them cannot be formulated within the framework of finite-state systems [22].

For many designs, there exist models where the communication between the components is abstracted into transactions, without specifying the details of the protocols implementing these transactions. This modeling style is called transaction-level modeling (TLM) [14]. Transaction-level models are widely used for architecture exploration, in early SW development, and as reference models for verification. The basic advantage of TLM models is that they can be simulated much faster than any other model described here. To be synthesized, TLM models need to be refined by specifying an implementation protocol for the transactions. This naturally leads to latency-insensitive designs, but it is also possible to refine TLM models into cycle-accurate or cycle-delayed models.

The acronym TLM has gained wide acceptance, and it is often being used as a synonym for any high-level model to which HLS is applied. We use it in a more limited way for models that abstract away the details of communication protocols.

11.2 HIGH-LEVEL SYNTHESIS BY EXAMPLE

This section provides an illustration of the HLS flow by means of an example. On one side the example is based on a very simple C-code structure for the ease of understanding. On the other hand, it is rich enough to introduce the main HLS steps and to present nontrivial trade-offs to be considered in the design process.

Let us assume that a system has a clear bottleneck in performance. One of the options is to move part of the software functionality into hardware, bearing in mind that hardware could deliver two to three orders of magnitude of speedup over software implementation. One might expect from the past experience of using software compilers that this would be a simple matter of finding the right HLS tool, setting a few parameters and enjoying the result. Unfortunately, today (and there is little hope of future change) there is no automatic push-button tool that takes a piece of software code and makes a "reasonable" hardware representation out of it. The reason is that contrary to traditional software compilation (where a sequential code is compiled into a lower level sequential code), hardware compilers take an input targeted at one architecture (software) and produce an output in a very different architecture (hardware) that is parallel by nature and typically cannot add resources (computational or memories) on the fly. The following is a list of tasks that a designer needs to take care of in order to use an HLS flow.

1. Make sure that the software code satisfies conditions for hardware synthesizability.
   This requires avoiding recursion and dynamic memory allocation because hardware has a static architecture that must be defined at compile time. Some limited forms of recursion and dynamic allocation (when the depth of recursion stack or the memory size are known statically) might be feasible, depending on how capable the HLS tool is, but in general a designer is discouraged from using these features. Note that in the early days of HLS, the use of pointers was limited as well. Fortunately, with the advances in the field, this limitation was lifted, and now most commercial tools happily accept code with pointers and pointer arithmetic, with very few limitations.
2. Explicitly reveal the coarse grain parallelism.
   This task is somewhat similar to preparing a program for parallel computing. It requires splitting a sequential program into a set of concurrent threads with predefined interfaces.

Note that revealing low-level parallelism is not required, since this is a task that is well understood by HLS tools, for example, given statements x = b + c; y = b * c; the tool would see that these two operations could be done in any order or simultaneously.

3. Provide information about hardware platform and design constraints.

   There are two main hardware platforms that HLS tools target: ASICs and FPGAs. The information about them is captured by specifying an underlying ASIC technology library or FPGA family type. Design intent about the implementation performance is provided by setting the clock period and latency constraints, which specify (optionally) an upper bound on the number of clock cycles to execute a particular piece of code.

4. Define a desired microarchitecture.

   The quality requirements (whether it is area, power, or combination of both) are of great importance in an HLS flow. If an implementation does not fit into an area/power envelope, it cannot be used as a part of a system and it is useless. The optimization space in minimizing the cost of hardware is much bigger than in software compilation. The decisions about loop unrolling or pipelining, memory architecture, or input/output protocols have a crucial impact on implementation cost (see, e.g., [3] where an order of magnitude difference in area and power numbers were observed for IDCT code implementations under different microarchitecture choices). There is little hope of coming up with an automatic solution for this problem that would work well for all cases. Instead, HLS tools provide "knobs" for the user to specify his or her intent about the implementation. An essential part of the synthesis process is the design exploration phase, when the user explores architectural options playing *what-if* scenarios and observing the tool results.

Figure 11.2 captures the aforementioned differences between software and hardware compilation flows. One can see that software compilation is a push-button flow with the tool being almost a black box for the user, while targeting hardware calls for user intervention and interaction with the tool at many steps (the tool is no longer a black box).

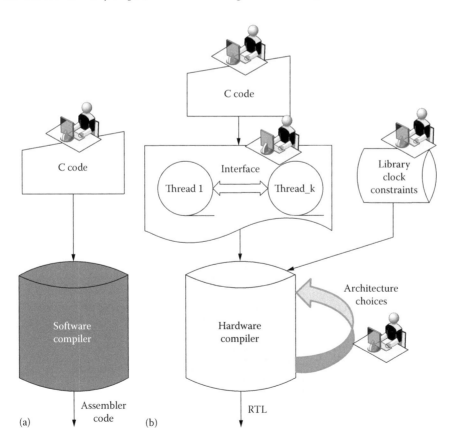

FIGURE 11.2   Compilation flows in software (a) and hardware (b).

The process of preparing sequential C-code for HLS is illustrated by the following example, showing the (simplified) conversion of the hue–saturation–light format into the more conventional red–green–blue (RGB) format (the code as follows generates the *red* component only).

```
int convertHSL2R(int hue,int light,int sat)
// function for the HLS->R conversion
// similar functions for green and blue are omitted for simplicity.
{
   int m1, m2, red;
   if (light < 64) m2 = light +light*sat/128;
   else m2 = light + sat - light*sat/128;
   m1 = 2*light - m2;
   red = m1;
   if (hue < 60) {
      red = m1 + (m2 - m1)*hue/64);
   } else if (hue < 180) {
      red = m2;
   } else if (hue < 240)
      red =(m1 + (m2 - m1)*(240 - hue)/64);
   }
   return red;
}
```

1. Threading and interface identification

   This function is a simple example and does not require splitting into several threads. The whole function becomes a thread. The function reads three input values *hue*, *light*, and *saturation* and produces a single output value *red*. Inputs and outputs in the function are conveniently specified as 32-bit integers. This is very typical in software because it does not incur large penalties (except for memory-critical applications) and keeps the functionality data agnostic. On the other hand, sizing the datapath to true bit widths is of paramount importance to achieve good quality of results (QoR) when designing hardware. In the aforementioned example, the ranges for inputs and outputs are the following: $0 \leq hue \leq 360$, $0 \leq light$, $sat \leq 128$, $0 \leq red \leq 256$. This is explicitly specified in the input specification for synthesis (see the following). Considerations similar to this must be made for the green and blue values, although our focus in this example is only on the red value computation. The computations on these different colors could be done in independent threads, or they could be all in a single thread. The designers decide the thread structure and write the functions accordingly.

2. Specifying design constraints

   Let us assume that the required throughput for the format conversion is 100 MHz, that is, a new pixel needs to be produced every 10 ns, and the clock cycle is 5 ns. For nonpipelined implementation, this defines the latency to be no greater than two cycles per pixel computation. The aforementioned latency constraint can be specified in different ways: (a) through latency annotations that an HLS tool understands or (b) by adding timing statements explicitly (wait() statements in SystemC).

The input specification in SystemC for the *convertHSL2R* is shown as follows:

```
class convertHSL2R: public sc_module {
    sc_in<bool>         clk, rst;
    sc_in<sc_uint<7>>   light, sat;
    sc_in<sc_uint<9>>   hue;
    sc_out<sc_uint<8>>  red;
    void                thread();
    ...
};
```

```
void convertHSL2R::thread() {
    while (true) {
      sc_uint<7> li, sa;
      sc_uint<9> hu;
      sc_uint<8> re;
      li = light.read();
      sa = sat.read();
      hu = hue.read();
      // From here the code is not changed
      int m1, m2;
      if (li < 64) m2 = li +li*sa/128;
      else m2 = li + sa - li*sa/128;
      m1 = 2*li - m2;
      re = m1;
      if (hu < 60) {
        re = m1 + (m2 - m1)*hu/64);
      } else if (hu < 180) {
        re = m2;
      } else if (hu < 240)
        re =(m1 + (m2 - m1)*(240 - hu)/64);
      }
      // Latency of computation is 2
      wait();
      wait();
      red.write(re);
    } // end while
} // end thread
```

Note that apart from explicit specification of interfaces, the computation part is almost unchanged in SystemC compared to the original code. In particular, the designer does not need to care about the true bit widths of internal variables (m1 and m2 are kept integer), because it is the job of the HLS tool itself to size them properly based on the I/O interface bit widths. The SystemC specification of *convertHSL2R* is used in this chapter to illustrate all the steps of the synthesis flow. This specification is intentionally simplified to ease the discussion. We will refine the flow while using more elaborated specification constructs in later sections.

Our example of *convertHSL2R* is limited in terms of

1. Control structures
   It represents a one-directional computation flow that does not contain loops or functions.
2. Explicit timing
   The timing intent of the designer is expressed by explicit *wait()* statements added to the input source. It assumes that there is no freedom in changing the latency of that computation and that the targeted RTL implementation should be IO cycle accurate with respect to the input source.
3. Data types and types of operations
   The example is solely restricted to integer data types (floating and fixed-point computations are significantly more complicated to handle in hardware). Representing software arrays in hardware is another source of synthesis complications. Typically, arrays are mapped onto memory modules with predefined interfaces for reading and writing. We postpone the discussion of this mapping until later and do not use arrays in this illustrative example.

11.2.1 HLS PROBLEM

Figure 11.3 shows the set of the tasks that typical HLS tools perform. We consider these tasks in more detail in the following.

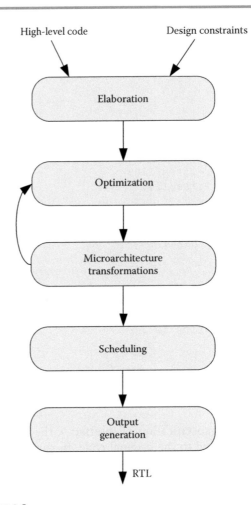

FIGURE 11.3   Typical HLS flow.

11.2.2 ELABORATION

In this step, the input software model and design constraints are elaborated into an internal repre-sentation that is suitable for hardware synthesis. Many representations have been proposed in the literature [2,6,7,9–11,27]. While they vary in details due to specific focuses on optimizations, they typically use graph structures to represent the design behavior to be synthesized, as well as the hard-ware resources and structures to be used for their implementation, so that the HLS design decisions described in Section 11.1.2 can be made efficiently. For the behavioral part, a good representation is given by the control and data flow graph (CDFG) model, which has become a de facto standard [9].

A CDFG can be formally represented as two flow graphs: control (CFG) and data (DFG) flow graphs. The nodes of the CFG represent control flow or latency. One can distinguish the following types of nodes in a CFG:

1. A unique origin node that represents the start of the thread. It has no input edges and a single output edge.
2. A fork node with a single input and multiple output edges to represent branching due to conditional statements in C code such as *if*, *switch*, or *loop exit*.
3. A join node with multiple input edges and a single output edge to merge conditional branches and to represent loop iterations (one input edge from the code before the loop, one from the body of the loop).
4. A simple node with a single input and a single output edge. These nodes come in two flavors: they either serve as a placeholder for a reference point in the control flow (e.g., a label in the C code) or correspond to "wait()" calls in SystemC (called state nodes).

From the aforementioned description of node types, it is easy to see that the CFG abstracts the computation and shows only the control paths and their latency. Sometimes, the CFG model is extended to include the so-called compound nodes [9,18] to form hierarchical structures such as if-then-else blocks, switch-case blocks, and loop nodes. The advantage is to gain in terms of scalability of the CFG representation. However, scalability comes at the expense of losing some optimization opportunities due to additional hierarchy levels, hence using this extension calls for cautious evaluation.

The DFG captures data transformations for a given behavior/thread. Its main objects are operations, which roughly correspond to simple (single operator) expressions in the source code. Variables in a DFG are represented in static single assignment (SSA) form in which all operation outputs (variables) are unique [4]. When a variable in the input source is assigned conditionally (say in two branches of an if-then-else statement), its value is derived from different computation branches using a special DFG operation MUX (also called Phi node).

Figure 11.4 shows the CFG and DFG for the *convertHSLToR* example. Variables m2 and re are assigned conditionally in the original specification. In the DFG, the assembling of their values from different conditional branches is performed by the corresponding DFG MUXes. The relationship between DFG and CFG is provided through a mapping from DFG operations to CFG edges. The first such mapping that is established after elaboration is called birthday mapping, which associates each DFG operation with the edge of the CFG where this operation is born according to the input source. For example, operation *SUB2 (m1 = sa − m2)* is associated with the out-edge of the if-bottom node in the CFG, while operations *SUB3 (240 − hu)* and *SUB4 (m1 − m2)* are associated with edges *c4* and *c2* of the CFG.

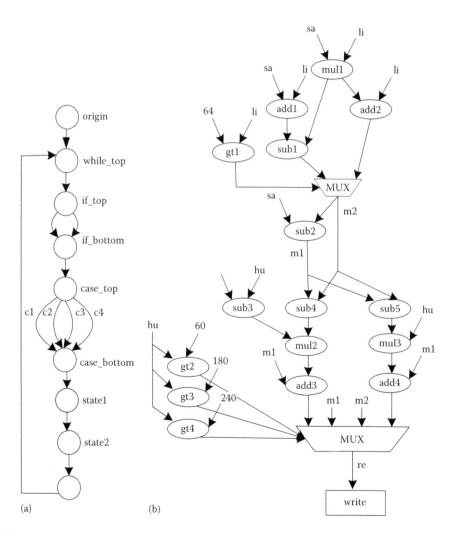

(a)   (b)

FIGURE 11.4   CFG (a) and DFG (b) for the *convertHSLToR* example.

11.2.3 OPTIMIZATION

We use the term "optimization" to specifically refer to the step of simplifying as much as possible the internal representation (CFG and DFG) before going to scheduling. The implications are much more important here than in software compilation, where getting the smallest possible size of the executable code is nice but not necessary. In HLS, redundant (or more complex than needed) operations in the DFG or a larger size of the CFG are directly translated into extra hardware, which in the end makes the implemented circuit too expensive and hence unusable in the highly competitive market of hardware IPs. An additional pressure comes from the fact that HLS technology often competes with manual RTL flows, where human designers have accumulated a lot of experience in squeezing gates out of their circuits. This is why optimization is very important for HLS. Optimization methods range from traditional compiler optimizations, such as dead code elimination, constant folding and propagation, and common subexpression extraction, to ones that are more specific to hardware and HLS. The most prominent example of hardware-specific optimization is bit trimming, which is based on careful range analysis of the operands, and then sizing the datapath to minimal bit widths. We will discuss HLS-specific optimizations (in particular those that simplify the CFG) in later sections.

An attentive reader may notice several aspects of our simple example from Figure 11.4b:

1. The application of common subexpression extraction for $li * si$ (see operation MUL that feeds both branches in the computation of m2)
2. The application of strength reduction to first replace division by 128 and 64 with shifts, which are further implemented by proper wiring of the operation outputs

In addition (though it is not shown in Figure 11.4b), the datapath for the computation is bit trimmed, resulting in operations that range from 7 to 11 bits.

11.2.4 MICROARCHITECTURE TRANSFORMATIONS

The quality of implementation after synthesis strongly depends upon the previously made architectural choices, which often require input from the designer. A good tool should provide flexibility for specifying design intent to achieve optimal results. The design decisions taken at this step fall into two main categories:

1. Restructuring control or design hierarchy
   Control restructuring typically chooses how to handle loops and/or functions in the computation. The designer has several options: loops can be unrolled or split, two or more loops can be merged into a single one, and loops can be pipelined letting iterations overlap. The design hierarchy coming from the source code is defined by functions and their calling graph. Functions in software are mainly used for two reasons (a) to reuse repeated functionality and (b) to encapsulate functionality in a single object, hiding internal details. While the goal of reuse is meaningful in hardware as well, encapsulation is less justifiable because it limits optimization opportunities, by creating a boundary between the caller and the called functions. When targeting hardware, the designer needs to reconsider the structure of the design as specified in the source code and decide which functions should be preserved (and implemented as a separate entity) and which ones should be merged (*inlined*) into the caller.
2. Specifying memory architecture
   Decisions on how to store data to be used in computations are of paramount importance for hardware synthesis. The most economical storage is provided by dedicated hardware memories. However, it has a negative impact on the amount of parallelism to access data (limited by memory bandwidth, i.e., by the number of ports and their widths). The opposite extreme is provided by flattening data (arrays or structs) to a set of scalar data elements, each of which could be accessed separately. This is more costly in terms of area

than memories, but it gives more flexibility in accessing the data elements. In addition, when implementing a memory, the designer needs to take into consideration whether this memory will be visible by other processes (and thus require multiple access support) or will be serving as a local scratchpad for a single process.

To help designers with making educated choices about microarchitecture, HLS tools should provide a reasonable set of defaults: for example, automatically inline a function when it is called only once and flatten an array when its size is smaller than a threshold. However, the solution space of microarchitecture choices is very large and decisions are highly interdependent (e.g., unrolling a loop that calls a function may make it too costly to inline this function afterward). Therefore, even with good defaults, specifying microarchitecture demands human interaction and it is a significant part of the design exploration process. After all, this is why designing with HLS is exciting—evaluating different architectures in a matter of minutes or hours is something that RTL designers can only dream of.

Our simple example of *convertHSLToR* does not contain internal loops or functions. For this example, the only possible designer choice is whether to implement the main loop "as is" or to pipeline it. This is fully defined by the required throughput and cycle time. With a simple back-of-the-envelope computation, one can quickly confirm that the computation fits in two cycles of 5 ns each, and therefore the desired throughput of 100 MHz is achieved without the need to pipeline the loop.

11.2.5 SCHEDULING

The final goal of HLS is to produce RTL. In RTL, the functionality is described by an explicit FSM, and the logic computing next states and outputs is associated with state transitions. The ultimate goal of the scheduling step is to make explicit which resources perform the computation and how the computational pieces are associated with states of the thread. Note that separation of control and computation in the form of the CFG and DFG helps significantly in formulating the scheduling problem. Indeed, one can clearly see a close similarity between the CFG and the FSM models. The main difference is that the CFG is explicit about mutually exclusive paths, while in an FSM all paths between identical source and destination states are merged in a single transition. For example, in Figure 11.4a the CFG shows paths between *state2* and *state1* around the loop explicitly (using two fork–join node pairs), while in the FSM there is a single transition *state2* → *state1* with an enabling condition *(if_true + if_false) & (case_c1 + case_c2 + case_c3 + case_c4)*.

With this observation, the scheduling problem can be formulated as finding of two mappings:

- Mapping *sched*: *dfg_operations* → *cfg_edges*, which defines the edge assigned to each operation as the result of its scheduling
- Mapping *res*: *dfg_operations* → *resources*, which defines which hardware instance implements an operation

The difficulty in finding these mappings is that their right-hand sides are not fixed: in general, there is freedom to both modify the CFG by adding states and/or to enlarge the set of resources, in order to satisfy scheduling constraints. Thus, scheduling can be thought of as a walk in the solution space defined by sets of resources and by CFG latencies (defined by the states currently present in CFG) with the goal of finding solutions in which none of the quality parameters (resource count or latency) can be improved without worsening the other (known as Pareto points) (see Figure 11.5a). At each step of this walk, a fixed latency/fixed resource scheduling problem is solved (if it is feasible), and when this succeeds, an RTL is produced. A failure to solve it shows that the problem is overconstrained and needs to be relaxed by increasing latency or adding resources. This is done by the *Analyzer*, which checks the reasons of failure and chooses the best relaxation to apply (see Figure 11.5b). The *Analyzer* can be implemented in many different ways, for example, as a domain-targeted expert system and as a general optimization engine (using, e.g., simulated annealing, genetic programming, integer linear programming [ILP], linear programming [LP], and satisfiability [SAT]), or rely on specific heuristics.

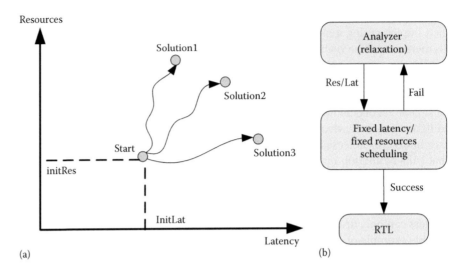

FIGURE 11.5 High-level view of scheduling process: walking optimization space (a) and relaxation (b).

Note that checking whether a given instance of the scheduling problem under fixed constraints (resources/latency) is feasible is known to be a difficult problem (NP-complete [5]). Hence, there is little hope to be able to solve it exactly, and the development of efficient heuristics is needed. The choice of the starting point when performing a scheduling walk (such as in Figure 11.5a) is very important: choosing it too conservatively increases scheduling time while overapproximating the number of needed resources or latency leads to lower implementation quality.

11.2.5.1 ESTIMATION OF MINIMAL RESOURCE AND LATENCY BOUNDS

Every DFG operation must be related to a resource that is capable of implementing its functionality. For many operations this mapping is one to one (Figure 11.6a) but some exceptions exist when several types of resources are capable of implementing the same type of operation (see Figure 11.6b where a "+" operation might be implemented by an adder or an adder/subtractor).

Resources have associated costs in terms of area and delay (see Figure 11.6b) that could be obtained by synthesizing them in the given technology library. Choosing the initial set of resources is a nontrivial problem in itself. The naive solution that provides a single resource for every operation type is too conservative. Indeed, it assumes that a DFG with, for example, n additions can be implemented with a single adder, while this is often not feasible under latency constraints. A tighter bound could be obtained by considering the distribution of operations within the CFG, taking into account their mobility. For example, if in a loop with latency m, n operations of type R are chained to produce a value used in every loop iteration, then clearly the lower bound on the number of resources of type R is given by $\lceil n/m \rceil$. This lower bound assumes "perfect sharing" of resources among loop operations, which may or may not be feasible.

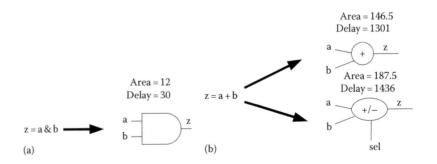

FIGURE 11.6 From operation to implementation resources: single (a) and multiple (b) maps.

Computation of a tighter (than naive) lower bound is illustrated by the estimation of the number of resources in the *convertHSLToR* example. Typically, only arithmetic operations (multiplications, additions, and subtractions) and user-defined operations (noninlined functions) are worth sharing. The rest of the DFG operations are simple (MUXes, logical, and comparisons to constants); hence, the cost of MUXes when sharing these operations would be excessive. In this case, the HLS tool would create an implementation using dedicated resources. The SystemC code for this example is a loop with a latency of two states, containing three multiplications (remember that common subexpression extraction merged two multiplications together), with two multiplications being mutually exclusive (as they are residing in opposite branches of an if-then-else statement). This gives a lower bound of $\lceil 2/2 \rceil = 1$ multiplier to implement this algorithm. Assuming additions and subtractions to be implemented by universal adder–subtractor resources (addsub), we find nine operations altogether; considering mutual exclusivity, we have at most five of them to be executed in two states, which gives $\lceil 5/2 \rceil = 3$ addsub resources as a minimal lower bound.

The generalization of this problem is more complicated because of the following reasons:

1. The effect of mutual exclusivity among operations is not straightforward. This information can be approximated by analyzing the conditions under which operations occur in the CDFG. Note that this is only an approximation, because originally mutually exclusive operations might be scheduled concurrently due to code motion (e.g., speculation).
2. Bit widths of operations must be taken into account in addition to operation types. Suppose that a loop with two states contains a 32-bit addition and a 4-bit addition. There is a trade-off in deciding whether a 4-bit addition should be mapped (bound) to a 32-bit adder (which is slow when measured by its worst-case delay) or a dedicated faster 4-bit adder needs to be created (possibly increasing area but reducing estimated delay).
3. Loops may provide too coarse a granularity to reason about resource contention. The mobility of operations might be restricted to a single state or a sequence of states. Formal reasoning about resource contention must be done on connected subgraphs within the CFG that correspond to operation mobility intervals. A trade-off between the number of considered subgraphs and their ability to contribute to the lower bound needs to be made, in order to keep the problem manageable.

Ideas similar to resource estimations can be exploited for deriving a minimal bound on the initial latency of each innermost loop body in a thread to start scheduling it. In this case, the tool needs to take into account multicycle and fixed-cardinality operations (e.g., memory accesses) and to compute how many cycles are required to fit these operations within the loop body. For example, if a specification contains a loop with two accesses to a single-port single-cycle memory, then the latency of the loop cannot be less than 2.

Note that due to heuristic nature of scheduling procedures, choosing a proper starting point for exploring the solution space is important not only to converge more quickly but for the final QoR as well. This suggests that investing into tightening the minimal bounds for resources and latency before scheduling is definitely worthwhile.

11.2.5.2 SCHEDULING REQUIREMENTS

Research on scheduling is very rich and summarizing it briefly is challenging (a good collection of pointers can be found in Reference 26). Instead, we choose to provide readers with a map that helps in navigating through hundreds of papers on the topic. We start from the most important requirements for scheduling algorithms that any successful approach must satisfy.

11.2.5.2.1 Scalability

This requirement is mostly driven by the size of the graphs that the scheduler needs to deal with. In industrial applications, it is quite possible to encounter a DFG with more than 10,000 operations (especially after extensive use of unrolling and inlining) and a CFG with hundreds of states. A typical arithmetic operation in a large design can be implemented with 2–20 types of different resources, for example, considering bit widths and architectures (e.g., ripple vs. carry-look-ahead). Let us assume 10 types of resources on average. One may need to solve 100 fixed latency/fixed resource problems (called scheduling steps) before reaching a fixed point in the scheduling walk discussed earlier. This gives a problem complexity in the order of 10^9 choices to be made and demands for algorithms with linear or pseudolinear complexity at most. In particular, it makes clear that scheduling approaches based on ILP, which were quite popular at some point, are of theoretical value only.

11.2.5.2.2 Accuracy of Timing Modeling

The scalability requirement imposes constraints on how complicated the modeling of resource delays should be. In logic synthesis, a 16-bit adder is characterized by many timing arcs corresponding to combinational paths from inputs to outputs (32 for the output MSB only), but using the same approach in HLS might affect the scalability requirement. Different approaches have been proposed in the literature to address this issue, ranging from characterizing resources by deterministic delays (single or multiple) [28], to statistical characterization [29], to trading off between accuracy and speed of timing computations.

Regardless of the HLS approaches for delay characterization, there are several important aspects to be considered, based on the nature of resource delays in state-of-the-art hardware designs. Consider the following table, which shows the maximum pin-to-pin delays obtained by logic synthesis using the TSMC_45nm_typical library for several resources, when logic synthesis is constrained to produce the fastest possible implementations.
Several observations can be made from this table:

1. With a clock cycle of 2 ns, commonly seen in consumer electronic devices (500 MHz clock rate), three multipliers or nine adders can perform their computation in a single clock cycle. Hence, any realistic approach must consider scheduling multiple operations in a single clock cycle (also called operation chaining), unlike a large number of theoretical approaches proposing scheduling a single operation on a resource in each control step.
2. Delays of resources vary widely and any assumption on their uniformity (such as unit delays) is not practical.
3. For high-frequency design (e.g., with a 2 GHz clock rate), the MUX delay is noticeable (takes about 8% of the clock cycle) and cannot be neglected.

As we mentioned earlier, timing modeling in HLS is less accurate than in logic synthesis because scalability requirements prevent us from exploiting pin-to-pin delays. By reducing the number of timing arcs when characterizing resources, the timing modeling is performed in a conservative way. This is needed to ensure timing convergence in the HLS-based design flow, that is, once timing is met in HLS, it should also be met in logic synthesis. It also helps to simplify HLS timing modeling by ignoring second-order timing effects like wire loads, which are very difficult to estimate up front, when the netlist is still under construction. However, the observation about the significance of MUX delays (see Table 11.1) prevents us from overly exploiting these conservative margins. In particular, it is not possible to ignore the impact of sharing on timing: resource and register sharing MUXes must be modeled explicitly to avoid timing violations in logic synthesis due to MUX delays and false paths (see the discussion in Section 11.2.5.4 for more details).

11.2.5.2.3 Controllability

The huge size of the optimization space in HLS is both a blessing and a curse. On one side it presents a lot of possibilities to achieve better implementation, but on the other (bearing in

TABLE 11.1   Maximum Delays Obtained after Logic Synthesis under Constraints to Produce Fastest Implementations

Resource	Mul (16 bit)	Add (16 bit)	MUX/Control	MUX/Data
Delay (ps)	576	209	26	38

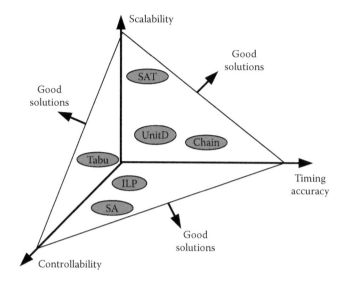

FIGURE 11.7   Scheduling landscape.

mind the heuristic nature of synthesis) it makes it easy for an automated optimization tool to end up being stuck in local minima. Some form of designer guidance is needed and inevitable. A designer needs to have a way to inform the tool about the design intent, provided in the form of constraints. Information such as "this loop is critical and should be synthesized separately from the rest of the thread" or "at most N multipliers should be used," or "for this design low power is the most critical constraint," will steer scheduling in very different directions. The other part of synthesis control provides capabilities to exactly replay the same synthesis decisions between different runs.

The controllability requirements make it difficult to build HLS synthesis based on general optimization engines such as simulated annealing (SA), genetic algorithms (GA, Tabu), or SAT-based approaches. These engines might be still helpful in providing quick and dirty solutions (to be refined by a more specific synthesis flow) but they are clearly not self-sufficient.

Figure 11.7 summarizes the aforementioned requirements graphically. The acceptable scheduling approaches should occupy the space outside the triangle created by requirements 1–3. The space under the triangle shows some of the approaches that are impractical because they violate one of the requirements. This map provides a quick sanity check when choosing a scheduling method.

11.2.5.3 HOW TO IMPLEMENT SCHEDULING

Each scheduling iteration (which solves a fixed latency/resource problem—see Figure 11.5) is represented conceptually as a bin packing problem instance [9], where bins are CFG edges and items to pack are DFG operations. Bin packing proceeds until either the timing constraint (clock period) or a resource bound (number of available resources of a given type) is violated for a clock cycle. Timing is checked by building a netlist corresponding to the set of operations that are scheduled in the current clock cycle (possibly from several CFG edges belonging to the same state).

The pseudocode for a typical scheduling procedure is shown in Figure 11.8. The priority function takes into account the mobility of the operations, their complexity, the size of their fanout cones, etc. Examples of different definitions of this function are list scheduling [9], force-directed scheduling [8], and other approaches.

Applying this procedure to the CFG/DFG of the *convertHSLToR* example gives the schedule in Figure 11.9, where the binding of operations to resources is shown with dotted arrows. Note that the original bound on the number of multipliers turned out to be precise, while scheduling had to relax the bound for addsubs, by adding two more resources (five versus the initially

```
SCHEDULE_(CFG C, DFG D, clock period T_clk, Library L, User Constraints U) {
forall edges in CFG {
    Ready ← operations ready to schedule;
    Compute_op_priorities(Ready);
    Op_best ← highest priority op;
    Op_best_success = false;
    Op_res ← resources compatible with op_best;
        forall r in Op_res {
            if (r is available at edge e
                && binding op_best on r satisfies timing)
                Op_best_success = true;
                break;
            }
        }
    if(!op_best_success and e is the last edge to which op_best can be scheduled ){
        Failed_ops ← op_best;
    } else {Update(op_best, Ready);}
} // end of forall edges in CFG
if (Failed_ops ! = ∅) {return failure;}
}
```

FIGURE 11.8   Scheduling procedure.

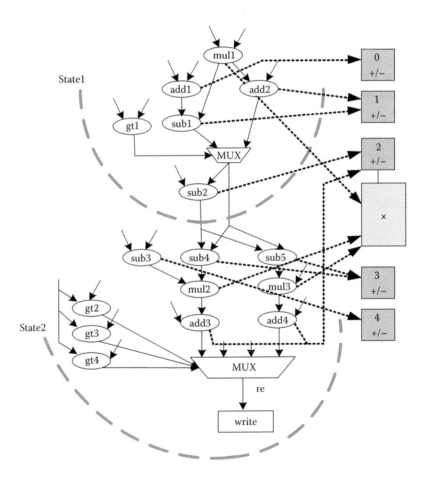

FIGURE 11.9   Schedule for *convertHSLToR* example.

estimated three). The explanation comes from restrictions to avoid structural combinational cycles during scheduling. Indeed, *addsub0* and *addsub1* are scheduled in *state1* in the fanout cone of the multiplier, while operations *sub3*, *sub4*, and *sub5* are in the fanin cone of this multiplier in *state2*. Sharing *sub3*, *sub4*, or *sub5* on *addsub0* and *addsub1* will create a (false) combinational cycle between the multiplier and the addsubs. These cycles must be accounted for in scheduling and avoided by adding more resources (*addsub3* and *addsub4*).

11.2.5.4 REGISTER ALLOCATION

When operation *op* is scheduled in state s_i and its result is consumed in a later state s_j, the output values of *op* must be stored in all states between s_i and s_j. The task of assigning values to be stored into registers is called *register allocation*. This task is somewhat simpler than the general task of assigning operations to resources for two reasons:

1. Even though the types of registers may vary (with different reset and enabling functionality), their timing characteristics are very close and the problem can be reduced to binding operation values to a single type of resource: a "universal" register.
2. Sharing of registers is relatively easy to account for, by adding to each register the delay of a sharing MUX. This is because during register sharing only the fanin logic needs to be considered, which creates a one-sided timing constraint that is easy to satisfy. Figure 11.10a shows timing before and after sharing of registers. Control and datapaths end up on maximal delays (denoted as fractions of clock cycle T) but they are still within the clock period. This is very different from the case of operations sharing the same resource, where both fanin and fanout cones matter and timing can be easily violated (see Figure 11.10b). The reason for this difference is that registers mark the start and end of all timing paths.

A scheduled DFG uniquely defines values to store and their lifespan intervals, that is, the states during which they should be stored. Two values can share the same register if they belong to mutually exclusive branches of computation or their lifespans do not overlap. Finding a minimum number of registers for a given set of states (defined by the CFG) can be reduced to applying the left-edge algorithm taking into account mutual exclusivity and lifespans. This problem can be solved in polynomial time [9].

Targeting a minimal bound on the number of registers might not provide the best result, because achieving this goal typically results in some values moving from register to register in different states. Moving values has a negative impact on both power and the size of MUXes, and it should be avoided whenever possible.

Although (as was stated earlier) there are good arguments for decoupling scheduling and register allocation, combining these steps may provide some benefits, as Figure 11.11 illustrates. Let us assume that operations are scheduled in their birthday positions, as in the SystemC code in Figure 11.11a, and we are about to schedule operation *y = a + b* in state *s3*. When deciding about sharing this operation with the previously scheduled *x = t1 + t2*, a conservative view on operand locations must be taken and sharing MUXes must be inserted into the netlist (see Figure 11.11b). These MUXes disappear if after scheduling we share registers for operands *t1,a* and *t2,b*. But in the scheduling step we do need to take into account the delays of sharing MUXes when considering state *s3*. When scheduling and allocation are done simultaneously, it is possible to reduce pessimism in timing analysis and achieve the simplified datapath of Figure 11.11c from the beginning.

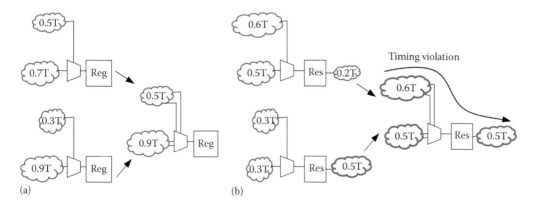

FIGURE 11.10 Impact of sharing on timing: registers (a) and resources (b).

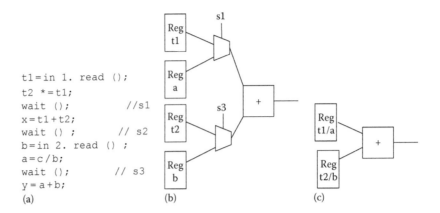

```
t1 = in 1. read ();
t2 * = t1;
wait ();           //s1
x = t1 + t2;
wait () ;          // s2
b = in 2. read () ;
a = c / b;
wait ();           // s3
y = a + b;
(a)
```

FIGURE 11.11   Register allocation: initial specification (a) separate from scheduling (b) and combined (c) approaches.

This however significantly increases the complexity of the scheduling task, which may overweigh the advantages of merging these synthesis steps.

11.2.6 PRODUCING RTL

A design in which all operations are scheduled and the values are assigned to registers is ready for RTL generation. There are several ways of generating RTL. A typical RTL model consists of the following parts:

1. Sequential part
   It contains the reset actions for state, output, and variable registers (initialization part) and their updates during normal functioning, where combinational "d-values" of register inputs are written to "q-values" of register outputs.
2. Combinational part that describes
   a. Control in the form of an FSM
      The FSM is extracted from the CFG, with a state encoding strategy (e.g., one hot) specified by the user.
   b. Datapath
      It specifies resources introduced by scheduling and their connections. For shared resources, inputs are taken from sharing MUXes. The latter can be specified through case statements controlled by the FSM.

In addition to the RTL model output, reports about the results of synthesis, as well as simulation-only models, can be generated. The latter are behavioral Verilog models that are neither synthesizable nor register and resource accurate but simulate faster than RTL and are used for debug and verification.

We have now walked through the main steps of the HLS process with an illustrative example. In the subsequent sections, we will describe details on some key aspects that need to be considered when devising HLS algorithms in practice.

11.3 SIMPLIFYING CONTROL AND DATA FLOW GRAPHS

11.3.1 DFG OPTIMIZATIONS

Optimizations reducing size and/or complexity of DFG operations are translated directly into smaller implementations. They could be divided into two categories: common compiler optimizations, equally applicable to software and hardware, and hardware-specific optimizations. Typically, an optimization step is implemented as a fixed-point computation, because applying one simplification may trigger another. Among compiler optimizations, the most prominent ones are (with examples)

1. Constant propagation

```
a + 0                     → a,
if (a == 128) {b = a - 1;} → b = 127;
a + 1 + 2                 → a + 3
```

2. Operation strength reduction

```
a/32                  → a >> 5,
a * 3                 → a + a << 1;
sc_uint<1> a, b; a + b; → a ^ b;
sc_uint a; a > 0      → OR(all_bits_of_a);
```

3. Common subexpression elimination

```
(a + b) * (a + b)              → tmp = a + b; tmp * tmp
a = M[i+1]; i++; b = M[i] + d; → a = M[i+1]; i++; b = a + d;
```

4. Dead code elimination

```
if (a > 128) sum = 128;
else if (b > 4 && a > 128) sum--; → if (a > 128) sum = 128;
```

A very powerful hardware-specific optimization is range minimization. The problem is posed as minimizing the ranges of operations based on (a) ranges of primary inputs, (b) ranges of primary outputs, and (c) user-specified ranges of some key variables (e.g., DFG feedback variables). Range analysis is implemented as a fixed-point forward and backward traversal of the DFG. In the forward traversal, the outputs of the operations are pruned based on information about their input ranges, while in the backward traversal the inputs of the operations are pruned based on the ranges of their outputs. Affine arithmetic (multiplications by constants and additions) is known to be well suited for range analysis, while nonaffine operations like nonconstant multiplications do not provide many of these opportunities. Range analysis is done by defining *transfer functions* for every DFG operation [12]. These functions come in two flavors:

1. Purely bit range based, say for $a = b + c$, with $b\_range = [0,4]$ (meaning that b has five bits) and $c\_range = [0,3]$ by forward propagation implies $a\_range = [0,5]$.
2. Value based, say for $a = b + c$, with $b\_value = [1,10]$ (meaning that b can take any value between 1 and 10) and $c\_value[1,5]$ by forward propagation implies $a\_value = [2,15]$, which requires one less bit than in the previous case. In addition to being more precise about maximum and minimum value reasoning, value-based analysis is also capable of providing insights about internal "holes" in the range to prove the redundancy of some internal bits.

Loops clearly present difficulties in range analysis due to values needing to be carried through the backward edge of the loop for the next iteration. For bounded loops, this difficulty could be avoided by virtually unrolling the loop during range analysis (similar to the bounded model checking approach). The other option is to use inductive reasoning between loop iterations, but its applicability is limited if the loop contains nonaffine operations [12]. In that case, the only option is a user-specified range for these feedback variables. Range analysis is a very important optimization technique that is particularly useful when the initial specification comes from legacy software with integral data types.

Timing is another major constraint in HLS; hence, there are many hardware-specific optimizations that are targeted particularly at timing:

1. Chain-to-tree conversion, where a long chain of operations is converted into a shallower (faster) tree structure. Figure 11.12a shows an example of this optimization, where a chain of five adders is converted into a tree form.
2. Retiming of MUXes across functional units (see Figure 11.12b).

To appreciate the power of timing optimizations, let us consider an example based on the following behavioral description, where c is an input Boolean signal and the loop is fully unrolled.

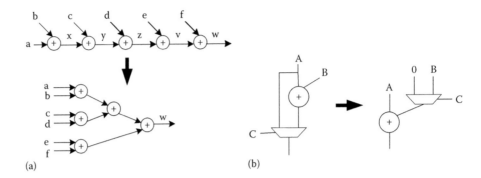

FIGURE 11.12   Timing optimizations: tree-conversion (a) and mux retiming (b).

```
for (i = 0; i < 4; i++) {
    if (c.read()) sum++;
}
```

The simplifications applied to the aforementioned timing transformations result in a structure that is far more suitable for hardware than the starting one (as illustrated by Figure 11.13).

11.3.2 CFG OPTIMIZATIONS

The optimizations presented so far are performed on the DFG and are targeted at datapath simplifications. Equally important is to simplify control, which is naturally formulated as a CFG structure simplification. The two main source code constructs that are responsible for complicating the CFG structure are conditional statements (*if, case*) and loops. We will consider loops in detail in the next section, while for now we will focus on conditional statements. Consider a SystemC code example as in Figure 11.14a and its CFG, where a conditional if-then-else statement is translated into a fork–join structure in the CFG. Fork–join structures in the CFG limit the mobility of operations, due to the need to synchronize computation in corresponding branches (MUXes are needed to produce a value for x, y, z). Typically, these merging MUXes (SSA Phi nodes) are fixed at the join nodes of the CFG. Even though it is possible to relax this condition and let them move to later nodes, this relaxation has limited applicability, especially when encountering a new fork–join structure. When these MUXes are considered fixed, the whole computation for x needs to happen in a single state, which requires at least two multipliers. A simple rewrite of the original specification in predicate form is shown in Figure 11.14b. It eliminates a conditional statement and

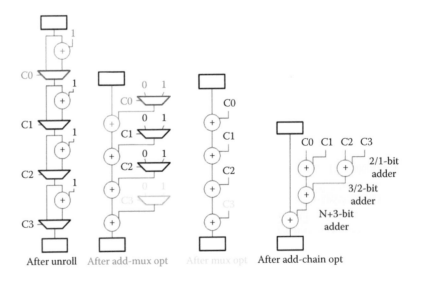

After unroll   After add-mux opt   After mux opt   After add-chain opt

FIGURE 11.13   Power of timing optimizations.

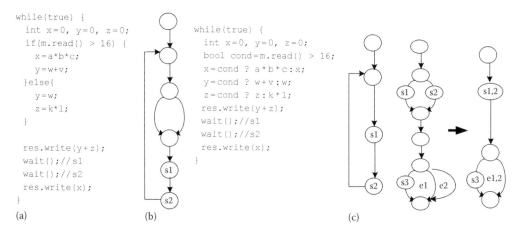

```
while(true) {
  int x=0, y=0, z=0;
  if(m.read() > 16) {
    x=a*b*c;
    y=w+v;
  }else{
    y=w;
    z=k*l;
  }

  res.write(y+z);
  wait();//s1
  wait();//s2
  res.write(x);
}
(a)
```

```
while(true) {
  int x=0, y=0, z=0;
  bool cond=m.read() > 16;
  x=cond ? a*b*c:x;
  y=cond ? w+v:w;
  z=cond ? z:k*l;
  res.write(y+z);
  wait();//s1
  wait();//s2
  res.write(x);
}
```

(b)   (c)

FIGURE 11.14 Simplifying conditional structures in CFG: initial specification (a) full (b) and partial (c) if-conversion.

simplifies the CFG. Instead, some operations get an additional attribute, called predicate condition, representing the condition under which they should be executed (e.g., w + v is executed when the variable *"cond"* = *true*). The advantage is the elimination of a synchronization point in the form of a join node, letting different computations span a different number of states. For example, computations for z and y are simple enough and can be completed before reaching state *s1*, while the computation for x may span both states, reducing the number of required multipliers to 1.

Reducing the CFG from the form shown in Figure 11.14a to that in Figure 11.14b automatically is called if-conversion (the fork–join structure is reduced to a single edge and predicates are added to DFG operations). The general case of if-conversion is represented in Figure 11.14c, which shows the extension of this transformation to branches with states (see states *s1* and *s2* merged into *s1,2*) and partial conversion (see edges *e1* and *e2* merged to *e1,2*).

If-conversion helps simplifying complicated CFG structures by eliminating fork–join constructs, and it relaxes many mobility constraints on operations, resulting in a richer solution space and potentially better QoR. A small downside is that mutual exclusivity of operations needs to be deduced not only from the CFG structure but from the predicate information as well. This, however, is relatively straightforward.

11.4 HANDLING LOOPS

Loops are commonly used in SystemC input code. Even our small *convertHSLToR* example has an embracing *while(true)* loop to describe the reactive nature of the algorithm infinitely repeating the transformation from HLS format to RGB. This loop however represents the simplest case because no values are carried from the outside loop to its internals and no values are carried between loop iterations. A more general example is represented by the following *FilterLoop* code snippet:

```
while(true)   {
  // setting a parameter to be used in the loop
  int smooth = param.read() + offset1.read() + offset2.read();
  wait();  //s1
  for (int i = 0; i < N; i++) {
      int aver = point[i] * scale[i] + point[i+2] * scale[i+2];
      out[i] = aver / smooth;
      smooth++;
      wait();  //s2
  }
}
```

The DFG and CFG for this example are shown in Figure 11.15a and b, respectively. One can notice that the DFG for *FilterLoop* is cyclic. Cycles occur when computing values i and *smooth* and are represented in the DFG by backward edges coming to MUX operations. These MUXes merge the computations from outside the loop (see the initialization of *smooth*) with the loop updates

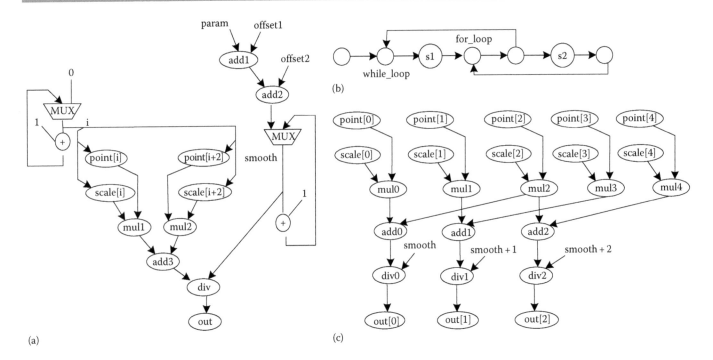

FIGURE 11.15   Data (a) and control flow graph (b) for Filter example with further unrolling (c).

that need to be passed to the next iteration (*smooth++*). Similar to merge MUXes for fork–join structures (as described in the previous section), the mobility of these operations is limited to the out-edge of the join node of the CFG. The presence of computational cycles in the DFG significantly complicates the problem of time budgeting during scheduling.

Indeed, operations from outside the loop that are scheduled on the in-edge of the loop share the same state as loop operations, and the more operations are scheduled before the loop, the less room remains in the bin for packing loop operations. This may result in operations from the loop body not fitting the current loop latency, calling for a loop latency increase. However, increasing the latency of the loop has a much bigger effect on the computation performance than increasing the latency outside the loop. We would rather add a state before the loop to separate operations outside the loop from those within the loop. The problem is that when scheduling operations before the loop, it is hard to anticipate the timing complications in the loop body because the scheduler has not seen it yet. Similar considerations could be applied for loop operations scheduled on the out-edge of the loop (i.e., on the out-edge of the *for_loop* node). These operations are in the fanout of the loop merge MUXes, and therefore they belong to the same timing path as operations from the last state of the loop feeding these MUXes backward. When scheduling the out-edge, it is as yet unknown which operations will end up in the last state and this presents a two-sided timing constraint that is nontrivial to solve. It typically requires a complicated reasoning to decide which relaxation to apply when scheduling fails to satisfy loop timing.

Simplifying or eliminating loops altogether is helpful in making the synthesis task easier. A common technique to do this is "loop unrolling." In its simplest form it results in *full unrolling*, which replaces a loop that is known to iterate at most N times with N copies of its loop body. The advantages of full unrolling could be appreciated by looking at the *FilterLoop* example. In its body, memories *point* and *scale* are accessed two times each, which for a single-port memory imposes a constraint of minimal latency 2 for executing a single loop iteration with four memory reads. The full unrolling of this loop (for N = 3) is shown in Figure 11.15c. Unrolling completely eliminates the hardware for loop control and reduces significantly the number of memory accesses and multiplications (doing a common subexpression extraction and noticing that *point[i + 2] * scale[i + 2]* for iteration i is the same as *point[i] * scale[i]* for iteration (i + 2)). The quantitative evaluations of the original and the unrolled implementations are summarized in Table 11.2, which clearly shows drastic improvements coming from unrolling.

TABLE 11.2 Evaluating Quality of Original and Unrolled Implementations

	Computation Time	No. of Memory Reads	No. of Multiplications
Original	$2 \times N \times Tclk$	$4 \times N$	$2 \times N$
Unrolled	$N \times Tclk + 2$	$2 \times (N + 2)$	$N + 2$

Unrolling might also be applied in a "partial" fashion when the *loop(Body)* is replaced by k copies *loop(Body$_1$, Body$_2$,..., Body$_k$)* and the loop is executed N/k times rather than N times.

This does not eliminate the loop but it helps in performing optimizations across different loop iterations, which are now explicitly expressed in the *Body$_1$, Body$_2$, ...* code.

Note that unrolling is beneficial only if the unrolled iterations are enjoying significant reductions due to optimization. If making iterations explicit does not create optimization opportunities and unrolling is not needed for performance reasons, then it is not advised because it simply bloats the DFG size and masks the structure of the original specification. There is no guarantee that different iterations would be scheduled uniformly and most likely the regularity of the datapath with N iterations will be lost. A careful evaluation of unrolling consequences should be done during microarchitecture exploration. When the increase in complexity overweighs the advantages given by the optimizations, the designer might want to keep the original loop structure. In the latter case, to make a loop synthesizable, one needs to make sure that it does not contain combinational cycles, that is, every computational cycle must be broken by a state. Breaking combinational cycles in loops that are not unrolled is typically performed automatically before starting scheduling, and it is up to the tool to find the best place for state insertion.

Besides unrolling and breaking loops, there are several well-known compiler techniques that rewrite loops in a simpler way that is more suitable for implementation (both in hardware and in software). The most popular ones are as follows:

1. *Loop inversion* that changes a standard *while* loop into a *do/while*. This one is particularly effective when one can statically infer that the loop is always executed at least once.
2. *Loop-invariant code motion* that moves out of the loop assignments to variables whose values are the same for each iteration.
3. *Loop fusion* that combines two adjacent loops that iterate the same number of times into a single one when there are no dependencies between their computations.

The advantages of loop fusion are illustrated by the following example when adjacent loops LOOP1 and LOOP2 are combined in a single loop LOOP1_2 reducing the number of reads from memory A and B by a factor of 2:

```
LOOP1 : for (int i = 0; i < N; i++) {
  CENTER[i].x = A[i].x + B[i].x;
  CENTER[i].y = A[i].y + B[i].y;
}

LOOP2: for (int i = 0; i < N; i++) {
  int x_dist = A[i].x - B[i].x;
  int y_dist = A[i].y - B[i].y;
  DIST_2[i] = x_dist * x_dist + y_dist * y_dist;
}

LOOP1_2:for (int i = 0; i < N; i++) {
  CENTER[i].x = A[i].x + B[i].x;
  CENTER[i].y = A[i].y + B[i].y;

  int x_dist = A[i].x - B[i].x;
  int y_dist = A[i].y - B[i].y;
  DIST[i] = x_dist * x_dist + y_dist * y_dist;
}
```

Finally, we would like to introduce one more loop transformation that is not common for software compilation but is very important for HLS. This is an important technique to reveal parallelism between loop iterations and is called loop pipelining. We will describe loop pipelining in detail in Section 11.7, but it is important to note here that it can only be applied to loops that do not have any other nested loops (apart from the special case of "stall loops," which wait until, e.g., input data or some shared resource becomes available). Therefore, when the user wants to pipeline a loop that does have a nested loop, the inner loop needs to be merged with the outer loop for pipelining to proceed. This transformation is illustrated in Figure 11.16. The following example shows the application of the merge loop transformation, which converts the *while_loop* with a nested *for_loop* into a single *while_loop*:

```
// Nested loops
while(1) {
    for (int j = 0; j < 10; j ++) {
        a[j] = b[j] + c[j];
    }
}
```

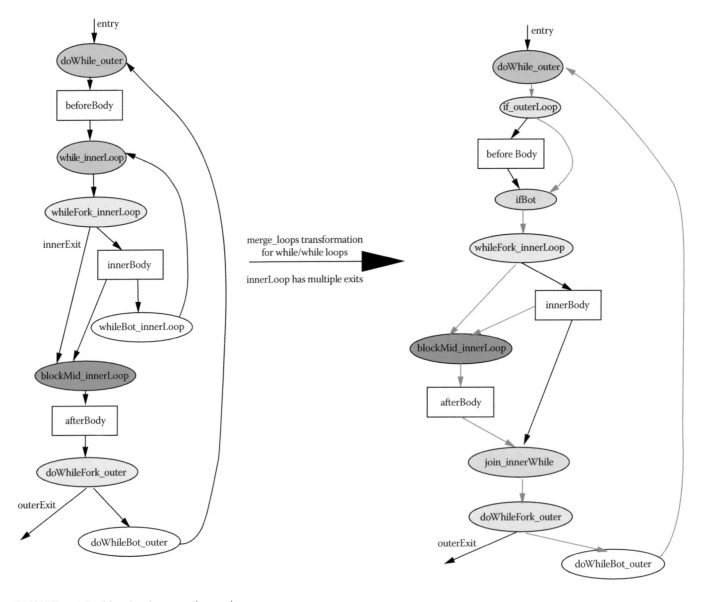

FIGURE 11.16   Merging inner and outer loops.

```
// Merged loops
bool start = true;
int j;
while(1) {
   if (start) {
      j = 0;
   }
   if (j < 10) {
     a[j] = b[j] + c[j];
     j++;
     start = false;
   } else {
      start = true;
   }
}
```

11.5 HANDLING FUNCTIONS

Almost every nontrivial computer program consists of multiple functions that call each other. Structuring a program in this way enables modularization, decomposition, and layering of levels of abstraction, all of which are key in designing complex systems. This program structure is often a useful guide when deciding the optimal hardware architecture. However, since the decision to separate some functionality into a function is driven by many reasons, and not only by the intended hardware structure, sometimes a better implementation can be achieved if function boundaries are ignored or modified. Thus, a part of the microarchitecture selection by the designer is to choose which functions should be implemented as separate units in hardware and which ones should be implemented as a part of the calling function. In the latter case, we say that the function is *inlined*, and in the former case, we say (for the lack of a better term) that the function is *noninlined*.

Inlining a function is equivalent to replacing the call to the function with a copy of the function's body. If a function is called at many places, multiple copies are created, and the DFG can grow significantly, sometimes testing the capacity limits of tools. On the positive side, each copy can be separately optimized in the context of its caller. This will enable optimizations like constant propagation that are specific to a call and cannot be applied to all calls.

Noninlined functions are implemented as stand-alone hardware design components. In general, such implementations can be treated similarly to resources, and function calls can be treated as operations. Multiple calls can share a single implementation, or the tool may decide to instantiate multiple copies of the implementation to improve performance.

One difference between a function call and other types of operations is that the latency of function calls (i.e., the number of cycles between receiving the inputs and generating the outputs) may be different for different values of input data. To deal with that, HLS tools typically implement some kind of protocol between the hardware implementation of the caller and the implementation of the function. For example, one can introduce "GO" and "RETURN" events and implement a function call as follows:

```
... // prepare arguments
notify(GO);
do {wait();} while (! notified(RETURN));
... // collect results
```

The block implementing the function then should behave like this:

```
while(true) {
 do {wait();} while (!notified(GO))
 ... // collect input argument
 ... // compute the function
 ... // prepare results
 notify(RETURN);
}
```

Similar implementations can be obtained if a function is explicitly modeled as a separate SystemC thread. However, according to the semantics of SystemC, every communication between threads requires at least one clock cycle, while the aforementioned scheme allows combinational communication between the caller and the function block.

This implementation still follows the sequential nature of the C++ code. At any given time, either the calling block or the function block may be doing some useful work, but the other is idling, waiting for an event to be notified. A more efficient implementation can be obtained if something is known about the latency of the block implementing the function. For example, if it is known that the function implementation takes at least N cycles to execute before generating RETURN, then the preparation of arguments and notification of GO on one side and waiting for RETURN and collection of results on the other side can be scheduled up to N cycles apart, and the caller can perform other useful work while the function implementation is computing the results.

Further optimization is possible if the latency of the function implementation is constant. Then, the RETURN event can simply be eliminated, and the collection of results can be scheduled a known constant number of cycles after notifying GO. Such implementations can be treated by the HLS tool simply as any other resource: either combinational (if latency is 0) or multicycle (otherwise).

Implementations of noninlined functions with constant latency may also be *pipelined*. Such an implementation may start processing a new set of inputs before it has computed the final results for the previous set. The minimum number of cycles between two consecutive input sets is called *the initiation interval*. Most commonly it is 1, but in general it can be any number of cycles smaller than the latency.

To create an implementation of a function, HLS tools typically use the same HLS algorithms they use for the caller. However, some tools also support an option where the usual HLS steps are skipped and a pipelined implementation is created by an embedded logic synthesis tool with retiming enabled. Logic synthesis can sometimes produce implementation with smaller latency and hence fewer storage elements, resulting in a smaller overall area.

11.6 HANDLING ARRAYS

Arrays are widely used in C and their hardware implementation requires special attention. First let us distinguish between arrays that are accessed only by a single thread and arrays that are accessed by multiple threads. We will first focus on the former, because it represents the base case. The latter introduces additional issues about managing concurrent accesses from two threads that go beyond the scope of this chapter.

HLS tools are typically capable of handling arrays in three ways:

1. Implementing them with memories created by a memory generator.
2. Implementing them as register banks. Such banks are defined in the generated RTL and they are then synthesized by a logic synthesis tool just like any other piece of generated RTL.
3. *Flattening* them, that is, treating them like any other variable and not applying any special rules to them. They will become values in the DFG, subject to the normal register allocation process.

The three options provide three different trade-off points in terms of design complexity and performance. The user usually has control over the implementation of each array.

There are two basic ways to flatten an array. For example, if an array is declared by "**char a[4]**," it could be treated as a 32-bit variable **a**, or it could be treated as four 8-bit variables **a_0**, ..., **a_3**. The latter approach increases the number of objects that need to be processed and thus increases the complexity, but it does make it easier to track dependencies between operations. For example, assume that the following loop is unrolled:

```
for (i=0; i<4; i++) a[i]++;
```

If array **a** is treated as a single variable, then the four increment operations need to be done in sequence, because they all modify the same variable **a**. However, if **a** is treated as four separate variables, after simple constant propagation, the tool may easily understand that there are no dependencies between the four increment operations and that they can be executed concurrently.

If an array is not flattened, then read and write accesses to the array are recorded as distinct operations that must be scheduled. However, scheduling these operations is subject to special constraints that do not apply to other kinds of operations. Consider, for example, the following code segment:

```
while(…) {
  …
  a[ind1] = x;    // operation W1
  y = a[ind2];    // operation R2
  out.write(y);   // operation O
  a[ind3] = z;    // operation W3
}
```

It contains two array write operations (W1, W3) and one array read operation (R2). If the array is implemented by a memory or a register bank, then executing each one of these operations occupies a port for one cycle. This creates a resource conflict in the sense that while one of the operations accesses the port, the other operations cannot access it. Normally, HLS tools can resolve resource conflicts by instantiating another copy of a resource. However, the resource in this case is a port, and the number of ports on a memory is typically fixed and cannot be increased. So, for example, if the array is implemented by a memory with a single read/write port, the body of the loop cannot be scheduled in less than three cycles.

In addition to resource constraints, scheduling of array operations is also subject to constraints derived from *array dependencies*. Consider operations W1 and R2, and assume that the array is implemented by a memory that has at least two write ports and one read port (hence there are no resource constraints). There are no data dependencies between W1 and R2; however, W1 will affect the result of R2, in case indices **ind1** and **ind2** are the same (this is often called a read after write [RAW] hazard). To cope with this kind of dependency, an HLS tool may require that R2 be scheduled at least one cycle after W1. This is a conservative rule that will generate correct implementations for all memories. It could be relaxed, for example, if the memory has a forwarding capability between a write port and a read port when the two addresses are the same or if forwarding logic is automatically generated [20].

Similarly, there is an array dependency (write after read [WAR] hazard) between R2 and W3. If W3 is not scheduled after R2 and **ind2** and **ind3** are the same, the value returned by R2 may be wrong. To satisfy this dependency, an HLS tool may require that W3 is scheduled at least one cycle after R2. Again, this is a conservative rule, but ironically it is needed exactly for the memories supporting the forwarding capability defined earlier. Finally, there is an array dependency between W1 and W3 (write after write [WAW] hazard), where W3 must be scheduled after W1 to avoid the possibility of W1 overwriting W3. There are no array dependencies between two read operations. In summary, even if the memory has plenty of ports, the body of the loop cannot be implemented in less than three cycles (unless forwarding is automatically supported by the memory or the tool), because R2 must be scheduled at least one cycle after W1, W3 must be scheduled at least one cycle after R2, and W1 in the next iteration must be scheduled at least one cycle after W3. This last dependency is an example of *interiteration array dependency*, which we will examine in more details in the section about pipelining.

HLS tools may analyze index expressions and not create array dependencies when two index expressions can be proven never to be equal. Analyzing array index expressions has been extensively studied, particularly for affine indices [23,24]. However, this kind of analysis is hard in practice, and it often misses cases for which designers are convinced that the dependency is not real. For these cases, most HLS tools allow the designers to break the dependencies that the tool could not prove to be false. However, if the designer is wrong and breaks a true dependency, then the generated hardware may not operate correctly.

In practice, reading from a memory often has a latency of one or more cycles. If the latency is N, then any operation that depends on the data returned by an array read needs to be scheduled at least N cycles after the read operation itself is scheduled. For example, the aforementioned operation O would have to be scheduled at least one cycle after the operation R1 if R1's latency is one cycle. Registers banks can be built with 0 latency, so that results are available in the same cycle in which the operation is scheduled. If this is the case, then only the usual data dependency rules need to be followed when scheduling array reads.

11.7 PIPELINING

Pipelining is a widely used technique for designing hardware that needs to meet stringent throughput requirements but has more relaxed latency requirements. In HLS, pipelined designs are created by pipelining loops. Without pipelining, an iteration of a loop will start executing only after the previous iteration is completely finished. When the loop is pipelined, computation for a single iteration is divided into stages and an iteration starts as soon as the previous iteration completes its first stage. Thus, at any given point in time, several iterations of the loop may be executing concurrently, each in a different stage. In a well-balanced design, all the stages take the same number of cycles, and henceforth we will only consider such designs. The number of cycles needed to execute one stage is called the "initiation interval." The number of cycles needed to complete all stages is called the "latency" of the pipelined loop. In practice, for a majority of designs, the initiation interval is required to be 1, and the latency varies widely from a few cycles to hundreds of cycles. Unless otherwise noted, we will assume that for all the examples in this section the initiation interval is 1. However, all the definitions and techniques are also valid for larger initiation intervals.

The user must specify the initiation interval for the loop to be pipelined. Most tools will accept some bounds on the latency, but most often the tools are searching for the optimal latency for the given initiation interval. A loop can be pipelined only if it meets certain conditions. The details vary from tool to tool, but typical conditions include the following:

- The loop can have no nested loops, except for *stall loops*. Stall loops in SystemC model conditions under which the pipeline needs to be stalled. Tools may accept different kind of loops as stall loops, but most tools require a stall loop to be of very special form, for example, **while(cond) wait();** or **do wait(); while(cond);**. To deal with nested loops, a tool may provide a transformation where multiple nested loops are merged into a single one (see Figure 11.16).
- It must be possible to schedule an iteration of the loop so that every execution in which stall loops are not taken takes the same number of cycles. Loops with such a property are said to be *balanced*. Every HLS tool has a set of transformations aimed at balancing unbalanced loops, but there can be no guarantee that these transformations will always succeed starting from a SystemC model with arbitrarily interspersed **wait()** calls.
- If a function call appears in a pipelined loop, it must either be inlined, or have a latency that is less than or equal to the initiation interval, or be pipelined. Furthermore, if it is pipelined, the initiation interval must be *compatible* with the initiation interval of the loop, and its latency must be less than the latency of the loop. A function pipelined with initiation interval II is compatible with loop initiations intervals that are multiples of II. In particular, a function pipelined with II = 1 is compatible with any pipelined loop. Note that pipelined functions can be called outside of a pipelined loop, but in that case, at most one of their stages is performing useful computations in any given clock cycle.

Thus, to schedule a pipelined loop means to choose the number of stages and then assign all operations in one iteration to their stages. This schedule is subject to the usual data dependency constraints, but it must also satisfy additional constraints specific to pipelining. Before we introduce these constraints, let us first introduce an example and some notation.

In the rest of this section, we will use the loop in Figure 11.17 as an example. Each iteration of the loop has four input operations (I1,..., I4), two arithmetic operation (>, +1), and two array accesses (R,W). We assume that array **arr** is implemented by a memory with one read and one

```
while(...) {
  while( ! a1_valid.read()) wait();   // input read I1, stall loop S1
  x = arr[a1.read()];                  // input read I2, array read R
  if (f(x) > 0) y = y+1;               // function f, greater than 0, increment +1,
                                       // multiplexor M
  while(! a2_valid.read()) wait();     // input read I3, stall loop S2
  arr[a2.read()] = y;                  // input read I4, array write W
}
```

FIGURE 11.17 *An example of pipeline.*

write port and read latency of 1. The loop also contains a call to function **f**. We assume that it is not inlined, but that it has a pipelined implementation with initiation interval 1 and latency 1. The multiplexor operation M computes the value of variable **y** at the bottom of the loop. The data inputs to M are the value of variable **y** from the previous iteration and the result of operation +1, while its control input is the result of operation >. Finally, the loop has two nested *stall loops*. They can be ignored for now, until we discuss stalling in detail.

To represent the schedule of the loop, we use expressions such as

(I1, S1, I2, R) | (f, +1) | (>, M) | (I3, S3, I4, W)

This expression indicates that

- ■ I1, S1, I2, and R are scheduled in stage 1
- ■ **f** and +1 are scheduled in stage 2
- ■ > and M are scheduled in stage 3
- ■ The rest of the operations are scheduled in stage 4

For the moment, we avoid the question if this is a valid schedule. By the end of this section, it should become clear that under certain conditions it indeed is.

To denote the state of the pipeline at any given point in time, we annotate the schedule with the iteration that is being executed in each stage. For example, we use

3(I1, S1, I2, R) | 2(f, +1) | b | 1(I3, S3, I4, W),

to denote the state where the first iteration is in stage 4, the second in stage 2, and the third in stage 1. Note that no iteration is in stage 3. In this case, we say that there is a *bubble* in stage 3 and use "b" to denote it. At times, we will omit some or all of the operations. So, to denote the same state as earlier, we may simply use

3 | 2 | b | 1

or if we want to focus only on array operations, we may use

3(R) | 2 | b | 1(W)

For the loop and the schedule as earlier, let us now examine the states through which the pipeline goes. Initially, the pipeline starts in the state

b | b | b | b

Assuming no stall loops are taken, the pipeline execution proceeds as follows:

1 | b | b | b
2 | 1 | b | b
3 | 2 | 1 | b
4 | 3 | 2 | 1
5 | 4 | 3 | 2
...

When pipelining a loop, one has to be careful that externally visible operations are executed in the same order as without pipelining. Consider the following execution trace focused on I/O operation I2 and I4:

4(I2) | 3 | 2 | 1(I4)
5(I2) | 4 | 3 | 2(I4)

...

Clearly, 4(I2) is executed before 2(I4), while without pipelining all operations in iteration 4 are executed after all operations in iteration 2. In this case, both operations are reads, which may not raise any concerns, but if I4 was a write and there was some external dependency between the output written by I4 and the input read by I2, the design may not be working properly after pipelining. The designer needs to make sure that these kinds of changes do not break the functionality, and some conservative tools may refuse to do such reordering without an explicit permission from the designer.

In the aforementioned loop, operation M depends on the result of operation +1, which in turn depends on the result of M in the previous iteration. So, in the data flow graph representing dependencies between operations, there is an edge from +1 to M and an edge from M to +1. In graph terminology, M and +1 belong to a *strongly connected component (SCC)* of the DFG. By definition, an SCC of a directed graph is a maximal subgraph such that there exists a path between any two nodes in the subgraph.

The dependency from M to +1 is between operations in different iterations of the loop, and hence it is called a "loop-carried dependency" *(LCD)*. In general, every SCC of the DFG must include some LCDs. Scheduling operations connected by LCDs requires special attention [3]. For example, the following chain of operations needs to be executed in sequence:

1(+1) -> 1(M) -> 2(+1)

But, 1(+1) and 2(+1) will be executed exactly one cycle apart (because the initiation interval is 1), posing significant restrictions on the scheduling of M. Some tools require that all the data from one iteration to the next be communicated through registers. If this is the case, then M must be scheduled in the same cycle as +1. More generally, all the operations in an SCC must be scheduled within the same initiation interval. Some tools use the technique known as *pipeline bypassing* [20] to relax this rule somewhat and allow M to be scheduled one stage after +1 without violating causality. For example, consider again the schedule

(I1, S1, I2, R) | (f, +1) | (>, M) | (I3, S3, I4, W)

where M is scheduled one cycle after +1. Consider a typical execution of the pipeline

2 | 1(+1) | b | b
3 | 2(+1) | 1(M) | b
4 | 3(+1) | 2(M) | 1

Operation 2(+1) depends on 1(M), which is being executed in the same cycle. To satisfy this dependency, the tool will create a combinational path from the hardware implementation of stage 3 to the implementation of stage 2, bypassing the registers that are normally used for such communication. Without pipeline bypassing, it may be impossible to schedule SCCs that contain sequential operations like memory reads.

From any operation in an SCC, there is a path to any other SCC member and a path back to itself. Thus, every operation in an SCC belongs to at least one *dependency cycle*. The sum of delays for all operations in such a cycle poses a fundamental limit on the throughput of the design. The timing path along any dependency cycle in an SCC must be less than the clock period times the initiation interval. This property holds regardless of the presence or absence of pipeline bypassing. Given a fixed initiation interval, this poses a fundamental limit on the maximal frequency of the design. Note that increasing pipeline latency does not affect this property. Changing the initiation interval does relax the frequency constraint, but this is often not an option because it may violate strict throughput requirements.

Let us now consider an execution of the pipeline while focusing on the array operations in the loop:

4(R) | 3 | 2 | 1(W)
5(R) | 4 | 3 | 2(W)
6(R) | 5 | 4 | 3(W)

Operation 4(R) is executed before 2(W) and 3(W), and it is executed in the same cycle as 1(W), while without pipelining it would be executed after all of them. This is not a problem as long as the address used by 4(R) is different from those used by all three write operations. Analyzing these interiteration array dependencies is the subject of intensive research [23,24], but in practice the tools rely on a combination of analysis and information provided by the user.

For this schedule to be legal, the user needs to specify that operations R and W have *distance* 3. This indicates that in any iteration the address of W will not be equal to the address of R in *at least* any of the three subsequent iterations. Based on this assurance from the user, the tool will allow W to be scheduled *up to* (but no more than) three stages after R. Note that if the user cannot give such a guarantee, the only option that would not violate such constraints between iterations would be to schedule R and W in the same stage, but this is not possible with II 1, because of the data dependency within a single iteration. Thus, in this case, the loop is not pipelineable, unless the user relaxes the distance constraint between array ops or increases the II.

Let us now examine how multicycle operations affect scheduling. The same rules as in nonpipelined case apply. Function call f depends on operation R, which has latency 1, so the earliest it can be scheduled is stage 2. But since function f is pipelined with latency 1, the earliest stage in which its successors can be scheduled is stage 3. For example, operation > cannot be scheduled sooner than that stage.

Finally, let us consider stall loops. Assume that the pipeline is in state

5 | 4 | 3 |2

and that **a2_valid** is 0. Stall loop 2(S2) will be executed, and we say that stage 4 has *stalled*. Since iteration 3 cannot proceed to stage 4 until iteration 2 has completed, iteration 3 will also stall in stage 3. The stalling will propagate backward and the pipeline will remain in this state until **a2_valid** becomes 1 again.

Let us now consider the same state but assume that **a2_valid** is 1 and **a1_valid** is 0. Now, stall loop 5(S1) is executed and stage 1 stalls. Many HLS tools are capable of dealing with such stalls in two ways: *hard stall* and *soft stall*. In the hard stall approach, the pipeline will be kept in the current state until **a1_valid** becomes 1. In the soft stall approach, stage 1 is stalled, but the later stages are allowed to proceed. Thus, in the next cycle the state will be

5 | b | 4 | 3

Note that a bubble has been created in stage 2. If **a1_valid** becomes 1, the pipeline will proceed, but the bubble will remain and will move down the pipe:

6 | 5 | b | 4

Now assume that in this state **a2_valid** is 0. Tools that support soft stall typically also support *bubble squashing*, that is, they will stall stage 4 but not stages 2 and 1 because there is a bubble in front of them. Thus, with bubble squashing the next state will be

7 | 6 | 5 | 4

and the pipeline will be fully filled again.

In general, soft stall ensures better average performance and avoids many deadlocks that may happen with hard stall. However, hard stall requires simpler logic, with shorter timing paths, and its simplicity makes it easier to coordinate multiple independent pipelines. Hard stall implementations also very often have a lower power profile than soft stall implementations. Therefore, both approaches are used in practice.

As indicated earlier, scheduling of pipelined loops is similar to scheduling of nonpipelined ones, with additional constraints due to LCDs, interiteration array dependencies, and I/O reordering. Another difference from the nonpipelined case is that resources cannot be shared between stages, since they may all be active in any given cycle. However, if the initiation interval is larger than one, then every stage consists of several cycles, and it is possible to share resources between different cycles.

Once a feasible schedule is found, HLS tools still need to take a few special steps before generating RTL. Registers to carry data from one stage to another need to be allocated. The tools need to take special care of data that need to be carried across multiple stages, as multiple versions of the data may exist at any given time, one for each active iteration. Pipeline control logic must be generated, including the stalling signals. Stalling poses a particular problem for multicycle operations like pipelined functions. For example, if a five-stage pipelined function is filled with data when the calling pipelined loop stalls, steps need to be taken not to lose data that are already in flight. The tools typically take one of two approaches. They either generate stalling control when the function is pipelined, so that internal computation of the function can be stalled at the same time as the calling loop stalls, or, if this is not possible, they add stall buffers to the function outputs to temporarily store the data that would otherwise be lost.

Pipelining changes the cycle-by-cycle behavior of the design, so if one wants to build a verification environment that could be used for both the source code and the generated RTL, then these changes need to be dealt with. In Section 11.1.4, we discussed verification approaches for cycle-delayed and latency-insensitive cases. The best way to deal with pipeline verification is to make sure that the design falls in one of these two categories. Pipelining without stalling is the basic tool to create cycle-delayed designs. Latency-insensitive designs based on handshaking protocols naturally lead to the use of stall loops in pipelined loops.

11.8 MANIPULATING LATENCY

Finding the optimal RTL implementation latency during synthesis is a nontrivial task. Let us start from the simplest reactive specification:

```
// Example: reactive_simple
while (true) {
      read_inputs();
      compute();
      write_outputs();
}
```

Typically, design requirements come in the form of a throughput constraint (Thr), defining the rate at which I/O operations take place, say 100 MHz, and of a clock period $Tclk$ (say 2 ns), derived from the environment in which the hardware block will be integrated. Latency is measured as a delay (in the number of clock cycles) between reading inputs and producing outputs and very often designers are flexible about it (as long as the throughput is satisfied). In this case, the latency of the computation is a parameter that needs to be determined during synthesis, where its upper bound is defined by $Lat_{max} = \lfloor 1/(Thr * Tclk) \rfloor = 1/(10^8 * 2 * 10^{-9}) = 5$. If the computations inside the loop fit into five clock cycles, then the synthesis task is over. This gives a first approximation for the procedure of synthesis under a latency constraint when the upper bound on Lat_{max} is known (see Figure 11.18a).

The procedure in Figure 11.18a is simple but it is effective only in very limited cases:

1. An implementation with maximal latency might not be optimal. Indeed if in the previous example the computation could fit three cycles, an implementation with latency = 5 is redundant in terms of both timing and area (a five-state FSM is more complicated than a three-state FSM).
2. Typically, computations are data dependent and working under a single worst-case latency significantly impacts the performance of the system.

```
Given upper bound Lat_max and thread T

 Lat_c = compute_init_latency ();
  //Lat_c=max (Lat_i) for all CFG paths in T

 add_states (Lat_max-Lat_c);

 success = schedule (T, 0);
  //("0" arg shows that latency is fixed)

 if (!success) report_overconstraint ();
```
(a)

```
Given upper bound Lat_max and thread T with
latency-insensitive IO protocols

  P_IO = specify_protocols ();

   set_max_latency (region_CFG, Lat_max);

   slack = compute_slack(CFG);

   success = schedule(T, P_IO, slack)

   if (!success) report_overconstraint();
```
(b)

FIGURE 11.18   Handling latency during scheduling: fixed (a) and latency-insensitive (b) approaches.

3. A computation within a thread is rarely represented as straight-line code, but it may contain several loops. Budgeting latency across loops requires careful consideration.
4. SystemC specification and RTL implementation may have different throughputs, which complicates the construction of testbenches. In our example, the specification needs a new input every two cycles, while the implementation requires an input every five cycles.

Addressing the data-dependent duration of the computations performed by different threads requires one to specify IO interactions among threads and with the environment in a latency-insensitive way. This can be addressed by using latency-insensitive I/O protocols that directly identify when an I/O transaction completes successfully. A simple example of such a protocol is given as follows:

```
//read_input protocol
while (!in_valid.read()) { wait();}
val = in_data.read();
in_accept = 0;
wait();
in_accept = 1;
```

In *read_input protocol* the design waits for its environment to notify it that a new input value is available by setting the *in_valid* signal to high. Before this happens, the design is stalled, simply checking the *in_valid* value from the environment. Once *in_valid* becomes true, the design reads the input value from the environment (*in_data.read()*) and informs the environment about the read completion by lowering *in_accept*. *in_accept* is asserted again in the next clock cycle, in preparation for consuming the next input. A similar protocol can be designed for *write_output*. Clearly, input and output protocols must preserve the same cycle-accurate behavior throughout synthesis, which implies that adding states (increasing latency) is not allowed inside the sub-CFG defining each protocol. Coming back to our *reactive_simple* example and substituting the *read_inputs()* and *write_outputs()* functions by the corresponding latency-insensitive protocols (with minimal latency 1 each), we can formulate the synthesis problem as implementing the *compute()* functionality by adding not more than three states to the *reactive_simple* thread. Note that this formulation is very different from our first setting that defined the latency of *reactive_simple* to be fixed at 5. Not only with a new setting the testbench can be used "AS IS" for both the input behavioral description and the implementation RTL (due to its latency-insensitive property), but the optimization space has increased, by exploring also the implementation solutions with latency 2, 3, 4, 5 in the same HLS run. The procedure for synthesis with latency constraints is refined as in Figure 11.18b.

The new function *specify_protocols()* aims at marking CFG regions (connected subgraphs with a single start node) as protocols. There are different ways of implementing this function: it can identify CFG regions based on protocol directives that are placed in the SystemC code or the user may want to specify protocols directly in the CFG, by pointing at the start and the end nodes of the region. Similarly, *set_max_latency()* specifies a maximal latency for a CFG region (for the whole CFG in particular) that is defined as the maximal latency for all acyclic paths in the region.

The goal of function *compute_slack()* is to provide a quantitative estimation of how many states can be legally added to a CFG to keep all latency constraints satisfied. This metric is conveniently represented by the notion of *sequential slack* on CFG edges. The sequential slack of an edge defines the number of states that could be legally added to this edge without violating any latency constraints. From the discussion about protocols, it is clear that any edge in the protocol region has slack 0 (no states are allowed to be added in protocols). To illustrate the notion of sequential slack, we refer to the example of CFG in Figure 11.14c (after if-conversion). Suppose that this CFG fragment is constrained to have maximum latency 3. Then for edge *e1,2* sequential slack is 2 because we can add two states to this edge and still meet the latency constraint. For the rest of the CFG edges, the slack is 1.

The function `schedule(T, Pro, slack)` performs scheduling and may add states to the edges that have positive sequential slack. The decision about adding a state can be taken, for example, by an expert system (see Figure 11.5b) by comparing the cost of this action versus other actions like adding resources and speculating an operation. Adding a state to an edge requires recomputing the slack for the rest of the CFG because the slacks of other edges are affected. Adding a state to any edge before the fork of the CFG in Figure 11.14c (assuming latency constraint 3) nullifies the sequential slack of all edges but *e1,2*, whose slack becomes 2−1 = 1.

To make the procedure from Figure 11.18b effective, the designer needs to translate throughput requirements down to latency constraints in the CFG. Sometimes it is difficult to do this explicitly. When several IO interfaces exist with complicated dependencies between them, expressing throughput as a single static upper bound might not be possible for practically meaningful bounds. Even when such a bound exists (as in our *reactive_simple* example), the computation could have loops with a number of iterations that is not statically known. This would make it impossible to derive latency constraints from the desired throughput. In general, latency constraints either are well defined only for loop-free regions of the CFG or must be interpreted as feedforward (ignoring backward edges). In the latter case, they constrain the number of states in a loop body but ignore the number of loop repetitions. Due to these reasons solving the task of converting throughput requirements into latency constraints is typically left to the user. This is called "latency budgeting" and its automated solution is an open problem. The typical way of approaching it is to let the tool run without latency constraints, when any number of states can be added outside the protocol regions. Then the designer analyzes the obtained solution (either manually or through simulation) and provides latency constraints to move in the desired direction, until a satisfactory implementation is obtained.

11.9 HLS PARADIGM AND FUTURE OPPORTUNITIES

HLS technology has advanced significantly, thanks to broad and persistent efforts in both academia and industry over the last several decades. Hence, when commercial HLS tools are evaluated today for production use, the question of whether the tool can produce high-quality implementations or not is no longer the central concern. Of course, HLS must meet stringent requirements of quality implementations, but it is widely recognized that the technology can be used to match or often outperform manual RTL in terms of QoR. People also know that the behavioral descriptions taken as input by HLS provide far more configurability than RTL, and this offers significant advantages to address the needs of reusability of design models across different design projects and across multiple generations of a given project with different design requirements.

This shifts the central concern for adopting HLS from whether it could be actually used as a design tool to how it could be integrated into the existing design flow, which starts from RTL. Specifically, there are three kinds of integration concerns: with the verification flow, with legacy RTL in the rest of the design, and with the backend implementation flow.

For verification there are two main issues. The first is how to extend state-of-the-art verification methodologies such as UVM (Universal Verification Methodology) [13] for verifying the behavioral description used as input to HLS. The second is how to reduce the verification effort currently required at RTL, when the RTL models are produced by HLS tools and not manually written by a human. For the first one, commercial verification tools have been extended to support functional verification of behavioral descriptions, especially when they are written in SystemC [25]. SystemC is a widely used language for modeling hardware behavior, not just for HLS but also for other use cases such as virtual platforms for software integration or architecture characterization. For this reason, extension efforts in terms of verification support of SystemC design models are active, and standardization of these extensions is also in progress [14,15]. On the other hand, other types of verification such as formal property checking or hardware-oriented code coverage are not yet fully adopted for these behavioral descriptions. Tools are available in this category for software engineering, but special requirements exist for hardware models, for example, interface protocol verification for concurrent systems or value coverage analysis for variables used in the models, and existing technologies are still not adopted in practice for industrial use with hardware models. For the second concern, the most desired solution is formal equivalence checking, which automatically confirms that the functionality of the synthesized RTL is equivalent to that of the original behavioral description under the specified synthesis transformation, such as pipelining or latency alteration. This area has made significant advances in the last decade. Nevertheless, the application domains for which HLS is used are expanding even at a higher rate. Furthermore, the capabilities of HLS technology are advancing to provide more aggressive transformations between the input behavioral descriptions and the resulting RTL. For these reasons, simulation-based verification is still required in practice for HLS-generated RTL in many commercial design projects.

Whether verification is done formally or by simulation, support for easy debugging of the generated RTL is extremely important, because the RTL models are generated by tools and are not always easy to understand for humans. What is needed is not only the debugging of the RTL itself, as would be done traditionally with manually written RTL code, but also the ability to correlate signals and design constructs in the RTL to the original behavioral description or the underlying design semantics. Modern HLS tools are advancing the correlation features for this purpose, thus helping humans to easily analyze the synthesis results. For example, if there is an adder resource in the RTL, designers can query the HLS tool to identify which addition operations of the original behavioral description are implemented by the RTL adder, which sequence those addition operations are executed, and where they are scheduled within the control data flow graph. To accomplish this, the HLS tool needs an internal representation to efficiently keep track of synthesis transformations made from the input descriptions to RTL and record this information in a database, so that it can be retrieved as needed.

The second type of integration is with the RTL-based design flow. In typical design projects, it is rare that all the design components required for a project are newly designed. Usually there are design models that can be reused or revised easily to implement some functionality, and many of such existing models are written in RTL. When HLS is used for the rest of the components, it is important that the HLS flow does not affect the design steps executed for the whole design. First of all, this means that when design components adjacent to the HLS components are written in RTL, they need to be connected seamlessly, despite the fact that they might be written in different languages and using different abstraction levels for the interfaces. Modern HLS tools provide features to support this need. For example, they automatically produce interlanguage interface adapters between the behavioral and RTL blocks. Sometimes, the boundary between the HLS components and RTL components does not exactly coincide with that of the design blocks specified in the block diagram of the design specification. There are cases where some part of the functionality of an HLS component must be implemented using existing RTL code. For example, a given function called in the behavioral description may have an RTL implementation that has been used for many years in multiple production projects. The designer knows that the RTL code works and there is no need to change it. Some HLS tools support this kind of fine-grain integration, by letting the user specify the binding of the behavioral function to the RTL design, with additional information on its input and output interfaces and timing characteristics, so that the tools can automatically reuse the RTL code, properly interfacing with the rest of the design synthesized by HLS.

Another requirement in the integration of HLS in the existing design flow is support for engineering change orders (ECOs). In the context of HLS, the main concern is support for top-down ECO, where at a late stage, when design components have been already implemented to the logic or layout level and verification has been done, the need arises to introduce small changes in the design functionality. In the RTL-based design flow, the designers carefully examine the RTL code and find a way to introduce the changes with minimal and localized modification of the code. If the designer tries to do the same with HLS, by introducing small changes in the behavioral description, when HLS is applied to the new description, the generated RTL often becomes very different from the original one. The logic implemented in RTL may change very significantly even if the functionality is very similar to the original one. Some HLS tools provide an incremental synthesis feature to address this issue [16]. In this flow, HLS saves information about synthesis of the original design in the first place, and when ECO happens, it takes as input this information together with the newly revised behavioral description. It then uses design similarity as the main cost metric during synthesis and produces RTL code with minimal differences from the original RTL code while meeting the specified functionality change.

In terms of integration with the backend implementation flow, a typical problem encountered with RTL produced by HLS is wire congestion. In behavioral descriptions used as input to HLS, it is not easy to anticipate how wires will be interconnected in the resulting implementations. A large array in the behavioral description may be accessed by many operations, but these operations could be partitioned in such a way that each subset accesses only a particular part of the array. In this case, decomposing the array into several small arrays and implementing each one of them separately could lead to better use of wires. But it is not easy to identify this up front in the input description. In fact, this single large array might be used in a concise manner in a loop, and this could look better than having multiple arrays. The designer might then instruct an HLS tool to implement this large array as a monolithic memory, and the tool would not be able to produce the better implementation with decomposed arrays. There are also cases where the HLS tool introduces artificial congestion. As we saw earlier in this chapter, HLS can effectively explore opportunities of sharing resources to implement arithmetic operations. It can also explore opportunities to share registers for storing results of different operations. These are good in terms of reducing the number of those resources required in the implementation, but they might create artificial wire congestion, because multiple wires are forced to be connected with the shared resources even though they might not be functionally related with each other and thus could be placed in a more distributed and more routable manner if resources were not shared.

The effect of wiring cannot be fully identified during the HLS process, because sufficient information on wiring is not available in that phase. It is therefore imperative to provide a mechanism that feeds the results of the analysis of wires made in the placement phase back to the HLS phase. Some companies that adopt HLS for production designs implement such a flow, where the reports of wire congestion obtained in the backend are correlated to resources used in the RTL generated by the HLS tool, which in turn are analyzed to identify the characteristics of the behavior accessing the resources associated with those resources [17]. With tool support for providing such correlations, engineers can develop an iterative design and analysis flow to cope with the congestion issue. A similar opportunity exists for power analysis and low-power design, in that effective feedback from the backend implementation flow to HLS can lead to better exploration by HLS tools or by their users of the design space, while considering power consumption. We anticipate that tighter integration of HLS with backend tool chains will be provided more and more in the future, so that design bottlenecks can be identified quickly, and HLS technology can be efficiently used to produce implementations that lead to smooth design closure in the downstream implementation process.

REFERENCES

1. R. Camposano, W. Wolf, *High-Level VLSI Synthesis*, Kluwer Academic Publishers, 1991.
2. R. Camposano, From behavior to structure: High-level synthesis, *IEEE Design & Test of Computers*, 7(5), 8–18, October 1990.
3. A. Kondratyev, M. Meyer, L. Lavagno, Y. Watanabe, Realistic performance-constrained pipelining in high-level synthesis, *Design, Automation & Test in Conference*, Grenoble, France, 2011, pp. 1382–1387.
4. B. Alpern, M. Wegman, F. Zadeck, Detecting equality of variables in programs, in *Conference Record of the 15th ACM Symposium on Principles of Programming Languages*, ACM, New York, January 1988, pp. 1–11.
5. C. Mandal, P. Chakrabarti, S. Ghose, Complexity of scheduling is high level synthesis, *VLSI Design*, 7(4), 337–346, 1998.

6. H. De Man, J. Rabaey, P. Six, L. Claesen, Cathedral II: Silicon compiler for digital signal processing, IEEE Design & Test of Computers, 3(6), 13–25, December 1986.

7. M.R.K. Patel, A design representation for high level synthesis, *Design Automation Conference*, Orlando, FL, 1990.

8. P. Paulin and J. Knight, Force-directed scheduling for the behavioral synthesis of ASICs, *IEEE Transaction on Computer-Aided Design of Integrated Circuits and Systems*, New York, NY, 1989.

9. G.D. Micheli, *Synthesis and Optimization of Digital Circuits*, McGraw-Hill, Inc., 1994.

10. M.C. McFarland, A.C. Parker, R. Camposano, Tutorial on high-level synthesis, *25th Design Automation Conference*, Anaheim, CA, 1988.

11. M.J.M. Heijligers, H.A. Hilderink, A.H. Timmer, J.A.G. Jess, NEAT: An object oriented high-level synthesis interface, *International Symposium on Circuits and Systems*, London, UK, 1994.

12. M. Stephenson, J. Babb, S. Amarasinghe, Bitwidth analysis with application to silicon compilation, *ACM PLDI*, 2000, pp. 108–120.

13. Salemi, R., *The UVM Primer: A Step-by-Step Introduction to the Universal Verification Methodology*, Boston Light Press, 2013.

14. B. Bailey, F. Balarin, M. McNamara, G. Mosenson, M. Stellfox, Y. Watanabe, *TLM-Driven Design and Verification Methodology*, Cadence Design Systems, Inc., San Jose, CA, 2010.

15. Sherer, A., Accellera's UVM in SystemC standardization: Going universal for ESL, available at http://accellera.org/resources/articles/accelleras-uvm-in-systemc-standardization-going-universal-for-esl, 2015.

16. L. Lavagno, A. Kondratyev, Y. Watanabe, Q. Zhu, M. Fujii, M. Tatesawa, Incremental high-level synthesis, *Asia South Pacific Design Automation Conference*, Taipei, Taiwan, 2010.

17. Q. Zhu, M. Tatsuoka, R. Watanabe, T. Otsuka, T. Hasegawa, R. Okamura, X. Li, T. Takabatake, Physically aware high-level synthesis design flow, *Design Automation Conference*, San Francisco, CA, 2015.

18. S. Gupta, N. Dutt, R. Gupta, A. Nicolau, SPARK: A high-level synthesis framework for applying parallelizing compiler transformations, *Proceedings of the 16th International Conference on VLSI Design*, New Delhi, India, January 4–8, 2003, p. 461.

19. L.P. Carloni, K.L. McMillan, A.L. Sangiovanni-Vincentelli, Theory of latency-insensitive design, *IEEE Transactions on Computer-Aided Design of Integrated Circuits and Systems*, 20(9), 1059–1076, 2001.

20. T. Kam, M. Kishinevsky, J. Cortadella, M. Galceran Oms, Correct-by-construction microarchitectural pipelining, *IEEE/ACM International Conference on Computer-Aided Design, ICCAD 2008*, San Jose, CA, 2008, pp. 434–441.

21. E.A. Lee, T. Parks, Dataflow process networks, *Proceedings of the IEEE*, 83(5), 773–799, 1995.

22. S. Krstic, J. Cortadella, M. Kishinevsky, J. O'Leary. Synchronous elastic networks, *Proceedings of the Formal Methods in Computer-Aided Design*, San Jose, CA, 2006, pp. 19–30.

23. T. Stefanov, B. Kienhuis, E.F. Deprettere, Algorithmic transformation techniques for efficient exploration alternative application instances, *Proceedings of the 10th International Symposium on Hardware/Software Codesign, CODES 2002*, 2002, pp. 7–12.

24. F. Balasa, F. Catthoor, H. De Man, Exact evaluation of memory size for multi-dimensional signal processing systems, *Proceedings of the 1993 IEEE/ACM International Conference on Computer-Aided Design*, 1993, pp. 669–672.

25. T. Grotker, S. Liao, G. Martin, S. Swan, *System Design with SystemC*, Kluwer Academic Publications, Boston, MA, 2002.

26. P. Coussy, D.D. Gajski, M. Meredith, A. Takach, An introduction to high-level synthesis. *IEEE Design & Test of Computers*, 26(4), 8–17, 2009.

27. D. Knapp and M. Winslett, A prescriptive formal model for data-path hardware, *IEEE Transactions on CAD*, 11(2), 158–184, February 1992.

28. A. Kondratyev, M. Meyer, L. Lavagno, Y. Watanabe, Exploiting area/delay tradeoffs in high-level synthesis, *Design, Automation & Test in Conference*, Dresden, Germany, 2012, pp. 1024–1029.

29. F. Wang, A. Takach, Y. Xie, Variation-aware resource sharing and binding in behavioral synthesis, *Proceedings of the Asia and South Pacific Design Automation Conference (ASP-DAC 09)*, Yokohama, Japan, 2009, pp. 79–84.

Microarchitecture Design

Back-Annotating System-Level Models

Miltos D. Grammatikakis, Antonis Papagrigoriou,
Polydoros Petrakis, and Marcello Coppola

CONTENTS

12.1 INTRODUCTION

The number of transistors on a chip grows exponentially, pushing technology toward highly integrated systems on a chip (SoCs) [1]. Existing design tools fail to exploit all the possibilities offered by this technological leap, and shorter-than-ever time-to-market trends drive the need for innovative design methodology and tools. It is expected that a productivity leap can only be achieved by focusing on higher levels of abstraction, enabling optimization of the top part of the design where important algorithmic and architectural decisions are made and massive reuse of predesigned system and block components are applied.

Detailed VHDL (VHSIC hardware definition language) or Verilog models are inadequate for system-level description due to poor simulation performance. Advanced system-level modeling may lead from a high-level system model derived from initial specifications, through successive functional decomposition and refinement, to implementing an optimized, functionally correct, unambiguous protocol, that is, without deadlock conditions or race hazards. A popular open-source C++ system-level modeling and simulation library that allows clock-accurate (also called cycle-accurate) modeling is Accellera Systems Initiative SystemC (first standardized by OSCI) [2]. SystemC (now IEEE standard 1666) consists of a collection of C++ classes describing mainly hardware concepts and a simulation kernel implementing the runtime semantics. It provides all basic concepts used by HDLs, such as fixed-point data types, modules, ports, signals, time, and abstract concepts such as interfaces, communication channels, and events. SystemC 2.3 allows the development and exchange of fast system models, providing seamless integration of tools from a variety of vendors [3]. A detailed system-level modeling framework can rely on a SystemC-based C++ Intellectual Property (IP) modeling library, a powerful simulation engine, a runtime and test environment, and refinement methodology. Two commercial tools for domain-independent SystemC-based design were developed in the early days: VCC from Cadence [4] and Synopsys Cocentric System Studio (extending RTWH-Aachen's COSSAP tool) [5] and Synopsys SystemC Compiler [6].

Despite current system design efforts, there is not yet a complete and efficient SystemC-based development environment capable of

- Providing parameterized architecture libraries, synthesis, and compilation tools for fast user-defined creation and integration of concise, precise, and consistent SystemC models

- Enabling IP reuse at the system and block level, which results in significant savings in project costs, time scale, and design risk [7,8]
- Supporting efficient design space exploration

In this chapter, we focus on general methodologies for clock-accurate system-level modeling and performance evaluation and design space exploration, a key process to enhancing design quality.

More specifically, in Section 12.2, we consider system modeling, design flow, and design space exploration. We first introduce abstraction levels and outline general system-level design methodology, including

- Top-down and bottom-up protocol refinement
- Hierarchical modeling, that is, decomposing system functionality into subordinate modules (blocks and subblocks)
- Orthogonalization of concerns (i.e., separating module specification from architecture and behavior from communication)
- Communication protocol layering

In Section 12.3, we outline SystemC-based system-level modeling objects, such as *module, clock*, intramodule memory, intramodule synchronization, and *intermodule communication channels*, and examine back-annotation of different operational characteristics of (predefined and customized) system-level models, focusing on time annotation.

Then, in Section 12.4, we discuss examples of system-level modeling and related back-annotation issues. We consider time annotation of clock-accurate models by providing an example based on processor architectures. We also examine a more complex example of system-level modeling and time annotation on a user-defined SystemC-based hierarchical finite-state machine (HFsm). This object provides two descriptions of model behavior: using *states, conditions, actions*, and *delays* as in traditional (flat) finite-state machines and using *states and events*. We also examine intermodule communication channels in SystemC and phase-locked loop (PLL) design in SystemC-AMS, the SystemC extension for analog mixed-signal (AMS) modeling.

In Section 12.5, we focus on system-level modeling methodologies for collecting performance statistics from modeling objects, including automatic extraction of statistical properties. Advanced high-level performance modeling environments may be based on advanced system-level monitoring activities. Thus, we examine the design of integrated system-level tools, *generating, processing, presenting*, and *disseminating* system monitoring information.

In Section 12.6, we discuss open system-level modeling issues, such as asynchronous processing, parallel and distributed system-level simulation, and interoperability with other design tools. By resolving these issues, SystemC-based system-level modeling can achieve a higher degree of productivity.

12.2 DESIGN METHODOLOGY

We examine fundamental concepts in system-level design, including abstraction levels, system-level modeling methodology, and design exploration.

12.2.1 ABSTRACTION LEVELS

A fundamental issue in system design is model creation. A model is a concrete representation of IP functionality. In contrast to component IP models, a *virtual platform prototype* refers to system modeling. A virtual platform enables integration, simulation, and validation of system functionality, reuse at various levels of abstraction, and design space exploration for various implementations and appropriate hardware/software partitioning. A virtual platform prototype consists of

- *Models of hardware components*, including peripheral IP block models (e.g., I/O, timers, audio, video code, or DMA), processor emulator via instruction set simulator (ISS) (e.g., ARM V4, PowerPC, ST20, or Stanford DLX), and communication network with internal or external memory (e.g., bus, crossbar, or network on chip)

■ *System software*, including hardware-dependent software, models of RTOS, device drivers, and middleware
■ *Environment simulation*, including application software, benchmarks, and stochastic models

Notice that a virtual prototype may hide, modify, or omit system properties. As shown in Figure 12.1, abstraction levels in system modeling span multiple levels of accuracy ranging from functional to transactional to clock and bit accurate to gate level. Each level introduces new model details [9]. We now provide an intuitive description for the most used abstraction levels from the most abstract to the most specific:

■ *Functional models* are appropriate for concept validation and partitioning between control and data. However, they have no notion of resource sharing or time, that is, functionality is executed instantaneously and the model may or may not be bit accurate. This includes definition of abstract data types, specification of hardware or software (possibly RTOS) computation, communication, and synchronization mechanisms and algorithm integration to high-level simulation tools and languages, for example, MATLAB® or UML.
■ *Transactional behavioral models* (simply noted as transactional) are functional models mapped to a discrete-time domain. Transactions are atomic operations with their duration stochastically determined, that is, a number of clock cycles in a synchronous model. Although detailed timing characteristics on buses cannot be modeled, transactional models are fit for modeling pipelining, RTOS scheduling, basic communication protocols, test-bench realization, and preliminary performance estimation. Transactional models are usually hundreds of times faster than lower-level models and are therefore appropriate for concept validation, virtual platform prototyping, design space exploration, and early software development [1]. SystemC currently integrates the TLM2 standard for transaction-level modeling that supports blocking and nonblocking interface functionality and aims at model interoperability.
■ *Clock-accurate models* (denoted CA) enforce accurate timing on all system transactions. Thus, synchronous protocols, wire delays, and device access times can be accurately modeled. Using discrete-event systems, this layer allows for simple, generic, and efficient clock-accurate performance modeling of abstract processor core wrappers (called bus functional models), bus protocols, signal interfaces, peripheral IP blocks, ISS, and test benches. Time delays are usually back-annotated from register-transfer-level (RTL) models, since clock-accurate models are not always synthesizable.

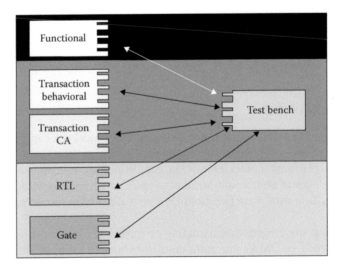

FIGURE 12.1   IP modeling in various abstraction levels.

- *Clock-accurate and bit-accurate* (CABA) *models* are clock-accurate models that are also bit accurate internally and at interface level. CABA models offer an efficient, attractive path to implementation via automatic hardware/RTOS synthesis (high-level synthesis), freeing the designer from timing and performance decisions while allowing for high reusability.
- *RTL models* correspond to the abstraction level from which synthesis tools can generate gate-level descriptions (or netlists). Register-transfer-level systems are usually visualized as having two components: data and control. The data part consists of registers, operators, and data paths, while the control part provides the time sequence of signals that stimulate activities in the data part. Data types are bit accurate, interfaces are pin accurate, and register transfer is clock accurate.
- *Gate models* are described in terms of primitives, such as Boolean logic with timing data and layout configuration. For simulation reasons, gate models may be internally mapped to a continuous-time domain, including currents, voltages, noise, clock rise, and fall times. Storage and operators are broken down into logic implementing the digital functions corresponding to these operators, while timing for individual signal paths can be obtained. Thus, according to these, an embedded physical SRAM memory model may be defined as
  - A high-level functional model described in a programming language, such as C or C++
  - A clock-accurate (and possibly bit-accurate) model, allowing validation of its integration with other components
  - An implementation-independent RTL logic described in VHDL or Verilog
  - A vendor gate library described using NAND, flip-flop schematics
  - A detailed and fully characterized mask layout at the physical level, depicting rectangles on chip layers and geometrical arrangement of I/O and power locations

12.2.2 SYSTEM MODELING METHODOLOGY

System modeling methodology is a combination of stepwise protocol refinement, hierarchical modeling, orthogonalization of concerns, and communication layering techniques. Stepwise protocol refinement is achieved through a combination of top-down and/or bottom-up approaches:

- In *bottom-up refinement*, IP reuse–oriented integration with optimal evaluation, composition, and deployment of prefabricated or predesigned IP block and system components drive the process.
- In *top-down refinement*, emphasis is placed on specifying unambiguous semantics, capturing desired system requirements, optimal partitioning of system behavior into simpler behaviors, and gradually refining the abstraction level down to a concrete low-level architecture by adding details and constraints in a narrower context, while preserving desired properties. Top-down refinement allows the designer to explore modeling at different levels of abstraction, thus trading model accuracy with simulation speed. This continuous rearrangement of existing IP in ever-new composites is a key process to new product ideas. It also allows for extensive and systematic reuse of design knowledge and application of formal correctness techniques. The concrete architecture obtained through formal refinement must satisfy the following properties:
  - *Relative correctness*, that is, it must logically imply the abstract architecture by conforming to system specifications
  - *Faithfulness*, that is, no new rules can be derived during the refinement process

While formal refinement is hard to achieve since there are no automated proof techniques, relative refinement is based on *patterns* consisting of a pair of architectural schemas that are relatively correct with respect to a given mapping. By applying refinement patterns, for example, a state transformation in a control flow graph [10], we can systematically and incrementally transform an abstract architecture to an equivalent lower-level form. Notice that formal architectural transformation is related to providing an augmented calculus with annotations for properties such as correctness, reliability, and performance [11]. If architectural components eventually become explicit formal semantic entities, then architectural compatibility can be checked in a similar way to type checking in high-level programming languages [12].

Hierarchy is a fundamental ingredient for modeling conceptual and physical processes in systems of organized complexity using either a top-down analytical approach, that is, the divide and conquer paradigm, or a bottom-up synthesis approach, that is, the design-reuse paradigm. Hierarchical modeling based on simpler, subordinate models controlled by high-level system models enables a better encapsulation of the design's unique properties, resulting in efficient system design with simple, clean, pure, self-contained, and efficient modular interfaces. For example, hierarchical modeling of control, memory, network, and test bench enables the system model to become transparent to changes in subblock behavior and communication components. A systematic framework supporting hierarchical modeling can be based on the SystemC module object, as well as the inheritance and composition features of the C++ language [13,14]. Thus, SystemC modules consist of SystemC submodules in nested structures. Moreover, SystemC facilitates the description of hierarchical systems by supporting module class hierarchy very efficiently since simulation of large, complex hierarchical models imposes *limited* impact on performance over the corresponding nonhierarchical models.

Stepwise refinement of a high-level behavioral model of an embedded system into actual implementation can be based on the concept of *orthogonalization of concerns* [15]:

- Separation of functional specification from final architecture, that is, *what* the basic system functions are for system-level modeling versus how the system organizes hardware and software resources in order to implement these functions. This concept enables system partitioning and hardware/software codesign by focusing on progressively lower levels of abstraction.
- Separation of communication from computation (called behavior) enables plug-and-play system design using communication blocks that encapsulate and protect IP cores. This concept reduces ambiguity among designers and enables efficient design space exploration.

Layering simplifies individual component design by defining functional entities at various abstraction levels and implementing protocols to perform each entity's task. Advanced intermodule communication refinement may be based on establishing distinct communication layers, thus greatly simplifying the design and maintenance of communication systems [16,17]. We usually adopt two layers:

1. The *communication layer* that provides a generic message API, abstracting away the fact that there may be a point-to-point channel, a bus, or an arbitrarily complex network on chip.
2. The *driver layer* that builds the necessary communication channel adaptation for describing high-level protocol functionality, including compatible protocol syntax (packet structure) and semantics (temporal ordering).

12.2.3 DESIGN SPACE EXPLORATION

In order to evaluate the vast number of complex architectural and technological alternatives, the architect must be equipped with a highly parameterized, user-friendly, and flexible design space exploration methodology. This methodology is used to construct an initial implementation from system requirements, mapping modules to appropriate system resources. This solution is subsequently refined through an iterative improvement strategy based on reliability, power, performance, and resource contention metrics obtained through domain- or application-specific performance evaluation using *stochastic analysis models and tools* and *application benchmarks*, for example, networking or multimedia. The proposed solution provides values for all system parameters, including configuration options for sophisticated multiprocessing, multithreading, prefetching, and cache hierarchy components.

After generating an optimized mapping of behavior onto architecture, the designer may either manually decompose hardware components to the RTL of abstraction or load system-level configuration parameters onto available behavioral synthesis tools, such as the Synopsys Synphony C Compiler or Xilinx Vivado. These tools integrate a preloaded library of configurable high-level (soft) IPs, for example, processor models, memories, peripherals, and interconnects, such as STMicroelectronics' STbus [18,19], and other generic or commercial system interfaces, such as VSIA's VCI [20], OCP [9], and ARM's AMBA bus [21]. The target is to parameterize, configure, and interface these IPs in order to generate the most appropriate synthesis strategy automatically. Now, the design flow can proceed normally with place and route, simulation, and optimization by interacting with tools, such as Synopsys IC Compiler, VCS, and PrimeTime PX for power analyses.

12.3 ANNOTATION OF SYSTEM-LEVEL MODELS

SystemC allows both CABA modeling and clock-approximate transaction-level modeling. In this section, we consider back-annotation of system-level models, focusing especially on hardware modeling objects.

12.3.1 SYSTEM-LEVEL MODELING OBJECTS

As a prelude to back-annotation, we examine a high-level modeling environment consisting of a SystemC-based C++ modeling library and associated simulation kernel. Proprietary runtime and test environment may be provided externally. Thus, our high-level modeling environment provides user-defined C++ building blocks (macro functions) that simplify system-level modeling, enable concurrent design flow (including synthesis), and provide the means for efficient design space exploration and hardware/software partitioning.

High-level SystemC-based modeling involves hardware objects as well as system and application software components. For models of complex safety critical systems, software components may include calls to an RTOS model [22]. An RTOS model abstracts a real-time operating system by providing generic system calls for operating system management (RTOS kernel and multitask scheduling initialization), task management (fork, join, create, sleep, activate), event handling (wait, signal, notify), and real-time modeling. Future SystemC versions are expected to include user-defined scheduling constructs on top of the core SystemC scheduler providing RTOS features, such as thread creation, interrupt, and abort. During software synthesis, an RTOS model may be replaced with a commercial RTOS.

Focusing on hardware components and on-chip communication components, we will describe system-level modeling objects (components) common in SystemC models. Most objects fall into two main categories: *active* and *passive modeling objects*. While active objects include at least one thread of execution, initiating actions that control system activity during the course of simulation, passive objects do not have this ability and only perform standard actions required by active objects. Notice that communication objects can be both active and passive depending on the modeling level. Each hardware block is instantiated within a SystemC class called sc_module. This class is a container of a user-selected collection of SystemC clock domains, and active, passive, and communication modeling objects including the following:

- *SystemC and user-defined data types* providing low-level access, such as bits and bytes, allowing for bit-accurate, platform-independent modeling.
- *User-defined passive memory objects* such as register, FIFO, LIFO, circular FIFO, memory, cache, as well as user-defined collections of externally addressable hierarchical memory objects.
- *SystemC and user-defined intramodule communication and synchronization objects* based on message passing, such as monitor, or concurrent shared memory such as

mutex, semaphore, conditional variables, event flag, and mailbox. This includes timer and watchdog timer for generating periodic and nonperiodic time-out events.

■ *Active control flow objects* such as SystemC processes: C-like functions that execute without delay (SC_METHOD), asynchronous threads (SC_THREAD), and clocked threads (SC_CTHREAD), and HFsm. HFsm encompasses abstract processor models based on independent ISS or even using custom techniques, that is, cross compiling the (bare metal) application code, mapping macro- to microoperations, and annotating with timing data based on a technical reference manual (e.g., timing delays for ARM Cortex-A9 [23]), ISS, and/or hardware implementation. Intramodule interaction, including computation, communication, and synchronization, is performed either by local processes or by remote control flow objects defined in other modules.

■ *Intermodule communication object*, for example, based on message passing, cf. [2].

■ *User-defined intermodule point-to-point and multipoint communication channels and corresponding interfaces*, for example, peripheral, basic, and advanced VCI [20], AMBA bus (AHB, APB) [21], ST Microelectronics' proprietary Request Acknowledge (RA), Request Valid (RV), and STBus (type 1, type 2, and type 3) [18,19]. These objects are built on top of SystemC in a straightforward manner. For example, on-chip communication network (OCCN) provides an open-source, object-oriented library of objects built on top of SystemC with the necessary semantics for modeling on-chip communication infrastructure [17], while heterogeneous system on chip (HSoC) provides a similar library for SoC modeling [24]. Notice that SystemC components can be integrated with a large number of SystemC-based libraries and tools already available in the market, such as a variety of open processor architectures models (in OVP suite) from Imperas [25].

12.3.2 TIME ANNOTATION

When high-level views and increased simulation speed are desirable, systems may be modeled at the transactional level using an appropriate abstract data type, for example, processor core, video line, or network communication protocol-specific data structure. With transactional modeling, the designer is able to focus on IP functionality rather than on details of the physical interface, including data flows, FIFO sizes, or time constraints. For efficient design space exploration, transactional models (including application benchmarks) must be annotated with a number of architectural parameters:

■ Realizability parameters capture system design issues that control system concurrency, such as processor, memory, peripheral, network interface, router, and their interactions, for example, packetization, arbitration, packet transfer, or instruction execution delays. RTOS models, for example, context switch or operation delays, and VLSI layout, for example, time–area trade-offs, clock speed, bisection bandwidth, power-consumption models, pin count, or signal delay models may also be considered.

■ Serviceability parameters refer to reliability, availability, performance, and fault-recovery models for transient, intermittent, and permanent hardware/software errors. When repairs are feasible, fault recovery is usually based on detection (through checkpoints and diagnostics), isolation, rollback, and reconfiguration. While reliability, availability, and fault-recovery processes are based on two-state component characterization (faulty or good), performability metrics evaluate degraded system operation in the presence of faults, for example, increased latency due to packet congestion when there is limited loss of network connectivity.

Timing information is only one of many possible annotations. In order to provide accurate system performance measurements, for example, power consumption, throughput rates, packet loss, latency statistics, or QoS requirements, all transactional models involving computation, communication, and synchronization components must be back-annotated with an abstract notion of time. Thus, timing characteristics, obtained either through hardware synthesis of clock-accurate hardware models or by profiling software models on an RTOS-based embedded processor with clock-accurate delays associated with each operation in its instruction set, can be back-annotated to the initial TLM model for transactional-level modeling. The clock-accurate system architecture model can be used for system-level performance evaluation, design space exploration, and also as an executable specification for the hardware and software implementation.

However, architectural delays are not always clock accurate but rather clock approximate. For example, for computational components mapped to a particular RTOS, or large communication transactions mapped to a particular shared bus, it is difficult to estimate accurately thread delays, which depend on precise system configuration and load. Similarly, for deep submicron technology, wire delays that dominate protocol timings cannot be determined until layout time. Thus, in order to avoid system revalidation for deadlock and data race hazards, and ensure correct behavior independent of computation, communication, and synchronization delays, one must include all necessary synchronization points and interface logic in the transactional models. Analysis using parameter sweeps helps estimate sensitivity of system-level design due to perturbations in the architecture, and thus examine the possibility of adding new features in derivative products [1,26].

For most SystemC-based hardware objects, time annotation is performed statically during instantiation time. Note that time delays also occur due to the underlying protocol semantics. As an example, we now consider memory and communication channel objects; HFsms are discussed in the next section:

- Time is annotated during instantiation of memory objects by providing the number of clock cycles (or absolute time) for respective read/write memory operations. The timing semantics of the memory access protocol implemented within an active control flow object may contribute additional dynamic delays. For example, a *congestion delay* may be related to allowing a single-memory operation to proceed within a particular time interval, or a *large packet delay* may be due to transferring only a small amount of information per time unit.
- Time is annotated during instantiation of the processes connected to the communication channels through timing constraints implemented within an active control flow object, blocking calls to intramodule communication and synchronization objects, and SystemC wait or wait_until statements. Blocking calls within intermodule communication objects instantiated within the SystemC channel object or adherence to specific protocol semantics may imply additional dynamic delays, for example, compatibility in terms of the size of communicated tokens; this parameter may be annotated during communication channel instantiation.

12.4 SYSTEM-LEVEL MODELING USING SYSTEMC

Design space exploration of complex embedded systems that combine a number of CPUs, memories, and dedicated devices is a tedious task. Thus, system designers must be equipped with an architecture simulation tool capable of converting quickly and without much effort high-level concepts to a specific model. In this context, a variety of design space exploration approaches based on interoperable system-level design tools entail different simulation efficiencies versus accuracy trade-offs that enable exploring a subset of the design space.

Next, we consider four short case studies related to system-level modeling and back-annotation: an abstract processor model, a hierarchical final-state machine and an asynchronous channel in SystemC, and a PLL in SystemC-AMS.

12.4.1 ABSTRACT PROCESSOR MODEL

In this section, we present a cosimulation approach based on an abstract processor model for an existing multicore architecture (e.g., focusing on ARM Cortex-A9). Using this approach, we are able to simulate the performance of a bare metal application, that is, without the use of an operating system, in a clock-approximate manner. Notice that full system simulation (e.g., under Linux) is not supported by SystemC. Although there is limited multiprocessor support in other high-level design space exploration frameworks, such as Imperas OVPSIM [25] and (open source) Gem5 [27–29], interoperability with SystemC is at a very early stage.

To start with, we need to model each ARM processor instruction (e.g., add, subtract, and multiply, compare and branch, move and shift, and local load/store) in SystemC at transaction-level and then statically back-annotate available clock-approximate timing information from the assembly instruction cycle delays. Accurate timing has been published in the technical architecture reference manual for Cortex-A9 [23]. Hence, each processor instruction is modeled as a SystemC function that contains a wait(x); statement, where x is the number of processor cycles that are back-annotated from the previously mentioned reference manual.

The desired application should be written in ANSI C and compiled using an ARM cross compiler [30] with flags suitable for generating assembly code. As a second step, this assembly code must be selectively replaced by the corresponding processor instructions modeled in our abstract processor model. In other words, each assembly instruction is converted to a SystemC function that is back-annotated with the corresponding delay. Notice that in order to actually follow the exact trace of the original program (when it branches), the equivalent C++ instruction is also executed by the simulator.

An example showing the conversion steps for an application (from source code to assembly and from assembly to SystemC function calls) is shown in Figure 12.2.

Architecture modeling of instruction and data caches requires knowledge of the architecture of the memory hierarchy (organization, cache line size, etc.) [31]. Certain model inaccuracies for instruction cache arise due to indirect addressing and the associated branch prediction model.

12.4.2 HIERARCHICAL FINITE-STATE MACHINES USING SYSTEMC

Finite-state machines are critical for modeling dynamic aspects of hardware systems. Most hardware systems are highly state dependent, since their actions depend not only on their inputs but also on what has previously happened. Thus, states represent a well-defined system status over an interval of time, for example, a clock cycle. In clock-accurate simulation, we assume that system events occur at the beginning (or the end) of a clock cycle. In this way, we can capture the right behavior semantics and back-annotate hardware objects in a very detailed way.

```
int sum = 0;
...
end = times(&tms2);
...
pthread_barrier_wait(&bar1);
```

☐ Multicore application
☐ ARM assembly
▨ Virtual platform application

```
263 0160  0030A0E3  mov  r3, #0 @ tmp155,
264 0164  1C300BE5  str  r3, [fp, #-28] @ tmp155, sum
...
756 0604  A0304BE2  sub  r3, fp, #160   @ tmp194,,
757 0608  0300A0E1  mov  r0, r3 @, tmp194
758 060c  FEFFFFEB  bl   times  @
759 0610  0030A0E1  mov  r3, r0 @ D.7289,
...
1404 0ca8 A0009FE5  ldr  r0, .L61+52 @,
1405 0cac FEFFFFEB  bl   pthread_barrier_wait @
```

```
mov_com(worker_id,...);
local_str(worker_id,...);
...
end = sc_simulation_time();
...
cpu_barrier(bar1, bar1_max, thread_id);
```

FIGURE 12.2 Program conversion steps for ARM cosimulation.

Hierarchical finite-state models based on Harel statecharts and variants [32,33] define complex, control-dominated systems as a set of smaller subsystems, by allowing the model developer to consider each subsystem in a modular way. Scoping is used for types, variable declaration, and procedure calls. Hierarchy is usually disguised into two forms:

1. Structural hierarchy that enables the realization of new components based on existing ones. A system is a set of interconnected components, existing at various abstraction levels.
2. *Behavioral hierarchy* that enables system modeling using either *sequential*, for example, procedure or recursion, or *concurrent computation*, for example, parallel or pipelined decomposition.

Similar to asynchronous and clocked threads (SC_THREAD and SC_CTHREAD constructs) defined in SystemC, an HFsm modeled in C++/SystemC is a powerful mechanism for capturing a description of dynamic behavior within a SystemC model. HFsm may be used to capture formally the dynamic behavior of a hardware module, for example, a circuit representing an embedded system controller or a network protocol. For control tasks, asynchronous or clocked threads in SystemC and HFsm implementations can be made equivalent. Although clocked threads are usually simpler and more efficient from the simulation time point of view, since the sensitivity list is just the clock edge, an HFsm implementation is more general and improves clarity and efficiency when specifying complex control tasks. It also allows for HFsm back-annotation in a more detailed way, thus making clock accuracy easier to obtain. Since outer HFsm states are typically behavioral generalizations of inner ones, HFsm facilitates the implementation of policy mechanisms such as signaling exceptions to a higher scope and preemptive events. It also avoids pitfalls in implementations of traditional (flat) finite-state machine implementations, such as

- Exponential explosion in the number of states or transitions when composing substates, hence characterizing the system-level model using less redundant timing information
- Difficulty in modeling complex hierarchical control flows, with transitions spanning different levels of the behavioral hierarchy
- Difficulty in handling hierarchical group transitions or global control signals specified for a set of states, for example, reset and halt
- No support for concurrent, multithreaded computation

12.4.2.1 FLAT HFsm DESCRIPTION: STATES, CONDITIONS, ACTIONS, AND DELAYS

With simple, commonly used modeling constructs, a flat HFsm is represented using states, conditions, delays, and actions. States are used to identify HFsm status, conditions are used to drive actions, and delay times are used to postpone actions. Composite transitions are defined as a quadruple formed by a condition, an action, a delay, and the next state. A flat HFsm represented graphically is shown in Figure 12.3.

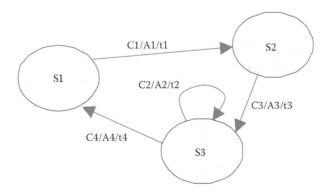

FIGURE 12.3   Graphical illustration of a flat, clocked finite-state machine.

An equivalent representation can be given in tabular form:

- State S1 = {{C1, A1, t1, S2}};
- State S2 = {{C3, A3, t3, S3}};
- State S3 = {{C4, A4, t4, S1}, {C3, A3, t5, S3}};

Let us consider the state S3. In this state, we have two transitions. In the first one, S1 is the next state, C4 is the condition, A4 is the action, and t4 is a constant compile-time delay. The semantics of this transition are as follows. If the HFsm is in state S3 and condition C4 is true, then HFsm performs action A4 and transfers control to state S1 after a delay of t4 clock cycles. Delay annotation during state changes is common in HFsm specifications, for example, when defining the dynamic behavior of a processor or memory controller. Thus, the HFsm designer could redefine this order, for example, condition–action–delay instead of condition–delay–action.

Furthermore, if HFsm conditions were *not* allowed to make blocking function calls, then the delay associated to performing the corresponding action would be *deterministic* and could simply be estimated and back-annotated directly by the model designer. *Dynamic delays*, for example, delays due to concurrent accesses to shared resources, cause nondeterministic delays. These are much harder to model accurately and difficult to replicate during testing. They could be handled at the beginning of the action routine by placing appropriate blocking function calls to time-annotated intramodule communication and synchronization objects, such as user-defined channels.

12.4.2.2 HFsm DESCRIPTION USING STATES AND EVENTS

For more complex examples, hierarchical states may have "parents" or "children," placed at different levels. This approach is common with machines providing different operating modes, that is, machine features that define multiple, alternative configurations. Then, HFsm is described conveniently using discrete-event systems, that is, a finite set of states, a finite set of event categories, start-up and regular state transitions mapping (state, event) pairs to other states, and actions associated with transitions. Events are triggered by changes in conditions that (similar to actions) can be time annotated. This description is natural, abstract, powerful, and efficient:

- *Initial start-up transitions* originate during initial transitions from a top superstate. These transitions invoke entry and start-up actions (e.g., setting state parameters) for the state associated with the transition as well as for intermediate superstates, if a transition crosses two or more levels of hierarchy.
- *Regular transitions* occur at the current state during normal HFsm processing. They are triggered by events selected either randomly or through external stimuli from a user-defined enumerated event list. These transitions cause entry (and exit) actions upon entering (respectively, exiting) an HFsm state. Depending on the position in the state hierarchy of the new versus the current state, entry and exit actions are executed in order, either from the least deeply nested to the most deeply nested state or vice versa. Entry and exit actions always precede processing actions possibly associated with the newly entered hierarchical state.

Hierarchical finite-state machines are common when describing operation control in digital systems, where, for example, a *top menu* superstate is usually specialized to one or more *system setting* modes (e.g., time and date keeping for a digital watch) and one or more *normal operating modes* (e.g., for Internet, text, or video content for an on-screen display of a digital TV). These modes of operation can be considered as independent superstates.

Within this hierarchical system specification context, we present a SystemC-related implementation of HFsm that clarifies the previous definition. A SystemC wait or wait_until can be associated with each processed event; notice that quite similarly to time annotation, power dissipation data can be annotated to each HFsm state transition. Moreover, blocking calls to intramodule communication and synchronization objects may account for additional dynamic delays. These delays can be implemented in start-up transitions (START_EVT) from a superstate and subsequent state entry (ENTRY_EVT) or state exit (EXIT_EVT) transitions. Note that state entry or exit transitions may occur within a given hierarchical level or between different levels; in fact, in order to discover which entry or exit actions to execute, it is necessary to compute dynamically, that is, during state transitions, the least common ancestor from the least deeply nested (source or destination) to the most deeply nested state (destination or source).

12.4.3 ASYNCHRONOUS INTERMODULE COMMUNICATION IN SYSTEMC

Unlike synchronous communication channels where individual modules share a clock, asynchronous blocking communication channels use local handshake protocols to exchange data among nodes connected to memory-less wires. Globally asynchronous locally synchronous (GALS) techniques relax the synchrony condition by employing multiple nonsynchronized clocks, achieving improved power consumption and reduced electromagnetic interference [34]. GALS systems may behave deterministically, if system output cannot differ for the same input sequence. Deterministic GALS systems must provide a unique output independent of frequency/phase variations, interconnect delays, and clock skew.

As an example, in asynchronous and GALS intermodule communication, we consider two SystemC modules exchanging information over an asynchronous pipeline. Information is transmitted only if one is ready to send and the other ready to receive. Many kinds of channels can be used to represent in SystemC this type of handshake. For CABA communication, the SystemC signal data channel provides the necessary event synchronization functionality, for example, wait(signal.event). However, as indicated in Sections 12.4.3.1 and 12.4.3.2, special care must be taken when implementing asynchronous or GALS communication using signals in SystemC, not only from the point of an annotation perspective but also from a viewpoint of correctness. In Section 12.4.3.3, we provide a possible solution to both problems based on the four-phase asynchronous handshake protocol that allows for back-annotation.

12.4.3.1 ASYNCHRONOUS INTERMODULE COMMUNICATION USING SYSTEMC SIGNALS

We first consider typical communication between two modules connected together via a SystemC signal of type data_type, that is, the data_in port of one is connected to the data_out port of the other and vice versa. Without loss of generality, we assume that both modules have similar functionality and focus on their communication interface. The following code snippet helps make this case familiar to SystemC developers.

```
class device: public sc_module {
        public:
        SC_HAS_PROCESS(device);
        device(sc_module_name nm): sc_module(nm) {
                SC_THREAD(in_action);
                        sensitive << data_in;

        }
        sc_in<data_type> data_in;
        sc_out<data_type> data_out;
        void in_action();
};
void device::in_action(){
        while(1){
                wait(); //waits an event in data_in port
                data_type module_data = data_in.read();
                // process received data
```

```
                           . . .
                          // Write data to the other module
                          data_out.write();


                   }
};
int sc_main (int argc, char* argv[]) {
...
          sc_signal<data_type> signal_1;
          sc_signal<data_type> signal_2;
          device module1("Module1");
          module1.data_in (signal_1);
          module1.data_out (signal_2);
          device module2("Module2");
          module2.data_in (signal_2);
          module2.data_out (signal_1);
...
          sc_start();
}
```

In the aforementioned implementation, we present a two-way intermodule communication that is asynchronous, since there is no clock synchronization. More specifically, when the sending module has data ready in its data_out port, the data_in port of the second module receives the data through the appropriate signal_x signal channel, where x = 1, 2.

This implementation is easy to use and has no clock dependency, so it supports intermodule communication in both synchronous and asynchronous systems, that is, systems that use the same or different clocks. However, it does not work properly in all cases, since there is a limitation in event generation related to SystemC signals. More specifically, a signal generates an event if and only if the signal changes its value. Thus, if a sender repeatedly executes the write method on its output port, but writes the same value, only one event in the event list of the listening port of the signal will be generated, that is, the second write is essentially lost.

12.4.3.2 INTERMODULE COMMUNICATION WITH CLOCK-SENSITIVE MODULES

Next, we consider two modules, each with a clock-sensitive thread. The modules use identical or different clocks and interact with each other as before through asynchronous signals. As in Section 12.4.3.1, we focus on communication interface definitions rather than the module functionality. The modules are again connected together via a SystemC signal of type data_type, as shown next. The following code extract helps make this example more concrete.

```
class device: public sc_module {
        public:
        SC_HAS_PROCESS(device);
        device(sc_module_name nm):sc_module(nm) {
            SC_THREAD(in_action);
                    sensitive << pkt_in << CLK;
        }
        sc_in_clk CLK;
        sc_in<data_type> data_in;
        sc_out<data_type> data_out;
        void in_action();
        };
void device::in_action(){
        while(1){
                //wait for event in pkt_in port
                wait(data_in.value_changed_event());
                data_type module_data = data_in.read();
```

```
          // process received data
          ...
          wait(x); //module can wait for x cycles
          // Write data to the other module
          data_out.write();
      }
};
int sc_main (int argc, char* argv[]) {
...
sc_clock clock1("CLOCK1", ZZ, SC_NS, X.X, Y.Y, SC_NS);
sc_clock clock2("CLOCK1", ZZ, SC_NS, .X, Y.Y, SC_NS);
...
      sc_signal<data_type> signal_1;
      sc_signal<data_type> signal_2;
      device module1("Module1");
      module1.data_out (signal_1);
      module1.data_in (signal_2);
      module1.CLK (clock1);
      device module2("Module2");
      module2.data_out (signal_2);
      module2.data_in (signal_1);
      module2.CLK (clock2);
...
      sc_start();
}
```

This implementation not only suffers from the same SystemC limitations as the previous example (see Section 12.4.3.1) requiring uniqueness of the signal values, but since the listening thread in_action is sensitive to both the data signal and the clock, it is possible that while waiting for a clock event in the in_action method, events are lost (signal values overwritten) or duplicated (signal values read twice) from the asynchronous data channel. Thus, correctness of this handshake protocol heavily depends on the relative speed of modules 1 and 2.

12.4.3.3 ASYNCHRONOUS INTERMODULE COMMUNICATION BETWEEN CLOCK-SENSITIVE MODULES

In order to resolve both previous problems, we consider a safe asynchronous intermodule communication channel for SystemC modules with clock-sensitive listening threads. A common asynchronous protocol for GALS communication is the *four-phase asynchronous handshake*. This protocol applies backpressure when the sender sends data that the receiver cannot accommodate. As shown in Figure 12.4, the protocol starts at an initial idle state. This state is entered again whenever the sender is not ready to send. The protocol is defined at the network level through a request (s_ready), an acknowledge signal (d_ready), and one or more data signals. The request is asserted by the sender, while the receiver responds by asserting the acknowledge; then both signals are deasserted in turn, that is, first the sender withdraws its request and then the receiver withdraws its acknowledge signal.

This fully interlocked asynchronous handshake allows correct operation without loss or duplicate information. We simplify things and focus (without loss of generality) on the channel implementation of two interacting modules: one sender and one receiver. The module connections are shown in Figure 12.4.

FIGURE 12.4 Sender and receiver connections.

```
class sender: public sc_module {
        public:
        SC_HAS_PROCESS(device);
        sender(sc_module_name nm):sc_module(nm) {
                SC_THREAD(out_action);
                        sensitive << CLK;
        }
        sc_in_clk CLK;
        sc_in<bool> d_ready;
        sc_out<bool> s_ready;
        sc_out<data_type> data_port;
        void out_action();
};
class receiver: public sc_module {
        public:
        SC_HAS_PROCESS(device);
        receiver(sc_module_name nm): sc_module(nm) {
                SC_THREAD(in_action);
                        sensitive << CLK;
        }
        sc_in<data_type> data_port;
        sc_in_clk CLK;
        sc_in<bool> d_ready;
        sc_out<bool> s_ready;
        void in_action();
};
void sender::out_action(){
        while(1){
                wait();
                s_ready.write(false);
                if (!d_ready.read()){
                        data_port.write(module_data);
                        s_ready.write(true); // ready to send
                        wait(d_ready.value_changed_event());
                }
        }
}
void receiver:: in_action() {
        while(1){
                wait();
                s_ready.write(false);
                if (d_ready.read()){
                        module_data = data_port.read();
                        s_ready.write(true);
                        wait(d_ready.value_changed_event());
                }
        }
}
int sc_main (int argc, char* argv[]) {
        ...
sc_clock clock1("CLOCK1", ZZ, SC_NS, X.X, Y.Y, SC_NS);
sc_clock clock2("CLOCK1", ZZ, SC_NS, X.X, Y.Y, SC_NS);
        ...
        sender sender0("SENDER");
        sender0.data_port(data_signal);
        sender0.m_s_ready(signal1);
        sender0.m_d_ready(signal2);
        sender0.CLK(clock1);
        receiver receiver0("RECEIVER");
        receiver0.data_port(data_signal);
        receiver0.m_d_ready(signal1);
        receiver0.m_s_ready(signal2);
```

```
        receiver0.CLK(clock2);
        sc_start();
}
```

As shown in this SystemC code, we use two extra signal ports of type bool. Each port has an s_ready (request) output port and a d_ready (acknowledgment) input port. The s_ready output port of the sender module is connected to the d_ready listening port of the receiver and vice versa. This implementation works as follows.

When the receiver sets its s_ready output signal to false (ready to receive), then the d_ready input signal of the sender module becomes true, and consequently, the out_action thread in the sender module is able to send data to the receiver via its data_out port. Then, it sets its s_ready signal to true and waits until its d_ready signal changes value, that is, until it becomes true.

The change in the s_ready signal of the sender allows the receiver module, which listens to the corresponding d_ready signal, to read data from its data_port. Subsequently, it sets its s_ready output signal to true and waits until its d_ready signal changes value, that is, until it becomes true. The change in the s_ready signal of the receiver unblocks the sender who listens to the corresponding d_ready port via the d_ready.value_changed_event().

Notice that both sender and receiver calls to the value_changed_event() are necessary to avoid duplicating/overwriting of the transmitted data, when the relative speeds of the sender and receiver differ. Moreover, by using Boolean signals, no messages can be lost, resolving the issue already discussed in Section 12.4.3.1.

This SystemC implementation seems to allow further back-annotation, for example, by calling either a simple wait(x, SC_NS) for a given number of ns or wait(x, SC_NS, ready.value_changed_event()), which waits for a specified time of x nanoseconds, or event ready to occur, whichever comes first. However, the semantics (earliest of the two) can lead to duplicates. Also, notice that a call to wait(x, SC_NS) (for an arbitrary x) could lead to synchronization hazards (deadlock, message loss, duplicates), since events generated at specific instances may never be consumed.

As an extension, the aforementioned handshake protocol can be rewritten based on a push instead of a pull functionality, that is, the sender (instead of the receiver) initiates the transaction. Such an implementation is described in Reference 35. It can also be utilized to design multiaccess channels (e.g., an asynchronous router) through the use of an sc_spawn() and an appropriately designed arbiter.

12.4.4 BEHAVIORAL PLL MODEL IN SYSTEMC-AMS

In the most embedded SoC devices, such as a wireless radio chip, digital circuits are combined with analog circuits. Most analog systems have a small transistor count, such as operational amplifiers, data converters (ADC or DAC), filters, PLLs, sensors, and power management chips, while complex analog blocks include analog controllers in transportation and consumer electronic appliances, such as motor controllers. AMS design must address design complexity, power efficiency, verification, and signal integrity at all levels, from single cell device to high-level components and from discrete- to continuous-time domain. Although digital design is highly automated, a very small portion of analog design is currently automated through EDA tools and is often described as *flat nonhierarchical design*. As a result, although digital circuits are increasingly larger parts in SoC design, critical analog circuits require a disproportionately large amount of chip area and require (comparatively) immense effort in interface design, validation, manufacturing, design analysis, and IP reuse.

In this context, system-level modeling of analog circuits may use an innovative approach based on the recently open SystemC-AMS 2.0 standard proposed by Accellera; a corresponding implementation is available from Fraunhofer Institut [36–39]. The SystemC-AMS extension complements the SystemC language with circuit-level simulation for critical AMS or RF circuitry (i.e., similar to Spice tools) and behavioral modeling (i.e., similar to Verilog-AMS, SystemVerilog, or VHDL-AMS), providing a high-level abstraction to support fast simulation of next-generation complex multimillion transistor SoCs.

The SystemC-AMS standard defines a User Guide and Language Reference Manual that define execution semantics and extensions to SystemC language constructs (classes, interfaces, analog kernel, and modeling of continuous-time analysis) for developing executable specifications of embedded AMS systems at system-level, analog behavioral and netlist level in the same simulation environment. This extension made SystemC the first AMS language offering frequency, together with time-domain analysis, enhancing system modeling and hw/sw codesign in many complex applications, such as electromechanical system control in robotics and simulation of analog RF and digital blocks of wireless networks, for example, Bluetooth.

More specifically, SystemC-AMS extensions define new language constructs identified by the prefix sca_. Depending on the underlying semantics, AMS modules can be declared in dedicated namespaces sca_tdf (timed data flow semantics that includes modeling systems in which activation periods or frequencies are either statically defined or dynamically changing), sca_eln (electrical linear networks), and sca_lsf (linear signal flow) [37]. By using namespaces, similar primitives to SystemC are defined to denote ports, interfaces, signals, and modules. For example, a timed data flow input port is an object of class sca_tdf::sca_in<type>.

12.4.4.1 SYSTEMC-AMS PARAMETER ANNOTATION

SystemC-AMS behavioral models focus on functionality, rather than on details of the physical implementation, enabling fast simulation and design space exploration early in the design phase, that is, before focusing on implementation details of low-level electrical circuit models and process technology. In contrast, low-level models allow precise evaluation of parasitic effects and electrical parameters of the analog circuit, such as wire delay parameters. This low-level characterization allows back-annotation of the initial behavioral models (at object instantiation time in SystemC-AMS) with a number of architecture-related performance, power, and reliability characteristics for clock-approximate or clock-accurate modeling.

Then, a back-annotated behavioral model can be used to do the following:

- Ensure correctness in the presence of deadlock and data race hazards.
- Enable interfacing to new hardware logic or software models.
- Perform accurate system-level design space exploration and analysis. For example, analysis using parameter sweeps helps estimate sensitivity of system-level design due to perturbations in the architecture in different operational environments and allows examination of possible new features in derivative products.

12.4.4.2 PLL DEFINITION IN SYSTEMC-AMS

As a case study, we consider the use of SystemC-AMS timed data flow for behavioral modeling of self-clocking structures for skew reduction based on PLL [40,41]. We explain how to annotate model parameters from existing open or freeware low-level electrical circuit designs in order to perform accurate system-level simulation, operational characterization, consistency validation, and sensitivity analysis. In addition, by understanding the sources and characteristics of different types of noise, distortion, and interference effects, it is possible to accurately model and validate a PLL's performance, reliability, and signal-to-noise ratio with SystemC-AMS.

For validating and characterizing PLL operation, we can use behavioral-level simulation in SystemC-AMS with input/output *timed data flow* (sca_tdf) signals. The PLL module described in SystemC-AMS models all circuit blocks, that is, phase detector, low-pass filter (which can be implemented in the frequency domain using sca_lsf), and analog/digital voltage–controlled oscillator (VCO). The block will generate an output signal that has a fixed (mathematically proven) relationship to the phase of an input reference signal.

The PLL module can also be integrated with other modules that induce perturbations to parameter and input signals (e.g., the reference oscillator). These modules, also modeled as generic SystemC-AMS classes with input and output time data flow signals, can cause variation due to nonlinear effects and thermal Gaussian noise [42,43], for example, due to component aging; block specification and binding details of these instantiated modules are omitted.

The PLL class definition in SystemC-AMS is as follows:

```
pll(sc_core::sc_module_name n,
      double ph_gain, // phase detector gain
      double lp_ampl, // low pass amplitude
      double lp_fc, // low pass frequency
      double vco_fc, // vco frequency
      double kvco, // vco sensitivity (Hz/V)
      double vco_gain, // vco gain
      double vco_vc) // vco voltage control
  Its timed data flow signals at the I/O interface are:
sca_tdf::sca_in<double> ref; // reference signal
sca_tdf::sca_out<double> lpo; // low pass filter
sca_tdf::sca_out<double> vco; // vco
```

The PLL model design parameters are as follows (specification and binding details, which are similar to SystemC, are omitted):

Parameter	Default	Description
sc_module_name n	—	
double ph_gain	3.7	Gain of phase detector
double lf_ampl	1	Amplitude of low-pass filter
double lf_fc	1	Cutoff frequency of low-pass filter (Hz)—small to remove high-frequency clock noise
double vco_fc	10^9	Central frequency of vco output (Hz)
double kvco	10^7	Sensitivity of vco (Hz/V)
double vco_gain	1	Gain of vco output
double vco_vc	0	Voltage control of vco
integer rate	1	Data rate of input port (no multirate channels)

Loop filter parameters (e.g., gain), as well as phase detector and VCO tuning characteristics, have a dramatic effect on PLL behavior. For example, abrupt small perturbations (e.g., ±1%) in the reference frequency or phase or filter dynamics (loop bandwidth, damping factor and frequency) influence PLL operational characteristics and stability, for example, lock time, cycle slipping and damping behavior, reference spurs, and phase noise. For instance, a lower bandwidth filter causes the loop to respond slower to the noise being injected by the VCO, while a higher bandwidth allows the loop to respond quickly to the noise and compensate for it.

12.4.4.3 PLL PARAMETER ANNOTATION

For a realistic, accurate, and efficient simulation of PLL operational characteristics, the previously described behavioral-level SystemC-AMS models must be annotated with different system state parameters from corresponding low-level electrical implementations. More specifically, it is necessary to modify SystemC-AMS model parameters, including initialization of constant system parameters (related to timing and power characteristics) and dynamic modification of parameter values during simulation, for example, when modeling specific stimuli or perturbations due to component interactions or technology process–related effects, such as noise.

Low-level implementations at electrical network and circuit level are already available as library models in many commercial and open or free PLL design toolkits and allow reuse of system performance metrics and test-bench scenarios:

- In addition to popular commercial signal processing toolkits, such as Virtuoso AMS from Cadence, LabView from National Instruments, and SimPLL from Analog Radio Systems [44], which provide basic PLL device component characteristics in a more general analog model environment, there are several free tools, as outlined next.
- CPPSIM, developed by Prof. M. Perrott at MIT, is designed mostly for teaching analog design [45]. This is a complete analog modeling environment, with simulator, model

design, and synthesis components. It is noteworthy that CPPSIM contains useful open-source C++ classes for AMS modeling; however, these classes are hidden within other free but not open-source code, so it is usually difficult to directly extract and reuse.

■ Several open-source or free (but limited license) tools providing real PLL operational characterization are very useful and easy to use for configuration and annotation between behavioral models and electrical network implementations. These are usually provided for free in executable form by engineers working at different analog design companies. For example, we mention EasyPLL from National Instruments [46], PLL from Michael Ellis [47], and PLLSim (with a nice GUI) from Michael Chan from the University of Queensland [48].

Unfortunately, back-annotation from low-level AMS models is not seamlessly integrated with the SystemC-AMS modeling environment, and circuit-level parameters can rarely be used without adaptation, a time-consuming and error-prone manual process even for the same process technology. In order to reduce time to market and increase the chance of first-time-right silicon, a systematic interoperable framework is necessary to automatically capture the available design knowledge information from low-level design schematics based on simulation results and technology-related parasitic models in a way that is transparent to the designer. In addition, after fast and efficient device simulation using the refined (annotated or calibrated) SystemC-AMS behavioral models, generic time series analysis and wave visualization tools, such as Fourier and oscillator analyzers, can be customized for analysis and correlation of the operational characteristics. For example, Figure 12.5 shows a typical frequency analysis after simulation, which shows a peak at the required frequency of 1 GHz. Spurs have much smaller amplitude by ~20 dB and reduce PLL capture range and lock time [49,50]. For a refined analysis, we can compute peak-to-peak and rms phase jitter and perform spur reduction using more sophisticated filters [51,52].

Apart from certain classes of filters, there is currently no automated model translation or synthesis algorithm for generating a circuit from top-level HDL-AMS specifications. New EDA tools are necessary to perform plug-and-play calibration, integration, verification, analysis, and optimization in terms of performance, fault tolerance, and power-efficiency characterization, extending and standardizing existing support for rapid development of reliable, configurable,

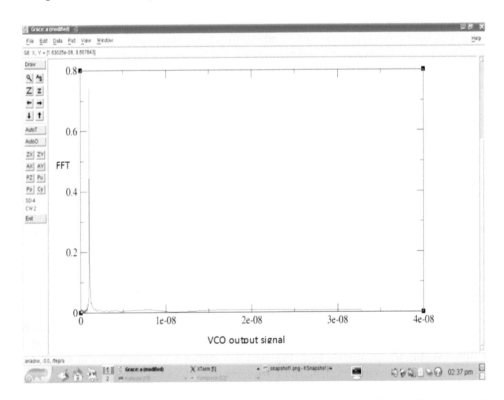

FIGURE 12.5   Fast Fourier transform of phase-locked loop voltage–controlled oscillator output signal.

and programmable AMS SoCs. These tools can also perform back-annotation automatically by applying metamodels, for example, symbolic analysis or graph rewriting principles in a hierarchical design, linking lower-level data to higher-level design entities.

12.5 AUTOMATIC EXTRACTION OF STATISTICAL INFO

System-level modeling is an essential ingredient of system design flow. Data and control flow abstraction of the system hardware and software components express not only functionality but also system performance characteristics that are necessary to identify system bottlenecks [53]. For example, virtual channels are used to avoid both network and protocol deadlock and also to improve performance and provide quality of service. While for software components it is usually the responsibility of the user to provide appropriate key performance indicators, for hardware components and interfaces, it is necessary to provide a statistical package that hides internal access to the modeling objects. The generated statistical data may be analyzed using visualization software, for example, the open-source Grace tool [54] or dumped to a file for subsequent data processing, for example, via a spreadsheet or a specialized text editor.

12.5.1 STATISTICAL CLASSES FOR SYSTEM-LEVEL MODELS

Considering the hardware system modeling objects previously proposed, we observe that dynamic performance characteristics, for example, latency, throughput, packet loss, resource utilization, and possibly power consumption (switching activity) are definitely required from the following:

- *Intermodule communication and synchronization objects* (message) representing communication channel performance metrics for throughput, latency, or packet loss.
- *Intramodule passive memory objects* (register, FIFO, LIFO, circular FIFO, memory, cache) reflecting memory performance metrics for throughput, latency, packet loss, buffer size, and hit ratio. Although similar metrics for certain joint intramodule communication and synchronization objects, for example, mailboxes, are possible, these objects may be implemented in hardware in various ways, for example, using control logic and static memory.

Assuming a common 2-D graphic representation, a statistical API for the aforementioned metrics can be based on a function enable_stat(args) enabling the data-capturing activity. Its arguments specify a distinct name of the modeling object; the absolute start and end time for statistics collection; the title and legends for the x and y axes; the time window for windowed statistics, that is, the number of consecutive points averaged for generating a single statistical point; and a Boolean flag for stopping or restarting statistics during simulation.

Since complex systems involve both time- (instant) and event-driven (duration) statistics, we may provide two general monitoring classes, collecting instant and duration measurements from system components with the following functionality.

In *time-driven simulation*, signals usually have instantaneous values. During simulation, these values can be captured at a specific time by calling the function:

stat_write (double time, double value).

In *event-driven simulation*, recorded statistics for events must include arrival and departure time (or duration). Since the departure time is known later, the interface must be based on two functions: stat_event_start and stat_event_end. Thus, first the user invokes an operation

int a = stat_event_start(double arrival_time)

to record the arrival time and save the unique location of the event within the internal table of values in a local variable a. Then, when the event's departure time is known, this time is recorded within the internal table of values at the correct location, by calling the stat_event_end function with the appropriate departure time.

void stat_event_end(double departure_time, a).

The stat_event_start/end operations may take into account memory addresses in order to compute duration statistics for consecutive operations, for example, consecutive read/write accesses (or enqueue/dequeue) operations corresponding to the same memory block (or packet) for a register, FIFO, memory, cache, or intermodule communication object, that is, message. Thus, a modified StatDurationLib class performs the basic stat_event_start/end operations using (transparently to the user) internal library object pointers. Then, upon an event arrival, we invoke the command

void stat_event_start(long int MAddr, double arrival_time)

to record the following in the next available location within the internal table:

- The current memory address (MAddr)
- The arrival time (arrival_time)
- An undefined (–1) departure time

Then, upon a corresponding departure event, we invoke the command

void stat_event_end(long int MAddr, double departure_time)

to search for the current MAddr in the internal table of values and update the entry corresponding to the formerly undefined, but now defined departure time.

For duration statistics in 2-D graphic form, the y-axis point may indicate time of occurrence of a read operation performed on the same address as a previous cache write operation whose time of occurrence is shown at the corresponding point on the x-axis. Event duration, for example, latency for data access from the same memory block can be obtained by subtracting these two values. For example, using Grace this computation can be performed very efficiently using the Edit data sets: create_new (using Formula) option. Furthermore, notice that the write–read mapping is one to one, that is, data are first written and then they are read, while a *reverse access*, that is, read before write, causes an error. Such conditions must be thoroughly checked in the code.

As explained before, statistics collection may be performed either by the user or directly by the hardware modeling objects, for example, bus channels, such as AMBA APB, AHB, AXI, or STBus, using library-internal object pointers. In the latter case, software probes are inserted into the *source code of library routines*, either manually by setting sensors and actuators or more efficiently through the use of a monitoring segment that automatically compiles the necessary probes. Software probes share resources with the system model, thus offering low cost, simplicity, flexibility, portability, and precise application performance measurement in a timely, frictionless manner.

Furthermore, based on the previous statistical functions, we can derive application-specific statistics for the following system modeling objects (similar to instant and duration classes, enable_stat(args) functions are provided for initializing parameters):

- *Average throughput over a specified time window* of register, FIFO objects, memory, cache, and intermodule communication objects (message)
- *Cumulative average throughput over consecutive time windows* of register, FIFO objects, memory, cache, and intermodule communication objects (message)
- *Instant value of counter-based objects* such as FIFO, LIFO, and circular FIFO objects; this class can also be used to compute the instant cache hit ratio (for read/write accesses) and cell loss with binary instantaneous counter values
- *Average instant value over a specified time window of counter-based objects*; this class can be used to compute average cache hit ratio and cell loss probability
- *Cumulative average value over consecutive time windows* of counter-based objects; this class can also be used to compute the cumulative average cache hit ratio and cell loss probability
- *Latency* statistics for read/write (or enqueue/dequeue) on register, FIFO, memory, cache, and intermodule communication objects (message)

In addition to the previously described classes that cover all basic cases, it is sometimes necessary to combine statistical data from different modeling objects, for example, from hierarchical memory units, in order to compare average read versus write access times, or for computing cell loss in network layer communication. For this reason, we need new user-defined *joint or*

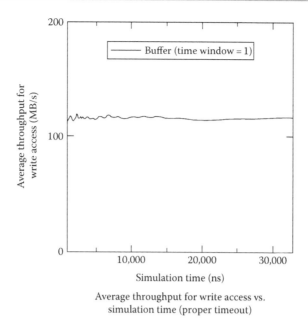

Average throughput for write access vs.
simulation time (proper timeout)

FIGURE 12.6   Performance results using Grace for a transport layer protocol.

merged statistics built on top of time- and event-driven statistics. Special parameters, for example, Boolean flags, for these joint statistic classes can lead to more detailed statistics.

We have discussed automatic extraction of statistical properties from hardware modeling objects. This statistical approach may also be extended to software modeling. As an example, the enable_stat_ construct for obtaining write throughput statistics (read throughput is similar) from a software FIFO object buffer is listed next. This function call will generate Figure 12.6.

```
// enable stats in [0,500], with time window = 1 sample
enable_stat_throughput_read("buffer", 0, 500, 1,
"Simulation Time", "Average Throughput (Writes)");
```

For more details on the implementation of statistical classes and the use of online or offline Grace-based statistical graphs, the reader is referred to the OCCN user manual and the statistical test benches that accompany the OCCN and HSoC libraries [17,24]. These test benches are compatible with both Solaris and Linux operating systems.

12.5.2 ADVANCED SYSTEM MONITORING

Conventional text output and source-code debugging are inadequate for monitoring and debugging complex and inherently parallel system models. Current system-level design tools, such as the Synopsys System Studio, generate vcd files for signal tracing, or build relational databases in the form of tables for data recording, visualization, and analysis. Efficient high-level performance modeling environments may be based on advanced monitoring activities in back-annotated system-level models. Although these activities may correspond to distinct monitoring phases occurring in time sequence, potentially there is a partial overlap between them:

- *Generation* refers to detecting events and providing *status and event reports* containing monitoring traces (or histories) of system activity.
- *Processing* refers to functions that process monitoring data, for example, merging of traces, filtering, correlation, analysis, validation, and updating. These functions convert low-level monitoring info to the required format and level of detail.
- *Presentation* refers to displaying monitoring information in appropriate form.
- *Dissemination* concerns the distribution of selected monitoring reports to system-level developers and external processing entities.

In addition to these main activities, implementation issues relating to intrusiveness and synchronization are crucial to efficient design and evaluation of monitoring activities.

12.5.2.1 GENERATION OF MONITORING REPORTS: STATUS AND EVENT REPORTS

In order to describe the dynamic behavior of an IP model instance (or group of objects) over a period of time, status and event reports are recorded in time order as *monitoring traces*. A *complete monitoring trace* contains all monitoring reports generated by the system since the start of the monitoring session, while a *segmented trace* is a sequence of reports collected during a limited period of time, for example, due to overflow of a trace buffer or deliberate halting of trace generation. A monitoring trace may also be used to generate nonindependent traces based on various logical views of objects or system activity. For each trace, we need to identify the reporting entity, the monitored object, the type of the report, as well as user-defined parameters, for example, start and end time, time window, priority, and size. We also provide browsing or querying facilities (by name or content) or runtime adjustments, for example, examining the order of event occurrences or readjusting the interval between event occurrences.

A *status report* contains a subset of system state information, including object properties, such as time stamp, status, and identity. Appropriate criteria define the sampling rate, the reporting scheme, and the contents of each report. For example, the report may be generated either periodically, that is, based on a predetermined finite-state machine or thread schedule, or on demand, that is, upon receiving a request for solicited reporting. The request may itself be periodic, that is, via polling or on a random basis.

System events may be detected immediately upon occurrence or after a delay. For example, signals on an internal system bus may be monitored in real time, while alternatively, status reports may be generated, stored, and processed in order to detect events at a later time. Event detection may be internal to the monitored object, that is, typically performed as a function of the object itself, or external, that is, performed by an external agent who receives status reports and detects changes in the state of the object. Once the occurrence of an event is detected, an *event report* is generated. In general, an event report contains a variable number of attributes such as reporting entity, monitored object, event identifier, type, priority, time of occurrence, state of the object immediately before and after event occurrence, application-specific state variables, time stamps, text messages, and possibly pointers to detailed information.

12.5.2.2 PROCESSING OF MONITORING INFORMATION

A system-level model may generate large amounts of monitoring information. Design exploration is successful if the data can be used to identify design problems and provide corrective measures. Thus, the data collection and analysis stage is split into four different phases.

Validation of system information provides consistency checks, possibly specified in a formal language, ensuring correct, nonintrusive monitoring, and harmonized operation. This includes

- *Sanity tests* based on the validity of individual monitoring traces, for example, by checking for correct token values in event fields, such as an identity or time stamp
- *Validation of monitoring reports against each other*, for example, by checking against known system properties, including temporal ordering

Filtering reduces the amount of monitoring data to a suitable rate and level of detail. For example, filter mechanisms reduce the complexity of displayed process communications by

- Incorporating a variable report structure
- Displaying processes down to a specified level in the module hierarchy
- Masking communication signals and data using filter dialogs
- Providing advanced filter functionality for displaying only tokens with predetermined values

Analysis processes monitoring traces based on special user-selected criteria. Since analysis is application dependent, it relies on sophisticated stochastic models involving combinatorics, probability theory, and Markov chain theory. *Analysis* techniques enable

- Simple processing, for example, average, maxima, variance statistics of state variables
- Statistical trend analysis for forecasting using data analysis packages such as MATLAB, Grace or SAS, and *correlation* (merging, combination, or composition) of monitoring traces, which raises the abstraction level

Together with filtering, correlation prevents the users from being overwhelmed by an immense amount of detailed information and helps identify system bottlenecks. Thus, for example, events generated by sensors or probes may be combined using AND, OR, and NOT operators to provide appropriate high-level reliability metrics. Since correlation of system-level monitoring information is a very challenging task, a relational database, such as mini-SQL, which includes selection, projection, and join operators, is sometimes useful.

12.5.2.3 PRESENTATION OF MONITORING INFORMATION

Various visualization techniques such as simple textual representation, time-process diagrams, and animation may be provided.

Textual representation (ascii) increases its expressive power by providing appropriate indentation, color, and highlighting to distinguish information at different abstraction levels. Events may be displayed in a causal rather than temporal order by including parameters such as event type, name of process initiating the event, name of process(es) handling the event, and contents of transmitted messages.

A *time-process diagram* is a 2-D diagram illustrating the current system state and the sequence of events leading to that state. The horizontal axis represents events corresponding to different processes, while the vertical one represents time. In synchronous systems, the unit of time corresponds to actual time, while in asynchronous systems, it corresponds to the occurrence of an event. In the latter case, the diagram is called a *concurrency map*, with time dependencies between events shown as arrows. An important advantage of time-process diagrams is that monitoring information may be presented on a simple text screen or a graphical one.

An *animation* captures a snapshot of the current system state. Both textual and graphical event representations, for example, input, output, and processing events, can be arranged in a 2-D display window. Graphical representations use formats such as icons, boxes, Kiviat diagrams, histograms, bar charts, dials, X–Y plots, matrix views, curves, pie charts, and performance meters. Subsequent changes in the display occur in single-step or continuous fashion and provide an animated view of system evolution. For online animation, the effective rates at which monitoring information is produced and presented to the display must be matched. For each abstraction level, animation parameters include enable/disable event monitoring or visualization, clock precision, monitoring interval, or level of detail, and viewing/printing statistics.

12.5.2.4 DISSEMINATION OF MONITORING INFORMATION

Monitoring reports would have to reach designers, managers, or processing entities. Thus, dissemination schemes range from very simple and fixed, to very complex and specialized. Selection criteria contained within the subscription request are used by the dissemination system to determine which reports (and with what contents, format, and frequency) should be delivered. Depending on the frequency of the queries and system events, the user may resort to a hybrid model combining both push (data duplication in small area around the server) and pull approaches (client contacting the server) for efficient on-demand dissemination of the required system monitoring information.

12.5.2.5 OTHER IMPORTANT MONITORING DESIGN ISSUES

Specifications for the actual graphical user interface of the monitoring tool depend on the designer's imagination, experience, and time. We present here desirable features of a system-level monitoring interface:

- *Visualization at different abstraction levels* that reduces the learning curve and enables the user to observe behavior at various abstraction levels on a per object, per block, or per system basis. Thus, the user may start observation at a coarse level and progressively (or simultaneously) focus on lower levels. At each level, system cost, application performance, and platform-specific metrics may be presented in appropriate easy-to-read charts and graphic visualization techniques. The communication volume (reliability or power consumption) may be visualized by adjusting the width or color of the lines interconnecting (respectively) the boxes representing the corresponding modules.
- A *history function* that visualizes inherent system parallelism by permitting the user to perform several functions, such as
  - Scrolling the display of events forward or backward in time by effectively changing a simulated system clock
  - Controlling the speed at which system behavior is observed using special functions, for example, to start, stop, pause, or restart an event display, perform single-step or continuous animation, and allow for real-time or slow-motion animation
- *Visibility of interactions* that enables the user to visualize dynamically contents of a particular communication message, object data structure, module or driver configuration, or general system data such as log files or filtering results.
- *Placement of monitoring information* that greatly enhances visibility and aids human comprehension. Placement may either be automatic, that is, using computational geometry algorithms, or manual by providing interface functions, for example, for moving or resizing boxes representing objects, system events, or coupled links representing process communication.
- *Multiple views* use multiple windows representing system activities from different viewpoints, thus providing a comprehensive picture of system behavior.
- *Scalable monitoring* focuses on monitoring large-scale models, with tens or hundreds of thousands of objects. Large-scale monitoring could benefit from efficient message queues or nonblocking and blocking concurrent queues [55,56] that achieve a high degree of concurrency at low implementation cost compared to other general methods [57,58].

In addition, the monitoring system designer should focus on the issues of intrusiveness and synchronization occurring in distributed system design.

- Intrusiveness refers to the effect that monitoring and diagnostics may have on the monitored system due to sharing common resources, for example, processing power, communication channels, and storage space. Intrusive monitors may lead not only to system performance degradation, for example, due to increased memory access, but also to possible deadlock conditions and data races, for example, when evaluating symmetric conditions that result in globally inconsistent actions.
- Distributed simulation can support time- and space-efficient modeling of large modular systems at various abstraction levels, with performance orders of magnitude better than existing commercial simulation tools [59]. Distributed systems are more difficult to monitor due to increased parallelism among processes or processors, random and nonnegligible communication delays, possible process failures, and unavailable global synchronized time. These features cause interleavings of monitoring events that might result in different output data from repeated runs of deterministic distributed algorithms or different views of execution events from various objects [60–63]. However, these problems do not occur with SystemC scheduling (and early parallel SystemC versions), since current versions of the kernel are sequential, offering simulated parallelism and limited event interleaving.

12.6 OPEN ISSUES IN SYSTEM-LEVEL MODELING

Clock-accurate transactional system-level modeling is performed at an abstraction level higher than traditional hardware description languages, thus improving design flow and enhancing design quality by offering an increased design space, efficient simulation, and block- and system-level reuse. We have described efforts toward a general, rich, and efficient system-level performance evaluation methodology that allows multiple levels of abstraction, enables hardware/software codesign, and provides interoperability between several state-of-the-art tools. There are activities that could enhance this methodology:

- System reliability, fault-tolerance, and validation/verification tools [64]; available tools in the latter area are usually external to the system-level modeling library and tightly linked to actual implementation. New system-level verification tools could rely on refinement techniques to test the functional equivalence between the original high-level design (e.g., SystemC) and the derived RTL design.
- Wider use of automated C/C++ or SystemC to RTL behavioral synthesis tools, such as Forte (now Cadence), Calypto Catapult, Synopsys Synphony, and Xilinx Vivado, can reduce time to market while providing area, speed, and latency trade-offs for complex designs, for example, signal and image processing or VLIW designs interfacing with standard system interconnects. Unlike in the past, automated design of virtual platforms that integrate AMS and mechanical components and system drivers should be considered as part of the system-level design flow.
- Efficient system-level modeling constructs, for example,
  - *Parallelizing the SystemC kernel* to execute efficiently in parallel and distributed systems
  - *Developing an asynchronous SystemC scheduler* to improve test coverage during the validation phase [65,66]
  - *Supporting system-level power-consumption* models; for example, unlike memory models, high-level network-on-chip power estimation methodology offers good accuracy [67–69]
  - *Providing full support of real-time operating system models* providing task management, context saving and restoring, preemptive scheduling, high-resolution timer, and task synchronization facility [70,71], for example, integration of embedded software stacks and applications, including user interfaces
  - *Handling asynchronous modeling* [72], for example, by modifying the SystemC kernel and providing waves, concurrency maps, and system snapshots
- Graphical visualization [73–75], including appropriate GUIs for
  - *Interactive model design* based on importing ready-to-use library modules, for example, HFsms, or reusable IP block or system components; these libraries would support best practices in *simulation configuration and control*, such as saving into files, starting, pausing, and restarting simulation as well as displaying waveforms by linking to graphical libraries and dumping or changing model parameters during initialization or runtime [76]
  - *Advanced monitoring features*, based on generation, processing, presentation, and dissemination of monitoring info
  - *Platform-specific performance metrics*, such as simulation efficiency, and computation and communication load, which help improve simulation performance in parallel platforms, for example, through automatic data allocation, latency hiding, or dynamic load balancing
- Documenting executable specifications by customizing tools, such as Doxygen, Doc++, or mkdoc.
- Interoperability between libraries and tools, for example, ISS, system and hardware description languages, and simulation interfaces, can provide long-term reuse opportunities. Although current efforts (e.g., for linking SystemC with Gem5, OPNET, or OMNET simulators) are limited to supporting one-to-many thread communication patterns and use ad hoc event-based synchronization and TCP sockets (e.g., MPI) or

shared memory to exchange information among the simulators, this could be achieved in a more organized way using middleware, for example, by inventing a generic interface definition language.

ACKNOWLEDGMENTS

This research has been cofinanced through the Operational Programme "Education and Lifelong Learning," Action Archimedes III by the European Union (European Social Fund), and Greek national funds (NSRF 2007–2013).

REFERENCES

1. Carloni, L.P. and Sangiovanni-Vincentelli, A.L., Coping with latency in SoC design, *IEEE Micro* (Special Issue on Systems on Chip), 2002, **22(5)**, 24–35.
2. SystemC, available from http://www.accellera.org/downloads/standards/systemc (Accessed on November, 2015).
3. *SystemC 2.0 User Guide*, 2001, available from http://www.accellera.org/downloads/standards/systemc (Accessed on November, 2015).
4. Krolikoski, S., Schirrmeister, F., Salefski, B., Rowson, J., and Martin, G., Methodology and technology for virtual component driven hardware/software co-design on the system level, *Int. Symp. Circuits Systems*, 1999, **6**, 456–459.
5. Synopsys, SystemC modeling with the synopsys system studio, 2002, available from http://www.synopsys.com/Prototyping/VirtualPrototyping/DigitalSignalProcessing/Pages/system-studio.aspx (Accessed on November, 2015).
6. Synthesis with the Synopsys Cocentric SystemC Compiler, User manual and presentations, 2002, available from http://www2.cic.org.tw/DSDWS2003/pdf/A6.pdf (Accessed on November, 2015).
7. Ferrari, A. and Sangiovanni-Vincentelli, A., System design: Traditional concepts and new paradigms, *Proceedings International Conference on Computer Design*, 1999, Austin, TX, October 10–13, 1999, pp. 2–13.
8. Rowson, J.A. and Sangiovanni-Vincentelli, A.L., Interface-based design, *Proceedings of the Design Automation Conference*, 1997, pp. 178–183.
9. Haverinen, A., Leclercq, M., Weyrich, N., and Wingard, D., SystemC-based SoC communication modeling for the OCP protocol, White Paper submitted to SystemC, 2002.
10. Moriconi, M., Qian, X., and Riemenschneider, R.A., Correct architecture refinement, *IEEE Trans. Software Eng.*, 1995, **SE-21**, pp. 356–372.
11. Medvidovic, N. and Taylor, R.N., A framework for classifying and comparing architecture description languages, *Proc. Software Eng. Conf.*, Lecture Notes in Comp. Sci., Vol. 1301, Springer, Berlin, Germany, 1997, pp. 60–76.
12. Allen, R. and Garlan, D., A formal basis for architectural connection, *ACM Trans. Software Eng. Method.*, 1997, **6**, 213–249.
13. Charest, L., Aboulhamid, E.M., and Tsikhanovich, A., Designing with SystemC: Multi-paradigm modeling and simulation performance evaluation, *Proceedings of the HDL Conference*, 2001, pp. 33–45.
14. Virtanen, S., Truscan, D., and Lilius, J., SystemC based object oriented system design, *Proceedings* of the *Forum on Design Languages*, Lyon, France, 2001, pp. 1–4.
15. Gajski, D.D., Zhu, J., Doemer, A., Gerstlauer, S., and Zhao, S., *SpecC: Specification Language and Methodology*, Kluwer, Dordrecht, the Netherlands, 2000. Also see http://www.cecs.uci.edu/~specc (Accessed on November, 2015).
16. Brunel, J.-Y., Kruijtzer, W.M., Kenter, H.J.H.N. et al., Cosy communication IP's, *Proceedings of the Design Automation Conference*, 2000, pp. 406–409.
17. Coppola, M., Curaba, S., Grammatikakis et al., OCCN: A NoC modeling framework for design exploration, *J. Syst. Arch.—Special Issue on Network-on-Chip*, 2004, **50**, 129–163. available from http://occn.sourceforge.net (Accessed on November 2015).
18. Carey, J., *STbus Superhighway: Type 3*, internal document, ST Microelectronics, March 2001.
19. Scandurra, A., *STbus C++ Class*, internal document, ST Microelectronics, December 2000.
20. VSI Alliance, http://www.vsia.org (Accessed on November, 2015).
21. ARM, AMBA Bus specifications, available from http://www.arm.com (Accessed on November, 2015).
22. Melkonian, M., Get by without an RTOS, *Embed. Syst.*, **13(10)**, 2000, pp. 1–7.
23. ARM, *Cortex-A9 Technical Reference Manual (Appendix B: Cycle Timings)*, 2012, available from http://www.arm.com.

24. Heterogeneous SoC Library, available from http://hsoc.sourceforge.net (Accessed on November, 2015).

25. Carloni, L.P., McMillan, K.L., and Sangiovanni-Vincentelli, A.L., Theory of latency-insensitive design, *IEEE Trans. Comp. Aided Design Integr. Circuits Syst.*, 2001, **20–29**, 1059–1076.

26. Imperas OVPSim, http://www.ovpworld.org.

27. Binkert, N. and Dreslinski, R.G., Black, G., et al. The M5 simulator: Modeling networked systems, *IEEE Micro*, 2006, **26(4)**, 52–60.

28. Martin, M.M.K, Sorin, D.J., Beckmann, B.M. et al., Multifacet's general execution-driven multiprocessor simulator (GEMS) toolset, *SIGARCH CAN*, 2005, **33(4)**, 92–99.

29. Binkert, N., Beckmann, B., Black, G. et al., The Gem5 simulator, *SIGARCH Comput. Archit. News*, 2011, **39(2)**, 1–7.

30. ARM Cross-Compilers frameworks, available from http://www.ibm.com/developerworks/linux/library/l-arm-toolchain (Accessed on November, 2015).

31. Gutierrez, A., Pusdersris, J., Dreslinksi, T. et al., Source of error in full-system simulation, *Proceedings International Symposium on Performance Analysis of Systems and Software*, Monterey, CA, 2014, pp. 13–22.

32. Harel, D., Statecharts: A visual formalism for complex systems, *Sci. Comput. Program.*, 1987, **8**, 231–274.

33. von der Beeck, M., A comparison of statecharts variants, *Proceedings of the Formal Techniques in Real Time and Fault Tolerant Systems*, Lecture Notes in Computer Science, Vol. 863, Springer, Berlin, Germany, 1994, pp. 128–148.

34. Sheibanyrad, A., Greiner, A., and Panades, I.M., Multisynchronous and fully asynchronous NoCs for GALS architectures, *IEEE Design and Test of Computers*, 2008, **25(6)**, pp. 572–580.

35. Crews, M. and Yuenyongsgool, Y., Practical design for transferring signals between clock domains, *EDN Network*, February 2003, **48(4)**, 65–70.

36. Fraunhoffer, SystemC-AMS 1.0 and 2.0 implementations, available from http://www. fraunhofer.de/en/business_areas/microelectronic_systems/system_development.html (Accessed on November, 2015).

37. SystemC-AMS publications (white papers, etc.), http://www.systemc-ams.org (Accessed on November, 2015).

38. Vachoux, A., Grimm, C., and Einwich, K., Towards analog and mixed-signal SoC design with SystemC-AMS, *Proceedings IEEE Workshop on Electronic Design, Test and Application*, 2004, pp. 97–102.

39. Vachoux, A., Grimm, C., and Einwich, K., SystemC-AMS requirements, design objectives and rationale, *Proceedings of the Conference on Design, Automation and Test in Europe*, 2003, pp. 388–393.

40. Antao, B., El-Turky, F., and Leonowich, R., Mixed-mode simulation of phase-locked loops, *Proceedings IEEE Custom Integrated Circuits Conference*, 1993, pp. 841–844.

41. Best, R., *Phase-Locked Loops: Design, Simulation and Applications*, 4th edition, McGraw-Hill, New York, 1999.

42. Nallatamby, J.-C., Prigent, M., Camiade, M. et al., Phase noise in oscillators—Leeson formula revisited, *IEEE Trans. Microw. Theory Tech.*, 2003, **51(4)**, 1386–1394.

43. Razavi, B., A study of phase noise in CMOS oscillators, *IEEE J. Solid State Circuits*, 1996, **31(3)**, 331–343.

44. SimPLL, Analog radio systems, available from http://www.radio-labs.com (Accessed on November, 2015).

45. CPPSim, Michael Perrot, MIT, tool available from http://www.cppsim.com (Accessed on November, 2015).

46. EasyPLL Web, National instruments, tool available from http://www.ti.com/lsds/ti/analog/webench/easypll.page (Accessed on November, 2015).

47. PLL software analysis tool, available from Mike Ellis website: http://michaelgellis.tripod.com/software.html (Accessed on November, 2015).

48. Chan, M., Postula, A., and Ding, Y., PLLSim—An ultra fast bang-bang phase locked loop simulation tool, *Proceedings Asia South Pacific Design Automation Conference*, Yokohama, Japan, 2007, pp. 74–79. (Tool withdrawn from the site but available via email.)

49. Stensby, J., Lock detection in phase-locked loops, *SIAM J. Appl. Math.*, 1992, **52(5)**, 1469–1475.

50. Stensby, J., An improved lock detector for phase-locked communication receivers, *J. Franklin Institute*, 2005, **342(2)**, 149–159.

51. Press, W.H. and Rybicki, G.B., Fast algorithm for spectral analysis of unevenly sampled data, *Astrophys. J.*, 1989, **338**, 277–280.

52. Scargle, J.D., Studies in astronomical series analysis, *Astrophys. J.*, 1982, **263(Part 1)**, 835–853.

53. Zimmerman, C., *The Quantitative Macroeconomics and Real Business Cycle*, for statistical software see: http://dge.repec.org/software.html (Accessed on November, 2015).

54. Grace, available from http://plasma-gate.weizmann.ac.il/Grace (Accessed on November, 2015).

55. Hunt, G.C., Michael, M., Parthasarathy, S., and Scott, M.L., An efficient algorithm for concurrent priority queue heaps, *Inf. Proc. Lett.*, 1996, **60**, 151–157.

56. Michael, M. and Scott, M.L., Simple, fast and practical non-blocking and blocking concurrent queue algorithms, *Proceedings of the ACM Symposium on Principles of Distributed Computing*, Philadelphia, PA, 1996, pp. 267–275.

57. Prakash, S., Yann-Hang, L., and Johnson, T., A non-blocking algorithm for shared queues using compare-and-swap, *IEEE Trans. Comput.*, 1994, **43(5)**, 548–559.

58. Turek, J., Shasha, D., and Prakash, S., Locking without blocking: Making lock-based concurrent data structure algorithms nonblocking, *Proceedings of the ACM Symposium on Principles of Database Systems*, San Diego, CA, 1992, pp. 212–222.

59. Grammatikakis, M.D. and Liesche, S., Priority queues and sorting for parallel simulation, *IEEE Trans. Soft. Eng.*, 2000, **SE-26**, 401–422.

60. Chandy, K. M. and Lamport, L., Distributed snapshots: Determining global states of distributed systems, *ACM Trans. Comp. Syst.*, 3(1), 1985, pp. 63–75.

61. Christian, F., Probabilistic clock synchronization, *Distr. Comput.*, 1989, **3(3)**, 146–158.

62. Fidge, C.J., Partial orders for parallel debugging, *Proceedings of the ACM Workshop Parallel Distributed Debugging*, 1988, pp. 183–194.

63. Lamport, L., Time, clocks and the ordering of events in distributed systems, *Comm. ACM*, 1978, **21(7)**, 558–564.

64. Grammatikakis, M.D., Hsu, D.F., and Kraetzl, M., *Parallel System Interconnections and Communications*, CRC Press, Boca Raton, FL, 2000.

65. Schumacher, C., Weinstock, J.H., Leupers, R. et al., legaSCi: Legacy SystemC model integration into parallel SystemC simulators, *Proceeding IEEE Parallel Distributed. Processing Symposium*, Boston, MA, 2013, pp. 2188–2193.

66. Weinstock, J., Schumaher, C., Leupers, R., Ascheid, G., and Tosoratto, L., Time-decoupled parallel SystemC simulation, *Proceedings of the Design Automation and Test in Europe Conference*, Dresden, Germany, 2014, pp. 1–4.

67. Benini, L. and De Micheli, G., System-level power optimization: Techniques and tools, *ACM Trans. Autom. Electr. Syst.*, 2000, **5(2)**, 115–192.

68. Benini, L., Ye, T.T., and De Micheli, G., Packetized on-chip interconnect communication analysis for MPSoC, *Proceedings of the Design Automation and Test in Europe Conference*, Munich, Germany, 2003, pp. 344–349.

69. Ye, T.T., Benini, L., and De Micheli, G., Analysis of power consumption on switch fabrics in routers, *Proceedings of the Design Automation Conference*, New Orleans, LA, 2002, pp. 524–529.

70. Koch-Hofer, C., Renaudin, M., Thonnart, Y., and Vivet, P., ASC, a SystemC extension for modeling asynchronous systems, and its application to an asynchronous NoC, *Proceedings of the International Symposium on Network-on-Chip*, Princeton, NJ, 2007, pp. 295–306.

71. Miramond, B., Huck, E., Verdier, F. et al., OveRSoC: A framework for the exploration of RTOS for RSoC platforms, *Int. J. Reconfig. Comput.*, 2009, **11**, 1–26.

72. Mueller, W., Ruf, J., Hoffmann, D., Gerlach, J., Kropf, T., and Rosenstiehl, W., The simulation semantics of SystemC, *Proceedings of the Design, Automation and Test in Europe Conference*, Munich, Germany, 2001, pp. 64–71.

73. Koutsofios, E. and North, S.C., *Dot User Manual*, AT&T Bell Labs, Murray Hill, NJ, 1993. Available from http://www.cs.brown.edu/cgc/papers/dglpttvv-ddges-97.ps.gz (Accessed on November, 2015).

74. Reid, M., Charest, L., Tsikhanovich, A., Aboulhamid, E.M., and Bois, F., Implementing a graphical user interface for SystemC, *Proceedings of the HDL Conference*, 2001, pp. 224—231.

75. Sinha, V., Doucet, F., Siska, C., Gupta, R., Liao, S., and Ghosh, A., YAML: A tool for hardware design, visualization, and capture, *Proceedings of the International Symposium on Systems Synthesis*, Santa Clara, CA, 2000, pp. 9–17.

76. Charest, L., Reid, M., Aboulhamid, E.M., and Bois, G., A methodology for interfacing open source SystemC with a third party software, *Proceedings of the Design, Automation and Test in Europe Conference*, Munich, Germany, 2001, pp. 16–20.

Microarchitectural and System-Level Power Estimation and Optimization

Enrico Macii, Renu Mehra, Massimo Poncino, and Robert P. Dick

13

CONTENTS

Abstract

Power consumption has become one of the main design considerations in electronic systems. This chapter introduces innovative methodologies for successfully dealing with power estimation and optimization during early stages of the design process. In particular, the presentation offers an insight into state-of-the-art techniques for power estimation at the microarchitectural level, describing how power consumption of components like datapath macros, glue and steering logic, memory macros, buses, interconnect, and clock wires can be efficiently modeled for fast and accurate power estimation. Then, the focus shifts to power optimization, covering the most popular classes of techniques, such as those based on clock gating, on exploitation of common-case computation, and on threshold and supply voltage management. *Ad hoc* optimization solutions for specific components, such as on-chip memories and global buses, are also briefly discussed for the sake of completeness. Our attention then turns to system-level modeling and optimization of power consumption in embedded systems such as mobile phones.

Most of the aforementioned approaches to microarchitectural and system-level power estimation and optimization have now reached a significant level of maturity and are finding their way into commercial CAD tools and software for designers of mobile computing systems. Strengths and limitations of the design technology that is at the basis of such tools will be discussed in detail throughout this chapter.

13.1 INTRODUCTION

Minimizing integrated circuit power consumption can prolong battery life span, decrease cooling subsystem costs, and decrease operating costs of electronic devices. Power consumption is an important consideration for many electronic devices. This is not surprising, given the massive market for mobile and portable telecommunication and computing systems. As a consequence, techniques and tools that enable tight power consumption control during design are required.

Pioneering work on low-power design techniques has focused on gate and transistor levels where, due to the available information on the structure and the macroscopic parameters of the devices, accurate power estimates are expected and satisfactory methods for both estimation and optimization are available.

The increased complexity of modern designs, facilitated by the advent of aggressively scaled technologies and the pressure of time-to-market constraints, called for modifications to the way ICs are designed. This resulted in the use of tools at higher levels of the design process. Today, design techniques and tools are available to assist in estimating and optimizing power consumptions at the microarchitectural (register-transfer level [RTL]) and system levels, in addition to lower levels of the design process. At these levels, basic design entities are no longer elementary objects such as transistors or logic gates, but rather blocks capable of performing complex functions. Typical components at the microarchitectural level include datapath macros (such as adders and multipliers), storage elements (such as registers and memory banks), communication resources (e.g., buses), and steering elements (e.g., multiplexors and codecs). At the system level, entire microprocessors or wireless communication interfaces may be used as building blocks. As a result, it has been necessary to develop power modeling and optimization techniques well suited for the RTL and above, many of which are available for industrial use.

As electronic technology evolves, so does EDA technology; higher levels of abstraction are currently being proposed as possible starting points for new design development. For instance, in the system-on-chip (SoC) domain, hardware–software combined specification, design, and synthesis are becoming common practice, and new methodologies and tools are being investigated. Although RTL remains the highest level of abstraction for which extensive EDA support is guaranteed, tools are now becoming available at higher levels of the design flow.

The objective of this chapter is to provide a comprehensive overview of the most advanced, yet well-established EDA solutions for power estimation and optimization. We will consider both RTL and system-level tools for estimating and optimizing power consumption. The chapter will start by providing some background information about power consumption in CMOS circuits. It continues with a brief overview of the architectural template assumed by most of the power modeling, estimation, and optimization approaches, which constitute the core of the chapter. It concludes with two sections on system-level power estimation and optimization.

13.2 BACKGROUND

CMOS is, by far, the most common technology used for manufacturing digital circuits. There are three major sources of power dissipation in a CMOS circuit [1]:

$$(13.1) \qquad P = P_{Sw} + P_{SC} + P_L.$$

where
 P_{Sw}, called "switching or dynamic power," is due to charging and discharging capacitors driven by the gates in the circuit
 P_{SC}, called "short-circuit power," is caused by the short-circuit currents that arise when pairs of PMOS/NMOS transistors are conducting simultaneously

Finally, P_L, called "leakage or static" or "stand-by power," originates from subthreshold currents caused by transistors with low threshold voltages and from gate currents caused by reduced thickness of the gate oxide.

For older technologies (e.g., 0.25 mm), P_{Sw} was dominant. For deep-submicron processes, P_L becomes more important. For instance, in application-specific integrated circuit (ASIC) designs, leakage power accounts for around 5%–10% of the total power budget at 180 nm, and this fraction grows to 20%–25% at 130 nm and to 35%–50% at 90 nm. Therefore, leakage power minimization must be addressed from the design standpoint, and not just at the technology or process level, as was done in the past.

Design methods for leakage power control are currently the subject of intensive investigation: approaches based on variable-threshold, dual-threshold, and multithreshold CMOS devices; the insertion of (possibly distributed) sleep transistors; the adoption of multivoltage gates; and the application of reverse and forward body biasing are some examples of promising solutions for reducing the impact of leakage in nanometer circuits. Yet, most of the methods and tools for low-power design in use today are still primarily targeting the minimization of the dynamic component of the power; this is because research in this domain has been carried out for a much longer time and the available solutions have now reached a significant degree of maturity. As a consequence, a significant part of this chapter will be focused on techniques addressing switching power modeling, estimation, and optimization.

Switching power for a CMOS gate working in a synchronous environment is modeled by the following equation:

$$(13.2) \qquad P_{Sw} = \frac{1}{2} C_L V_{dd}^2 f_{Ck} E_{Sw},$$

where
 C_L is the output load of the gate
 V_{dd} is the supply voltage
 f_{Ck} is the clock frequency
 E_{Sw} is the *switching activity* of the gate, defined as the probability of the gate's output of making a logic transition during one clock cycle
 Reductions of P_{Sw} are achievable by combining the minimization of the four parameters in Equation 13.2

Historically, supply voltage scaling has been the most used approach to power optimization, since it normally yields considerable savings owing to the quadratic dependence of P_{Sw} on V_{dd}. The major shortcoming of this solution, however, is that lowering the supply voltage affects circuit speed. As a consequence, both design and technology solutions must compensate for reduced voltage. In other words, speed optimization is applied first, followed by supply voltage scaling, which brings the design back to its original timing but with a lower-power requirement.

A similar problem, that is, performance decrease, is encountered when power optimization is obtained through frequency scaling. Techniques that rely on reductions of the clock frequency to lower-power consumption are thus usable under the constraint that some performance slack does exist. Although this may seldom occur for designs considered in their entirety, it happens quite often that some specific units in a larger architecture do not require peak performance for some clock/machine cycles. Selective frequency scaling (as well as voltage scaling) on such units may thus be applied, at no penalty on the overall system speed.

Optimization approaches that have a lower impact on performance, yet allowing significant power savings, are those targeting the minimization of the *switched capacitance* (i.e., the product of the capacitive load and the switching activity). Static solutions (i.e., applicable at design time) handle switched capacitance minimization through area optimization (which corresponds to a decrease in the capacitive load) and switching activity reduction via exploitation of different kinds of signal correlations (temporal, spatial, and spatiotemporal). Dynamic techniques, on the other hand, aim at eliminating power waste that may be originated by the application of certain system workloads (i.e., the data being processed).

Static and dynamic optimizations can be achieved at different levels of design abstraction. Actually, addressing the power problem from the very early stages of design development offers enhanced opportunities to obtain significant reductions of the power budget and to avoid costly redesign steps. Power-conscious design flows must then be adopted; these require, at each level of the design hierarchy, the exploration of different alternatives, as well as the availability of power estimation tools that could provide accurate feedback on the quality of each design choice.

This chapter first describes the state of the art in tools and techniques for RTL power optimization. It then describes system-level power optimization techniques, many of which are now supported by EDA tools. More specifically, although RTL descriptions may contain components of different kinds (see the next section for more details on the architectural template), we will primarily concentrate on estimation and optimization of the portion of the architecture that is normally designed with the help of automatic tools (e.g., datapath, controller, interconnect, and clock tree). Modeling, estimation, and optimization techniques for components with peculiar characteristics, such as memories and buses, will also be surveyed in order to make the picture as complete as possible. At the system level, we will describe tools for power estimation and optimization that allow at least part of the process to be automated.

13.3 ARCHITECTURAL TEMPLATE

The definition of a "microarchitecture" generally assumes an *architectural template*, which serves as a reference for all microarchitectural descriptions. There are two main advantages in assuming such a template: First, it allows one to infer a fine-grained partition of the entire design into *objects* of manageable size, for which customized models and specific optimizations can be devised. Second, by assuming that all microarchitectural descriptions will map onto this template, the chance of reusing models and optimization techniques increases.

A traditional microarchitectural template views a design as the interaction of a *datapath* and a *controller*, which fits well to the so-called finite-state machine with datapath (FSMD) model [2]. Variants of the base FSMD model concern the structure of the controller (e.g., sparse logic implementation vs. wired logic), the structure of the interconnect (e.g., number, type, and size of the buses), or the supported arithmetic operations (e.g., number and type of the available datapath units).

Figure 13.1 shows a possible architecture of a typical FSMD that exposes its basic building blocks: the controller (shown on the left) and the datapath (right), consisting of a register file (or equivalently a set of sparse registers), a memory, various interconnection buses, and some datapath units (integer and/or floating point). Besides the blocks themselves, the interconnection between them (not explicitly depicted in the figure) is also a source of power consumption; these wires can be global or local, depending on whether they span the entire architecture or not. Examples of the former classes are the clock signal and the global buses; examples of the latter are the signals connecting the units to the global wires. The most important of all wires is the clock that constitutes a significant source of consumption, because of its large load and high activity.

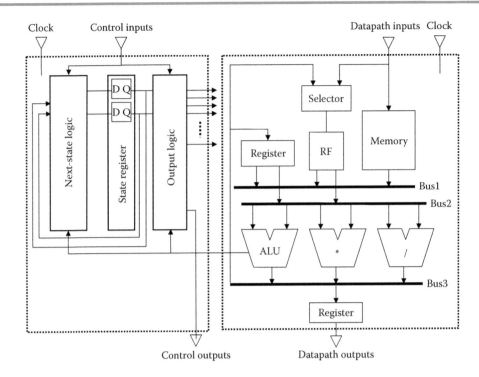

FIGURE 13.1   Architectural template for RTL power estimation.

While most of the local wires are usually hidden inside the primitive blocks, global wires need a special treatment because they cannot be captured otherwise.

The architectural template introduced earlier is sometimes defined as *structural RTL*, with the purpose of emphasizing the explicit notion of structure it contains, as well as the RT-level nature of the operations that take place during execution.

The key issue is that fitting a microarchitectural description to this template allows us to restrict the granularity of the power models and of the optimizations to that of the following main building blocks: *controller*, *datapath units*, *memory*, *buses*, and *wires*. Registers may fall either in the class of memory devices (for a register-file implementation style) or in that of data-path units (for "sparse" style). This approach is followed by most microarchitectural power estimation and optimization frameworks that have been proposed in the recent literature [3–10].

The FSMD template of Figure 13.1 also nicely fits microarchitectural descriptions specified using hardware description languages (e.g., VHDL or Verilog). They are in fact described as a state machine (the controller), whose states consist of a series of microarchitectural operations (e.g., assignments, arithmetic, or logic operations) corresponding to datapath operations. It is worth emphasizing, however, that such descriptions, after parsing by standard hardware description language (HDL) compilers, tend to lose some of their structural semantics and are partly flattened into a netlist of finer-grain primitives such as abstract logic gates [11], in which only memory macros are preserved.

13.4 MICROARCHITECTURAL POWER MODELING AND ESTIMATION

The problem of power estimation at the RTL amounts to building a *power model* that relates the power consumption of the target design to suitable quantities. In formula,

$$P = P(X_1, ..., X_n),$$

where X_i, $i = 1,..., n$, represent the *n model parameters*. The construction of the model shown here implies addressing several issues that call for the development of various modeling alternatives and are discussed in detail in the sequel.

13.4.1 MODELING ISSUES

This section summarizes the challenges in determining appropriate model granularity, parameters, semantics, structure, and location.

13.4.1.1 MODEL GRANULARITY

Section 13.3 shows that an architectural template determines the granularity of the power models; power estimation of a design requires thus the construction of power models for the following classes of building blocks: *controller, datapath units, memory, buses, and wires.*

13.4.1.2 MODEL PARAMETERS

Parameters included in the model must be observable at the abstraction level at which they are used. Under the architectural model of Section 13.3, the abstract model of switching power in Equation 13.2 translates into the following high-level expression:

$$(13.3) \qquad P_{\text{total}} = k \sum_{\forall \text{component } i} A_i C_i,$$

where

A_i and C_i denote the *switching activity* and the *physical capacitance* of component i, respectively

k lumps the $f V_{\text{dd}}^2$ terms, which can be considered as scaling factors at the RTL

This decouples a power model into a model for activity and a model for capacitance for each type of component. Activity and capacitance models may depend on different parameters because they are affected by different physical quantities.

An activity is easier to model, because at the RTL it is a well-defined quantity. Activity models rely on *activity parameters*, such as bit-wise (referred to specific input or output signals of a component) or word-wise (regarding input or output values of a component) transition and static probabilities. Most activity models proposed in the literature use these probability measures as parameters. Other choices for activity parameters may include variants of transition probabilities such as *transition density* [12], defined as an average (over time) switching rate or various *correlation measures*.

Modeling physical capacitance is a more problematic task than modeling activity. The term "physical" suggests the difficulty in linking capacitance to quantities observable at the RTL. In spite of that, RTL capacitance models can be derived with a reasonable degree of accuracy. They all rely on the intuitive observation that capacitance will be roughly related to the number of "objects" (gates or similar lower-level primitives) of the target component. In other words, physical capacitance at the RTL is approximated by *complexity*, and we thus speak of *complexity models*, based on *complexity parameters*.

Complexity parameters that are available at the RTL are restricted to the *width* of a component (i.e., its number of inputs and outputs) or the *number of states* (which is relevant only to the controller since notion of state is explicit). Any complexity parameter different from these would require some additional information derived from back annotation of physical information of previous implementations.

13.4.1.3 MODEL SEMANTICS

Models can be distinguished by interpretation of their return values. The most intuitive option is to assume that models express *average power*, which is commonly used as a metric to track battery lifetime or average heat dissipation. In this case, the semantics of the model is that of a single figure to represent the consumption of the target description. Average power models are called "cumulative power models" [13].

However, the notion of a *cycle* intrinsic to RTL descriptions allows us to obtain a power model with a richer semantic by simply changing the way we collect statistics. The first step in this direction consists of expanding the model of Equation 13.3 into

$$(13.4) \qquad\qquad P_{\text{total}} = k \sum_{\forall \text{cycle } j} P_j = k \sum_{\forall \text{cycle } j} \sum_{\forall \text{component } i} A_{ij} C_{ij},$$

where P_j denotes the power consumption at cycle j, which can be obtained by summing the power consumption for each component (as in Equation 13.3), this time using activities and capacitances of component i at each cycle j. The semantics of the model of Equation 13.4 is *cycle accurate* because it allows to track cycle-by-cycle (total) power.

The use of a cycle-accurate model affects the choice of the parameters. For example, transition or static probabilities are not suitable quantities anymore, since they are intrinsically "average." Conversely, cycle-accurate models should use cycle-based activity measures, such as the *number of bit toggles* between consecutive patterns (i.e., the *Hamming distance*) [14,15] or *the values* of consecutive input patterns [13,16].

A cycle-accurate model provides several advantages over a cumulative one. First, it allows one to go beyond the bare evaluation of average power and can be used to perform sophisticated analysis of power consumption over time, such as reliability, noise, or IR drop analysis. In addition, a cycle-accurate model is more accurate than a cumulative one, but not just because it provides a set of power values rather than a single one. In fact, the relation between input statistics and power is nonlinear: average consumption is usually different from the consumption associated to average input statistics, especially when power consumption varies significantly over time. Therefore, even when average power is the objective, averaging the series of cycle-by-cycle values will yield a more accurate estimate than a model of average power. On the negative side, cycle-accurate models require significantly larger storage than cumulative ones.

13.4.1.4 MODEL CONSTRUCTION

Concerning model construction, an RTL power model can be built *top-down* (or *analytical*) or *bottom-down* (or *empirical*) [3].

Top-down approaches relate activity and capacitance of an RTL component to the model parameters through a closed formula. The term "top-down" refers to the fact that the model is derived directly from the microarchitectural description and is not based on lower-level information. For this reason, such formula usually has a *physical interpretation*. Analytical models are particularly useful either when dealing with a newly designed circuit for which no information of previous implementations is available or when the implementation of the circuit follows some predictable template, which can be exploited to force some specific relation between the model parameters. A memory is a typical example of an entity for which an analytical model is suitable: its internal organization is quite fixed (cell array, bit and wordlines, decoders, MUXes, and sense amps), thus allowing accurate modeling based on *internal* parameters [17,18].

If we exclude these special cases, however, top-down models are not very accurate, since their links to the implementation (e.g., technology and synthesis constraints) are quite weak.

Bottom-up approaches, conversely, are based on "measuring" the power consumption of existing implementations, from which the actual power model is derived. Typically, the template of the power model (i.e., the parameters and a set of coefficients used to weigh the parameters) is defined up front; statistical techniques are then used to fit the model template to the measure of power values. This approach is known as *macromodeling* and has proved to be a very accurate and robust methodology for RTL power estimation and can be considered the state-of-the-art solution. Section 13.4.2 will be devoted to the detailed description of the macromodeling flow (Figure 13.2).

13.4.1.5 MODEL STORAGE

The issue of model storage is concerned with the *shape* of the model. Since models express a mathematical relation between power and a set of parameters, the problem amounts to that of representing such a relation. The two options are to store it (1) as an equation or

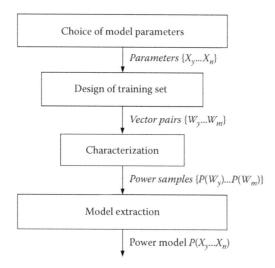

FIGURE 13.2 Macromodeling flow.

(2) as a lookup table, corresponding to the choice of representing a relation as a continuous function (equation-based models), or a discrete function approximated by points (lookup table–based models). These two types of models differ in their storage requirements and *robustness*, that is, the sensitivity of the model to the conditions (i.e., the experiments) used for its construction. In that sense, robustness is an issue only for empirical models. Concerning storage requirements, equation-based models are much more compact than table-based ones. In general, an equation will only require the storage of the coefficients of the model, as opposed to a full table. In addition, the accuracy of a table-based model is proportional to its size (the denser the table, the higher the accuracy), whereas the accuracy of an equation-based model is independent of the model size.

13.4.2 RTL POWER MODELS

In this section, we will review the most relevant results on power modeling of datapath units, controller, and wires. On the other hand, as already mentioned in Section 13.2, we will not consider components such as memories and communication buses.

13.4.2.1 POWER MODELING OF DATAPATH UNITS

There are three main reasons for which accurate datapath power estimates must resort to empirical models. First, because of the variety of available datapath units, a distinct analytical model for each type of component would be required. Second, analytical models, because of their functional meaning, would not be able to capture differences due to various implementation styles of a given component (e.g., a ripple-carry vs. a carry-lookahead adder). Third, accuracy of analytical models would be too low for units that have poorly regular structures (e.g., dividers or floating-point operators).

Besides coping with these issues, empirical models facilitate a single-model template that can be used for all types of datapath units. Empirical power models are commonly called "macromodels." The term has been borrowed from the statistical domain to denote the fact that such models have a "coarse" level of detail and are used to relate quantities pertaining to different abstraction levels, such as RTL parameters, to the actual measured power. We thus talk of "macromodeling" when defining the process of building an empirical power model for a generic component.

Before reviewing the most important results in datapath macromodeling, it is worth summarizing the flow that is normally used for building such macromodels.

13.4.2.2 MACROMODELING FLOW

Macromodeling involves the following four major steps:

1. *Choice of model parameters*: Although this step applies to nonempirical models also, it is especially important for macromodels, since it defines the parameter space and it affects the complexity of the following steps.

2. *Design of the training set*: The training set is a representative subset of the set of all possible input vector pairs that will be used to determine the model. Key points in the choice of the training set are its size (i.e., the total number of vector pairs) and its statistical distribution. The former is responsible for the simulation time, while the latter impacts accuracy; a bad distribution of the training set may offset the advantage of a large number of vector pairs. What defines a *good* distribution depends on the chosen parameters. A requirement for the training set is that it should span the domain of all the model parameters as much as possible. When one or more domains are not sufficiently covered by the training set, we say that the model is *insufficiently trained*. For instance, if the parameter of the model is the switching activity, the choice of random patterns as a training set would not be a good one, since only a very small portion of the activity domain (the one around a switching activity of 0.5) would be exercised.

3. *Characterization*: This step consists of using the training set to generate a set of points in the (*power, parameters*) space. For each element in the training set (a vector pair), a corresponding value of power is obtained by means of a low-level power simulator (gate or transistor level).

4. *Model extraction*: This step consists of deriving the model from the set of measurements obtained in the previous step. The actual calculation depends on how the model is stored. For equation-based models, a least-mean-square (LMS) regression engine is typically applied to the sample measurements; if a lookup table is used, extraction consists of collecting the power values for each of the discrete points of the parameters space.

13.4.2.3 MACROMODELS FOR DATAPATH UNITS

The literature on power macromodeling for datapath units is quite rich, and it includes solutions that cover all points of the power model space discussed in Section 13.4.1, with different accuracy/effort trade-offs. In spite of the vast amount of available material, it is quite easy to identify the key results in the power macromodeling domain.

The technique described by Powell and Chau [19] can be considered as the first proposal of a power macromodel for datapath units; it is actually just a capacitance macromodel, because it assumes fixed activity, that is, no activity parameters are included. Landman and Rabaey [20] recognized that modeling activity is essential, by observing that even "random" data do not exhibit true randomness on all bits, because of sign extension bits due to reduced dynamic ranges of typical values. Their model, called "dual-byte type" (DBT), separates data bits in two regions (sign and random), with distinct activity parameters. Since each unit may have different combinations of input and output regions depending on their functionalities, each unit must have a distinct model.

Gupta and Najm [21] proposed a power macromodel that can be considered as the state of the art; it consists of a 3D lookup table of power coefficients, whose dimensions are P_{in}, the average input probability; D_{in}, the average input *transition density*; and D_{out}, the average output transition density. Characterization is based on quantizing each dimension into equally spaced intervals.

Figure 13.3 shows the three dimensions of the lookup table in the case of a quantization interval of 0.2. Notice how the relation between D_{in} and P_{in} constrains the number of feasible points in the (D_{in}, P_{in}) parameters' space.

Under this model, characterization involves the generation of a number of input streams for each feasible value of parameters D_{in} and P_{in}. Values of parameter D_{out}, instead, are extracted from the resulting output stream (obtained through simulation) and are discretized using a binning technique.

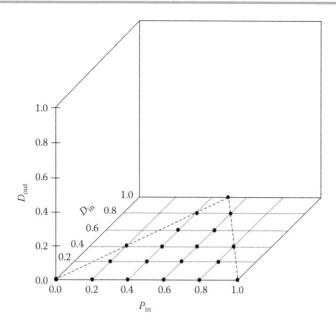

FIGURE 13.3 Three-dimensional lookup table macromodel.

The 3D lookup table model provides two main advantages with respect to the DBT model (or its variants). First, the use of a lookup table with discretized dimensions significantly improves the robustness of the estimates. Second, since parameters are normalized values, the same model template can apply to any datapath unit, without the need of (1) customizing it to its functionality and (2) including an explicit *capacitance* (i.e., complexity) parameter (thus, without the need of being parameterized).

Many macromodels published in the literature build around this 3D table-based model, either by improving some specific aspects that help to increase accuracy (e.g., interpolation schemes and local accuracy improvements) [22–25] or by generalizing it into a parameterized model (i.e., where the bit width of the operators becomes a parameter, as in the DBT model) [26].

13.4.2.4 POWER MODELING OF CONTROLLERS

Control logic typically has a smaller impact than the datapath; however, its dissipation is not negligible for several reasons. First, some designs are control intensive and may contain many interacting controllers whose impact on power may be sizable. Second, datapath units are often hand crafted while controllers are synthesized; then, they use silicon area less efficiently than datapath. Third, controllers are usually active for a large fraction of the operation time, while the datapath can be partially or completely idle.

What makes controller power modeling more challenging than datapath units is the fact that controllers are usually specified in a very abstract fashion even at the RT level and they are specific to each design. In contrast, datapath units are usually instantiated from a library that can be precharacterized once and for all. Some controller implementation styles employ regular structures, like PLAs or ROMs, which can be precharacterized with good accuracy. On the other hand, the most common choice for controllers is to synthesize them as sparse logic. Because of its irregular structure, this makes power modeling considerably harder.

One additional difficulty lies in the fact that the description of controllers is more abstract than that of datapath components because states are usually specified symbolically. Thus, controller synthesis goes through two steps, namely, state assignment and logic synthesis.

Power modeling of controllers at the microarchitectural level can be classified into *prestate assignment* [27,28] and *poststate assignment* [4,6,29] depending on the target description of the controller. The former is generally less accurate and can be used to provide upper and lower bounds on power consumption. The latter tries to give actual estimates of average power. Strictly speaking, however, RTL descriptions of controllers typically fall in the former category.

Due to the poor link between an abstract, state-based description and the actual implementation, power models for controllers that have some practical value are empirical. The analytical model presented in [27] provides in fact theoretical bounds for the switching in finite-state machines (FSMs), but the correlation to actual power figures is intrinsically very weak. For controllers, macromodeling is thus more than just an option. The difference with respect to datapath components, however, is that the model template is not so intuitive, because it is difficult to estimate the correlation between parameters and the actual power consumption. Therefore, the choice of the parameters is even more critical than in the case of datapath units. Possible parameters for controllers include

- *Static behavioral* parameters (i.e., parameters that can be obtained from the behavioral specification): the number of inputs, the number of outputs, and the number of states
- *Dynamic behavioral* parameters (i.e., parameters that can be obtained from functional simulation): the average input and output signal and transition probabilities
- *Static structural* parameters (i.e., parameters that are available after state encoding): the number of state variables
- *Dynamic structural* parameters (i.e., parameters that can be obtained from RTL simulation): the average transition probability of the state variables

A framework for the exploration of all possible models of order M with up to N terms is described in [30]; it defines a third-order model with four terms that yields an average relative error of about 30% and relative standard deviation of the error of about the same magnitude. The model uses only a subset of the aforementioned parameters, but from all of the four categories. This model has been shown to be far more accurate and robust than other models, including the *intuitive* model, that is, the one that uses a linear combination of switching activities (for inputs, outputs, and state variables) weighted by their corresponding cardinalities (number of inputs, outputs, and state variables).

13.4.2.5 POWER MODELING OF MEMORIES

As for any other component, memory power can be fitted into the template of Equation 13.2:

$$P_{\mathrm{mem}} = \frac{1}{2}\, C_{\mathrm{mem}} V_{\mathrm{dd}}^2 f_{\mathrm{Ck}} E_{\mathrm{Sw}},$$

where C_{mem} denotes the capacitance that is switched on a memory access. Since the capacitance for a read access is, in general, different from that for a write access, C_{mem} should be considered as an average capacitance.

For this reason, it is more accurate to resort to a cycle-accurate model and express average power as follows:

$$(13.5) \qquad P_{\mathrm{mem}} = \frac{1}{2}\, \frac{1}{N_{\mathrm{cycles}}}\, (N_{\mathrm{read}} C_{\mathrm{read}} + N_{\mathrm{write}} C_{\mathrm{write}}) f_{Ck} V_{\mathrm{dd}}^2,$$

where
  N_{cycles} is the total number of cycles
  N_{read} (N_{write}) is the number of read (write) accesses
  C_{read} (C_{write}) is the corresponding capacitances

Memories and in particular SRAM arrays, unlike generic RTL blocks, lend themselves to a relatively easy modeling of capacitance, owing to their well-defined internal structure, which is characterized by high regularity. In other words, it is feasible to adopt an analytical model that expresses the total memory power as the sum of the various components: read/write circuitry, decoders, cells, bitlines, wordlines, MUXes, and sense amplifiers [17,31,32]. These models are very accurate but suffer from being strongly dependent on technology; in fact, they require the knowledge of parameters such as the capacitance of the bitline or the wordline or the capacitance of a minimum-sized transistor, which cannot always be easily accessible.

A simpler, yet effective model for microarchitectural estimation is an empirical one, where *all* model parameters are chosen in order to be easily available at a higher level of abstraction (e.g., the number of words) and the relation to the technology is established in a characterization run. For example, a typical capacitance model can be expressed as

$$C_{mem} = a + bW + cN + dW N,$$

where
  W is the number of rows
  N is the number of columns of the memory array [4]

A similar model that also included the number of words as parameters was used in [33]. The characterization proceeds very similarly to the conventional macromodeling flow, in which the values of a, b, c, and d are determined by means of LMS regression. If we stick to the template of Equation 13.5, separate characterizations for read and write capacitances are required.

13.4.2.6 POWER MODELING OF WIRES

If we assume that the power consumption of datapath units and controller is based on empirical macromodels, their power estimates include the contribution of (internal) wires. There are then two categories of wires that need a custom power model: *global* wires, such as reset and clock, and *intercomponent* wires, such as those connecting RTL blocks.

For both types of wires similar considerations apply: in fact, power is consumed by wires when charging and discharging the corresponding capacitances. Therefore, the model of Equation 13.3, where A_i and C_i are the switching probability and the parasitic capacitance of the ith net, respectively, well fits wires also. Since factor A_i is available for both global wires and intercomponent wires, the problem of modeling wiring power reduces to that of modeling wiring capacitance, or, equivalently, wiring length. Wiring capacitances are unknown at the RTL, but realistic estimates can be obtained based on structural information, area estimates, and low-level wiring models. The problem of estimating the total wire length at high levels of abstraction is quite well understood; it is based on variants of Rent's rule [34–36], which relate the length of the interconnect to macroscopic parameters that can be easily inferred from a high-level specification, such as the number of I/O pins.

The main difference between the power models for global wires and intercomponent wires lies in their *context*. Area-based estimates work reasonably well for global wires, since they span the entire design. As a matter of fact, typical power models for global wires directly relate wire length to a power k of design area (often $k = 0.5$) and the latter to a power q of the number of I/O pins [20].

For intercomponent wires, conversely, their length is more weakly correlated to the entire design area. The solution adopted is very similar, yet on a smaller scale—that is, rather than referring to the entire design, each wire refers to the components it connects. In this case, wire length is made proportional to a power of the component's area, which is related to a power of the number of I/O pins of the component [37].

Figure 13.4 shows the conceptual topology of an intercomponent wire. The output capacitance C_{out} of a component is implicitly accounted for by its power model. Conversely, the input

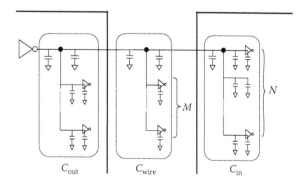

FIGURE 13.4 Hierarchical topology of an intercomponent wire.

capacitance C_{in} is not included in the power model, since power estimates provided by low-level simulation represent the power drawn from the supply net, while input capacitors are directly charged by the primary input lines. The total wire capacitance should thus take into account the fanout of a wire (known at the RTL), both externally (C_{wire}) and internally (C_{in}), by splitting each contribution into a fixed term (the "stem" capacitance) and a quantity proportional to the number of fanouts (the "branch" capacitance).

It is worth emphasizing that in ultra-deep-submicron technologies, where wires dominate chip area, the estimation of wire length requires some early floorplanning information in order to be reliable. Such information can then be back-annotated into the RTL description to guide wire length estimation based on the empirical models discussed earlier.

13.4.2.7 POWER MODELING OF BUSES

At the microarchitectural level, buses are relatively straightforward entities consisting of a set of wires. However, we believe they deserve a separate treatment from the generic wires we have considered in the previous section, since they are usually seen as a single-interconnection resource that will eventually be routed together as *a set* of wires; this physical *grouping* of bus wires is also reflected by typical HDL descriptions, where buses are represented as arrays of signals.

Buses may be shared (i.e., they may not just be point-to-point connections) and, mostly, are single-master buses relying on a simple protocol that defines the physical (i.e., the signaling scheme) and the *data-link* (i.e., the binary representation of the transmitted *values*) layers. More complex on-chip bus architectures that include features such as support for multimaster and complex protocols (i.e., those required by modern multiprocessor SoC architectures) tend to be categorized as hardware blocks rather than as a wiring infrastructure, because of the high amount of hardware control they require [38]. This trend is witnessed by the wide availability of synthesizable IP blocks that can be modularly used by RTL designers and automatically synthesized (e.g., Synopsys' DesignWare IPs for AMBA buses [39]). In this section, we will focus on single-master buses that are typical in non-core-based designs.

Specifically, with reference to the template of Figure 13.1, buses are the resources that are used to connect blocks of two types (possibly in a shared manner): computational units and memories.

A conceptual model of the average power consumed by a bus consisting of n lines can be obtained by simply summing up the contribution of each wire, according to the model of Equation 13.2:

$$P_{bus} = \frac{1}{2} \sum_{i=1}^{n} (C_L V_{dd}^2 f_{Ck} E_{Sw}^i),$$

where
C_L is, in this case, the capacitance of a single bus line
E_{Sw}^i is the switching activity of the ith bus line

The equation is based on the assumption that all lines have roughly the same capacitance. This model would be reasonably accurate if the bus lines were routed independently of each other, possibly on different metal layers. Conversely, the *grouped* nature of buses makes the model unrealistic because it considers each wire as if it were isolated from the others, thus completely ignoring the contribution due to *coupling capacitances*, that is, the mutual capacitive effect of two neighboring wires.

Figure 13.5 shows a simplified view of the capacitances switched between two adjacent wires. C_l is the ground capacitance considered in the model (also called *self*-capacitance), while C_C denotes the coupling capacitance. In deep-submicron technologies, the magnitude of C_C far exceeds the self-capacitance: Electrical-level simulations for 0.13 mm technologies have shown that C_C is more than three times the value of C_l [40] and that this factor will further increase in future technologies.

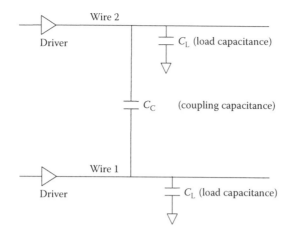

FIGURE 13.5 Capacitances involved in bus line switching.

In view of the aforementioned discussion, accounting for the contribution of C_C in on-chip buses is mandatory; this can be done by augmenting the basic power model as follows [41]:

$$(13.6) \qquad P_{\text{bus}} = \frac{1}{2} \left(\sum_{i=1}^{n} E_{\text{Sw}}^i C_L + E_{\text{Sw}}^i C_C \right) f_{\text{Ck}} V_{\text{dd}}^2,$$

where E_{Sw}^i denotes the *coupling switching activity*, that is, a quantity related to the simultaneous switching of two adjacent lines. In fact, when the transitions on two adjacent lines, a and b, are aligned in time, there are only two transition pairs that cause C_C to switch: (1) when both a and b switch to different final values and (2) when one of the two lines switches, while the other one does not, and their final values are different.

Table 13.1 shows the normalized switched capacitance for a two-line bus, when all capacitive effects are considered. The ratio C_c/C_l is represented by l. In the table, only $0 \rightarrow 1$ transitions are counted as power-dissipating transitions on C_l.

The table clearly shows that increasingly larger values of l will tend to emphasize the importance of the energy due to switching of the coupling capacitance.

13.4.3 POWER ESTIMATION

The modeling technology discussed in the previous sections can be successfully used to enhance state-of-the-art RTL-to-physical design flows with power estimation capabilities.

Assuming that the design to be estimated is described by an FSMD, as defined in Section 13.3, the estimation procedure consists of three basic steps [8], as shown in Figure 13.6.

The first operation implies identifying and separating the datapath components from each other and from the FSM that represents the control. This is needed in order to enable the power estimator to generate the power models for each component in the FSMD.

TABLE 13.1 Normalized Switched Capacitance on Two Adjacent Bus Lines

		(b^t, b^{t+1})			
		$0 \rightarrow 0$	$0 \rightarrow 1$	$1 \rightarrow 0$	$1 \rightarrow 1$
(a^t, a^{t+1})	$0 \rightarrow 0$	0	$1 + \lambda$	0	0
	$0 \rightarrow 1$	$1 + \lambda$	2	$1 + 2\lambda$	1
	$1 \rightarrow 0$	0	$1 + 2\lambda$	0	λ
	$1 \rightarrow 1$	0	1	λ	0

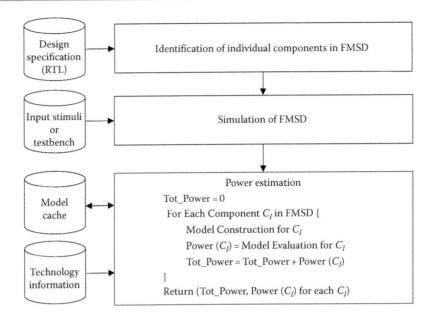

FIGURE 13.6 Microarchitectural power estimation flow.

The FSMD then needs to be simulated. To this purpose, an RTL simulator is used to trace all the internal signals that define the boundaries between the various components in the FSMD. This is of fundamental importance for the evaluation of the power models that is needed to complete the estimation procedure.

Finally, the actual power estimation must take place. The design hierarchy is traversed at the top level, from the inputs toward the outputs, and for each component, model construction and model evaluation are carried out. Model construction entails building the proper model according to the type of component being considered. A caching strategy may be used to limit the number of times models are built; more specifically, after the model for a given component is built, it is stored in a cache, so that it can possibly be reused at later times. Model evaluation, on the other hand, requires that the parameter values obtained during the RTL simulation phase are plugged into each model to get the actual power values.

The total power information for the design is then obtained by summing up the contribution of the model of each component; both a total power budget and a power breakdown can thus be reported to the user.

13.5 MICROARCHITECTURAL POWER OPTIMIZATION

This section presents some of the most popular microarchitectural power optimization techniques used in designs today. We first deal with datapath components and controllers; in this case, we classify the existing solutions into three categories:

1. Those based on clock gating, whose objective is to stop the clock for some cycles of operation in order to achieve a reduction of the switching component of the power.
2. Those based on exploitation of common-case computation, whose objective is to optimize switching power consumption for the most common operation conditions.
3. Those based on dynamic management of threshold and supply voltages, whose goal is that of reducing either leakage or switching power or both by operating the logic in a multi-V_{th}/multi-V_{dd} regime.

For each class of approaches, we discuss the basic idea as well as the algorithms used and challenges faced in applying these techniques in automated flows. We then move to techniques applicable to memories and buses. Owing to the peculiarity of these components, their optimization is treated separately toward the end of this section.

13.5.1 CLOCK GATING

Clock gating dynamically shuts off the clock to portions of a design that are *idle* for some cycles. The theory of this technique has been investigated extensively in the past [42–44], and clock gating is now considered the most successful and widely adopted solution for switching power reduction in real designs [45–48].

In some cases, it may be possible to shut off the clock to an entire block in the design, thereby saving large amounts of power when the block is not functioning. Perfect cases for this are when a block is used only for a specific mode of operation, for example, the receiver and transmitter parts in a transceiver may not be active at the same time, and the receiver can be shut off during transmit stages, or vice versa.

It is also possible to gate the clock of a single register or set of registers. For instance, synchronous, load-enabled (LE) registers are usually implemented using a clocked D-type flip-flop and a recirculating multiplexor, with the flip-flop being clocked at every cycle as shown in Figure 13.7a.

In the gated clock version of Figure 13.7b, the register does not get the clock signal in the cycles when no new data are loaded, therefore reducing switching power. Savings are further enhanced by the removal of the multiplexor. Gating a single-bit register, however, has the associated penalty of power consumption in the clock-gating logic. The key is then to amortize such a penalty over a large number of registers, by saving the flip-flop clocking power and the multiplexor power of all of them using a single clock-gating circuit.

The transparent latch is used to guarantee that spurious glitches in the enable signal occurring when the clock is high are not propagated to the clock input of the register. The latch freezes the output at the rising edge of the clock and ensures that the new enable signal, EN1, at the AND gate is stable when the clock is high. Meanwhile, the enable signal can time-borrow from the latch, so that it has the entire clock period available to propagate.

We cover advanced methods for detecting clock-gating opportunities and present issues associated with implementing clock gating in a typical automatic flow, along with possible solutions in this section.

13.5.1.1 ADVANCED CLOCK GATING

The clock-gating conditions that depend on the enable signal of the register bank, which can be identified by means of topological inspection and analysis of the RTL description, can be extended by considering the functional behavior of the circuit. In particular, it is possible to

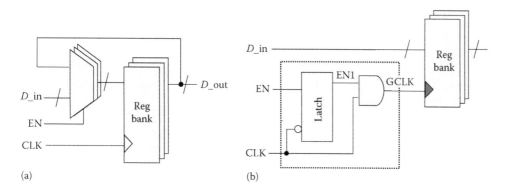

(a)   (b)

FIGURE 13.7 Clock gating: (a) load-enabled register implementation and (b) gated clock implementation.

augment the opportunities for stopping the clock that feeds a given register anytime the output of such a register is not observable, that is, the value of the output is not used by the gates in the register's fanout cone. Calculating these additional options for clock gating entails finding the *observability don't care* (ODC) conditions for the output of the register; Babighian et al. [49] propose to extract idle conditions from a RTL netlist, by focusing on control signals that drive steering modules (e.g., multiplexors, tristate buffers, and LE registers). Referring, for instance, to a multiplexor, all but one data inputs are unobservable at any given time depending on the value(s) that are carried out by the control signal(s), so that the logic values of the unobservable branches become irrelevant to the correct operation of the circuit. Hence, the clock signals of the registers in the fanin cone of an unobservable multiplexor branch can be gated without compromising the functionality.

The goal of the approach of Babighian et al. [49] is to improve the effectiveness of clock gating by creating an activation function that can stop the clock of a set of registers for a significant fraction of cycles when the register outputs are unobservable. This is done in two steps: first, by performing ODC computation, which is based on a backward traversal of the datapath in decreasing topological order, considering only ODCs created by steering modules, and second, by generating the activation function, which implies the synthesis of the ODC function and the addition of clock-gating logic to the RTL netlist. Even though ODC computation is simple for a single-steering module, ODC expressions can become quite large if the netlist has many levels of steering modules and many fanout points. Therefore, traversal of only a limited number of levels in the netlist is allowed.

Once the ODC expression is computed, the corresponding logic must be instantiated in order to drive properly the clock-gating logic. The main difficulty in this step is due to the fact that ODC conditions masking register in clock cycle T may be used to gate their clock in cycle T–1. In other words, the clock-gating logic may need to be active in the clock cycle immediately before the register becomes unobservable. Unfortunately, the control signals at the inputs of the ODC functions are generated one clock cycle too late.

If the control signals are available directly as outputs of the registers, the instantiation of the clock-gating logic is relatively straightforward. Logic gates implementing the ODC expressions are inserted and their inputs are connected to the inputs of the registers. In real-life designs, however, the control inputs of the steering modules seldom come directly from the registers; instead, they are often generated by additional logic. In this case, the entire cone of logic between registers and control signals should be duplicated and connected at the inputs of the registers, and ODC computation gates should then be connected at the outputs of the duplicated cones. Clearly, the addition of this extra logic may represent a nonnegligible overhead.

Most of today's commercial synthesis tools deal with this issue by restricting the type of activation function used to gate the clock. In practice, they just detect ODCs generated locally to the registers by assuming the register's outputs to be always observable by the environment. This corresponds to considering, as ODC function, the complement of the enable signal, which feeds the clock-gating logic.

By detecting clock-gating conditions only when enable signals are present, no precomputation of clock-gating conditions is required and thus no duplication is needed. Notice that ODC-based clock-gating subsumes traditional automatic clock gating as a very special limit case (i.e., backward traversal is completely avoided). Thus, the clock-gating conditions computed by the ODC-based approach are guaranteed to be a superset of those targeted by tools that introduce clock gating only for LE registers (see Figure 13.7).

13.5.1.2 CLOCK-SKEW ISSUES

A problem in latch-based architectures comes from the fact that clock skew between the latch and the AND gate can result in glitches at the gated clock output. This is explained in Figure 13.8. In particular, Figure 13.8a shows the case when the clock arrives much earlier at the AND gate than at the latch. Here, the clock skew between the latch and the AND gate should be less than the clock-to-output delay of the latch for the circuit to function properly. Figure 13.8b depicts the situation when the clock arrives earlier at the latch. In this case, the clock skew between the AND gate and the latch should be less than the sum of the setup time

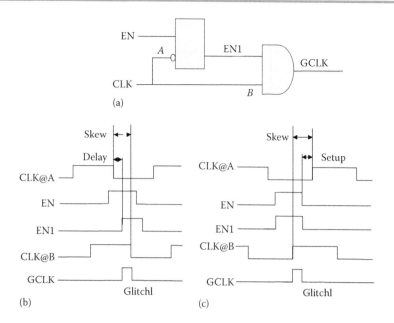

FIGURE 13.8 Clock skew within clock-gating logic: (a) clock-gating logic, (b) glitches due to positive clock skew between latch and AND gate, and (c) glitches due to negative clock skew between latch and AND gate.

of the latch, and the input-to-output delay of the latch, to function properly. Therefore, the clock skew between the latch and the AND gate, C_s, should be carefully controlled according to the following equation:

$$- (s + d_{in}) < C_s < d_{clk},$$

where
  s is the setup time of the latch
  d_{in} the input-to-output delay of the latch
  C_s the difference in clock arrival time between the latch and the AND gate (the clock arrival time at the AND gate minus the clock arrival time at the latch), and
  d_{clk} the clock-to-output delay of the latch.

Depending on the relative placement of the latch and the AND gate, these requirements may pose very stringent constraints on the clock-tree synthesis (CTS) tool.

The best way to control the relative timing of the two clock signals is to keep the entire structure in a single cell, called the "integrated clock-gating" (ICG) cell. The cell should be designed specifically for clock gating, with the explicit requirements discussed earlier. Most technology libraries today do include the ICG cell as part of their primitives. Another way to address this issue is to ensure that the latch and the AND gate are close to each other during the placement phase of the design by placing hard constraints on the distance between them. This makes it simpler for CTS tools to reduce the clock skew between them during the clock routing phase.

13.5.1.3 CLOCK LATENCY ISSUES

To maximize power savings, a single clock-gating cell may be used to gate several registers, if the activation function is common to all of them. But the gated clock may not have enough strength to drive all these registers, calling for a clock tree at its output. If a clock tree is introduced between the ICG cell and the registers it controls, the clock signal at the gating logic arrives much before the clock signal at the registers, and the activation signal must be ready before the clock arrives at the gating logic. This applies strict timing constraints on the activation signal, which must be addressed during synthesis.

13.5.1.4 CLOCK-TREE SYNTHESIS

In the presence of massively gated clocks, CTS tools must automatically address the presence of gated clocks, both combinational and sequential, and also the ICG cell on the clock line. CTS tools must support different relative latency requirements at different points in the clock tree, since the clock latency at the gated clocks can be very different from the latency at the registers. If ICG cell is not used, the CTS tool would need to provide stringent control of clock skew between the latch and the AND gate.

13.5.1.5 PHYSICAL CLOCK GATING

Physical clock gating simultaneously takes into account the factors mentioned in the previous subsections, namely, skew, latency, and CTS issues. There exists a spectrum of clock-gating approaches with regard to the placement of clock-gating cells into a clock tree. Designers often opt to place the clock-gating cells as close as possible to the final placement of their corresponding registers, as shown in Figure 13.9a. This placement can be enforced during physical synthesis by specifying a bound for the proximity of the clock-gating cells to the registers. Some advantages of this approach are that it makes it easier to estimate the latency from the clock-gating cell and it also increases the amount of available slack for the arrival of the activation signal. The impact on the clock tree is fairly minimal since the clock-gating cells are placed close to the registers and can eliminate the need for buffer insertion after clock-gating cell insertion. A disadvantage to this approach is that it leaves most of the clock-tree switching even when branches are leading to registers that will have the clock blocked by a gated clock. In order to save as much power as possible, it is desirable to gate as many buffers on the clock tree as possible. This is difficult for the designer to do without knowledge of the actual physically induced timing constraints.

In a physically aware clock-gating system, CTS works in conjunction with placement and clock gating to determine an optimized placement and insertion of the clock-gating cells into the clock tree. This information is used to balance the delay on the activation signal with the amount of potential power saved by placing the clock-gating cell closer to the root of the clock tree, as shown in Figure 13.9b.

13.5.1.6 CLOCK-TREE PLANNING

Cell placement, clock-gating cell insertion, and CTS can also be decoupled, provided that the process of clock-tree planning is started after cell placement and it does not interfere with clock routing. The objective of the approach in [50,51] is to build a power-optimal gated clock-tree structure fully compatible with state-of-the-art physical design tools to perform detailed clock routing and buffering. Thus, the output of the proposed clock-tree planning methodology is not a completely routed clock tree; instead, it is a clock netlist (including clock-gating cells and related control logic) and constraints that, provided as input to CTS tools, lead to a low-power gated clock tree, while still accounting for all non-power-related requirements (e.g., controlled skew and low crosstalk-induced noise).

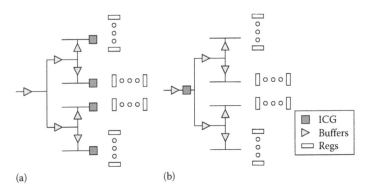

(a)   (b)

FIGURE 13.9   Integrated clock-gating cell placement: (a) Close to registers. (b) Buffering after ICG cell insertion.

The methodology consists of three steps:

1. Calculation of the clock-gating activation functions
2. Generation of the clock-tree logical topology
3. Instantiation and propagation of the clock-gating cells and related logic

The gated clock activation functions for all the RTL modules are computed first. Next, according to the activation functions and the physical position of the registers, the logical topology of the clock tree is planned. This entails balancing the reduction in clock-switching activity against clock and activation function capacitive loads. Clock-gating cells are then inserted into the clock-tree topology and propagated upward in the tree whenever this is convenient, thus balancing the clock power consumption against the power of the gated clock subtree. The information about the gated clock tree is finally passed to the back-end portion of the flow, which will take care of clock-tree routing and buffering.

13.5.1.7 TESTABILITY ISSUES

Clock gating reduces test coverage of the circuit since gated clock registers are not clocked unless the activation signal is high. During test or scan modes, test vectors need to be loaded into the registers, and hence they must be clocked irrespective of the value of the activation signal. One way to address this, investigated from the theoretical standpoint in [52], is to include a control point or control gate at the activation signal, as shown in Figure 13.10a. This allows clock-gating signal, EN, to be overridden during the scanning in or out of vectors by the test-mode signal. In this way, during the test clock cycles, the clock signal is not gated by the activation signal, EN, and the register can be tested to see if it holds the correct state. Further, the test-mode signal is held at logic "1" during test mode, making any stuck-at fault on the activation signal unobservable. If full observability is required, this signal must be explicitly made observable by tapping it into an observability exclusive or (XOR) tree, as shown in Figure 13.10b.

A growing concern around scan-based testing is the power consumed during the scanning in and out of test vectors [53,54]. The changing register values during scanning can create activity levels that are much higher than those experienced during *normal* operation and lead to *good* chips failing during testing.

13.5.2 EXPLOITATION OF COMMON-CASE COMPUTATION

It is well known that in complex digital architectures, some functionalities are exercised far more than others; this is due, mainly, to the fact that the data to be processed (i.e., the workload) may not have an equiprobable distribution. Optimizing the common-case computation has thus

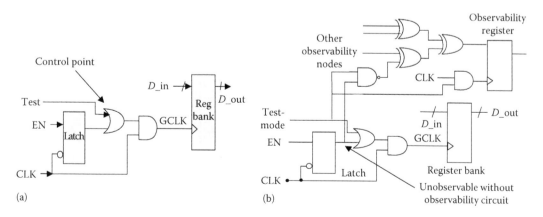

FIGURE 13.10 Testability issues with clock gating: (a) adding controllability and (b) improving observability.

become an established practice in high-throughput design [55], where variable-latency units replace fixed-latency ones to improve the overall system performance by adapting the latency of the datapath to the length of the computation to be executed. Architectural retiming [56], speculative completion [57], and telescopic units [58,59] are examples of application, even in an automatic fashion, of the optimization paradigm based on exploitation of common-case computation.

The concept of common-case computation has been extended recently to the case of power minimization; more specifically, as switching power depends on what the units of a design are, which are activated by the input data as well as on the type of data to be processed, it may be possible to come up with variable-power architectural solutions that guarantee minimum power demand for the most probable execution conditions [60]. Figure 13.11 shows a possible architecture implementing the common-case optimization approach.

Block A supports the full functionality of the design, while block B only covers the most common subset of it. A and B work in mutual exclusion, owing to the latch-enabled registers placed on the primary inputs of the two blocks. Based on the next datum to be processed, block SEL is in charge of selecting whether A or B should be activated in the next clock cycle. As B is much smaller than A, any time B is active, power is reduced. On the other hand, when A has to compute, there may be a penalty in power, as the whole design is larger (thus, there are more gates that have to switch) if compared to block A alone. Let q be the probability of block B to be active; let P_A and P_B be the average power consumed by blocks A and B, respectively, when they are computing; and let P_{Ov} be the power overhead due to block SEL and the additional logic needed to make the architecture working. Actual power savings are achieved if

$$(1 - q)P_A + qP_B + P_{Ov} < P_a. \tag{13.7}$$

This tells us that the architecture exploiting the common-case computation obtains power reductions over block A alone for high values of probability q. Clearly, high q usually implies a larger SEL and a larger B (intuitively, what happens is that as q increases, some of the functionality of A is incorporated also into B). The challenge is thus that of designing the smallest possible SEL and B blocks that maximize the value of q and limit the impact on area and delay penalty.

In the following, we review some solutions proposed in the recent literature that can be considered as practical actuation, with a slightly different flavor, of the design framework discussed earlier.

13.5.2.1 OPERAND ISOLATION

The concept at the basis of *operand isolation* [61] is illustrated in Figure 13.12. In Figure 13.12a, we observe that the output of the multiplier is only used when the control signals to the multiplexors, SEL_0 and SEL_1, are both high. In cycles when either of the control signals is low, if the multiplier inputs change, the multiplier performs computation but its result is not used. The wasted

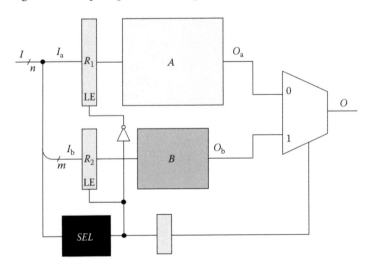

FIGURE 13.11   Exploitation of common-case computation.

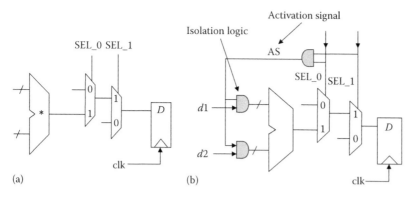

FIGURE 13.12 Operand isolation: (a) Original circuit. (b) After operand isolation.

power may be substantial if these idle cycles occur for long periods of time. Figure 13.12b shows operand isolation applied to the multiplier. First, the activation signal AS is created to detect the idle cycles of the multiplier. AS is high in the *active* clock cycles when the multiplier output is being used, otherwise low. This signal is used to isolate the multiplier by freezing its inputs during idle cycles using a set of gates, called "isolation logic." In Figure 13.12b, the isolation logic consists of AND gates, but OR gates or latches may also be used. Using AND/OR gates avoids the introduction of new sequential elements and reduces the impact on the rest of the flow. Also, AND/OR gates are cheaper and tend to give better power savings overall.

Operand isolation saves power by reducing switching in the operator being isolated, but it also introduces timing, area, and power overhead from the additional circuitry for the activation signal and the isolation logic. This overhead must be carefully evaluated against the power savings obtained to ensure a net power saving without too much delay or area penalty.

13.5.2.2 PRECOMPUTATION

Precomputation [62,63] relies on duplication of part of the logic with the purpose of precomputing the circuit output values one clock cycle before they are required and then using these values to reduce the total amount of switching in the circuit during the next clock cycle. Knowing the output values one clock cycle in advance allows the original logic to be turned off during the next time frame, thus eliminating any charging/discharging of internal capacitances.

The size of the logic that precalculates the output values must obviously be kept under control, since its contribution to the total power balance may offset the savings achieved by blocking the switching inside the original circuit. Several variants to the basic architecture can be adopted to take care of this problem. In particular, it may sometimes be convenient to resort to partial, rather than global, shutdown, that is, to select for precomputation only a (possibly small) subset of the circuit inputs.

As an example, consider Figure 13.13a; the combinational block, A, implements an N-input, single-output Boolean function, f, and it has the I/O pins connected to registers R_1 and R_2. A possible precomputation architecture is depicted in Figure 13.13b.

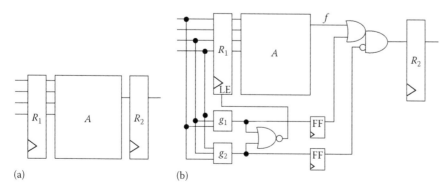

FIGURE 13.13 Precomputation: (a) Original circuit. (b) Precomputed architecture.

Key elements of the precomputation architecture are the two N-input, single-output predictor functions, called g_1 and g_0, whose behavior is required to satisfy the following constraints:

$$g_1 = 1 \Rightarrow f = 1$$

$$g_0 = 1 \Rightarrow f = 0$$

The consequence is that if at the present clock cycle either g_1 or g_0 evaluates to 1, the LE signal goes to 0, and the inputs to block A at the next clock cycle are forced to retain the current values. Hence, no gate output transitions inside block A occur, while the correct output value for the next time frame is provided by the two registers located on the outputs of g_1 and g_0.

As mentioned earlier, the choice of the predictor functions is a difficult task. Perfect prediction requires $g_1 \equiv f$ and $g_0 \equiv f9$. However, this solution would not give any advantage in terms of power consumption over the original circuit, since it would entail the triplication of block A, and thus it would cause the same number of switchings as before, but with an area three times as large as the original network. Consequently, the objective to be reached is the realization of two functions for which the probability of their logical sum (i.e., $g_1 + g_0$) being 1 is as high as possible, but for which the area penalty due to their implementations is very limited. Also, the delay of the implementation of g_1 and g_0 should be given some attention, since the prediction circuitry may be on the critical path, and therefore, it may impact the performance of the optimized design.

One way of guaranteeing functions g_1 and g_2 to be much less complex than function f, thus implying a marginal area overhead in addition to a remarkable power savings, consists of making the two predictor functions depend on a limited number of inputs as compared to f.

Precomputation-based power optimization has been shown to be effective in the case of designs with pipelined structure. On the contrary, it seems to be hardly applicable to the case of sequential circuits with feedback. The reason for this is that the precomputation functions never attempt to stop the present-state inputs, which represent the majority of the inputs to the combinational logic for sequential circuits with a realistic number of memory elements (i.e., flip-flops).

13.5.2.3 COMPUTATIONAL KERNEL EXTRACTION

It is known that, when in their steady state, complex sequential circuits tend to run through a limited set of states. Once such a set, called a "computational kernel" [64], is extracted from a given circuit specification, it can be successfully used for various types of optimization, including power minimization.

Given a sequential circuit with the traditional topology shown in Figure 13.14a, the paradigm proposed in [64] for improving its power dissipation is based on the architecture depicted in Figure 13.14b.

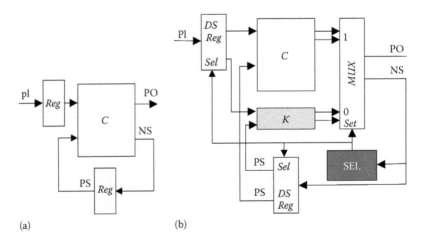

FIGURE 13.14 Computational kernel extraction: (a) Original circuit. (b) Kernel-based architecture.

The essential elements of the architecture are the following:

- The combinational portion of the original circuit (block C)
- The computational kernel (block K)
- The selector function (block SEL)
- The dual-state registers (block $DS\ Reg$)
- The output multiplexor (block MUX)

The computational kernel can be seen as a *dense* implementation of the circuit it has been extracted from. In other terms, K implements the core functions of the original circuit, and because of its reduced complexity, it usually implements such functions in a faster and less power consuming way.

The purpose of the selector function SEL is that of deciding what logic block, between C and K, will provide the output value and the next state in the following clock cycle. To take a decision, SEL examines the values of the next-state outputs at clock cycle n. If the output and next-state values in clock cycle $n + 1$ can be computed by the kernel K, then SEL generates the value 1. Otherwise, it generates the value 0. The output of block SEL is connected to the multiplexor that selects that block produces the output and the next state, as well as to the control input of the dual-state registers, which steer the primary and state inputs to the appropriate combinational block (either C or K). The optimized implementation is functionally equivalent to the original one.

The scheme in Figure 13.14b is just one among several possible architectures. The peculiar feature of this solution concerns the topology of the selection logic. In particular, the choice of having a selection function that only depends on the next-state outputs is dictated by the need of obtaining a small implementation. Reducing the size of the support of SEL, that is, not including the primary inputs, is one way of pursuing this objective.

Fundamental for a successful application of the kernel-based optimization paradigm is the procedure adopted for kernel extraction. If the circuit is described by means of its state transition graph, the kernel can be determined exactly through symbolic (i.e., binary decision diagram [BDD] based) procedures similar to those employed for FSM reachability analysis. On the other hand, when the state transition graph is too large to be managed, there are two options. If a gate-level description of the circuit is available, then block K can be iteratively synthesized by means of implication analysis followed by redundancy removal. Otherwise (i.e., if only a functional description of the circuit is available), kernel identification and synthesis can be performed by resorting to simulation of typical input traces and by then running probabilistic analysis and resynthesis of the state transition graph based on the results of the simulation. Obviously, in case approximate kernel extraction is adopted, the savings that can be obtained are usually more limited.

13.5.3 MANAGING VOLTAGES

As discussed in Section 13.2, power consumption in CMOS designs heavily depends both on the operating supply voltage (switching power) and on the transistors' threshold voltage (leakage power).

The two variables are not independent, and switching speed constitutes the link. Starting from the 0.5 mm technology node, supply voltage levels have been scaling at approximately 1 V/0.1 μm. As we are now below the 100 nm feature size, the reduced operating voltages are forcing threshold voltages down to 0.25 V and below, in order to preserve speed. This has had a major impact on the leakage current of the transistors built into these technologies. As Equation 13.8 shows, the subthreshold leakage current grows exponentially as the threshold voltage decreases:

$$(13.8) \qquad I_{\mathrm{sub}} = I_0(\mathrm{e}^{(-V\mathrm{th}/S)}(1 - \mathrm{e}^{-qV\mathrm{ds}/kT})) \ (\text{at } V_{\mathrm{gs}} = 0).$$

For approximately every 65–85 mV decrease in threshold voltage (depending on temperature), there is an order of magnitude increase in subthreshold leakage current.

13.5.3.1 MANAGING THRESHOLD VOLTAGES

Silicon foundries have started to offer multiple threshold devices at the same process node to address the need to control leakage current and enabling designers to trade leakage and performance [65]. Along with the standard V_{th}, a low- and high-V_{th} transistor may be offered where the low-V_{th} device may have an order of magnitude higher leakage than the standard V_{th} device and the high-V_{th} device may have an order of magnitude lower leakage than the standard. This reduction in leakage is not free though, and it comes at the expense of the speed of the device—there could be a 20% to 2x delay penalty between the standard and the high-V_{th} devices—and an increase in the cost of the fabrication process.

Synthesis algorithms can be used to optimize leakage by using high-threshold voltages while still meeting the timing requirements using low-V_{th} devices on the more critical paths. The threshold voltage of a transistor can be dynamically changed by varying its back bias. The change in V_{th} is roughly proportional to the square root of the back-bias voltage. As threshold voltages drop below 0.25 V, variable back biasing may gain more appeal. A distinct advantage of this approach is that during periods when heavy processing is needed, V_{th} can be reduced, thus speeding up the cells. When the cells are in a slower drowsy or idle mode, the threshold voltage can be raised, thus lowering the leakage. One significant impact of using variable back biasing is that there are two new terminals for each cell that now need to be routed. A common ASIC design practice is to create cells that tie the *n-well* regions to V_{dd} and the *p-well* regions to ground. In the physical implementation, these are simply predefined contacts designed into the cell that are connected as part of the power and ground routes. To enable back biasing, new voltage lines are routed to control the bias. These can be to individual cells or, more likely, to regions that contain multiple cells sharing the same *well* and a common tie-cell to control the *well bias* [66,67].

13.5.3.2 POWER GATING–BASED SUPPLY VOLTAGE MANAGEMENT

For applications like cell phones that have long shutdown periods, the power consumption is dominated by the leakage power consumption during off periods. *Power gating* addresses this issue by shutting off the power supply to blocks that are not in use. This not only eliminates the switching power dissipation in the block but also hugely saves the leakage power dissipation when the block is shut down. In the simplest case of power gating, the voltage level across the chip is the same, but different power supply grids are used in different parts of the chip. Besides both leakage and switching power savings, this helps to control the IR drop on the power grids. These design techniques create interesting switching on the chip. When one block is shut down, the voltage on its output ports may drift to undefined values causing large leakage and unexpected functionality in the gates that it drives. Therefore, it is necessary to isolate all the output ports of a block that is shut down. Further, if a block is shut down, it might be important for it to *remember* some of its previous state values. The registers used to store these values need to be specially designed to be powered by a secondary supply that will allow them to retain their values during shut down and restore it when the primary supply is up [68,69].

Figure 13.15 shows a typical floorplan of a power-gated chip with isolation cells and retention registers.

EDA tools need to insert automatically isolation cells and retention registers for powered-down blocks. To complete the entire flow, it is important to be able to represent all this information in the source description (the RTL) and infer both isolation and retention logic during RTL synthesis. Since the power-down behavior directly impacts the functionality of the chip, it is important to be able to verify this functionality both with simulation tools and formal verification tools.

To enable power gating in a complete automated design flow, place, route, and back-end optimization tools need to be aware of the different power supply regions. Cells powered by each supply need to be placed in separate areas, and signals should be routed so that they do not traverse areas serviced by a different power supply, since subsequent repeater/buffer insertion on these signals will not be legal.

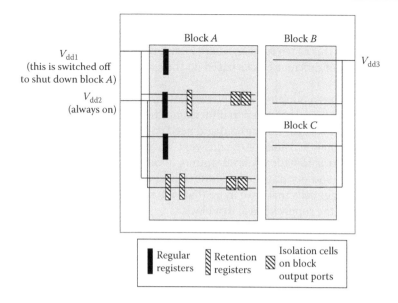

FIGURE 13.15   Applying power gating.

With complex power-gating schemes, the role of power and power network analysis tools becomes more important. Turning the voltage ON or OFF to a block can cause large transients on the power grid, affecting other blocks on the chip. These effects should be accounted for by the power analysis tools.

13.5.3.3 MULTI-V_{dd} SUPPLY VOLTAGE MANAGEMENT

Integrated circuit designs commonly support multiple voltages in one form or the other, and EDA tools need to support the various special needs of these design styles [70,71]. For instance, the power-gating technique discussed earlier can be thought of as the simplest version of multivoltage design, where the ON voltage level is the same for all the blocks, but can be changed to 0 V during shutdown.

A more complex methodology uses different voltages for different blocks on the chip, which are powered with separate power supplies that can be independently shut down [72]. An example of a multivoltage chip is shown in Figure 13.16.

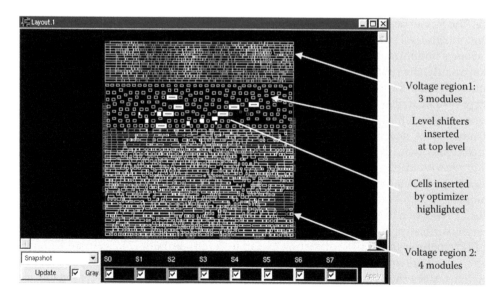

FIGURE 13.16   Example of multivoltage chip.

Variations of multivoltage design styles include dynamic V_{dd} scaling, where on-chip controls are used to switch the voltage on the different parts of the chip to different predefined levels. Adaptive scaling takes this a step further, where high-level control feedback loops can be used to change the V_{dd} values of different blocks based on the speed/power requirements for the chip at any given time.

Along with the issues associated with power gating, multivoltage design has other requirements from EDA tools. First, tools need to understand the impact of voltage levels on timing and power of library cells (hence, on the design) and need to provide both analysis and optimization capabilities that accurately account for the different V_{dd} levels. Second, the implementation tools should be able to insert automatically level shifters to be able to adapt signals from one level to another. Instantiation and optimization of level shifter cells is key to multivoltage design [73]. The tools have to manage cells that have more than one supply rail and circuitry that can vary or completely shut down the supply voltage to a block.

Clock-tree generators need to account for buffers that operate at different voltages to provide clock signals to each block, and the router needs to account for buffer placement in the context of different voltage regions on the chip. Routing a feed-through signal via a region may now require the insertion of level shifters in order to drive the signal adequately. Analysis tools need to understand these different situations—tracking all of these new voltage-based modes—and provide useful feedback to the designer.

An additional major impact on the design flow is the need to treat the supply line as another variable. For most previous mainstream designs, logical netlists only specified the input and output connections between gates. V_{dd} pins for all the cells (as also the V_{ss} pins) were attached to a single power (ground) network after the place and route step was done. In a multivoltage design, different cells are connected to different power (ground) networks and this information needs to be managed in the entire design flow starting from the RT level, to be able to simulate correctly and verify the design right from the start.

A final design implication that optimization and analysis tools must account for is the signal integrity impact of driving some lines at higher voltages than others. The higher-voltage lines can cause larger spikes in neighboring low-voltage lines than other lower-voltage aggressors, which impacts timing analysis, power, and the routing of lines on the chip.

13.5.4 MEMORY POWER OPTIMIZATION

The power model of Equation 13.5 exposes the two quantities that can be targeted to reduce memory power, namely, the number of (read or write) accesses and the (read or write) capacitance. The former are actually dictated by the scheduling of the operations resulting from RTL design or high-level synthesis and should be considered as fixed. Attention should then focus on capacitance minimization.

Apparently, the read or write capacitances are related to physical parameters and are also uniquely defined for a given technology. In practice, C_{read} and C_{write} should be regarded as *average* read or write capacitances, that is, they should represent the cost of an *average* read or write access. This subtle difference opens the way to a class of optimization techniques that modify the *organization* of the memory, which can be seen as the RTL counterpart of techniques that, in a more general macroarchitectural context, aim at optimizing the *memory hierarchy* of a system. The most well-known example is the use of a cache between processor and memory, whose net effect is that of reducing the *average cost* of accessing memory.

Such a reduction of average cost (i.e., capacitance) can be achieved either by augmenting the hierarchy vertically (i.e., adding levels to the hierarchy) or by growing it horizontally (i.e., adding support structures in parallel to existing hierarchy levels), or both [74,75]. In an RTL context, the cycle-accurate timing information contained in the description rules out the vertical transformations, since they affect the average cycle time of a memory. We can reasonably assume that, in a microarchitecture, only one level of memory hierarchy does (typically) exist. The latter option is therefore the only viable one that allows a relatively seamless integration in a

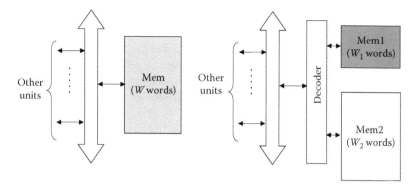

FIGURE 13.17 Common-case optimization applied to memory.

microarchitecture. Horizontal modification of a memory can be restated in terms of the usual common-case paradigm discussed in Section 13.5.2.

In this specific situation, the common case is represented by the memory cells that are accessed most. The idea is that of "isolating" this common case from the average case; since both are identified by memory cells, this amounts to instantiating a memory block "in parallel" to the main memory. This is shown in the example in Figure 13.17.

The left side of the figure depicts the original memory, consisting of $W = 1000$ words, while on the right-hand side, the W memory locations are mapped onto two different memory blocks of sizes $W_1 = 200$ and $W_2 = 800$. Assume that for a given workload the memory cells in the monolithic block Mem are accessed a total of $N = 10,000$ times. For the partitioned architecture, the common case is that the most frequently accessed memory cells are placed into the smaller block Mem$_1$ (with $N_1 = 8000$ accesses) and the least frequently accessed locations are mapped to the larger block Mem$_2$ (with $N_2 = 2000$ accesses).

We realistically assume that the power cost for accessing a cell in a memory block is directly related to the size of the block itself (the larger the block, the higher the power cost). Specifically, for our example, we assume the power access cost for cells in Mem to be $C_{\text{Mem}} = 0.5$, the cost for cells in Mem$_1$ to be $C_{\text{Mem1}} = 0.1$, and the cost for cells in Mem$_2$ to be $C_{\text{Mem2}} = 0.4$.

The average power for the monolithic memory is $P_{\text{mono}} = P_{\text{Mem}} = \dfrac{N C_{\text{men}}}{N} = 0.5$. For the partitioned memory, neglecting the decoder overhead, the power is

$$P_{\text{part}} = P_{\text{Mem1}} + P_{\text{Mem2}} = \frac{N_1 C_{\text{Mem1}} + N_2 C_{\text{Mem2}}}{N} = \frac{8,000 \cdot 0.1 + 2,000 \cdot 0.4}{10,000} = 0.16.$$

The partitioned scheme is more convenient as N_1 gets larger with respect to N_2, meaning that the most frequently accessed locations should be mapped as much as possible onto smaller blocks.

The generic template illustrated in Figure 13.17 lends itself to numerous variants, depending, for instance, on whether the subblocks are overlapping or not, on the choice of the number and of the size of the subblocks, on the possibility of having noncontiguous partitions (i.e., addresses could be *relocated* within blocks), on the architecture and implementation style of the decoder, on the organization and routing of the address and data buses, and on the placement of the subblocks.

The common-case principle applied to memories requires proper extra logic to drive the accesses to the correct memory bank (the block generically denoted as "encoder" in the figure) whose power and performance overhead constrains the type of partitioning scheme allowed. Some memory partitioning variants, such as those proposed in [76], provide significant power reductions with a very limited hardware overhead that can be easily tolerated into an RTL design.

It is important to mention that although the delay of the decoder is on the critical path, it usually does not affect the overall cycle time, since the access times of the subblocks are smaller than that of the monolithic memory; it suffices that $(t_{\text{decoder}} + \max_i(t_{\text{block}i})) \leq t_{\text{mem,mono}}$ in order to guarantee a seamless integration of the partitioned scheme ($t_{\text{mem,mono}}$ is the access time of the monolithic memory, and $t_{\text{block}i}$ is the access time of the generic subblock after partitioning).

13.5.5 BUS POWER OPTIMIZATION

Among the various parameters that appear in the model of Equation 13.6, only switching activity can be exploited at the microarchitectural level in order to reduce power. While supply voltage and frequency are somehow assigned up front, capacitance values (both switching and coupling) can only be reduced during physical design through proper wire sizing (for self-capacitance), spacing, or shielding (for coupling capacitance).

Reducing the activity factors amounts to modifying the binary values that are transmitted to the bus, in other terms, *encoding* the values. For correct operations, clearly, both encoding and decoding are required at each end of the bus.

Since we are dealing with hardware components, encoding/decoding must be implemented through specialized hardware blocks. This obvious requirement poses serious limitations on the encoding that can be applied on an on-chip bus: codecs will in fact consume power, which if not kept under control may easily offset the power gain achieved from the reduction of switching and coupling activities. Moreover, the codec also adds delay to the corresponding paths, which must be limited since the cycle time is fixed up front at the RTL. There is, therefore, a clear trade-off between codec complexity (power, but also delay) and the reduction in the number of transitions. Since the latter are weighted by the switched capacitance, this trade-off can be cleanly expressed as a minimum value (switching and coupling) of capacitance that represents the break-even point of the power cost function; beyond this minimum value, the encoding scheme becomes advantageous.

This trade-off drastically limits the type of encoding to be chosen. The existing literature on low-power bus encoding is vast and it consists, in most of the cases, of quite complex encoding schemes that are more suitable for off-chip buses or for long global on-chip buses rather than for short datapath buses. Moreover, technology scaling plays against the application of encoding to on-chip buses: as wire capacitance progressively increases with respect to that of the cells, the average length of the wires must be decreased, thus making it more and more difficult to amortize the cost of the codec.

13.5.5.1 BUS-ENCODING SCHEMES

Bus encoding can be applied at the physical level (*signal encoding*) or at the data-link level (*data encoding*). The former consists of modifying (in time or space) the way the binary 0s and 1s are represented. For instance, a 1 could be encoded as a $0 \rightarrow 1$ transition. Data encoding, instead, consists of modifying the way the binary patterns are represented. For example, a word could be represented by adding one parity bit. The application of the two types of encoding is not mutually exclusive. However, signal encoding may have an impact on the technology used to implement the design. For instance, a typical signal-encoding technique consists of reducing the voltage swing on (possibly some) of the bus wires. To make this possible, the technology on which the system will be implemented must be able to support that voltage level. For this reason, signal-encoding techniques are more related to the physical level of abstraction and will not be analyzed further. Conversely, data encoding is more general, since the only underlying assumption is that of tolerating the insertion of additional hardware to perform the encoding.

The problem of defining a code that minimizes the number of (self and/or coupling) transitions enables a theoretical formulation that has led to solutions that mix results from information and probability theory. Although surveying these techniques is out of the scope of this chapter, a rough classification may help in understanding the problem better. Bus-encoding techniques can be categorized based on two dimensions:

1. *The amount of redundancy allowed.* Some encoding schemes rely on spatial or temporal redundancy. Spatial redundancy implies the addition of extra bus lines, whereas temporal redundancy implies the addition of extra cycles to the bus transfers.
2. *The amount of knowledge on the statistics of the transmitted data.* Some schemes assume *a priori* knowledge of the statistical properties of the information transmitted on the bus, which can be exploited to customize the encoding functions to the most typical

behavior. An example is that of address buses, in which, although the specific patterns sent on the bus are not known, there exists a high degree of correlation between them, because of the sequential execution of a program.

Redundant codes are very popular in literature. One of the most referenced ones is the *bus-invert* (BI) scheme [77], in which the transmitter computes the Hamming distance between the word to be sent and the previously transmitted one. If the distance is larger than half of the bus width, the word to be transmitted is inverted, that is, complemented. The information about inversion is carried by an additional wire that is used at the receiver end to restore the original data.

The BI scheme has some interesting properties. First, the worst-case number of transitions of an n-bit bus is $n/2$ at each cycle. Second, if we assume that data are uniformly randomly distributed, it is possible to show that the average number of transitions with this code is lower than that of any other encoding scheme with just one redundant line. Moreover, the basic 1-bit redundant BI code has the property that the average number of transitions per line increases as the bus gets wider and asymptotically converges to 0.5, which is also the average switching per line of an unencoded bus and is already close to this value for 32-bit buses. This shortcoming has spun a number of variants of the basic BI scheme, based on the partitioning of the bus into smaller blocks and on the use of bus inversion on each block independently. Since the trivial application of this partitioned variant on an m-block bus would require m control lines, these methods have tried to reduce this additional complexity (e.g., see the work in [78,79]).

Addition of redundancy is not very desirable at the RTL. Temporal redundancy obviously alters the timing of the operations, thus giving rise to performance issues, while spatial redundancy may require the modification of the bus interface to support the extra connections, which may not be feasible when connecting synthesizable IP blocks with predefined I/O.

Discarding redundancy drastically limits the spectrum of applicability of bus encoding. In particular, if no assumption on the statistical properties of the transmitted data can be made, results from information theory show that it is not possible to reduce the number of transitions [80]. Some irredundant codes proposed in the literature bypass this theoretical limitation by building statistical information online over a given timing window [81,82]. These adaptive schemes require, however, a significant hardware overhead that is not generally affordable at the RTL.

When some knowledge of the statistical properties is available *a priori*, more effective encodings can be devised. The most realistic option at the RTL is the case of address buses, for which there exists a high degree of correlation between consecutive addresses; in particular, addresses generated by processors typically exhibit a high degree of *sequentiality*; this is particularly true for data-dominated applications, where the few control structures only occasionally break the sequentiality of the address stream.

Some authors have suggested the adoption of Gray coding [83] for address buses. This code achieves its asymptotic best performance of a single transition per emitted address when infinite streams of consecutive addresses are considered. This average can be lowered to asymptotic zero transitions, at the price of adding some spatial redundancy [84]. Alternatively, the asymptotic zero-transition behavior can be achieved without any redundancy by exploiting the decorrelating characteristics of the XOR function, when applied to consecutive bus patterns. In this way, the values on the bus are encoded using a transition signaling scheme [85]. Notice that the decorrelation implies an operation of a signal with its previous copy and may not be feasible at the RTL for timing reasons. Conversely, Gray encoding is a fully combinational transformation and it is potentially feasible to insert Gray codecs into an RTL description, should the cycle time constraints allow it.

So far, we have discussed techniques that aim at reducing switching activity. When addressing coupling activity, things are even more critical, because the encoding involves pairs of wires, thus making the hardware overhead required by the codec more complex by construction. In addition, solutions based on encoding the data to reduce coupling activity tend to ignore the most important implication of coupling, that is, its impact on timing due to *crosstalk*. Minimizing the number of simultaneous transitions on adjacent wires may reduce power consumption of coupling capacitance, yet it does not reduce crosstalk by itself.

Crosstalk mainly affects signal integrity, and even a significant reduction of crosstalk-induced power is of little interest for designers, if it does not guarantee the proper functionality of the circuit. Therefore, a solution to the problem must be consistent with performance-oriented crosstalk reduction techniques. In other terms, since crosstalk is mainly a capacitive effect, the only way to reduce it is by reducing the capacitance that causes it, and let power reduction come as a by-product. This makes the power optimization an issue to be dealt with during the physical design step, with the application of techniques ranging from (static or dynamic) wire permutation [86–88] to nonuniform wire spacing [89], from the insertion of shielding lines [90–92] to different combinations of the aforementioned approaches [93,94]. Solutions based on encoding of the data, however, may still be applied on top of capacitance reduction techniques, should the timing constraint allow it.

13.6 SYSTEM-LEVEL POWER MODELING

This section describes three commonly used methods of system-level power modeling. In all three, macromodels of components are combined to produce a macromodel of the entire system. The primary challenge is constructing these subcomponent macromodels. The first approach to solving this problem makes heavy use of the macromodeling techniques described in Section 13.4.2, but operates at a higher level of abstraction, with entire microprocessors or wireless communication subsystems being treated as components. The second approach makes use of measured power consumptions of components, although the system may not yet exist. The third approach is based on direct or indirect measurement of power consumption in a working system subject to carefully monitored, and possibly controlled, workloads. The appropriate method of model construction depends on whether the model is intended for use during, or after, the hardware design process.

13.6.1 SUBCOMPONENT MACROMODEL–BASED MODELING

We now describe how a system-level macromodel can be developed based on component-level macromodels.

If a system-level power model is intended for use in refining components within the system-level architecture, it is necessary to construct the component-level models before it is possible to measure their power consumptions. Therefore, a bottom-up approach is used, in which the component power model is constructed from circuit structure–based or historical macromodels of its components. In Figure 13.18, these macromodels are represented by the "macromodel built from circuit-level models" blocks.

Figure 13.18 illustrates a system-level macromodel for an embedded computing system containing several component macromodels: an organic light-emitting diode (OLED) display, a microcontroller, and several other components. Based on these macromodels, it is possible to construct a higher-level model in which power consumption depends on the sequence of the activity and power management states of the component. In the case of an OLED display, this state might depend on the distribution of pixel intensity RGB values and the display power management state. In the case of a microcontroller, this state might depend on the operating voltage, frequency, and utilization (percent of time not idle). More detailed models might also consider instruction and operand properties as well as cache state. In the case of a wireless communication interface, this state may depend on whether the interface is idle, ready, receiving, or transmitting and the states it has been in recently. Macromodels of lower-level subcomponents can be used to determine the relationships between activity and power management states on component power consumption. For example, given an appropriate subcomponent macromodel and an understanding of the impact of a particular instruction on switching activity in subcomponents such as ALUs, caches, and buses, it is possible to estimate the power implications of particular instructions on the higher-level component, such as a microprocessor.

The subcomponent macromodel-based modeling approach has some disadvantages. Typically, substantial errors are introduced due to inaccuracies in the subcomponent models. This approach

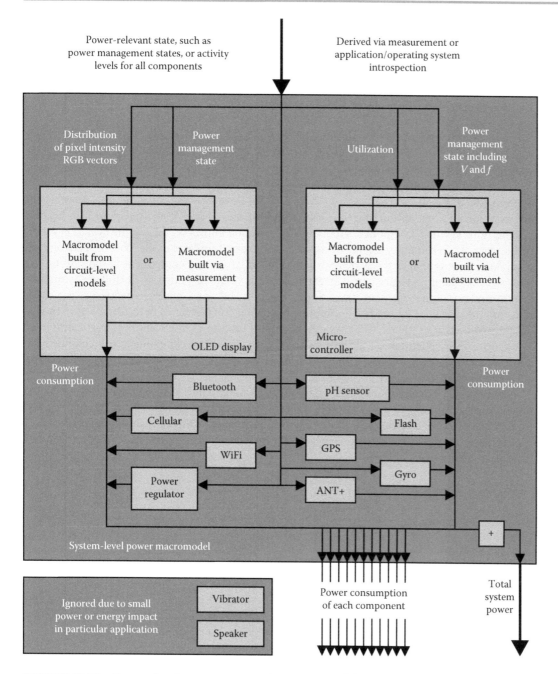

FIGURE 13.18 System-level power consumption macromodel.

can also be computationally expensive to evaluate because it depends on potentially complex subcomponent macromodels. However, if knowledge of the impact of system-level workloads on the power consumptions of individual components is needed before the designs of these subcomponents are finalized, there is little alternative.

13.6.2 COMPONENT MEASUREMENT–BASED MODELING

If the designs of components within the system have already been finalized and have been used in prior designs, an alternative modeling approach with some advantages in accuracy and modeling complexity can be used. In this case, it is possible to measure the impact of component model parameters such as input and power management state and use the resulting component-level power models to construct a system-level model. The macromodels produced by this modeling method are represented by the "macromodels built via measurement" blocks in Figure 13.18.

13.6.3 SYSTEM MEASUREMENT–BASED MODELING

We now describe the process of developing a system-level power model based on direct or indirect measurement of component power consumptions. The end result will be a system-level macromodel such as that shown in Figure 13.18, though the model can be produced without access to macromodels for, or direct measurements of, individual components.

Even after a system-level hardware architecture has been finalized, a system-level power model can be valuable when optimizing the designs and policies of operating system and applications or determining which components to focus optimization efforts on during the design of related future systems. In this case, instead of constructing the system-level model bottom-up from subcomponent macromodels, there are advantages to constructing a component-based system-level power model using measurements in carefully observed, and perhaps controlled, component activity and power management states. Such models are based on measurements of real systems, reducing the potential for error. This approach also has the benefit of permitting a system-level designer to construct the necessary component-level models even if their circuit structures are not known. Even when circuit structures are known, for many components the complexity of constructing measurement-based models is lower than that of building upon circuit structure subcomponent macromodels.

Despite the benefits of using measurement-based system-level macromodels, there are some potential risks. Foremost among these is the risk of failing to exercise some activity and power management state of a component that is later encountered during real-world use. Fortunately, it is possible to construct such models for complex systems such as smartphones with single-digit error percentages [95].

13.6.3.1 SIMPLIFIED PROBLEM DEFINITION

Let us now consider the model construction process for a fairly complex and heterogeneous system: the smartphone. A similar approach can be used for many other classes of electronic systems. The simplest version of the problem can be stated as follows.

Abstract system-level power modeling problem definition: Given knowledge of the possible power management and activity states of all components within an electronic system, as well as the ability to control these states and measure the resulting component-level power consumptions, construct a power model capable of estimating the power consumption of the entire system as a function of the current and historical states of its individual components.

Starting from this definition, a naïve approach to modeling the entire system would be to transition the system through each possible combination of component states and measure the resulting component power consumptions. This process could be quite expensive, as the number of system-wide states would increase exponentially in the number of components. Fortunately, the system-level power consumption is the sum of the power consumptions of its components and, in most real-world systems, the power consumptions of most components are mostly independent of the states of other components. Therefore, the expedient of varying the state of each component while holding the states of other components constant is generally sufficient. We found that this simplification introduced 6% error when used during power modeling of a smartphone [95].

13.6.3.2 PRACTICAL PROBLEM DEFINITION

Unfortunately, the power modeling problem faced by real-world system designers is harder and more complex, for the following reasons:

"Given knowledge of the possible power management and activity states of all components…": In fact, this information is commonly unavailable from component vendors, so the system

designer must often play the role of sleuth, combining knowledge of component operation with comprehensive testing to determine the states possible for each component. This is not a fundamental problem. It exists only because component designers, whose customers are system designers, do not provide enough information about their components to enable their most power-efficient use.

"…as well as the ability to control these states…": Although it is generally possible to exercise control over the states of components, this control is imprecise. For example, precisely controlling the percent of time a processor spends idle may be difficult without burdensome changes to its operating system. Precisely controlling the power management state of a wireless interface may depend on time-consuming reverse engineering of its firmware or having control of a cellular base station. Fortunately, combining approximate control of component states with more precise measurement of these states offers a solution. Consider, for example, the problem of modeling the dependence of microprocessor power consumption as a function of duty cycle, the proportion of time spent actively executing instructions. It may be difficult to precisely control the percent of cycles spent idle due to operating system interference. However, rough control of this parameter can be combined with operating system introspection to ensure that each power measurement is associated with its corresponding duty cycle. A similar approach can be used when modeling the dependence of wireless communication interface power consumption on parameters such as transmission power.

"… measure the resulting component-level power consumptions…": In many systems, measuring the power consumptions of individual components is a challenge because distributed current meters are not available and accessing the appropriate interconnects is physically challenging or would cause damage to the system. As a result, the system designer must often make due with measuring the power consumption of the entire system during the modeling process.

This leaves us with the following problem definition:

Practical system-level power modeling problem definition: Given the ability to (1) determine all possible power management and activity states of all components within an electronic system, (2) control these states with some probability of success, (3) measure the current states precisely, and (4) measure the resulting system-level power consumption, construct a power model capable of estimating the power consumption of the entire system as a function of the current and historical states of its individual components.

13.6.3.3 MODEL CONSTRUCTION

This problem of constructing a system-level power model can be solved using the procedure illustrated in Figure 13.19. The first step of determining the major parameters on which component power consumptions depend can be challenging due to the variety of possible dependencies and because understanding each requires some component-specific knowledge. For example, backlit display power consumption depends primarily on backlight output level; OLED power consumption is a function of pixel color intensities; microprocessor power depends primarily on the percent of active cycles and power management state including operating frequency and voltage; and wireless communication interface power consumption depends on the recent history of data transmission and reception rates, which influence internally controlled power management states that are generally not externally visible. Determining which components and states must be considered can be more challenging than exercising those states.

In the second step, methods are developed to determine the states of components in the system. These methods may use software introspection or measurement with an external data acquisition card to determine the state.

The third step in Figure 13.19 can be challenging, as well, because it can be impractical to precisely control the states of individual components. Sometimes, only approximate control, or control that has some limited probability of success on a given attempt, is possible.

In the fourth step, built-in or external sensors are used to measure system power consumption.

In the fifth step, transitioning through the power-relevant states of a single component at a time, while holding the states of other components constant, can simplify the model construction process, although it is not strictly necessary. This leaves open the possibility of building a

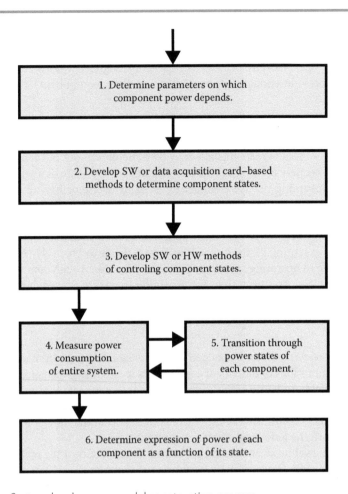

FIGURE 13.19   System-level power model construction process.

model without explicitly controlling component states, provided that each component enters all power-relevant states during the model construction process.

Step six is generally easiest when regression on the data measured in step five is used to determine component macromodel parameters and potentially structure.

There are several papers describing this process, or particular steps within it, in detail [95–101].

13.6.3.4 MODEL EVALUATION

The process of evaluating a system-level power model such as that described in the prior sections is generally simpler than that of constructing it. The model is generally inexpensive to evaluate function of the power-relevant states of all power-relevant components. This opens the possibility of hosting the system-level power model on the physical electronic system being modeled with low overhead, making it practical to estimate the power consumptions of the system and its individual components even if the system does not have built-in current sensors or has only one such sensor. There are several tools based on this concept [95,96].

Although automation plays a central role in the generation and evaluation of system-level power models, it is common for several steps of the process to require designer involvement and knowledge. In most cases, this is the result of component designers neglecting to provide relevant component-level power models that, if available, would support automated construction of system-level power models. As a result, there are major opportunities for component designers catering to designers of low-power systems to differentiate their products from their competitors. It is now fairly common for independent design groups around the world to build power models for components without a detailed understanding of their circuit structures, when the cost could have instead been paid once by the better-equipped component designer.

13.7 SYSTEM-LEVEL POWER OPTIMIZATION

This section describes three methods by which system-level power consumption is commonly optimized. Although the first is generally manual, it is included because it is one of the most important uses of the results yielded by the automated power modeling tools described in Section 13.6. It may also be possible to automate in the future. In the second approach, automated compiler optimizations and transformations are used to improve the energy efficiency of software. In the third approach, the activity and power management states of individual components are adjusted to simultaneously optimize system power and other objectives, such as functionality or user satisfaction.

13.7.1 APPLICATION REDESIGN

System-level power modeling often yields surprising information to system-level hardware and software designers. For example, a designer might find that a particular component has much higher power or energy consumption than expected, leading to the identification of an application design flaw that can be corrected through redesign. The process of finding such design flaws is based on the use of automated power modeling and estimation tools, but can depend on painstaking manual analysis.

There has been some recent work on automating the analysis process. For example, researchers have used virtual machine–based tracking of information flow through applications and platform (operating system) to identify operations that consume energy but never have any impact on the output of the system [102]. One can determine when an energy-consuming operation, such as receipt of a packet via a wireless interface, is useless by tracking all of the computations transitively depending on the product of the operation and determining whether any change the system state in a way that can become visible to the system's user. If not, the operation could be eliminated without any harmful impact on the system. Such operations exist in real, widely used, smartphone applications.

13.7.2 COMPILER OPTIMIZATIONS

Some of the power optimization mechanisms can be explicitly or implicitly controlled by compile-time code transformations. For example, the sequence of instructions can influence the resulting energy consumption and multiple sequences are often permissible, allowing compilers to reorder instructions, for example, to reduce switching activities on the instruction bus, thereby reducing energy consumption [103]. Although such approaches can significantly reduce power consumption in some circumstances, the benefits of such low-level techniques are often low when compared with related optimization techniques that consider performance, alone.

Researchers have analyzed the performance and energy implications of enabling and disabling particular compiler optimizing transformations. Although this can sometimes greatly reduce energy consumption, in most work most of the benefit comes from reducing execution time, that is, performance-oriented compiler optimizations (a fairly mature research area) would have yielded a similar result. In a few cases, operations that decreased energy consumption increased execution time, but these generally had only a few percent impact on each.

Techniques in which compilers intelligently control the temporally fine-grained power management states of microprocessors have generally yielded substantial energy savings (e.g., 50%) [104,105]. Compiler control of memory layout including the software-directed use of scratchpad memory and compiler-assisted cache control have also resulted in energy reductions of more than 25% [106–108].

13.7.3 COMPONENT STATE CONTROL

It is common for operating systems and firmware to control the power management states of system components with the goal of minimizing energy consumption while maintaining good quality of service. More precisely, the goal is to transition each component to the appropriate state, at the right time, to minimize system-level energy consumption while offering an adequate user experience. This is a hard problem for two reasons.

First, changing power management state, for example, changing the voltage and frequency of a microprocessor or transitioning a wireless communication interface to a lower-power mode offering less functionality, often imposes a substantial time and energy cost. Circuit designers are working to reduce these penalties, but they remain substantial for many components. Perfectly solving the component power management state scheduling problem would require knowledge of future component use. The most commonly used strategy is to assume that future component use will be similar to that in the recent past, and transition to lower-power, lower-functionality component states after a certain amount of time has elapsed since the last intensive use of the component. Researchers have developed power management policies based on more sophisticated predictive models, some of which are appropriate for use in automated online power management system software [109].

Second, the only performance-related metric that ultimately matters for a computer system is the satisfaction of its user or owner. Commonly used proxies for this metric, such as instructions processed per second, do not relate in any simple way to user satisfaction [109]. Researchers have considered metrics that better approximate user satisfaction, such as application response time [110], that are suitable for use in evaluating and improving automated online system-level power management techniques.

13.8 CONCLUSIONS

This chapter has reviewed the basic principles of power modeling, estimation, and optimization for digital CMOS circuits described at the microarchitectural and system levels. The most common and successful modeling solutions for various types of components, such as datapath macros, controllers, memories, wires, and buses, have been discussed in detail, offering a comprehensive overview of the state of the art in this domain. We have also described methods of constructing power models for complex, heterogeneous digital systems. Power estimation methods making use of the various models have been illustrated.

Building on the foundation of power estimation, we have described power optimization technologies such as clock gating, exploitation of common-case computation and dynamic V_{th}/V_{dd} management, memory partitioning, bus encoding, and careful control of component activity and power management states at the system level. Emphasis has been put on design automation aspects of most of the techniques considered, with the objective of making this chapter of practical use not only to IC designers and architects but also to circuit-level and system-level EDA engineers.

REFERENCES

1. A.P. Chandrakasan, S. Sheng, and R.W. Brodersen, Low-power CMOS digital design, *IEEE J. Solid State Circ.*, 27, 473–484, 1992.
2. D.D. Gajski, N.D. Dutt, A.C.-H. Wu, and S.Y.-L. Lin, *High-Level Synthesis: Introduction to Chip and System Design*, Kluwer Academic Publishers, Dordrecht, the Netherlands, 1992.
3. P. Landman, High-level power estimation, *ISLPED-96: ACM/IEEE International Symposium on Low Power Electronics and Design*, Monterey, CA, August 1996, pp. 29–35.
4. P. Landman and J. Rabaey, Activity-sensitive architectural power analysis, *IEEE Trans. Comput. Aided Des. Integr. Circ. Syst.*, 15, 571–587, 1996.
5. P. Landman, R. Mehra, and J. Rabaey, An integrated CAD environment for low-power design, *IEEE Des. Test Comput.*, 13, 72–82, 1996.
6. A. Raghunathan, S. Dey, and N. Jha, Register-transfer level estimation techniques for switching activity and power consumption, *ICCAD-96: IEEE/ACM International Conference in Computer-Aided Design*, San Jose, CA, November 1996, pp. 158–165.
7. S. Katkoori and R. Vemuri, Architectural power estimation based on behavioral profiling, *J. VLSI Des.*, 7, 255–270, 1998.
8. A. Bogliolo, I. Colonescu, R. Corgnati, E. Macii, and M. Poncino, An RTL power estimation tool with on-line model building capabilities, *PATMOS-01: IEEE International Workshop on Power and Timing Modeling, Optimization and Simulation*, Yverdon-les-Bains, Switzerland, September 2001, pp. 2.3.1–2.3.10.

9. S. Ravi, A. Raghunathan, and S. Chakradhar, Efficient RTL power estimation for large designs, *IEEE International Conference on VLSI Design*, New Delhi, India, January 2003, pp. 431–439.

10. D. Helms, E. Schmidt, A. Schulz, A. Stammermann, and W. Nebel, An improved power macro-model for arithmetic datapath components, *PATMOS-02: IEEE International Workshop on Power and Timing Modeling, Optimization and Simulation*, Sevilla, Spain, September 2002, pp. 16–24.

11. M. Bruno, A. Macii, and M. Poncino, A statistical power model for non-synthetic RTL operators, *PATMOS-03: IEEE International Workshop on Power and Timing Modeling, Optimization and Simulation*, Torino, Italy, September 2003, pp. 208–218.

12. F. Najm, Transition density: A new measure of activity in digital circuits, *IEEE Trans. Comput. Aided Des. Integr. Circ. Syst.*, 12, 310–323, 1993.

13. Q. Qiu, Q. Wu, C.-S. Ding, and M. Pedram, Cycle-accurate macro-models for RT-level power analysis, *IEEE Trans. VLSI Syst.*, 6, 520–528, 1998.

14. H. Mehta, R.M. Owens, and M.J. Irwin, Energy characterization based on clustering, *DAC-33: ACM/IEEE Design Automation Conference*, Las Vegas, NV, June 1996, pp. 702–707.

15. S. Gupta and F. Najm, Energy-per-cycle estimation at RTL, *ISLPED-99: ACM/IEEE International Symposium on Low-Power Electronics and Design*, San Diego, CA, August 1999, pp. 16–17.

16. L. Benini, A. Bogliolo, M. Favalli, and G. De Micheli, Regression models for behavioral power estimation, *PATMOS-96: IEEE International Workshop on Power and Timing Modeling, Optimization and Simulation*, Bologna, Italy, October 1996, pp. 125–130.

17. D. Liu and C. Svensson, Power consumption estimation in CMOS VLSI chips, *IEEE J. Solid State Circ.*, 29, 663–671, 1994.

18. E. Schmidt, G. Jochens, L. Kruse, F. Theeuwen, and W. Nebel, Memory power models for multilevel power estimation and optimization, *IEEE Trans. VLSI Syst.*, 10, 106–109, 2002.

19. S. Powell and P. Chau, Estimating power dissipation of VLSI signal processing chips: The PFA technique, *VLSI Signal Process.*, 4, 250–259, 1990.

20. P. Landman and J. Rabaey, Architectural power analysis: The dual-bit type model, *IEEE Trans. VLSI Syst.*, 3, 173–187, 1995.

21. S. Gupta and F. Najm, Power macromodeling for high level power estimation, *DAC-34: ACM/IEEE Design Automation Conference*, Anaheim, CA, June 1997, pp. 365–370.

22. M. Barocci, L. Benini, A. Bogliolo, B. Ricco, and G. De Micheli, Look-up table power macro-models for behavioral library components, *IEEE Alessandro Volta Memorial Workshop on Low-Power Design*, Como, Italy, March 1999, pp. 173–181.

23. R. Corgnati, E. Macii, and M. Poncino, Clustered table-based macromodels for RTL power estimation, *GLS-VLSI-99: IEEE/ACM Great Lakes Symposium on VLSI*, Ann Arbor, MI, March 1999, pp. 354–357.

24. A. Bogliolo, E. Macii, V. Mihailovici, and M. Poncino, Combinational characterization-based power macro-models for sequential macros, *PATMOS-99: IEEE International Workshop on Power and Timing Modeling, Optimization and Simulation*, Kos, Greece, October 1999, pp. 293–302.

25. M. Anton, I. Colonescu, E. Macii, and M. Poncino, Fast characterization of RTL power macromodels, *ICECS-01: IEEE International Conference on Electronics, Circuits and Systems*, La Valletta, Malta, September 2001, pp. 1591–1594.

26. A. Bogliolo, R. Corgnati, E. Macii, and M. Poncino, Parameterized RTL power models for combinational soft macros, *IEEE Trans. VLSI Syst.*, 9, 880–887, 2001.

27. A. Tyagi, Entropic bounds on fSM switching, *ISLPED-96: ACM/IEEE International Symposium on Low Power Electronics and Design*, Monterey, CA, August 1996, pp. 323–327.

28. D. Marculescu, R. Marculescu, and M. Pedram, Theoretical bounds for switching activity analysis in finite-state machines, *ISLPED-98: ACM/IEEE International Symposium on Low Power Electronics and Design*, Monterey, CA, August 1998, pp. 36–41.

29. S. Katkoori and R. Vemuri, Simulation-based architectural power estimation for PLA-based controllers, *ISLPED-96: ACM/IEEE International Symposium on Low Power Electronics and Design*, Monterey, CA, August 1996, pp. 121–124.

30. L. Benini, A. Bogliolo, E. Macii, M. Poncino, and M. Surmei, Regression-based RTL power models for controllers, *GLS-VLSI-00: ACM/IEEE Great Lakes Symposium on VLSI*, Evanston, IL, March 2000, pp. 147–152.

31. U. Ko and P.T. Balsara, Characterization and design of a low-power, high-performance cache architecture, *IEEE International Symposium on VLSI Technology, Systems and Applications*, Taipei, Taiwan, May 1995, pp. 235–238.

32. M.B. Kamble and K. Ghose, Analytical energy dissipation models for low power caches, *ISLPED-97: ACM/IEEE International Symposium on Low Power Design*, Monterey, CA, August 1997, pp. 143–148.

33. S.L. Coumeri and D.E. Thomas, Memory modeling for system synthesis, *IEEE Trans. VLSI Syst.*, 8, 327–334, 2000.

34. B.S. Landman and R.L. Russo, On a pin vs. block relationship for partitions of logic graphs, *IEEE Trans. Comput.*, 20, 1469–1479, 1971.

35. S. Sastry and A.C. Parker, Stochastic models for wireability of analysis of gate arrays, *IEEE Trans. Comput. Aided Des. Integr. Circ. Syst.*, 5, 52–65, 1986.

36. F.J. Kurdahi and A.C. Parker, Techniques for area estimation of VLSI layouts, *IEEE Trans. Comput. Aided Des. Integr. Circ. Syst.*, 8, 81–92, 1989.

37. C. Anton, A. Bogliolo, P. Civera, I. Colonescu, E. Macii, and M. Poncino, RTL estimation of steering logic power, *PATMOS-00: IEEE International Workshop on Power and Timing Modeling, Optimization and Simulation*, Gottingen, Germany, September 2000, pp. 36–45.

38. L. Benini and G. De Micheli, Networks on chips: A new SoC paradigm, *IEEE Comput.*, 35, 70–78, 2002.

39. Synopsys Design Ware Library, http://www.synopsys.com/IP/SOCInfrastructureIP/DesignWare/Pages/default.aspx.

40. R. Ho, K.W. Mai, and M.A. Horowitz, The future of wires, *Proc. IEEE*, 89(4), 490–504, April 2001.

41. P. Sotiriadis and A. Chandrakasan, A bus energy model for deep submicron technology, *IEEE Trans. VLSI Syst.*, 10, 341–350, 2002.

42. L. Benini, P. Siegel, and G. De Micheli, Automatic synthesis of gated clocks for power reduction in sequential circuits, *IEEE Des. Test Comput.*, 11, 32–40, 1994.

43. L. Benini and G. De Micheli, Transformation and synthesis of FSMs for low power gated clock implementation, *IEEE Trans. Comput. Aided Des. Integr. Circ. Syst.*, 15, 630–643, 1996.

44. L. Benini, G. De Micheli, E. Macii, M. Poncino, and R. Scarsi, Symbolic synthesis of clock-gating logic for power optimization of synchronous controllers, *ACM Trans. Des. Automat. Electron. Syst.*, 4, 351–375, 1999.

45. M. Gowan, L.L. Biro, and D.B. Jackson, Power considerations in the design of the Alpha 21264 microprocessor, *DAC-35: ACM/IEEE Design Automation Conference*, San Francisco, CA, June 1998, pp. 726–731.

46. A. Correale, Overview of the power minimization techniques employed in IBM powerPC 4xx embedded controllers, *ISLPD-95: ACM International Symposium on Low Power Design*, Dana Point, CA, August 1995, pp. 75–80.

47. V. Tiwari, D. Singh, S. Rajgopal, G. Mehta, R. Patel, and F. Baez, Reducing power in high-performance micro-processors, *DAC-35: ACM/IEEE Design Automation Conference*, San Francisco, CA, June 1998, pp. 732–737.

48. Z. Khan, and G. Mehta, Automatic clock gating for power reduction, *SNUG-99: Synopsys Users Group*, San Jose, CA, March 1999.

49. P. Babighian, L. Benini, and E. Macii, A scalable algorithm for RTL insertion of gated clocks based on observability don't cares computation, *IEEE Trans. Comput. Aided Des. Integr. Circ. Syst.*, 24, 29–42, 2005.

50. L. Benini, M. Donno, A. Ivaldi, and E. Macii, Clock-tree power optimization based on RTL clock-gating, *DAC-40: ACM/IEEE Design Automation Conference*, Anaheim, CA, June 2003, pp. 622–627.

51. M. Donno, E. Macii, and L. Mazzoni, Power-aware clock tree planning, *ISPD-04: ACM/IEEE International Symposium on Physical Design*, Phoenix, AZ, April 2004, pp. 138–147.

52. L. Benini, M. Favalli, and G. De Micheli, Design for testability of gated-clock FSMs, *EDTC-96: IEEE European Design and Test Conference*, Paris, France, March 1996, pp. 589–596.

53. B. Pouya and A. Crouch, Optimization for vector volume and test power, *ITC-00: IEEE International Test Conference*, Atlantic City, NJ, October 2000, pp. 873–881.

54. J. Saxena, K. Butler, and L. Whetsel, An analysis of power reduction techniques in scan testing, *ITC-01: IEEE International Test Conference*, Baltimore, MD, October 2001, pp. 670–677.

55. J.L. Hennessy and D.A. Patterson, *Computer Architecture: A Quantitative Approach*, Morgan Kaufmann Publishers, San Francisco, CA, 1996.

56. S. Hassoun and C. Ebeling, Architectural retiming: Pipelining latency-constrained circuits, *DAC-33: ACM/IEEE Design Automation Conference*, Las Vegas, NV, June 1996, pp. 708–713.

57. S. Nowick, Design of a low-latency asynchronous adder using speculative completion, *IEE Proc. Comput. Digit. Tech.*, 143, 301–307, 1996.

58. L. Benini, G. De Micheli, E. Macii, and M. Poncino, Telescopic units: A new paradigm for performance optimization of VLSI designs, *IEEE Trans. Comput. Aided Des. Integr. Circ. Syst.*, 17, 220–232, 1998.

59. L. Benini, G. De Micheli, A. Lioy, E. Macii, G. Odasso, and M. Poncino, Automatic synthesis of large telescopic units based on near-minimum timed supersetting, *IEEE Trans. Comput.*, 48, 769–779, 1999.

60. G. Lakshminarayana, A. Raghunathan, K.S. Khouri, N.K. Jha, and S. Dey, Common case computation: A new paradigm for energy and performance optimization, *IEEE Trans. Comput. Aided Des. Integr. Circ. Syst.*, 23, 33–49, 2004.

61. M. Munch, B. Wurth, R. Mehra, J. Sproch, and N. Wehn, Automating RT-level operand isolation to minimize power consumption in datapaths, *DATE-00: IEEE Design Automation and Test in Europe*, Paris, France, March 2000, pp. 624–631.

62. M. Alidina, J. Monteiro, S. Devadas, A. Ghosh, and M. Papaefthymiou, Precomputation-based sequential logic optimization for low power, *IEEE Trans. VLSI Syst.*, 2, 426–436, 1994.

63. J. Monteiro, S. Devadas, and A. Ghosh, Sequential logic optimization for low power using input-disabling precomputation architectures, *IEEE Trans. Comput. Aided Des. Integr. Circ. Syst.*, 17, 279–284, 1998.

64. L. Benini, G. De Micheli, A. Lioy, E. Macii, G. Odasso, and M. Poncino, Synthesis of power-managed sequential components based on computational kernel extraction, *IEEE Trans. Comput. Aided Des. Integr. Circ. Syst.*, 20, 1118–1131, 2001.

65. S. Svilan, J.B. Burr, and G.L. Tyler, Effects of elevated temperature on tunable near-zero threshold CMOS, *ISLPED-01: ACM/IEEE International Symposium on Low Power Electronics and Design*, Huntington Beach, CA, August 2001, pp. 255–258.

66. K. Flautner, D. Flynn, and M. Rives, A combined hardware–software approach for low-power SoCs: Applying adaptive voltage scaling and the vertigo performance-setting algorithms, *DesignCon*, Paper SA2-3, Santa Clara, CA, January 2003.

67. S. Martin, K. Flautner, T. Mudge, and D. Blaauw, Combined dynamic voltage scaling and adaptive body biasing for lower power microprocessors under dynamic workloads, *ICCAD-02: ACM/IEEE International Conference on Computer Aided Design*, San Jose, CA, November 2002, pp. 721–725.

68. S. Shigematsu, S. Mutoh, Y. Matsuya, Y. Tanabe, and J. Yamada, A 1-V high-speed MTCMOS circuit scheme for power-down application circuits, *IEEE J. Solid State Circuits*, 32, 861–869, 1997.

69. V. Zyuban and S. Kosonocky, low power integrated scan-retention mechanism, *ISLPED-02: ACM/IEEE International Symposium on Low Power Electronics and Design*, Monterey, CA, August 2002, pp. 98–102.

70. K. Usami, M. Igarashi, F. Minami, T. Ishikawa, M. Kawakawa, M. Ichida, and K. Nogami, Automated low-power technique exploiting multiple supply voltages applied to media processor, *IEEE J. Solid State Circuits*, 33, 463–472, 1998.

71. L. Wei, K. Roy, and V. De, Low power low voltage CMOS design techniques for deep submicron ICs, *VLSI-00: International Conference on VLSI Design*, Calcutta, India, January 2000, pp. 24–29.

72. D.E. Lackey, S. Gould, T.R. Bednar, J. Cohn, and P.S. Zuchowski, Managing power and performance for system-on-chip designs using voltage islands, *ICCAD-02: ACM/IEEE International Conference on Computer Aided Design*, San Jose, CA, November 2002, pp. 195–202.

73. F. Ishihara, F. Sheikh, and B. Nikolic, Level conversion for dual supply systems, *ISLPED-03: ACM/IEEE International Symposium on Low Power Electronics and Design*, Seoul, Korea, August 2003, pp. 164–167.

74. A. Macii, L. Benini, and M. Poncino, *Memory Design Techniques for Low-Energy Embedded Systems*, Kluwer Academic Publishers, Dordrecht, the Netherlands, 2002.

75. P. Panda and N. Dutt, *Memory Issues in Embedded Systems-on-Chip Optimization and Exploration*, Kluwer Academic Publishers, Dordrecht, the Netherlands, 1999.

76. L. Benini, L. Macchiarulo, A. Macii, and M. Poncino, Layout-driven memory synthesis for embedded systems-on-chip, *IEEE Trans. VLSI Syst.*, 10, 96–105, 2002.

77. M. Stan, and W.P. Burleson, Bus-invert coding for low-power I/O, *IEEE Trans. VLSI Syst.* 3, 49–58, 1995.

78. Y. Shin, S.-I. Chae, and K. Choi, Partial bus-invert coding for power optimization of application-specific systems, *IEEE Trans. VLSI Syst.* 9, 377–383, 2001.

79. U. Narayanan, K.-S. Chung, and K. Taewhan, Enhanced bus invert encodings for low-power, *ISCAS-02: IEEE International Symposium on Circuits and Systems*, Vol. 5, Scottsdale, AZ, May 2002, pp. 25–28.

80. S. Ramprasad, N. Shanbhag, and I. Hajj, Signal coding for low power: Fundamental limits and practical realizations, *IEEE Trans. Circ. Syst. II: Analog Digit. Signal Process.*, 46, 923–929, 1999.

81. L. Benini, A. Macii, E. Macii, M. Poncino, and R. Scarsi, Architectures and synthesis algorithms for power-efficient bus interfaces, *IEEE Trans. Comput. Aided Des. Integr. Circ. Syst.*, 19, 969–980, 2000.

82. M. Mamidipaka, D. Hirschberg, and N. Dutt, Low power address encoding using self-organizing lists, *ISLPED-01: ACM/IEEE International Symposium on Low Power Electronics and Design*, Huntington Beach, CA, August 2001, pp. 188–193.

83. H. Mehta, R.M. Owens, and M.J. Irwin, Some issues in gray code addressing, *GLS-VLSI-96: IEEE/ACM Great Lakes Symposium on VLSI*, Ames, IA, March 1996, pp. 178–180.

84. L. Benini, G. De Micheli, E. Macii, D. Sciuto, and C. Silvano, Asymptotic zero-transition activity encoding for address busses in low-power microprocessor-based systems, *GLS-VLSI-97: IEEE/ACM Great Lakes Symposium on VLSI*, Urbana-Champaign, IL, March 1997, pp. 77–82.

85. Y. Aghaghiri, F. Fallah, and M. Pedram, Irredundant address bus encoding for low power, *ISLPED-01: ACM/IEEE International Symposium on Low Power Electronics and Design*, Huntington Beach, CA, August 2001, pp. 182–187.

86. Y. Shin and T. Sakurai, Coupling-driven bus design for low-power application-specific systems, *DAC-38: ACM/IEEE Design Automation Conference*, Las Vegas, NV, June 2001, pp. 750–753.

87. L. Macchiarulo, E. Macii, and M. Poncino, Low-energy encoding for deep-submicron address buses, *ISLPED-01: ACM/IEEE International Symposium on Low Power Electronics and Design*, Huntington Beach, CA, August 2001, pp. 176–181.

88. J. Henkel and H. Lekatsas, A2BC: Adaptive address bus coding for low-power deep sub-micron designs, *DAC-38: ACM/IEEE Design Automation Conference*, Las Vegas, NV, June 2001, pp. 744–749.

89. L. Macchiarulo, E. Macii, and M. Poncino, Wire placement for crosstalk energy minimization in address buses, *DATE-02: IEEE Design Automation and Test in Europe*, Paris, France, March 2002, pp. 158–162.

90. A. Vittal and M. Sadowska, Crosstalk reduction for VLSI, *IEEE Trans. Comput. Aided Des. Integr. Circ. Syst.*, 16, 290–298, 1997.

91. H. Kaul, D. Sylvester, and D. Blauuw, Active shields: A new approach to shielding global wires, *GLS-VLSI-02: ACM/IEEE Great Lakes Symposium on VLSI*, New York, March 2002, pp. 112–117.

92. S. Salerno, E. Macii, and M. Poncino, Crosstalk energy reduction by temporal shielding, *ISCAS-04: IEEE International Symposium on Circuits and Systems*, Vol. 2, Vancouver, British Columbia, Canada, May 2004, pp. 749–752.

93. R. Arunachalam, E. Acar, and S. Nassif, Optimal shielding/spacing metrics for low power design, *ISVLSI-03: IEEE Annual Symposium on VLSI*, Tampa, FL, February 2003, pp. 167–172.

94. S. Salerno, E. Macii, and M. Poncino, Combining wire swapping and spacing for low-power deep-submicron buses, *GLS-VLSI-03: IEEE/ACM Great Lakes Symposium on VLSI*, Washington, DC, April 2003, pp. 198–202.

95. L. Zhang, B. Tiwana, Z. Qian, Z. Wang, R.P. Dick, Z.M. Mao, and L. Yang, Accurate online power estimation and automatic battery behavior based power model generation for smartphones, *Proceedings of International Conference on Hardware/Software Codesign and System Synthesis*, Scottsdale, AZ, October 2010, pp. 105–114.

96. M. Dong and L. Zhong, Self-constructive high-rate system energy modeling for battery-powered mobile systems, *Proceedings of International Conference on Mobile Systems, Applications, and Services*, Washington, DC, June 2011, pp. 335–348.

97. S. Gurun and C. Krintz, A run-time, feedback-based energy estimation model for embedded devices, *Proceedings of International Conference on Hardware/Software Codesign and System Synthesis*, Seoul, Korea, October 2006, pp. 28–33.

98. T. Cignetti, K. Komarov, and C. Ellis, Energy estimation tools for the Palm, *Proceedings of International Conference on Modeling, Analysis and Simulation of Wireless and Mobile Systems*, Boston, MA, 2000, pp. 96–103.

99. C. Isci and M. Martonosi, Runtime power monitoring in high-end processors: Methodology and empirical data, *Proceedings of International Symposium Microarchitecture*, San Diego, CA, December 2003, pp. 93–104.

100. G. Contreras et al., XTREM: A power simulator for the Intel XScale, *Proceedings of Conference Languages, Compilers, and Tools for Embedded Systems*, Washington, DC, June 2004, pp. 115–125.

101. J. Flinn and M. Satyanarayanan, PowerScope: A tool for profiling the energy usage of mobile applications, *Proceedings of Workshop on Mobile Computer Systems and Applications*, New Orleans, LA, 1999, pp. 2–10.

102. L. Zhang, M.S. Gordon, R.P. Dick, Z.M. Mao, P. Dinda, and L. Yang, ADEL: An automatic detector of energy leaks for Smartphone applications, *Proceedings of International Conference on Hardware/Software Codesign and System Synthesis*, Tampere, Finland, October 2012, pp. 363–373.

103. C. Lee, J.K. Lee, T. Hwang, and S.-C. Tsai, Compiler optimization on instruction scheduling for low power, *Proceedings of International Symposium System Synthesis*, Madrid, Spain, September 2000, pp. 55–60.

104. Q. Wu, V. Reddi, Y. Wu, J. Lee, D. Connors, D. Brooks, M. Martonosi, and D.W. Clark. Dynamic compilation framework for controlling microprocessor energy and performance, *Proceedings of International Symposium on Microarchitecture*, Barcelona, Spain, November 2005, pp. 271–282.

105. R. Banakar, S. Steinke, B.-S. Lee, M. Balakrishnan, and P. Marwedel, Scratchpad memory: Design alternative for cache on-chip memory in embedded systems, *Proceedings of International Symposium. Hardware/Software Codesign*, Estes Park, CO, May 2002, pp. 73–78.

106. O. Avissar, R. Barua, and D. Stewart, An optimal memory allocation scheme for scratch-pad-based embedded systems, *ACM Trans. Embed. Comput. Syst.*, 1(1), 6–26, November 2002.

107. R.A. Ravindran, P.D. Nagarkar, G.S. Dasika, E.D. Marsman, R.M. Senger, S.A. Mahlke, and R.B. Brown, Compiler managed dynamic instruction placement in a low-power code cache, *Proceedings of International Symposium on Code Generation and Optimization*, March 2005, pp. 179–190.

108. S. Irani, S. Shukla, and R. Gupta, Online strategies for dynamic power management in systems with multiple power-saving states, *ACM Trans. Embed. Comput. Syst.*, 2(3), 325–346, August 2003.

109. L. Yang, R.P. Dick, P.A. Dinda, G. Memik, and X. Chen, HAPPE: Human and application driven frequency scaling for processor power efficiency, *IEEE Trans. Mobile Comput.*, 12(8), 1546–1557, August 2013.

110. L. Zhang, D.R. Bild, R.P. Dick, Z.M. Mao, and P. Dinda, Panappticon: Event-based tracing to optimize mobile application and platform performance, *Proceedings of International Conference on Hardware/Software Codesign and System Synthesis*, Montreal, Canada, September 2013.

Design Planning

Ralph H.J.M. Otten

CONTENTS

14.1 INTRODUCTION

In the past decade, *design automation* faced a sequence of what were called *closure problems.* "How to achieve wireability in placement of components or modules on a chip," and "how to allocate resources in order to optimize schedules" were among the early ones. In the 1990s, timing closure was a dominant challenge; that is "how to ensure timing convergence with minimal size." Toward the end of that decade so many additional characteristics and constraints, such as (static and active) power, signal integrity, and electromagnetic compliance, had to be considered that industry started to speak of the ever elusive *design closure* (For an analysis of these characteristics and constraints, see Chapter 13 of *Electronic Design Automation for IC Implementation, Circuit Design, and Process Technology*). Yet, that is today's challenge of design automation: "how to specify a function to be implemented on a chip, feed it to an *EDA* tool, and get, without further interaction, a design that meets all requirements of functionality, speed, size, power, yield and other costs."

14.1.1 WIRING CLOSURE

Around 1980, designers realized that the complexity of chips in the decade to come forced them to use more than just a router and occasionally a placer to find a starting point. At the same time, it was obvious that the two tasks were heavily dependent on each other. Routing without a placement was inconceivable, while at the same time a placement might render any routing infeasible. This "phase problem" between placement and routing was tackled with error-prone spacing repairs and wasteful reservations, which turned out to be inadequate. Thus the quest for *wiring closure* was born.

The answer came with the introduction of *floorplans*, data structures that capture relative positions of objects rather than co-ordinates. Once such a "topology" was generated and stored, optimizations could convert it into geometry, often a dissection of the rectangular chip into rectangular regions for the modules. The restriction to rectangular shapes was convenient because the optimization could be organized as trading off dimensions in order to reserve an adequate shape for each module while achieving an optimal overall contour. Under a mild structural constraint this required only polynomial time.

In a sense, floorplanning is simply a generalization of placement. Whereas placement was the manipulation of geometrically fixed objects to arrange them in a configuration without overlaps, floorplanning handled objects for which the shape did not have to be predetermined. It lent itself much better to top-down approaches where the shape and sometimes even the precise size of the modules were unknown. Certain restraints enabled *stepwise refinement* in that the tree representation of the floorplan was forced to be a refinement of a given hierarchy tree. These features caused a strong association of the floorplan concept with hierarchical design.

The salient feature of floorplans is that they allow designers to perform early analysis on their design decisions so that performance can be improved without resorting to lengthy iterations (in its original context floorplanning was presented as an iteration-free approach to chip design!). It thus enables *silicon virtual prototyping*, where sophisticated assessment procedures are integrated with hierarchical floorplan design to adapt modules in a timely manner to their (preliminary) environments, while respecting intellectual property reuse.

Today it can safely be stated that placement has been replaced by floorplanning, followed by a legalization step. Many modern back-end tools start with generating point configurations (with or without size-dependent spacing) or overlapping geometrical objects, mostly by analytic or stochastic optimizers that can handle complex but flat incidence structures with library elements of fixed shape. Once the optimum for this configuration is established, legalizers ensure that all overlap is removed, and that routability is enhanced. Ironically, this was exactly what early fully automatic floorplanners did [1]. Other back-end tools start by partitioning a netlist while giving relative positions to the blocks. Details about the algorithms can be found in Chapter 5 of *Electronic Design Automation for IC Implementation, Circuit Design, and Process Technology*.

In Section 14.2, we work from the definition of a floorplan to the many models and configurations that have served to capture relative positions in one way or another. We emphasize the importance of meaningful and easy to handle metrics for evaluating floorplans, although the in-depth treatment and comparison is left to other chapters in this book.

14.1.2 TIMING CLOSURE

With wiring as the main closure problem of the 1980s, timing closure became the target of the next decade. Two timing closure approaches emerged in the middle of the 1990s, both founded on the observation that gate delay can be kept constant under load variation by sizing [2]. Applying this rule to the whole design flat is one of them, and it has been commercialized successfully. There is however a flaw in this approach: as interconnects get longer, wire resistance cannot be neglected anymore. Unfortunately, keeping the delay constant by sizing is a model valid only when there is no resistance between the driving gate and the capacitive load(s). Line buffering is a patch that often helps, but cannot rescue the method. The other approach is by *wire planning.* It allocates delay to *global* interconnect first, then assigns budgets to modules, and applies the constant delay method to each of these modules. The assumptions of this method are that interconnect delay is linear with its length, and the size of the individual models is small compared to the whole design. The latter is used in two ways: interconnect delay within the modules cannot be reduced by buffer insertion, and modules can be treated as points during the global delay allocation.

If interconnect delay is proportional to interconnect length, path delay is insensitive to module position as long as detours are avoided. Interconnect linear with length is achieved by optimal segmentation of wires with buffers. To achieve minimal delay, buffers should be optimized in size. This fixes line impedance, thus giving up degrees of freedom to keep module delay within the budget. Adapting line impedance changes the delay per unit length.

Fortunately, delay change by size variation around the optimum is small, and by minimizing area during the budget assignment, slack is created that can be used to compensate for additional interconnect delay.

What remains is to implement an algorithm for time budget assignment. This implies trading module size for speed: the smaller the module, the slower the signal propagation through it. If we know the relation between the two performance characteristics for every module, a minimization of the total area under timing constraints is easily formulated. Every path imposes a constraint on the budget assignment, and every budget assignment to a module implies a certain amount of area. The mathematical program obtained in this way can be efficiently solved if the trade-off function is convex, i.e., there is an algorithm with runtime polynomial in the size of the tableau.

There may however be exponentially many paths in a graph, and the size of the tableau could be exponential in the size of the graph. What is needed is a formulation that is polynomial in the size of the graph. Such a formulation will be presented in Section 14.3.3. Before that, we will review constant delay synthesis, and summarize the assumptions of wire planning. We conclude with a section that shows how mild violations of these assumptions can be accommodated.

14.1.3 DESIGN CLOSURE

Up until 2000, the EDA industry had concentrated on tools and techniques for achieving closure with respect to one target design parameter. Moreover, these parameters were mainly related to that part of the trajectory that synthesized mask level data from an RTL description or even lower levels. After wiring and timing, power is likely to become the closure problem of this decade.

Today's chip synthesis requires tools and methodologies for manipulating designs at higher levels of abstraction to meet a large variety of performance constraints. What is needed is a formal system to handle trade-offs between performance characteristics over many levels of abstraction. Such a system will be indispensable to achieve *design closure*, in the sense alluded to at the beginning of this chapter. It should free design trajectories from the need for iterations and interactions to which users resort when their tools fail in one or more aspects to achieve closure, not knowing whether the process will converge or not, and in the latter case not even being sure whether a solution within specifications exists.

The approach hinges on a generalization of the concept of a performance characteristic, and a set of operations over such characteristics that extend and reduce performance spaces without ever losing candidates for optimal solutions with respect to any monotonic cost function. Of course, the key is the effectiveness in keeping the search space small, and the efficiency of the algorithms which do that.

After analyzing successes of the past in achieving closure, and pointing to the efforts of today, we formulate a number of observations that are the basis for the construction of such a formal system, an algebra for trade-offs, the topic of Section 14.4.

14.2 FLOORPLANS

Formally, a floorplan is a data structure that captures the relative positions of objects which get their final shape by optimizing some objective without violating constraints.

This definition does not imply any underlying hierarchical structure. It simply generalizes the notion of placement: the manipulation of objects with fixed geometrical features in order to allocate them without any overlap in an enclosing plane region. The generalization refers to both the uncertainty or flexibility of the shape of some or all of the objects, and the possibility of overlap when these objects have a shape or area associated with them.

It is evident that hierarchical approaches that want to find regions for allocating subsets of modules almost always must resort to floorplanning in some way, because it is neither always wise nor often feasible to predetermine the shape of all modules in a hierarchy. This obvious need for floorplanning in approaches that want to preserve, completely or partly, properties of earlier partitioning, caused an almost complete identification of floorplanning with hierarchical layout design. The elegant formulation of stepwise refinement for layout synthesis, where preserving the hierarchical structure of a functional design occurs automatically by adoptation of a structural restraint that expanded, ordered and labeled the original hierarchy tree, planted this idea firmly in the minds of back-end tool developers of the late 1980s.

This is easily understood if we see that a supermodule's floorplan (that is the relative positions of its modules) was derived from:

■ A preliminary shape of the supermodule
■ A preliminary environment (e.g., pin positions)
■ Shape information for the modules (e.g., shape constraints)
■ An incidence structure (i.e., a netlist)
   and possibly
■ Timing information of external signals
■ Module delay information
■ Path delay information.

The result was subsequently used in

■ Creating a preliminary environment for the supermodule's modules
■ Assessing wire space requirements
■ Estimating the delays.

All these are controlled by the (given or derived) hierarchy.

Figure 14.1 illustrates this, and essentially invites the recursive application of the central floorplan function. The terminology, the creation, and the use of module environments, and the explicit mention of hierarchy in the last item, seems to bind floorplanning exclusively to hierarchical design. However, a flat design with a single supermodule and a single level of leaf cells (undivided modules) is perfectly consistent, and was always the most important practical application of the floorplanning concept. It has rendered classical constructive placement obsolete, because of its flexibility and relative efficiency.

14.2.1 MODELS

In this section, we look at three classes of models for floorplans, not necessarily in chronological order of origin, but fairly representative of what has been conceived over three decades for capturing relative positions of modules.

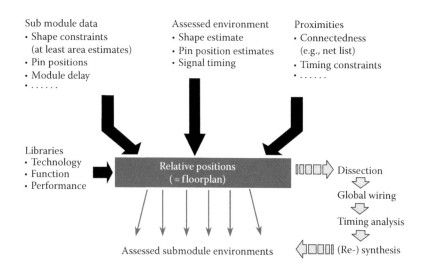

FIGURE 14.1   Floorplan design.

14.2.1.1 GRAPHS

The oldest floorplans are graphs. Drawing a rectangle dissection already produces a graph, where line segments are interpreted as edges and the points where these come together (mostly *T-junctions*, sometimes *crossings*) as vertices. It should be noted that different dissections may lead to the same graph. Special properties that such a graph should have is that its edges can be divided into two classes, *h-* and *v-edges*, and it should possess a plane representation where these edges are represented by orthogonal line segments, one orientation per class. This graph is called the *floorplan graph* (Figure 14.2a).

The first formal floorplanning procedure started with a graph in which the edges represented desired adjacencies. The vertices represent the objects, that is modules, but in the first applications, *rooms*. Obviously, not every such graph can be converted into a rectangle dissection. They should be planar, but that is not enough. In addition, they should have a dual that has a plane representation with the two orthogonal sets of line segments, as described above. Graphs with such a dual are called *Grason graphs* in honor of the first scientist to publish a floorplanning procedure [3]. This pioneer also gave a characterization of Grason graphs, but efficient algorithms for recognizing Grason graphs and constructing an associated rectangle dissection came some 15 years later [4–7].

On the basis of the division into h- and v-edges, we can also divide the edges of the Grason graph, yielding two graphs: the *Grason-h-graph* and the *Grason-v-graph*. By directing these graphs consistently with the relative positions of the rectangles of their vertices, we obtain the *Grason-h-digraph* (Figure 14.2b) and the *Grason-v-digraph*.

Equivalent to this digraph pair (or the colored Grason graph), and also to the colored floorplan graph (i.e., with explicitly given h- and v-edgesets) is the *channel digraph* (Figure 14.2d) where each channel (i.e., maximal line segment) has a vertex and each *T*-junction has an arc pointing

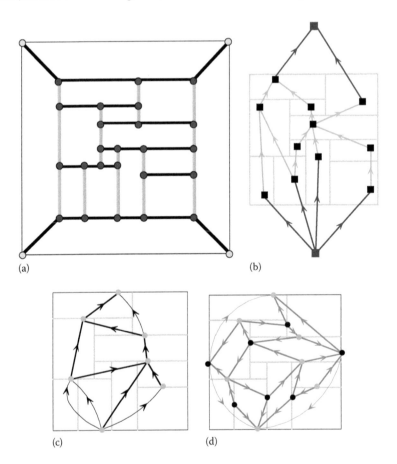

(a)   (b)

(c)   (d)

FIGURE 14.2 Graphs: (a) Floorplan graph with cube extension, (b) Grason *h*-digraph, (c) polar *h*-digraph, and (d) channel digraph.

to the "bar of the *T*" [8]. This graph has rarely been used though it has some characteristic properties and retains generality, where the much more popular *polar digraphs* are not capable of capturing all adjacencies.

Although known already in the first half of the last century, these polar graphs were introduced [9] into layout synthesis in 1970 by Tatsuo Ohtsuki. The *polar h-digraph* (Figure 14.2c) takes the maximal horizontal line segments as its vertices, and has an arc for every rectangle. It does capture incidence of rectangles to line segments, but adjacency between rectangles is not exactly covered. Consequently, the *Grason h-digraph* is a subgraph of the line graph of the *polar h-digraph*.

The relations between the graphs acting as floorplans for the same rectangle dissection are brought together in the table of Figure 14.3, where "D" stands for dualization and "L" for forming the line digraph. The upward arrow toward the polar digraphs and superset signs indicate the loss of information when going to polar graphs. Polar graphs are nevertheless the most popular graph models because they can be transformed into sets of linear equalities that must be satisfied by the rectangle dimensions of any compatible dissection.* This forms a good starting point for sizing using mathematical programming.

14.2.1.2 POINT CONFIGURATIONS

Plane point sets are by now the most commonly used floorplans. If the only information that is used is the sequence of the points (representing the modules) after projecting them on an axis, then it is better to speak of a sequence pair, because all distance information is left unused.

Methods that try to embed a distance space into a euclidean space, and finally into a two-dimensional euclidean space often use eigenvalues to find the plane with maximum spread and to lose as little distance information as possible when projecting on that plane. Depending on the distance metric certain statements about optimality can be made. However, the euclidean distance is not really a useful metric and projection may force modules apart that should and can be kept together. Eigenvalue methods are good mainly for recovering structure in low-dimensional configurations.

With the rise of annealing as a competitive method for improving complex configurations, it was soon considered for floorplan design [10]. The question was how to represent relative positions in such a way that total wiring length was calculated very quickly. The first answer was a pair of sequences, which meant that the moves were transpositions of two elements in one such sequence, but the points were then spaced according to their area (not the square root of area!).

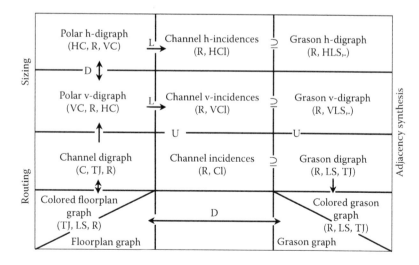

FIGURE 14.3   Relation table.

---

* In fact, these equalities are the Kirchhoff equations from basic electric circuit theory, with e.g., the widths as current intensities, and the heights as voltage differences between the nodes.

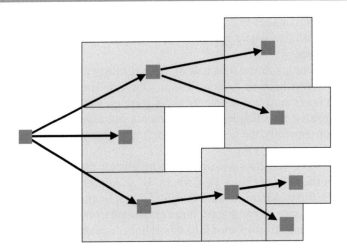

FIGURE 14.4 Labeled ordered tree.

Sequence pairs obtained a more specific meaning when the two permutations of modules implied the following meaning for relative positions:

1. A module m_i is to the left of module m_j if it precedes that module in both sequences.
2. A module m_i is above a module m_j if it precedes that module in the first sequence, and if it follows that module in the second sequence.

Till now sequence pairs are only generated by stochastic algorithms such as annealing. Evaluations are time consuming, although polynomial or even linear, which causes long runtimes, because these evaluations are placed in the inner loop. This makes such algorithms impractical. In favor of sequence pairs, it should be noted that they can easily handle empty rooms without explicitly encoding them in the sequences. This is not possible, for example, with polar graphs!

14.2.1.3 TREES

Trees for representing rectangle configurations have become popular recently, as shown by the appearance of *O-*, *bi-*, *star* and *stair trees*. These representations compete in their efficiency to solve the packing problem for their compatible rectangle configurations. These packing problems are easy to solve for rectangles with fixed shape and orientation. An example is the *labeled ordered tree* (Figure 14.4) from which a packing can be computed in linear time [11]. Among all possible trees with the adequate number of vertices there is one that produces the smallest possible packing. There are however very many possible trees though their number is bounded by the *Catalan number* times the number of permutations, and this number is asymptotically smaller than $(n!)^2$, the number of sequence pairs.

14.2.2 DESIGN

In many ways, one can argue that floorplan design has not been solved adequately. The various algorithms for generating floorplans modeled by graphs are either confined to subspaces that do not have to contain optimum or even near-optimum plans, or produce in essence a number of configurations that cannot be bounded by a polynomial in the number of modules.

14.2.2.1 ADJACENCY-BASED FLOORPLAN DESIGN

The first thoughts on floorplan design were directed toward realizing the required or desired adjacencies. These adjacencies form a graph, and if this graph is a Grason graph, a rectangle

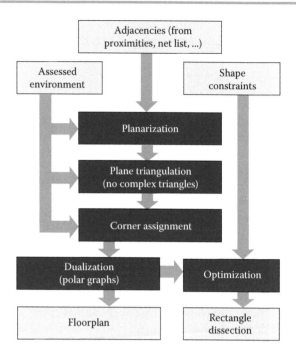

FIGURE 14.5   Adjacency-based floorplan design.

dissection with these adjacencies exists. Linear-time algorithms* for testing graphs for this property exist. In general, these adjacencies do not form a Grason graph, and seldom a subgraph of a Grason graph. The adjacency requirements therefore have to be pruned to such a subgraph and subsequently extended to a suitable Grason graph, either explicitly or implicitly. A generic procedure is given in Figure 14.5.

Graphs with rectangular duals (even graphs with duals, when combinatorially defined) have to be planar. Procedures that try to realize adjacencies therefore often start with heuristics for planarizing the corresponding graph. A straightforward way for doing that is by removing "weak" adjacencies. Removing enough edges will planarize a graph sooner or later. But adding nodes also can have this effect. These "phony" rectangles (because nodes are supposed to represent rectangles) are called wiring nodes in [12]. Other heuristics can be found, but after the first stage in Figure 14.5 the graph is supposed to be planar, so it has a dual and of course a plane representation. But does it have a rectangular dual? Not for a representation with *complex triangles*—that is, representations with circuits of three edges with other edges inside. If that is the case there are a number of options:

- Modify the representation
- Remove the complex triangles
- Triangulate avoiding complex triangles.

In [13], all options are exploited although in the reverse sequence. The remaining freedom can be used for assigning the corners.

The resulting graph has a rectangular dual, and therefore a corresponding rectangle dissection can be found. A constructive procedure for generating such a dissection in an efficient way was published in [5]. An elegant method, based on the observation that construction is in essence assigning T-junctions, was *complete matching* in [14].

A flow such as in Figure 14.5 does not consider shape requirements (not even size requirements) in its floorplan generation. Shape constraints are taken into account only afterward in a floorplan optimization step, mostly a tableau of inequalities holding the floorplan (e.g., by the equations derived from polar graphs) and the shape constraints. This often leads to unacceptable

---

* Do not trust the textbooks on this! Consult the original articles.

results, the main reason why pure adjacency-based floor-planning was never successful. The approach in the next section is much better equipped to take shapes into account.

14.2.2.2 PROXIMITY-BASED FLOORPLAN DESIGN

Point configurations can be generated efficiently depending on the objective. Eigenvalue methods are very fast but are bound to use a euclidean metric. Besides, they usually use projection to lower the dimensionality of the solution, which may reduce distances, leaving other modules unnecessarily far apart. Repeated probing based on eigenvalues makes the generation slow.

Probabilistic techniques, such as annealing, genetic algorithms, and evolution, need very fast evaluations, which makes them resort to crude measures. For fixed rectangles, techniques such as in Section 14.2.1.3 can be of advantage, but they cannot handle more flexible shape constraints.

Currently the best results seem to be obtained with analytic optimizers of the quadratically convergent Newton-type, followed by legalizers. They are usually applied to flat designs, that is designs without any hierarchical structure, regardless of the number of objects to be treated. Since these analytic optimizers do not produce nonoverlapping configurations of the objects to be allocated, they are classical examples of floorplanners, although the literature often calls them (global) placement procedures, which is only justified when combined with a valid legalizer.

In any way, critical for floorplanners is the metric by which the results are evaluated. In Chapter 5 of EDA for IC Implementation, Circuit Design, and Process Technology, this issue will be addressed.

14.2.3 CONSTRAINED FLOORPLANS

Much more can be said about floorplanning, especially about the misconceptions and mis-interpretations surrounding it. A particular topic that attracts many scientists (but few tool developers) is the avoidance of special constraints. Early in its history the restraint of preserving sliceability was recommended [1]. It was exactly that constraint that made floorplanning fit as a glove with stepwise refinement from a given hierarchy. Floorplans with that property could be represented as a tree, but there were other and more important properties. The fact that slicing structures have the minimum number of "channels" may have had its day as well as the conflict-free routing sequence of those channels. But the fact that many algorithms that are at least *NP*-hard for general floorplan structures become tractable for slicing structures, remains of paramount importance. Among these algorithms is of course the floorplan optimization problem [15]. Later it was proven by Stockmeyer [16] that even the simple special case of fixed module geometries and free orientations was *NP*-hard! Given a slicing tree and the more general shape constraints defined in [17] makes floorplan optimization efficiently solvable. Also, labeling a given tree to find the optimum slicing tree with that topology can be done in polynomial time [18]. And last but not the least, given a point configuration (but actually a pair of sequences), the optimal compatible slicing structure can be found in polynomial time. "Compatible" means that every slice in the floorplan corresponds exactly with a rectangular set in the point configuration [19].

Considering the place and the nature of floorplanning capable of handling uncertainties, flexibility, and preliminary positions, there is not much sense in paying any price for also including nonslicing floorplan topologies [20,21].

14.3 WIREPLANS

A wireplan is an incidence structure of modules and global wires. A global wire is one that can be sped up by buffer insertion. The latter implies that its (minimum) length and layer are known. For the modules, a trade-off between speed and size should be available. If timing constraints are given for the inputs and outputs, time budgets can be assigned to the modules so that the total size is minimal. The modules can then be designed using the concept of constant delay so that they do not exceed their budget.

14.3.1 CONSTANT DELAY

14.3.1.1 DELAY MODELS

In [2], the observation was made that sizing a gate proportional to its load keeps the gate delay constant. This confirms the delay model introduced in [22] that reads as

$$\tau = g \frac{C_L}{C_{in}} + p$$

where
   p is the inherent parasitic delay
   g is the computational effort

The latter depends on the function, the gate topology, and the relative dimensioning of the transistors. Both p and g are size-independent, i.e., increasing the gate size with an arbitrary factor does not affect these parameters. The quotient C_L/C_{in} is called the *restoring effort* of the gate.

14.3.1.2 SIZING FORMULATION

Constant delay synthesis starts from a network of gates where each gate has a fixed delay assigned to it. In order to realize this delay, its *restoring effort* has to have the appropriate value. Satisfying these requirements for all gates in the network simultaneously leads to a set of linear equations, as can be seen from Figure 14.6.

Using the notations in that figure the set can be written as

(14.1)
$$c = q + N f^D c \quad \text{or} \quad (I - N f^D) c = q$$

The algebraic and numerical aspects of this set have been analyzed in [23] where a detailed derivation can be found as well.

The sizing procedure is summarized in Figure 14.7. It shows that if synthesis produces, beside a netlist, also a vector f of reciprocal values of restoring efforts (thereby fixing the delay of each gate), solving set Equation 14.1 yields the vector c of input capacitances (and implicitly the gate sizes). Placement of these sized gates may reveal changes in the wire capacitances, and consequently in the vector q. Iterating the calculation converges to a set of sizes and imposed capacitances consistent with the placement.

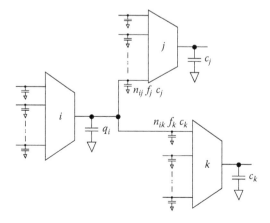

FIGURE 14.6   Composing the load capacitance of gate i: it consists of an "imposed" capacitance q_i, and the input capacitances in its fanout, which are proportional to the respective load capacitances.

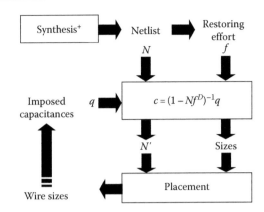

FIGURE 14.7 Computational procedure for sizing the gates.

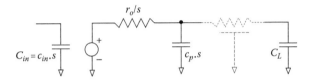

FIGURE 14.8 Wire resistance, here de-emphasized, is neglected in the model underlying the constant delay synthesis.

14.3.1.3 LIMITS

The gate model of Section 14.3.1.1, which underlies the procedure in Section 14.3.1.2, assumes negligible resistance between the gate and its load. For relatively short wires in the network this assumption is certainly acceptable. Modules of up to 100,000 gates can be safely dimensioned in this way. For larger modules wires may get so long that wire resistance is no longer negligible. In today's chips, it is even said that interconnect dominates performance. It is not possible to derive an equally simple and numerically robust iteration method when resistance has to be accounted for. Moreover, improving speed by sizing does not work when there is line resistance. Line segmentation and buffering is effective up to a certain level. This limits the validity of constant delay synthesis (Figure 14.8).

14.3.2 WIRE PLANNING

14.3.2.1 PROBLEM STATEMENT

Larger networks will have wires with nonnegligible wire resistance; if some wires have to connect widely separated points this can only be avoided in very regular designs by only-neighbor connections. To extend the methodology of the previous section to these designs, such wires have to be treated appropriately. Not uncommon in complex designs is dividing the whole into parts that can be handled by the sizing procedure. Such a partitioning often comes naturally in functional design. The time plan can then identify connections that allow for (or have to have) delays that exceed a critical value. If this value is taken to be the delay of a segment in an optimally buffered interconnection, the thus identified interconnect can be characterized as wires with a length that enables delay reduction by segmentation and buffering. It has been shown that the critical value only depends on the chosen technology and not, for example, on the layer to which the wire is going to be assigned [23].*

---

* The delay of a segment in an optimally buffered line is indeed a process constant while the length of a segment depends on the layer. That length does not depend on the buffer sizes!

The approach is to allocate the delay of such identified interconnect first, and then divide the remaining time budget over the modules. These modules do not contain such interconnect and can be sized with the procedure described in Section 14.3.1.2, and thus be held within the budget assigned to that module. The problem to be solved is therefore: given a network of modules connected by *global* interconnect and timing requirements on paths in that network, assign size and delay to each module so that a network can be produced that satisfies the timing requirements. Of course, the relation between size and delay of each module has to be known. This relation will be discussed in Section 14.3.2.3. But first, possible properties of global interconnect will be exploited.

14.3.2.2 WIRE DELAY

Optimally segmented wires have a delay linear in the length [23]. And if wire delay is linear with its length, its contribution to the path delay between two points on a chip does not depend on the position of the modules on this path. Even stronger, if the path is detour-free, the wire delay on a path is fixed: it only depends on the coordinates of the two endpoints.

Of course, it may in general not be possible to realize the paths of a network in a detour-free manner when the coordinates of these endpoints are given.* An easy net-by-net criterion has been formulated in [23], answering the question whether for a given endpoint placement, a detour-free realization exists. Even a point placement can be derived straightforwardly if that criterion is tested (and satisfied) for all nets. The given endpoint placement is called *monotonic* in that case. Also it has been demonstrated there that every acyclic functional network has a realization that makes every endpoint placement monotonic. For logic networks a synthesis procedure has been implemented that preserves monotonicity [25].

Here we assume that an acyclic network is given with the coordinates of its endpoints and that it can be realized without path detours. In Section 14.3.4, we discuss what can be done if detours cannot be avoided and extra wire delay is incurred. The endpoints can be the location of connections to the outside, entries and exits of a supermodule containing all modules in the network, the positions of preplaced registers, etc.

14.3.2.3 MODULE SIZE AND DELAY

The relation between the size and delay of a module is essentially a trade-off: speeding up a module is usually paid for with additional area. Plotted in an area versus delay quadrant, the realization values will be in a region bounded below for both performance characteristics by a monotonically decreasing function with the points of interest on that function. In practice, that function can be well approximated by a convex curve. In [26] this curve is approximated by finding a best fit for $(delay - c)\,(area + b) = a$ through the identified *Pareto* points among the design points.

14.3.3 TIME BUDGETTING

14.3.3.1 PROBLEM FORMULATION

Assume that a wireplan is given with a monotonic placement of the endpoints (in the sense of Section 14.3.2.2). This means that there exists a point placement for its modules enabling simultaneous detour-free routing of all paths between inputs and outputs. The Manhattan or L_1 length of any input–output path ($L_{i,o}$) can therefore be made equal to the L_1 distance between the input and output terminal pins of that path. The wire delay on such a "monotonic" path is proportional to this length, regardless of the position of the modules on this path, because the delay of optimally buffered wires is linear with length, and the summed delay of the wire segments between the modules is the summed length of the segments multiplied by some

---

* In practice, the overwhelming majority of two-point routes are without detours as experiments in [24] show: only 1.37% were detoured!

constant. The main conclusion is that we can calculate the wire delay of a monotonic path directly from the input and output pin positions. Because module positions are not needed to do that delay calculation, it is possible to do time budgeting even before relative module positions are known.

To obtain the total path delay, the delay of its modules has to be added to the wire delay. That delay depends on the implementation of the modules. Let P^{jk} be the set of all paths from input j to output k; in a given wire plan and p a single path in this set. Then the constraints imposed by the timing requirements read as follows:

$$\forall_{j,k} \forall_{p \in P_{jk}} \left[D_{W_{jk}} + \sum_{m \in p} D_m(A_m) \le T_{req_{jk}} \right]$$

These constraints simply state that for each single path between two pins, the wire delay plus the summed module delay should not exceed a given timing constraint. $T_{req_{jk}}$ is an upper bound on the delay between pins j and k. $D_{W_{jk}}$ is the wire delay on the path, and thus equal to L_1 distance between both pins multiplied by a (technology/layer dependent) constant. The minimum (Pareto) delay of a module m is approximated by $D_m(A_m)$, a function of its size (area) A_m.

The first question that can be posed is whether the wire plan can be realized in a footprint of a given size, while satisfying the timing requirements. It can be answered by minimizing total area, ΣA_i, subject to the given constraints. If the obtained minimal area is larger than the area of the footprint, the wire plan does not fit. Otherwise, not only is the answer to the question obtained, but also the available "slack" area that can be used to adapt modules to the wire characteristics.

The method to use for solving the above optimization problem depends on the delay model, that is on the module delay as a function of module area. Linearizing the D_m curves, and using linear program solvers yields a method which can be polynomial in the size of the program, but the linearizations need to be sufficiently coarse. Other time budgeting techniques have appeared in the literature. Among the more popular techniques, we have to count the *zero-slack algorithm* of Nair et al. [27] and its variants, the *knapsack of approach* of Karkowski [28] and the convex optimization of Sapatnekar et al. [29].

This optimization problem can be solved by *geometric program* solvers if both object function and constraints have the form $f(t) = \Sigma c_j \prod t_i^{aij}$. Choosing for the module delay

$$D(s) = d/s + r.$$

achieves that. The function captures the essence of common area-delay trade-offs. D is the delay as a function of a size factor s, making the area of a module $A(s) = a_0 s$. For each module d, r and a_0 are given. Now, the resulting program for n modules is

$$\text{minimize} \sum_{i=1}^{n} a_{0i} s_i$$

with constraints of the following form:

$$\sum_{j \in p} \left[d_j s_j^{-1} + r_j \right] + D_{W_{jk}} \le T_{req_{jk}}$$

A certain class of geometric programs (including the example above) has *posynomial* functions for both object function and constraints $C_j > 0$. By applying the transformation $t_i = e^{xi}$, both object function and constraints of the posynomial formulation become convex, and can be solved in polynomial time. Further, general convex solvers can be used. The number of constraints depends on the number of paths. Unfortunately, the latter is exponential in the number of edges, resulting in a program of unmanageable size.

14.3.3.2 COMPLEXITY REDUCTION

The number of *constraints* is determined by the total number of input–output paths. The number of terms in a constraint is the number of modules in the associated path. Note that the number of terms is far greater than the number of variables (module sizes), since modules may be on several paths. The number of terms is of interest here, and renders the straightforward formulation impractical already for relatively small designs.

It is easy to see that the number of paths in a graph can be exponential with regard to the number of edges, but there may be some hope that in practical designs the number of paths is small. This is not the case, and an approach based on considering all paths is out of the question.

However, all the necessary information can be captured by maintaining a number of variables at each node. The delay from a primary input i to an internal node n depends on the *maximum* delay of the paths from i to n. If s of those subpaths exists, s constraints appear for each subpath from n to an output. Since only maximum delays count, only the delay of the slowest (sub)path is of interest. To use that observation, we introduce at each node, for each primary input in its fanin cone, a variable representing the maximum module delay on the subpath from the input to this node. Using the fact that we have a directed, acyclic graph, this value can easily be calculated as the sum of the delay of this node and the *maximum* of the values of its predecessors.

For example, in Figure 14.9, there are two paths between I_2 and O_2. With w_x denoting the wire delay of segment x, and M_y the module delay of module y, we find (in the straight forward formulation) as constraints for these path

$$\omega_2 + M_2 + \omega_3 + M_1 + \omega_6 + M_4 + \omega_8 \leq T_{req, I_2 O_2}$$

$$\omega_2 + M_2 + \omega_4 + M_3 + \omega_7 + M_4 + \omega_8 \leq T_{req, I_2 O_2}$$

Using monotonicity and linear wire delay

$$\omega_2 + \omega_3 + \omega_6 + \omega_8 = \omega_2 + \omega_4 + \omega_7 + \omega_8 = Dw_{I_2, O_2}$$

we obtain the form of Section 14.3.3.1

$$M_2 + M_3 + M_4 + Dw_{I_2, O_2} \leq T_{req, I_2 O_2}$$

$$M_2 + M_3 + M_4 + Dw_{I_2, O_2} \leq T_{req, I_2 O_2}$$

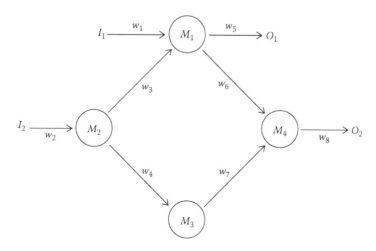

FIGURE 14.9   A circuit of modules.

In the formulation of this section, introducing the variable AT_n^1 to denote the maximum module delay from input I to node n, we find

$$AT_{M_4}^{I_2} + Dw_{I_2,O_2} \leq T_{req,I_2O_2}$$

$$AT_{M_4}^{I_2} = M_4 + \max\left(AT_{M_4}^{I_2}, AT_{M_3}^{I_2}\right)$$

$$AT_{M_3}^{I_2} = M_3 + \max\left(AT_{M_2}^{I_2}\right)$$

$$\vdots$$

$$AT_{M_2}^{I_2} = M_2 + 0$$

Although the savings in efficiency are not noticeable in such a small example, it is obvious that they are tremendous in large examples. In fact, the number of constraints becomes equal to the number of nodes plus the product of the number of edges and the number of primary inputs. Each node has for each primary input in the fanin cone a variable representing the module delay up to here. The number of these variables is bound by the number of primary inputs. For each edge, a constraint is generated for each of those variables at the source node, linking the delay of a node with the delay of its successor. For each node, an additional constraint representing its own delay is generated. Both kinds of constraints are simple and have only two terms.

This shift from paths to nodes makes the *timing graph* from [30] a more convenient model. It is in essence the line digraph of the original: the arrows become nodes, and two arrows that have a head–tail connection at a module node yield an arrow in the timing graph. The conversion is illustrated in Figure 14.10.

In this model, we can distinguish different delays between different module pin pairs and associate them with the corresponding arrow. Figure 14.11 shows the timing graph that corresponds with the wire plan of Figure 14.9.

As an alternative to [30], we have for each node as many variables AT as there are primary inputs in its input cone. In Figure 14.11, node n_3 has two variables associated, $AT_{n_3}^{I_2}$ and $AT_{n_3}^{I_2}$. In this way, all input–output delays can be taken into account. The complexity of the problem is now, as is the case with static timing analysis, polynomial in the size of the graph.

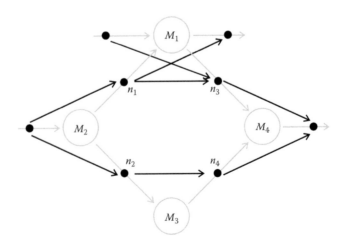

FIGURE 14.10   The line graph of the digraph in Figure 14.9.

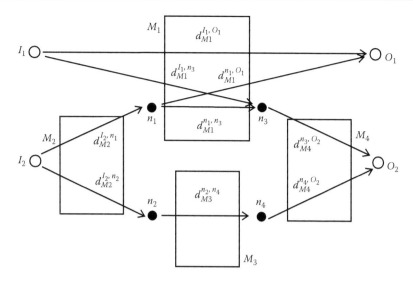

FIGURE 14.11   The timing graph of the digraph in Figure 14.9.

The problem formulation after this complexity reduction becomes

$$\text{minimize} \sum_M A_M$$

subject to

(14.2)
$$\forall I \in C(m) \; \forall (m,n) \in E\left[AT_n^I \geq AT_m^I + d_M^{mn}(A_M) \right] \qquad (E)$$

$$\forall I \in P_I \; \forall O \in P_O \left[AT_n^I \leq T_{req,IO} - Dw_{IO} \right]$$

where
   PI and PO are the sets of primary inputs and outputs, respectively
   E is the set of edges in the timing graph
   $C(m)$ is the set of primary inputs in the input cone of m

Moving from the straightforward path-based formulation of Section 14.3.3.1 to the node-based approach of this section enables an efficient area optimization under timing constraints. But there are some more techniques to reduce the problem even further and save on runtime.

A simple reduction in the number of constraints is obtained by not restricting the formulation to variables derived from looking at maximum delays from the inputs. The first set of constraints in (E) contains

$$\sum_m \gamma(m) |C(m)|$$

constraints, $\gamma(m)$ being the outdegree of node m. If the generation of constraints is done with reference to the primary outputs, the total number of constraints is determined by output cones and indegrees. This may yield a quite different and sometimes considerably lower number of constraints. Although a graph may be analyzed for this comparison in a straightforward manner, the tableau of constraints may be generated for both cases after which the smaller one is used. The computation time for generating such a tableau is small compared to the time taken by the actual computation.

A more intricate improvement in efficiency is obtained by using a technique called *pruning* [31]. It reduces problem size, degeneracy, and redundancy without sacrificing accuracy. As illustrated by Figure 14.12, the basic pruning operation is a graph transformation that replaces a node

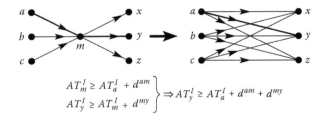

$$AT_m^I \geq AT_a^I + d^{am} \atop AT_y^I \geq AT_m^I + d^{my} \Bigg\} \Rightarrow AT_y^I \geq AT_a^I + d^{am} + d^{my}$$

FIGURE 14.12   The basic pruning operation with constraints for the bold path.

with touching arcs, and replaces it with other arcs. The variables on the arcs are such that the associated optimization problem is equivalent to the original one.

Generally, solver performance depends on the number of constraints, the number of variables, and the total number of terms in the constraints. Pruning affects these numbers. It is possible to assign to each node a *gain*, a measure of the benefit in case this node was pruned, taking these effects into account. If, for example, two constraints with four variables are replaced by one constraint with six variables, this may or may not be beneficial, depending on the used solver. This is reflected by a positive or negative gain. Nodes are greedily pruned until no nodes with positive gains exist anymore.

In [31], only the numbers of variables and constraints are taken into account, and given equal weights, but one can also consider the number of terms, and tailor the associated weights to the solver used. In the original formulation, only one *AT* variable resides at each node, while one for each primary input may also exist. Therefore, the pruning procedure has to be adjusted to calculate gains for *AT variables* rather than nodes. In this way, an *AT* at node *n* may be pruned for primary input I_1, but not for primary input I_2. Another way of looking at this is that *for each primary input*, a timing graph is being constructed, consisting of the input and its fanout cone, and to which the original pruning procedure can be applied. Then, the resulting optimization program is simply the sum of object functions and the concatenation of the associated constraints. The pruning procedure requires only one graph traversal, and results not only in a dramatically more compact formulation, but also one that is numerically better conditioned.

14.3.4 ROBUSTNESS

The time budgets are based on assumptions concerning path lengths, monotonicity, and trade-offs. In real life, however, some of these assumptions may be hard to meet later on: one may, for instance, sometimes need to deviate from monotonicity. Another uncertainty is the accuracy of the trade-off models. More generally, one can say that in a noniterative design flow, one needs a certain amount of "slack separation"* in order to obtain "robustness" with respect to uncertainties later on. Fortunately, slack on the majority of paths can be introduced at very low cost.

First, the formulation itself ensures a certain amount of slack: wire delay is calculated as the path length times some constant. When the modules are realized, however, they will occupy space, effectively reducing wire length, and thus delay. Second, the formulation ensures that all input–output pairs will have a critical path: a path whose delay equals the constraint. This implies that every module is on a critical path. It does not however imply that every wire segment is. This offers the possibility of detouring these wires. Third, if simple bounds on area are used, this may also result in slack. Finally, wire delay is calculated for ideally buffered wires. If logic is pulled out of the module, and spread over the wires, buffers may be replaced by "useful" gates. Therefore, wire delay estimation is kept conservative.

If the slack inherent in the formulation is still not sufficient, Westra et al. [26] introduces a way to provide additional slack at little area cost. It also has the advantage that *truely critical paths* that limit circuit performance most are revealed, making it clear where the main thrust of design effort should focus.

---

* *Negative slack* means constraints are violated, (*positive*) *slack* means there is room to tighten a constraint, and *slack separation* is the difference between the most critical path and other paths.

14.4 A FORMAL SYSTEM FOR TRADE-OFFS

It is interesting to look back on how wiring closure was achieved, though keeping in mind that in those days wiring was realized in two or three layers and allocated to channels between modules. Critical was the generalization of placement, concerned with the arrangement of geometrically fixed objects without overlap in the plane (possible bounded or preferably small) into floorplan design, where only relative positions of objects are being fixed. Accepting slicing as a structural restraint then guaranteed sufficient wiring space that could be treated uniformly with the modules during sizing, and yielded a conflict-free sequence of channel routing problems. In addition, it rendered several optimizations tractable, including sizing which was in essence a trade-off between the dimensions of height and width. These dimension pairs are partially ordered by a *dominance relation:* one pair dominates another if neither of its dimensions is larger than the corresponding dimension of the other pair.

Similar observations can be made with respect to timing closure. The essential shift there is from sizing with speed as an arbitrary outcome to timing constraints with size beyond control. For smaller modules the concept of constant delay enabled closure. However, the underlying assumption that the delay of a gate can be kept constant by varying its size linearly with its load is only valid when resistance between the gate and its load can be neglected. This required planning of all wires for which resistance could not be neglected. They take a considerable but constant part of the timing budget leaving the remainder to be divided over the modules. Again this comes down to trade-offs, now between size and speed, and a corresponding *dominance relation.*

After timing, the primary concern for many of today's *SoC* devices is power consumption, which if appropriate paradigm shifts can be found, will lead to similar trade-offs between speed and power. The problem however is that performance characteristics can no longer be dealt with by considering them in isolated pairs or triples. What is needed is a formal system that allows us to manipulate a design without losing potentially optimal final solutions. Such a system is the topic of this section, but before introducing it we observe the following.

Closure has up to now been achieved by Pareto-style trade-offs, where every realization implies a combination of values of performance characteristics. These values are linearly ordered, though the realization points are partially ordered by dominance. Only points that are not dominated by others can be optimal under monotone objectives. At the end, the best one with respect to such an objective is chosen. The two examples above enable hierarchical application of the methods.

Today, it is of paramount importance to avoid premature commitment while being unaware of the consequences, because of the many relevant performance characteristics that interact with different levels of design. Choosing the best one at a certain level may preclude optimal ones in later stages. Yet, multidimensional Pareto analysis may be, and mostly is, very complex.

14.4.1 REALIZATION SPACES

A performance characteristic is a set of realization values that indicate a quality aspect of a product or design step. Performance characteristics that quickly come to mind in chip design are *size*, *delay*, and *power*. Considered by themselves they are of the type "the smaller the better." Without loss of generality we will assume all characteristics to be of that type. Another property of these classic performance characteristics is that they are totally ordered.[*] That is, for every pair of realizations we can say whether one is better than the other with respect to such a parameter. The performance characteristics measurable in the final product often share this property. Although performance characteristics without that property do not easily come to mind, we will not adopt such a constraint for performance characteristics. We assume that they are ordered, i.e., there is a reflexive, transitive, and anti-symmetric relation associated with any performance characteristic, but not all of its values are comparable. Any trade-off between totally ordered performance characteristics may produce such

---

[*] A performance characteristic whose ordering is total (or linear) is called a basic performance characteristic here.

a relation. Power and delay, for example, form pairs that are so ordered by dominance: such a pair dominates another pair if it is at least as good in both aspects. Obviously, two pairs each of which improves upon the other in some aspect are not comparable. If we consider a number of realizations of a given product specification and compare them on the basis of power consumption and speed, we are in fact looking at a number of pairs that are *partially ordered*. We call such a partially ordered set of realizations with respect to some performance characteristic a *realization space*, and its ordering relation *dominance*, i.e., a realization *dominates* another realization if it is at least as good with respect to every concerned performance characteristic. Abstractly, a *realization space* is simply a set Q with a partial order $\preceq Q$, and we say $q \in Q$ *dominates* $q' \in Q$ whenever $q \preceq Q q'$.

Let us consider all performance characteristics that might be relevant for a design. Some of these characteristics will be measurable in the final result. Others might be absorbed during the design process after they have played a role in certain stages of that process. It may even be so that certain characteristics have not been encountered at all upon completion because early decisions precluded them. That is, at each stage of the design process, realization spaces concerned with subsets of the performance characteristics are being manipulated and transformed into other realization spaces. It is important to note here that when realization spaces are combined to form a new space they induce a new partial ordering. In other words, the result is also a realization space. We no longer distinguish spaces concerned with basic performance characteristics and results that are associated with partial orders!

Each element (or point) in a realization space S represents a (partial) realization of a system. For convenience, we write them as vectors, $\bar{c} = (c_1, c_2, \ldots, c_n)$, where each c_i is a realization value of a characteristic Q_i. Then $\bar{a}$ in S *dominates* $\bar{b}$ in S ($\bar{a} \preceq s \bar{b}$) if $a_i \preceq_{Qi} b_i$ for all i. Dominance still expresses the fact that the underlying realization is in no aspect worse than the other. It can be raised to subsets of a realization space to say that any point in the dominated subset can be matched or improved upon by some point in the dominating subset: a subset A of a realization space dominates another subset B if for every element $\bar{b} \in B$ there is an element $\bar{a} \in A$ such that $\bar{a} \preceq c \bar{b}$; this is written as $A \preceq B$. If A and B dominate each other, they are called *Pareto-equivalent* which is denoted as $A \equiv B$. Pareto-equivalence between two subsets means that neither subset contains a point that cannot be matched by a point in the other subset.

Realization points that are strictly dominated by other points in the same set can be removed without losing interesting realization options. In general, we do not want to maintain larger sets than necessary. This brings about the notion of minimality: a subset C of a realization space is *Pareto-minimal* if the removal of any of its elements yields a subset that is not Pareto-equivalent to C. That is, no element dominates another element in a Pareto-minimal subset. Or in other words, the Pareto-minimal subsets are *anti-chains* of the partial order.

Note that two Pareto-equivalent subsets need not be the same. However, when they are Pareto-minimal they have to be identical! More disturbing is the fact that not all realization spaces have a unique Pareto-equivalent subset that is Pareto-minimal. Whenever it exists for a realization space S we denote it by $\min(S)$. A sufficient requirement for such a *Pareto-frontier* is that every chain in the space has a smallest element (i.e., the space is *well-ordered*). This implies that *every finite realization space has a Pareto-equivalent subset that is Pareto-minimal.* In the present context, we mostly have only discrete points obtained by profiling in the realization spaces. From that perspective, the implication is certainly satisfactory. We assume in the rest of this section that all realization spaces are well-ordered.

14.4.2 OPERATORS

Of course, the unbridled combination of the performance characteristics yields immense, unmanageable spaces. Essential, therefore, is an efficient algorithm that takes a realization space and returns a so-called *Pareto frontier*. This potentially reduces realization spaces, and might, if applied intelligently, keep a design flow manageable. Other operations may be defined provided that no potentially optimal candidate realizations are lost. Operations should preferably *preserve minimality*, i.e., applying the operation on minimized sets produces a minimal resulting set. However, a number of obviously necessary operations do not have that property. In such cases, we have to settle for *optimum preserving*, meaning that minimizing the result of such an operation on minimized operands, is the same as minimizing the result of that same operation on the unminimized operands.

14.4.2.1 MINIMIZATION

We denote by min(C) the unique Pareto-minimal set of a realization spaces C. It contains all realizations from that space that we would be interested in practice because all other realizations are dominated by some realization in min(C).

We call the operation to obtain that subset *minimization*. It is the key operation in the system. It would be nice if other operations when applied to minimized realization spaces would not require minimization to obtain the Pareto-minimal equivalent space. More precisely, if we have an operator Ψ that takes n spaces as its operands, we would like

(14.3) $$\min(\Psi(C_1,..., C_n)) = \Psi(\min(C_1),..., \min(C_n))$$

We say then that Ψ *preserves minimality*. In that case minimization after applying operator Ψ is unnecessary. The result is always minimal when the operands have been minimized, which saves time in computing the results of the operation.

Unfortunately, some indispensable operations will not preserve minimality. That does not mean that it is not advantageous to apply the operation on minimized spaces. The computational effort spent on these generally smaller spaces is almost always considerably less than when the unreduced spaces are the arguments. If only a Pareto-equivalent subset is obtained, that is

(14.4) $$\Psi(C_1,..., C_n) \equiv \Psi(\min(C_1),...,\min C_n))$$

Subsequent minimization then yields the same realization space as when the operation was applied to C_1, ..., C_n. So by minimizing operands, which are often intermediate results in the design process, before applying Ψ, no optimal realizations are lost. Every practical operator should *support minimization* in that sense, for otherwise all realizations have to be analyzed to identify the optimal ones.

If Pareto-equivalence is a congruence with respect to the operator Ψ, that is;

$$\forall_{1 \leq i \leq n} \left[C_i' \equiv C_i \right] \rightarrow \left[\Psi(C_i',...,C_n') \equiv \Psi(C_i,...,C_n) \right]$$

then the operation necessarily supports minimization. This means that to prove that minimization support it suffices to show that the operator preserves dominance:

$$\forall_{1 \leq i \leq n} \left[C_i' \preceq C_i \right] \rightarrow \left[\Psi(C_i',...,C_n') \preceq \Psi(C_i,...,C_n) \right]$$

14.4.2.2 FREE PRODUCT

When designing subsystems independently with their own realization spaces and tradeoffs, they have to be put together later, which means that their realization spaces have to be combined. The corresponding operator is called *free product*.

Let C_1 and C_2 be realization spaces each with their own performance characteristics. The free product $C_1 \cdot C_2$ consists of the points $\bar{c}_1 \cdot \bar{c}_2$, the concatenation of the "vectors" of each point $\bar{c}_1 \in C_1$ and each point $\bar{c}_2 \in C_2$

Obviously, not only is Pareto-equivalence a congruence with respect to free product, the operator even preserves minimality, which was to be expected when its purpose is considered.

14.4.2.3 UNION

To enable a choice between realizations resulting from different design paths, the operator *union* is introduced. It can be thought of as the realization points of both spaces being brought into one and the same space spanned by the same set of performance characteristics as the two individual spaces. The notation is $C_1 \sqcup C_2$.

The operator does not preserve minimality. Obviously, a design trajectory may produce a realization that is dominated by one resulting from the other design path. So, minimization may result in a subset strictly included in its operand. But minimization is supported, because it can be seen as easily that $C_1 \sqcup C_2 \equiv \min(C_1) \sqcup \min(C_2)$. It is even true that Pareto-equivalence is a congruence with respect to the union operator.

14.4.2.4 CONSTRAINTS

When we have a set of realizations, some of which are invalid because of additional constraints, then we can apply a constraint to filter out only those realizations that satisfy the constraint. Such constraints can be formally expressed as a set D of acceptable realizations or equivalently as a proposition on the realization space, identifying the acceptable realizations. Application of the constraint D to a realization space C leaves those realizations of C satisfying the constraint. We denote the new space as $C \sqcap D$.

When defined in that general way, constraints present a problem. It may easily happen that a constraint filters out a realization that dominated one that was not filtered out. Minimization cannot be supported in this way, because that operator may have removed such dominated realization points, while they may be part of the minimized result of the constraint applied to the unminimized space. The constraints in our algebra—we will call them *safe constraints*—will never have a dominating realization passing while the dominating realization fails the test. More precisely, a constraint D is called a safe constraint when for every $\bar{a}$ and $\bar{b}$ with $\bar{a} \preceq \bar{b}, \bar{b} \in D$ implies that $\bar{a} \in D$. The notation for applying a safe constraint D on a realization space C is $C \stackrel{\vee}{\sqcap} D$.

14.4.2.5 PROJECTION

In certain design stages, certain performance characteristics may no longer be relevant for future decisions. Projection can then be used to remove such a dimension of the realization space. In general, minimality is not preserved by projection, but dominance is. Therefore congruence and hence minimization support follow.

14.4.3 COST FUNCTIONS

The selection of a final realization point often happens with a cost function. Such a cost function should be one which never selects a point that is dominated under the space's partial order by another point. If that were possible, the whole formal system constructed here would be pointless, because the purpose was to manipulate and form realization spaces without losing any candidates for optimal realizations. In the process, strictly dominated realizations are being removed when reducing a space to the Pareto frontier. To ensure that cost functions are in concord with the motivation for building this formal system, we require *monotonicity*, i.e., whenever $\bar{c}_1 \preceq \bar{c}_2$ the cost function f has to satisfy $f(\bar{c}_1) \leq f(\bar{c}_2)$.

Thus defined, it requires cost functions to assign real numbers to every point in the final realization space. This is perfectly in order when the real objective is like that, as for example the chip size (or area). Often the selection is not so unambiguous, and designers resort to weighted combinations. Setting these weights is more often than not a trial-and-error affair. Yet the outcome may depend heavily on the exact setting. In essence, it is a consequence of restricting the range of cost functions to a basic performance characteristic. This is not necessary. If we allow the range to be any performance characteristic, that is any partially ordered set, then a simple adaptation of the definition of monotonicity preserves the essential property cost functions should have.

Monotone functions from a realization space C into a performance characteristic are then defined to be functions satisfying

$$\forall_{c_1, c_2 \in C} \left[\overline{c}_1 \preceq \overline{c}_2 \rightarrow \overline{c}_2 f\left(\overline{c}_1 \preceq f(\overline{c}_2) \right) \right]$$

Thus defined they still can only select realizations in the Pareto-minimal set of the space as optimal. Moreover, any member of that set can be selected by such a function. That is, for every point in the Pareto-minimal set of a realization space, there is a monotone cost function that will select that point as minimal. This means that further reduction of spaces is not possible without giving up the guarantee that no potentially optimal candidate realizations are lost, if the cost function is not known. With the generalized cost function however a partial selection of candidates is possible, leaving fewer realizations to consider in subsequent steps.

14.4.4 CONCLUDING REMARKS

This section describes the algebra $(C, \min, \bullet, \sqcup, \downarrow, \check{\sqcap})$, where C is the set of all possible realization spaces that can be encountered during the design of a chip. The algebra was introduced in [32] where it mainly served as a formal system for runtime reconfiguration. The manipulations and calculations then have to take place on mostly resource-constrained devices. Here it is used for design-time exploration with a wide choice in computer power and lesser constraints on compute time. Not surprisingly, different algorithms may be chosen in these two application areas.

In the latter application, the algebra is meant for supporting the design while it forms realization spaces by combining subsystems and their realization spaces. The number of realizations, also when restricted to Pareto-minimal sets, can grow very fast, and it is therefore crucial to keep the space sizes small. Reducing them to their Pareto-minimal spaces is the most important means here, but pruning the space may still be necessary. The challenge is then to find the best approximation of realization space by one with a limited number of points [33,34].

To date the most efficient minimization algorithm has time complexity $O(N (\log N)^d)$ [35], where d is the number of dimensions of the realization space and N the number of realizations. The number of points for which the algorithm is faster than simpler algorithms such as the *simple cull* in [36], having a complexity of $O(N^2)$, is quite high. This makes the former algorithm the choice only for large spaces, while *simple cull* and hybrid algorithms [36] are recommendable for smaller numbers of realizations. A similar effect is observed for data structures for storing sets of realizations. The quad-tree data structure [37,38] is shown to be more efficient than linear lists in [38], but here also, the gain is achieved for large numbers of data points. Yukish [36] shows that when keeping points lexicographically sorted, normalized linear tables can be advantageous for computing the union and intersection of sets.

REFERENCES

1. R.H.J.M. Otten, Automatic floorplan design, *Proceedings of the 19th Design Automation Conference*, Las Vegas, NV, 1982, pp. 261–267.
2. J. Grodstein, E. Lehman, H. Harkness, B. Grundmann, and Y. Watanabe, A delay model for logic synthesis of continuously-sized networks, *Proceedings of ICCAD*, San Jose, California, November 1995.
3. J. Grason, *A Dual Linear Graph Representation for Space-Filling Location Problems of the Floor-Planning Type*, MIT Press, Cambridge, MA, 1970.
4. K. Kozminsky and E. Kinnen, Rectangular duals of planar graphs, *Networks*, 15, 145–157, 1985.
5. J. Bhasker and S. Sahni, Representation and generation of rectangular dissections, *Proceedings of the 23rd Design Automation Conference*, Las Vegas, NV, 1986, pp. 108–114.
6. J. Bhasker and S. Sahni, A linear time algorithm to check for the existence of a rectangular dual of a planar triangulated graph, *Networks*, 17, 307–317, 1987.
7. R. Schmid, Synthese von Floorplans auf der Basis von Dualgraphen mit rechtwinkliger Einbettung, PhD thesis, Universität Karlsruhe, Karlsruhe, Germany, 1987.

8. U. Flemming, Representation and generation of rectangular dissections, *Proceedings of the 15th Design Automation Conference*, Las Vegas, NV, 1978, pp. 138–144.

9. T. Ohtsuki, N. Sugiyama, and H. Kawanishi, An optimization technique for integrated circuit layout design, *Proceedings of ICCST*, Kyoto, Japan, 1970, pp. 67–68.

10. R.H.J.M. Otten and L.P.P.P. van Ginneken, Floorplan design using annealing, *Proceedings of ICCAD*, Santa Clara, CA, 1984.

11. T. Takahashi, A new encoding scheme for rectangle packing problem, *Proceedings ASP-DAC 2000*, Yokohama, Japan, 2000, pp. 175–178.

12. W.R. Heller, G. Sorkin, and K. Maling, The planar package planner for system designers, *Proceedings of the 19th Design Automation Conference*, Las Vegas, NV, 1982, pp. 663–670.

13. S. Tsukiyama, K. Koike, and I. Shirakawa, An algorithm to eliminate alle complex triangles in a maximal planar graph for use in VLSI floorplan, *Proceedings ISCAS*, Philadelphia, PA, 1986, pp. 321–324.

14. S.M. Leinwand and Y.-T. Lai, An algorithm for building rectangular floor-plans, *Proceedings of the 21st Design Automation Conference*, Albuquerque, NM, 1984, pp. 663–664.

15. R.H.J.M. Otten and P.S. Wolfe, Algorithm for wiring space assignment in slicing floorplans, Technical Report Y0882-0682, IBM Thomas J. Watson Research Center, Yorktown Heights, NY, 1982.

16. L. Stockmeyer, Optimal orientations of cells in slicing floorplan designs, *Inform. Control*, 57, 91–101, 1983.

17. R.H.J.M. Otten, Efficient floorplan optimization, *Proceedings of ICCD*, Port Chester, NY, 1983, pp. 499–502.

18. W. Keller, F. Schreiner, B. Schurman, E. Siepmann, and G. Zimmerman, Hierarchisches top down chip planning, Report of the University of Kaiserslautern, Kaiserslautern, Germany, 1987.

19. L.P.P.P. van Ginneken and R.H.J.M. Otten, Optimal slicing of plane point placements, *Proceeding of the European Design Automation Conference*, Glasgow, U.K., 1990, pp. 322–326.

20. F.Y. Young and D. F. Wong, How good are slicing floorplans? *Integration: VLSI J.* 23, 61–73, 1997.

21. M. Lai and D.F. Wong, Slicing Tree is a complete floorplan representation, *Proceedings of the Design, Automation and Test in Europe Conference (DATE)*, Munich, Germany, March 2001.

22. I. Sutherland and R. Sproull, The theory of logical effort: Designing for speed on the back of an envelope, in *Advanced Research in VLSI*, C. Séquin, Ed., MIT Press, Cambridge, U.K., 1991.

23. R.H.J.M. Otten, A design flow for performance planning, in *Architecture Design and Validation Methods*, E. Borger, Ed., Springer, Berlin, Germany, 2000.

24. J. Westra, C. Bartels, and P.R. Groeneveld, Probabilistic congestion prediction, *Proceedings of the International Symposium on Physical Design*, Phoenix, AZ, 2004, pp. 204–209.

25. W. Gosti, A. Narayan, R. K. Brayton, and A. L. Sangiovanni-Vincentelli, Wireplanning in logic synthesis, *Proceedings of ICCAD*, San Jose, California, 1998.

26. J. Westra, D.-J. Jongeneel, R.H.J.M. Otten, and C. Visweswariah, Time budgeting in a wire planning context, *Proceedings of DATE*, Munich, Germany, 2003.

27. R. Nair, C.L. Berman, P.S. Hauge, and E.J. Yoffa, Generation of performance constraints for layout, *IEEE Trans. CAD*, 8, August 1989, pp. 860–874.

28. I. Karkowski, Circuit delay optimization as a multiple choice linear knapsack problem, *Proceedings of the European Design Automation Conference (EDAC)*, Paris, France, March 1993, pp. 419–423.

29. S.S. Sapatnekar, V.B. Rao, P.M. Vaidya, and S.M. Kang, An exact Solution to the transistor sizing problem for CMOS circuits using convex optimization, *IEEE Trans. CAD*, 12, 1621–1634, 1993.

30. A.R. Conn, I.M. Elfadel, W.W. Molzen, Jr., P.R. O'Brien, P.N. Strenski, C. Visweswariah, and C.B. Whan, Gradient-based optimization of custom circuits using a static-timing Formulation, *Proceedings of the 36th Design Automation Conference*, New Orleans, LA, June 1999, pp. 452–459.

31. C. Visweswariah and A.R. Conn, Formulation of static circuit optimization with reduced size, degeneracy and redundancy by timing graph manipulation, *Proceedings of the International Conference on Computer-Aided Design (ICCAD)*, San Jose, CA, 1999, p. 244.

32. M. Geilen, A.A. Basten, B. Theelen, and R.H.J.M. Otten, An algebra of Pareto points, *Proceedings of ASCD*, St. Malo, France, 2005, pp. 88–97.

33. C. Mattson, A. Mullur, and A. Messac, Minimal representations of multiobjective design space using a smart Pareto filter, *Proceedings of the Ninth AIAA/ISSMO Symposium on Multidisciplinary Analysis and Optimization*, Atlanta, GA, 2002.

34. E. Zitzler, L. Thiele, M. Laumanns, C. Fonseca, and V.G.D.Fonseca, Performance assessment of multiobjective optimizers: An analysis and a review, *IEEE Trans. Evol. Comput.*, 7, 117–132, 2003.

35. J. Bentley, Multidimensional command-and-conquer, *Commun. ACM*, 23, 214–229, 1980.

36. M. Yukish, Algorithms to identify pareto points in multidimensional data sets, PhD thesis, Pennsylvania State University, University Park, PA, August 2004.

37. R. Finkel and J. Bentley, Quad trees: A data structure for retrieval on composite keys, *Acta Inform.*, 4, 1–9, 1974.

38. M. Sun and R.E. Steuer, Quad trees and linear lists for identifying nondominated criterion vectors, *ORSA J. Comput.*, 8, 367–375, 1996.

39. L. Cheng, L. Deng, and M.D.F. Wong, Floorplan design for 3-D VLSI design, *Proceedings of the IEEE Asia South Pacific Design Automation Conference (ASP-DAC)*, Shanghai, China, January 2005, Vol. 1, pp. 405–411.

Logic Verification

Design and Verification Languages

Stephen A. Edwards

15

CONTENTS

Abstract

After a few decades of research and experimentation, register-transfer dialects of two languages—SystemVerilog and VHDL—have emerged as the industry standard starting point for automatic large-scale digital integrated circuit synthesis. Writing register-transfer-level descriptions of hardware designs remains a largely human process, and hence the clarity, precision, and ease with which such descriptions can be coded correctly have a profound impact on the quality of the final product and the speed with which the design can be created.

While the efficiency of a design (e.g., the speed at which it can run or the power it consumes) is obviously important, its correctness is paramount, consuming the majority of the time (and hence money) spent during the design process. In response to this challenge, a number of so-called verification languages have arisen. These have been designed to assist in a simulation-based or formal verification process by providing mechanisms for checking temporal properties, generating pseudorandom test cases, and for checking how much of a design's behavior has been exercised by the test cases.

Through examples and discussion, this chapter describes the two main design languages—VHDL and SystemVerilog—and SystemC, a language currently used to build large simulation models.

15.1 INTRODUCTION

Hardware description languages (HDLs) have been the preferred way to design an integrated circuit since about the mid-1990s, when they supplanted graphical schematic capture programs. A typical design methodology in 2014 starts with back-of-the-envelope calculations that lead to a rough architectural design. This design is refined and tested for functional correctness by coding a simulator for it in a software language such as C or C++. Once this high-level model is satisfactory, it is passed to designers who implement it in a register-transfer-level (RTL) dialect of VHDL or SystemVerilog—the two industry-dominant HDLs. This new model is then simulated to verify that it behaves equivalently to the high-level reference model and then fed to a logic synthesis system such as Synopsys' Design Compiler, which translates the RTL into an efficient gate-level netlist. Finally, this netlist is given to an automated place-and-route system that ultimately generates the polygons that will become wires and transistors on the chip.

None of these steps is at all simple. Translating a C model of a system into RTL involves adding many details, ranging from protocols to cycle-level scheduling. Despite many years of research, this step remains stubbornly manual, although automatic translation has become feasible in certain narrow domains, such as signal processing. Synthesizing and optimizing a netlist from an RTL dialect of an HDL has been automated, but is the result of many years of university and industrial research, as are all the automated steps after it.

15.2 HISTORY

Many credit Reed [46] with the first HDL. His formalism, simply a list of Boolean functions that define the inputs to a block of flip-flops driven by a single clock (i.e., a synchronous digital system), captures the essence of an HDL: a formal method for modeling systems at a higher level of abstraction. Reed's formalism does not mention the wires and vacuum tubes that would actually implement his systems, yet it makes clear how these components should be assembled.

In the six decades since Reed, both the number and the need for HDLs have increased. In 1973, Omohundro and Tracey [41] could list nine languages and dozens more have been proposed since.

The main focus of HDLs has shifted as the cost of digital hardware has dropped. In the 1950s and 1960s, the cost of digital hardware remained high and was used primarily for general-purpose computers. Chu's CDL [14] is representative of languages of this era: it uses a programming language–like syntax, has a heavy bias toward processor design, and includes the notions of arithmetic, registers and register transfer, conditionals, concurrency, and even microprograms. Bell and Newell's influential ISP (described in their 1971 book [6]) was also biased toward processor design.

The 1970s saw the rise of many more design languages [11,13]. One of the more successful was ISP. Developed by Charles Rose and his student Paul Drongowski at Case Western Reserve in 1975–1976, ISP was based on Bell and Newell's ISP and used in a design environment for multiprocessor systems called N.mPc [44]. Commercialized in 1980 (and since owned by a variety of companies), it enjoyed some success, but starting in 1985, the Verilog simulator (and accompanying language) began to dominate the market.

The 1980s brought Verilog and VHDL, whose descendants remain the dominant HDLs to this day (2014). Initially successful because of its superior gate-level simulation speed, Verilog started life in 1984 as a proprietary language in a commercial product, while VHDL, the very high-speed integrated circuit (VHSIC) HDL, was designed at the behest of the US Department of Defense as a unifying representation for electronic design [18].

While the 1980s was the decade of widespread commercial adoption of HDLs for simulation, the 1990s brought them an additional role as input languages for logic synthesis. While the idea of automatically synthesizing logic from an HDL dates back to the 1960s, it was only the development of effective multilevel logic synthesis algorithms in the 1980s [12] that made HDLs practical for specifying hardware, much as compilers for software require optimization to produce competitive results. Synopsys was one of the first to release a commercially successful logic synthesis system that could generate efficient hardware from RTL Verilog specifications. By the end of the 1990s, parts of virtually every large integrated circuit were designed this way.

HDLs continue to be important for providing inputs for synthesis and modeling for simulation, but their importance as aids to validation continues to grow. Long an important part of the design process, the use of simulation to check the correctness of a design is now absolutely critical, and languages have evolved to better perform simulation quickly, correctly, and judiciously.

Language features in SystemVerilog now directly support an approach to verification that is grown in popularity: it is now common to automatically generate biased random test cases (e.g., input sequences in which a system's *reset* signal is asserted very rarely), check how thoroughly these cases exercise the design (e.g., by checking whether certain values or transitions have been overlooked), and by checking whether invariants have been violated during the simulation process (e.g., making sure that each request is followed by an acknowledgment).

15.3 VERIFICATION

Verification—making sure that C and RTL models are functionally correct—is the most serious, time-consuming challenge in digital hardware design. At the moment, simulation remains the dominant way of raising confidence in the correctness of these models, but has many drawbacks. One of the more serious is the need for simulation to be driven by appropriate test cases. These need to exercise the design, preferably the difficult cases that expose bugs, and be both comprehensive and relatively short since simulation takes time. The so-called formal verification techniques, which amount to efficient exhaustive simulation, have been gaining ground, but suffer from capacity problems.

Knowing when simulation has exposed a bug and estimating how complete a set of test cases is are the two other major issues in a simulation-based functional verification methodology. SystemVerilog has recently adopted a wide variety of constructs that can generate biased, constrained random test cases, check temporal properties, and check functional coverage. VHDL has also adopted many verification-centered constructs, but currently lags behind SystemVerilog.

Simulation applies a stimulus to a model of a design to predict the behavior of the fabricated system. Naturally, there is a trade-off between highly detailed models that can predict many attributes, say, logical values, timing, and power consumption, and simplified models that can only predict logical behavior but execute much faster.

Because the size of the typical design has grown exponentially over time, functional simulation, which only predicts the logical behavior of a synchronous circuit at clock-cycle boundaries, has become the preferred form of simulation because of its superior speed. Of course, timing still matters, but verifying whether a design meets timing is more frequently checked using a static timing analyzer, which is far quicker and more reliable than simulation-driven timing analysis, anyway. Furthermore, designers have shied away from more timing-sensitive circuitry such as transparent latches and other "asynchronous" design styles because they require more detailed simulation models and are therefore more costly to validate. Finally, checking a design's power consumption usually demands simulation. While a detailed timing simulation of a design should provide a very precise estimate of power consumption (e.g., by accounting for glitches), it is faster and nearly as effective to run a functional simulation to estimate activity factors. Even static power consumption (e.g., leakage) can be affected by state, something conveniently analyzed by functional simulation.

Simulation-based validation raises three important questions: what the stimulus should be, whether it exposes any design errors, and whether the stimulus is long and varied enough to have exposed "all" design errors. Historically, these three questions have been answered manually, that is, by having a test engineer write test cases, check the results of simulation, and make some informed guesses about how comprehensive the test suite actually is.

A manual approach has many shortcomings. Writing test cases is tedious, and the number that needs to be written for "complete" verification grows faster than the size of the system description. Manually checking the output of simulation is similarly tedious and subtle errors can be easily overlooked. Finally, it is difficult to judge quantitatively how much of a design has really been tested.

More automated methodologies, and ultimately languages, have evolved to address some of these challenges, but the verification problem remains one of the most difficult. Biased random test case generation is now typical, joining manual test case generation as a standard practice in verification. Designer-inserted assertions, long standard practice for software development, have also become standard for hardware, although the sort of assertions needed in hardware are more complicated than the typical "the argument must be nonzero" sort of checking that works well in software. Finally, automated "coverage" checking, which attempts to quantify how much of a design's behavior has been exercised in simulation, has also become standard.

All of these techniques, while an improvement, are not a panacea. While biased random test case generation can quickly generate many interesting tests, it provides no guarantees of completeness, meaning bugs may go unnoticed. Because they must often check a temporal property (e.g., "acknowledge arrives within three cycles of every request"), good assertions in hardware systems are much more difficult to write than those for software (which most often check data structure consistency), and again, there is no way to know when enough assertions have been added, and it is possible that the assertions themselves have flaws (e.g., they let bugs by). Finally, test cases that achieve 100% coverage can also let bugs by because the criterion for coverage is necessarily weak. Coverage typically checks what states particular variables have been in, but it cannot consider all combinations because their number grows exponentially quickly with design size. As a result, certain important combinations may not be checked even though coverage checks report "complete coverage".

While the utility of biased random test generation and coverage metrics is mostly limited to simulation, assertion specification techniques are useful for, and have been heavily influenced by, formal verification. Pure formal techniques consider all the possible behaviors by definition and therefore do not require explicit test cases (implicitly, they consider all possible test cases) and also do not need to consider coverage. But knowing what behavior is unwanted is crucial for formal techniques, whose purpose is to either expose unwanted behavior or formally prove it cannot occur.

In the early 2000s, a sort of renaissance occurred in verification languages. Temporal logics, specifically linear temporal logic (LTL) and computation tree logic (CTL), form the mathematical basis for most assertion checking, but their traditional mathematical syntax is awkward for hardware designers. Instead, a number of more traditional computer languages, which combine a more human-readable syntax for the bare logic with a lot of "syntactic sugar" for more naturally expressing common properties, were proposed for expressing properties in these logics. Two industrial efforts from Intel (ForSpec) and IBM (Sugar) emerged as the most complete and were later adopted in part by SystemVerilog and VHDL.

Meanwhile, some EDA companies were producing languages designed for writing test benches and checking simulation coverage. Vera, originally designed by Systems Science and since acquired by Synopsys, and e, designed and sold by Verisity, were the two most commercially successful. Bergeron [7] discusses how to use the two languages.

All four of these languages underwent extensive crossbreeding. Vera was made public, was rechristened OpenVera, had Intel's ForSpec assertions grafted onto it, and was added almost in its entirety to SystemVerilog. Sugar, meanwhile, has been adopted by the Accellera standards committee, rechristened the Property Specification Language (PSL), and also added in part to SystemVerilog and VHDL. Verisity's e language has changed the least, but was eventually made public.

15.3.1 PSL

The property specification language (PSL) evolved from the proprietary Sugar language developed at IBM and has since been adopted as an IEEE standard [27] and grafted onto both SystemVerilog and VHDL.

Beer et al. [5] provide a nice introduction to an earlier version of the language, which they explain evolved over many years. It has been used within IBM in the RuleBase formal verification system since 1994 and was also pressed into service as a checker generator for simulators in 1997. Accellera, an EDA standards group, adopted it as their formal property language in 2002. Cohen et al. [17] provides instruction on the language.

PSL is based on CTL [15], a powerful but rather cryptic temporal logic for specifying properties of finite-state systems. It is able to specify both safety properties ("this bad thing will never happen") as well as liveness properties ("this good thing will eventually happen"). Liveness properties can only be checked formally because it makes a statement about all the possible behaviors of a system, while safety properties can also be tested in simulation. LTL, a subset of CTL, expresses only safety properties and can therefore be turned into checking automata meant to be run in concert with a simulation to look for unwanted behavior. PSL carefully defines which subsets of its properties are purely LTL and are therefore candidates for use in simulation-based checking.

PSL is divided into four layers. The lowest, Boolean, consists of instantaneous Boolean expressions on signals in the design under test. The syntax of this layer follows that of the HDL to which PSL is being applied and can be Verilog, SystemVerilog, VHDL, or others. For example, `a[0:3] & b[0:3]` and `a(0 to 3) and b(0 to 3)` represent the bitwise *and* of the four most significant bits of vectors a and b in the Verilog and VHDL flavors, respectively.

The second layer, temporal, is where PSL gets interesting. It allows a designer to state properties that hold across multiple clock cycles. The `always` operator, which states that a Boolean expression holds in every clock cycle, is one of the most basic. For example, `always !(ena & enb)` states that the signals `ena` and `enb` will never be true simultaneously in any clock cycle.

More interesting are operators that specify delays. The `next` operator is the simplest. The property `always(req -> next ack)` states that in every cycle that the `req` signal is true, the `ack` signal is true in the next cycle. The -> symbol denotes implication, that is, if the expression to the left is true, that on the right must also be true. The `next` operator can also take an argument, for example, `always req -> next[2] ack` means that `ack` must be true two cycles after each cycle in which `req` is true.

PSL provides an extended form of regular expressions convenient for specifying more complex behaviors. Although it would be possible to write `always (req -> next (ack -> next !cancel))` to indicate that `ack` must be true after any cycle in which `req` is true, and `cancel` must be false in the cycle after that, it is much easier to instead write `always {req; ack; !cancel}`. This illustrates a basic principle of PSL: most operators are actually just "syntactic sugar"; the set of fundamental operators is quite small.

PSL draws a clear distinction between "weak" operators, which can be checked in simulation (i.e., safety properties) and "strong" operators, which express liveness properties and can only be checked formally. Strong operators are marked with a trailing exclamation point (!), and some operators come in both strong and weak varieties.

The eventually! operator illustrates the meaning of strong operators. The property always (req -> eventually! ack) states that after req is asserted, ack will always be asserted eventually. This is not something that can be checked in simulation: if a particular simulation saw req but did not see ack, it would be incorrect to report that this property failed because running that particular simulation longer might have produced ack. This is the fundamental difference between safety and liveness properties: safety states something bad never happens; liveness states something good eventually happens.

Another subtlety is that it is possible to express properties in which times moves backward through a property. A simple example is always ((a && next[3] (b)) -> c, which states that when a is true and b is true three clock cycles later, c is true in the first cycle, that is, when a was true. While it is possible to check this in simulation (for each cycle in which a is true, remember whether c is true and look three clock cycles later for b), it is more difficult to build automata that check such properties.

The third layer of PSL, the verification layer, instructs a verification tool what tests to perform on a particular design. It amounts to a binding between properties defined with expressions from the Boolean and temporal layer, and modules in the design under test. The following simple example

```
vunit ex1a(top_block.i1.i2) {
      A1: assert never (ena && enb);
}
```

declares a "verification unit" called ex1a, binds it to the instance named top_block.i1.i2 in the design under test, and declares (the assertion named A1) that the signals ena and enb in that instance are never true simultaneously.

In addition to assert, verification units may also include assume directives, which state the tool may assume a given property; assume_guarantee, which both assumes and tests a particular property; restrict, which constrains the tool to only consider inputs that have a given property; cover, which asks the tool to check whether a certain property was ever observed; and fairness, which instructs the tool to only consider paths in which the given property occurs infinitely often, for example, only when the system does not wait indefinitely.

The fourth, modeling layer of PSL essentially allows Verilog, SystemVerilog, VHDL, or other code to be included inline in a PSL specification. The intention here is to allow addition details about the system under test to be included in the PSL source file.

15.3.2 The "e" LANGUAGE

The e language was developed by Verisity as part of its Specman product as a tool for efficiently writing test benches. In this sense, it is quite different than PSL, which acts as an add-on to existing languages such as SystemVerilog or VHDL. It is an imperative object-oriented language with concurrency, the ability to generate constrained random values, mechanisms for checking functional (variable value) coverage, and a way to check temporal properties (assertions). Books on e include Palnitkar [43] and Iman and Joshi [32].

The syntax of e is a little unusual. First, all code must be enclosed in <' and '> symbols, otherwise it is considered a comment. Unlike C, e declarations are written "name: type." The syntax for fields in compound types (e.g., structs) includes particles such as % and !, which indicate when a field is to be driven on the device under test and not randomly computed, respectively.

Figure 15.1 shows a fragment of an e program that defines an abstract test strategy for a very simple microprocessor, specifically how to generate instructions for it. It illustrates the type system of the language as well as the utility of constraints. It defines two enumerated scalar types, opcode and reg, and defines the width of each. The struct instr defines a new compound type (instr) that represents a single instructions. First is the op field, which is one of the opcodes defined earlier. They, the op1 and the op2 fields, are marked with %, indicating that they should be considered by the pack built-in procedure, which marshals data to send to the simulation.

The kind field is also a enumerated scalar, but is used here as a type tag. It is not marked with %, which means that its value will not be included when the structure is packed and sent to the simulation. The two when directives define two subtypes, that is, "reg instr" and "imm instr."

```
Instruction encoding for a very simple processor
<'
type opcode : [ ADD, SUB, ADDI, JMP, CALL ] (bits: 4);
type reg : [ REG0, REG1, REG2, REG3 ] (bits: 2);

struct instr {
  %op  : opcode;         // Four-bit opcode
  %op1 : reg;            // Two-bit operand
  kind : [imm, reg];     // Whether instruction is immediate or register

  when reg instr { %op2 : reg; }   // Second operand register
  when imm instr { %op2 : byte; }  // Second operand and immediate byte

  // Constrain certain instructions to be register, immediate
  keep op in [ ADD, SUB ] => kind == reg;
  keep op in [ ADDI, JMP, CALL ] => kind == imm;

  // Constrain the second operand for JMP and CALL instructions
  when imm instr {
    keep opcode in [JMP, CALL] => op2 < 16;
  }
};

extend sys {
  !instrs : list of instr; // Add a non-generated field called instrs
};
'>
```

FIGURE 15.1 e code defining instruction encoding for a simple 8-bit microprocessor. An example from the Specman tutorial.

Such subtypes are similar to derived classes in object-oriented programming languages. Here, the value of the kind field, which can be either imm or reg, determines the subtype.

The two keep directives impose constraints between the kind field and the opcode, ensuring, for example, that ADD and SUB instructions are of the reg type. Although these constraints are simple, e is able to impose much more complicated constraints on the values of fields in a struct.

The final when directive further constrains the JMP and CALL instructions, that is, by restricting what values the op2 field may take for these instructions.

The extend sys directive adds a field named instrs to the sys built-in structure, which is the basic environment. The leading ! makes the system create an empty list of instructions, which will be filled in later.

Figure 15.2 illustrates how the definition of Figure 15.1 can be used to generate tests that exercise the ADD and ADDI instructions. It first adds constraints to the instr class (the template for instructions defined in Figure 15.1) that restrict the opcodes to either ADD or ADDI, then imposes a constraint on the top level (sys) that makes it generate exactly 10 instructions. Running the source code of Figures 15.1 and 15.2 together makes the system generate a sequence of 10 pseudorandom instructions.

```
<'
extend instr {
  keep opcode in [ADD, ADDI];
  keep op1 == REG0;
  when reg instr { keep op2 == REG1; }
  when imm instr { keep op2 == 0x3; }
};

extend sys {
  keep instrs.size() == 10;
};
'>
```

FIGURE 15.2 e code that uses the instruction encoding of Figure 15.1 to randomly generate 10 instructions. An example from the Specman tutorial.

15.4 SYSTEMVERILOG

SystemVerilog began as the Verilog HDL [1,28,29], designed and implemented by Phil Moorby at Gateway Design Automation in 1983–1984 (see Moorby's history of the language [11] for more details). The Verilog product was very successful, buoyed largely by the speed of its "XL" gate-level simulation algorithm. Cadence bought Gateway in 1989 and largely because of pressure from the competing, open VHDL language, made the language public in 1990. Open Verilog International was formed shortly thereafter to maintain and promote the standard, and IEEE adopted it in 1995 and then ANSI in 1996. Extensions were added in 2001 and 2005, leading to new standards. Meanwhile, Superlog, a Verilog dialect extended with software-like constructs, was released in 1999, grew in popularity, and was ultimately donated to the Verilog standardization body in 2002. In the same timeframe, the Vera language began life around 1995 as a commercial language for describing Verilog test benches. In 1998, Synopsys bought its creator, System Science, Inc., and opened the language to the public in 2001 as OpenVera. In 2005, Verilog, Superlog, and OpenVera were merged to produce the SystemVerilog standard. As of this writing (2014), SystemVerilog is a superset of "classical" Verilog (although it retains its own IEEE standard number 1800) and seems poised to supplant Verilog.

The first Verilog simulator was event driven and efficient for gate-level circuits, the fashion of the time, but the opening of the Verilog language in the early 1990s paved the way for other companies to develop more efficient compiled simulators, which traded upfront compilation time for higher simulation speed.

Like tree rings, the syntax and semantics of SystemVerilog language now embodies a history of simulation technologies and design methodologies. At its conception, gate- and switch-level simulations were in fashion, and Verilog contains extensive support for these now rarely used modeling styles. Moorby had worked with others on this problem before designing Verilog [22]. Since then, it has acquired constructs from object-oriented software languages, biased random test generation techniques, and test bench coverage tools.

Like many HDLs, Verilog supports hierarchy for structural modeling, but was originally designed assuming modules would have at most tens of connections. Hundreds or thousands of connections are now common, and Verilog-2001 [29] added a more succinct connection syntax to address this problem.

Procedural or behavioral modeling, once intended mainly for specifying test benches, was pressed into service first for RTL specifications and later for the so-called behavioral specifications. Again, Verilog-2001 added some facilities to enable this (e.g., `always @*` to simplify the procedural modeling of combinational logic) and SystemVerilog has added even more support explicitly for RTL modeling, for example, `always_comb` and `always_ff`.

The syntax and semantics of Verilog have always been a compromise between modeling clarity and simulation efficiency. A "reg" in Verilog, which used to be the main storage class for behavioral modeling, is exactly a shared variable. This means it simulates very efficiently (e.g., writing to a reg is just an assignment to memory), but also means that it can be misused (e.g., when written to by two concurrently running processes) and misinterpreted (e.g., its name suggests a memory element such as a flip-flop, but it often represents purely combinational logic). In SystemVerilog, the use of "reg" is deprecated in favor of the more flexible "logic" type, which subsumes both "reg" and "wire" types, allowing the compiler to choose the most appropriate implementation.

Thomas and Moorby [52] has long been the main text on the Verilog language (Moorby was the main designer). Unlike many standards documents, the Verilog standard [28] is quite readable, since it was adopted from the original Verilog simulator user manual, a tradition that has been carried over to the SystemVerilog standard [31]. Other references include Palnitkar [42] for an overall description of the language, and Mittra [39] and Sutherland [50] for the programming language interface (PLI). Smith [48] compares Verilog and VHDL. French et al. [23] present a clever way of compiling Verilog simulations and also discuss more traditional ways. Sutherland et al. [51] describe SystemVerilog relative to Verilog; Spear and Tumbush [49] focus on SystemVerilog's verification features.

15.4.1 CODING IN SYSTEMVERILOG

A SystemVerilog model is a list of modules. Each module has a name; an interface consisting of named, typed, directional ports; a list of local "variables"; and a body that can contain instances of primitive gates such as ANDs and ORs, instances of other modules (allowing hierarchical structural modeling), continuous assignment statements, which can be used to model combinational datapaths, and concurrent processes written in an imperative style.

Figure 15.3 illustrates some of SystemVerilog's modeling styles. Shown are various ways to model a two-input multiplexer: primitive gates in Figure 15.3a, a user-defined primitive—a truth table—in Figure 15.3b, a continuous assignment in Figure 15.3c, and a concurrent process in Figure 15.3d. All of these models exhibit roughly the same behavior (minor differences occur when some inputs are undefined) and can be mixed freely within a design.

```
module mux_struct(
  output logic f,
  input logic a, b, sel);

  logic nsel, f1, f2;

  and g1(f1, a, nsel),
      g2(f2, b, sel);
  or  g3(f, f1, f2);
  not g4(nsel, sel);

endmodule
(a)
```

```
primitive mux_prim(
  output logic f,
  input logic a, b, sel);

table
  1?0 : 1;
  0?0 : 0;
  ?11 : 1;
  ?01 : 0;
  11? : 1;
  00? : 0;
endtable

endprimitive
(b)
```

```
module mux_cont(
  output logic f,
  input logic a, b, sel);

  assign f = sel ? b : a;
endmodule
(c)
```

```
module mux_imp(
  output logic f,
  input logic a, b, sel);

  always_comb
    if (sel) f = b;
    else f = a;

endmodule
(d)
```

```
module testbench;
  logic a, b, sel, f;

  mux dut(f, a, b, sel);

  initial begin
    $display("a,b,sel -> f");
    $monitor($time,,
            "%b%b%b -> ",
            a, b, sel, f);
    a = 0; b = 0; sel = 0;
    #10 a = 1;
    #10 sel = 1;
    #10 b = 1;
    #10 a = 0;
    #10 sel = 0;
    #10 b = 0;
  end
endmodule
(e)
```

```
a,b,sel -> f
 0 000 -> 0
10 100 -> 1
20 101 -> 0
30 111 -> 1
40 011 -> 1
50 010 -> 0
60 000 -> 0
(f)
```

FIGURE 15.3 SystemVerilog examples. (a) A two-input multiplexer described with a structural model. (b) A user-defined primitive for the multiplexer. (c) The multiplexer described using a continuous assignment. (d) The multiplexer described with imperative code. (e) A test bench for the multiplexer. (f) The output from the test bench.

One of SystemVerilog's strengths remains its ability to represent test benches along with the model being tested. Figure 15.3e illustrates a test bench for this simple mux, which applies a sequence of inputs over time and prints a report of the observed behavior. Figure 15.3f shows its output.

SystemVerilog modules communicate through *logic* variables (Figure 15.4), which can be set both continually (e.g., by an *assign*) and imperatively through assignment statements in *initial* and the various *always* blocks. (These replace Verilog's more error-prone *reg* and *net* types, which had to be used for imperative and continuous assignments, respectively.)

Figure 15.5 illustrates the syntax for defining and instantiating models. Each module has a name and a list of named ports, each of which has a direction and a width. Instantiating such a module consists of giving the instance a name and listing the signals or expressions to which it is connected. Connections can be made positionally or by port name, the latter being preferred for modules with many (perhaps 10 or more) connections.

Continuous assignments are a simple way to model both Boolean and arithmetic datapaths. A continuous assignment uses SystemVerilog's comprehensive expression syntax to define a function to be computed and its semantics are such that the value of the expression on the right of a continuous expression is always copied to the net on the left (regs are not allowed on the left of a continuous assignment). Practically, SystemVerilog simulators implement this by recomputing the expression on the right whenever any variable it references changes. Figure 15.6 illustrates some continuous assignments.

```
logic a;                     // Wire: four-valued (0,1,X,Z)
bit b;                       // Bit: two-valued (0,1)
byte by;                     // Eight-bit, two-valued
int i;                       // Thirty-two bit, two-valued
wire c;                      // Wire with weak to strong drivers
tri [15:0] dbus;             // 16-bit tristate bus
tri #(5,4,8) b;              // Wire with delay
logic [5:0] vec;             // Six-bit register
trireg (small) q;            // Wire stores a small charge
integer imem[0:1023];        // Array of 1024 integers
logic [31:0] dcache [0:63];  // A 32-bit memory
real d;                      // Double-precision floating point
```

FIGURE 15.4   A sampling of SystemVerilog variable definitions.

```
module mymod(
  output logic out1,        // Outputs first by convention
  output logic [3:0] out2,  // four-bit vector
  input logic in1,
  input logic [2:0] in2);

// Module body: instances,
// continuous assignments,
// initial and always blocks

endmodule

module usemymod;
logic a;
logic [2:0] b;
logic c, e, g;
logic [3:0] d, f, h;

mymod m1(c, d, a, b);            // simple instance
mymod m2(e, f, c, d[2:0]),       // instance with part-select input
      m3(.in1(e), .in2(f[2:0]), // connect-by-name
         .out1(g), .out2(h));

endmodule
```

FIGURE 15.5   SystemVerilog structure: an example of a module definition and another module containing three instances of it.

```
module add8(
  output logic [8:0] sum,
  input logic [7:0] a, b,
  input logic carryin);

// unsigned arithmetic
assign sum = a + b + {8'b0, carryin};
endmodule

module datapath(
  output logic [2:0] addr_2_0,
  output logic       icu_hit,
  input logic        psr_bm8,
  input logic        hit);

logic [31:0]  addr_qw_align;
logic [3:0]   addr_qw_align_int;
logic [31:0]  addr_d1;
logic         powerdown;
logic         pwdn_d1;
logic [1:0]   addr_offset;

// part select, vector concatenation is {}
assign addr_qw_align = { addr_d1[31:4], addr_qw_align_int[3:0] };

// if-then-else operator
assign addr_offset = psr_bm8 ? addr_2_0[1:0] : 2'b00;

// Boolean operators
assign icu_hit = hit & !powerdown & !pwdn_d1;

// ...

endmodule
```

FIGURE 15.6 SystemVerilog modules illustrating continuous assignment. The first is a simple 8 bit full adder producing a 9 bit result. The second is an excerpt from a processor datapath.

Behavioral modeling in Verilog uses imperative code enclosed in *initial* and *always* blocks that write to variables to maintain state. Each block effectively introduces a concurrent process that is awakened by an event and runs until it hits a delay or a wait statement. The example in Figure 15.7 illustrates basic behavioral usage.

Figure 15.8 shows a more complicated behavioral model, in this case a simple state machine. This example is written in a common style where the combinational and sequential parts of a

```
module behavioral;
logic [1:0] a, b;

initial begin
  a = `b1;
  b = `b0;
end

always begin
  #50 a = ~a; // Toggle a every 50 time units
end

always begin
  #100 b = ~b; // Toggle b every 100 time units
end

endmodule
```

FIGURE 15.7 A simple SystemVerilog behavioral model. The code in the *initial* block runs once at the beginning of simulation to initialize the two registers. The code in the two *always* blocks runs periodically: once every 50 and 100 time units, respectively.

```
module FSM(
  output logic o,
  input logic a, b, reset, clk);
logic [1:0] state, nextState;

// Combinational logic block: sensitive to changes on all inputs;
// outputs o and nextState always assigned

always_comb
  case (state)
    2'b00: begin
      o = a & b;
      nextState = a ? 2'b00 : 2'b01;
    end
    2'b01: begin
      o = 0; nextState = 2'b10;
    end
    default: begin
      o = 0; nextState = 2'b00;
    end
  endcase

// Sequential block: sensitive to clock edge and reset signal

always_ff @(posedge clk or posedge reset)
  if (reset)
    state <= 2'b00;
  else
    state <= nextState;

endmodule
```

FIGURE 15.8   A SystemVerilog behavioral model for a state machine illustrating the common practice of separating combinational and sequential blocks.

state machine are written as two separate processes. The first process is purely combinational, as indicated by the `always_comb` directive. The process executes on any change on signals a, b, or state. The code consists of a multiway choice—a *case* statement—and performs procedural assignments to the o and nextState variables.

The second process models a pair of flip-flops that holds the state between cycles. The `@(posedge clk or posedge reset)` directive makes the process sensitive to the rising edge of the clock or a change in the reset signal. At the positive edge of the clock, the process captures the value of the `nextState` variable and copies it to `state`.

The example in Figure 15.8 illustrates the two types of behavioral assignments. The assignments used in the first process are the so-called blocking assignments, written =, and take effect immediately. Nonblocking assignments are written <= and have somewhat subtle semantics. Instead of taking effect immediately, the right-hand sides of nonblocking assignments are evaluated when they are executed, but the assignment itself does not take place until the end of the current time instant. Such behavior effectively isolates the effect of nonblocking assignments to the next clock cycle, much like the output of a flip-flop is only visible after a clock edge. In general, nonblocking assignments are preferred when writing to state-holding elements for exactly this reason. See Figure 15.12 and the next section for a more extensive discussion of blocking versus nonblocking assignments.

15.4.2 VERIFICATION WITH SYSTEMVERILOG

One of Verilog's strengths has long been its ability to model both systems and test benches, two very different tasks. System modeling demands both accuracy (i.e., the model behaves like the system) and precision (i.e., the model supports sufficient details in the model); test engineers, by contrast, are concerned mostly about how easily they can code a test bench that can produce the desired stimulus and check for the desired responses. How the test bench code does its job is nearly immaterial.

The majority of the additions that turned Verilog into SystemVerilog focused on simplifying the verification task. The Superlog additions [21] were mostly software-inspired constructs from

C and C++, including enumerated types, record types (*struct*s), *typedef*s, type casting, a variety of operators such as +=, operator overloading, control-flow statements such as `break` and `continue`, as well as object-oriented programming constructs such as classes, inheritance, and dynamic object creation and deletion. At the very highest level, it also adds strings, associative arrays, concurrent process control (e.g., fork/join), semaphores, and mailboxes, giving it features only found in concurrent programming languages such as Java. All of these had the goal of making it easier to express more complicated test bench behavior more succinctly using software-inspired modeling techniques.

Perhaps the most interesting additions, however, directly support biased random test generation. In this methodology, which has been growing in popularity and effectiveness, a test engineer writes a program that generates many "random" inputs for a particular module along with some sort of checker that can verify the system under test behaves as desired when given the inputs. The idea is that it is easier to generate a comprehensive test suite by leaving certain elements to chance rather than insisting a human conceive of all the corner cases.

But purely randomly generated tests are likely to be terribly inefficient, making biased, constrained random tests preferred. Consider a module with a *reset* signal. Unbiased test vectors might assert *reset* half the time, making long runs of the module quite unlikely. Instead, we want biased random test vectors that assert *reset* perhaps 1/100th or 1/1000th of the time.

To these ends, SystemVerilog adopted constructs pioneered in the (Open)Vera, Sugar, and ForSpec verification languages that facilitate biased constrained random test generation, functional coverage checks, and temporal assertions. Biased, constrained random test generation allows the user to control the generation of "random" test inputs, constraining them to regions of interest, avoiding illegal or don't-care inputs. Functional coverage checks provide a mechanism for the user to classify events, such as being in a certain state or making a specific, expected transition between states, and then easily count the number of such events while the system under test is being subjected to the test vectors. This provides a way to look for gaps in test coverage, for example, when a particular state transition should be possible but was never observed.

Temporal assertions allow a test engineer to specify both desirable and undesirable properties that take place over one, two, or more clock cycles, such as entering an illegal state or asserting a particular output for too many cycles. The idea here is to provide the user with the ability to easily look for unwanted behavior (i.e., bugs) while a system is being subjected to random test vectors. Such assertions came originally from the formal verification community, which has long wanted to test systems for all possible inputs using more analytical techniques. Which formal verification techniques grow in quality, mostly the ability to analyze large designs quickly, simulation remains a key component in any test engineers' arsenal because it is always able to answer questions about the largest of designs.

Figure 15.9 illustrates some of the random test generation constructs in SystemVerilog, which were largely taken from the Vera language.

Figure 15.10 illustrates some of the coverage constructs in SystemVerilog. In general, one defines "covergroups," which are collections of bins that sample values on a given event, typically a clock edge. Each covergroup defines the sorts of values it will be observing (e.g., values of a single variable, combinations of multiple variables, and sequences of values on a single variable) and rules that define the "bins" each of these values will be placed in. In the end, the simulator reports which bins were empty, indicating that none of the matching behavior was observed. Again, much of this machinery was taken from Vera.

Figure 15.11 shows some of SystemVerilog's assertion constructs. In addition to signaling an error when an "instantaneous" condition does not hold (e.g., a set of variables are taking on mutually incompatible values), SystemVerilog has the ability to describe temporal sequences such as "`ack` must rise between one and five cycles after `req` rises" and check whether they appear during simulation. Much of the syntax comes from PSL/Sugar.

15.4.3 VERILOG SHORTCOMINGS

Compared to VHDL, SystemVerilog does a poor job at protecting users from themselves. Verilog's variables are shared variables and the language permits all the standard pitfalls associated with them, such as races and nondeterministic behavior. Most users avoid such behavior by following

```
class Bus;
  rand bit[15:0] addr;
  rand bit[31:0] data;

  constraint world_align { addr[1:0] = 2'b0; }
endclass

initial begin
  Bus bus = new;

  repeat (50) begin
    if (bus.randomize() == 1)
      $display("addr = %16h data = %h\n",
                bus.addr, bus.data);

    else
      $display("overconstrained: no satisfying values exist\n");
  end
end

typdef enum { low, mid, high } AddrType;

class MyBus extends Bus;
  rand AddrType atype; // Additional random variable

  // Additional constraint on address: still word-aligned}
  constraint addr_range {
    (atype == low ) -> addr inside { [0:15] };
    (atype == mid ) -> addr inside { [16:127] };
    (atype == high) -> addr inside { [128:255] };
  }
endclass

task exercise_bus;
  int res;

  // Restrict to low addresses
  res = bus.randomize() with { atype == low; };

  // Restrict to particular address range
  res = bus.randomize()
        with { 10 <= addr && addr <= 20 };

  // Restrict data to powers of two
  res = bus.randomize() with { data & (data - 1) == 0 };

  // Disable word alignment
  bus.word_align.constraint_mode(0);

  res = bus.randomize with { addr[0] || addr[1] };

  // Re-enable word alignment
  bus.word_align.constraint_mode(1);
endtask
```

FIGURE 15.9 Constrained random variable constructs in SystemVerilog. The example starts with a simple definition of a Bus class that constraints the two least significant bits of the address to be zero and then invokes the **randomize** method to randomly generate address/data pairs and print the result. Next is a refined version of the Bus class that adds a field taken from an enumerated type that further constrains the address depending on its value. The example ends with a task that illustrates various ways to control the constraints, after examples in the SystemVerilog LRM. (From Accelera, SystemVerilog 3.1a language reference manual: Accellera's extensions to Verilog, Napa, CA, May 2004.)

```
enum { red, green, blue } color;

bit [3:0] adr, offset;

covergroup g2 @(posedge clk);
  Hue:    coverpoint color;
  Offset: coverpoint offset;

  // Consider (color, adr) pairs, e.g.,
  // (red, 3'b000), (red, 3'b001), ..., (blue, 3'b111)
  AxC:    cross color, adr;

  // Consider (color, adr, offset) triplets
  // Creates 3 * 16 * 16 = 768 bins
  all:    cross color, adr, Offset;
endgroup

g2 g2_inst = new; // Create a watcher

bit [9:0] a; // Takes values 0--1023

covergroup cg @(posedge clk);

  coverpoint a {
    // place values 0--63 and 65 in bin a
    bins a = { [0:63], 65 };

    // create 65 bins, one for 127, 128, ..., 191
    bins b[] = { [127:150], [148:191] };

    // create three bins: 200, 201, and 202
    bins c[] = { 200, 201, 202 };

    // place values 1000--1023 in bin d
    bins d = { [1000:$] };

    // place all other values (e.g., 64, 66, .., 126, 192, ...) in their own bin
    bins others[] = default;
  }

endgroup

bit [3:0] a;

covergroup cg @(posedge clk);
  coverpoint a {
    // Place any of the sequences 4 -> 5 -> 6, 7 -> 11, 8 -> 11, 9 -> 11, 10 ->11,
    // 7 -> 12, 8 -> 12, 9 -> 12, and 10 -> 12 into bin sa.
    bins sa = (4 => 5 => 6), ([7:9],10 => 11,12);

    // Create separate bins for 4 -> 5 -> 6, 7 -> 10, 8 -> 10, and 9 -> 10
    bins sb[] = (4 => 5 => 6), ([7:9] => 10);

    // Look for the sequence 3 -> 3 -> 3 -> 3
    bins sc = 3 [* 4];

    // Look for any of the sequences 5 -> 5, 5 -> 5 -> 5, or 5 -> 5 -> 5 -> 5
    bins sd = 5 [* 2:4];

    // Look for any sequence of the form 6 -> ... -> 6 -> ... -> 6
    // where "..." represents any sequence that excludes 6
    bins se = 6 [-> 3];
  }
endgroup
```

FIGURE 15.10   SystemVerilog coverage constructs. The example begins with a definition of a *covergroup* that considers the values taken by the color and offset variables as well as combinations. Next is a covergroup illustrating the variety of ways *bins* may be defined to classify values for coverage. The final covergroup illustrates SystemVerilog's ability to look for and classify sequences of values, not just simple values. Examples from the SystemVerilog LRM. (From Accelera, SystemVerilog 3.1a language reference manual: Accellera's extensions to Verilog, Napa, CA, May 2004.)

```
                        // Make sure req1 or req2 is true if we are in the REQ state}
                        always @(posedge clk)
                          if (state == REQ)
                            assert (req1 || req2);

                        // Same, but report the error ourselves
                        always @(posedge clk)
                          if (state == REQ)
                            assert (req1 || req2)
                            else
                              $error("In REQ; req1 || req2 failed (\%0t)", $time);

                        property req_ack;
                          @(posedge clk) // Sample req, ack at rising clock edge
                            // After req is true, between one and three cycles later,
                            // ack must have risen.
                            req ##[1:3] $rose(ack);
                        endproperty

                        // Assert that this property holds, i.e., create a checker
                        as_req_ack: assert property (req_ack);

                        // The own_bus signal goes high in 1 to 5 cycles,
                        // then the breq signal goes low one cycle later.
                        sequence own_then_release_breq;
                          ##[1:5] own_bus ##1 !breq
                        endsequence

                        property legal_breq_handshake;
                          @(posedge clk)        // On every clock,
                          disable iff (reset)   // unless reset is true,
                          // once breq has risen, own_bus should rise and breq should fall.
                          $rose(breq) |-> own_then_release_breq;
                        endproperty

                        assert property (legal_breq_handshake);
```

FIGURE 15.11 SystemVerilog assertions. The first two *always* blocks check simple safety properties, that is, that **req1** and **req2** are never true at the positive edge of the clock. The next property checks a temporal property: that **ack** must rise between one and three cycles after each time **req** is true. The final example shows a more complex property: when reset is not true, a rising **breq** signal must be followed by **own_bus** rising between one and five cycles later and **breq** falling.

certain rules (e.g., by restricting assignments to a shared variable to a single concurrent process), but SystemVerilog allows more dangerous usage. Tellingly, a number of EDA companies exist solely to provide lint-like tools for SystemVerilog that report such poor coding practices. Gordon [24] provides a more detailed discussion of the semantic challenges of SystemVerilog.

Nonblocking assignments are one way to ameliorate most problems with nondeterminism caused by shared variables, but they, too, can lead to bizarre behavior. To illustrate the use of shared variables, consider a three-stage shift register. The implementation in Figure 15.12a appears to be correct, but in fact may not behave as expected because the language says the simulator is free to execute the three *always* blocks in any order when they are triggered. If the processes execute top to bottom, the module becomes a one-stage shift register, but if they execute bottom to top, the behavior is as intended. The real danger here is that the simulation *might* work as desired but the synthesized circuit may behave differently, defeating the value of the simulation.

Figure 15.12b shows a correct implementation of the shift register that uses nonblocking assignments to avoid this problem. The semantics of these assignments are such that the value on the right-hand side of the assignment is captured when the statement runs, but the actual assignment of values is only done at the "end" of each instant in time, that is, after all three blocks have finished executing. As a result, the order in which the three assignments are executed does not matter, and therefore the code always behaves like a three-stage shift register.

The behavior of nonblocking assignments can be unexpected to software programmers. In most programming languages, the effect of an assignment can be felt by the instruction

```
module bad_sr(                              module good_sr(
  output logic o,                             output logic o,
  input logic i, clk);                        input logic i, clk);
logic a, b, o;                              logic a, b, o;

always_ff @(posedge clk) a = i;             always_ff @(posedge clk) a <= i;
always_ff @(posedge clk) b = a;             always_ff @(posedge clk) b <= a;
always_ff @(posedge clk) o = b;             always_ff @(posedge clk) o <= b;

endmodule                                   endmodule
(a)                                         (b)
```

FIGURE 15.12 SystemVerilog examples illustrating the difference between blocking and nonblocking assignments. (a) An erroneous implementation of a three-stage shift register that may or may not work depending on the order in which the simulator chooses to execute the three *always* blocks. (b) A correct implementation using nonblocking assignments, which make the variables take on their new values after all three blocks are done for the instant.

```
                                       module good_counter(
module bad_counter(                      output logic [3:0] o,
  output logic [3:0] o,                  input logic clk);
  input logic clk);                    logic [3:0] count;
logic [3:0] o;
                                       always @(posedge clk) begin
always @(posedge clk) begin              count = count + 1;
  o <= o + 1;                            if (count == 10)
  if (o == 10)                             count = 0;
    o <= 0;                              o <= count
end                                    end

endmodule                              endmodule
(a)                                    (b)
```

FIGURE 15.13 Verilog examples illustrating a pitfall with nonblocking assignments. (a) An erroneous implementation of a counter, which counts to 10, not 9. (b) A correct implementation using a combination of blocking and nonblocking assignments.

immediately following it, but the delayed-assignment semantics of a nonblocking assignment violates this rule. Consider the erroneous decimal counter in Figure 15.13a. Without knowing the subtle semantics of Verilog nonblocking assignments, the counter would appear to count from 0 to 9, but in fact it counts to 10 before being reset because test of o by the *if* statement gets the value of o from the previous clock cycle, not the results of the o <= o + 1 statement. A corrected version is shown in Figure 15.13b, which uses a local variable count to maintain the count, blocking assignments to touch it, and finally a nonblocking assignment o to send the count outside the module.

Coupled with the rules for register inference, the circuit implied by the counter in Figure 15.13b is fine. Only the count variable will actually become a state-holding element.

15.5 VHDL

Although VHDL and SystemVerilog are often used interchangeably for specifying RTL hardware, they could not have had more different histories. Unlike Verilog, VHDL was deliberately designed to be a standard HDL. As Dewey explains [18], VHDL was created at the behest of the US Department of Defense in response to the desire to incorporate integrated circuits (specifically VHSIC, hence the name of the program from which VHDL evolved) in military hardware. Starting with a summer study at Woods Hole, Massachusetts, in 1981, requirements for and the scope of the language were first established, then after a bidding process, a contract to develop the language was awarded in 1983 to three companies: Intermetrics, which was the prime contractor for Ada, the software programming language developed for the US military in the early 1980s; Texas Instruments; and IBM. Dewey and de Geus describe this history in more detail [19].

The VHDL language was created in 1983 and 1984, essentially concurrently with Verilog, and first released publicly in 1985. Interest in an IEEE standard HDL was high at the time, and VHDL was eventually adopted as IEEE standard 1076 in 1987 [33] and revised in 1993, 2002, and most recently 2008 [30]. Verilog, meanwhile, remained proprietary until 1990. The standardization and growing popularity of VHDL at the time was certainly instrumental in Cadence's decision to make Verilog public.

The original objectives of the VHDL language [18] were to provide a means of documenting hardware (i.e., as an alternative to imprecise English descriptions) and of verifying it through simulation. As such, a VHDL simulator was developed along with the language.

The VHDL language is vast, complicated, and has a verbose syntax derived from Ada. For many years, it had a more robust type system than Verilog, but recent additions to SystemVerilog have narrowed this gap. While VHDL's popularity as a means of formal documentation is questionable, it has succeeded as a modeling language for hardware simulation and, like Verilog, as a specification language for RTL logic synthesis.

While much of the VHDL and Verilog languages are very different, in practice the subsets designers use to describe hardware are similar, in large part because most synthesis tools accept similar subsets of both languages. The way the languages are employed to write test benches varies far more. Many of the features targeting biased random test generation recently added to SystemVerilog do not have VHDL analogs.

The 2008 revision of the language [30] fixes many infelicities in the language that had increased its verbosity (e.g., VHDL 2008 now allows logic types in conditionals and provides a matching set of relational operators that can eliminate the former plague of `='1'` operators) and adds the IEEE PSL [27] as a mechanism for specifying temporal correctness properties.

VHDL has spawned many books discussing its proper usage. Basic texts include Lipsett et al. [37] (one of the earliest), Dewey [20], Bhasker [8], Perry [45], Rushton [47], and Ashenden [3,4]. More advanced is Cohen [16], which suggests preferred idioms in VHDL, and Harr and Stanculescu [26], which discusses using VHDL for a variety of modeling tasks, not just RTL.

15.5.1 CODING IN VHDL

Like SystemVerilog, VHDL describes designs as a collection of hierarchical modules. But unlike SystemVerilog, VHDL splits them syntactically into interfaces—called entities—and their implementations—architectures. In addition to named input and output ports, entities also define compile-time parameters (generics), types, constants, attributes, use directives, and others.

Figure 15.14 shows code for the same two-input multiplexer roughly equivalent to the SystemVerilog examples in Figure 15.3. Figure 15.14a is the entity declaration for the multiplexer, which defines its input and output ports. Figure 15.14b is a purely structural description of the multiplexer: it defines internal signals, the interface to the Inverter, AndGate, and OrGate components, and instantiates four of these gates. The name of the architecture, "`structural`," is arbitrary; it is used to distinguish among different architectures. Furthermore, the Verilog example used the built-in gate-level primitives; VHDL itself does not know about logic gates, but can be taught about them.

Figure 15.14c illustrates a dataflow model for the multiplexer with each logic gate made explicit. VHDL does have built-in logical operators. Figure 15.14d shows an even more succinct implementation, which uses the multiway *when* conditional operator.

Finally, Figure 15.14e shows a behavioral implementation of the mux. It defines a concurrently running process sensitive to the three mux inputs (a, b, and c) and uses an if–then–else statement (VHDL provides most of the usual control-flow statements) to select between copying the a signal and the b signal to the output d.

One of the design philosophies behind the VHDL language was to maximize its flexibility by making most things user definable. As a result, unlike Verilog, it has only the most rudimentary

```
entity mux2 is
  port (a, b, c : in Bit; d : out Bit);
end;
(a)

architecture structural of mux2 is

  signal cbar, ai, bi : Bit;      -- internal signals

  component Inverter              -- component interfaces
    port (a:in Bit; y: out Bit);
  end component;
  component AndGate
    port (a1, a2:in Bit; y: out Bit);
  end component;
  component OrGate
    port (a1, a2:in Bit; y: out Bit);
  end component;

begin
  I1: Inverter port map(a => c, y => cbar); -- connect-by-name
  A1: AndGate  port map(a, c, ai);          -- connect-by-position
  A2: AndGate  port map(a1 => b, a2 => cbar, y => bi);
  O1: OrGate   port map(a1 => ai, a2 => bi, y => d);
end;
(b)

architecture dataflow1 of mux2 is
  signal cbar, ai, bi : Bit;
begin
  cbar <= not c;                 architecture dataflow2 of mux2 is
  ai   <= a and c;               begin
  bi   <= b and cbar;              d <= a when c else -- Allowed in VHDL 2008
  d    <= ai or bi;                    b;
end;                             end;
(c)                              (d)

architecture behavioral of mux2 is
begin
  process(all) -- Sensitive to all inputs
  begin
    if c then -- Allowed in VHDL 2008
      d <= a;
    else
      d <= b;
    end if;
  end process;
end;
(e)
```

FIGURE 15.14 VHDL 2008 code for a two-input multiplexer. (a) The entity definition for the multiplexer specifies its interface. (b) A structural implementation instantiating primitive gates. (c) A dataflow implementation with an expression for each gate. (d) A direct dataflow implementation. (e) A behavioral implementation.

built-in types (e.g., Boolean variables, but nothing to model four-valued logic), but has a much more powerful type system that allows such types to be defined. The Bit used in the examples in Figure 15.14 is actually a predefined part of the standard environment, that is,

```
type BIT is ('0', '1');
```

which is a character enumeration type whose two values are the characters 0 and 1. VHDL is case insensitive; Bit and BIT are equivalent.

Figure 15.15 is a more elaborate example showing an implementation of the classic traffic light controller from Mead and Conway [38]. This is written in a synthesizeable dialect, using the common practice of separating the output and next-state logic from the state-holding element.

```vhdl
library ieee;
use ieee.std_logic_1164.all;

entity tlc is
  port (
    clk, reset, cars, short, long : in std_ulogic;
    highway_yellow, highway_red, farm_yellow, farm_red
    start_timer                   :    out std_ulogic);
end tlc;

architecture imp of tlc is
signal current_state, next_state : std_ulogic_vector(1 downto 0);
constant HG : std_ulogic_vector := "00";
constant HY : std_ulogic_vector := "01";
constant FY : std_ulogic_vector := "10";
constant FG : std_ulogic_vector := "11";
begin

P1: process (clk) -- Sequential process
begin
  if (clk'event and clk = '1') then
    current_state <= next_state;
  end if;
end process P1;

-- Combinational process: sensitive to input changes
P2: process (all)
begin
  if reset then
    next_state <= HG;
    start_timer <= '1';
  else
    case current_state is
      when HG =>
        highway_yellow <= '0'; highway_red <= '0';
        farm_yellow    <= '0'; farm_red    <= '1';
        if cars and long then
          next_state <= HY; start_timer <= '1';
        else
          next_state <= HG; start_timer <= '0';
        end if;
      when HY =>
        highway_yellow <= '1'; highway_red <= '0';
        farm_yellow    <= '0'; farm_red    <= '1';
        if short then next_state <= FG; start_timer <= '1';
        else          next_state <= HY; start_timer <= '0';
        end if;
      when FG =>
        highway_yellow <= '0'; highway_red <= '1';
        farm_yellow    <= '0'; farm_red    <= '0';
        if not cars or long then
          next_state <= FY; start_timer <= '1';
        else
          next_state <= FG; start_timer <= '0';
        end if;
      when FY =>
        highway_yellow <= '0'; highway_red <= '1';
        farm_yellow    <= '1'; farm_red    <= '0';
        if short then next_state <= HG; start_timer <= '1';
        else          next_state <= FY; start_timer <= '0';
        end if;
      when others =>
        next_state     <= "XX"; start_timer <= 'X';
        highway_yellow <= 'X'; highway_red  <= 'X';
        farm_yellow    <= 'X'; farm_red     <= 'X';
    end case;
  end if;
end process P2;
end imp;
```

FIGURE 15.15   The traffic light controller from Mead and Conway [38] implemented in VHDL, illustrating the common practice of separating combinational and state-holding processes.

Specifically, the first process is sensitive only to the clock signal. The *if* statement in the first process checks for an event on the clock (VHDL signals have a variety of attributes; *event* is true whenever the value has changed) and the clock being high, that is, the rising edge of the clock. The second process is sensitive only to the inputs and present state of the machine, not the clock, and is meant to model combinational logic. It illustrates the multiway conditional *case* statement, constants, and bit vectors. It employs types (i.e., `std_ulogic` and `std_ulogic_vector`) and operators from the ieee.std_logic_1164 library, an IEEE standard library [34] for modeling logic with unknown values (X).

15.5.2 VERIFICATION IN VHDL

Simple assertions that check variable values at a single point in time have long been part of VHDL. The assert construct may appear in both concurrent and sequential contexts:

```
assert a < b;

assert state /= BAD
  report "bad state encountered"
  severity "FAILURE";

process

begin

  for i in 0 to numvector-1 loop
    inputs <= vectors(i).in;
    wait for shortdelay;

    assert outputs = vectors(i).out
      report "incorrect output"
      severity error;

    wait for longdelay;
  end loop;
  wait;
end process;
```

However, many interesting properties of hardware designs are temporal, meaning that desired behavior changes over time. Standard VHDL assertions can be used in processes to check temporal properties. For example, the following process verifies that an acknowledge signal is always asserted in the cycle following a cycle in which *request* is asserted.

```
process begin
  wait until clk'event and clk = '1' and request = '1';
  wait until clk'event and clk = '1';
  assert acknowledge = '1';
end process;
```

However, VHDL 2008 adds PSL constructs to the `assert` construct, allowing this to be coded much more simply:

```
assert always (request -> next acknowledge);
```

Here, *always* means this holds in all cycles (i.e., not just the first); *next* indicates the given property (here, *acknowledge* is true) must hold in the next clock cycle (i.e., after the next rising edge of the clock).

15.5.3 VHDL SHORTCOMINGS

One shortcoming of VHDL is its obvious verbosity: the use of begin/end pairs instead of braces, the need to separate entities and their architectures, the need to spell out things like ports, its lengthy names for standard logic types (e.g., `std_ulogic_vector`), and its requirement of enclosing Boolean values and vectors in quotes. Some of these issues have been addressed in VHDL 2008 (e.g., many formerly mandatory = '1' constructs are no longer needed), but not all. Many of these are artifacts of its roots in the Ada language, another fairly verbose language commissioned by the US Department of Defense, but others are due to questionable design decisions. Consider the separation of entity/architecture pairs. While separating these concepts is a boon to abstraction and simplifies the construction of simulations of the same system in different configurations (e.g., to run a simulation using a gate-level architecture in place of a behavioral one for more precise timing estimation), in practice most designers only ever write a single architecture for a given entity and such pairs are usually written together.

The flexibility of VHDL also has advantages and disadvantages. Its type system was for a long time more flexible than Verilog's (although SystemVerilog has added features that close the gap), providing things such as aggregate types and overloaded functions and operators, but this flexibility also comes with a need for standardization and also tends to increase the verbosity of the language. Some of the need for standardization was recognized early, resulting in libraries such as the widely supported IEEE 1164 library for multivalued logic modeling. However, a standard for signed and unsigned arithmetic on logic vectors was slower in coming (it was eventually standardized in 1997 [35]), prompting both Synopsys and Mentor to each introduce similar but incompatible and incomplete versions of a similar library.

Fundamentally, many of the problems stem from a desire to make the language too general. Aspects of the type system suffer from this as well. While the ability to define new enumerated types for multivalued logic modeling is powerful, it seems a little odd to require virtually every VHDL program (since its main use has long been specification for RTL synthesis) to include one or more standard libraries. For many years, this also led to the need to be constantly comparing signals to the literal '1' instead of just using a signal's value directly and requiring a user to carefully watch the types of subexpressions. VHDL 2008 has addressed many of these issues, but many more remain.

15.6 SYSTEMC

SystemC is a relative latecomer to the HDL wars. Developed at Synopsys in the late 1990s, primarily by Stan Liao, SystemC was originally called Scenic [36] and was intended to replace Verilog and VHDL as the main system description language for synthesis (see Arnout [2] for some of the arguments for SystemC). SystemC is not so much a language as a C++ library along with a set of coding rules, but this is exactly its strength. It evolved from the common practice of first writing a high-level simulation model in C or C++, refining it, and finally recoding it in RTL Verilog or VHDL. SystemC was intended to smooth the refinement process by removing the need for a separate HDL.

SystemC can be thought of as a dialect of C++ for modeling digital hardware. Like Verilog and VHDL, it supports hierarchical models whose blocks consist of input/output ports, internal signals, concurrently running imperative processes, and instances of other blocks. The SystemC libraries make two main contributions: an inexpensive mechanism for running many processes concurrently (based on a lightweight thread package; see Liao et al. [36]) and an extensive set of types for modeling hardware systems, including bit vectors and fixed-point numbers. A SystemC model consists of a series of class definitions, each of which define a block. Methods defined for such a class become concurrently running processes, and the constructor for each class starts these processes running by passing them to the simulation kernel. Simulating a SystemC model starts by calling the constructors for all blocks in the design and then invoking the scheduler, which is responsible for executing each of the concurrent processes as needed.

The computational model behind earlier versions of SystemC was cycle based instead of the event-driven model of Verilog and VHDL. This meant that the simulation was driven by a collection of potentially asynchronous, but periodic clocks. Later versions (SystemC 2.0 and higher) adopted an event-driven model much like VHDL's.

SystemC can also be viewed as a complementary approach to the test bench problem, which Verilog and VHDL have long grappled with. While both Verilog and VHDL have long had facilities for writing test benches, they have never been as advanced, efficient, flexible, or as interoperable as mainstream software languages such as C or C++. To address this, both Verilog and VHDL have long provided application programming interfaces that allow C/C++ code to link with their simulators (Verilog's PLI and VHDL's Procedural Interface, respectively). These can be used to develop very sophisticated test benches, link in simulation models written in C/C++, and communicate with real hardware.

These interfaces, however, are clumsy compared to simply coding in C/C++ to begin with, the approach SystemC takes. Instead, SystemC pushes the awkwardness to modeling concurrency and structure, things which Verilog and VHDL do better. For writing large functional models, SystemC definitely has the advantage, but the few commercial products that have provided hardware synthesis from SystemC have met with little commercial success.

A number of SystemC books are now available. Black and Donovan [10] is popular. Grötker et al. [25] provide a nice introduction to SystemC 2.0. Bhasker [9] is also an introduction. The volume edited by Muller et al. [40] surveys more advanced SystemC modeling techniques.

15.6.1 CODING IN SYSTEMC

Figure 15.16 shows a small SystemC model for a 0–99 counter driving a pair of seven-segment displays. It defines two modules (the decoder and counter *struct*s) and an sc_main function that defines some internal signals, instantiates two decoders and a counter, and runs the simulation while printing out what it does.

The two modules in Figure 15.16 illustrate two of the three types of processes possible in SystemC. The decoder module is the simpler one: it models a purely combinational process by defining a method (called, arbitrarily, "compute") that will be invoked by the scheduler every time the number input changes, as indicated by the sensitive << number; statement beneath the definition of compute as an SC_METHOD.

The second module, counter is an SC_CTHREAD process: a method (here called "tick") that is invoked in response to a clock edge (here, the positive edge of the clk input, as defined by the SC_CTHREAD(tick, clk.pos()); statement) and can suspend itself with the wait() statement. Specifically, the scheduler resumes the method when the clock edge occurs, and the method runs until it encounters a wait() statement, at which point its state is saved and control passes back to the scheduler.

This example illustrates only a very small fraction of the SystemC type libraries. It uses unsigned integers (sc_uint), bit vectors (sc_bv), and a clock (sc_clock). The nonclock types are wrapped in sc_signals, which behave like VHDL signals. In particular, when an SC_CTHREAD method assigns a value to a signal, the effect of this assignment is felt only after all the processes triggered by the same clock edge have been run. Thus, such assignments behave like blocking assignments in Verilog to ensure that the nondeterministic order in which such processes are invoked (the scheduler is allowed to invoke them in any order) does not affect the ultimate outcome of simulating the system.

15.6.2 SYSTEMC SHORTCOMINGS

Like many languages, the most common use of SystemC has diverged from its designers' original intentions—an input for hardware synthesis in the case of SystemC. A number of synthesis tools for the language do exist, but SystemC is now used primarily (and quite successfully) for system modeling. This does mean, however, that it does not solve the "separate language for synthesis problem."

A big disadvantage of SystemC is that C++ was never intended for modeling digital hardware and as a result is even more lax about enforcing rules than Verilog. The syntax, similarly, is

```
#include "systemc.h"
#include <stdio.h>

struct decoder : sc_module {
  sc_in<sc_uint<4> > number;
  sc_out<sc_bv<7> > segments;

  void compute() {
    static sc_bv<7> codes[10] = {
      0x7e, 0x30, 0x6d, 0x79, 0x33,
      0x5b, 0x5f, 0x70, 0x7f, 0x7b };
    if (number.read() < 10)
      segments = codes[number.read()];
  }

  SC_CTOR(decoder) {
    SC_METHOD(compute);
    sensitive << number;
  }
};

struct counter : sc_module {
  sc_out<sc_uint<4> > tens;
  sc_out<sc_uint<4> > ones;
  sc_in_clk clk;

  void tick() {
    int one = 0, ten = 0;
    for (;;) {
      if (++one == 10) {
        one = 0;
        if (++ten == 10) ten = 0;
      }
      ones = one;
      tens = ten;
      wait();
    }
  }

  SC_CTOR(counter) {
    SC_CTHREAD(tick, clk.pos());
  }
};

int sc_main(int argc, char *argv[])
{
  sc_signal<sc_uint<4> > ones, tens;
  sc_signal<sc_bv<7> > ones_segments, tens_segments;
  sc_clock clk;

  decoder decoder1("decoder1");
  decoder1(ones, ones_segments);
  decoder decoder2("decoder2");
  decoder2(tens, tens_segments);

  counter counter1("counter1");
  counter1(tens, ones, clk);

  for (int i = 0 ; i < 12 ; i++) {
    sc_start(clk, 1);
    printf("%d %d %x %x\n",
           (int)tens.read(), (int)ones.read(),
           (int)(sc_uint<7>)tens_segments.read(),
           (int)(sc_uint<7>)ones_segments.read());
  }
```

FIGURE 15.16 A SystemC model for a two-digit decimal counter driving two seven-segment displays.

somewhat awkward and relies on some very tricky macro preprocessor definitions. On detailed models, the simulation speed of a good compiled-code Verilog or VHDL simulator may be better, although SystemC is much faster for higher-level models. For such systems, which consist of complex processes, SystemC should be superior since the simulation becomes nearly a normal C++ program. However, the context-switching cost in SystemC is higher than that of a good Verilog or VHDL simulator when running a more detailed model, so a system with many small processes would not simulate as quickly.

Another issue is the ease with which a SystemC model can inadvertently be made nondeterministic. Although carefully following a discipline of only communicating among processes through signals will ensure the simulation is deterministic, any slight deviation from this will cause problems. For example, library functions that use a hidden global variable may cause nondeterminism if called from different processes. Accidentally holding state in an `SC_METHOD` process (e.g., writing code that stores values in class or global variables) can also cause problems since such methods are invoked in an undefined order.

Many argue that nondeterministic behavior in a language can be desirable for modeling nondeterministic systems, which certainly exist and need to be modeled. However, the sort of nondeterminism in a language such as SystemC or Verilog creeps in unexpectedly and is difficult to use as a modeling tool. For the simulation of a nondeterministic model to be interesting, there needs to be some way of seeing the different possible behaviors, yet a nondeterministic artifact such as an `SC_METHOD` process that holds state provides no mechanism for ensuring that it is not, in fact, predictable. As a result, a designer has a hard time answering whether a model of a nondeterministic system can exhibit undesired behavior, even through a careful selection of test cases.

15.7 CONCLUSIONS

VHDL and SystemVerilog remain the dominant HDLs and are likely to be with us for a long time, although perhaps they will become like assembly language has become to programming: a part of the compilation chain, but not generally written manually. Both have deep semantic flaws, but these can be largely avoided by adhering to coding conventions, and in practice are quite practical design entry vehicles.

SystemC offers an alternative approach that starts from a "stock" (and very well supported) software programming language and addresses the high-level modeling problem directly, but it has not succeeded as a HDL, despite multiple attempts to make it bridge the gap.

When this chapter was first written 10 years ago, there were a plethora of competing verification-focused languages designed to interface with VHDL and Verilog models. Since then, SystemVerilog and, to a lesser extent, VHDL have adopted the better ideas from these languages. More than most other languages, Verilog and VHDL have evolved significantly over time as infelicities are fixed and constructs supporting new methodologies have been introduced.

The fundamental burdens of specifying digital hardware and verifying its correctness will continue to fall on design and verification languages. Even if those in the future bear little resemblance to those described here, the current crop forms a strong foundation on which to build.

REFERENCES

1. Accelera. SystemVerilog 3.1a language reference manual: Accellera's extensions to Verilog, Napa, CA, May 2004.
2. G. Arnout. SystemC standard. In *Proceedings of the Asia South Pacific Design Automation Conference (ASP-DAC)*, pp. 573–578, Yokohama, Japan, January 2000.
3. P.J. Ashenden. *The Designer's Guide to VHDL*, 3rd edn. Morgan Kaufmann, San Francisco, CA, 2008.
4. P.J. Ashenden. *The Student's Guide to VHDL*, 2nd edn. Morgan Kaufmann, San Francisco, CA, 2008.
5. I. Beer, S. Ben-David, C. Eisner, D. Fisman, A. Gringauze, and Y. Rodeh. The temporal logic Sugar. In *Proceedings of the 13th International Conference on Computer-Aided Verification (CAV)*, Lecture Notes in Computer Science, Vol. 2102, pp. 363–367, Paris, France, 2001. Springer, Berlin, Germany.
6. C. Gordon Bell and A. Newell. *Computer Structures: Readings and Examples*. McGraw-Hill, New York, 1971.

7. J. Bergeron. *Writing Testbenches: Function Verification of HDL Models*, 2nd edn. Kluwer, Boston, MA, 2003.
8. J. Bhasker. *A VHDL Synthesis Primer*, 2nd edn. Star Galaxy Publishing, Allentown, PA, 1998.
9. J. Bhasker. *A SystemC Primer*, 2nd edn. Star Galaxy Publishing, Allentown, PA, 2004.
10. D.C. Black and J. Donovan. *SystemC: From the Ground Up*. Springer, Berlin, Germany, 2005.
11. D. Borrione, R. Piloty, D. Hill, K.J. Lieberherr, and P. Moorby. Three decades of HDLs: Part II, Colan through Verilog. *IEEE Design & Test of Computers*, 9(3):54–63, September 1992.
12. R.K. Brayton, G.D. Hachtel, and A.L. Sangiovanni-Vincentelli. Multilevel logic synthesis. *Proceedings of the IEEE*, 78(2):264–300, February 1990.
13. Y. Chu, D.L. Dietmeyer, J.R. Duley, F.J. Hill, M.R. Barbacci, C.W. Rose, G. Ordy, B. Johnson, and M. Roberts. Three decades of HDLs: Part I, CDL through TI-HDL. *IEEE Design & Test of Computers*, 9(2):69–81, June 1992.
14. Y. Chu. An ALGOL-like computer design language. *Communications of the ACM*, 8(10):607–615, October 1965.
15. E.M. Clarke and E.A. Emerson. Design and synthesis of synchronization skeletons using branching time temporal logic. In *Proceedings of the Workshop on Logic of Programs*, Lecture Notes in Computer Science, Vol. 131, pp. 52–71, Yorktown Heights, New York, May 1981. Springer, Berlin, Germany.
16. B. Cohen. *VHDL Coding Styles and Methodologies*, 2nd edn. Kluwer, Boston, MA, 1999.
17. B. Cohen, S. Venkataramanan, and A. Kumari. *Using PSL/Sugar for Formal Verification*. VhdlCohen Publishing, Palos Verdes Peninsula, CA, 2004.
18. A. Dewey. VHSIC hardware description (VHDL) development program. In *Proceedings of the 20th Design Automation Conference*, pp. 625–628, Miami Beach, FL, June 1983.
19. A. Dewey and A.J. de Geus. VHDL: Toward a unified view of design. *IEEE Design & Test of Computers*, 9(2):8–17, April 1992.
20. A.M. Dewey. *Analysis and Design of Digital Systems with VHDL*. Brooks/Cole Publishing (Formerly PWS), Pacific Grove, CA, 1997.
21. P.L. Flake and S.J. Davidmann. Superlog, a unified design language for system-on-chip. In *Proceedings of the Asia South Pacific Design Automation Conference (ASP-DAC)*, pp. 583–586, Yokohama, Japan, January 2000.
22. P.L. Flake, P.R. Moorby, and G. Musgrave. An algebra for logic strength simulation. In *Proceedings of the 20th Design Automation Conference*, pp. 615–618, Miami Beach, FL, June 1983.
23. R.S. French, M.S. Lam, J.R. Levitt, and K. Olukotun. A general method for compiling event-driven simulations. In *Proceedings of the 32nd Design Automation Conference*, pp. 151–156, San Francisco, CA, June 1995.
24. M.J.C. Gordon. The semantic challenge of Verilog HDL. In *Proceedings of the 10th Annual IEEE Symposium on Logic in Computer Science (LICS)*, San Diego, CA, June 1995.
25. T. Grötker, S. Liao, G. Martin, and S. Swan. *System Design with SystemC*. Kluwer, Boston, MA, 2002.
26. R.E. Harr and A.G. Stanculescu, editors. *Applications of VHDL to Circuit Design*. Kluwer, Boston, MA, 1991.
27. IEEE. *Standard for Property Specification Language (PSL) (IEEE 1850–2010)*, New York, 2010.
28. IEEE Computer Society. *IEEE Standard Hardware Description Language Based on the Verilog Hardware Description Language (1364–1995)*, New York, 1996.
29. IEEE Computer Society. *IEEE Standard Verilog Hardware Description Language (1364–2001)*, New York, September 2001.
30. IEEE Computer Society. *IEEE Standard VHDL Language Reference Manual (1076–2008)*, New York, 2008.
31. IEEE Computer Society. *IEEE Standard for SystemVerilog—Unified Hardware Design, Specification, and Verification Language (1800–2012)*, New York, 2012.
32. S. Iman and S. Joshi. *The e Hardware Verification Language*. Springer, Berlin, Germany, 2004.
33. The Institute of Electrical and Electronics Engineers (IEEE). *IEEE Standard VHDL Reference Manual (1076–1987)*, New York, 1988.
34. The Institute of Electrical and Electronics Engineers (IEEE). *IEEE Standard Multivalue Logic System for VHDL Model Interoperability (Std_logic_1164)*, New York, 1993.
35. The Institute of Electrical and Electronics Engineers (IEEE). *IEEE Standard VHDL Synthesis Packages (1076.3–1997)*, New York, 1997.
36. S. Liao, S. Tjiang, and R. Gupta. An efficient implementation of reactivity for modeling hardware in the Scenic design environment. In *Proceedings of the 34th Design Automation Conference*, pp. 70–75, Anaheim, CA, June 1997.
37. R. Lipsett, C.F. Schaefer, and C. Ussery. *VHDL: Hardware Description and Design*. Kluwer, Boston, MA, 1989.
38. C. Mead and L. Conway. *Introduction to VLSI Systems*. Addison-Wesley, Reading, MA, 1980.
39. S. Mittra. *Principles of Verilog PLI*. Kluwer, Boston, MA, 1999.

40. W. Muller, W. Rosenstiel, and J. Ruf, editors. *SystemC: Methodologies and Applications*. Kluwer, Boston, MA, 2003.

41. W.E. Omohundro and J.H. Tracey. Flowware—A flow charting procedure to describe digital networks. In *Proceedings of the First International Conference on Computer Architecture* (*ISCA*), pp. 91–97, Gainesville, FL, December 1973.

42. S. Palnitkar. *Verilog HDL: A Guide to Digital Design and Synthesis*. Prentice Hall, Upper Saddle River, NJ, 1996.

43. S. Palnitkar. *Design Verification with e*. Prentice Hall, Upper Saddle River, NJ, 2003.

44. F.I. Parke. An introduction to the N.mPc design environment. In *Proceedings of the 16th Design Automation Conference*, pp. 513–519, San Diego, CA, June 1979.

45. D.L. Perry. *VHDL*, 3rd edn. McGraw-Hill, New York, 1998.

46. I.S. Reed. Symbolic synthesis of digital computers. In *Proceedings of the ACM National Meeting*, pp. 90–94, Toronto, Ontario, Canada, September 1952.

47. A. Rushton. *VHDL for Logic Synthesis*, 3rd edn. John Wiley & Sons, New York, 2011.

48. D.J. Smith. VHDL & Verilog compared & contrasted—Plus modeled examples written in VHDL, Verilog, and C. In *Proceedings of the 33rd Design Automation Conference*, pp. 771–776, Las Vegas, NV, June 1996.

49. C. Spear and G. Tumbush. *SystemVerilog for Verification*, 3rd edn. Springer, 2012.

50. S. Sutherland. *The Verilog PLI Handbook*. Kluwer, Boston, MA, 1999.

51. S. Sutherland, S. Davidmann, and P. Flake. *SystemVerilog for Design: A Guide to Using SystemVerilog for Hardware Design and Modeling*, 2nd edn. Springer, Berlin, Germany, 2006.

52. D.E. Thomas and P.R. Moorby. *The Verilog Hardware Description Language*, 5th edn. Kluwer, Boston, MA, 2002.

Digital Simulation

John Sanguinetti

CONTENTS

16.1 INTRODUCTION

Logic simulation is the primary tool used for verifying the logical correctness of a hardware design. In many cases, logic simulation is the first activity performed in the process of taking a hardware design from concept to realization. Modern hardware description languages are both simulatable and synthesizable. Designing hardware today is actually writing a program in the hardware description language. Performing a simulation is just running that program. When the program (or model) runs correctly, then one can be reasonably assured that the logic of the design is correct, *for the cases that have been tested in the simulation.*

Simulation is the key activity in the design verification process. That is not to say that it is an ideal process. It has some very positive attributes:

1. It is a natural way for the designer to get feedback about his design. Because it is just running a program—the design itself—the designer interacts with it using the vocabulary and abstractions of the design. There is no layer of translation to obscure the behavior of the design.
2. The level of effort required to debug and then verify the design is proportional to the maturity of the design. That is, early in the design's life, bugs and incorrect behavior are usually found more quickly. As the design matures, it takes longer to find the errors.
3. Simulation is completely general. Any hardware design can be simulated. The only limits are time and computer resources.

On the negative side, simulation has two drawbacks, one of which is glaring:

1. There is (usually) no way to know when you are done. It is not feasible to completely test, via simulation, all possible states and inputs of any nontrivial system.
2. Simulation can take an inordinately large amount of computing resources, since typically it uses a single processor to reproduce the behavior of many (perhaps millions of) parallel hardware processes.

Every design project must answer the question "have we simulated enough to find all the bugs?" and every project manager has taped out his design knowing that the truthful answer to that question is either "no" or "I don't know." It is this fundamental problem with simulation that has caused so much effort to be spent looking for both tools to help answer the question, and formal alternatives to simulation.

Code coverage, functional coverage, and logic coverage tools have all been developed to help gauge the completeness of simulation testing. None are complete solutions, though they all help. Formal alternatives have been less successful. Just as in the general software world where proving programs correct has proven intractable, formal methods for verifying hardware designs have still not proven general enough to replace simulation. That is not surprising, since it is the same problem.

The second drawback motivates much of the material in this chapter. That is, simulation is always orders of magnitude slower than the system being simulated. If a hardware system runs at 1 GHz, a simulation of that system might run at 10–1000 Hz, depending on the level of the simulation and the size of the system. That is a slowdown from 10^6 to 10^8! Consequently, many people have spent considerable time and effort finding ways to speed up logic simulation.

Considering both the advantages and disadvantages of logic simulation, it is really quite a good tool for verifying the correctness of a hardware design. Despite its drawbacks, simulation remains the first choice for demonstrating correctness of a design before fabrication, and its value has been well established.

16.1.1 LEVELS OF ABSTRACTION

Because simulation is a general technique, a hardware design can be simulated at different levels of abstraction. Often it is useful to simulate a model at several levels of abstraction in the same simulation run. The commonly used levels of abstraction are gate level, register transfer level (RTL), and behavioral (or algorithmic) level. However, it is possible to incorporate lower levels like transistor level or even lower physical levels as well as higher levels such as transaction level or domain-specific levels. For this discussion, we will restrict our attention to behavioral, RTL, and gate levels, with the understanding that other levels are completely compatible with the techniques described here.

16.1.2 DISCRETE EVENT SIMULATION

we decided to focus on simulation ..., because that's the only really interesting thing to do with a computer

Alan Kay, Second West Coast Computer Faire (1978)

Logic simulation is a special case of the more general discrete event simulation methods, which were initially developed in the 1960s [5]. Discrete event simulation is a method of representing the behavior of a system over time, using a computer. The system may be either real or hypothetical, but its state is assumed to change over time due to some combination of external stimulus and internal state. The fundamental idea is that the behavior of any system can be decomposed into a set of discrete instants of time at which things happen. Those instants are called events, and the "things that happen" are state changes. This is very analogous to the way we digitize continuous physical phenomena, like audio sampling. In essence, we digitize a time-based process by dividing it up into discrete events. It is easy to see that with a fine enough granularity, one can get an adequately accurate representation for just about any purpose.

The basic operation of a discrete event simulation is given in Figure 16.1.

Each event in the system is represented by an event routine. An event routine is some code to be executed to represent the action at that event, which usually amounts to a state change and a determination of the occurrence of one or more future events. The set of possible events is usually (though not always) fixed, but the number of occurrences of each type of event can be variable.

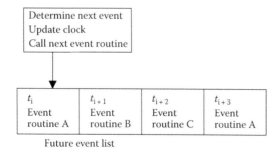

FIGURE 16.1 Discrete event simulation basic operation.

The list of events shown in Figure 16.1 is a list of those events that are scheduled to happen at known times in the future (t_{i+1}, t_{i+2}, ...). The entries on this list contain the time at which the event will occur, and the type of event it is; that is, what event routine should be called to realize that event's behavior. It is common to have more than one instance of any particular type of event scheduled to occur at various times on this list. This list is called the future event list.

16.2 EVENT- VS. PROCESS-ORIENTED SIMULATION

The basic discrete event simulation paradigm was in use in the early 1960s (and probably earlier), and several subroutine libraries were developed for FORTRAN and Algol to support it. It was soon realized that there were more natural ways to write simulation models of many systems, and that led to the development of specialized simulation languages, particularly GPSS, SIMSCRIPT, and Simula [3]. Of these, GPSS is still used today to model systems as extensive as the U.S. air traffic control system. Simula was the first object-oriented programming language and introduced the concepts of classes and objects.

Both GPSS and Simula took a process-oriented view of modeling. That is, instead of writing a separate routine for each event, one would write a routine that represented a process which might include several events within it taking place over a nonzero period of simulated time. The routine would have wait statements in it to indicate that some time would pass between one event and the next. This did not affect the underlying simulation mechanism, but it did change the way the simulation model was written. A process orientation is particularly useful for hardware simulation at a behavioral level, while an event orientation works for RTL and gate levels. The most popular hardware description languages, Verilog [6,7], VHDL [2], and SystemC [4], are process-oriented simulation languages which are indirectly descended from Simula.

16.3 LOGIC SIMULATION METHODS AND ALGORITHMS

Logic simulation is essentially a process of computing a state trajectory of the system over time. The system's state is defined by the state variables, which are the storage elements in the circuit. We usually think of these as the registers and latches in the design. How the state, taken as a whole, changes over time is called the state trajectory.

A simple state trajectory as a function of time can be written as follows:

$$f(m, n, t + 1) = m_t + n_t$$

where
    m is a state variable
    n is an input

It is obvious how to compute this function:

$$\text{Do } (i = 1, \text{ endtime})$$
$$t = t + 1$$
$$m = m + n$$
$$n = \text{new input}$$
$$\text{enddo}$$

This example illustrates how a function is simulated over time. There is a loop, which consists of updating the system time or clock, evaluating the logic components, and updating the state variables. Compare this loop to the process of Figure 16.1. There is only one event routine, and it gets executed repeatedly at a regular time interval.

Now we can expand this example to two functions,

$$f(m,n,t+1) = m_t + n_t$$
$$g(x,n,t+1) = x_t \wedge n_t$$

Evaluating these two functions together would look like this:

$$\text{Do } (i = 1, \text{ endtime})$$
$$t = t + 1$$
$$m = m + n$$
$$x = x \wedge n$$
$$n = \text{new input}$$
$$\text{enddo}$$

This set of state functions when taken together, make up a system with a state variable, which is a duple (m, x). Running this simulation computes the trajectory of the state vector (m, x) over the simulation interval.

Notice that the way we have defined the functions f and g, the value of n used in the computation is the value of n at time t, even though the time of the evaluation is $t + 1$. This leads to a further refinement of the simulation process. In real hardware, state variables (registers or latches) do not get updated instantaneously. That is, the computation even if it is a simple assignment, takes some nonzero amount of time. So the simulation loop could be rewritten as in Figure 16.2.

While we have not represented the actual delays involved in the computation, we have abstracted them into a behavior that takes into account their effect. We now have the typical event-processing loop in a digital logic simulation: advance the clock, evaluate all the logic functions, and update the variable values.

Note that this has incorporated an abstraction of physical behavior over time, since we are moving time from one discrete moment to another, and we are assuming that nothing interesting happens in between those two instants. We also assumed that m, x, and n do not change their values instantaneously, but they do change before the next time instant $(t + 1)$. Note that they all change together after all the evaluation has been done. This is usually called a "delta cycle," meaning it is a set of events that happen at time t, but after all the evaluation events. That is, the new-value assignments happen at $t + \delta t$.

A further complication arises when time delays must be introduced into a computation. That is, the function being computed may look like this:

$$f(m,n,t) = g(m_{t-k}, n_{t-k})$$

```
do (i=1, endtime)
    t = t + 1
    // evaluate the functions
    t_m = m + n
    t_x = x ^ n
    t_n = new input
    // update the variable values
    m = t_m
    x = t_x
    n = t_n
enddo
```

FIGURE 16.2   Basic simulation loop.

```
do (i=1, endtime)
    t = t + 1
    // evaluate the functions
    t_m(t+k) = m + n
    t_n = new input
    // update the variable values
    m = t_m(t)
    n = t_n
enddo
```

FIGURE 16.3   Simulation loop with update step.

```
do (i=1, endtime)
    t = t + 1
    // evaluate the functions
    t_m = m + n
    put_on_future_update_list(t_m,t+k)
    t_n = new input
    // update the variable values
    m = take_off_future_update_list(t)
    n = t_n
enddo
```

FIGURE 16.4   Simulation loop with update list.

That is, the new value does not get updated into the state variable until k time units after the evaluation. To handle this, the simulation loop would look like Figure 16.3.

It is common practice to replace the array $t\_m()$ in Figure 16.3 with a linked list, which is ordered by the index t. In order to make this work, the delay is usually incorporated into the new-value computation, and the new value is put on a future update list at time $t + k$. The future update list is analogous to the future event list, and in fact does not need to be separate from it. The simulation loop would then look like Figure 16.4.

Here we have assumed that the future update list has values only for the variable m. Generally, there may be many different state variables whose new values may be saved on the future update list, so each entry on the list must identify the state variable that the value is associated with. Note also that the new value is always at the front of the list, if the list is ordered by increasing time t. That is, at any given time t, there are no values on the list with time $< t$ (those would be values in the past), so all the variable updates can be found at the front of the list. Of course, there may be no values on the list associated with time t, that is, the first entry on the list may have time $> t$, in which case no variable would be updated at time t.

16.3.1 SYNCHRONOUS AND ASYNCHRONOUS LOGIC

The example above shows synchronous, or sequential, logic. That is, each variable's new value is computed at a regular interval $(t, t + 1, …)$. There is also asynchronous, or combinational, logic that might be present in a system to be simulated. The salient characteristic of asynchronous logic is that the function is computed continuously. That is, when the inputs change, the output changes immediately, at least as compared to the granularity of the simulation time. For example, in an *and* gate followed by an inverter, the two inputs are *and*ed into the inverter whose output is created with no (apparent) delay (Figure 16.5).

FIGURE 16.5   Example asynchronous logic.

This would be represented by $d = \sim(c\&b)$. In terms of our simulation loop, asynchronous logic functions have their outputs updated during the evaluation phase, not delayed until the update phase. That is, there is no temporary variable created to hold the new value.

It now is apparent that we need some way of distinguishing between the kind of assignment that is immediate and the previous, synchronous, kind, which is delayed. A simulation language typically has the ability to represent instantaneous change as well as delayed change.

16.3.2 PROPAGATION

Combinational logic leads to a further complication of our simulation loop, namely, where do we do the evaluation of a combinational expression? Combinational logic must be evaluated whenever one of its constituent inputs changes. Sequential logic can be evaluated regularly, at time intervals corresponding to a clock signal. That is why it was natural to write the simulation loop with the evaluations happening immediately after updating the time variable. But where do we put combinational logic update?

The answer can be found by looking at what would cause a combinational expression to be evaluated. That is, when would its inputs change? If we look at a typical circuit that has both combinational and sequential logic, it might look like Figure 16.6.

Looking at this, we see that the input to a combinational expression is either the value of a state variable (a register) or the output of another combinational expression. Going back to the simulation loop, the logical place to put the evaluation of combinational expressions is after the update of state variables. If our combinational logic consisted of

$$c = m \mid a$$
$$d = \sim (c \,\&\, b)$$

Then the loop would look like Figure 16.7.

The simulation loop as it is described in Figure 16.7 is now sufficient to simulate circuits that have the form of Figure 16.6, but with an important caveat. That caveat is that the combinational expressions must be ordered such that each one is evaluated only after all of its inputs have been updated. If a circuit does not contain any combinational feedback loops, then it is possible to satisfy this ordering requirement. Ordering the expressions is called *levelizing*.

If the delays for the sequential variables are all zero, which is a common case, then this simulation loop is quite efficient, and in fact many logic simulators have been created with just this simulation loop. The drawback of this approach however, is that all of the expressions in the circuit must be evaluated on every loop iteration, i.e., at each time instant. When very much of the circuit remains unchanging for long periods, this can be quite wasteful, as expressions that do not change get continually reevaluated.

To improve the simulation efficiency in this common case, we can return to the basic event scheduling idea of discrete event simulation, and only evaluate an expression when one or more of its inputs has changed. This is called *propagation* of a value from an output to an input, or in

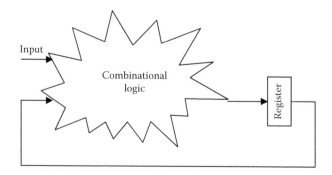

FIGURE 16.6 Example with synchronous and asynchronous logic.

```
do (i=1,endtime)
    t = t + 1
    // evaluate the functions
    t_m = m + n
    put_on_future_update_list(t_m,t)
    t_n = new input
    // update the variable values
    m = take_off_future_update_list(t)
    n = t_n
    //update all combinational expressions
    c = m | a
    d = ~(c & b)
enddo
```

FIGURE 16.7   Simulation loop with asynchronous logic.

language terms, from a left-hand side of an assignment to a right-hand side. The simulation loop now would look like Figure 16.8.

Just as we introduced a future update list, we can also create a propagation list, where the entry on the list is the expression whose inputs have changed. Rewriting the simulation loop using that, we get Figure 16.9.

Notice here that the propagate list contains a pointer or other identifier of the expression that needs to be evaluated. Notice also that the expressions that depend on any variable are given by the function $c(v)$. The function $c(v)$ is static, since in logic simulation, the expressions that depend on any variable are known at compile time. Finally, notice that it is

```
do (i=1, endtime)
    t = t + 1
    // evaluate the state variables
    t_m = m + n
    put_on_future_update_list(t_m,t)
    t_n = new input
    // update the variable values
    m = take_off_update_event_list(t)
    n = t_n
    // propagate m and n to combinational expressions
    c = m | a
    // propagate c to combinational expressions
    b = ~(c & d)
    // propagate b to combinational expressions
    ...
enddo
```

FIGURE 16.8   Simulation loop with propagation.

```
do (i=1,endtime)
    t = t + 1
    for (all state variables m)
        t_m= f( . ) // evaluate the state variables
        put_on_future_update_list(m, t_m, t)
    endfor
    // update the variable values
    while (u = take_off_future_update_list(t) != empty)
        u.v = u.t_v
        put_on_propagate_list(c(u.v), t+u.d)
    endwhile
    // propagate changed values to combinational expressions
    while (c = take_off_propagate_list(t) != empty)
        c.v = eval(c.expr)
        put_on_propagate_list(c(c.v), t+c.d)
    endwhile
enddo
```

FIGURE 16.9   Simulation loop with propagation list.

```
do (while t < endtime)
   t = min(future_event_list, future_update_list, propagate_list)
   while (m = take_off_future_event_list(t) != empty)
     t_m= eval(m.expr) // evaluate the state variable in this event
     put_on_future_update_list(m.v, t_m, t+m.d)
     put_on_future_event_list(m.v, m.expr, t+m.next)
   endwhile
   // update the variable values
   while (u = take_off_future_update_list(t) != empty)
     u.v = u.t_v
     put_on_propagate_list(c(u.v), t+u.d)
   endwhile
   // propagate changed values to combinational expressions
   while (c = take_off_propagate_list(t) != empty)
     c.v = eval(c.expr)
     put_on_propagate_list(c(c.v), t+c.d)
   endwhile
enddo
```

FIGURE 16.10   Simulation loop with future event, update, and propagate lists.

a simple efficiency improvement to only propagate those values that have changed. That is, if the new value of a variable is the same as the old value, there is no need to reevaluate any dependent expressions. With that enhancement, we have a complete simulation loop for a logic simulator.

The only remaining necessary generalization is time. We have dealt with time as simply an ordered set of instants, t_1, t_2, …. This would be sufficient if each time instant was able to be mapped onto a real time line, for example, if each t_i corresponded to clock cycle i. However, in the general case, there may be nonuniform time intervals at which events happen. Just as there could be a delay in updating the state variables with new values, there could be delays updating combinational variables with new values (this would represent gate delays). So both the update event list and the propagate list would be ordered by time, and each event on either list would include a time at which it was to occur. Now, when all the events at the current time t are finished, the main loop would be iterated upon and time would be advanced to the nearest time in the future from either of the two lists (Figure 16.10).

The entry which is put on the future event list is the sequential expression, which will be evaluated at a given time in the future (often a clock-cycle boundary). This now has all the elements of a logic simulation loop, sometimes also called the scheduling loop or the simulation kernel. People often talk about a "simulation engine" in an attempt to make simulators sound more sophisticated than they really are. The scheduler is about the only thing that could be called a "simulation engine," and as we have seen, conceptually it is pretty simple.

16.3.3 PROCESSES

We have described a logic simulation model as a collection of synchronous state variables (registers) and asynchronous combinational expressions. This corresponds to a typical RTL-style description of a logic circuit. We might write a description of such a circuit as follows:

at rising edge of clock:

$\quad$ $reg_i = newval_i$

$\quad$ $reg_j = newval_j$

$\quad$ …

asynchronous:

$\quad$ $newval_i = f(reg_i, reg_j, …)$

$\quad$ $newval_j = g(reg_i, reg_j, …)$

$\quad$ …

```
Process
    while (forever)
        while (!valid)
            wait (rising edge of clock)
        d <= data.in
        ready <= 0
        wait (rising edge of clock)
        for (i=0; i<10; i++)
            wait (rising edge of clock)
        dataout <= f(d)
        ready <= 1
    endwhile
endprocess
```

FIGURE 16.11   An example process.

This maps directly onto the simulation loop described in Figure 16.10. However, this is a pretty low level of description, and it would be laborious to write circuit models this way routinely. By introducing the idea of a process, we can gain expressive power and still use the same simulation mechanics. A process would look like this:

$$process$$
$$wait\ (rising\ edge\ of\ clock)$$
$$reg_i <= f(reg_i, reg_j, ...)$$
$$reg_j <= g(reg_i, reg_j, ...)$$
$$...$$
$$endprocess$$

Here we have combined the combinational logic with the sequential logic. Note that we have used a special symbol <=, to indicate that the assignment is a delayed assignment. This is not strictly necessary, as we could infer that any assignment to a state variable should be a delayed assignment. Verilog uses a special assignment symbol, VHDL and SystemC do not.

We could extend the process by allowing time to pass during the execution of the process, as in Figure 16.11.

This process represents a communication protocol with another process whereby data are exchanged using a ready/valid handshake, and the output happens 10 clock-cycles after the input. It is straightforward to map this process onto our simulation loop by decomposing the process into events that get put onto the future event list. Each of the events corresponds to a wait in the process. This is left as an exercise for the reader.

The notion of a process becomes more powerful when a system has multiple processes, which can operate on common variables. Processes are inherently independent. That is, the events in two processes may or may not have any ordering relationship between them. Typically, all processes are assumed to begin at the beginning of simulation (time 0), but the events they wait on may be different and unrelated. Events within processes may be synchronized by means of common events or variables, but they do not have to be.

It is easy to see how arbitrarily complicated behavior can be described with multiple processes, and it is straightforward to map the events in each process into our simulation structure. Indeed, it is fairly easy to construct models that cannot be physically realized using these structures. While it may not be apparent what the value of describing unrealizable systems is, this power can be very useful when modeling the environment that the system is subjected to. This is usually called the *testbench*, while the target system is called the *design under test*. It is very useful to be able to describe both in the same simulation environment.

16.3.4 RACE CONDITIONS

A race condition can occur in a concurrent system when the behavior of the system depends on the order of execution of two events that are logically unordered. The most common cause of this is when one process modifies a variable and another reads the same variable at the same simulated time. This will not happen with state variables when delayed assignment is used, but it can happen with combinational variables, or with state variables if delayed assignment is not used. VHDL took the approach of making all assignments to state variables delayed, while Verilog did not. Thus, it is easier to write a model with race conditions in Verilog than in VHDL. There is an efficiency cost to delayed assignment of course, which is one of the reasons that VHDL simulators are typically slower than Verilog simulators.

16.4 IMPACT OF LANGUAGES ON LOGIC SIMULATION

The three major hardware description languages, Verilog, VHDL, and SystemC, are the primary languages used for hardware simulation today. All of them have the richness required to represent the vast majority of hardware designs. Verilog has more low-level capabilities than VHDL or SystemC, while SystemC has more high-level capabilities than VHDL or Verilog. They have common features that enable efficient simulation of hardware constructs, primarily hardware data types, hardware-oriented hierarchy, and hardware-oriented timing and synchronization.

16.4.1 DATA TYPES

One of the ways that hardware description languages differ from other programming languages is the data types they offer. Logic signals are either 1 or 0, or in some cases neither (i.e., undriven or floating). But real hardware can be more complicated, so it is often convenient to be able to model signals as a range of strengths. That is, a strong signal can override a weak signal. It is also convenient for simulation purposes to include an unknown value (x), which indicates that a value is either uninitialized, or driven by conflicting values. This range of possible values is most useful when modeling at low levels, which is why Verilog has the richest set of signal values and SystemC has the least.

All hardware description languages have data types that allow the explicit specification of bit widths. That is, it can be specified that a variable is n bits wide, where the maximum value of n is usually some large number. This is useful when describing buses and collections of signals that are to be treated as a single variable. There are also operations to go along with these data types, like concatenation and subset selection. Operations on these data types cause the simulator to do more work than would be done in a normal C program, since the underlying computer must use several instructions to accomplish them, rather than a single native instruction.

16.4.2 VARIABLES

Verilog, VHDL, and SystemC make a distinction between variables which are state variables and those which are combinational variables. This makes the simulator's job easier because the classification is made by the programmer. In fact, a simulator could determine which variables are state variables and which are not by context. Verilog goes even further and requires the programmer to indicate which assignments are delayed assignments and which are not.

16.4.3 HIERARCHY

The organization of a model in a hardware description language consists of a collection of modules in a tree structure. This corresponds with the way hardware is built. Signals are passed between modules in the tree by means of input and output ports. As far as the simulation goes, the port connections between parent and child are the same as combinational logic assignments. That is, for an input port, the left-hand side of the assignment is the port variable in the child module, and the right-hand side is the port variable in the parent module. By including these constructs in the language, the user does not have to write so much code, but the simulator still has to do the same amount of work.

```
reg clk;
                                  reg [7:0] state, newstate;
clk = 0;
                                  always @ (posedge clk)
always #10
                                      state <= newstate;
   clk = ~clk;
(a)                               (b)
always begin
   while (go == 0)
      @(posedge clk);    //wait for go
   count = 0;
   while (count <10)      // wait for 10 cycles
      @(posedge clk) count = count + 1;
   done <= 1;            // raise flag
   @(posedge clk)
      done <= 0;    // drop flag after one cycle
end
(c)
```

FIGURE 16.12 Description and use of a clock signal in Verilog (a) clock signal; (b) register using clock signal; (c) process using clock signal.

16.4.4 TIME CONTROL

Hardware description languages offer a clock-based model of synchronization and event control. This makes synchronous logic circuits easy to write. Again, it does not make the simulator do any less work. The main characteristic is the definition of events in relation to one or more common signals, which are usually interpreted as a system clock signal, or a set of clock signals. Thus, a process can wait on the rising or falling edge of a signal, or either edge. The common way of indicating that two processes execute at the same time is to have both wait on the same edge of a common signal. At the RTL, where registers are explicitly instantiated, it is common to have hundreds or thousands of registers all triggered by the rising edge of a clock. This can be taken advantage of by the simulator to reduce the overhead of these events. Note however, that there is nothing inherently different about a clock signal from any other signal. The only difference is how it is used.

A Verilog description of a clock signal is shown in Figure 16.12a, a register using it is shown in Figure 16.12b, and a process using it is shown in Figure 16.12c.

16.4.5 COMBINATIONAL LOGIC

A distinctive feature of hardware description languages from a simulation point of view, is the inclusion of a separate construct to describe combinational or continuous logic. As previously described, combinational logic is composed of assignments to variables that are done reactively. That is, the assignment is evaluated and performed whenever any of its constituent variables changes. We saw earlier how propagation is handled in the simulator. Combinational expressions are evaluated only as a result of propagation. Their use is a natural way to represent hardware at the RTL, but their use imposes complications for the simulator, since a poor choice of when to evaluate them can have dramatic consequences on the running time of the simulation. Figure 16.13a shows a Verilog continuous assignment, and Figure 16.13b shows the same variable assignment written as a process. Note that in both cases, the expression evaluation and assignment will be performed whenever one of the right-hand side variables changes.

16.5 LOGIC SIMULATION TECHNIQUES

Simulation speed is defined as the ratio of simulation time to simulated time. Simulation time is the real time required to execute the simulation model, while simulated time is the time represented in the model. Because an essentially unbounded amount of simulation is required to verify assign var = a + (b ^ c);

always @(a or b or c)

var = a + (b ^ c);

(a)

(b)

FIGURE 16.13 Logic written as (a) continuous assignment; (b) process.

the correct behavior of a complex digital design, simulation speed is very important. It is interesting to note that in logic simulation, the simulation speed ratio is often called the slow-down, since the simulation takes longer than the real system, usually by orders of magnitude. However, in general system simulation, the simulation speed ratio is often called a speedup, since the simulated time scale may be very large.

The underlying mechanics of logic simulation are as described above. Except in the case of cycle-based simulation as described below, all simulators use a scheduler that functions pretty much the same. The scheduler is responsible for selecting the next event to execute and transferring control to it. There are well-known algorithms for implementing schedulers, and all mainstream logic simulators have reasonably well-optimized schedulers.

Nevertheless, different simulators can have vastly different performances on the same model written in the same language (or the same model written in a different language). The difference in execution efficiency between simulators is due to how the event routines are executed, how operations are executed, and especially how events are scheduled. It is not uncommon for two different simulators to execute a different number of events—differing by a factor of 2 or 3—for the same model, and still yield the same results.

16.5.1 INTERPRETED SIMULATION

Most early logic simulators were interpreters. That is, the simulator read in the model source, built some internal data structures, and encoded the event routine operations in a unique instruction set. Then, as the simulation was run, the event routines were interpreted by a special piece of code, the interpreter, which "executed" those custom instructions. This is a technique that has been used for many years to translate and execute programs in a variety of languages. In general, interpreted execution offers a good opportunity for debugging the program, since the interpreter can relate any errors directly to the source of the program. However, there is an efficiency cost. Interpreted execution is slow, because the interpreter has to do a lot of work for each instruction. As a result, there are few modern logic simulators that are interpreters.

16.5.2 COMPILED CODE SIMULATION

Logic simulators have been categorized as "compiled code" simulators to distinguish them from interpreters. A better description would be simply "compiler" [1]. A compiled code simulator is nothing more than a compiler for the simulation language it implements. Fundamentally, it differs from an interpreter in how the event routines are executed. While an interpreter executes the event routines by executing a sequence of operations represented as higher-level instructions, a compiler prepares the event routines so that they can be executed directly by the host machine. That is, an event routine is compiled into machine code so that it can be called as a subroutine. Essentially, the compiler has to do more work initially, but it produces event routines that execute much faster. It is common for compiled code to execute one to two orders of magnitude faster than interpreted code. That difference in efficiency has been observed across many different programming languages and many different underlying machine architectures. It is no surprise that compiled logic simulators typically run between 10 and 20 times faster than interpreted logic simulators on the same models.

It is common for compiled logic simulators to emit event routine code in C and then use the host machine's C compiler to produce machine code. The C code emitted is nothing more than an intermediate form of the compiled program, with the C compiler serving as the last phase of the compiling process. In general, there is little difference in efficiency between compiled simulators that use the host C compiler to produce machine instructions and those that emit machine instructions directly.

It is worth noting that the relative efficiency of interpreters and compilers is dependent on the level of abstraction of the model being simulated. The lower the level, the smaller the difference. The fastest gate-level simulator, even today (2006), is still an interpreted Verilog simulator.

16.5.3 CYCLE-BASED SIMULATION

In the general case, updating a variable value may have a time delay which results in an event being scheduled and then executed by the scheduling loop. In a typical synchronous logic model, every clock cycle could have many intermediate events within the cycle when different variables are updated. In most cases however, the same state behavior will be observed at the clock boundaries if those intermediate events are collapsed to two events, one at the clock edge and one immediately after. This is generally called cycle-based simulation. Since the state behavior is all that is important when verifying logical correctness, this simplification is appealing. The question then becomes, how much faster can you simulate a model using this abstraction?

A number of different cycle-based techniques have been tried. The obvious way to do it was mentioned previously, where every expression in the model is ordered and executed once in every cycle. This is sometimes called ubiquitous cycle-based simulation. The big advantage this has is that there is no overhead of putting events on the future event list or taking them off. In an RTL logic simulation, this overhead can amount to as much as 40% of the execution time. For some designs this static scheduling has proven effective, but the big drawback it has is that if the model has few state variables changing in each cycle, there is a lot of wasted work done. This technique is seldom used for commercial logic simulators now.

However, the ubiquitous cycle-based technique is amenable to hardware acceleration. The scheduling loop is simple enough in that it is relatively easy to program an FPGA-based hardware device to perform the simulation. Products that do this are called emulators, and they can perform logic simulations several orders of magnitude faster than software simulators.

16.5.4 LEVEL OF ABSTRACTION AND SPEED

As noted above, while not limited to these levels, the primary levels of abstraction that logic simulation is concerned with are gate level, RTL, and behavioral level. From a simulation perspective, the main difference between these levels is the number of events that are executed. Between RTL and gate level, the primary abstraction is the width of operands and results. At the RTL, multiple-bit variables, or *vectors* in Verilog terminology, are operated on as a unit. At gate level, vectors are typically split into their individual bits. Thus, at gate level, every logical operation is performed by a hardware element, which requires an event to produce its output. If the timing of the design is important for the simulation, then simulating each physical component is required, since each component may have a different delay.

At the RTL, operations are typically performed as aggregates on their input variables, producing vectors as results. It is easy to see that *and*ing two 16-bit vectors to produce a 16-bit result can be done with one event at RTL, while it will take 16 events at gate level. Since the event routine itself takes about the same amount of time to *and* two 1-bit inputs as two 16-bit inputs, the gate-level simulation will run 16 times slower than the RTL simulation. It is common for the same design to simulate an order of magnitude faster at RTL than at gate level.

Moving up to behavioral level, the primary abstraction is the reduction of clocked events. That is, in behavioral code, variable values are computed without regard to the mechanics of the computation. A computation that may take several cycles in the ultimate hardware will be done in just one event. Synchronization in behavioral code is typically done via communication signals rather than fixed-cycle counts. Consequently, the simulation can be done with significantly fewer events at behavioral level than at RTL. Again, the difference between simulation time for a behavioral model and the same design at the RTL can be an order of magnitude.

16.5.5 CO-SIMULATION METHODS

There are a variety of hardware simulation languages that have been invented and used over the years. In addition to the primary HDLs, Verilog, VHDL, and SystemC, other languages have been created to write testbench code in, most prominently e and Vera. In addition, code written in other more general-purpose languages (C, PERL, etc.) is often incorporated into a simulation model. While these other languages are mainly used for modeling the parts of the system that make up the environment for the hardware design, there is no difference between them and the hardware design as far as the simulator is concerned. An event is an event, whether in the hardware model or the environment model.

Integrating this multiplicity of languages into a single simulation model can be a problem. Integrating a general-purpose programming language with a simulation language is not especially hard, since there is no concept of time in general-purpose languages. Thus, all one really needs to do is allow an escape mechanism so that code written in the general-purpose language can be called in an event routine. Providing an API so that the general-purpose language code can schedule events and get called as an event routine is pretty straightforward. All mainstream logic simulators provide this capability.

Integrating a simulator with another logic simulator is a more difficult problem. Ideally, a logic simulator would be able to understand all the required simulation languages and compile a model written in all of them into a single model. Practically, logic simulators are created to handle one language, and integration with other languages is done at a coarser level of granularity.

The simplest way to integrate models written in two or more languages is to compile the components with separate compilers, and then put them together by coordinating their schedulers. This is called *co-simulation*. That is, each simulator runs as it normally would, keeping its own scheduler and future event list as well as other local data, and synchronizing through the schedulers. This can be done as shown in Figure 16.14. The points marked as <==n==> are points at which the schedulers must synchronize with each other. All schedulers must have finished the preceding section before any can proceed to the next section.

The only remaining difficulty is that propagation may be required between parts of the model that reside in different simulators. That is, a Verilog module could have a variable whose value depends on a variable in a VHDL module. It is easy to see how to do this—just provide a way for the VHDL model to propagate a value onto the Verilog model's propagate list. Actually doing it in a concise and automatic way is more problematic, and this problem has been solved in a variety of ways through a combination of language features (to identify a variable as external) and an API. Sometimes, this ability is simply restricted to module ports.

16.5.6 SINGLE-KERNEL SIMULATORS

Co-simulation with multiple schedulers works, but the coarse granularity of the synchronization can impose a substantial overhead if there are many propagation events across the simulator boundary. In general, the API calls between simulators involve a fair amount of overhead as

```
do (forever)                              do (forever)
  t = next t               <==1==>          t = next t
  while (m = future_event_list(t))          while (m = future_event_list(t))
    execute event routine m                   execute event routine m
  endwhile                 <==2==>          endwhile
  // update the variable values             // update the variable values
  while (u = update_list(t))                while (u = update_list(t))
    u.v = u.t_v                               u.v = u.t_v
    put_on_propagate_list(c(u.v), t+u.d)      put_on_propagate_list(c(u.v), t+u.d)
  endwhile                 <==3==>          endwhile
  // propagate changed values               // propagate changed values
  while (c = propagatelist(t))              while (c = propagatelist(t))
    c.v = eval(c.expr)                        c.v = eval(c.expr)
    put_on_propagate_list(c(c.v), t+c.d)      put_on_propagate_list(c(c.v), t+c.d)
  endwhile                 <==4==>          endwhile
enddo                                     enddo
```

FIGURE 16.14 Synchronization points in co-simulation.

each one must establish its environment on every call. It would be more efficient if the model components written in all the languages could use the same scheduler. This is the approach taken by the so-called single-kernel simulators.

In essence, simulators that handle multiple languages with a single scheduler have separate compilers for each language but they all use the same scheduler. That avoids the need for synchronization between different schedulers, and can save a substantial amount of overhead. There are commercial single-kernel simulators for Verilog and VHDL, and Verilog, VHDL, and SystemC.

16.6 IMPACT OF HVLs ON SIMULATION

Hardware verification languages (HVLs) came into vogue as special-purpose languages for writing testbenches. They provide convenient means of generating stimulus for a hardware model, and also provide useful abstractions for modeling the hardware model's environment. For instance, it is generally easier to write a communication protocol that provides correct, variable input in an HVL than it is to write the same thing in an HDL. From the simulator's point of view, the mechanics of dealing with an HVL are the same as co-simulation. Indeed, an HVL is just another flavor of simulation language. Simulators that integrate an HVL and an HDL into a single scheduler are becoming more common.

16.7 SUMMARY

In this chapter, we have covered many of the details of digital logic simulation. Logic simulation is simply a special case of discrete event simulation, which has a long history in general system modeling. The speed of simulation is proportional to the level of detail in the simulation model, which in turn is determined by the level of abstraction at which the model is expressed. In logic simulation, the three common levels of abstraction are gate level, RTL, and behavioral level. There are techniques like cycle simulation used to speed up simulation, as well as techniques like co-simulation used to improve the hardware modeling capability. Ultimately, logic simulation is the most general technique available to verify that a hardware design does what it is intended to do.

REFERENCES

1. A. Aho, R. Sethi, and J. Ullman, *Compilers Principles, Techniques, and Tools*, Addison-Wesley, Reading, MA, 1986.
2. P. Ashenden, *The Designer's Guide to VHDL*, Morgan Kaufmann Publishers, San Francisco, CA, 1995.
3. G.M. Birtwistle, O.-J. Dahl, B. Myhrhaug, and K. Nygaard, *Simula Begin*, Auerbach Publishers, Philadelphia, PA, 1973.
4. T. Grötker, S. Liao, G. Martin, and S. Swan, *System Design with SystemC*, Kluwer Academic Publishers, Dordrecht, the Netherlands, 2002.
5. M.H. McDougall, Computer system simulation: An introduction. *Comput. Surv.*, 2, 191–209, 1970.
6. S. Palnitkar, *Verilog HDL: A Guide to Digital Design and Synthesis*, 2nd edn., Sunsoft Press, Mountain View, CA, 2003.
7. D.E. Thomas and P. Moorby, *The Verilog Hardware Description Language*, Kluwer Academic Publishers, Dordrecht, the Netherlands, 1991.

Leveraging Transaction-Level Models in an SoC Design Flow*

Laurent Maillet-Contoz, Jérôme Cornet, Alain Clouard,
Eric Paire, Antoine Perrin, and Jean-Philippe Strassen

CONTENTS

* This chapter is a replacement for the chapter "Using Transactional-Level Models in an SoC Design Flow," by A. Clouard, F. Ghenassia, L. Maillet-Contoz, and J. P. Strassen from the first edition of the *EDA Handbook* (2006).

17.1 INTRODUCTION

Multimillion gate circuits currently under design with the latest CMOS technologies include not only hardwired functionalities but also embedded software running most often on multiple processors. This complexity drives the need for extensions of the traditional RTL-to-layout design and verification flow.

Systems on chip (SoCs), as the name implies, are complete systems composed of processors, buses, hardware accelerators, input/output (I/O) peripherals, analog/RF devices, memories, and embedded software. Fifteen years ago, these components were assembled on boards; nowadays, they can be embedded into a single circuit. This skyrocketing complexity has two major consequences: (1) mandatory reuse of many existing IPs to avoid redesigning the entire chip from scratch for each new generation and (2) employment of embedded software to provide major parts of the expected functionality of the chip. Allowing software development to start very early in the development cycle is, therefore, of paramount importance to reduce time to market. Meanwhile, real-time requirements are key parameters of the specifications, especially in the application domains targeted by semiconductor companies such as STMicroelectronics (e.g., automotive, multimedia). It is, therefore, equally important to be able to analyze the expected real-time behavior of a given SoC architecture. Another crucial issue is the functional verification of IPs composing the system as well as their integration. The design flow must support an efficient verification process to reduce the development time and avoid silicon respins that could jeopardize the return on investment of the product under design.

The guidelines and methodology discussed in this chapter are based on more than a decade of industrial experience in electronic system-level (ESL) design with transaction-level modeling (TLM) work that the team of authors did at STMicroelectronics, ranging from base technology development, to industrial deployment in product groups, to ESL standardization with IP-XACT and SystemC/TLM [1]. The team started investigations on high-level modeling in the late 1990s and identified about 15 proprietary languages at that time, but there is no commercial product available to fully cover the needs. STMicroelectronics decided to invest in the development of a modeling kit to support the needs related to the complex and fast-moving architectures of our

SoCs. Model portability across different simulation environments was a key concern to enable early adoption and, later, model reuse—to serve internal and external customers.

During the 1990s, more and more designs were embedding several processors, bus masters (such as DMAs), and a number of peripherals and subsystems. The need for models with fewer details than RTL that would be available earlier and would simulate faster became obvious in the SoC area. As the number of processors increased in SoCs, so did the need to easily execute and debug software during presilicon verification and validation phases. This involved test software (used for driving functional verification of RTL-level hardware), as well as production software that would be shipped with the SoC to the customer.

The kind of SoC model needed to address this problem must provide a view of the hardware tailored for software execution, typically capturing data movement to and from memory and registers, as well as interrupts from peripherals. As software programmers most often do not require visibility at the cycle level, but rather need simulation fast enough for interactive debugging, there was an opportunity to create a level of modeling that would not be cycle accurate (CA) but would transfer data as single block read/write (data transaction) whenever possible: TLM. While this concept existed for some time, its implementation on top of SystemC in the early 2000s made TLM a widely recognized level of modeling in the system flow [2]. The Open SystemC Initiative TLM Working Group, founded in 2003, contributed to its standardization as an Open SystemC Initiative* and, later on, as an integral part of IEEE 1666-2011 [2]. TLM also appeared as a modeling level on top of other languages, such as SystemVerilog, to help build testbenches. SystemC TLM models (C++ based) are used along with software (most often C based) in a natural fit, all along the system design flow, from architecture to hardware functional verification, to firmware pre- and postsilicon debug, and to software-on-hardware presilicon validation [3,4].

The structure of this chapter is as follows: first, we introduce the types of presilicon platforms in use today and their benefits in the flow. Then, we discuss the key factors required to ensure the successful development and reuse of such platforms. We insist on the importance of modeling standards in system-level design and identify missing pieces that would further increase model interoperability and facilitate platform integration. System design automation is then discussed, as well as procurement requirements. Before concluding, we summarize the main emerging needs identified from intensive usage of the virtual platforms in product groups: representation of clock trees, power supplies, and reset, software-in-the-loop early power estimates, capture of system synchronization aspects, etc.

17.2 RISE OF PRESILICON PLATFORMS

An SoC flow exploiting transactional platforms can be summarized by (somewhat overlapping) steps such as architecture, design, joint verification of hardware (including firmware) and software, as well as system pre- and postsilicon validation.

This section describes the different transactional platforms exploited by the various categories of users during these steps. We first review the major platform types and introduce key concepts such as *transactors*. We then explore how the different user needs are addressed by the various platform types.

In the following sections, we focus on modeling the digital parts of SoCs. We also acknowledge the possibility and interest in modeling analog blocks using the AMS extensions of the SystemC language and supporting simulators [5]. From our experience, very detailed AMS models execute too slowly to support interactive software debug sessions, so caution should be taken to ensure a balance between accuracy of analog parts and simulation speed requirements.

17.2.1 TRANSACTIONAL PLATFORM TYPES AND USAGES

The simplest type of transaction-level platform contains only transaction-level models (i.e., SystemC/ TLM models) without using other abstraction levels (such as RTL). These platforms are typically used for software development activities such as firmware, OS kernel porting and drivers, middleware,

___
* Now Accellera Systems Initiative.

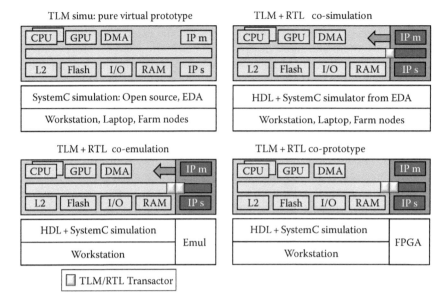

FIGURE 17.1   Types of transaction-level platforms.

and some hardware-related application features. In addition, these platforms are used by engineers preparing software-driven tests (typically in the C language) for the hardware functional verification and validation. In this chapter, we refer to such platforms as "pure virtual prototypes."

In other situations, it is necessary to combine a virtual prototype with some RTL blocks (see Figure 17.1). Depending on the need, VHDL or Verilog RTL models may be included for one or several IPs or subsystems. In some cases, the whole SoC is simulated with RTL models, except for the processor or processor subsystem (to leverage the speedup of a TLM abstraction of the CPU). To interface the TLM part and the RTL part, a special simulation component is used: a *transactor* [6,7], which performs the translation between TLM abstract transactions and corresponding RTL signal sequences. Transactors are available from third parties, such as Bluespec,[*] Cadence,[†] Mentor,[‡] and Synopsys,[§] for many standard protocols, such as AMBA, USB, and PCIe. They may also be developed on purpose for design-specific protocols, such as direct IP-to-IP links, if needed by the verification plan. We call the simulations that integrate both TLM and RTL parts as "TLM cosimulations."

When greater simulation speed is required for the RTL parts of a TLM cosimulation, these RTL models can be moved to a hardware emulator machine. These machines are an easy means to speed up simulation, but they are generally available in limited quantity at a given company due to their cost. We call "TLM coemulation" the corresponding TLM/RTL platform.

As an alternative or complement to hardware emulation, it is possible to use field-programmable gate array (FPGA) boards to host the RTL portions. In that case, we call the resulting TLM/RTL mix a "TLM coprototyping platform." The main advantage is the wider deployment of such platforms as FPGA boards are less costly than a hardware emulator.[¶]

Next, we cover the various user categories: developers and validation engineers, designers, and architects. For each category, we present the most useful platform types.

17.2.1.1 MIDDLEWARE AND APPLICATION DEVELOPMENT

Development of these software layers requires satisfying two important properties: simulation speed and observability. Indeed, as developers spend their time doing interactive debugging, it is important that they do not wait when stepping through the execution of their embedded C code.

---

[*] http://www.bluespec.com/.

[†] http://www.cadence.com/.

[‡] http://s3.mentor.com/.

[§] http://www.synopsys.com/.

[¶] The interested reader may refer to Chapter 19 for more information.

This implies efficient infrastructure to execute the embedded software, as well as abstract enough TLM models (TLM loosely timed [2]), for the rest of the hardware. Regarding observability, users require at least the same level of observability as they would expect from a debugger on the physical target: contents of memory and registers accessible through address map, possibility to suspend upon interrupt, etc.

Running the software requires a processor model, together with the same address map and interrupt network as the future silicon. Several processor modeling technologies exist as third-party or in-house solutions. One option is to integrate an instruction set simulator (ISS), exposing the targeted instruction set. In this case, the functional effect of the instructions within the embedded software binary is simulated by the ISS. An alternate solution is to execute the embedded software on a processor model relying on host processor instructions, while keeping complete functional accuracy for the rest of targeted processor specifications; this is usually called native wrappers/host code execution [8,9]. Here, it is the functional effect of the actual source code (C or C++) that is reproduced, by compiling the embedded software for the instruction set of the machine executing the simulation (host), based on the fact that compilation to a different target will not change its functionality. Obviously, this latter way of executing software brings some minor restrictions (assembly code needs to be rewritten in C), but in our experience, this methodology brings unrivaled software execution performance. It is used successfully in production not only for simple C tests but also for operating systems like Linux or FreeRTOS.

These software execution technologies available for TLM platforms satisfy the simulation speed requirement. The other part of the equation regarding efficient software debug with virtual platforms relies on the simulation speed of TLM models for the rest of the hardware. Our experience in that area is that both the level of detail and proper modeling practices are crucial to guarantee efficiency. It is tempting to target cycle accuracy or even precise timing when writing TLM models for embedded software execution. However, this inevitably impedes simulation speed without providing significant benefits. We realized that only functional timing was necessary, together with a way to capture uncertainties about the other time values (see Section 17.4.4 for more details). Finally, modeling practices also play a key role in determining simulation speed. Causes of inefficiency range from bad software practices (naïve loops with repeated memory allocations/deallocations, unused string setup, polling instead of explicit synchronization) to clumsy modeling (capturing a timer with a process that awakes at each clock cycle).

Regarding observability, TLM platforms actually provide more information than execution on the real chip. It is possible to inspect details of the hardware behavior during the embedded software execution, without the traditional limitations of silicon debugging. In addition, TLM platforms can raise crucial warnings about hardware misuses, which are generally symptoms of software bugs. The advanced user can set breakpoints in both their own software and the C++ code of the various TLM models, which is a powerful codebug capability in order to help understand the hardware–software interactions. The complexity of current SoCs, with multiple cores and cache coherency, not to mention OS virtualization and security, makes these capabilities even more important, as the developer strongly needs to understand what is happening in the SoC architecture. Another trend is also developing, due to the same observability advantages: postsilicon use of TLM platforms. Developers tend to use the virtual prototype to better understand the root causes behind corner-case situations experienced on the final chip.

17.2.1.2 FIRMWARE AND LOW-LEVEL DRIVER DEVELOPMENT

Developers for these layers of software benefit from pure virtual prototypes in the same way as upper-layer developers do (see previous section). It is important to mention that even at that level of the software stack, the TLM platform captures enough detail to allow the actual execution of the software/firmware and also to provide interesting observability features.

17.2.1.3 IP AND SUBSYSTEM DESIGN FUNCTIONAL VERIFICATION

The success of an SoC is not only determined by its processor (efficient implementation, use of a popular instruction set, etc.) but also based on its integration of specialized subsystems

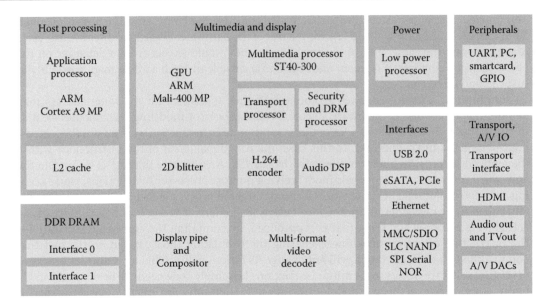

FIGURE 17.2   Block diagram of a complex system on chip. (From www.st.com/web/en/catalog/mmc/FM131/SC2058.)

(see Figure 17.2). A subsystem is a set of hardware IPs that, together with a dedicated processor (or set of processors) running firmware, provide the most efficient implementation of a given functionality. It is controlled by the main processor of the SoC and may exchange data and interrupts with other subsystems. In addition, a subsystem may communicate with the outside world through I/Os. For instance, a TVOUT subsystem will communicate with an external display through HDMI connections.

Our strategy is to rely on SystemC/TLM platforms to develop C tests for verification purposes, which will be reused for validation later in the project. Such platforms require a basic backbone to execute the test software, together with a TLM reference model of the subsystem to be verified. This approach separates test development from actual RTL debug: this results in more mature tests (as they have been developed using a reference model) and eases RTL verification.

Other methods for functional verification exist, such as universal verification methodology (UVM) [10], assertion-based verification, or formal verification techniques. They do not address this strategy and are not covered here. Interested readers may refer to [11] or Chapter 18 for more information.

There are two types of complementary verification tests. Internal subsystem tests typically replace the actual firmware, when the goal is to verify the subsystem's hardware as seen from the subsystem's dedicated processor. In this case, a *white box* subsystem model is used, describing the several IPs and internal buses involved (see Figure 17.3).

External subsystem tests, running on the main processor of the SoC, aim at verifying the functionality of the subsystem as a whole. The model of the subsystem may be here a *black box* model, exposing only the system-level-relevant functional interfaces (mainly bit-true registers, data, and interrupts) and having a more abstract model for the subsystem internals. This kind of model typically does not reflect the separation between hardware and firmware internal to the subsystem (see Figure 17.4).

For instance, for a video decoder, the black box model internals may be the reference algorithm coming from the corresponding standard committee. When the subsystem is a simple variant of a previous generation, the choice may be to skip the white box and go directly to a black box verification platform.

Once the verification tests are developed, RTL debug is performed by replacing the TLM reference model by the RTL of the IP to be verified, using transactors, yielding a TLM cosimulation platform. It allows debugging the RTL in a very efficient way, benefitting from the speed and observability of the TLM subset.

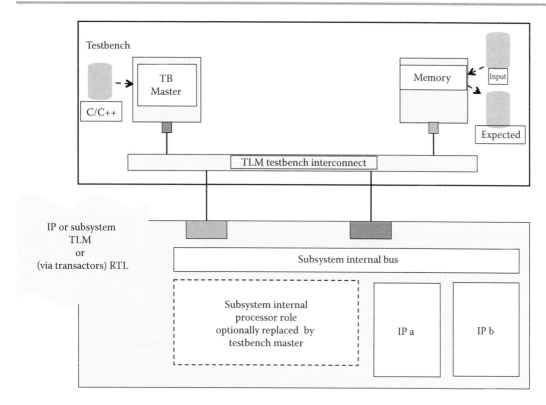

FIGURE 17.3   White box verification.

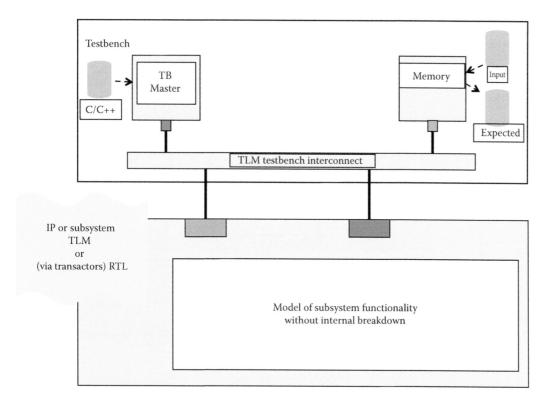

FIGURE 17.4   Black box verification.

When the RTL part is too slow to simulate (e.g., video streams taking days or weeks to be decoded), faster execution is achievable by exploiting a TLM coemulation or coprototyping platform. The TLM part remains unchanged, while the RTL moves to a hardware emulator or an FPGA board.

17.2.1.4 SoC VERIFICATION AND SYSTEM VALIDATION

SoC RTL verification and SoC validation benefit primarily from mixed TLM/RTL cosimulation platforms. They are used by SoC RTL integration verification, where observability is crucial. The validation team benefits from TLM coemulation platforms as speed is essential to run long, data-intensive test suites.

To build a platform suited to SoC verification, the TLM reference models of the various subsystems are integrated with an SoC backbone using the memory map of the final chip (Figure 17.5). At this stage, IP verification C tests can be directly reused in SoC verification test suites, saving a lot of time and effort during the SoC project and providing a golden reference (stimuli and expected data) across IP and SoC phases and teams. When doing the actual RTL debug, the RTL part represents most of the SoC within the simulation. Only the processor subsystem is kept at the TLM abstraction level for faster embedded software execution (Figure 17.6).

17.2.1.5 ARCHITECTURE EXPLORATION

With today's complex designs, it is no longer possible to predict and estimate the performance of the final system using spreadsheet formulas. Such spreadsheets typically include hundreds of formulas that are very hard to maintain over the project. Moreover, formulas cannot represent the dynamic behavior of software-rich systems (such as on-the-fly mode changes generated by user interactions).

SoC system architects, hardware architects, and software architects need a simulation platform with the following characteristics:

- Available early, from the beginning of the project
- Easy to modify, for quick what-if architecture study loops
- Easy to get performance estimates from, to help the SoC hardware and software organizations commit to guarantees of performance for the future product
- Easy to understand by other stakeholders of the system design flow, making the platform a common reference

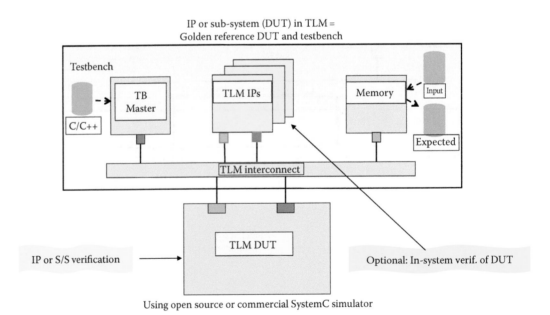

FIGURE 17.5 From IP to system-on-chip verification—step 1.

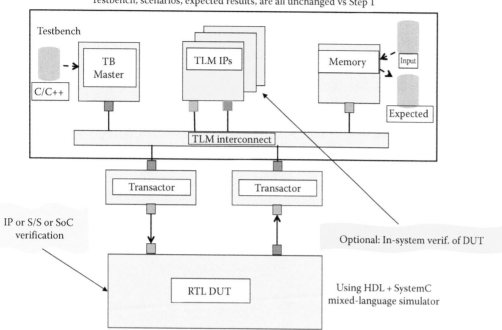

FIGURE 17.6   From IP to system-on-chip verification—step 2.

These requirements are somewhat contradictory: for instance, it may be difficult to get estimates that are precise enough to take SoC architecture decisions when using a model intended for the beginning of the project. In practice, the availability of models is the driving factor for an architect to select the virtual platform that is the closest to his or her needs.

For the SoC architecture phase, the RTL of the various IPs may exist (direct reuse from a previous generation) or may be obtained easily (existing configurable IP). In these two cases, the IP can be integrated directly in the simulation. If the simulation speed of the RTL IP is not satisfactory, it is possible to use RTL-to-SystemC translator tools to produce SystemC cycle accurate (CA) models. Commercial solutions such as Carbon Model Studio* or HIFSuite[†] can be used; open-source tools like Verilator[‡] are also available. These models are identical in accuracy to the RTL (at the boundaries of the IP) but simulate an order of magnitude faster. It is also possible to rely on third-party libraries, which generally include SystemC CA models (processors, hardware accelerators, etc.). The use of a hardware emulator can also be envisaged for fast execution. However, in the latter case, the project time to be spent on the first mapping and for analyzing the results needs to be considered before betting on this approach for the architecture phase.

When no RTL of the IP is available, a TLM model may already exist or could be possible in terms of effort. Such models can be used as part of the performance study platform, with some care. They need to be approximately timed (TLM-AT [2]) to be useful. Depending on the goal of the architecture study, it may be necessary to model all IP internals (including TLM-AT submodules) or an abstraction could be made, exposing only the interface of the IP model to the rest of the platform. Such models are notoriously time consuming to create and are not 100% accurate. Furthermore, it is difficult to assess their contribution to the global accuracy of the platform. Hence, these models do not help architects to commit on performances of the future silicon, except in simple architecture cases. However, they may be better than nothing, because they provide a way to actually run simulations and help architects to compare different architecture options, in a relative way.

---

* http://www.carbondesignsystems.com/.
† http://www.hifsuite.com/index.php/home.
‡ http://www.veripool.org/wiki/verilator.

Last, it is also possible that neither RTL models nor TLM functional models are available for a given IP (for example, new hardware blocks at early stages of a project). The new blocks may, in this frequent case, be represented by IP traffic generators (IPTGs) [12]. An IPTG typically generates functional traffic (read/write, send/receive transactions, etc.) at the TLM level, with a configurable profile (such as sizes of block transfers or bursts) corresponding to the IP being modeled.

When the aim of the study is to estimate SoC interconnect (number of nodes for a network on chip, internode widths, frequencies, FIFO sizes, etc.), the IPTG TLM-level traffic is connected to a CA model of the interconnect with transactors.

When the aim of the study is a software performance estimate, a simpler, TLM-AT model of the interconnect may suffice. Again, the actual level of accuracy obtained in such a setup is still the subject of ongoing research. In practice, interconnects are most often built from design kits offering parameterized preexisting modules (such as network-on-chip routing nodes) and related assembly tools. Such assembly makes it possible to quickly build CA or RTL models of interconnect variants. For this reason, CA/RTL models of the interconnect are generally used, with their TLM approximately timed counterparts being employed for software-intensive benchmarks or reference applications used in what-if architecture loops.

17.2.2 IMPACTS ON THE DESIGN CHAIN

The benefits of presilicon debug of software for the SoC under development (time to market, early and powerful codebug of software with hardware) can be propagated to the final customer, who has to develop or port their own software. Pure virtual prototypes are the transactional platforms of choice for this activity. In some cases, the virtual prototypes can be delivered very quickly, as only new IPs or IP variants would require extra modeling effort. It is also possible that the missing TLM model IPs are developed by a third-party SoC or IP provider (who is sometimes the final customer of the SoC itself), speeding up the availability of the platform. The benefits are obvious: the final customer is exposed early to a prototype of the product; these virtual prototypes can be easily replicated and upgraded worldwide without per-board return or onsite upgrade by field application engineers.

The same advantages are available in other contexts, for example, when the SoC is codeveloped by several companies. In this case, the virtual prototype provides a live, simulatable reference across companies. As shown earlier, this reference can be leveraged in many areas: from presilicon OS driver development to verification test reuse between the various teams. In addition, it provides a tool to help quickly spot integration or interoperability issues, hence improving the time to market and quality of the end product, as well as intercompany cooperation.

17.2.3 PLATFORMS AT A GLANCE

The types of virtual platforms and types of uses described in the previous paragraphs (see Figure 17.1) can be summarized in Table 17.1.

TABLE 17.1   Summary of Types of Virtual Platforms and Uses

Virtual Platform Types and Uses (Pre- and Postsilicon)	SoC Architect Phase	IP Design and Verification	SoC Design and Verification	SoC Validation	Software
Traffic generator platform	Yes				
Pure virtual prototype		Yes	Yes	Yes	Yes
TLM cosimulation with RTL		Yes	Yes	Yes	
TLM coemulation		Yes	Yes	Yes	Yes
TLM coprototype				Yes	Yes

The ecosystem keeps growing. All major EDA providers, as well as a number of specialized companies, have built an offer of tools, models, and services. A good indicator of the vitality of the ecosystem is the large number of organizations participating to the Accellera Systems Initiative (ASI)* organization, in charge of the standardization of this area.

17.3 PLATFORM DEVELOPMENT

Virtual prototype development has triggered new activities related to IP modeling and platform integration. In this section, we devise the importance of standards, how platform automation helps to increase productivity, and finally the procurement strategy.

17.3.1 USAGE-DRIVEN MODELING GUIDELINES

In an industrial context, it is crucial to ensure model reuse over time in various simulation environments. Therefore, special attention is paid to define and use the appropriate standards, as described next.

17.3.1.1 STANDARDS AND THE PLATFORMS ECOSYSTEM

In this section, we give an overview of the importance of standards in the system-level design area. It is possible to categorize the various types of standards as follows:

- Design and modeling language standards, such as VHDL/Verilog/SystemVerilog and SystemC: From an industrial perspective, they are important in order to ensure model usability over a multiyear period and to secure corresponding investments.
- Methodology standard: UVM [10] is emerging as a recognized standard to rationalize the structure of RTL verification environments and ease reuse and sharing of verification platforms.
- Interoperability standards, which we describe in more detail later on in this section: One can list model-to-model interoperability standards like SystemC/TLM [13], as well as model-to-tool interoperability standards like APIs for transaction recording as defined in the SystemC Verification Library [14] or parameter definitions as defined in the upcoming cci_param standard [15].

The availability of interoperability standards is fundamental, as they impact model developers and platform integrators as well as end users.

Model-to-tool interoperability standards enable models to interact with various tools through well-defined interfaces. This is of utmost importance to ensure portability of models across CAD environments. Indeed, when such standards are not available, model developers have to include proprietary libraries in their models to support a given tool (e.g., relying on a given modeling object to allow a register to be seen in a register introspection tool). One proprietary library from one vendor will rarely work in environments provided by other vendors. This creates extra implementation and validation effort on the shoulders of model developers and makes it difficult to reuse models across teams or organizations. Getting standard model-to-tool interfaces also improves the end user experience. With them, the user can choose the tool environment they are most comfortable with, without asking for any changes in the models.

Model-to-model interoperability standards facilitate integration of TLM models to create virtual prototypes. In the absence of such standards, platform integrators receive models based on different and often incompatible modeling technologies. The integration of foreign models then requires the following steps:

- Understanding of the concepts of the foreign modeling technology
- Identification of the integration strategy
- Implementation and validation of the appropriate adapter or wrapper

---

* http://www.accellera.org/.

All these steps must be performed for all foreign technologies and are obviously not cost effective. Moreover, adding adapters is a cause of integration bugs and negatively impacts the overall simulation performance. The SystemC TLM-2 standard [13] is an example of model-to-model interoperability standard for memory-mapped bus interfaces.

Modeling standards offer a great opportunity to help developers create their models, considering the standard as a guideline, instead of building models from the ground up. This gives a framework where examples are usually available and provides guidance to implement the right interfaces. Additionally, modeling standards usually include a set of rules to comply with, which is crucial to ensure interoperability. Accompanying the standards, release of supplemental material, like a reference implementation, significantly helps model developers to ramp up.

Standards are an *ecosystem enabler*. The growth rate of the ESL ecosystem has been quite disappointing during the last decade. One of the major reasons for the slow ramp-up of the market has been the absence of standards. As long as proprietary solutions were dominating the technical space, it was difficult to get interoperable commercial offers for models and tools. As a consequence, customers were reluctant to significantly invest in these solutions. When standards become available, IP providers can extend their offerings with models of their IPs that can interoperate with models from other parties, facilitating virtual platform integration. Likewise, platform integrators can adopt tools from various providers without creating unnecessary dependencies or constraints on models. The availability of standards also creates the opportunity for new, dynamic participants like start-ups to penetrate the market and offer innovative products with high added value. Standards are therefore the seed to initiate a rich and interoperable ecosystem.

As far as the community is concerned, the existence of ESL standards is a great opportunity to educate people along the design chain: IP and tool suppliers, IP sourcing teams, and end customers can adopt and spread good practices. From the academic perspective, educational programs can be built in engineering schools to provide the industry with highly skilled modeling engineers.

ESL standards are delivered as a normative document, usually called a Language Reference Manual. When it comes to model implementation, they must also be implemented as pieces of C++/SystemC.

Tool vendors* are offering *commercial implementations* of the standards, in their simulators or associated tools. Benefits of these implementations include an easy installation process and availability of documentation and support. The integration of various tools into ESL frameworks can offer a good added value, as vendors can propose performance-optimized implementations, extended functionalities, and coupling between complementary tools. However, this can put technical constraints on the environment: version of supported compilers, ability to link models with specific libraries, etc.

Some ESL standards, such as SystemC, also come with an *open-source implementation*. From the model validation perspective, such an implementation is fundamental to provide a *reference* point. Considering that the underlying technology is C++, it is critical that this point is neutral with respect to any commercial implementation. As an example, as IP providers deliver their models as binary code, for obvious reasons, this introduces a dependency on the version of SystemC used during the compilation process. When the platform integrator receives models from several providers, alignment to a single configuration needs to be ensured. From our experience, it is very difficult to have an IP provider release a binary package for a SystemC implementation provided by a competing vendor. The open-source implementation fully plays its *neutral ground* role, as the IP model can be compiled against this freely available version and released without simulator-related licensing issues.

There is a strong benefit of keeping an open-source implementation of the SystemC simulation kernel. This strategy helps a wide community of participants to develop added-value tools that are built on top and beside simulation kernels, as well as connections to other simulators. In addition, there is a strong interest to share maintenance costs of the kernel, seen as an enabler of added-value technologies, through an open-source approach.

---

* Several companies, including major EDA and more specialized providers, are offering commercial SystemC simulators.

17.3.1.2 MISSING PIECES

Modeling standards are still in their infancy. SystemC has been standardized as IEEE 1666 in 2005 and revised in 2011, integrating especially the TLM-1 and TLM-2 model-to-model standards [2]. However, this only covers memory-mapped bus interfaces, whereas the industry needs to integrate IPs with additional connections. Not every interface in this domain needs a standard TLM definition. Focus should be put on interfaces actually required when connecting IPs from different suppliers. In particular, some point-to-point protocols or very specific communication mechanisms have less need to be standardized. ST, Cadence, and ARM have joined forces to create several technical proposals, both in model-to-model and model-to-tool areas, which have been donated to the ASI to contribute to further standardization. We have listed here the various needs, sorted by perceived urgency and underlined the ones that are addressed as part of ST/Cadence/ARM joint proposals [16] and shared with ecosystem partners.

Improving model-to-model interoperability requires standard interfaces in the following areas:

- *Wire modeling*: This is the most common type of interface encountered in IP integration, after memory-mapped bus interfaces. SystemC already provides a class for modeling such wires, called sc_signal. But it has proven to be too focused on the RTL abstraction level, with several drawbacks when used in TLM modeling. In particular, changes in value are not seen immediately, but at the next delta cycle, which is at odds with TLM instantaneous interface method calls and can create situations where the event order is difficult to predict or too constrained with respect to the real chip. In addition, in TLM, multiple different processes often end up driving changes on a given wire line, which causes a multiple driver issue with sc_signal. Finally, the actual data type used with the signal is left to the choice of the model developer, causing model-to-model interoperability issues. Therefore, the ST–Cadence–ARM proposals advocated the standardization of a dedicated TLM protocol fixing the data type to use (to C++ bool data type) and integrating, among others, immediate propagation of value-change calls.
- *Frequency and voltage*: Functional and extrafunctional properties of complex SoCs are more and more intertwined. Capturing operating frequency and voltage in the models will soon become mandatory, to validate advanced embedded software, as described in more detail in Section 17.4.1 below. Consequently, standard interfaces for these facets should be defined to facilitate model integration.
- *Serial communication*: SoCs are exposing more and more I/O capabilities. In the automotive domain and other application areas, it is desirable to integrate several models communicating through serial protocols like CAN, SPI, or FlexRay to validate complex interactions occurring in the final product. The definition of standard modeling interfaces for serial communication would greatly facilitate platform integration for system houses.

Model-to-tools interoperability would be significantly improved with the standardization of the following elements:

- *Configuration*: While parameters might be defined in IP-XACT descriptions, there is no standard API to define parameters in SystemC models and easily set/update their values. Each model provider has its own configuration infrastructure implemented in C++, which makes the integration of models from various sources very painful, having several (potentially incompatible) configuration mechanisms to be supported in a single platform. Started several years ago, this standardization effort [34] is still ongoing in the Configuration, Control, and Inspection (CCI) Working Group of the ASI.
- *Memory region interface*: No standard is defined to support introspection in SystemC. This dramatically limits debugging capabilities, as each CAD vendor has implemented proprietary hooks to visualize register or memory values in their tools. This usually ends up with the instantiation of vendor-specific modeling objects in the models, which negatively impacts model portability across CAD environments. To address this issue, our

proposal is to define an introspection API at the boundary between models and tools that enables any model to work with any tool, without the need to rely on proprietary modeling objects.

■ *Address map*: Debugging embedded software on a multicore platform with complex hierarchical interconnects becomes very difficult. Due to security features and memory management strategies, address translation mechanisms are becoming more and more complex, making the memory regions actually reachable from one core very difficult to understand. Moreover, the reconstruction of the system address map cannot be computed statically due to interconnect remapping capabilities. System-level tools fail today to reconstruct the system address map, as seen from one specific initiator, due to the lack of standard API to collect this information dynamically at runtime. Our standard proposal here is a companion API based on TLM-2, which allows discovering dynamically the address map as seen from a given point in the system.

Methodological aspects should not be neglected as well as divergent choices could impede the correct integration of models and execution of virtual prototypes:

■ *System synchronization*: In many cases, virtual prototypes are adopting a loosely timed approach, to benefit from associated simulation speed. However, the inherent issue of system synchronization coming with this modeling style is neither well understood nor addressed in a sound and consistent manner. This can lead to integration problems (one block taking over the simulation kernel or degrading performance) or even worse may render the virtual prototype *unrealistic*, that is, omitting real-world behaviors that would trigger bugs in the embedded software. We provide more details on this topic in Section 17.4.4.

■ *Conformance to standards*: Defining standards is a first step to solve the interoperability problem; it is also necessary for stakeholders of the ecosystem to adopt and *comply* with them. We have unfortunately observed some SystemC simulator implementations being late to conform to IEEE 1666 evolutions. Similarly, model providers viewing TLM-2 as a simple API casually ignore the 500+ rules associated with this standard. It is crucial for the ecosystem to improve in this area, in order to globally increase interoperability.

■ *Avoid proprietary or exotic languages*: The natural language used to develop transaction-level models to be integrated in virtual prototypes is SystemC. However, IP providers are sometimes adopting other proprietary languages, such as LISA+ [17] or MATLAB®/Simulink® [18]. In some cases, an intermediate compilation flow is needed to generate SystemC code from a model described in another language, or a SystemC wrapper is required to connect the model to the platform. The binary form of the resulting SystemC model might cause integration issues when integrated with other binary models into a virtual prototype, due to dependencies on external proprietary libraries. Similar situations occur when models are written in a scripting language, such as Python, to initially support other needs (for instance, generation of data for verification purposes). Integrating such models in SystemC requires executing them in different OS processes and an exchange of data with the virtual prototype using IPC channels like sockets or shared memory. In summary, these exotic practices make integration more difficult: they cause extra effort, negatively impact the simulation speed, and create maintenance issues.

■ *Facilitate integration and configuration*: Model developers should avoid items that cause integration or configuration issues. For instance, they should not use environment variables to pass parameter values to the models, when they can use either constructor parameters or a more general configuration API. The definition of project-level or company-level coding guidelines can help us to enforce the adoption of the right practices. Likewise, developers should avoid using APIs from software packages obtained from third parties when they are undocumented, poorly supported, or not portable from one environment to the other. Special attention should be paid to the licensing scheme of such packages to prevent legal issues [19] or license contamination, for example, with the GPL license [20].

Last but not least, availability of vendor-agnostic modeling building blocks should facilitate the ramp-up of the industry in the ESL area:

- *Register bank*: Register modeling is a recurrent question when developing TLM models. A minimal approach consists in representing registers as scalar variables and private data members of the SystemC module. A huge case statement must then be implemented to respond to read/write accesses and update the variables accordingly. This is very error prone and tedious for the model developer. A good practice consists of using a generic register bank object that is configured at runtime, exploiting an IP-XACT [21] definition when available. SystemRDL [22], although new and not yet widely used, might also address such requirements in the future. It improves the readability of models and offers built-in modeling capabilities and services like configurable verbosity of the reporting mechanisms. EDA vendors usually provide such capabilities through proprietary modeling objects, but they unfortunately include tool-specific dependencies that are incompatible with competing environments. A vendor/tool agnostic offering exposing a standardized introspection interface is consequently highly needed.

- *Convenience layer*: The primary goal of model developers is to focus on the right modeling of their IPs. They are usually not experts in the underlying simulation infrastructure and should not spend their efforts in dealing with low-level details of standards. Therefore, a convenience layer providing user-friendly APIs should be available. It improves the productivity of model developers (as simpler read/write API calls are easier to understand than the low-level intricate details of some standards). It ensures standard compliance by construction (as it is the responsibility of the implementers of the convenience layer to be compliant). It can also provide extra services on top of standard APIs such as endianness management, address alignment, error checking, unified messaging, and logging mechanism. Finally, it facilitates alignment of models with the evolution of the standards (because the user code is isolated from the standards' APIs).

- *Reporting/messaging/monitoring*: SystemC offers built-in basic reporting primitives that are not complete enough to address all the needs of virtual prototyping. Messaging capabilities should offer flexibility in verbosity level. Several configurable categories of messages should be defined (information, warning, error, fatal error, etc.). In order to facilitate postprocessing of simulation traces, messages should be systematically formatted, with timestamp, name of the module, message category, etc.

17.3.1.3 MODELS AND TOOLS ORTHOGONALITY

In the past, a tight and unexpected dependency of models on simulation tools has often been observed, due to the usage of vendor-specific modeling objects within models: add-ons for debugging features, proprietary introspection classes, etc. Such practices have drawbacks that become the most visible quite late, at integration time. Indeed, platform integrators often face integration issues due to incompatible dependencies triggered by the models.

The models can also be reused in a variety of contexts. For instance, a given SystemC/TLM model might be reused for functional verification purposes in cosimulation platforms in conjunction with HDL models. It could also sometimes be delivered to partners or customers, who have their own constraints. Due to the fast evolution of CAD tools and their respective feature set, it may be necessary for a user to switch from one vendor to another depending on the project requirements. Migrating models is almost impossible if they have dependencies on proprietary capabilities of a certain vendor. Considering this context, no assumption should be made on the CAD environment used to simulate the models. A robust policy should be adopted to ensure the portability of models across CAD environments. This can be implemented through a strict adherence to the standards, consequently making the models portable. In addition, model developers should ensure independence from proprietary modeling elements and check that the models can execute correctly on any standard-compliant simulator by running appropriate nonregression test suites (functional tests to check model correctness when it is evolved).

17.3.2 SYSTEM DESIGN AUTOMATION

A standardized format to describe the externally visible information of an IP block is also important to facilitate IP reuse. To address this need, IP-XACT has been defined to provide a standardized XML-based machine-readable format that captures IP metadata. Initially, it was thought to ease IP and subsystems integration in platforms (both at RTL and TLM levels), by providing all the information required to instantiate and bind IPs together into a design. Tools like netlisters can draw benefit from this description independent of the IP provider, as the format is vendor independent. The intention is that IP providers should deliver IP-XACT descriptions of their IPs, to facilitate the growth of the IP ecosystem.

An alternate usage of this standard has emerged, relating to model generation flow. An IP-XACT description of an IP contains a lot of information that can be seen as a machine-readable subset of the IP specification:

- List of registers and bit fields that are defined with associated offsets, potentially tagged with predefined behaviors (attributes)
- Connection points for buses and wires
- Instance parameters*

This information present in the IP-XACT file can be used to support the automatic generation of a wide variety of files as depicted in Figure 17.7, ranging from TLM and/or RTL model skeletons to header files for embedded software or register tests or UVM sequences for functional verification [23].

17.3.3 PROCUREMENT REQUIREMENTS

Availability of models is a prerequisite to create virtual prototypes. This calls for new practices in the supply chain, as well as from the project perspective.

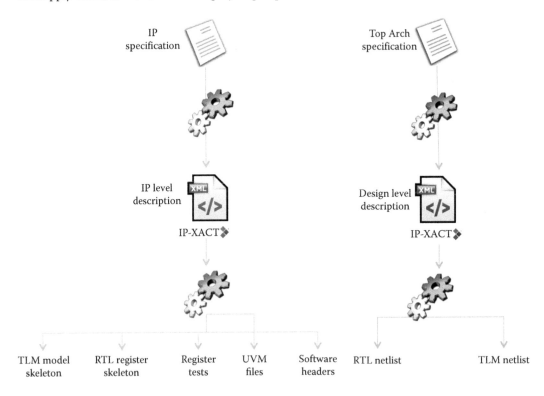

FIGURE 17.7   IP-XACT-based generation flow.

---

* It is, however, required to define on the SystemC model side the corresponding standard to avoid in-house parameter implementations and facilitate model exchange and reuse.

17.3.3.1 SUPPLY CHAIN ECOSYSTEM

The situation has progressed well since the first edition of this handbook in 2006. The leading hardware IP vendors nowadays provide IEEE 1666 TLM and IEEE 1685 IP-XACT views for their IP. The most advanced ones use these views for their own internal needs (see previous sections on software, verification, and validation), in a specification to TLM to RTL flow, and then offer them optionally, in addition to their IP RTL. We are, however, not yet at a stage where IP vendors provide these TLM views in advance of RTL when a new IP or a new variant is introduced. Fortunately, this is the case between an internal IP supplier team and the SoC team inside the same company, where incremental and iterative releases are better accepted.

As another emerging trend, leading customers of SoC companies now increasingly request TLM platforms prior to silicon, to make progress in the validation of their own software integration and, thus, have less to do after silicon is available. They consequently arrive earlier to the market with their complete product. Here, an understanding of the type of company partnership is needed to maximize the schedule gain; without physical devices (silicon mask, FPGA, etc.), releases can start very soon with a jointly agreed reduced set of features, evolving afterward incrementally until silicon is available.

The expanding use of TLM models across the supply chain also means a larger number of potential users for tools to help create and to use transaction-level platforms: SystemC-aware C++ debuggers, IP-XACT to TLM generators, and platform-aware profilers, to name a few.

17.3.3.2 PROJECT-LEVEL PLATFORM STRATEGY

From SoC specification to silicon, we have seen that the various categories of users (e.g., developers and hardware verification engineers) need different subsets of the SoC, one most relevant to each task. Many of these subsets include the initial, common *transactional backbone* that represents the memory path from the main processor(s) of the SoC to the RAM (e.g., DDR controller model) via cache levels and interconnect. It is typically created by the SoC integration team or a specialized modeling team working with a specification from the SoC team. Around this backbone, a given IP verification team may plug the model of the IP to be verified, to create expected results data for a given set of stimuli. Afterward, more complex use cases require the simultaneous presence in the TLM virtual platform of several subsystems. The backbone is incrementally complemented by subsystem models, to the benefit of all teams, even if they do not use all of the contents in the virtual platform. It simplifies the version and configuration management of the platform and keeps a single reference at any given point in project time. This is possible with the pure virtual prototype, provided models are correctly written to keep acceptable speed of the complete platform. It is also possible with the TLM coemulation or coprototyping platform, but with a smaller number of incremental steps during the SoC project, as each step requires the time to map for emulation of the corresponding subsystem(s). Given the very high cost of large-capacity hardware emulators, it may be relevant to make available to a given team only the subplatform that is needed by this team. More simultaneous platforms might be loaded in the emulator, hence more teams working simultaneously for a given emulator budget. This multiple-platform approach is mandatory for mixed TLM/RTL platform usages, such as IP verification, as a close-to-complete RTL SoC platform would be too slow to simulate (typically hours to days per run) compared to the verification IP team needs.

17.4 EMERGING NEEDS

In the early days of virtual prototyping, the focus was on the core functionality of the system. This subject is now well understood and addressed in the industry (IP companies providing models, standards being in use, etc.). However, the ever-growing complexity of on-chip systems and the wider deployment of virtual prototypes demand new modeling capabilities, as described in Sections 17.4.1 to 17.4.6 below.

17.4.1 NEXT STEPS IN PLATFORM-ASSISTED SOFTWARE VALIDATION

Presilicon development of embedded software modules requires extending the virtual prototypes to support development related to clock trees, power supplies, and resets. These capabilities are also beneficial in postsilicon use.

17.4.1.1 CLOCKS

The bring-up of complex systems is quite often impacted by poor management of clock trees. It is therefore very useful to verify upfront what the reachable frequencies are for each IP, considering the multiple frequency multipliers and dividers and detecting errors in programming sequences during the development of bring-up sequences and software drivers.

This cannot be checked using hardware prototypes on FPGAs or hardware emulators. On these platforms, clock trees are not accurately captured and are at best visible as waveforms.

A TLM model captures this information in a different, more understandable way. Clocks can be modeled by their frequency value (rather than signals), which is then transformed by various multipliers, dividers, and clock gating blocks. On the receiving end, IPs can trigger warnings or errors if they are used outside of specified frequency ranges. In this way, it is possible to validate software driving such clocks.

17.4.1.2 VOLTAGE

In the power area, more and more IPs such as DC/DC converters are software programmable and voltage information should be captured in the virtual prototypes accordingly. Working or retention voltages may also be needed to support validation of the programming sequences. Finally, voltage information is also required when implementing low-power strategies,* as explained in Section 17.4.1.4.

17.4.1.3 RESET

Functional verification activities require modeling of reset capabilities in the IPs to be able to restart each IP independently and validate their behavior. This is also useful in the wake-up phase.

Embedding reset management in IP TLM models has often become a necessity today. Generally, there is one reset condition specified for each IP block: an external agent asks for reset assertion/deassertion of the IP, to put it in a stable state. This is usually called "hardware" reset, as controlled by external hardware connections, and always occurs at the start-up of the platform. Hardware reset may also be triggered during the lifetime of the platform, particularly when IPs are put in low-power mode and need to be reset to wake up. As a first approximation, it can be easy to ignore the reset functionality at the beginning of platform execution, considering that all IP resets are deasserted at start-up time. However, this excludes low-power wake-up from the simulation environment, missing the opportunity to observe and check the interactions between software drivers and IPs restarting operation.

There is a second case of reset, which is less widespread: *software* reset. IPs may sometimes crash (e.g., generating erroneous traffic), requiring a way to *restart* them independently. Such software resets are generally triggered by internal commands (such as a register bit). Generally, complex IPs always provide these capabilities, to allow quickly putting back IPs into a stable state in emergency or error situations.

17.4.1.4 VALIDATION OF LOW-POWER STRATEGIES

Energy management is becoming a key differentiating factor for complex SoCs. This is true for nomadic devices not only as they are operating under battery constraints but also more generally as energy efficiency is now seriously considered by the end customer on top of new regulations.

---

* An interested reader may also refer to Chapters 7 and 13 for more information.

A large variety of low-power strategies exists today, such as clock and power gating or dynamic voltage and frequency scaling. Advanced low-power policies are controlled by specific embedded software usually running on a dedicated core. If appropriately extended, virtual prototypes can help to anticipate the functional validation of these policies. This may require integrating new dedicated TLM protocols and offering corresponding introspection capabilities. Each IP model should expose a mechanism to update operating frequency, voltage, and apply reset. The resulting modeling infrastructure should also provision execution modes allowing these features to be disabled: this enables gradual adoption in the models, without impacting them when no policy is implemented yet.

The key benefits expected from such activity are the following:

- Being able to check reset management and correctness of programming sequences for frequency and voltage update
- Allowing operating point validity checks: combined values of frequency and voltage that must be valid from the IP, as well as from the SoC perspective

17.4.2 EARLY POWER ESTIMATION

The next step to support the development of energy-efficient devices is the ability to estimate the power consumption of the system operating in real conditions, early in the design cycle. This should be done taking into account all hardware features and executing the full software stack. An additional interesting estimation is that of the thermal behavior of the system (which is closely linked to its power consumption). Some experimental activities to connect virtual prototypes to dedicated power and thermal models have already been conducted [24–27], but we are still far from a production-level deployment. From our experience, it is not realistic to expect detailed and accurate figures using virtual prototypes, due to their high level of abstraction. However, it might be possible to extract trends or orders of magnitude to compare the relative impacts of several low-power policies* that could be applied to the system. Depending on their characteristics, we consider three categories of systems, for which the power estimate might be impacted by different factors:

- Systems significantly composed of analog IPs, such as complex microcontrollers: the presence of analog IPs complicates the overall power estimation.
- CPU-based architectures composed of multiple clusters of cores: it is necessary to model the various levels of cache, their size, and refill policies.
- Complex SoCs, mixing multiple cores, and hardware accelerators: interconnect is difficult to model, and subsystems might be very complex and heterogeneous. The embedded application software is required to provide system estimates, whereas it is usually available only after the architectural studies.

While it is desirable to get early power and thermal estimates, a certain number of challenges still need to be tackled:

- It is difficult to assess the level of accuracy of such estimates.
- A floorplan is usually not available for early thermal estimates.
- The timing accuracy currently available in virtual prototypes is probably not sufficient to accurately assess power consumption, including leakage and temperature-related effects.
- The modeling effort is huge during the architectural step, and models might be available far too late to make system architecture decisions, which need to be made early.
- Realistic embedded software needs to be available to obtain valid estimations, which is usually not the case when performing architectural studies.

---

* Interested readers may also refer to Chapter 13 for more information.

17.4.3 SCALABILITY OF PLATFORMS

Despite having a simulation speed of multiple orders of magnitude faster than traditional RTL simulations, TLM simulation performance is more and more affected by the growing complexity of the systems. For example, current virtual prototypes representing complex multimedia systems embed several hundred IPs or subsystems. They execute several hundreds of SystemC processes running more than one million lines of code. On top of the hardware models, several millions lines of embedded software are commonly executed.

Current SystemC simulators only exploit a single processor core due to multithreading primitives (coroutines) that were designed to both simplify description of model parallelism (no need for mutexes or locks) and allow reproducibility of the simulations from one run to another. On today's machines, not being able to exploit the many cores/processors available on a host machine is an important bottleneck, and the cooperative semantics of SystemC will clearly become a major concern on tomorrow's machines. It is therefore crucial to conduct research in this area [28–30].

17.4.4 SYSTEM SYNCHRONIZATION

One of the most overlooked issues in today's embedded systems is their potential sensitivity to *functional event order*. A *functional event* is a concrete event happening on the real chip, whose occurrence can influence the behavior of other components in the system and therefore the embedded software.

Simple examples of functional events include an interrupt request (IRQ) being raised or a value change in a register whose content is read by a CPU for synchronization. Functional events are often perceived as independent from one another; however, there are many situations where this is not the case. For instance, consider two IRQs, one signaling the end of a video-processing activity and the other originating from a timer. The second IRQ could be used as a *watchdog* to indicate that the video processing did not complete on time and trigger a special behavior (such as cancelling further processing or skipping a frame). With this simple example, it is obvious that there are several possibilities to consider for the designer: either the video processing completes on time (i.e., the end-of-video-processing IRQ occurs before the watchdog IRQ) or it does not complete on time (i.e., the watchdog IRQ occurs before the end-of-video-processing IRQ). By definition, the embedded software is sensitive to the order of these functional events and should be tested in the various situations. Even two apparently unrelated functional events can exhibit bugs in the software. We also saw real-world bugs occurring when two functional events were happening *simultaneously* from the CPU point of view.

It is important to anticipate such behaviors with a virtual prototype. To achieve this, two important elements must be present:

1. The simulation must explicitly capture the possibility for various orders or *interleavings* of the functional events.
2. The embedded software (as well as the system itself) must be tested under these various possibilities. This requires multiple runs of the simulation.

In the context of SystemC simulations, the first point is achieved by ensuring that the various components are modeled with separate threads and that these threads have enough *degree of freedom* between them. More precisely, degrees of freedom are achieved by yielding back control to the SystemC scheduler at some given points in each process. The locations where each process should yield are dependent on the presence of functional events and hence on the actual functionality being captured. We call these locations *synchronization points*.

Yielding back control means that the simulation scheduler will be free to choose which process to select for execution, resulting in multiple interleaving possibilities. However, there are two remaining issues: (1) how to explore the various possibilities (to test the embedded software on multiple runs) and (2) how to ensure that the resulting interleavings are realistic.

One possibility would be to have some "orchestra conductor," which would decide which process should be resumed first. Such an orchestra master would also take into account the exploration of the various possibilities, by changing its decisions based on the simulation run iteration. It would also have to generate realistic simulation runs, by not choosing to elect a process "too often" or "not often enough." This notion of orchestra master is present whenever there is parallelism. It has been used in the past to capture models of computation [31] or study the notion of fairness [32]. In our case, it is quite difficult to write such an orchestra master, because we have to reconcile both interleaving exploration and some notion of realism that is difficult to precisely define.

The pragmatic solution to this modeling problem is to reintroduce "timing" in the model. There are two types of timing situations to be considered:

- Some of them are *functional* in the sense that the hardware will follow them as written in the specification: timers, video timings, etc. Whatever *legal* value is programmed by the software will cause a timer to raise its IRQ accordingly.
- The others are not precisely known or are highly dependent on data or other factors. For these cases, using fixed values can actually remove valid degrees of freedom in the model and be equivalent to making wrong assumptions about the hardware (cache causing delays, new hardware revision of the IP being faster, unexpected bus contention, etc.).

For instance, taking the previous example, saying that the video processing takes 200 ns and the timer is programmed at 300 ns will result in the first IRQ always happening before the second one, and thus the "does not complete in time" scenario will not occur. This could be fine if it is never possible that the situation happens, by construction (in which case, it means the software would not have to support it). If the video processing is actually not guaranteed to take a given fixed duration (complex dependency on actual data, variability, etc.), then it is necessary to capture that lack of knowledge. In order to do that, we propose to use a classical tool to deal with fuzziness: intervals of values. Such intervals allow expressing an order of magnitude and injecting a notion of realism in the model. The wider these timing intervals are, the more behaviors can exist around the degrees of freedom; conversely, narrowing the intervals allows selection of a limited set of realistic situations. Exploring the corresponding sets of interleaving can be done using sophisticated techniques [33], but it can also be simply done by choosing random values in the intervals for each simulation run (and changing the random seed). The approach can be extended by providing more information, such as constraints between timing intervals, which could then be explored with constrained random number generation. The only limitation in this area is the actual information available on timing relations at this stage of the design flow.

In the TLM-2 standard, various and somewhat conflicting synchronization/simulation techniques have been standardized, to fulfill the broadest set of views on the subject. The technique that we are mentioning is covered by the standard under the term "explicit synchronization" or "synchronization on demand." Another technique proposed by TLM-2 is the "time quantum," which is often wrongly associated with the term "temporal decoupling." Actually, decoupling time is inherent in the modeling style associated with TLM (at least for loosely timed models). By definition, multiple actions are taken without any interactions with the SystemC scheduler, both because the time is not known precisely at this level of detail and because this would generate simulation overhead with context switches. The term synchronization on demand indeed mentions synchronization with something: it is resynchronizing local decoupled time with the global SystemC simulation time. This resynchronization is done with the SystemC wait() statement, which will put the current process on pause and yield back control to the scheduler (which will resume execution of the process once the wait() delay is elapsed). The difference between synchronization on demand and time quantum is the way in which the decision to resynchronize is taken. With the former, the locations where time should be resynchronized are linked to actual functionality of the real system. With the latter, the decision is based purely on the comparison between the time accumulated during temporal decoupling and some arbitrary global threshold: the time quantum.

There is absolutely no guarantee that the time quantum will indeed reach the threshold for interesting locations that would need to be degrees of freedom. Furthermore, the locations depend on the actual timing values, which are assumed to be fixed (no intervals there). Not only such simulations would not exhibit the proper set of functional event *interleavings* required to validate correctly the embedded software, but sometimes, due to insufficient degrees of freedom, the simulation can just break, that is, it would not exhibit any functionally correct behavior. In other words, with this technique, the user is left with trying different arbitrary values for the global quantum, hoping that the simulation will run correctly and exhibit interesting behaviors.

On the other hand, synchronization on demand, combined with loose timing, allows a reliable definition and exploration of a set of realistic behaviors. The degrees of freedom are caused by resynchronization occurring at functionally defined locations: system synchronizations. The functional timing values, together with the knowledge of orders of magnitudes (defined with the loose timings), constrain the set of behavior to yield realistic situations.

Given the complexity of current systems, it is crucial that the embedded software is developed to be robust to the various events occurring on the real chip (which may change with a different environment or the next generation of the product). To achieve this, the software must be tested on virtual prototypes that provide a reliable way of exposing varied, yet realistic, behaviors. The modeling techniques presented in this section have successfully been used in production, allowing, among others, discovery of corner-case bugs in critical sections of various kinds of embedded software.

17.4.5 VIRTUAL PLATFORM INTEGRATION ISSUES

The complexity of large SoC integration has brought similar levels of complexity to the integration of their presilicon virtual prototypes. Of course, all well-known problems linked to large application integration do occur, and standards (IEEE 1666 TLM, IEEE 1685 IP-XACT, upcoming CCI WG, etc.) are created to solve some of them. However, it becomes more and more difficult to guarantee proper alignment of the various technical parameters in a given virtual prototype. For instance, upgrading to the latest version of a given standard may require revalidation of every IP in this new environment. This can be time consuming or prove difficult in the case of legacy IPs. Another example of technical parameter decision during integration is the choice of a SystemC simulator for the whole virtual platform. As of today, the only solution to set up a virtual platform is to select one SystemC simulator and to try to compile and integrate all IPs on top of it. As we mentioned previously, the availability of an open-source implementation of SystemC is really beneficial to the virtual platform integrators, as it is a neutral point allowing integration from different IP model providers. In the absence of this open-source version, would all SystemC/TLM suppliers provide compatibility between their different implementations to allow this mandatory heterogeneous integration?

17.4.6 VIRTUAL PLATFORM COMMUNICATION ISSUES

While virtual platforms may require huge integration efforts, software developers need to be able to inject dynamic data inside their virtual platforms and get results from the execution of their software on the virtual platform. At a minimum, this can be done through files, but this solution quickly shows unacceptable limitations. Embedding internal traffic generators inside the virtual platform is just another way to hide the need to perform real I/O communication with the virtual platform.

To generalize this interaction of the virtual platform with the external world, it is required to have actual communication between two or more virtual platforms, as software becomes more and more distributed between multiple independent SoCs. While it is already difficult for a single virtual platform, the integration of multiple platforms would become a real nightmare, and it will mandate another level of standardization, namely, to be able to connect multiple virtual platforms together through standard communication channels (e.g., one virtual platform producing a video stream from a video file out of an HDMI transmitter IP and another decoding from an HDMI receiver IP and displaying the decoded images, both being connected only through their respective HDMI IP models).

Undoubtedly, this need will expand to complex communicating heterogeneous systems, connecting multiple virtual platforms together in order to run complex distributed software. This will probably start with point-to-point serial communication protocols (RS232, I2S, CAN, LIN, etc.), but will quickly evolve to multipoint and bidirectional ones (I2C, USB, Ethernet, etc.). This is clearly one important requirement brought by the network cloud or even the Internet of Things: a huge number of small devices connected and interacting together. Testing interoperability will be much easier if using connectable virtual prototypes, thanks to their introspection capabilities.

17.5 CONCLUSION

Since the first edition of this handbook, using transaction-level models in an SoC design flow has become a recognized and efficient methodology. Use cases and associated virtual platforms are now well understood. In addition, stronger standards are now available: SystemC/TLM IEEE 1666-2011, and IP-XACT IEEE 1685-2014 (which was upgraded with TLM additions to match the former). These standards enabled the creation of a TLM model market; many IP suppliers now provide a TLM view of their IPs. Benefits of TLM platforms across most phases of an SoC design project are proven. Several months are saved in the design cycle, enabling shorter time to market, early detection, and easy debug of hardware/software integration complex bugs and reduced schedule risks. Customers of SoC companies have also started to leverage the benefits for their own presilicon software developments. Standardization efforts are continuing to add a next level of third-party model interoperability and further reduce the expertise level needed by a company to fully benefit from these virtual platforms. The technical progress on virtual prototypes is far from finished: it is necessary to capture even more aspects (clocks, power supply, etc.) while keeping up with the evolution in complexity (scalability, integration issues, etc.). With the complex features coming with new architectures (e.g., cache coherency with peripherals, virtualization, security, power), one can be sure that the impact of transaction-level platforms will go beyond productivity improvement: their unique hardware/software observability will simply be mandatory to succeed in the design of such SoCs.

One of the best rewards of our 15 years of efforts (including concept exploration, industrial deployment, and standardization activities) has been the shift of perception from software developers concerning virtual prototypes. They switched from complete skepticism about their advantages to a deep sense of confidence that running software on them provides an important step toward a higher level of quality for their software. What we observe now is that people who try it become eager to adopt it.

ACKNOWLEDGMENTS

This chapter is an overview of the system-level design flow developed at STMicroelectronics. We earnestly appreciate the contributions of all our colleagues to the developments described in this chapter. Our gratitude goes to Frank Ghenassia, who initiated this activity and was the coauthor of the first edition of this chapter. We also thank our many colleagues who have contributed to the TLM deployment in STMicroelectronics.

REFERENCES

1. F. Ghenassia (ed.), *Transaction-Level Modeling with SystemC: TLM Concepts and Applications for Embedded Systems*, Dordrecht, The Netherlands, Springer, 2005.
2. IEEE Standards Association, IEEE 1666™: Open SystemC Language Reference Manual [Online]. Available at http://standards.ieee.org/getieee/1666/download/1666-2011.pdf, accessed Nov 20, 2015.
3. B. Bailey and G. Martin, *ESL Models and Their Application*, Dordrecht, The Netherlands, Springer, 2010.
4. W. Chen, *Out-of-Order Parallel Discrete Event Simulation for Electronic System-Level Design*, Dordrecht, The Netherlands, Springer, 2015.
5. K. Einwich, Introduction to the SystemC AMS extension standard: Tutorial, in *International Symposium on Design and Diagnostics of Electronic Circuits and Systems*, Cottbus, Germany, 2011.
6. F. Balarin and R. Passerone, Specification, synthesis, and simulation of transactor processes, (pp. 1749–1762), *IEEE Transactions on Computer-Aided Design of Integrated Circuits and Systems*, 26(10), 2007.

7. N. Bombieri, N. Deganello, and F. Fummi, *Integrating RTL IPs into TLM Designs through Automatic Transactor Generation*, Munich, Germany, 2008.

8. J. Kenney, Firmware driven OVM testbench, Mentor Graphics [Online]. Available at http://www.mentor.com/products/fv/verificationhorizons/firmware-driven-ovm-testbench, accessed Nov 20, 2015.

9. H. Shen and M. Hamayun, and F. Pétrot, Native simulation of MPSoC using hardware-assisted virtualization, (pp 1074–1087), *IEEE Transactions on Computer-Aided Design of Integrated Circuits and Systems*, 31(7), 2012.

10. Accellera Systems Initiative, Standard Universal Verification Methodology Class Reference 1.2, June 2014 [Online]. Available at http://www.accellera.org/downloads/standards/uvm, accessed Nov 20, 2015.

11. B. Bailey, *The Functional Verification of Electronic Systems: An Overview from Various Points of View*, IEC, Chicago, IL, 2005.

12. A. Perrin and G. Poivre, Architecture analysis and system debugging—A transactional debugging environment, in *Transaction-Level Modeling with SystemC: TLM Concepts and Applications for Embedded Systems*, F. Ghenassia (ed.), Dordrecht, The Netherlands, Springer, 2005.

13. Accellera Systems Initiative, TLM transaction-level modeling library, Release 2.0.1, July 2009 [Online]. Available at http://www.accellera.org/downloads/standards/systemc, accessed Nov 20, 2015.

14. Accellera Systems Initiative, SystemC verification library (SCV), Release 2.0, April 2014 [Online]. Available at http://www.accellera.org/downloads/standards/systemc, accessed Nov 20, 2015.

15. U. Sisodia, *Standard Methodology for Configuration, Control & Inspection of Models*, San Diego, CA, 2011.

16. STMicroelectronics, STMicroelectronics, ARM and Cadence improve tool and model interoperability with three joint contributions to Accellera systems initiative, press release, July 2013 [Online]. Available at http://www.st.com/web/en/press/t3433, accessed Nov 20, 2015.

17. ARM Ltd., *LISA+ Language for Fast Models Reference Manual* [Online]. Available at http://infocenter.arm.com/help/topic/com.arm.doc.dui0372i/DUI0372I_sg_lisa_rm.pdf, accessed Nov 20, 2015.

18. D. Xue and Y. Chen, *System Simulation Techniques with MATLAB and Simulink*, John Wiley & Sons, Inc., Chichester, UK, 2013.

19. T. A. Lipinski, *The Librarian's Legal Companion for Licensing Information Resources and Services*, Neal-Schuman Publishers, Chicago, IL, 2012.

20. GNU General Public License, June 29, 2007 [Online]. Available at http://www.gnu.org/copyleft/gpl.html.

21. IEEE Standards Association, IEEE 1685™: IP-Xact, Standard Structure for Packaging, Integrating, and Reusing IP Within Tool Flows, 2014 [Online]. Available at http://standards.ieee.org/getieee/1685/download/1685-2014.pdf, accessed Nov 20, 2015.

22. Accellera Systems Initiative [Online]. Available at http://www.accellera.org/activities/committees/systemrdl/, accessed Nov 20, 2015.

23. Video tutorial: Verification and automation improvement using IP-XACT, Accellera Systems Initiative, February 2, 2012 [Online]. Available at http://www.accellera.org/resources/videos/ipxactverif, accessed Nov 20, 2015.

24. T. Bouhadiba, M. Moy, F. Maraninchi, J. Cornet, L. Maillet-Contoz, and I. Materic, Co-simulation of functional SystemC/TLM models with power/thermal solvers, in *Workshop on Virtual Prototyping of Parallel and Embedded Systems*, Boston, MA, 2013.

25. L. Maillet-Contoz, Using virtual platform for application level power estimation, in *Designer Track, DAC*, San Francisco, CA, 2014.

26. L. Maillet-Contoz, Integration of functional and extra-functional properties in a system level modeling kit, in *Workshop on System-to-Silicon Performance Modeling and Analysis, DAC*, San Francisco, CA, 2014.

27. O. Mbarek, A. Pegatoquet, and M. Auguin, A methodology for power-aware transaction-level models of systems-on-chip using UPF standard concepts, in *PATMOS*, Madrid, Spain, 2011.

28. J. H. Weinstock, C. Schumacher, R. Leupers, G. Ascheid, and L. Tosoratto, Time-decoupled parallel SystemC simulation, In *Design, Automation and Test in Europe Conference and Exhibition (DATE)*, (pp. 1–4), Dresden, Germany, 2014.

29. A. V. d. Mello, I. M. Pessoa, A. Greiner and F. Pêcheux, Parallel simulation of systemc TLM 2.0 compliant MPSoC on SMP workstations, In *Design, Automation & Test in Europe Conference & Exhibition (DATE)*, (pp. 606–609), Dresden, Germany, 2010.

30. M. Moy, Parallel programming with SystemC for loosely timed models: A non-intrusive approach, In *Design, Automation & Test in Europe Conference & Exhibition (DATE)*, (pp. 9–14), Grenoble, France, 2013.

31. C. Ptolemaeus (ed.), *System Design, Modeling, and Simulation using Ptolemy II*, Berkeley, CA: Ptolemy.org, 2014.

32. R. Milner, Processes: A mathematical model of computing agents, in *Logic Colloquium*, Bristol, U.K., 1973.

33. C. Helmstetter, F. Maraninchi, L. Maillet-Contoz, and M. Moy, Automatic generation of schedulings for improving the test coverage of systems-on-a-chip, in *Formal Methods in Computer Aided Design (FMCAD)*, San Jose, CA, 2006.

34. G. Verma, Standard methodology for configuration, control & inspection of models, in *16th North America SystemC User Group*, San Diego, CA, June 2011.

Assertion-Based Verification

Harry Foster and Erich Marschner

18

CONTENTS

18.1 INTRODUCTION

Functional verification is a process of confirming that the *intent* of the design was preserved during its implementation. Hence, this process requires two key components: a *specification of design intent* and a *design implementation*. Yet historically, describing design intent in a fashion useful to the verification process has been problematic. For example, typical forms of specification are based on natural languages, which certainly do not lend themselves to any form of automation during the verification process. Furthermore, ambiguities in the specification often lead to misinterpretation in the design and verification environments. The problem is compounded when a verification environment cannot be shared across multiple verification processes (i.e., the lack of interoperability between the specification and the various verification environments for simulation, acceleration, emulation, or formal verification).

In this chapter, we introduce an approach to addressing the functional verification challenge, known as *assertion-based verification* (ABV), which provides a unified methodology for unambiguously specifying design intent across multiple verification processes using assertions. Informally, an assertion is a statement of design intent that can be used to specify design behavior [1]. Assertions may specify internal design behaviors (such as a specific FIFO structure) or external design behavior (such as protocol rules for a design's interfaces or even higher-level requirements that span across design blocks). One key characteristic of assertions is that they allow the engineer to specify *what* the design is supposed to do at a high level of abstraction without having to describe the details of *how* the design intent is to be implemented. Thus, this abstract view of the design intent is ideal for the verification process—whether we are specifying high-level requirements or lower-level internal design behavior.

18.1.1 OBSERVABILITY AND CONTROLLABILITY

Fundamental to the discussion of ABV is understanding the concepts of *controllability* and *observability* [2,3]. Controllability refers to the ability to influence an embedded finite state machine, structure, or specific line of code within the design by stimulating various input ports. Note that while in theory a simulation testbench has high controllability of the design model's input ports during verification, it can have very low controllability of an internal structure within the model. Observability, in contrast, refers to the ability to observe the effects of a specific internal finite state machine, structure, or stimulated line of code. Thus, a testbench generally has limited observability if it only observes the external ports of the design model (because the internal signals and structures are often hidden from the testbench).

To identify a design error using the testbench approach, the following conditions must hold (i.e., evaluate true):

1. The testbench must generate proper input stimulus to activate (i.e., sensitize) a bug.
2. The testbench must generate proper input stimulus to propagate all effects resulting from the bug to an output port.

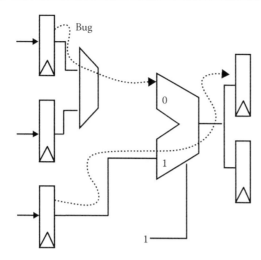

FIGURE 18.1   Poor observability misses bugs.

It is possible, however, to set up a condition where the input stimulus activates a design error that does not propagate to an observable output port (Figure 18.1). In these cases, the first condition cited earlier applies; however, the second condition is absent.

Embedding assertions in the design model increases observability. In this way, the verification environment no longer depends on generating proper input stimulus to propagate a bug to an observable port. Thus, any improper or unexpected behavior can be caught closer to the source of the bug, in terms of both time and location in the design intent.

While assertions help solve the observability challenge in simulation, they do not help with the controllability challenge. However, by adopting an assertion-based, constraint-driven simulation environment or applying formal property checking techniques to the design assertions, we are able to address the controllability challenge.

18.2 HISTORY

In this section, we discuss different approaches to solving the functional specification challenge. We begin by briefly introducing propositional temporal logic, which forms the basis for modern property specification languages. Building on this foundation, we then present a historical perspective for various forms of assertion specification.

18.2.1 REASONING ABOUT BEHAVIOR

Logic, whose origins can be traced back to ancient Greek philosophers, is a branch of philosophy (and today mathematics) concerned with reasoning about behavior. In a classical logic system, we state a proposition and then deduce (or infer) if a given model satisfies our proposition, as illustrated in Figure 18.2.

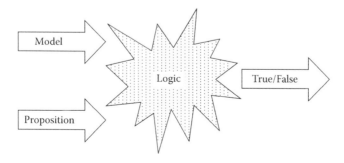

FIGURE 18.2   Classical logic system.

For example, consider the following set of propositions:

- *The moon is a satellite of the earth.*
- *The moon is rising (now).*

If we let the universe be our model, then using classical logic we can check whether our set of propositions hold for the given model.

Classical logic is good for describing *static* situations. However, classical logic is unsuitable for describing dynamic behavior (i.e., situations involving time). Returning to the previous example, it would not be possible to express the following proposition since it involves time:

- *The moon will rise again and again.*

Note that our interest in functional verification of hardware systems requires that we use a logic that is expressive enough to describe properties of reactive systems. For a reactive system, components of the system concurrently maintain ongoing interaction with their environment, as well as other concurrent components of the system. Hence, in the next section, we discuss a more expressive logic that involves time.

18.2.1.1 PROPOSITIONAL TEMPORAL LOGIC

In this section, we build a foundation of understanding by introducing a few basic concepts of propositional temporal logic. The advantage of using temporal logic to specifying properties of reactive systems is that it enables us to reason about these systems in a very simple way. That is, temporal logic eliminates the need to explicitly specify the time relationships between system events, which are represented as Boolean formulas that describe states of the design. For example, instead of writing the property expression involving time explicitly, such as

```
∀t.!(grant1(t) & grant2(t))
```

which specifies for all values of time t, `grant1` and `grant2` are mutually exclusive, we simply write in a temporal property language, such as Property Specification Language (PSL):

```
always !(grant1 & grant2)
```

which states that `grant1` and `grant2` should not hold at the same time.

In temporal logic, we define a computational path π as an infinite sequence of states

$$\pi = s_0, s_1, s_2, s_3, \ldots,$$

which represents a forward progression of time and a succession of states, s_i. When proving a temporal assertion, we may assume that a point in time or given state along the path has a unique future or successor state (e.g., $s_0 \rightarrow s_1$), in which case the assertion is proven on a given path or trace of execution (e.g., a single simulation trace). Thus, each possible computational path of a system is considered separately. We refer to this form of logic as *linear-time temporal logic*, and linear temporal logic (LTL) is one example [4]. Alternatively, we may assume that each point in time or given state along a path may split into multiple futures or successor states (e.g., $s_0 \rightarrow s_1$ and $s_0 \rightarrow s_5$). In that case, all computational paths are considered concurrently, which is usually represented as a tree of infinite computational paths. We refer to this form of logic as *branching-time temporal logic*, and computational tree logic (CTL) is one example [5].

CTL and LTL play essential roles in formal hardware verification for branching-time temporal logic and linear-time temporal logic, respectively. To show how these two types of logic differ, we first introduce CTL* [6]. CTL* contains operators for reasoning about paths of computation, such as the *path formula* operators **G** (always), **F** (eventually), **U** (until), and **X** (next) and operators for reasoning on branching paths of execution, that is, the *state formula* operators **A** (for all paths of execution) and **E** (for any path of execution). In addition to these quantifiers, any Boolean compositions of CTL* formulas are CTL* formulas as well.

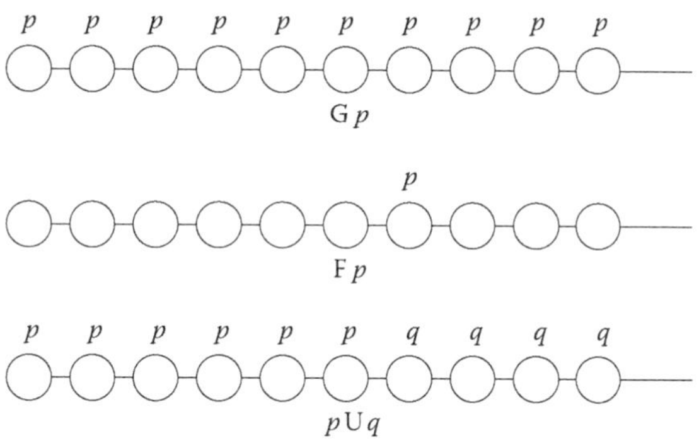

FIGURE 18.3 Temporal formula path operator semantics.

For any temporal formulas p and q, as illustrated in Figure 18.3, the temporal formula **G** p specifies that the temporal formula p holds for all states of a path π (or simply, p always holds). The temporal formula **F** p specifies that p holds for some future state of a path π. The temporal formula p **U** q specifies that the temporal formula p holds in all states of a path π until q holds in some future state of π.

The temporal formula **A** f specifies that for all paths π starting from the current state, f holds. The temporal formula **E** f means that there is a path π starting from the current state for which f holds.

As seen in the previous paragraphs, CTL* can be separated into *state formulas* (involving **A** and **E**) and *path formulas* (involving **G**, **F**, **X**, and **U**). Any atomic proposition p (or Boolean expression) over *state formulas* is by definition a *state formula*. In addition, existential quantification over path formulas (e.g., **E** f, where **E** is the existential quantifier and f is a *path formula*) is also by definition a *state formula*.

Any *state formula* is a *path formula*, as are Boolean compositions of *path formulas*. In addition, path formulas can be composed using the temporal operators **X** f and f_1 **U** f_2.

Note that the **F** operator can be thought of as an alias for the unary form of the *until* operator (e.g., **F** $p = true$ **U** p), and that **G** and **F** are dual (i.e., **G** p is equivalent to $\neg$**F**$\neg p$). The rationale behind the first alias is that *eventually* (**F**) is equivalent to waiting vacuously until p is valid; and the rationale behind the second formula is that saying "p always valid" is equivalent to saying that "it is not the case that ¬p will be valid in the future." Similarly, it is not difficult to show that **E** and **A** are dual.

Note that in CTL* we do not make any restriction on the order in which temporal and branching operators appear in a valid formula. As a result, **FG** and **AG** are valid CTL* formulas. The first formula (**FG**) states that *eventually our proposition will be valid forever*. The second formula (AG) states that *for all paths starting from the current state, our proposition will always be true*.

Now that we have presented CTL*, we can restrict this logic to CTL and LTL:

- *CTL*: A CTL formula is a CTL* formula beginning with a branch quantifier (**A** and **E**), restricting that temporal operators (**F**, **G**, **U**, and **X**) be proceeded by branch quantifiers. For example, the formula **AG** p is a valid CTL formula, but **FG** p is not.
- *LTL*: An LTL formula is the subset of CTL* formulas obtained by simply restricting the valid formulas to path formulas.

For example, the formula **FG** p is a valid LTL formula, but **EG** p is not.

If we consider an implicit universal quantifier for all paths (i.e., **A**) in front of an LTL formula, we can see that certain behaviors, such as **A**(**GF** p), cannot be represented by CTL, although it is a valid LTL formula. Similarly, **AG**(**EF** p) is a valid CTL formula, but not a valid LTL formula. While the first formula says that for all states of all paths, eventually p holds (a fairness constraint), the second formula says that for all branches of all states, at least in one of the paths, eventually p holds.

Note that neither LTL nor CTL can express a property that involves counting, such as

- *p is asserted in every even cycle*

In the next section, we introduce extended regular expressions, which allow us to express behavior that involves counting.

We do not thoroughly discuss the semantics of CTL*, CTL, and LTL in this chapter, but we give a short introduction on this subject to subsidize the remainder of this chapter. We refer the reader to [7] for a more complete definition of these logics' semantics and complexities.

18.2.1.2 EXTENDED REGULAR EXPRESSIONS

Regular expressions provide a convenient way to define sets of computations (i.e., a temporal pattern used to match various *sequences* of states). Extended regular expressions are regular expressions extended with conjunctions and negation. Hence, the computational path defined by an extended regular expression can be combined to form the building blocks used to specify hardware design assertions. Regular expressions can express counting *modulo n* type behavior through the use of the* operator. Hence, regular expressions allow us to specify properties that cannot be described by either CTL or LTL, such as {`true;!p}*`, which states that *p is asserted in every even cycle*.

Note that not all properties can be expressed with extended regular expressions. For example, the property *eventually p holds forever* cannot be expressed using extended regular expressions, yet this property can be expressed in LTL (i.e., **FG**p).

18.2.2 ASSERTION LANGUAGES

Assertions may be expressed either declaratively or procedurally. A declarative assertion is always active and is evaluated concurrently with other components in the design. A procedural assertion, on the other hand, is a statement within procedural code and is executed sequentially in its turn within the procedural description. Hence, declarative properties are natural for specifying block-level interfaces, as well as systems. Similarly, a procedural assertion is convenient for expressing implementation-level properties that must hold in the context of procedural code. A key difference between the declarative assertion and the procedural assertion is that the declarative assertion concurrently monitors the assertion expression, while the procedural assertion only validates the assertion expression during sequential visits through the procedural code.

In the following sections, we discuss various languages and techniques in use today for expressing assertions, which includes *VHDL*, the *Open Verification Library* (OVL), and *SystemVerilog Assertion* (SVA). Note that a comprehensive discussion of the Accellera PSL follows in Section 18.3.1.

18.2.2.1 VHDL ASSERTIONS

The concept of assertions was introduced quite early in the development of the VHSIC Hardware Description Language, better known as VHDL. VHDL was originally developed in the early 1980s by a team consisting of Intermetrics, Inc., IBM, and Texas Instruments, under a DoD-sponsored effort to create a design and documentation language for VHSIC-class electronic designs. The initial requirements for this language were defined in June 1981 during a workshop in Woods Hole, MA. The requirements published in the proceedings of that workshop [8] included a requirement to support exception handling, but they did not explicitly mention assertions. However, by January 1983, when these requirements were included in the request for proposal for the VHDL effort [9], they had been extended to include many of the so-called "Steelman" requirements developed earlier for the Ada programming language [10], including a requirement for assertions that was almost identical to the assertion requirement for Ada (Section 3.8.6):

> It shall be possible to include assertions in programs. If an assertion is false when encountered during simulation, it shall raise an exception. It shall also be possible to include assertions, such as the expected frequency for selection of a conditional path, that cannot be verified. [Note that assertions can be used to aid optimization and maintenance.]

Ironically, the original Ada requirement did not lead to an explicit assertion construct in Ada, yet the derivative requirement for VHDL resulted in the creation of not just one, but two kinds of assertion in VHDL: a sequential assertion statement, which can appear within sequential (i.e., procedural) code in a process or subprogram; and a concurrent assertion statement, which can appear in a block and acts like an independent process.

VHDL assertions involve only combinational expressions; no temporal operators are available in VHDL, other than the very limited capability provided by certain signal-valued attributes such as S'delayed. Even so, VHDL assertions quickly proved to be very useful mechanism for expressing invariants that are expected to hold within a design, such as a relationship among variables that is expected to hold at a given point in the execution of algorithmic code (expressed with a sequential assertion statement) or a relationship among signals that is expected to hold at all times (expressed with a concurrent assertion statement).

The simulation semantics for VHDL assertions, which are defined in the VHDL Language Reference Manual [11], ensure that an assertion will issue an error message any time it is executed and the asserted expression evaluates to false. However, error detection during simulation is not the only value of such assertions. They also can play a significant role in documentation of code and therefore maintenance of the code over time. Furthermore, as formal equivalence checking developed as an alternative to simulation, VHDL assertions were adapted to that verification method as well. For example, in the early-to-mid-1990s, VHDL assertions were used in one formal verification tool [12] to specify both *axioms* or assumptions about the environment and assertions about the design that needed to be verified.

The IEEE 1076-2008 VHDL standard extended the language by incorporating PSL (see Section 18.3.1) directly into the language. This extension enables PSL temporal operators, properties, sequences, and directives to be used alongside the existing VHDL concurrent assertion statements to express much more complex behavioral requirements.

18.2.2.2 OVL

One of the challenges in creating an assertion-based methodology is ensuring consistency within a design project. Any inconsistencies between multiple stakeholders involved in the design and verification process can become so problematic that the benefits the assertions provide during the verification process are overshadowed by an unmanageable methodology. For example, an assertion-based methodology needs to provide a consistent manner of

- Controlling assertions (e.g., enabling and disabling assertions)
- Reporting assertion violations
- Specifying reset

Aside from methodology consistency, another challenge related to assertion adoption has been the effort required to teach engineers new languages that specify assertions.

The OVL [13] was developed as a means to overcome these challenges within an assertion-based methodology [14]. Actually, the OVL is not an assertion language and lacks the expressiveness of languages such as PSL or SVA. In contrast, the OVL is a library of simulation monitors written in both Verilog and VHDL. Hence, the OVL could be classified as a declarative form of assertion specification. During the register-transfer level (RTL) development process, engineers select and then instantiate OVL monitors into the RTL model. Each assertion monitor is then used to check for a specific Boolean and temporal violation during the verification process. The following is an example of a Verilog-instantiated OVL monitor that checks for the case when `grant1` and `grant2` are not mutually exclusive:

```
assert_always mutex (clk, reset_n, !(grant1 & grant2));
```

The OVL monitors address assertion-based methodology considerations by encapsulating a unified and systematic method of reporting and a common mechanism for enabling and disabling assertions during the verification process. The reporting and enable/disable features use a consistent process, which provides uniformity and predictability within an assertion-based methodology.

The OVL offers a wide set of monitors, enabling the engineer to capture a large class of assertions, for example, the simple `assert _ always` Boolean invariant shown in the previous example and one-hot checking, as well as multicycle temporal checks, such as `assert _ next` and `assert _ change`.

18.2.2.3 SYSTEMVERILOG ASSERTIONS

Unlike VHDL, the IEEE-1364 Verilog language never contained a Boolean assertion construct. It was not uncommon for designers to capture assertions in their Verilog RTL in an ad hoc fashion using the Verilog `$display()` system task. The problem with this approach is that not all `$display()` calls could be treated or identified as assertions, which means that any ad hoc assertions specified using a `$display()` could not be verified by formal verification tools. In addition, this ad hoc approach to specifying assertions required a significant amount of extra modeling to express sequences or other complex temporal behaviors.

SVA [15] was developed to provide engineers the means to describe complex behaviors about their designs in a clear and concise manner. SVA is based on LTL built over sublanguages of regular expressions and supports two types of assertions: *procedural* (or immediate) and *declarative* (or concurrent). Both assertion types are intended to convey the intent of the design engineer and to identify the source of a problem as quickly and directly as possible. In addition, engineers can use SVA to define temporal correctness properties and coverage events.

The following is an example of a SystemVerilog concurrent assertion used to check for mutual exclusion of grant signals:

```
property mutex;
      @(posedge clk) disable iff (!reset_n) (!(grant1 & grant2));
endproperty
assert property (mutex);
```

In 2004, the Accellera 3.1a SystemVerilog LRM [16] was moved to the IEEE for standardization under the auspices of the IEEE's Corporate Standards Group (CAG) as IEEE P1800. Great effort was taken within Accellera to ensure semantics alignment between SVA with PSL. However, there are a few syntactical differences between SVA and PSL due to the diverse objectives of these two languages. Most notably, SVA was designed to provide an embedded assertion capability directly within the SystemVerilog language, while PSL was designed as a stand-alone assertion language that harmoniously works across multiple hardware description languages (HDLs) (such as SystemVeriliog, Verilog, and VHDL).

18.3 LANGUAGE PRINCIPLES AND CONCEPTS

The following section introduces principles and concepts behind the IEEE PSL [17].

18.3.1 PSL PRINCIPLES

PSL is a comprehensive language that includes both LTL and CTL constructs. This enables PSL to support various kinds of formal verification flows, including event-driven and cycle-driven simulation and various algorithms for formal verification (model checking).

Most tools are based on a single algorithm and therefore cannot support all of PSL. In particular, most simulation tools are characterized by a single trace and monotonically advancing time, which preclude support for many CTL-based properties and make it difficult to support some LTL-based properties efficiently. The subset of PSL that can be supported efficiently in both simulation and formal verification is known as the "Simple Subset" and is defined by a short list of simple syntactic restrictions.

The temporal semantics of PSL are formally defined. This formal definition ensures precision; it is possible to understand precisely what a given PSL construct means and does not mean.

The formal definition also enables careful reasoning about the interaction of PSL constructs, or about their equivalence or difference. The formal definition has enabled validation of the PSL semantic definition, either manually or through the use of automated reasoning [18].

However, for practical application, it is imperative that PSL statements be grounded in the domain to which they apply. This is accomplished in PSL by building on expressions in the HDL used to describe the design that is the domain of interest. At the bottom level, PSL deals with Boolean conditions that represent states of the design. By using HDL expressions to represent those Boolean conditions, PSL temporal semantics are connected to, and smoothly extend, the semantics of the underlying HDL.

18.3.2 BASIC PSL CONCEPTS

PSL is defined in *layers* (Figure 18.4). The Boolean layer is the foundation; it supports the temporal layer, which uses Boolean expressions to specify behavior over time. The verification layer consists of directives that specify how tools should use temporal layer specifications to verify functionality of a design. Finally, the modeling layer defines the environment within which verification is to occur.

PSL is also defined in various *flavors* (Figure 18.4). Each flavor corresponds to a particular underlying HDL. Currently defined flavors include VHDL, Verilog, SystemVerilog, and SystemC.

Since PSL works with various underlying HDLs, it must adapt to the various forms of expression allowed in each language and the various notions of datatype that exist in each language. To that end, PSL defines a number of expression type classes—Bit, Boolean, BitVector, Numeric, and String—and specifies how native and user-defined datatypes in VHDL, Verilog, SystemVerilog, and SystemC all map to those type classes.

18.3.2.1 BOOLEAN LAYER

The Boolean layer consists of expressions that belong to the Boolean-type class, that is, those whose values are, or map to, true or false. These expressions represent conditions within the design—for example, the state of the design, or the values of inputs, or a relationship among control signals. Typically, this includes any expression that is allowed as the condition in an *if* statement in a given HDL. PSL extends this set of expressions with a collection of built-in functions that check for rising or falling signals, signal stability, and other useful utilities.

18.3.2.2 TEMPORAL LAYER

The temporal layer involves time and Boolean expressions that *hold* (i.e., are true) at various points in time. This includes expressions that hold pseudo-continuously (i.e., at every point in time considered by the verification tool) as well as expressions that hold at selected points in time, such as those points at which a clock edge occurs, or an enabling condition is true. Temporal operators enable the specification of complex behaviors that involve multiple conditions over time.

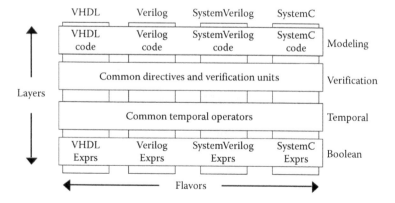

FIGURE 18.4 Property Specification Language layers and flavors.

One class of temporal operators is used to construct Sequential Extended Regular Expressions, or *sequences*. A sequence describes behavior in the form of a series of conditions that hold in succession. A sequence may specify that a given condition repeats for a minimum, maximum, or even unbounded number of times before the next condition holds, and it may specify that two or more subordinate sequences overlap or hold in parallel. If the behavior described by a sequence matches the behavior of the design starting at a given time, then the sequence holds at that time. PSL also supports *endpoints*, which are identical to sequences except that they hold at the time in which the behavior completes.

For example, consider the following sequence declarations, which describe a simple handshake-based bus protocol. In this case, a default clock declaration specifies the times at which successive expressions are evaluated:

```
default clock = (posedge clk);
sequence GetBus = { req[+]; (req && ack) };
sequence HoldBus = { (req && ack)[*] };
sequence RlsBus = { (!req && ack); !ack };
sequence ReadOp = { rwb && ardy; !drdy[*0:2]; drdy };
sequence ReadT = { GetBus; {HoldBus && ReadOp}; RlsBus };
```

The first three define the series of control conditions involved in arbitrating for control of the bus. Sequence GetBus describes behavior in which a request signal is high for one or more cycles, followed by a cycle in which the request signal is still high and an acknowledge signal goes high. Sequence HoldBus describes behavior in which both request and acknowledge signals stay high for an indefinite length of time. Sequence RlsBus describes behavior in which the request signal drops, and in the next cycle the acknowledge signal drops. Similarly, sequence ReadOp describes the control signals involved in a read operation on the bus: in the first cycle, a read operation is indicated and *address ready* signal is asserted; following that, *data ready* is asserted after a delay of 0, 1, or 2 cycles. Sequence ReadT combines all of these to represent a read transaction in which the GetBus sequence occurs first; then the ReadOp sequence occurs in parallel with the HoldBus sequence; and finally, the RlsBus sequence occurs.

A second class of temporal operators provides a means of expressing conditional behavior using implication. This includes Boolean implication (a → b) as well as suffix implication, in which the antecedent may be a sequence rather than a simple Boolean. In the latter case, the consequent holds at the end of the initial sequence. A third class of temporal operators consists of a set of English keywords that describe temporal relationships. These include **always**, **never**, **next**, **eventually!**, **until**, **before**, and slight variations thereof. Temporal operators may be combined with Booleans and sequences to describe behaviors, or *properties*, of a design.

For example, consider the following property declarations, which describe certain characteristics of the same simple bus protocol:

```
property BusArbitrationCompletes =
    always GetBus |=> {HoldBus; RlsBus};

property ReqSteady =
    always rose(req) -> next req until ack;

property AckSteady =
    always rose(ack) -> next !req before !ack;

property AckWithin (constant N) =
    always rose(req) -> next {{ack} within [*N]};
```

The first declaration defines a property BusArbitrationCompletes, which says that if Sequence GetBus occurs, then it will always be followed by Sequence HoldBus and then Sequence RlsBus. The second declaration defines property ReqSteady, which says that if signal req rises, then it will stay high until signal ack is high. Similarly, the third declaration defines property AckSteady, which says that if signal ack rises, then signal req will go low before ack is low. The final declaration defines parameterized property AckWithin, which says that after req rises, ack will occur within *N* cycles, where *N* is to be provided later.

18.3.2.3 VERIFICATION LAYER

The verification layer specifies how sequences and properties are expected to apply to the design and therefore how verification tools should attempt to verify the design using those sequences and properties. PSL *directives* indicate what to do with a given property or sequence. The **assert** directive says that a given property is expected to hold for the design, and that this should be checked during verification. The **assume** directive says that a given property should be taken for granted—it can be used in verifying other properties, but it need not be checked. The **restrict** directive is similar; it says that that a given sequence should be taken for granted. Typically, **assert** directives apply to the design being verified, while **assume** and **restrict** directives are used to constrain the environment within which the design is being verified. Additional directives (in particular, the **cover** directive) provide additional capabilities.

The verification layer also includes *verification units*, which enable packaging a collection of related PSL declarations and directives so they can be applied as a group to a particular part of the design. A verification unit may be bound to a particular instance within the design, or to a module used within the design; this causes the PSL directives to apply to that instance, or to all instances of that module, respectively. Unbound verification units may be defined to create reusable packages of PSL definitions. One verification unit may *inherit* another verification unit, which makes the definitions of the latter available for use in the former.

18.3.2.4 MODELING LAYER

The modeling layer consists of HDL code used to model the environment of the design and/or to build auxiliary state machines that simplify the construction of PSL sequences or properties. Modeling the environment of design under verification is primarily of interest for formal verification, which is addressed elsewhere. Building auxiliary state machines applies to both formal verification and simulation, and is addressed here.

Counters are one class of auxiliary state machines that are often necessary in specifying behavior. The modeling layer of PSL allows the underlying HDL to be used for constructing such a counter within a verification unit, so that it is kept separate from the design itself, yet is still available for use in a PSL assertion. For example, consider a byte-serial bus protocol that specifies the number of bytes being transferred as the first byte following the packet header. Assuming the Verilog flavor of PSL, the following verification unit might be written to specify the behavior of this protocol:

```
vunit SerialProtocol (serial_interface_module) {
    integer count, crc;

    always @(posedge clk)
      if (header)
        begin
        count = currentbyte;
        crc = 0;
        end
      else
        begin
        if (count > 0) crc = compute_crc(crc, currentbyte);
        count = count - 1;
        end

    endpoint header = {currentbyte==`PKT_HDR; 1};
    sequence data = {(count>0)[*]; crc==currentbyte};

    assert always header -> next data;
}
```

In this example, verification unit SerialProtocol contains Verilog code (in Courier font) that acts as an auxiliary state machine supporting the assertion at the end of the verification unit. The Verilog code maintains a counter, loaded from the first byte after the header, and decrements the counter as successive data bytes appear. In parallel, it computes a cyclic redundancy check (CRC) for the packet, to compare against the CRC embedded at the end of the packet.

This example illustrates reference to a PSL endpoint within the HDL modeling layer code. Endpoint *header* is a sequence consisting of the header byte of a packet followed by the packet size. When this sequence has been recognized, it triggers the loading of the packet size into the counter in the Verilog code. The assertion also uses this endpoint as the indication that a packet has started, which implies that the data must follow (and that the CRC computed by the Verilog code must match the byte following the data in the packet).

Although PSL is defined as a separate language, it is also possible to embed PSL directly into HDL code. One approach involves a convention adopted by many PSL tools in which PSL code is embedded in HDL comments, following the comment delimiter and the keyword *psl.* Another approach involves incorporation of PSL directly into the underlying HDL, as recently occurred in the IEEE 1076-2008 VHDL standard. Embedded assertions are of most use for designers who want to add assertions as they go, to document their assumptions. External assertions (in verification units) are often of more use to verification engineers, who want to verify blocks without modifying the source code of those blocks.

18.4 INDUSTRY ADOPTION OF ABV

While the process of writing assertions is fairly well understood by those skilled in the art—or whose skill can be easily acquired though a wealth of published papers and books that focus on language syntax and semantics—the process of creating a repeatable ABV methodology that integrates into an existing verification flow is not. Hence, this section focuses on practical process considerations that must be implemented to ensure an organization's successful adoption of assertion-based techniques.

18.4.1 WHO SHOULD CREATE THE ASSERTIONS?

One question often asked by many engineers just beginning the process of adopting an ABV methodology within their flow is: "Who should write the assertions? Should it be the design engineers or the verification engineers?" In fact, confusion about who should write the assertions has prevented many projects from adopting assertion-based techniques. However, in organizations that have matured their ABV process, both design and verification engineers generally create assertions. Although the ABV objectives are similar for both the design and verification engineer (i.e., to reduce debugging time while clarifying the design intent), the types of assertions each of these stakeholders write are generally different, as illustrated in Figure 18.5.

The assertions that a verification engineer writes are generally derived from the project's verification testplan. These are often referred to as high-level assertions, and they are used to check for compliance of the design against the specification. For example, a verification engineer might write a set of assertions to check compliance for a specific bus protocol. This set of assertions could be bundled together to form an assertion-based IP component, which then could be reused to check multiple similar interfaces. The key concept here is that the types of assertions that the verification engineer writes are typically specification-focused black box in nature (i.e., they have no internal implementation knowledge of the design), and they should be accounted for in the verification testplan.

In contrast, the types of assertions that a design engineer writes are focused on some low-level aspects of the design implementation that they want to monitor. This is particularly useful for checking design assumptions as well as checking for some characteristic of the design implementation where there are potential concerns due to complexity. The type of low-level assertions that a design engineer writes is analogous to a C software programmer who writes an assertion to monitor that a pointer that was passed to a function is valid before it is used within the function— thus reducing debugging time if the function was used incorrectly.

Verification engineer

➤ High-level assertions
  ✓ Requirement focused
➤ Black-box assertions
➤ Accounted for in testplan
➤ Creating reusable assertion IP
➤ Compliance traceability

Design engineer

➤ Low-level assertions
  ✓ Implementation focused
➤ White-box assertions
➤ Not accounted for in testplan
➤ Improve observability
➤ Reduce debugging time

FIGURE 18.5   Assertion-based verification stakeholders and types of assertions.

As an example, the design engineer might decide to add low-level assertions to monitor a particular FIFO for overflow or underflow conditions. Or the design engineer might add a low-level assertion to monitor for multiple, simultaneous grants that might be erroneously issued by an arbiter. The key concept here is that the types of assertions that the design engineer writes are focused on some aspect of the implementation, white box in nature, and they are not accounted for in the verification testplan. In addition, these low-level assertions should not be verification targets for the verification engineer. The objective of creating these low-level assertions by the design engineer is to reduce debugging time by monitoring potential incorrect behavior related to some aspect of the implementation.

For large projects (where the engineering team has matured its ABV process), the verification engineering team typically creates a few hundred high-level, specification-focused assertions. In contrast, for the same project, it is not unusual to find that the design engineering team (when combining all their RTL) will have contributed 10s to 100s of thousands of low-level assertions [19].

18.4.2 MATURING AN ORGANIZATION'S ABV PROCESS CAPABILITIES

Successful adoption and integration of ABV within a project's design and verification flow involves much more than (1) learning an assertion language, (2) creating a few assertions, and (3) purchasing verification tools that support assertions. The successful adoption of ABV within these organizations does not happen by accident; that is, process issues must be addressed.

Why is process important? When an engineer conducts a set of new tasks in an ad hoc fashion, they are often not repeatable and can prevent effective adoption of these new tasks by other engineers throughout the organization. Furthermore, if no metrics are gathered related to the new tasks that were introduced into the flow, then it becomes difficult to quantify any productivity gains or identify areas that need to be addressed.

So what do we mean by process? A process (Figure 18.6) obviously consists of tools and technology. Yet, it also consists of a clearly defined set of conventions and procedures. This makes the process repeatable throughout an organization. Furthermore, for successful adoption to occur, the organization must develop or acquire the required skills required by the process, as well as motivation to develop those skills. Finally, metrics are fundamental to quantitatively determine the effectiveness of a process, as well as identify opportunities for process optimizations or modifications. Without metrics, the organization lacks visibility into a process.

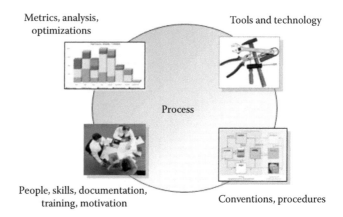

Metrics, analysis,
optimizations

Tools and technology

Process

People, skills, documentation,
training, motivation

Conventions, procedures

FIGURE 18.6   What is a process?

18.4.2.1 ABV TOOLS AND TECHNOLOGY

One of the benefits of assertions is that they can be specified once, and then the project can reuse the assertions across an entire suite of verification tools. In other words, assertions can be reused across a custom verification tool, a standard RTL testbench, a commercial simulator, a semiformal tool, or a formal verification tool. A number of projects today are even synthesizing assertions into hardware emulation, FPGA prototypes, and actual silicon for postsilicon debugging. In reality, ABV is not a process unto itself but an extension of existing verification techniques (e.g., simulation, emulation, and formal verification, FPGA prototyping, and even postsilicon validation).

Although the ABV tools, languages, and technology are necessary components of the ABV process, there is a wealth of material published that narrowly focuses on this aspect of the ABV process. Hence, this topic is not addressed in this chapter.

In terms of reusing assertions across multiple tools, although it is possible, this is something that must be carefully considered within a project. For example, it might be unproductive to formally verify every low-level, implementation assertion created by a design engineer for a given block—particularly if the block has been identified as low risk. Similarly, it is likely be unproductive to synthesize every assertion from every design block into emulation.

18.4.2.2 ABV CONVENTIONS AND PROCEDURES

To ensure that the ABV process is repeatable and consistent, the design project must establish conventions for the creation and use of assertions. For example, the project team must provide a consistent manner of controlling assertions (i.e., enabling and disabling assertions), reporting assertion violations, and specifying reset. If these conventions are not established early in the project's planning phase, then the ABV process can become unwieldy and even overwhelming as multiple stakeholders get involved.

18.4.2.2.1 Assertion Control   As previously stated, it is a good idea to establish a consistent mechanism for enabling and disabling assertions within the verification process. SystemVerilog provides constructs to enable and disable assertions, but the conventions used for that purpose must be decided by the project (e.g., `$assertoff`, `$asserton`, `$assertkill`). Some organization use either the Verilog `generate` or `` `ifdef `` constructs for various reasons to enable or disable assertions. However, it is important to choose a convention that integrates well with other processes within the design teams specific verification flow. Hence, there is no single correct convention for all organizations; however, a project convention should be established.

Other process considerations that the project might need to address are the grouping of sets of assertions and the ability to control specific groups based on the verification focus. For example, some groups of assertions might need to be disabled during error injection testing in simulation or emulation.

18.4.2.2.2 Assertion Reset   Similar to assertion control conventions, it is necessary to establish a consistent convention and mechanism for resetting the assertions. For example, some projects create a global or master reset that is controlled from the testbench, in addition to the local resets within the scope of the assertions. The decision to have a global reset is something that must be decided early in the verification planning process.

18.4.2.2.3 Reporting   If each engineer develops their own error reporting convention, it becomes problematic to create scripts that automatically process the various simulation log files. While today's assertion languages provide constructs for reporting, they leave it up to the engineer to decide what to report. Conventions should be established to ensure consistent looking assertion messages and their appropriate error or warning message levels. This simplifies the creation of any scripts used for postprocessing verification tool log files.

Some projects establish a convention for limiting the number of times each specific assertion can report an error within a given simulation run. This can be helpful at preventing any given assertion from overwhelming a regression log file for a previously identified error.

18.4.2.2.4 Naming Conventions   Naming conventions are something you might want to consider as part of your ABV process. For example, to ensure consistency across a project, many organizations require the engineers to add labels to their assertions and coverage properties. One convention that is often adopted is to add prefixes to the labels, or named properties and sequences, to simplify their identification. Table 18.1 provides one example of a prefix naming convention that is often used. The organization might decide to create different convention for your own project. The point is that naming conventions provide clarity when reviewed by multiple ABV multiple engineers and managers concerned with the ABV process.

18.4.2.2.5 Binding versus Embedding Assertions   Verification engineers often group their set of related assertions into a reusable verification component—often referred to as assertion-based IP [20]. However, for the design engineer, the question is often asked if the assertion should be directly embedded into the RTL code or maintained separately within a bind file. There are advantages and disadvantages to each approach. The advantage of embedding the assertions directly in the RTL is that if the RTL is reused at some future point in time, the embedded assertions travel with the code, which can simplify maintenance. In addition, assertions directly embedded in the RTL often help clarify the original design intent, which is helpful if the RTL code is modified by someone other than the original creator in the future.

However, there are times when an RTL module intended for reuse is considered golden and cannot be modified. Yet, it is still often desirable to create a set of assertions for that reused module. Under these circumstances, binding a set of assertions (located in a separate file from the RTL) preserves the integrity of the original golden RTL code, while permitting the addition of assertions [21].

Another consideration when establishing an assertion bind versus embed convention relates to the synthesis (and other) processes, which might be under the control of a makefile. For example, if there is a bug in an assertion that needs to be fixed, or if the engineer decides to add a new assertion directly into the RTL code, a timestamp for the RTL file would be updated, which would unnecessarily trigger a costly synthesis run and affect the physical flow. However, if the assertions are maintained in a separate bind file, the maintenance of those assertions has no effect on the timestamp on the RTL file, which prevents an unnecessary make from executing.

TABLE 18.1   Suggested Assertion Prefix Naming Convention

Prefix	Definition
`s_`	Prefix used to identify a named sequence, e.g., `s_req_gnt`
`p_`	Prefix used to identify a named property, e.g., `p_mutex`
`a_`	Prefix used on labels to identify an assertion, e.g., `a_mutex`
`c_`	Prefix used on labels to identify coverage properties, e.g., `c_write_burst`

18.4.2.3 LEARNING HOW TO CREATE ASSERTIONS EFFECTIVELY

It is generally easy for an engineer to learn the syntax and semantics of an assertion language, and there are many resources available that focus on this subject [1,19–25]. Yet, skills must be developed beyond the *how* aspect of writing an assertion. In fact, many engineers getting started with the ABV process are often confused on the *what* that assertions should address. This is less of a problem for verification engineers since the verification testplan identifies what should be checked. However, for design engineers, understanding where to add assertions in the RTL can be frustrating until they gain sufficient experience.

18.4.2.3.1 Simple Assertion Creation Principles   Following a few simple principles and guidelines greatly simplifies the process of creating assertions for design engineers. The most important principle is to keep the assertions simple and short. However, design engineers often think too hard about the type of assertion they want to create. For example, many engineers are thinking in terms of high-level, end-to-end checks—similar to the check they might consider when creating a directed test. Or they might think in terms of a complex sequence they want to check as defined by a specific requirement. One problem with this approach is that many of these complex assertions often require advanced assertion language skills to create. But more importantly, they will not provide the design engineer with the simulation debugging benefits that low-level assertions provide.

Hence, when getting started, our first recommendation is that the design engineers avoid thinking about assertions in the same way as they would when creating directed end-to-end tests. Instead, the design engineers should write simple assertions to check low-level discrete behavior—for example, a FIFO will not underflow or overflow, or an input packet's tag has a legal value, or the grants from an arbiter are mutually exclusive. The advantage of keeping the assertions simple and short is that the design engineer will find it much easier to write these assertions. Furthermore, these simple assertions will reduce the design engineers' debugging effort in simulation. In contrast, high-level, end-to-end assertions (although useful to verification engineers for creating reusable verification components) generally do not reduce the debugging effort.

Our second recommendation is that the design engineers write their assertions in place of a comment whenever possible. For example, many engineers write a comment about some assumption they have made about the design or some aspect of the design that is concerning them—such as the two control signals must never be high at the same time or the design will not function correctly. Adding an assertion (in addition to a comment) is a great way to validate your assumptions during simulation and monitor specific design concerns.

Our final recommendation is, whenever it is possible, the design engineer should create assertions on the interface ports of their design block to validate basic control signals or legal command values. For example, if a block has two input control signals (e.g., `en1` and `en2`) that must never be active at the same time, then a simple assertion can check this condition. Or, if there is a specific timing relationship or restriction between interface control signals, such as a grant must follow a request within 10 cycles, then a simple assertion can quickly identify problems when the timing restriction is violated. For design engineers, it is recommended that they evolve their assertion skills first before they try to create more complex assertions that verification engineers would typically create, such as a set of assertions checking a bus protocol.

18.4.2.3.2 Assertion Reviews   An effective method to build skills within an organization is to conduct assertion reviews with the team at different stages of RTL development. This is not as painful or as time consuming as traditional code reviews. The goal is for the various team members to describe (from a high level) the various assertions they have added to their design and what they are checking. What happens through this process is that engineers who are inexperienced at creating assertions will learn from the engineers who are more experienced at creating assertions. For example, an experienced engineer might share with the team a set of assertions that were created to check a complex arbiter. The inexperienced engineers will often realize that they have a similar complex arbiter and could create a set of similar assertions for it.

Another benefit with assertion reviews is that the process often exposes incorrect assumptions made between multiple engineers working on neighboring blocks. These bugs can then be fixed prior to any form of verification, which increases productivity. For example, one engineer might share with the team a set of assertions written to check the interface of a block. The engineer responsible for the neighboring block might then point out an incorrect assumption that was made on the neighboring interface.

18.4.2.3.3 Learning from Bug Escapes   Another useful method of maturing a project's ABV skills is to carefully analyze bug escapes, and then learn how to improve the team's assertion-creating skills from these mistakes. For instance, when a bug slips through simulation and into emulation, FPGA prototyping, or even postsilicon validation—the design and verification team should try to identify whether some low-level, RTL error (which manifested itself into the larger issue identified at a later stage of verification) could have been caught with an assertion. As an example, if a router block periodically drops a packet as a result of a fair arbitration error, then a good candidate assertion might be one that checks to ensure the arbiter is fair. The key point is that to build skills with the design engineers, you want to focus on the root cause(s) of the bug and the low-level assertions that the design engineer could have created to reduce the overall debugging effort (e.g., the arbiter error in our case)—versus focusing on writing a single assertion that would detect the high-level, end-to-end failure. Focusing on low-level assertions does not mean that the high-level assertions are of little value, but a larger set of low-level assertions will dramatically reduce debugging time by identifying the root cause of a higher-level failure.

By applying this approach, the project team matures their ABV skills over time, and increases the quality and density of assertions in the design.

18.4.2.4 METRICS IN THE ABV PROCESS

Finally, metrics are a fundamental component for any process a design team creates. Without quantifiably measuring the effectiveness of a process, the organization has no real visibility into its effectiveness—nor the ability to identify aspects of the process that need to be optimized, modified, or eliminated [26].

The following are just a few examples of metrics that can be introduced into an ABV process:

18.4.2.4.1 Bug Identification   Mature organizations generally log the bugs they identified during verification regressions, which enables them to observe bug rates and other trends associated with bugs (Figure 18.7). These process metrics are useful for decision making by the design and verification team, as well as project managers. For example, when bug rates level off, then either the design is stabilizing in terms of quality, or the metrics have identified a lack of effectiveness with the current verification approach, and an alternative strategy might need to be considered.

In addition to logging specific bugs, mature organizations often log the technique in the verification environment that found the bug, for example, a linting tool, a scoreboard within a testbench, a VIP protocol checker, and an assertion in simulation. By logging the verification technique that identified each specific bug, it is then easy to calculate the number of bugs found by the various techniques (e.g., assertions), which can quantify the effectiveness of various verification techniques.

18.4.2.4.2 Debugging Effort   If a project logs the effort spent on debugging each specific bug, as well as the verification approach used to identify the bug, then it is possible to measure the impact that a new verification technique has on reducing debugging time. Managing the effort spent in debugging is an effective way to improve productivity since debugging consumes the most effort spent in verification.

18.4.2.4.3 Assertion Density   This metric calculates the number of assertions for each block in the design, and it is useful for identifying potential verification risks. Experience has shown that, in general, complex blocks with a low assertion density have more bugs and are more difficult to debug [27]. For designers who have low assertion density metric, and have expended high effort

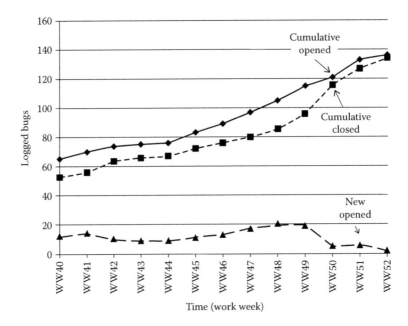

FIGURE 18.7 Bug rates measure design stability.

in debugging, publishing assertion density within an organization, along with debugging effort, can be a great motivation to improve assertion density numbers.

Note that an alternative approach to calculating assertion density is to calculate the ratio of the number of assertions per lines of code. In general, simply calculating the number of assertions in a design block is sufficient. Also, some commercial tools calculate assertion density as the minimum sequential distance between each design element and an assertion [28]. The objective is to identify portions of the design with poor observability.

18.5 SUMMARY AND FUTURE DIRECTIONS

Today's assertion language standards, such as PSL and SVA, have become a driving force in industry adoption of advanced functional verification techniques, bridging the gap between theory and practice. For example, recent industry studies indicate that debugging, on average, has grown to consume the largest amount of time within today's overall SoC verification effort [29,30]. These same studies indicate that assertion use in industry has grown from a 37% adoption rate in 2007 to 68% by 2012. The significance of this proliferation is that as numerous published case studies indicate, organizations that have adopted an ABV methodology have seen a significant reduction in verification debugging time (as much as 50% [1,31]) due to improved observability. In addition, adopting ABV provides an integration path for more advanced forms of verification into the design flow (such as formal property checking).

The current interest in ABV is focused on the use of assertions in simulation. Assertions are usually added after the fact, to increase observability of events within the design so that bugs can be detected closer to their source. But this approach is only beginning to take advantage of the power of assertion languages.

Many engineers consider the biggest challenge in adoption of assertion-based techniques is related to the manual effort and time required to generate assertions—with no guarantees about the quality or coverage of the assertions produced. To address this challenge, researchers are exploring assertion synthesis techniques to automatically generate assertions by data mining simulation traces in order to identify both design and environment properties [32]. This research has recently led to multiple commercial assertion synthesis solutions, as listed in Table 18.2.

Over time, the industry has witnessed an increased use of the coverage capabilities of assertion languages to enable coverage-driven verification [33]. At the same time, by taking advantage of

TABLE 18.2   Commercial Tools Supporting Assertions

Company	Assertions in Simulation	Assertions in Formal	Assertion Synthesis
ALDEC [35]	✓		
Atrenta [36]			✓
Cadence Design Systems [37]	✓	✓	
Jasper Design Automation [38]		✓	✓
Mentor Graphics [39]	✓	✓	✓
OneSpin Solutions [40]		✓	
Synopsys [41]	✓	✓	

the formal semantics of assertion languages such as PSL and SVA, formal verification of designs has become common. Eventually, we can expect to see assertion languages used to thoroughly specify the behavior of design IP blocks and their interfaces [34] with such specifications developed before, rather than after, the design is done. This practice may even lead to use of assertion languages as a vehicle for design if appropriate methods for assertion-based design synthesis or component selection can be developed.

REFERENCES

1. H. Foster, A. Krolnik, and D. Lacey, *Assertion-Based Design*, 2nd edn., Kluwer Academic Publishers, Berlin, Germany, 2004.
2. S. Devadas, A. Ghosh, and K. Keutzer. An observability-based code coverage metric for functional simulation. In *Proceedings of the 33rd Design Automation Conference*, Las Vegas, NV, pp. 418–425, 1996.
3. F. Fallah, S. Devadas, and K. Keutzer. OCCOM: Efficient computation of observability-based code coverage metrics for functional simulation. In *Proceedings of the 35th Design Automation Conference*, San Francisco, CA, pp. 152–157, 1998.
4. A. Pnueli. The temporal logic of programs. In *Proceedings of the 18th IEEE Symposium on Foundations of Computer Science*, Providence, RI, pp. 46–57, 1977.
5. E.M. Clarke and E.A. Emerson. Design and synthesis of synchronization skeletons using branching time temporal logic. In *Proceedings of the Workshop on Logic of Programs*, Lecture Notes in Computer Science, Yorktown Heights, NY, Vol. 131, pp. 52–71, 1981.
6. E.A. Emerson and J.Y. Halpern. "Sometimes" and "not never" revisited: On branching vs. linear time. In *Proceedings of 10th ACM Symposium on Principles of Programming Languages*, Austin, TX, pp. 127–140, 1983.
7. T. Kropf. *Introduction to Formal Hardware Verification*, Springer, Berlin, Germany, 1998.
8. G.W. Preston. Report of IDA summer study on hardware description language (IDA Paper P-1595), Institute of Defense Analyses, Alexandria, VA, October 1981.
9. Department of Defense Requirements for Hardware Description Languages, January 1983.
10. Department of Defense requirements for high order computer programming languages ("Steelman"), June 1978, http://www.adahome.com/History/Steelman.
11. IEEE Standard 1076-2002. IEEE standard VHDL language reference manual, IEEE, May 17, 2002.
12. V Formal User Guide, COMPASS design automation, August 1995.
13. Accellera standard OVL library reference manual, www.accellera.org, 2013.
14. H. Foster, K. Larsen, and M. Turpin. Introducing the new Accellera open verification library standard. In *Proceedings of DVCon*, San Jose, CA, 2006.
15. IEEE Standard 1800-2012. SystemVerilog: Unified hardware design, specification and verification language, IEEE, Inc., New York, 2005.
16. Accellera SystemVerilog language reference manual, www.accellera.org, 2004.
17. IEEE Standard 1850-2010. Property Specification Language (PSL), IEEE, Inc., New York, 2010.
18. M.J.C. Gordon. Validating the PSL/sugar semantics using automated reasoning. *Formal Aspects of Computing*, 15(4), December 2003, 406–421, Springer-Verlag, London, U.K.
19. B. Turumella and M. Sharma. Assertion-based verification of a 32 thread SPARC™ CMT microprocessor. In *Proceedings of the 45th Design Automation Conference, DAC 2008*, Anaheim, CA, pp. 256–261, 2008.
20. H. Foster and A. Krolnik. *Creating Assertion-Based IP*, Springer, Berlin, Germany, 2008.

21. C. Cummings. SystemVerilog Assertions design tricks and SVA bind files, 2009, http://www.sunburst-design.com/papers/CummingsSNUG2009SJ_SVA_Bind.pdf. (Accessed on October, 2013).

22. S. Dellacherie, H. Foster, E. Marschner, S. Ruah, and S. Smith. Tutorial: Assertion-based verification. In *40th Design Automation Conference*, Anaheim, CA, 2003.

23. C. Eisner and D. Fisman. *A Practical Introduction to PSL*, Springer, Berlin, Germany, 2006.

24. H. Foster and C. Coelho. Assertions targeting a diverse set of verification tools. In *Proceedings of 11th Annual International HDL Conference*, San Jose, CA, March 2001.

25. J. Long, A. Seawright, and H. Foster. SVA local variable coding guidelines for efficient use. In *Proceedings of DVCon*, San Jose, CA, 2007.

26. A. Meyer and H. Foster. Metrics in SoC verification: Not just for coverage anymore. In *Proceedings of DVCon*, San Jose, CA, 2012.

27. G. Kudrijavets, N. Nagappan, and T. Ball. Assessing the relationships between software assertions and code quality: An empirical investigation. Microsoft Research, October 2013.

28. R. Sathianathan. Are you missing assertions? *Verification Horizon*, February 2008, http://www.mentor.com/products/fv/verificationhorizons/are-you-missing-assertions.

29. FarWest Research, 2007 Functional Verification Study.

30. Wilson Research Group, 2012 Functional Verification Study.

31. Y. Abarbanel, I. Beer, L. Gluhovsky, S. Keidar, and Y. Wolfsthal. FoCs—Automatic generation of simulation checkers from formal specifications. In *Proceedings of 12th International Conference Computer Aided Verification*, pp. 414–427, 2000.

32. S. Vasudevan, D. Sheridan, S. Patel, D. Tcheng, B. Tuohy, and D. Johnson. GoldMine: Automatic assertion generation using data mining and static analysis. In *Proceedings of the Design, Automation & Test in Europe Conference & Exhibition*, 2010.

33. F. Haque, J. Michelson, and K. Khan. *The Art of Verification with SystemVerilog Assertions*, 1st edn., Verification Central, 2006.

34. E. Marschner, B. Deadman, and G. Martin. IP reuse hardening via embedded sugar assertions. In *Proceedings of the IP-Based System-on-Chip Design Workshop*, Grenoble, France, October 30–31, 2002.

35. ALDEC. http://www.aldec.com/en.

36. Atrenta. http://www.atrenta.com/.

37. Cadence Design Systems. http://www.cadence.com.

38. Jasper Design Automation. http://www.jasper-da.com/.

39. Mentor Graphics. www.mentor.com/.

40. OneSpin Solutions. http://www.onespin-solutions.com/.

41. Synopsys. www.synopsys.com/.

Hardware-Assisted Verification and Software Development

Frank Schirrmeister, Mike Bershteyn, and Ray Turner

19.1 INTRODUCTION

The state of hardware-assisted development has changed fundamentally since the first edition in 2006. This chapter therefore is split into two parts. In the first part, we describe how a typical modern system-on-chip (SoC) design and the different development engines—from virtual prototyping through register-transfer-level (RTL) simulation, acceleration and emulation, and field-programmable gate array (FPGA)-based prototyping and bring-up on-chip prototypes—are being used in concert, outlining their individual strengths and weaknesses. The second part of the chapter delves further into the state-of-the-art emulation technology, contrasting processors and FPGA fabrics as underlying architectures.

19.1.1 DEVELOPMENT TRENDS OF THE LAST DECADE

Since the first edition of this book almost 10 years ago in 2006, the chip and system design characteristics that define the requirements for development and verification processes of hardware and software have changed significantly.

First, according to Reference 1, the number of design starts, while still hovering around 9000 per year in 2014 after a dip to a low of about 7000 in 2009, is expected to remain at the same order of magnitude until 2020. However, it has significantly changed with respect to the technology node and the complexity of each design. In 2004, only 3.7% of the design starts targeted 90 nm or smaller geometries. In 2014, that percentage has grown to 33.5% and is predicted to reach 41.5% in 2020. The growth in design complexity, thanks to available silicon area, is significant.

Second, at 90 nm, the industry experienced a shift in the distribution of cost as well. The overall development and verification effort for software exceeded that for hardware, making up about 48% of the effort for 65 nm designs. This effort is expected to grow by eight times from 65 nm nodes to 16/14 nm. The execution cycles required for verification of those complex designs cannot be provided by just faster execution engines but also require smarter use of verification cycles and even decisions as to which use cases may be neglected and do not need verification.

The third and possibly the most significant design trend that has impacted design technologies is that of hardware intellectual property (IP) reuse. According to Reference 2, in 2004, it was a modest percentage of just under 30% of the design, with just under 20 IP blocks on average, and it is expected to grow to about 70% reuse and an average of 120 IP blocks in 2014. In 2017, the same study predicts that, on average, about 180 IP blocks will contribute to 80% overall hardware reuse.

19.1.2 TYPICAL 2015 DEVELOPMENT PROJECT

Comprehensive functional verification of hardware itself, as well as its associated software, is key to reducing development costs and delivering a product on time and as specified. Figure 19.1 indicates a design flow from specification to silicon, including production and postsilicon validation. Hardware has to be developed using integration of IP into subsystems, subsystems into SoCs, and SoCs into systems. Complex software stacks, from bare-metal software (directly interfacing to the registers in hardware and forming abstraction layers to access the hardware) to operating systems (OSs) to middleware and applications, have to be able to execute on the various processors in the system.

More specifically, based on a study of 12 projects [3], a specification phase of 8–12 weeks is followed by a phase combining RTL design, integration, and verification, with a major factor being the qualification of IP. The overall duration from RTL to GDSII can range from 49 to 83 weeks. The key is that in the last 10–12 weeks only small gate-level changes and engineering change orders are allowed as the focus of development will have shifted to silicon implementation and verification. The actual tapeout is followed by an 11–17-week production phase and 14–18 weeks of postsilicon validation.

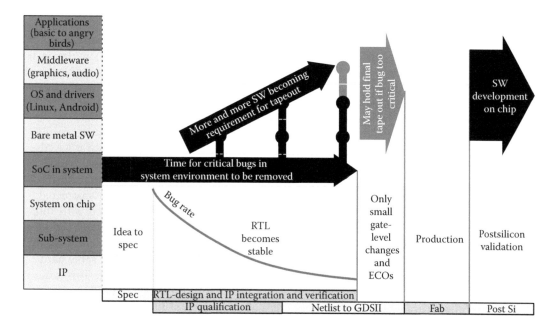

FIGURE 19.1   Hardware/software codevelopment.

The left axis of Figure 19.1 indicates the HW/SW (Hardware/Software) development stack. The SoC integrates subsystems and IP blocks and then operates within its system environment, that is, the PCB board and package. Different types of software are executing on the SoC, ranging from bare-metal software that, together with its associated hardware, actually defines the functionality of the chip, to drivers and OSs such as Linux, Android, iOS, and Windows or real-time OSs such as OSE or vxWorks. These OSs are hosting middleware for audio, video, graphics, and networking that in turn enable applications responsible for the end user experience.

A couple of key dependencies are also indicated in Figure 19.1. As RTL matures during the design flow, there comes a point at which hardware functionality has to be frozen as the focus shifts toward silicon implementation and only changes at the gate level can be easily implemented. At that point, all the aspects of the chip as to how it interacts with its system environment also have to be verified, posing unique challenges to execution platforms for the RTL at that stage of the project because either the system environment has to be modeled or virtualized or the engines of varying speeds that execute the design under development need to be connected to actual physical representations of the system environment.

The other dependency relates to software development. Generally, the interactions at the hardware/software interfaces need to be validated as early as possible, and today, proper boot-up of OSs has become a de facto requirement to allow tapeout. This poses unique challenges for the RTL execution engines as a large number of cycles needs to be executed—billions of instructions—to get OSs to boot. While the actual use of presilicon development platforms will continue during the final phase prior to tapeout, decisions to hold the tapeout at that point due to software issues found while functional verification proceeds need to be considered very carefully as the verification of physical implementation has to be restarted. Once the chip is back from production, software development can be finalized using the actual chip prototypes.

Many improvements in system design processes focused on engines that can execute hardware representations earlier in the design cycle. These engines must also be able to execute sufficient cycles for verification and for software development—which can reach billions of cycles just for an OS boot-up. We have been witnessing a shift to the left (as illustrated in Figure 19.1) that is still continuing [4]. More and more tests that used to be performed only post silicon are now attempted presilicon. As mentioned earlier, software development has become so complex that some of it has to be brought forward in the design flow as a requirement for tapeout, like booting to the prompt of an OS. Being able to start hardware/software codevelopment 4–6 weeks earlier can significantly impact the financial return for a product if a given target market window can be met.

19.1.3 CORE DEVELOPMENT ENGINES

During the aforementioned project phases, verification and software development are mainly done on four different types of core execution engines—virtual prototyping, RTL simulation, in-circuit emulation (ICE), and FPGA-based prototyping. A very common use model utilizing both RTL simulation and emulation in combination is called "simulation acceleration" and is further detailed in the following five definitions:

1. *Virtual prototypes* are transaction-level representations of the hardware, able to execute the same code that will be loaded on the actual hardware. They can execute at speeds well above hundreds of MIPS on x86-based hosts running Windows or Linux. To the software developer, virtual prototypes look just like the hardware because the registers are represented correctly, while functionality is accurately represented but abstracted. For example, processor pipelines and bus arbitrations are not represented with full accuracy, and video/audio decoding algorithms may functionally execute using a C/C++ implementation but will not be timing or bit accurate. For more information on virtual prototyping, please see Chapters 6, 8, and 12. *RTL simulation* executes the same hardware representation that is later fed into logic synthesis and implementation. This is the main vehicle for hardware verification, and it executes in the single-digit Hertz range but is fully accurate as the RTL becomes the golden model for implementation, allowing detailed hardware debugging. RTL simulation should be used early in the verification process when bugs and fixes are frequent.

2. *Simulation acceleration* executes a mix of RTL simulation and hardware-assisted verification. Simulation acceleration can address the performance shortcomings of RTL simulation when the design is mapped into a hardware accelerator to execute faster, while the testbench (and any behavioral design code) continues to execute in simulation on the workstation. A high-bandwidth, low-latency channel connects the workstation to the accelerator to exchange signal data between the testbench and the design. By Amdahl's law, the slowest device in the chain will determine the achievable speed. Normally, this is the testbench executing nonsynthesizable verification constructs (e.g., written using "e" or SystemVerilog) in the simulator. With a very efficient testbench (written in C or transaction based), the channel connecting the host and the hardware accelerator may become the bottleneck. As indicated by the name, the primary use case is acceleration of simulation. This combination allows engineers to utilize the advanced verification capabilities of language-based testbenches with a faster device under test (DUT) that is mapped into the hardware accelerator. Typical speedups over RTL simulation can reach or exceed 1000×, but the achievable speed is typically limited to tens of kilohertz.

3. *ICE* executes the design using specialized hardware verification computing platforms into which the RTL is mapped automatically and for which the hardware debug is as capable as in RTL simulation. Interfaces to the outside world (Ethernet, USB, etc.) can be made using rate adapters. ICE takes the full design including monitors and checkers (but typically excluding test benches unless they are synthesizable) and maps it into the hardware platform, allowing much higher speed up into the megahertz range and thus hardware/software codevelopment. ICE greatly reduces the long time required to implement and change designs typically seen with FPGA prototyping and provides a comprehensive, efficient debugging capability. While it takes weeks or months to implement an FPGA prototype, it takes only days to implement emulation. And design changes take a few hours or less. ICE does this at the expense of running speed and cost compared to FPGA prototypes. Looking at emulation from a different perspective, it improves acceleration's performance by substituting "live" stimulus for the testbench simulated on the host. This stimulus can come from a target system (the product being developed) or from test equipment. At 10,000–100,000 times the speed of simulation, ICE alone delivers the speed necessary to test application software while still providing a comprehensive hardware debug environment. Due to its fast compilation into processor arrays and since it does not require an expensive FPGA layout phase, ICE can be used much earlier in the design flow than FPGA-based prototyping.

4. *FPGA-based prototyping* uses a collection of FPGAs into which the design is mapped directly. Due to the need to partition the design, remap it to a different implementation technology, and reverify equivalence to the incoming RTL, the bring-up of an FPGA-based prototype can be cumbersome and takes months (as opposed to hours or minutes in ICE); debug is mostly an offline process. In exchange, speeds can reach tens of megahertz, making software development a realistic use case. The time required to map a large design into a collection of FPGAs can be very long and error prone. Changes to fix design flaws also take a long time to implement and may require board wiring changes. Since FPGA prototypes have little debugging capability, probing signals inside the FPGAs in real time is difficult and intrusive, leading to a reduction in execution speed, and recompiling FPGAs to move probes takes a long time. FPGA-based prototyping should be used toward the end of the development cycle when the design is basically complete, and speed is needed to get sufficient testing to uncover any remaining system-level bugs. Due to its speed, FPGA-based prototyping is still often the preferred software development vehicle, especially when accurate hardware representations are required (as compared to virtual prototyping on less accurate transaction-level models) and when cost is important and designs are small enough to fit into a single FPGA.

Notably, simulation and prototyping involve two different styles of execution. Simulation executes the RTL code serially while a prototype executes fully in parallel. This leads to differences in debugging. In simulation, one sets a breakpoint and stops the simulation to inspect the design state, interact with the design, and resume simulation. One can stop execution *midcycle* as it were, with only a part of the code executed. One can see any signal in the design and the contents of any memory location at any time. You can even back up (if you have saved a checkpoint) and rerun. With a prototype, one has to rely on a logic analyzer for visibility, so you can see only a limited number of signals, which you determine ahead of time (by clipping on probes). The target does not stop when the logic analyzer triggers, so each time one changes the probes or triggers condition, you have to reset the environment and start again from the beginning. Acceleration and emulation are more like prototyping and silicon in terms of RTL execution and debugging, since the entire design executes simultaneously as it will in the silicon. Since the same hardware is often used to provide both simulation acceleration and ICE, these systems provide a blend of these two very different debugging styles.

Another difference between simulation versus acceleration and emulation is a consequence of accelerators using hardware for implementation—they inherently use only two-valued logic (0/1)—acting the way the silicon will when fabricated. More values, such as X and Z, are used only in specific scenarios, like X-state handling during initialization or strength resolution, as will be discussed in Section 19.4. Accelerators also do not model precise circuit timing; hence, they may miss race conditions or setup and hold-time violations. Verification of those is better carried out during simulation or with static timing analysis tools. A key distinction between an emulator and an FPGA prototyping system is that the emulator provides a rich debug environment, while a prototyping system has little to no debug capability and is primarily used after the design is debugged to create multiple copies for system analysis and software development.

19.2 HARDWARE-ACCELERATED VERIFICATION SYSTEMS

Key tasks during the development include system modeling and trade-offs, early software development, IP selection and design verification, SoC and subsystem verification, gate level, timing and power sign-off, HW/SW validation for SoC and bare-metal software, software integration, and quality assurance, as well as system and silicon validation. The following sections will outline the main trade-offs, user concerns, requirements, and solutions.

19.2.1 HARDWARE VERIFICATION AND SOFTWARE DEVELOPMENT

Figure 19.2 outlines appropriate engine use areas as an overlay on the main user tasks.

Depending on whether models are available, virtual prototyping can enable software development as early as a couple of weeks after the specification is available. It is fast, allows good software debug insight and execution control, and is typically the quickest way to bring up software on a new design. By itself, it does not allow detailed hardware debug, which is the initial strength of RTL simulation. Used initially for RTL development, IP integration and design verification, RTL simulation can be used up to the complexity level of subsystems and certainly is a sign-off criterion for gate-level simulation and timing sign-off. It allows the fastest turnaround time for new RTL, offers excellent hardware debug, but it is typically too slow to execute meaningful amounts of software. With the advent of extending the use of graphical processing units (GPUs) to more general parallel compute tasks, the speed of RTL simulations can be increased using accelerators built into workstations.

To better extend to subsystems and the full SoC, verification acceleration uses specialized hardware attached to the simulation workstation and moves the DUT into hardware. In addition to faster hardware verification, this can allow enough speedup for bare-metal software development. With its in-circuit capabilities, emulation extends the verification to the full-chip and chip-in-system levels by enabling connections to real system environments such as PCI, USB, and Ethernet. The main advantage of processor-based emulation is fast turnaround time for bring-up, which makes it ideal for the project phase in which RTL is not quite yet mature. In addition, it allows multiuser access and excellent hardware debug insight in the context of real software that can be executed at megahertz speeds, resulting in very efficient hardware/software debug cycles. Standard software debuggers can be attached using JTAG adaptors or virtual connections. FPGA-based emulators are typically weaker with respect to debug efficiency and turnaround time.

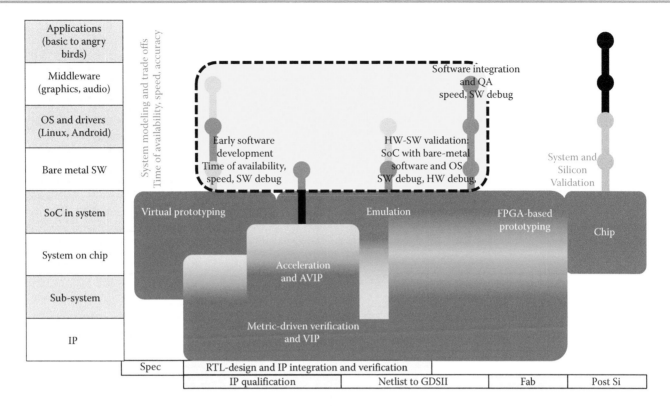

FIGURE 19.2   Sweet spots for execution engines to run verification and software development.

FPGA-based prototyping allows the extension of the speed range into the tens of megahertz range and often offers the best cost per gate per megahertz for software development and hardware regressions in the project phase when RTL has become stable enough so that fast turnaround time and hardware debug matter less. The downside to standard FPGA-based prototyping is capacity limitations as well as longer bring-up due to the changes that have to be made to map the RTL to FPGAs. Key characteristics of the different dynamic development engines are compared in Table 19.1.

19.2.2 USER CONCERNS AND REQUIREMENTS

The top requirements and concerns of customers when choosing a hardware-assisted verification and software development engine are execution speed, capacity, and cost:

- Execution speed is often looked at as a primary concern for software developers, which is not surprising given the number of cycles necessary to boot an OS. However, as presented by ARM® during CDNLive India 2013 [5] on how many cycles it took to verify the ARM big.LITTLE™ processor configuration, it is also an important parameter for verification given that trillions of cycles have to be executed per week. Traditionally, emulation is in the megahertz range and FPGA-based prototyping in the range of tens of megahertz.
- Capacity—the design size that can be mapped—is often a key criterion. In emulation, users can map billion-gate designs. For FPGA-based prototyping, the upper limit is typically in the area of 100 million gates.
- The marginal cost per unit is usually measured in dollars per gate, and it is of key interest in balancing the investment customers make. Emulation reflects its value of advanced debug, shorter bring-up, earlier time of availability, and, of course, its versatility of use models at higher cost compared to FPGA-based prototyping.

TABLE 19.1   Computational Comparison of Dynamic Execution Engines

	Computational Element	Granularity (# of Comp. Elements)	Speed per Comp. Element	Cycles/s	Examples
Software simulation, TLM (virtual prototyping)	X86 cores	Under 16	3 GHz	100+ MIPS	Cadence VSP, Synopsys Virtualizer, Mentor Visa, Intel Virtutech
Software simulation, RTL	X86 cores	Under 16	3 GHz	Under 1	Cadence Incisive, Synopsys VCS, Mentor Questa
Simulation acceleration using accelerators	GPU processing elements	Hundreds	1 GHz	10–1,000	Rocketick
Verification acceleration	Mix of host and emulator	Millions	Under 1 GHz	40–10,000	Cadence Palladium XP with Incisive Mentor Veloce with Questa Synopsys Zebu Server plus VCS
Processor-based emulation	Custom processors	Hundreds of thousands to millions	Under 1 GHz	100K–2M, processor-based scales better with design size than FPGA	Cadence Palladium Series, IBM Awan
FPGA-based emulation	FPGA gates	Millions	1–100 MHz	500K–4M, does not scale with design size, debug causes further slowdown	Mentor Veloce, Synopsys Zebu
FPGA-based prototyping	FPGA gates	Millions	1–100 MHz	2M–50M, sometimes up to 100 M	Synopsys HAPS, Cadence Protium, DINI, Aldec, S2C, ProDesign

Besides these top three requirements, additional requirements are growing in importance and help moderate the decision of which engine is the most appropriate at a specific point in a development project:

- The effort in addition to RTL is an often-overlooked component. Emulation can use RTL in an almost unmodified form in order to map it into the hardware. FPGA-based prototyping requires users to manually modify the RTL to map into the FPGA as opposed to the actual target technology for which the RTL was meant. Memories have to be remodeled, the design needs to be partitioned between FPGAs, clock domains must be managed, and so on. Both FPGA-based prototyping and FPGA-based emulation require an actual layout of the partitioned design into the individual FPGAs, a process that often does not achieve closure at first try and is very time consuming, even when using multiple cores and parallelizing across multiple multiprocessor servers. In contrast, processor-based emulation maps into the hardware at 75 million gates per hour on a single workstation. This impacts power and cost.
- Hardware debug, a key requirement for hardware verification, is almost as capable in emulation as it is in RTL simulation, with full visibility and interactivity while not slowing down the execution. In FPGA-based prototyping, hardware debug is less advanced, with probes inserted in advance and waveforms analyzed offline after execution. In both FPGA-based prototyping and FPGA-based emulation, the instrumentation for debug slows down the execution and modifies the actual RTL.
- Software debug is an area in which emulation and FPGA-based prototyping are similar in capabilities. Standard software debuggers such as the ARM DS-5, GDB, and Lauterbach's Trace32 can be attached via a JTAG port. Techniques are emerging that allow synchronized offline debug of software traces together with hardware, perhaps giving emulation a slight edge.
- Bring-up time is closely tied to the modifications that are necessary to map RTL into the hardware. Users of processor-based emulation have reported [6] that they can map new RTL several times a day. In traditional FPGA-based prototyping, users would have to wait several months until bring-up.
- Time of availability in the development cycle is a key difference, and it is tied to maturity of the RTL being mapped. Due to its fast compile-time, processor-based emulation can be used very early in the development cycle when RTL becomes available. Due to its

longer bring-up and efforts to modify RTL, FPGA-based systems are more suitable once RTL is stable, later in the design flow, but still prior to silicon availability.

- Hardware accuracy is not a big differentiator between emulation and FPGA, and it is in the diagram mostly to differentiate from other related technologies, such as virtual prototyping, for which the hardware is abstracted. Emulation has a slight edge because fewer modifications are required to the original RTL to enable bring-up.
- Finally, system connections represent how an engine connects to the actual chip environment. Both emulation and FPGA-based prototyping fare well here as they can be connected to the system environment using rate adaptors, like Cadence SpeedBridges™ [7] and Mentor iSolve™ [8]. FPGA-based prototyping has a slight edge due to the higher possible speed that allows some interfaces to be executed natively.

19.2.3 HISTORY OF HARDWARE-ACCELERATED VERIFICATION SYSTEMS

In the 1980s, IBM developed the engineering verification engine (EVE) family of simulation acceleration hardware for internal use. EVE later extended to the Awan architecture in 1998, AwanNG in 2004, and AwanStar in 2008 [9].

In 1988, Quickturn delivered the first commercial emulator when FPGA technology had reached sufficient capacity to allow their implementation. In 1997, Quickturn delivered a processor-based emulator based on custom silicon and was acquired by Cadence Design Systems in 1999. In 2002, Cadence delivered the Palladium product line with 128 million gates and up to 750 MHz. Axis announced Xtreme emulator based on reconfigurable computing technology in 2001, it was acquired by Verisity in 2004, and then it merged into Cadence in 2005. In 2010, Cadence subsequently introduced the Palladium XP Verification Computing technology, followed by Palladium XP II in 2013.

Mentor Graphics entered the emulation business by acquiring Meta Systems (founded in 1991) in 1995. After exiting the emulation business in 1996, they introduced the Celaro Pro technology in 2002 and in 2003 acquired IKOS, which had acquired Virtual machine Works in 1995. They introduced the Veloce product family in 2007, followed by Veloce 2 in 2012.

Eve had launched their Zebu product line in 2002 and was acquired by Synopsys in 2012. Subsequently, Zebu Server 3 was introduced in early 2014.

With growing complexity of FPGAs, custom FPGA prototypes found more and more usage as internally developed platforms in the 2000s. These systems either used the FPGA implementation tools delivered by FPGA providers like Xilinx and Altera, or commercial FPGA implementation tools from companies like Synplicity. By 2005, ProDesign Elektronik GmbH, HARDI Electronics, The Dini Group, and Aldec had introduced commercial FPGA-based prototyping systems, which prompted make versus buy discussions among designers, evaluating whether they should build in house prototypes by themselves versus relying on commercial offerings [10]. HARDI was later acquired by Synplicity and merged into Synopsys in 2008. In 2009, Synopsys acquired the ChipIt business unit of ProDesign. At the time of the revision of this chapter for the second edition, the main FPGA-based prototyping offerings in the market were provided by Synopsys (HAPS), The Dini Group, S2C (TAI), ProDesign (duo v7), Cadence (Protium™), and Aldec (Riviera PRO).

In addition, in 2015, the lines between FPGA-based prototyping and emulation are blurring. Differentiation between the two is mostly done by use model and support. The majority of users for FPGA-based prototyping is doing software development [11], while the majority of users for emulation are doing hardware verification. With software's growing importance for SoC development, emulation is used for software development too, as is FPGA-based prototyping for hardware verification.

The following sections will introduce in more detail the characteristics of processor-based and FPGA-based systems for emulation and prototyping. To emphasize the blurring lines between different solutions, Figure 19.3 outlines the capabilities of different solutions along the user concerns and requirements as introduced in Section 19.2.5. Generally, FPGA systems excel at higher speeds and lower marginal costs per unit, while processor-based systems excel at time of availability during a development project due to faster bring-up time and hardware debug (see also [12]).

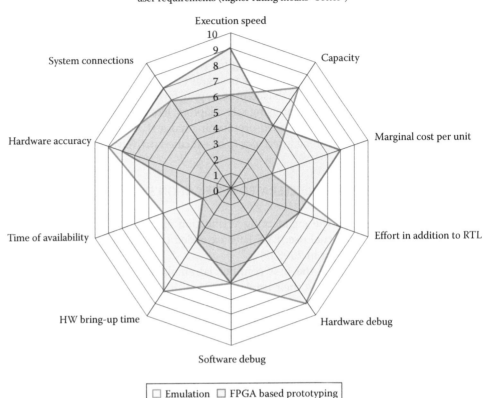

FIGURE 19.3 Processor-based emulation and field-programmable gate array–based prototyping.

19.2.4 FPGA-BASED SYSTEMS

FPGAs are an obvious choice for implementing an emulator, especially when the design fits into a single FPGA. The number of gates an FPGA can hold drops as a design is partitioned into multiple FPGAs. When a design is partitioned into many FPGAs, the FPGA I/O pin count rather than the vendor-rated gate can determine the capacity of the FPGA.

This problem of how many pins to provide for each partition of a system came up in the IBM 1401 project back in 1960. Ed Rent found an empirical rule for the relationship between pins per logic block and the number of gates in the block, by which the number of pins is equal to the product of a *Rent constant* K and the number of gates G, adjusted by the *Rent exponent* R [13]. While Rent's rule is empirical, it has had profound influence on system architecture and CAD/EDA tools and has been central to logic emulation system architecture [14,15]. Different Rent coefficients apply to different environments; the exponent is typically larger than 0.5 and determines global connectivity, while the constant, at values larger than 1, impacts net fan out.

System modules that have to be implemented in an emulator are typically richly connected inside, with fewer connections to other modules outside. Partitioning system modules into smaller portions requires cutting of many of the internal nets. Rent's rule applies to modules that get partitioned and predicts how many internal nets are cut depending on the size of the partition. Partitioning many modules across small FPGAs requires large interconnect hardware, which in exchange reduces speed and increases power. At the time of the writing of the first edition a decade ago, it looked like FPGAs will run out of pins long before their available capacity was utilized. Figure 19.4 shows Rent's rule pins calculated from Xilinx logic gate count capacities and illustrates how Rent's rule pins required have gone up with capacity much faster than real pins available (just as predicted in the first edition of this book).

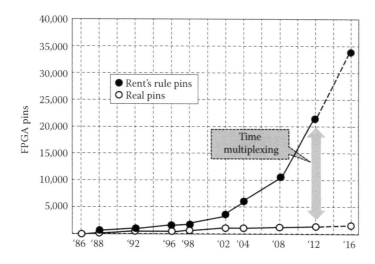

FIGURE 19.4   Rent's rule pins calculated from Xilinx logic gate count capacities.

As a result, FPGA-based emulators need very special attention when it comes to partitioning and routing the design to be mapped into them, specifically because the FPGA capacity grows much more quickly than the available bandwidth between them. Time-division multiplexing (TDM) of wires has been used, with several signals sharing the bandwidth of one wire. The less bandwidth is required per signal, the more signals per wire can be used at the same execution speed. It turned out that in reality small bandwidth per signal is hard to achieve in FPGAs because route delay unpredictability increases bandwidth requirements and timing constraints increase compilation time significantly. However, the lack of available pins today is compensated by time-multiplexing many design nets onto fewer pins, upward of 30× per pin, now at gigahertz rate.

As shown in Reference 16 in December 2014, Rent's rule applies when each major module gets partitioned into many FPGAs, but if entire modules fit in an emulation chip then only its system-level nets get cut as there are naturally much fewer of them. Mike Butts uses in Reference 16 the example of the 3.8 million gate "OOO" module of the Intel Nehalem CPU that fits into one 28 nm FPGA but needs to be partitioned if it is mapped into smaller FPGAs (see Figure 19.5). Using the example of a 6.6 million gates (MG) design, Butts shows using Rent's rule that when a large FPGA containing all 6.6 MG is used, it only requires us to cut 37% as many nets within the major modules as the same design mapped into 10 chips with 660,000 gates each (i.e., cutting 19,300 instead of 52,000 nets). With FPGAs getting large enough to fit major modules, the cut

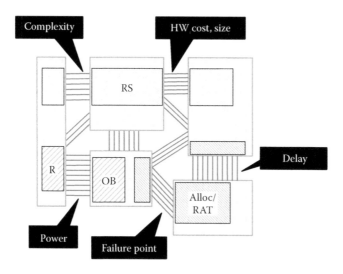

FIGURE 19.5   Partitioned "OOO" module of Intel Nehalem CPU.

nets are mostly system-level busses to which Rent's rule does not apply. Over the last decade, as indicated in Section 19.1.1, the number of major modules has grown in number. However, its growth has been easily matched by the growth in size of FPGAs, which have been following Moore's law. As a result, if the hierarchy of a design is known, then Rent's rule becomes less significant for modern FPGA-based emulators.

19.2.5 PROCESSOR-BASED SYSTEMS

Processor-based emulators consist of a massive array of Boolean processors able to share data with one another, running at very high speed. The software technology consists of partitioning a design among the processors and scheduling individual Boolean operations in the correct time sequence and in an optimal way. At their introduction, performance did not match FPGA-based emulators, but compile times of less than an hour and the elimination of timing problems that plague FPGA-based systems made the new technology appealing for many use models, especially simulation acceleration.

Later generations of this technology eventually surpassed FPGA systems in emulation speed, while retaining the large advantage in compilation times—and without a farm of a hundred PCs for compilation. Advances in software technology extended the application of processor-based emulators to handle asynchronous designs with any number of clocks. Other extensions supported 100% visibility of all signals in the design, visibility of all signals at any time from the beginning of the emulation run, and dynamic setting of logic analyzer trigger events without recompilation. At the same time that the emulation speed of FPGA-based systems was decreasing (due to heavily multiplexing pins), new generations of processor-based systems not only increased emulation speed but also proved scalable in capacity to several billions of gates (Table 19.2).

19.3 EMULATOR ARCHITECTURES

19.3.1 SMALL-SCALE EMULATION AND LOGIC PROTOTYPING WITH FPGA

FPGAs provide a convenient and inexpensive mechanism for prototyping low-complexity logic designs. The logic gate and memory capacity of FPGAs has recently grown so much that the high-end FPGA parts are capable of holding smaller application-specific integrated circuits (ASICs) up to 25 million gates. Similarly, modern FPGAs can operate at speeds of up to hundreds of megahertz, which is also close to an average ASIC clock rate. These factors have resulted in growing popularity of FPGA prototyping for small- to medium-size logic designs.

In the majority of cases, the designers of such prototypes create and customize them independently based on the requirements and characteristics of the systems in which their logic designs are intended to operate. The external inputs and outputs are connected either to a production board or to a prototype board specifically designed for that purpose. A minimal set of tools and infrastructure required for prototyping is supplied by FPGA vendors for whom such prototyping activities represent a significant market. Such tools typically consist of a compiler that converts the logic design into an FPGA programming bitstream and a debugger.

As much as FPGA vendors attempt to simplify prototyping, there are still some difficulties that the developers of such a prototype must overcome. For example, FPGA devices are volatile. Their configurations must be stored in nonvolatile off-chip memory and downloaded at power-up. The facilities for that purpose need to be created. In many cases, the verification environment for the design is implemented in software. For that, the FPGA needs to be interfaced with the workstation that runs the software. These issues create a business opportunity for vendors of small FPGA-based rapid prototyping systems that are typically implemented on boards, which contain the target FPGA and the facilities for design download and workstation interfacing. The use of such systems shortens the time required for prototype bring-up in comparison with the completely custom-designed setups.

TABLE 19.2   Comparison of Characteristics of Hardware-Assisted Verification Engines

	Cadence Palladium XP	Mentor Veloce 2	Synopsys Zebu	Synopsys HAPS, Cadence Protium, DINI, Aldec, S2C, ProDesign
Emulator architecture	Custom silicon, processor-based architecture, custom board, custom box, scalable memory architecture that eliminates need for backplanes or crossbars	Custom silicon, FPGA-based architecture, custom board, custom box, switching backplane and virtual wires	Off-the-shelf FPGA, custom board, custom box, cross bar and TDM	Off-the-shelf FPGA, off-the-shelf board, off-the-shelf box (2M–100+ M for Virtex-7-based systems); HAPS and RPP having automated software flows
Granularity	4M–2B	16M–2B	25M–200M	4M–300M
Design capacity	Up to 2 billion Typical 100M–1B	Up to 2 billion Typical 100M–1B	Up to 3B gates Typical 100M–1B	Up to 300M gates Typical 2M–100M
Typical utilization	90%–110%	60%–75%	60%–75%	Depends on partitioning
Primary target designs	SoCs 100M–1B gates; large CPUs, GPUs, multichip systems, and application processors; extending into IP and subsystems just fine, due to its fine granularity	SoCs 100M–1B gates; large CPUs, GPUs, multichip systems, and application processors; limited applicability to smaller designs due to lesser granularity	SoCs 25M–200M gates	IP blocks, subsystem, and SoCs 2M–100M
Speed range (cycles/s)	100K–2M scaling with design size	100K–1.5M degrading with design size	500K–5M degrading with design size and probes	2M–20M degrading with design size and probes
Compile time	70M gates/h single workstation (Palladium) including automated partitioning time no need to parallelize	Frontend partitioning 40M gates/h with PC farm plus time to layout the FPGAs, which is parallelizable	25M–100M gates/h for PC farm proprietary software for fast FPGA partitioning, synthesis, and place & route (P&R) (parallelizable)	1M–15M gates/h, constrained by FPGA vendor synthesis and P&R times; doesn't include partitioning time that is often mostly done manually
Partitioning	Automated	Automated; see Rent's rule	Automated; see Rent's rule	Cadence Protium having an automated flow adjacent to Palladium For others, partitioning depending on the # of FPGAs
Visibility	Full visibility at-speed probe capturing buffer for 1M cycles	Full visibility at-speed probe capture causing slowdown in execution speed; smaller buffer	Static and dynamic probes; at-speed probe capture causing significant slowdown in execution speed	Static probes (vendor dependent), always causing slowdown in execution speed
Debug	Breakpoints, assertions, unique simulation hot swap, SW debug	Breakpoints, some assertions, SW debug	Breakpoints, some assertions, SW debug	Breakpoints, little assertions, SW debug
Virtual platform API	Yes	Yes	Yes	Varies by vendor
Transactor availability	Standard/off the shelf	Standard/off the shelf	Standard/off the shelf	Little standard/off-the-shelf availability, developed ad hoc
Verification language—native support	C++, SystemC, Specman e, SystemVerilog, OVM, SVA, PSL, OVL	C++, SystemC, SystemVerilog, OVM, SVA, PSL, OVL	Synthesizable Verilog, VHDL, SystemVerilog	Synthesizable Verilog, VHDL, SystemVerilog
Memory	Up to 1 TB	Up to 125 GB	Up to 200 GB	Up to 32 GB
Users	1–512 users	1–128	1–49	1 user

Things rapidly become more complicated when the design size grows beyond the capacity of one FPGA. Although multiple FPGA prototyping systems are also available, they all suffer from difficult problems of partitioning, creating signal connectivity, and multi-FPGA timing closure.

19.3.2 LARGE-SCALE EMULATION WITH FPGA ARRAYS

As discussed earlier, when mapping very large-scale integrated circuits into FPGAs, often multiple FPGAs are required to fully emulate the design. The systems of the early 1990s had already identified the major difficulties of large-scale FPGA array emulation: partitioning, timing closure, and interconnect schemes. One distinctive approach used for FPGA interconnect was a regular pattern of direct FPGA to FPGA connections. Although moderately successful, this scheme significantly magnified the difficulties of timing closure and partitioning because many signals had to visit several FPGA parts in order to reach their destinations.

FPGA arrays with partial crossbar architecture, highly successful in the 1990s, implement the interconnection of FPGA parts using a number of special crossbar devices [17]. Each such device is a fully programmable crossbar with N terminals. Assuming that the FPGA array consists of K parts, each having M external inputs and outputs, every crossbar device will connect to at most N/K inputs or outputs of each FPGA. This ratio, commonly called the "richness" of a partial crossbar, determines the ability to route the interconnection scheme. The number of pins of a single crossbar device defines the limit to which a one-level partial crossbar can be scaled. Further capacity increase is accomplished using multilevel schemes, where the crossbar devices are again connected in a partial crossbar pattern. It has been shown that with appropriate use of the internal symmetry of FPGA pins, even low richness values allow most practical logic designs to be routed successfully. Hybrid schemes that combine direct connections with partial crossbar have also been implemented [18].

The problem of multi-FPGA timing closure has also been addressed. This problem in FPGA-based emulators primarily manifested itself in hold-time violations, due to the difficulty of managing clock skews in multi-FPGA designs. Unlike the insufficient setup time, which can be corrected simply by reducing the operating speed of an emulator, hold-time violations affect the functionality of the design and must be removed. A number of approaches have been proposed, which utilize partitioning constraints and clock signal duplications or special circuit modification [15].

As an answer to the challenge of insufficient FPGA pin count, several TDM schemes have been proposed, which divide the bandwidth available at an FPGA input/output pin among several signals that need to be transmitted between the parts.

A virtual wire scheme [14,18], though defined in general terms, has been primarily applied to direct FPGA connectivity architecture. It schedules the transmission of signals between FPGA parts as well as the storage of signal values at design flip-flops, in terms of the cycles T_V of a single-clock signal that is distributed throughout the emulator. After each logical signal that crosses an FPGA partition boundary is assigned to both the physical pin and the T_V cycle, the total number of the T_V cycles necessary to advance the circuit to the next state is determined. This number essentially constitutes a TDM factor M for this design, which will be equal to the ratio of the design clock cycle (under the simplifying assumption that it was a single-clock synchronous design to start with) to T_V. Circuits are later built in the FPGA programmable logic that implement and control TDM and signal evaluation schedules. It is important to note that virtual wire technology assumes that the design can be statically scheduled (although exceptions are made for asynchronous signals that are not to be multiplexed at all) and that the execution of exactly M T_V cycles assures its correct transition to the next state. This methodology effectively transforms the original circuit to a synchronous equivalent, which operates with a clock period T_V. It has to ensure that the signal setup time requirements for such synchronous circuits are satisfied, and therefore, no combinatorial path exists with a propagation time larger than T_V.

A different scheme, though again defined in general terms, has been primarily used with the partial crossbar architecture. This scheme is different in that it makes no attempt to transform the circuit to its synchronous equivalent, and it makes no assumptions about the signal propagation times inside the FPGA. All it does is the high-speed multiplexing of the FPGA inputs and outputs with the maximal speed permitted by the partial crossbar, printed circuit board, etc. The multiplexing period T_V does not depend at all on the design being emulated and is in fact constant, as is the multiplexing factor M. On the other hand, there are no guarantees that exactly M T_V cycles would be sufficient for advancing the design state. Signal transitions are propagated freely until they reach the FPGA boundary where they are synchronized to the TDM clock. As in the virtual wires technique, some signals can be excluded from multiplexing (and the clocks always are) so that the asynchronous behavior can be emulated more closely. The crossbar parts have the ability to demultiplex signals before switching them. Therefore, the fact that some signals share a physical wire does not impose any constraints on the way these signals should be routed.

For each solution, it has been argued that it has advantages over the other in terms of execution speed, support for asynchronous design styles, as well as the overall cost and ease of adoption. However, they encounter similar difficulties. The virtual wires technology allows the increase of effective bandwidth of the FPGA pins only to the extent that T_V is less than the design state evaluation time. In order to keep the emulation speed from going down, it is necessary therefore to keep T_V low. However, to satisfy setup time requirements, T_V must be larger than any combinational path delay. Controlling these delays in an FPGA, while keeping compilation time low, is notoriously difficult. Besides, the delays are known only after the FPGA compilation is finished while the signal schedule has to be determined before this compilation starts. As a result, pin bandwidths cannot be increased so that they keep up with the ever-growing gate/pin ratio. Although the technique by Sample et al. [19] does not explicitly require controlling setup time, in the end it suffers from similar limitations.

One can make an intriguing observation while considering multi-FPGA systems with TDM. A configurable logic structure (typically, a lookup table) that produces a signal subject to time multiplexing is severely underused. While its output needs to be valid only during one T_V period, this output is in fact being produced continuously. Thus, on average, only one out of M configurable logic structures could be in actual use at any given time. Conversely, we could use the configurable logic much more efficiently if it could compute different signals during different T_V cycles. Had this been the case, however, it would effectively turn a lookup table into a processor that executes instructions over time.

19.3.3 PROCESSOR ARRAYS FOR EMULATION

A processor-based hardware emulator [20] consists of bit-wide processors that execute a different Boolean function of N variables every T_V cycle. The input values on which the Boolean functions operate are supplied by sets of multiplexers that switch every T_V cycle as well. The Boolean functions and the control codes for the input multiplexers (addresses) together constitute instructions that are stored in the instruction memory.

The depth of such memory determines the TDM factor M that applies both to the processors and to the interconnect structures. In this approach, every processor effectively performs the work that M FPGA lookup tables formerly had, plus the multiplexing structures for one chip I/O pad and port multiplexing structure of a block memory array.

In the semiconductor implementations of a processor-based emulator (Figure 19.6), routing structures are significantly simplified compared with FPGAs. In place of a large variety of wire segments that could be arbitrarily combined with programmable switches, processor-based emulation chips contain a two-level routing structure [21] that recognizes processor cluster-level routing in addition to the uniform chip-level routing. Because of that, signal propagation delays are uniform. Processor-based emulators offer tighter control over the signal timing than an FPGA-based emulator's virtual logic can allow, at a correspondingly lower duration of the T_V cycle and higher values of M.

To better understand how a processor-based emulator works, it is useful to briefly review how a logic simulator works. Recall that a computer's arithmetic logic unit (ALU) can perform basic Boolean operations on variables, for example, AND, OR, and NOT, and that a language construct such as "always @ (posedge Clock) Q = D" forms the basis of a flip-flop. In the case of

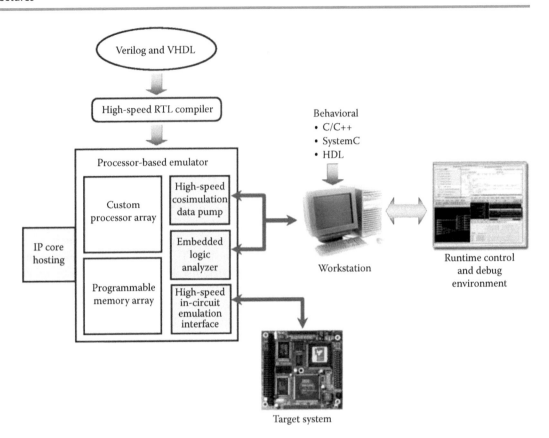

FIGURE 19.6 Processor-based emulator architecture.

gates (and transparent latches), simulation order is important. Signals race through a gate chain *left to right* so to speak, or *top to bottom* in RTL source code. Flip-flops (registers) break the gate chain for ordering purposes.

One type of simulator, a levelized compiled logic simulator, executes the Boolean equations one at a time in the correct order. (Time delays are not relevant for functional logic simulation.) If two ALUs were available, you can imagine breaking the design up into two independent logic chains and assigning each chain to an ALU, thus parallelizing the process and reducing the time required, perhaps to one half. A processor-based emulator has from tens of thousands to millions of ALUs, which are efficiently scheduled to perform all the Boolean equations in the design in the correct sequence. The following series of illustrations illustrates this process.

The first step is to reduce Boolean logic to four-input functions (Figure 19.7). The following sequencing constraint set applies:

- The flip-flops must be evaluated first.
- S must be calculated before M.
- M must be calculated before P.
- Primary inputs B, E, and F must be sampled before S is calculated.
- Primary inputs G, H, and J must be sampled before M is calculated.
- Primary inputs K, L, and N must be sampled before P is calculated.

Note that primary input A can be sampled at any time after the flip-flops.

The second step is to schedule logic operations among processors and time steps (Figure 19.7).

In the schedule as shown in Table 19.3, an emulation cycle consists of running all the steps for a complete modeling of the design. Large designs typically take 60–300 steps to schedule, depending on whether the compiler optimizes for speed or capacity. The resulting schedule is shown in Figure 19.8.

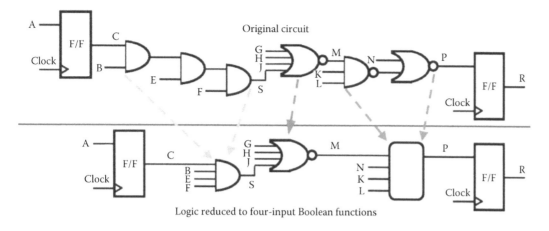

FIGURE 19.7   Reducing Boolean logic.

TABLE 19.3   Emulation Schedule

Time Step	Processor 1	Processor 2	Processor 3	Processor 4	Processor 5
1	Calculate C	Sample B	Sample G		Calculate R
2		Sample E	Sample H		Sample N
3		Sample F	Sample J		
4					Sample K
5		Calculate S			Sample L
6			Calculate M		
7					Receive M
8					
9					
10	Sample A				Calculate P

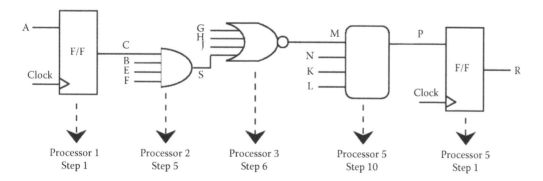

FIGURE 19.8   Logic as it is scheduled on processors.

In addition to the Boolean equations representing the design logic, the emulator must also efficiently implement memories and encrypted "soft" IP and support physical IP "bonded-out" cores. For simulation acceleration, it must also have a fast, low-latency channel to connect the emulator to a simulator on the workstation, since behavioral code cannot be synthesized into the emulator. For in-circuit operation, the emulator needs to support connections to a target system and test equipment. Finally, to provide visibility into the design for debugging, the emulator contains a logic analyzer. The logic analyzer will be described later.

19.4 DESIGN MODELING

As demonstrated earlier, the further we move along the line from a simple FPGA-based prototype to an advanced, highly scalable emulation system, the less the actual model execution mechanism resembles the target design implementation in silicon. In other words, for processor-based emulation, the user's design must be restructured to facilitate an efficient mapping onto the given emulation hardware. Next, we shall cover the major reasons and methods of design modeling.

19.4.1 TRI-STATE BUS MODELING

Tri-state busses are modeled with combinatorial logic. When none of the enables are on, emulators give the user a choice of pull-up, pull-down, or retain state. In the latter case, a latch is inserted into the design to hold the state of the bus when no drivers are enabled. In case multiple enables are on, the expected result is a value X, which emulators can represent as either 0 or 1 depending on compiler implementation. (Note that this is a good place to use assertions.)

19.4.2 BREAKING ASYNCHRONOUS LOOPS

Emulators do not model the precise gate-level timing of silicon. FPGA-based emulators have random gate delays and processor-based emulators have delays of 0. As processor-based emulators execute the operations sequentially in levelized order, a topological loop in the netlist leads to a problem of correct instruction ordering. If a wrong order is chosen, several rounds of evaluation may be required to compute the correct next state of a design. Several techniques for automatic instruction ordering (loop breaking) have been developed [22]. Application of these techniques typically assures design state computation in a single round of gate evaluations (no oversampling).

19.4.3 CLOCK HANDLING IN PROCESSOR-BASED EMULATORS

As described earlier, clocking was one of the prime sources of unreliability in FPGA-based emulators. Processor-based emulators completely avoid this problem by scheduling all combinatorial gate evaluations to complete before updating any storage elements. They can generate all the clocks necessary for a design or they can accept externally generated clocks. It is much more convenient to have the emulator generate all the design clocks needed, and it runs faster. To provide the maximum possible emulation speed while retaining the asynchronous accuracy required, processor-based emulators provide two methods of handling asynchronous design clocks: aligned edge and independent edge (Figure 19.9).

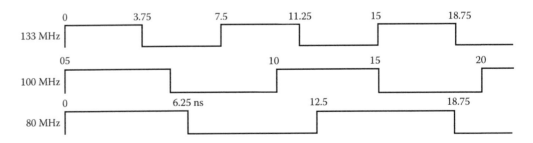

FIGURE 19.9   Clocking example—three asynchronous clocks.

19.4.4 CLOCKING WITH INDEPENDENT EDGES

Since emulators do not model design timing, but rather are functional equivalents, the exact timing between asynchronous clock edges is not relevant. One only needs to independently emulate clock edges that are not simultaneous in real life. With independent edge clocking (Figures 19.10 and 19.11), an emulation cycle is scheduled for every edge of every clock unless an edge is naturally coincident with an already scheduled edge. This is very similar to an event-driven simulator.

19.4.5 CLOCKING WITH ALIGNED EDGES

Aligned edge clocking is based on the fact that although many clocks in a design happen to have noncoincident edges because of their frequencies, proper circuit operation does not depend on the edges being independent (Figure 19.12). In this case, while proper frequency relationships are maintained, clock edges are aligned to the highest frequency clock, thereby reducing the number of emulation cycles required (compared to independent edge clocking) and increasing emulation speed.

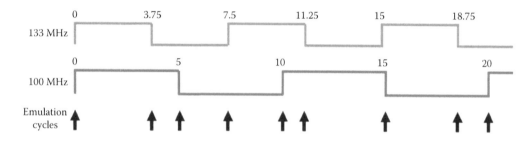

FIGURE 19.10   Independent edge clocking of two asynchronous clocks.

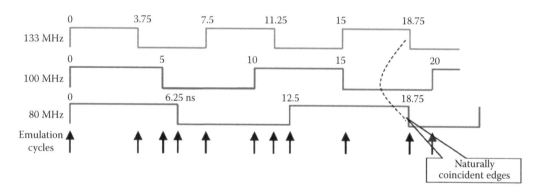

FIGURE 19.11   Adding a third asynchronous clock with independent edge clocking.

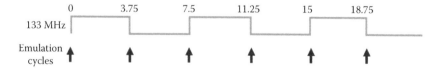

FIGURE 19.12   Aligned edge clocking starts by assigning an emulation cycle for each edge of the fastest clock.

In aligned edge clocking, an emulation cycle is first scheduled for each edge of the fastest clock in the design. Then, all other clocks are scheduled relative to this clock, with the slower clock edges *aligned* to the next scheduled emulation cycle. Note that no additional emulation cycles have been added for the second and third (slower) clocks (Figures 19.13 and 19.14). Thus, emulation speed is maintained.

Also note that while edges are moved to the following fastest clock edge, frequency relationships are maintained, which is essential for proper circuit operation. Giving the user the flexibility to switch between aligned edge clocking and independent edge clocking, processor-based emulators provide both high asynchronous accuracy and the fastest possible emulation speed for a wide variety of design styles as compared in Table 19.4.

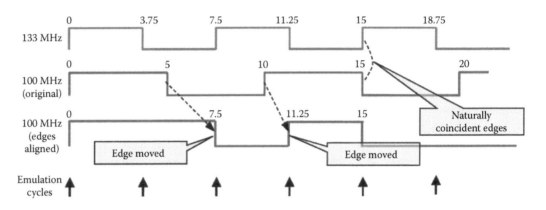

FIGURE 19.13 Add second clock aligning edges to the fastest clock.

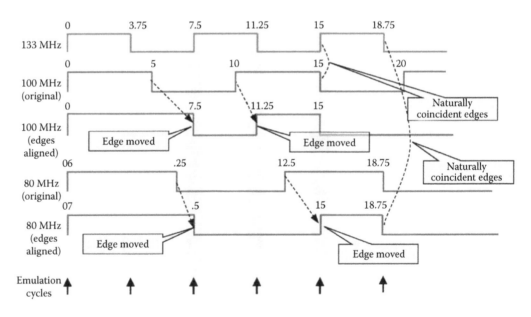

FIGURE 19.14 Add additional clocks by aligning edges to the fastest clock.

TABLE 19.4 Comparison of Clocking Techniques in Processor-Based Emulator

	Independent Edge Clocking	Aligned Edge Clocking
Preserves frequency relationships	Yes	Yes
Matches simulation	Yes	May not
Maintains speed with more clocks	No	Yes
Maintains edge independence	Yes	No

19.4.6 TIMING CONTROL ON OUTPUT

When interfacing to the real world, it is sometimes necessary to control the relative timing of output signals. A dynamic random access memory (DRAM) memory interface is one such example—all the address lines must be stable before asserting write enable. Since processor-based emulators schedule logic operations to occur in sequence, it is easy to add a constraint on the timing within the emulation cycle on individual (or groups of) output signals to control the timing to a very high resolution relative to other output signals (Figure 19.15). The compiler then schedules this output calculation at the appropriate point in the emulation cycle. This is not possible with FPGA-based emulators, since they have no control over timing within a design clock.

19.4.7 TIMING CONTROL ON INPUT

In a similar way, and for similar reasons, timing on input signals (e.g., sampling a pin when another pin rises or falls) may be controllable by the user to meet specific situations. Again, processor-based emulators can simply schedule specific input pins to be sampled before or after others within the emulation cycle. With FPGA-based emulators, the user must tweak timing by adding delays to certain signals—including a large guard band—because FPGA-based emulators cannot control absolute timing along different logic paths. For FPGA-based emulators, this is very hard to control and may even vary from compile to compile.

19.4.8 GENERATING HIGH-SPEED CLOCKS

Occasionally, it may be necessary to generate a clock several times faster than the fastest clock in the design. An example is a chip with an internal clock divider. Running the entire emulation at higher speed would reduce emulation performance when only a couple of flip-flops need this speed. Since there are usually over a hundred steps in an emulation cycle, it is easy for processor-based emulators to generate a higher-frequency clock by scheduling multiple changes on an output within a single emulation cycle (Figure 19.10). As shown later, the user may exercise control over the timing of all the changes and create fairly complex clock signals, if needed. FPGA-based emulators cannot do this because they have no control over timing within the emulation cycle.

19.4.9 REDUCING UNNECESSARY EVALUATIONS

It is well established that in a typical design, a large percentage of gates switch only infrequently relative to the fastest clock signal. Techniques have been developed to reduce the number of evaluation cycles in a processor-based emulator such that a larger share of logic signals is in fact changing their values in each evaluation.

Such techniques are based on compile-time proof of alignment of signal transitions with one of several successive edges of the fastest design clock. For example, if it can be proven that all

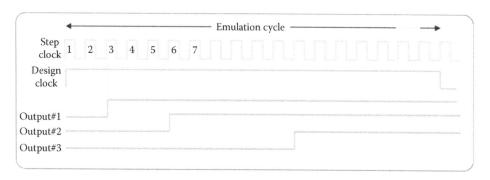

FIGURE 19.15   Processor-based emulators can adjust output timing with high precision.

signals change only with the positive edge of the clock, there would be no need to evaluate the design on its negative edge and the emulation speed can be effectively doubled. In practice, such proof can usually be achieved for the majority (but not for all) of signals. Those signal transitions for which alignment to a particular edge cannot be proven will need to be evaluated twice (duplicated), thus doubling the performance at the cost of some increase of model size.

19.4.10 MEMORY MODELING

Memory modeling is a critically important implementation task in emulation systems. Its effect upon overall performance and capacity utilization is at least as big as that of the modeling of logic. Typically, emulation systems have several levels of memory resources such as external off-chip memories (DRAM or SRAM), on-chip block RAM, and the configurable logic that can be utilized as memory (flip-flops, lookup tables). Each of these types of resources offers its own performance/capacity trade-off. Off-chip RAM is typically large but slow, configurable logic is fast but only appropriate for small memory sizes, and block RAM offers a compromise. Both off-chip RAM and block RAM come in a fixed set of form factors (width vs. depth) and have one or two ports accessible simultaneously. On the other hand, a model presented for emulation by the user commonly includes a wide variety of memory form factors, as well as memories with large numbers of ports. The goal of memory modeling is to map the memories presented in a design onto the available memory resources in such a way as to achieve the fastest performance with little to no effect on emulation capacity.

Several techniques have been developed to achieve this objective. Some of them (memory volume reduction, read port duplication, optimal type of memory selection) are generically applicable to most kinds of emulation systems. Some (compaction, merging) are specific to processor-based systems. For an overview of the most popular techniques, see [22].

19.5 DEBUGGING

Most of the time spent verifying designs is spent while debugging problems. Thus, productivity in debugging is paramount to reducing time to market. Key to this capability is being able to rapidly display any signal in the design and being able to diagnose the observed anomaly.

19.5.1 LOGIC ANALYZER AND SIGNAL TRACE DISPLAY

The built-in logic analyzer is the heart of an emulator's debugging capability. The debugger displays in an integrated way signal waveforms, RTL source code, generated schematic views, and the design hierarchy. Signal values at a specific point in time can be back-annotated onto the schematic and RTL source code displays to accelerate the user's analysis of the situation.

To give some context, consider that an emulator with a very large design with 20 million signals, running at 500 kHz, would generate 60 TB of signal trace data per minute. In any logic analyzer, there is a finite amount of memory for signal trace data storage. Emulators deal with this by presenting the user with trade-offs between width (number of signals) and depth (number of clock cycles). Emulators also use several techniques to *extend* the logic analyzer trace capacity greatly.

Emulators only need to capture design inputs and the outputs of flip-flops and memories. The values of other signals are Boolean combinations of these and can be quickly calculated, when needed. This capability would be necessary in any event since combinatorial logic in the design is optimized into n-input Boolean functions, and thus, some combinatorial signals in the design do not exist individually in the emulator.

Oftentimes, the engineer doing the debugging may not know ahead of time which signals will need to be probed to debug a problem. This is especially true when verification is performed by a separate team of engineers. In the course of debugging in such situations, the engineer may need to see any signal in the design to trace the problem back to the root cause. Emulators often allow the user to trade-off display speed, logic analyzer depth, and the support of dynamic target systems to get the best debugging capabilities for their particular needs.

In contrast to 100% signal visibility, dynamic probing provides a very deep logic analyzer with a smaller number of probes. Users must specify prior to running—and also prior to compilation with FPGA-based emulators—the signals they wish to see. Much as with simulation, adding a signal to the list allows the display of that signal going forward in time, but not backward in time. ("Forward" refers to clock cycles that have not yet been emulated, while "backward" refers to clock cycles that have already been emulated.) Processor-based emulators accomplish this rapid change of signals to be probed (<1 s to change 1000 signal probes) by utilizing otherwise *idle* time steps in the processor array to route signals to available logic analyzer memory channels. The compiler reserves a small percentage of the processor array bandwidth for this purpose when using dynamic probing, allowing new signals to be probed without changing the *program* for emulating the design. Note that in FPGA-based emulators, when not using 100% visibility, recompiling the entire design with the new probes is necessary—which changes the timing of the design model, sometimes resulting in a different functional behavior or reduced operating speed.

For any of the aforementioned debugging choices, emulators often also provide the ability to extend the depth of signal tracing (number of samples) to infinity—from the beginning of emulation, showing all the signals, all the time—but with some trade-offs. To support *infinite* trace data, the emulator is stopped periodically (as the physical logic analyzer becomes full) to upload the necessary data to the workstation. Then emulation continues. This approach also requires a use model, which tolerates periodic stopping of the clocks to the target system (typically a static target system). The advantage is that the logic analyzer now has infinite depth and is able to display signal trace data from the beginning of the emulation run for any or all signals in the design. The trade-off is that stopping the clocks to upload the logic analyzer data results in reduced net emulation speed—usually two to three times slower. However, the user can turn on and off the infinite trace recording at run time so that the reduced speed is only experienced during times of interest. For example, one could boot a PC at full speed and then turn on infinite tracing to study a particular problem. Thus, in spite of having a finite, practical amount of logic analyzer trace memory, emulators can provide a rich set of debugging choices from which the user can select a suitable set of trade-offs for their use model.

19.5.2 DEFINING TRIGGER CONDITIONS

Emulators allow the user to specify complex trigger conditions for the logic analyzer. In the course of debugging a problem, the need to change the trigger condition is often encountered. Processor-based emulators support changing the trigger condition instantly, without needing to recompile the entire design. FPGA-based emulators may require recompiling one or more FPGAs in order to change trigger conditions when the change involves new signals.

19.5.3 DESIGN EXPERIMENTS WITH SET, FORCE, AND RELEASE COMMANDS

Design experiments allow the user to make certain design changes instantly, without recompilation, and see the result. This can be useful

- To disable sections of the circuit that are interfering with the section you want to inspect
- To force signals on or off to enable/disable signal effects

This allows the user to experiment with ideas without having to recompile the design. Design experiments are accomplished using specific commands to set, force, and release signals, possibly included in a script for time sequencing. Set instantly sets a flip-flop to a state of 1 or 0, but then allows normal circuit operation to change it. Force and Release are complementary commands; Force will freeze a signal to 1 or 0 until a Release command is given for that signal.

19.6 USE MODELS

Emulators and accelerators can be used in a number of ways to accelerate functional verification and enable software development.

19.6.1 IN-CIRCUIT EMULATION

This model of operation connects the emulator with a user's target system—a prototype of the system the user is designing. The emulator typically replaces the ASIC(s) being designed for the target system, allowing system-level and software testing prior to silicon availability. Because the emulator runs slower than the ASIC being designed, slowdown solutions must be implemented to match the real-world interfaces to the emulator. Clocks may be derived in the target system or in the emulator. In some simple environments, the target system may be a piece of test equipment that generates and verifies test data—network testers are a typical example of this. As of 2015, ICE is the most common use model for emulation.

19.6.2 SIGNAL-BASED ACCELERATION

As described in Section 19.1.3, in this use model the synthesizable RTL is moved into the emulator, where it runs several orders of magnitude faster than on the workstation. Behavioral code (e.g., the test bench and sometimes the memory models) continues to be run on the workstation by the simulator. Overall performance is limited by the communication channel between emulator and workstation and also by the testbench execution on the simulator. Most users in this mode prefer to debug in the familiar simulation environment, providing visibility into the RTL signals evaluated in the emulator.

19.6.3 TRANSACTION-BASED ACCELERATION

Transaction-based acceleration (TBA) is another form of *simulation acceleration* using cosimulation in which transactions (a set of bundled signals, see also Chapter 17) are sent to the emulator rather than bit-by-bit signal exchanges. This reduces the traffic between workstation and emulator, allowing higher performances to be achieved (where the channel is the limiting factor). Transactors are designed in synthesizable RTL and put into the emulator with the design.

These transactors map the transactions from the workstation into bit-by-bit operations for the design in the emulator. TBA has the additional advantage of raising the debugging level of the design from individual logic levels to transactions that are more meaningful to the designer. There is a significant issue of acquiring a rich library of transactor IP that can be (or has been) synthesized. The Accellera committee (www.accellera.org) has standardized an interface between simulators and emulators as the Standard Co-Emulation Modeling Interface [23], which, if supported, allows mix-and-match use of emulators/accelerators with simulators for transaction-based testbenches.

As the second edition of this book is being written, a discussion about virtualization continues. With TBA, the remote access of emulation becomes easier than with the "hard" connected interfaces used by ICE. Virtualization can even be pushed so far as to represent some of the peripheral components with which the chip communicates on the workstation that is connected to the emulator. A prime example is a USB peripheral like a memory stick, which can be represented fully virtually. While increasing the ability to debug and to inject specific errors, due to the limitations in the channel between the emulator and the host, this use model typically results in slower execution speed than can be achieved by ICE.

19.6.4 TEST BENCHES IN HIGH-LEVEL PROGRAMMING LANGUAGES

Similar to cosimulation, this use model has a test bench (and possibly some submodels) written in a high-level programming language like C/C++ instead of a hardware description language (HDL) testbench. The C code is interfaced to the emulator through an application programming interface (API). This mode can provide higher performance than HDL cosimulation because high-level programming languages execute faster than HDL in a simulator. High-level test benches are becoming more popular because they are often used for system-level modeling before designing the ASIC and some of the test routines may be reusable for the ASIC testbench. Debugging in this mode involves using the emulator debug environment (which the user prefers to look like the simulator) as well as a debugger for the high-level programming language used.

19.6.5 VECTOR REGRESSION

This use model applies large sets of input vectors for regression testing of small changes in the design. The vectors may come from simulation, testers, or other sources. Typically, they are used close to tapeout when small changes are made to the logic and the user must ensure that nothing was broken by the change. Comparison of resulting vectors against known good results may be done either in the emulator or the workstation (depending on emulator capabilities). Emphasis is on fast execution and pass/fail results.

19.6.6 EMBEDDED TESTBENCH

In this use model, the testbench (and the entire design) is written in synthesizable RTL code and the design and testbench are run in the emulator at full emulation speeds. Few users are willing to write a testbench in synthesizable RTL because of the extra effort and the existing investment in behavioral testbench code. This use model is most often applied to high-performance processor development, because performance is valued so highly as to make the extra testbench effort worthwhile. When the ASIC contains a processor core, embedded testbench mode can be applied using a simple synthesizable testbench and processor object code executing out of emulator memory.

19.7 CONSIDERATIONS FOR SUCCESSFUL EMULATION

Creating a successful ICE environment involves a few additional considerations relative to simulation or acceleration.

19.7.1 CREATING AN IN-CIRCUIT EMULATION ENVIRONMENT

Emulation generally runs up to a few megahertz, while the actual system under design will run much faster. When the emulator must interface with real-world full-speed devices, an emulation environment that bridges the speed difference between the real-world speeds and emulation speeds must be created. There are a variety of ways in which this might be accomplished. Sometimes, it is as simple as using a tester that can be slowed down. For example, for Ethernet interfaces, usually the PHY is bypassed. A commercial Ethernet tester can slow down the bit stream to emulation speeds (both transmit and receive) and both drive traffic into the design and analyze protocol results. Alternatively, one can use an Ethernet bidirectional packet buffer, such as a Cadence SpeedBridge or Mentor iSolve, between multiple real-world connections and sloweddown Ethernet ports running at emulation speed.

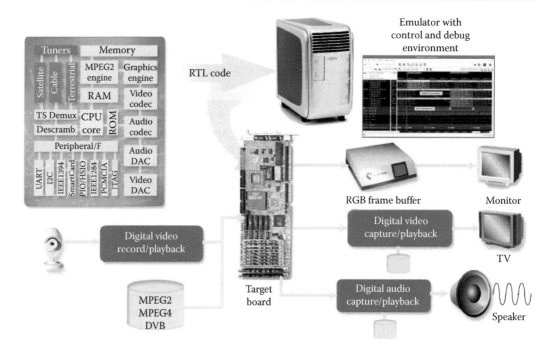

FIGURE 19.16 Emulation environment for set-top box design.

Another emulation environment example is the verification of an ASIC for a video graphics card. The red–green–blue (RGB) data will be produced at emulation speeds, but the monitor cannot sync the image at these speeds. You can use a video frame buffer, which accepts the RGB data at emulation speeds and puts the pixel data into a frame buffer. The frame buffer is then read out at normal scan rates and drives a monitor. In this way, the user views the image produced by the emulated design but the frame update rate is reduced. Such an environment is common as it allows users to *see* the results of their design.

A set-top box emulation environment is shown in Figure 19.16. The RF circuitry is dropped—emulation only performs functional verification of the digital portion of the design. The rest of the ASIC is compiled into the emulator and the emulator is plugged into the ASIC socket on the PC board. A digital video recorder/player is used to deliver the digital data stream representing the demodulated RF. This data stream could come from some standard data sets for Motion Picture Expert Group (MPEG) decompression testing or it could be data that the user recorded from a video camera and microphone. The video frame buffer is used to display (nearly) still images. A second digital video recorder/player captures video data at emulation speed, records it to hard disk, and then plays it back at full speed. Being able to view motion is a critically important factor in MPEG decoder work, since the MPEG process does not reproduce the exact input bit stream (it is "lossy"). Hence, decoder algorithms are evaluated subjectively. In a similar way, audio is produced via a digital audio recorder/player that buffers and replays audio at full speed so that one can hear the quality of reproduction.

19.7.2 DEBUG ISSUES WITH DYNAMIC TARGETS

Some emulators temporarily suspend the clocks during operation, either to capture or to calculate and display signal data. While the emulator is doing this, it is not available to maintain the connection to the target system. As long as the target system is static, there is no problem. Emulation resumes normally when the clocks are restarted. When components connected to emulation have dynamic characteristics, they may not recover automatically from having their clocks stopped for a few seconds. With dynamic target systems, for example, including a dynamic RAM, which must be refreshed, it may be necessary to select a different choice for capturing and/or displaying signal data to avoid this problem—if one is available. Note that some rate adapters solve this problem by keeping the target clocks running even when the emulator is not supplying clocks to it.

19.7.3 CONSIDERATIONS FOR SOFTWARE TESTING

As discussed in the opening section, most electronic products today have increasing software content. The traditional approach to software verification is to wait for (mostly) working silicon to begin software debugging. This makes the hardware and software debugging tasks largely sequential and increases the product's development time. It also means that a serious system bug may not be found until after first silicon, requiring a costly respin and delaying the project for several months.

The objective of hardware/software coverification is to make the hardware and software debugging tasks as concurrent as possible in order to start software debugging earlier. At a minimum, this means starting software debug as soon as the IC is taped out, rather than waiting for good silicon. But even greater concurrency is possible. Software debugging with the actual design can begin as soon as the hardware design achieves some level of correct functionality. Starting software debugging early can save up to 9 months of product development time. Early software testing can also uncover a bug in the ASIC early enough to fix it in the ASIC, thereby avoiding a software patch that may degrade usability or performance of the product. The increased level of software testing available, with the speed of ICE, also reduces the risk of a software bug going undetected and getting released with the product, and the expensive upgrades, recalls, or lost business that might result.

There are additional benefits to starting software verification prior to freezing the hardware design. If system problems or performance issues are found, designers can make intelligent trade-offs in deciding whether to change the hardware or software, possibly avoiding a degradation in product functionality, reduced performance, or an increase in product cost.

Whether a microprocessor in-circuit emulator (MP-ICE) or real time operating system (RTOS) debugger is used, the software debugging and hardware debugging environments can be synchronized so that hardware/software interface issues can be debugged conveniently. The breakpoint/trigger systems of the emulator and MP-ICE are cross-connected such that the emulator's logic analyzer trigger is one of the MP-ICE breakpoint conditions and the MP-ICE breakpoint trap signal is set as an emulator logic analyzer trigger condition. Therefore, if a software breakpoint is reached, the emulator captures the condition of the ASIC at the same moment. If an ASIC event triggers the logic analyzer, the software is stopped at that moment. This allows inspection of the hardware events that led to a software breakpoint or of the ASIC operation resulting from executing a set of software instructions. This kind of coordinated debugging is extremely valuable for understanding subtle problems that occur at the hardware/software interface.

There are a variety of ways to interface to a microprocessor for software testing. JTAG, RS-232, or Ethernet connections can control a resident debug monitor. The OS in the processor may provide its own debug environment.

The microprocessor itself could be a packaged part or a bonded-out core mounted on the target system, an RTL model that gets mapped into the emulator with the rest of the design, or an instruction set simulator model running on the workstation.

19.7.4 MULTIUSER ACCESS

Some emulators allow their total capacity to be dynamically partitioned among a number of simultaneous users. This increases the usage considerably, especially early in the design cycle for block-level acceleration and in environments where multiple projects are sharing the emulator. As the design progresses and full-chip and system-level testing is required, a larger percentage of the emulator capacity (perhaps all of it) is used to emulate it.

19.8 SUMMARY AND OUTLOOK

Hardware accelerators and emulators provide much higher verification performance than logic simulators, but require some additional effort to deploy. ICE provides the highest performance—often 10,000–1,000,000 times faster than a simulator—but requires an emulation environment to be built around it with speed buffering devices. Accelerators and emulators require the user to be aware of the differences between simulation and silicon (i.e., emulators and chips):

1. Simulation supports 12 values and strength levels per bit or more, whereas silicon has only 2 values.
2. Simulation generally executes RTL statements sequentially, whereas silicon *executes* RTL concurrently.
3. Simulation is highly interactive (observable and controllable), whereas silicon is less so.

FPGA-based emulators use commercial FPGAs and are smaller and consume less power, while processor-based emulators require custom silicon designs and consume more power, but they compile designs much faster and can deliver higher performance. The key benefit of using emulation for design verification is shorter time to market and higher product quality.

It should be clear at this point that no engine fits all requirements set forward by users. A continuum of engines is needed, from TLM-based virtual prototypes through RTL simulation, simulation acceleration, emulation, FPGA-based prototyping, and even the actual silicon when it comes back from fabrication. All major EDA vendors have announced a suite of connected engines (Cadence in May 2011, Mentor in March 2014, and Synopsys in September 2014 [24–26]), and it will be interesting to see how the continuum of available engines evolves over the next decade.

REFERENCES

1. H. Jones, Global system IC industry service report, *IBS*, 23(7), p. 25, July 2014.
2. R. Wawrzyniak, System(s)-on-a-chip: Changes in SoC design methodology, SC101-13, Semico, Phoenix, AZ, p. 22, March 2013.
3. F. Schirrmeister. Combining prototyping solutions to solve hardware/software integration challenges, *EETimes*, December 13, 2011, http://www.eetimes.com/document.asp?doc_id=1279285.
4. F. Schirrmeister. The great shift to the left, *SemiEngineering*, March 27, 2014, http://semiengineering.com/the-great-shift-to-the-left/.
5. F. Schirrmeister. How many cycles are needed to verify ARM's big.LITTLE on Palladium XP, http://community.cadence.com/cadence_blogs_8/b/sd/archive/2012/10/30/how-many-cycles-are-needed-to-verify-arm-s-big-little-on-palladium-xp.
6. W.C. Wong and R. Binnamangalum, T4240 DMA/PCIE performance analysis using Cadence Palladium, p.7, CDNLive Silicon Valley, Santa Clara, CA, March 2013, http://bit.ly/1KU80gu.
7. Cadence SpeedBridge Datasheet, http://www.cadence.com/.
8. Mentor Graphics. iSolve Datasheet, https://www.mentor.com/products/fv/emulation-systems/isolve.
9. M.D. Moffitt, G.E. Gunther, K.A.Pasnik. Place and route for massively parallel hardware-accelerated functional verification, in International Conference on Computer-Aided Design (ICCAD), San Jose, CA, 2013.
10. M. Santarini. ASIC prototyping: Make versus buy, *EDN*, 50, 30–40, 2005.
11. J. Blyler. What drives ASIC prototyping with FPGAs in 2012 and beyond?, *JB's Circuit*, http://chipdesignmag.com/sld/blyler/2012/09/27/what-drives-asic-prototyping-with-fpgas-in-2012-and-beyond/.
12. F. Schirrmeister. The agony of hardware-assisted development choices, *SemiEngineering*, http://semiengineering.com/the-agony-of-hardware-assisted-development-choices/.
13. B. Landman and R. Russo. On a pin versus block relationship for partitioning of logic graphs, *IEEE Trans. Comput.*, C20, 1469–1479, 1971.
14. J. Babb, R. Tessier, and A. Agarwal. Virtual wires: Overcoming pin limitations in FPGA-based logic emulators, in *Proceedings of IEEE Workshop on FPGA Based Custom Computing Machines*, Napa, CA, April 1993, pp. 142–151.
15. P. Christie and D. Stroobandt. The interpretation and application of Rent's rule, *IEEE Trans. VLSI Syst.*, 8, 639–648, December 2000.

16. M. Butts. Logic emulation in the MegaLUT era—Moore's law beats Rent's rule, in *ICFPT*, Hillsboro, OR, 2014, http://www.icfpt2014.org/uploadfiles/FPT14_Keynote_MButts_2Dec14.pdf.
17. J. Varghese, M. Butts, and J. Batcheller. An efficient logic emulation system, *IEEE Trans. VLSI Syst.*, 1, 171–174, 1993.
18. K. Kuusilinna et al. Designing BEE: A hardware emulation engine for signal processing in low-power wireless applications, *EURASIP J. Appl. Signal Process.*, 6, 502–513, 2003.
19. S.P. Sample et al. Emulation system with time-multiplexed interconnect, U.S. Patent 5,960,191.
20. W.F. Beausoleil et al. Multiprocessor for hardware emulation, U.S. Patent 5,551,013.
21. W.F. Beausoleil et al. Tightly coupled emulation processors, U.S. Patent 6,035,117.
22. P. Beletsky et al. Techniques and challenges of implementing large scale logic design models in massively parallel fine-grained multiprocessor systems, in *International Conference on Computer-Aided Design*, San Jose, CA, 2013.
23. Accelera. SCE-MI Reference Manual, http://accellera.org/images/downloads/standards/sce-mi/SCE_MI_v21-110112-final.pdf.
24. Cadence. Cadence announces breakthrough in system development to meet demands of "app-driven" electronics, http://www.cadence.com/cadence/newsroom/press_releases/pages/pr.aspx?xml=050311_sys_dev.
25. C. Dunn Mentor graphics enterprise verification platform unites Questa and Veloce for 1000× productivity gain, Mentor Graphics, 2014, https://www.mentor.com/company/news/mentor-questa-veloce-productivity-gain.
26. S. Gulizia and L. Gillette-Martin. Synopsys unveils verification continuum to enable next wave of industry innovation in software bring-up for complex SoCs, Synopsys, 2011, http://news.synopsys.com/2014-09-23-Synopsys-Unveils-Verification-Continuum-to-Enable-Next-Wave-of-Industry-Innovation-in-Software-Bring-Up-for-Complex-SoCs.

Formal Property Verification

Limor Fix, Ken McMillan, Norris Ip, and Leopold Haller

CONTENTS

20.1 INTRODUCTION

In 1994, Intel spent $500 million to recall the Pentium CPU due to a functional bug that produced erroneous floating-point division results that were off by as much as 61/1,000,000 [112]. The growing usage of hardware and software in our world and their insufficient quality have already cost enormously in terms of human life, time, and money [112,114,115]. Both hardware and software designs are experiencing a verification crisis since simulation-based logic verification is becoming less and less effective due to the growing complexity of the systems. To fight the low coverage in simulation, large systems are often decomposed into blocks and each of these blocks is simulated separately in its own environment. Only when the block-level simulations are stable will integration and full-chip/system simulation be carried out. This methodology caused a dramatic increase in the effort invested in simulation-based verification since a complete simulation environment, including tests, checkers, coverage monitors, and environment models, is developed and maintained for each block. Industrial reports show that half of the total design effort is devoted to verification today, the number of bugs is growing exponentially, and hundreds of billions of simulation cycles are consumed [1]. Despite the huge investment in verification, only a negligible percentage of all possible executions of the design is actually being tested.

Formal property verification enjoys major advantages over simulation. It offers an exhaustive verification technology, which is orders of magnitude more efficient than scalar exhaustive simulation. In formal verification, all possible executions of the design are analyzed and no test inputs are required. The main deficiency of formal verification is its limited capacity compared with simulation. Around the early 2000s, the state-of-the-art formal property verification engines could handle models on the order of 10K memory elements, and with complementary automatic model, reduction techniques properties were verified on hardware designs and software programs with 100–500K memory elements. Further advances in the techniques since then have enabled formal property verification even at the system-on-a-chip (SoC) level and also for postsilicon debugging, handling designs with millions of gates and memory elements.

To give the reader some initial intuition into the formal verification approach, let us consider the simple program:

```
Int x;
read (x);  if x < 0 then x = -x;  print(x);
```

Assume that x is a 32-bit integer with a range of 2^{32} different values. To verify fully that for every legal input the program prints a nonnegative number, simulation needs to cover the entire input space, that is, 2^{32} simulation runs. In a formal verification approach, a single symbolic simulation run provides full coverage of the input space and thus exhaustively verifies the program. In this example, the input to the symbolic simulation is a set of n Boolean variables representing the input bits and the result of the symbolic simulation is a Boolean expression that reflects the program computation over the input variables. The resulting Boolean expression is then checked to be nonnegative.

A *property* is a specification of some aspect of a system's behavior that is considered necessary, but perhaps not sufficient for correct operation. Properties can range from simple *sanity checks* to more complex functional requirements. For example, we might specify that a mutually exclusive pair of control signals is never asserted simultaneously or we might specify that a packet router chip connected to the PCI bus always conforms to the rules for correct signaling on the bus. In principle, we could even go so far as to specify that packets are delivered correctly, but full functional specifications of this sort are rare, because of the difficulty in verifying them. More often, formal verification is restricted to a collection of properties considered to be of crucial importance or difficult to check adequately by simulation.

In hardware design, formal property verification is used throughout the design cycle.

As Table 20.1 illustrates, a hardware design project consists of five major simultaneous activities (rows), that is, register-transfer level (RTL), timing, circuit, layout, and postsilicon. Over the duration of the project (columns), each activity has several consecutive subactivities. For example, the RTL activity starts with microarchitectural (uArch) development, which is followed by RTL development, and it ends with RTL validation. At the early stages of the project when the architecture and the microarchitecture are designed, formal models of new ideas and protocols are developed and formally verified against a set of requirements. For example, the formal verification of a new cache coherence protocol may be established at this stage. This verification is usually carried out by the validation team working closely with the architects. An important benefit from this activity is the intimate familiarity of the validation team, early at the design stage, with the new microarchitecture. Later, while the RTL is being developed, the formal property verification activity shifts to RTL verification. The properties to be verified at this stage are derived either from the high-level microarchitectural specification, for example, the IEEE specification of the arithmetic operations, or from the implementation constraints that need to be verified on the RTL, for example, verifying that all timing paths are within the required range. Constraints consumed by synthesis, static timing analysis, and power analysis are formally verified to be valid, and thus formal verification is offering not only better functional correctness but also debugging of implementation problems. Formal property verification of the RTL was mostly carried out by validation teams around the early 2000s. However, with the increased capacity of the formal verification technology and increased ease of use, RTL and circuit designers have recently started to use formal property verification products. Toward the end of the project and even after tape-out, the focus of formal property verification moves back to the verification of the high-level microarchitecture protocols—but this time, these properties are verified against the RTL and not against artificial formal models. It is still a big effort to verify high-level microarchitecture protocols against the RTL—only a few high-risk areas in the design are selected to be verified and the verification is carried out by the validation teams.

TABLE 20.1   Formal Property Verification in Hardware Design Flow

Project Phases	Design	Stabilization	Implementation	Convergence	Debug
RTL	uArch development	RTL development	RTL validation		
Timing	Timing specification	Full timing specification	Converging timing to project goals	Timing converged	
Circuit	Schematic entry	25% schematic	100% schematic	Only bug fix in schematic	
Layout			Layout clean at block level	Layout assembled and clean	
Postsilicon				A0 tape-out	Functional silicon
	Verify microarchitectural properties on FV models				
		Verify RTL properties			
FPV			Verify microarchitectural properties on the RTL		

In the electronic hardware industry, tools for formal property verification have been developed since the early 1990s in internal CAD groups of big VLSI companies like Intel, IBM, Motorola, Siemens [2,3], and Bell Labs [4]. In the late 1990s and early 2000s, several startup companies, like @ HDL, 0-in, Real Intent, Verplex, and Jasper, were offering similar tools. The large electronic design automation (EDA) companies are also offering products for formal property verification as of 2015.

In advanced electronic systems, software and hardware are becoming more and more interchangeable, interacting in different ways both in advanced electronic systems and in the system design process. Let us look at two examples. One common usage of software is in high-level modeling of complex systems. These systems are designed using high-level software languages like C and SystemC [5], and only later in the design cycle, parts of the system (or all of it) are designated to be implemented in hardware. Another example of common usage of software is in the design of embedded systems. These systems usually consist of embedded software that executes on top of an embedded microprocessor or, increasingly, processors. Modern CPUs may also include and run embedded software called "microcode" that is used to simplify the instruction set implemented in hardware.

Simulation is currently the main verification tool for software. However, the increased frequency of fatal and catastrophic software errors is driving the electronic design community to look for better, more exhaustive verification solutions. During the 5 years or so after 2000, intensive efforts have been directed to finding new and fully automatic ways to apply formal methods to software verification.

Many applications of formal verification to software have been analyzed in various research and development efforts, including two areas worthy of special recognition. One is *property verification of software*, which is aimed at establishing the functional correctness of the software with respect to properties developed by the programmer and/or automatically extracted sanity properties that check, for example, array bounds [6–11]. The second application of formal verification to software is *translation validation*, which is aimed at verifying the correct translation of the high-level software model into the corresponding hardware description, for example, Verilog, or into a lower-level software program [12–15].

In the early 2000s, formal property verification of software just started to move into the industry—most promising results were reported by the research team in Microsoft, verifying Windows NT drivers. As early as 2002, Bill Gates pointed out in a keynote speech: "Things like even software verification, this has been the Holy Grail of computer science for many decades but now in some very key areas, for example, driver verification we're building tools that can do actual proof about the software and how it works in order to guarantee the reliability" [16]. Since then, Microsoft has started shipping formal software verification tools as part of their Static Driver Verifier Platform for device vendors [88]. Formal verification tools have also been successfully used to prove the absence of critical runtime errors in large-scale avionics control software [105].

Similarly, in the early 2000s, formal validation of automatic translation of software programs was also at the initial transfer stage in the electronics industry. For example, Intel has developed a formal tool for verifying correspondence between microcode programs [15]. A few EDA startups, like Calypto, were offering C-to-Verilog translation validation in the early 2000s as well. In the same timeframe, results were reported for academic systems demonstrating the ability to verify 10,000 lines of source code in C against the compiler results [17] and being able to verify an academic microprocessor with 550 C lines against its RTL implementation with 1200 latches [14].

20.2 FORMAL PROPERTY VERIFICATION METHODS AND TECHNOLOGIES

The static nature of an RTL design provides a structured environment for development of highly automated formal verification engines and techniques. In this section, we introduce the concept of formal property specification and explain the underlying mathematical techniques for several commonly used approaches, including the automata-theoretic approach, symbolic model checking (SMC), satisfiability analysis, interpolation, symbolic simulation, and theorem proving.

20.2.1 FORMAL PROPERTY SPECIFICATION

Hardware designs and embedded software generally fall into a class known as *reactive systems*. As defined by Pnueli [18], these are systems that interact continuously with their environment, receiving input and producing output. This is in contrast to a computer program, such as a compiler, that receives input once, then executes to termination, producing output once. To specify a reactive system, we need to be able to specify the valid input/output *sequences*, not just the correct function from input to output. For this purpose, Pnueli proposed the use of a formalism known as *temporal logic* that had previously been used to give a logical account of how temporal relationships are expressed in natural language. Temporal logic provides operators that allow us to assert the truth of a proposition at certain times relative to the present time. For example, the formula *Fp* states that *p* is true at *some* time in the future, while *Gp* says that *p* is true at *all* times in the future. These operators allow us to express succinctly a variety of commonly occurring requirements of reactive systems. We can state, for example, that it is never the case that signals grant1 and grant2 are asserted at the same time or that if req1 is asserted now, eventually (at some time in the future) grant1 is asserted. The former is an example of a *safety* property. It says that some bad condition never occurs. The latter property is an example of a *liveness* property. It says that some good condition must eventually occur. To reason about liveness properties, we must consider infinite executions of a system, since the only way to violate the property is to execute infinitely without occurrence of the good condition.

Since Pnueli introduced the use of temporal logic for specification, a number of notations have been developed to make specifications either more expressive or more succinct. For example, we may want to specify that a condition must occur not just eventually, but within a given amount of time. In a real-time temporal logic [19], the formula $F_{(<5)}p$ might be used to specify that *p* must become true within five time units. Moreover, because hardware systems are typically pipelined, it is common to refer to the conditions occurring in consecutive clock cycles. For this purpose, it is more convenient to use a notation similar to regular expressions. For example, using the property specification language (PSL), a standard in the EDA industry, we can say "if req1 is followed by grant1, then eventually dav is asserted" like this:

```
(req1; gnt1) |-> F dav
```

The semicolon represents sequencing. Such notations also allow us to express some "counting" properties that have no expression in ordinary temporal logic. For example, we can state that p holds in every other clock cycle like this: (p; true)*.

Modern PSLs used in EDA provide some additional features that are convenient for hardware specifications. For example, one can specify a particular signal as the "clock" by which time is measured in a given property, and one can specify a "reset" condition that effectively cancels the requirements of the property. The latter is useful for dealing with reset and exception conditions.

20.2.1.1 HISTORY OF TEMPORAL SPECIFICATION LANGUAGES

Pnueli suggested the use of temporal logic in the late 1970s [20]. Starting in the early 1980s, a variety of logics and notations have been used in research on automatic verification, including linear temporal logic, computational tree logic, CCS, dynamic logic, and temporal logic of actions (TLAs) [20–29]. Some of these notations, notably CCS and TLA, are used for both modeling systems and specifying properties. In the 1990s, property languages specialized to EDA were developed. These include ForSpec, developed at Intel [30], Sugar from IBM [31], the temporal "e" language from Cadence (formerly Verisity), and CBV from Motorola. These languages were donated to the Accellera standards body and have been the basis for defining a new IEEE standard called PSL (IEEE standard 1850-2010) [32,79].

20.2.2 FORMAL VERIFICATION TECHNOLOGIES AND ENGINES FOR HARDWARE DESIGNS

Having specified properties of a system in a suitable notation, we would then like to be able to verify these properties formally, in as automated a manner as possible. Fully automated verification of temporal properties is generally referred to as "model checking," in reference to the first such technique developed by Clarke and Emerson [22]. Model checking can not only verify temporal properties but also provide *counterexamples* for properties that are false. Counterexamples are traces of incorrect system behavior that are valuable in diagnosing errors.

Since hardware systems are finite-state systems, they can be verified by model checking, at least in principle. Advances in model checking algorithms have allowed checking designs of significant size.

20.2.2.1 AUTOMATA-THEORETIC APPROACH

We can think of a property as specifying a set of acceptable input/output traces of a system. If all possible traces of a given system fall into this set, then the system satisfies the property. Most algorithms for model checking work by translating the property into a *finite automaton* that accepts exactly the set of traces accepted by the property [33]. The automaton is a state graph, whose edges are labeled with input/output pairs. The *language* of the automaton is the set of input/output traces observed along *accepting paths* in this graph. By defining the notion of an accepting path in various ways, we can define different kinds of languages. For example, if we specify the legal initial and final states of an accepting path, we obtain a set of finite traces. To obtain a language of infinite traces, we can specify the initial states, and the sets of states that may occur infinitely often on an accepting infinite path. This allows us to represent liveness properties with automata. Once the property has been translated into an equivalent automaton, we need no longer concern ourselves with the property language.

To check the property of a system, we also represent the system with an equivalent automaton. We now have only to check that every trace of the system automaton is a trace of the property automaton. This can be done by combining the automaton for the complement of the property with the system automaton to produce a *product automaton*. If this automaton accepts any traces, then the system does not satisfy the property. This test can be made, in the simplest case, by searching the state graph of the product automaton for an accepting path. For finite traces, this can be accomplished by a depth- or breadth-first search, starting from the initial states. This kind of search-based model checking (referred to as "explicit state") is exemplified by systems such as COSPAN [34], SPIN [6], and Murphi [7], which are quite effective at verifying software-based protocols, cache coherence protocols, and other systems with relatively small state spaces.

20.2.2.2 IN SEARCH OF GREATER CAPACITY

The main practical difficulty with the aforementioned approach is that the state graphs of the automata can be prohibitively large. The number of states of the system automaton is, in the worst case, the number of possible configurations of the registers (or other state-holding elements) in the system, which is exponential. For most property languages, the number of states in the automaton corresponding to a property is also exponential in the size of the property text. Thus, the brute-force approach described earlier is hopeless for large hardware designs. Instead, a number of heuristic methods have been developed that avoid explicit construction of the state graph. Here, we discuss a few of these methods that are currently used in the EDA.

20.2.2.3 BINARY DECISION DIAGRAM–BASED SYMBOLIC MODEL CHECKING

The SMC approach was introduced in the late 1980s [35,75] and was first implemented in a tool called SMV [36]. In this approach, automata are not explicitly constructed. Instead, a logical formula is used to characterize implicitly the set of possible transitions between states. In this formula, variable x represents the state of register x in the current state, while x' represents the state of register x at the next time. The possible transitions of a sequential machine are easily characterized in this way as a set of Boolean equations (see Figure 20.1).

For most property languages, the property automaton can also be characterized in this way, without an exponential expansion. A breadth-first search can then be accomplished purely by operations on Boolean formulas. To do this, we use a formula to stand for the set of states that satisfies it. The set of successors P' of a set of states P can then be obtained by the following

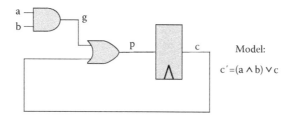

FIGURE 20.1   Characterizing transitions of a circuit.

equation, where V is the set of state variables, R the transition formula, and $\exists$ the existential quantifier over Boolean variables:

$$P'(V') = \exists V \cdot P(V) \wedge R(V, V')$$

A symbolic breadth-first search starts with a formula characterizing the initial states and iterates the aforementioned *image operation* until a fixed point occurs (i.e., no new states are obtained).

To use this idea effectively, we need a compact representation for Boolean formulas, on which Boolean operations and existential quantification can be efficiently applied. For this purpose, binary decision diagrams (BDDs) are commonly used. A BDD is a decision graph in which variables occur in the same order along all paths and common subgraphs are combined. In 1986, Bryant gave an efficient algorithm for Boolean operations on this structure [37]. SMC using BDDs provides a means of property checking for circuits of modest size (typically up to a few hundred registers) in cases where the state graph is far too large to be constructed explicitly. This method was used successfully to find errors in the cache coherence protocols of a commercial multiprocessor [38] and has since been used in a number of commercial EDA tools, with many improvements introduced in the 1990s.

20.2.2.4 SATISFIABILITY-BASED BOUNDED AND UNBOUNDED MODEL CHECKING

The Boolean satisfiability problem (SAT) seeks to determine whether there exists a truth assignment that makes a given Boolean formula true (or equivalently, whether there exists an input pattern that makes the output of a given Boolean circuit one). The first algorithm to solve this classic NP-complete problem was described in 1960 [76], and since then many improvements have been made. The success of SAT solvers such as Chaff [39], GRASP [40], and MiniSat [113] in solving very large problem instances prompted interest in using SAT solvers for model checking. In 1999, the notion of *bounded model checking* (BMC) using a SAT solver was introduced [41]. As in SMC, the transitions of the system and property automata are characterized by Boolean formulas. By *unfolding* this representation (i.e., making k consecutive copies, corresponding to consecutive time frames), the question of existence of a counterexample of k-steps can be posed as a SAT problem. This made it possible to exploit advances in SAT solvers to find counterexamples to properties for systems much larger than can be handled by SMC (typically up to a few thousand registers).

20.2.2.5 UNBOUNDED MODEL CHECKING USING A SAT SOLVER

The limitation of BMC is that it can only find counterexamples. It does not provide a practical way of proving that a property holds for all possible behaviors of a system. However, the success of BMC in handling large designs led to new methods for complete model checking that exploit SAT. One method, called k-induction [42], uses a SAT solver to test whether there is a path of k distinct good states leading to a bad state. If not and if there is no path of k-steps from an initial state to a bad state, then by induction the property holds at all times. This method can prove properties using just a SAT solver, though in practice the required k value can be quite large, making the BMC step prohibitive.

Another approach, called "localization," uses a SAT solver only to decide which components of a system are relevant to proving a particular property. This is done by using the SAT solver to refute either a partial counterexample or all counterexamples of a given length [43,44]. The components of the system used in this refutation are then used to attempt to verify the property using BDD-based SMC or other methods. Since modern SAT solvers are quite effective at ignoring irrelevant facts, this process can produce a significant reduction in the size of the system to be checked.

20.2.2.6 INTERPOLATION-BASED MODEL CHECKING

In 2003, a SAT-based algorithm for unbounded model checking based on *Craig interpolation* was introduced for hardware verification [45]. Interpolation-based model checking (IMC) succeeded BDD-based methods as the dominant method for unbounded verification. Craig interpolation can be used to approximate a symbolic image computation that directly operates over formulas. That is, given a formula representing a set of states, interpolation can generate a formula representing an approximation of the set of possible successor states. This technique allows a symbolic breadth-first search similar to BDDs, but unlike BDDs this search can be directly performed over logical formulas and is approximate rather than precise.

A Craig interpolant for an inconsistent pair of logical formulas A and B is a formula I that is implied by A and inconsistent with B and uses only variables that are common to A and B. Intuitively, we may think of I as an explanation for the inconsistency of the pair expressed in a common vocabulary.

We now detail how interpolation may be used for image computation starting from a BMC run. A formula generated in BMC consists of k linked copies of the transition system, together with constraints representing knowledge about the initial state of the system and constraints representing the property to be checked. We may split such a formula into two parts A and B as follows: The first part A contains the first copy of the transition system and the constraints imposed by the initial state. The second part B contains all other copies of the transition systems and the constraints representing the check for property violation. If no property is violated in k-steps, then the pair of formulas is inconsistent and the result of interpolation I will correspond to an approximate image operation. The interpolation I consists only of variables that are common to A and B, that is, variables that represent knowledge about states reachable in one step. It is implied by A and therefore represents at least all states reachable in one step, and it is inconsistent with B, that is, no errors can be reached from I in $k - 1$ steps.

Interpolants can be computed efficiently from refutation proofs generated by SAT solvers [89]. The ability of SAT solvers to focus on relevant parts of a problem leads to approximate images that, when used as part of a symbolic breadth-first search, are often precise enough to prove the property of interest, but imprecise enough to retain efficiency. If the images computed via interpolants are found to be too imprecise to yield a clear result during the search, the search procedure is restarted with a higher value for the k parameter.

20.2.2.7 SYMBOLIC SIMULATION

In 1990, the technique of *symbolic simulation* was introduced for hardware verification [46]. This technique resembles ordinary simulation in the sense that the user provides an input sequence to drive the design. However, in symbolic simulation, the input values can be symbolic variables as well as numeric zeros and ones. Thus, a single simulation run may in effect represent many runs of the design. The outputs of the simulation are Boolean formulas over the symbolic variables instead of numeric values. The outputs can then be compared to the desired functions. This comparison can be done using BDDs or a SAT solver. The complexity of this check can be reduced by introducing logical *unknown* (X) values at inputs that are considered irrelevant to the case being verified. By itself, symbolic simulation does not prove properties because, like BMC, it simulates only bounded runs of the system. However, complete property verification is possible using the *symbolic trajectory evaluation* methodology, based on symbolic simulation [47]. This method requires the user to provide part of the proof, but has the advantage that it can be applied to fairly large designs.

20.2.2.8 THEOREM-PROVING METHODS

From the 1950s into the 1980s, a great deal of progress was made in mechanizing mathematics and logical deduction. Classic papers in the field include [48–55]. Beginning in the 1970s with the Boyer–Moore theorem prover NQTHM [56] and LCF [57], many practical tools have been developed that assist the user in formalizing and proving mathematical theorems. Later systems include ACL2 [68], PVS [59], and HOL [60]. These systems usually provide some form of mechanized *proof search* to aid in the construction of proofs. The proof search mechanism may be fully automatic, as in NQTHM, or customizable by writing *tactics*, as in LCF and its descendants.

To use a theorem prover for formal property verification, we translate both the system model and the desired property into the logic of the prover. This can be done, for example, by automatically translating the program into a logical assertion that characterizes its transition behavior (a *shallow embedding*) or by treating the program itself as an object in the formalism and writing an interpreter for the programming language in the logic (a *deep embedding*). In either case, verifying the property reduces to proving an appropriate theorem in the logic.

To prove properties of complex systems with a theorem prover, a considerable amount of user guidance is required, in the form of lemmas, tactics, or manual guidance of the deduction process. The most common proof approach is to construct an *inductive invariant* of the system (or of the system augmented with auxiliary structures). An inductive invariant is a fact that is true initially, and implies itself at the next time, thus is true always. Once a suitable invariant is obtained, the proof that it is inductive can be aided by the use of a *ground decision procedure* [61]. This approach greatly reduces the required amount of proof guidance. However, manually constructing an inductive invariant is still a time-consuming and intellectually taxing process.

For this reason, a number of tools that combine a theorem prover and a model checker have been developed. In such systems, the overall proof of a property can be reduced to lemmas that can be discharged by a model checker. For example, the HOL-VOSS system [62] and its successor Forte [63] use a theorem prover to reduce the proof to lemmas that can be checked by symbolic simulation. The Cadence SMV system [64] can be used to reduce the proof to verification of temporal logic properties by SMC. Such systems have the potential to shift a substantial amount of the proof effort from the user onto automated tools. To make this possible, however, we require an overall proof strategy that reduces the problem to a scale that can be handled by model checking.

20.2.3 MODULARIZATION AND ABSTRACTION

The most common proof approaches for dealing with complex systems are *modularization* and *abstraction*. By modularization, we mean breaking a property to be proved (a *goal*) into two or more properties (or *subgoals*) that are more easily verified automatically, usually because they depend on fewer system components. We must show, in turn, that the subgoals imply the goal. In a strictly modular approach, such as the *assume-guarantee paradigm* [65], we verify a system consisting of two modules A and B by proving properties of A and B in isolation and then combining these to deduce a property of the system. Strictly modular proofs can be difficult to obtain, however, because of the need to capture in the subgoals all properties of one module that are required by the other. In practice, one can take a more relaxed approach. It is necessary only that the subgoals be provable in a coarser *abstraction* of the system than the original goal. The abstraction may be obtained by localization (i.e., considering only a subset of system components) or by various methods of abstracting data (including *predicate abstraction*, which is described later in this section).

In fact, the key to all known methods for verifying large, complex systems is *abstraction*. Abstraction may involve replacing a complex system with a simpler system that captures the required properties or focusing on certain relevant aspects of the system under analysis and ignoring others. An abstract system can be constructed manually, in which case we must prove that the abstract system in fact preserves the properties of interest. Alternatively, an abstraction can be constructed automatically, perhaps according to parameters given by the user. A general framework for this approach is provided by *abstract interpretation* [66], which was developed in the context of software verification. In this framework, the user chooses a suitable representation for facts about data in the system, and an automatic analysis computes the strongest inductive invariant of the system that can be expressed in this representation. As an example, we might use a representation that can express all linear affine relations between program variables. An important instance of abstract interpretation is *predicate abstraction* [67]. Here, the user provides a set of simple predicates, like $x < y$. The analysis constructs the strongest inductive invariant that can be expressed as a Boolean combination of these predicates.

Of course, the invariant constructed by abstract interpretation may or may not be strong enough to prove the desired property. The key to this approach is to choose an abstraction appropriate to the given property. Thus, in predicate abstraction, the predicates must be chosen carefully so that they are sufficient to prove the property, without being so numerous as to make the analysis intractable. Considerable attention has been given to automating the selection of predicates [10,68]. This has proved to be especially effective in software model checking as we will see. Abstract interpretation has also been effectively used to prove the subgoals arising in a modularization approach [64]. Here, the key is to choose subgoals that can be proved using a coarse abstraction. By this approach, much of the work of constructing invariants can be done automatically, thus simplifying the manually constructed part of the proof.

20.3 SOFTWARE FORMAL VERIFICATION

The dynamic nature of software poses a big challenge to the application of formal verification. Programs may contain dynamic memory or process allocations, dynamic loops, dynamic jump targets, aliases, and pointers. Whereas the state space of a discrete hardware system is finite (and every hardware model is therefore, at least in principle, amenable to automatic verification), the state space of a software program may be infinite and the respective formal verification problem is, in general, undecidable. In practice, this means that any reliable software verification algorithm will either target a restricted class of programs or will fail to produce a clear yes or no result in some cases. The following is a brief survey of automatic software verification techniques based on model checking.

20.3.1 EXPLICIT MODEL CHECKING OF PROTOCOLS

Around 1991–1992, protocols were verified using dedicated modeling languages like Promela [6] and Murphi [7] and explicit model checkers. This approach best fits concurrent asynchronous software systems. The explicit model checking algorithm performs a search on the state space of the system, and on the fly verifies for each newly visited state that the property holds for it. The algorithm stores all the states it encounters in a large hash table. When a state is generated that is already in the hash table, the search does not expand to its successor states. To allow efficient search of the state space, several optimization techniques have been developed. The most dominating one is the partial order reduction. Not all possible successors of a given state are generated and included in the search. However, states and flows of the program that are not visited in the search are proven to be redundant for establishing the correctness of the property. For example, if the program has two processes *P1:: a; b;* (in process *P1*, statement *a* is performed first and then statement *b*) and *P2:: c; d*. Then there are six possible execution flows (orders) *abcd*, *acbd*, *acdb*, *cdab*, *cabd*, and *cadb*. However, if the correctness of the property to be checked is only influenced by the order of execution of the statements *b* and *d*, the search can be limited to check only two flows: *acbd* and *acdb*. This technique has been proven to be successful for protocols that have high degree of concurrency and thus partial reduction can be very beneficial. In particular, industrial cache coherence protocols, cryptographic, and security-related protocols were successfully verified. Known academic and industrial model checkers based on this family of techniques are SPIN [6], Murphi [7], and TLA+ [69]. The main drawback of this approach is that the verification is not applied to the "golden model" of the design, that is, to the actual C and Java programs.

20.3.2 EXPLICIT AND SYMBOLIC MODEL CHECKING OF JAVA AND C PROGRAMS

In 1999–2001, Java programs were abstracted using abstract interpretation [66] and translated into finite-state models to be verified using explicit and symbolic model checkers [8,9,70,71]. Similar technology has been proposed for translating C programs using *predicate abstraction* [67] into Boolean programs and submitting them to a symbolic model checker [10]. The extracted Boolean program has the same control-flow structure as the original program; however, it has only Boolean variables, each representing a Boolean predicate over the variables of the original program. For example, if x and y are complex data structures with the same type, for example, structures, then a Boolean variable in the extracted program may represent the predicate $x = y$, t thus resulting in a much smaller state space in the Boolean program.

To give the reader some intuition into predicate abstraction, let x be an integer variable in the original C program and let the Boolean program have only two predicates that refer to x, "the predicate *b1* is defined as $x > 2$" and "the predicate *b2* is defined as $x = 5$."

Original program	Boolean program
pc1: x = 5;	*pc1: b1 = true; b2 = true;*
pc2:...	*pc2:...*

In this case, it is easy to see that if x is assigned 5 in the original program, then both the predicates *b1* and *b2* should become *true* in the corresponding program location, *pc1*, in the Boolean program. A more interesting case is as follows:

Original program	Boolean program
pc1: x = x+1;	*pc1: b1 = ?; b2 = ?;*
pc2:...	*pc2:...*

In this case, since *b1* represents $x > 2$, *b1* should be assigned *true* if $x + 1 > 2$, that is, if at program location *pc1*, $x > 1$ holds. The condition $x > 1$ is called the weakest precondition that would guarantee that $x > 2$ holds after the assignment $x = x + 1$. Similarly, since *b2* represents $x = 5$, *b2* should be assigned *true* if $x + 1 = 5$, that is, if at program location *pc1*, $x = 4$ holds. However, at location *pc1* in the Boolean program, we only have access to the truth value of *b1* and *b2* and not to the values of the predicates $x > 1$ and $x = 4$, which are not included in the set of predicates used for the generation of the Boolean program. In this case, an automated decision procedure is invoked to strengthen these predicates into an expression over the set of the chosen predicates. In our example, the predicate $x > 1$ will be strengthen to $x > 2$ and the predicate $x = 5$ will strengthen into *false*. Thus, we get the following:

Original program	Boolean program
pc1: x = x+1;	*pc1: b1 = if b1 then true else false;*
	b2 = false;
pc2:...	*pc2:...*

The predicate abstraction technology has been extended to handle pointers, procedures, and procedure calls [10]. For example, let the predicate *b1* be **p > 5*:

Original program	Boolean program
pc1: x = 3;	*pc1: b1 = ?;*
pc2:...	*pc2:...*

If *x* and **p* are aliases, then *b1* should be assigned *false* since **p* becomes *3*, else *b1* should retain its previous value. Thus, we get the following:

Original program	Boolean program
pc1: x = 3;	pc1: b1 = (if &x = p then false else retain);
pc2:...	pc2:...

The SMC algorithm is applied to the Boolean program to check the correctness of the property. If successful, we can conclude that the property holds also for the original program. If failed, the Boolean program is refined to more precisely represent the original program. Additional Boolean predicates are automatically identified from the counterexample and a new Boolean program is constructed. This iterative approach is called counterexample-guided abstraction refinement (CEGAR) [77]. The CEGAR methodology has proven highly influential in the field of formal verification. It was successfully applied to checking safety properties of Windows device drivers and to discovering invariants regarding array bounds. Academic and industrial model checkers based on this family of techniques include SLAM and SLAM2 by Microsoft [10,88], which became part of the Static Driver Verifier toolset used internally and distributed to device vendors, Bandera [70], Java Pathfinder [9], TVLA [72], Feaver [73], Java-2-SAL [74], and Blast [68].

20.3.3 BOUNDED MODEL CHECKING OF C PROGRAMS

Around 2003, C and SpecC programs were translated into propositional formulae and then formally verified using a bounded model checker [11,14]. This approach can handle the entire ANSI-C language consisting of pointers, dynamic memory allocations, and dynamic loops, that is, loops with conditions that cannot be evaluated statically. These programs are verified against user-defined safety properties and also against automatically generated properties about pointer safety and array bounds.

The C program is first translated into an equivalent program that uses only *while, if, go to,* and *assignment* statements. Then each loop of the form *while (e) inst* is translated into *if (e) inst; if (e) inst; ...; if (e) inst; {assertion !e}*, where *if (e) inst* is repeated *n* times. The assertion *!e* is later formally verified and if it does not hold, *n* is increased until the assertion holds. The resulting program that has no loops, and the properties to be checked are then translated into a propositional formula that represents the model after unwinding it *k* times. The resulting formula is submitted to a SAT solver, and if a satisfying assignment is found, it represents an error trace. During the *k*-steps unwinding of the program, pointers are handled. For example,

```
int a, b, p;
if (x)  p=&a; else p=&b;
*p=5;
```

The aforementioned program is first translated into

```
p1=(x ? &a: p0) ∧p2=(x? p1: &b)
*p2=5;
```

That is, three copies of the variable *p* are created: *p0, p1,* and *p2*. If *x* is true, then *p1=&a* and *p2=p1,* and if *x* is *false, p1=p0* and *p2=&b*. Then **p2=5* is replaced by

*(x? p1; &b)=5	that is replaced by
(x ? *p1; *&b)=5	that is replaced by
(x? *p1; b)=5	that is replaced by
(x? *(x? &a;p0); b)=5	that is replaced by
(x? *&a; b)=5	that is replaced by
(x? a; b)=5	

After all these automatic transformation, the pointers have been eliminated and the resulting statement is $(x? a; b) = 5$. For details on dynamic memory allocations and dynamic loops, see [11].

20.3.4 TRANSLATION VALIDATION OF SOFTWARE

Two technological directions are currently pursued for formally verifying the correct translation of software programs. One [12,17] that automatically establishes an abstraction mapping between the source program and the object code offers an alternative to the verification of synthesizers and compilers. The other direction [13,14] automatically translates the two programs given in C and Verilog into a BMC formula and submits it to a SAT solver. The value of each Verilog signal at every clock cycle is visible to the C program. Thus, the user can specify and formally verify the desired relation between the C variables and the Verilog signals. Both the C and Verilog programs are unwound for k-steps as we have described in the previous section.

20.4 RECENT DEVELOPMENTS (2004–2015)

In the years between 2004 and 2015, formal property verification has evolved significantly and has become an indispensible complement to traditional verification methods in the RTL functional verification domain. Ball et al. [88] provides good information from the 2008 perspective. Many design projects have adopted formal property verification as an integral part of the verification methodology—it is no longer a point tool and it no longer requires PhD-level expertise to perform formal property verification in commercial design projects.

Model checking algorithms using SAT solvers and interpolation techniques have become widely used in formal verification software, and new approaches such as IC3/property directed reachability (PDF) and satisfiability modulo theory (SMT) have made formal verification possible in large designs in a commercial setting. Academic interest in formal verification has shifted toward verification of software in recent years.

The availability of several standards has also contributed to the increased proliferation of formal property verification, including standards for property languages, SoC metadata, low-power metadata, and coverage measurements.

20.4.1 PROPERTY SPECIFICATION

Section 20.2.1.1 reviewed the early history of temporal specification languages as of 2005. Since then, there have been several major standardization efforts in this area:

- PSL [79]: Property specification language provides a formal notation for specifying behaviors of a design. It was initiated by Accellera with version 1.0 released in 2003 and eventually approved as IEEE 1850 in 2005 with a revision in 2010.
- SVA [80]: SystemVerilog assertion is part of the SystemVerilog standard, initially approved as IEEE 1800 in 2005, with a revision in 2012. It comprises a subset of the SystemVerilog language that permits specification of properties declaratively outside a procedural context or embedded in procedural code.
- OVL [81]: Open Verification Library is a library of property checkers written in hardware description languages including Verilog, VHDL, SystemVerilog, and PSL with Verilog flavor. Since the official release of OVL 1.0 in 2005 by Accellera, it has continued to be evolved by the committee, with OVL Version 2.8 released in December 2013.

While OVL provides the convenience of specifying assertions and properties using predetermined library elements, PSL and SVA provide comprehensive language constructs for the specification of design behaviors.

Assertion monitors in OVL are composed of an *event, message,* and *severity.* For example, the *assert_never* module in OVL takes a clock signal, a reset signal, and a Boolean expression as the event. A severity level is provided as the parameter for the instantiated module, and an optional message string is the string displayed in the case of an assertion failure. More information can be found in Chapter 18.

PSL is defined in four layers: the Boolean layer, the temporal layer, the verification layer, and the modeling layer. Figure 20.2 illustrates an assertion, specifying that it is always the case (i.e., for all time) that whenever the "req" signal is asserted, either the "read_ack" signal or the "write_ack" signal must be asserted in the next cycle (captured through the construct "next"). The PSL constructs "assert" and "always" belong to the verification and temporal layers, respectively, and the expression "read_ack || write_ack" belongs to the Boolean layer. For this example, to illustrate the concept of modeling, the "req" signal is trivially created from a combination of signals found in the design. PSL uses the concept of verification unit to group and organize PSL properties. Because of the modeling layers, two favors of PSL are available, one for Verilog and one for VHDL.

SVA contains several components: sequences, declarations, directives, and bindings. Figure 20.3 shows an example of an assertion similar to the one in Figure 20.2. The SVA construct "|=>" describes "next cycle"; the "property" construct provides the declaration; and the "assert" construct provides the *directive* regarding how to use the property in a verification setting. Binding (not shown in the example) is used to bind these properties to RTL modules in the design. Since SVA is designed to be used directly as a part of SystemVerilog designs and testbenches, it inherits the expression language of SystemVerilog, including its data types, expression syntax, and semantics.

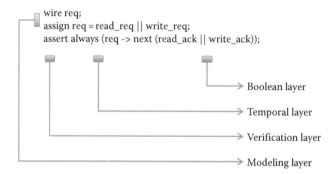

FIGURE 20.2 Property specification language assertion.

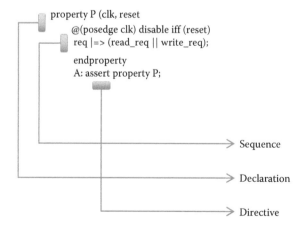

FIGURE 20.3 SystemVerilog assertion.

20.4.2 PROOF ENGINES AND ABSTRACTIONS

In this section, we highlight the recent evolutions of formal verification engines for hardware and software. In particular, the field of SMT has seen significant advances since its introduction from 2000.

20.4.2.1 FORMAL VERIFICATION ENGINES FOR HARDWARE DESIGNS

Although academic interest in formal verification has slowly shifted toward software verification over the past 10 years, several major breakthroughs occurred with regard to core model checking algorithms:

- IMC was introduced in 2003 [45] and succeeded BDDs as a dominant verification methodology for unbounded verification. Other applications of interpolation to verification were identified, such as interpolation for predicate selection as part of the predicate abstraction loop or for the approximation of transition relations [89].
- PDR, originally called IC3, was introduced in 2010 [90]. IC3/PDR is a SAT-based verification algorithm that incrementally constructs an inductive invariant for the system under analysis using a large number of easy-to-solve single-step SAT queries.
- The system under analysis is represented by a number of frames. Each frame is associated with a set of logical formulas that represent knowledge about reachable states at the corresponding depth of execution. In the main phase, the algorithm uses SAT queries to find counterexamples to induction (CTIs). A CTI is a set of states at a certain depth of execution that may lead to a property violation and that is not excluded by the current state of knowledge about the system.
- Once a CTI is generated, the algorithm attempts to prove that it does not correspond to a property violation. In the process of doing so, additional proof obligations may be generated at earlier execution depths, which are processed recursively. Once a CTI has been shown not to correspond to a property violation, additional knowledge is added to the system, which excludes the CTI from being considered again at a later point during the search. The algorithm is guaranteed to eventually find a counterexample or converge on an inductive invariant that proves safety of the system.
- Today, IC3/PDR is considered the strongest single-engine verification procedure in hardware verification and has spawned intense research interest, both in terms of low-level improvements and efficient implementation techniques [91] and with regard to the underlying paradigm of inductive, incremental verification.
- Work on modern proof engines such as BMC, IMC, and IC3/PDR has focused overwhelmingly on safety properties. In 2002, the Biere translation was introduced [92], which allows liveness properties to be translated to safety properties. The translation implicitly checks for the presence of an infinite counterexample by attempting to find looping parts of the state space and involves a duplication of the number of state-holding circuit elements. The k-liveness methodology was introduced in 2012 [93] and provides an alternative method for reducing liveness to a sequence of safety checks. The k-liveness method often works better than the Biere translation but cannot be used for finding counterexamples.

20.4.2.2 SATISFIABILITY MODULO THEORIES

Since the introduction of Chaff [39] and Grasp [40] in 2000, algorithmic and engineering advances have continued to increase the performance of SAT solvers at an exponential rate until at least 2007 [94].

The field of SMT attempts to generalize the advances of propositional solvers to fragments of first-order logic interpreted over a given first-order theory. An SMT solver can handle logical formulas that contain variables and operations over some nonpropositional target domain. Popular first-order theories supported by SMT solvers include linear and rational integer arithmetic, equality with uninterpreted functions, bit vectors, and theories for specific data types such as floating-point numbers or arrays. An informative overview over SMT technology is provided in [95]. Modern solvers include Z3 [96], MathSat [97], Yices [98], and CVC4 [99]. The SMT-LIB project [100] standardizes first-order theories handled by solvers as well as file formats.

One of the central developments in SMT is the DPLL(T) framework [101,102], which lifts the architecture of modern propositional SAT solvers to incorporate support for first-order theories. The core idea is to split the satisfiability task among two communicating solvers: a SAT solver is used to reason about the propositional structure of the problem and handles basic Boolean operators, while a theory solver reasons about conjunctions of atomic subformulas. The SAT solver represents theory atoms abstractly as Boolean variables and generates candidate truth assignments. The theory solver is used to check whether these candidate assignments are satisfied by some first-order model in the theory. Theory solvers may also communicate inferred truth values to the propositional solver or help learn new propositional information. SMT solvers may combine multiple different theory solvers into a single solver for the combined theory [103,104].

20.4.2.3 FORMAL VERIFICATION TECHNOLOGIES FOR SOFTWARE

Academic interest in formal verification has shifted toward verification of software in recent years. In part, this shift is driven by developments in logical decision procedures, which made available effective decision procedures for first-order theories, which are used to reason about programs. We now summarize some recent developments in software verification:

- Mature tools based on *abstract interpretation* were developed and used to analyze large-scale system. The Astrée abstract interpreter [105] has been used to prove absence of runtime errors in projects consisting of more than 100,000 lines of automatically generated C code. Abstract interpretation–based tools have also proven effective in the context of floating-point verification [106].
- Novel *software model checking* algorithms were developed, including the interpolation-based algorithm implemented in the Impact model checker [107] and extensions of the PDR procedure to software [108,109]. Modern software model checkers typically combine a variety of verification techniques such as predicate abstraction [67], abstract interpretation [66], BMC [41], and interpolation [107]. Significant advances were made also made in proving that programs terminate (an overview over these efforts is given in [110]) and in the analysis of heap-manipulating programs using tools based on separation logic [111].

20.4.3 NEW STANDARDS, APPLICATIONS, AND USE MODELS

Early formal verification approaches had a reputation of being difficult to use, and mostly requiring dedicated formal verification experts in the design project in order to attain its goals. However, between 2004 and 2014, new standards, new applications, and new use models have enabled the usage of formal verification by people with minimal formal verification background.

This section highlights some of these changes. The need to manually write properties for formal verification has been significantly reduced thanks to verification IPs (VIPs), property generation, metadata for SoC assembly, and metadata for low power. New use models have emerged, such as design comprehension and debugging. These use models hide the formal algorithm details from the users and provide benefit to people not interested in achieving exhaustive proofs of properties. Finally, coverage models for formal verification provide metrics of verification completeness even when exhaustive proofs are impractical.

20.4.3.1 VERIFICATION IPs AND PROPERTY GENERATION

The availability of verifiable properties is critical to the deployment of formal property verification techniques. Writing properties in standard languages such as SVA or PSL demands specialized knowledge. Even for experts it can be extremely time consuming to manually create a robust set of properties for a design.

Two complementary approaches address this need:

- *VIPs* that contain properties for standard interfaces
- *Property generation* techniques that extract properties from RTL and/or traces either manually or automatically

Specifications of standard protocols, such as AMBA (AHB, APB, AXI, ACE), PCI-Express, DDR, LPDDR, and OCP, can often be captured precisely as properties, including assertions, assumptions, and functional coverage points. These VIPs can be used both for formal property verification and for simulation.

VIPs intended for formal property verification are typically configurable into three modes: master, slave, and monitor. These modes determine which properties in the VIP should be interpreted as assumptions or assertions. For example, assertions from the master mode would be used as assumptions in the slave mode, and assumptions from the master mode would be used as assertions in the slave mode. These assertions and assumptions in the master and slave modes would all be used as assertions in the monitor mode.

Using simulation to verify compliance of a design with respect to standard interfaces is challenging and time consuming; with formal property verification, these VIPs help exhaustively verify the design with respect to the compliance of the protocol. The assumptions in these VIPs provide the necessary constraints for verification of other properties with respect to the design, by configuring the proof engines to avoid returning counterexamples that violate the protocol assumptions.

Property generation techniques help create properties directly from an RTL description of the design. Before a testbench is available, properties can be extracted directly from the logic in the RTL, using predefined functional checks such as dead code checks, finite-state machine (FSM) checks, and arithmetic overflow checks. Once simulation tests are running with the design, properties can be deduced from the collections of waveforms, identifying behaviors common to multiple waveforms, and coverage holes never exercised in any of the waveforms. The properties generated from the process could be focusing on multicycle temporal properties, cross-hierarchical properties, FIFOs, counters, FSMs, one-hot candidate, etc.

It is critical in the property generation flow to enable user guidance to control the generation process. Uncontrolled property generation creates numerous property candidates, and it is hard for the user to identify the important properties among the other candidates that can be considered as noise. Heuristics to rank and classify the property candidates enable the user to focus on the important items first, and the concept of waiver to pinpoint clearly illegal or impossible cases, is necessary to avoid generating similar properties during analysis in the future. Furthermore, user-provided templates allow the generation process to provide only properties of the form that the user is interested in.

20.4.3.2 METADATA FOR SoC AND LOW-POWER DESIGNS

As formal property verification became indispensible in many design projects, many automated or dedicated applications have been discovered along the way. For example, metadata for SoC and low-power designs is a good source of properties to be formally verified against the design.

In recent years, many standards for metadata format have been created:

- *IPXACT* [82]: Initiated by the SPIRIT Consortium in 2003 and approved as IEEE 1685-2009 in 2009, IPXACT enables the capture of metadata related to a design, its components, the related bus interfaces and connections, abstractions of those buses, and details of the components including address maps, register, and field descriptions.
- *SystemRDL* [83]: Initiated by the SPIRIT Consortium in 2009, SystemRDL focuses on the metadata related to the design and delivery of registers to be used in IP blocks in an SoC.
- *Common power format (CPF)* [84]: Initiated by Silicon Integration Initiative in 2007, CPF captures the designer's intent for power management, including information on

power modes, power domains, power switches, state retention registers, isolation cells, and level shifters.

- *Unified power format (UPF)* [85]: Initiated by Accellera in 2007 and approved as IEEE 1801-2009, UPF is also designed to capture the designer's intent for power management, similar to CPF. It continued to evolve into IEEE 1801-2013 and IEEE 1801a-2014, and an upcoming revision will significantly reduce the differences from CPF.

More information on IPXACT can be found in Chapter 5.

Given an IPXACT or SystemRDL description, formal property verification techniques can be used to exhaustively determine whether the register behaviors in the design description have been implemented according to the parameters specified in such metadata. While the fundamental read/write operations to the registers through an interface (known as software access) can be captured as properties easily, there are wide variations of register styles, potentially causing the resulting properties to be very complicated. Some of the complexity includes hardware modification of the register, modified write value semantics, register aliasing, and locking and security mechanisms. The complexity of such variations is the main reason why formal property verification has become more prominent in register behavior verification.

An IPXACT description also provides instantiation and connectivity information within an SoC. Instead of writing simulation tests for connectivity verification, formal property verification has become the main approach for verifying connectivity in many projects. While IPXACT enables specification of the simple use cases, such as direct point-to-point connections and interface-to-interface connections, vendor extensions have been made in an ad hoc fashion to capture more complicated connections, with conditional behavior, temporal latency, etc.

The availability of low-power metadata has enabled low-power-related formal verification on an RTL description that does not contain low-power implementation yet. After reading in the RTL description, the low-power metadata can be used to convert this RTL description to a power-aware formal model in accordance with power partitioning specification, such as

- Adding corruption to the nonretained data when power is off
- Adding extra logic to retain the value of a register if retention has been specified in the low-power metadata
- Adding extra logic to isolate the power-off logic from surrounding power domain if isolation has been specified in the low-power metadata

Using the power-aware formal model, formal property verification can be performed on

- The behavior of power-related properties extracted from the low-power metadata, which is impossible to analyze in the original RTL description
- Nonpower-related properties with the power-related artifacts introduced by the low-power transformation, which may behave differently from the RTL without the low-power artifacts

20.4.3.3 DESIGN COMPREHENSION AND DEBUGGING

Even though the primary purpose of formal property verification is to determine the validity of properties with respect to a design description, it has recently been used to aid design comprehension and debugging, as described in [86].

Given a functional coverage point, formal property verification can generate a trace exercising the coverage point if the coverage point is reachable from the reset state(s). Rearranging the problem statement so that the formal verification algorithm receives a set of targets with optional temporal relationships, an interesting waveform exercising the design can be created without setting up a simulation test bench. Instead of tuning the inputs generated from a simulation test bench to hit specific targets, the formal algorithm enforces such targets to be exercised in the generated waveform.

On the other hand, properties used for formal property verification can be used for comprehension of a waveform from simulation and from formal verification of other properties. Given a waveform, an assertion (or an assumption) can be evaluated to see whether it has been violated anywhere in the waveform and whether its triggers have been exercised anywhere in the waveform. For a functional cover point specified as a property, it can be evaluated to see if it has been exercised anywhere in the waveform. These properties can be displayed as transactions in the waveform display, and the specific clock cycles of the relevant signals can be highlighted to aid comprehension of the waveform.

20.4.3.4 COVERAGE METRICS FOR FORMAL VERIFICATION ACTIVITIES

With more verification projects utilizing both simulation and formal property verification as two complementary approaches to verify the same design, it is often important to combine results from both approaches into a coherent measurement of verification progress. Unified branch and statement code coverage models have been the first step toward such an effort. More information on this topic can be found in Chapter 18.

Finding a coherent metric for verification progress is one of the goals for the UCIS working group. This group was formed in November 2006 in Accellera for the purpose of defining an application interface [87]. The standard allows for interoperability of verification coverage data across multiple tools from multiple vendors, including symbolic simulation and formal tools. The UCIS 1.0 standard was released in June 2012.

The EDA industry has suggested several metrics of coverage with formal property verification:

- *Dead code and stimuli coverage*: Measure the stimuli coverage by checking what is reachable in the design under a given set of constraints, and determine dead code based on analysis in an underconstrained environment, that is, while considering all legal input sequences and some illegal input sequences.
- *Proof coverage and property completeness*: Determine the completeness of the set of properties and the effectiveness of each property. One of the techniques used in such measurement is to extract the proof core using the proof-based abstraction technique, that is, from the portion of the design that affects the validity of the proof.
- *Bounded proof coverage*: Measure quality of a bounded proof when exhaustive analysis is not available. A simple reachability bound is not a good measurement of the quality of a bounded proof—a simple counter within the design can drive a formal algorithm to reach a large reachability bound, but the activities analyzed within such a bound may be few and not useful. One of the techniques used in such measurement is to analyze other design cover points with respect to the reachability bound of a bounded proof. The more design cover points can be exercised with the current constraints and within the reachability bound, the more design scenarios would have been analyzed by such a bounded proof. This provides a functional interpretation of the bound with respect to what the design can do and cannot do within the reachability bound.

20.5 SUMMARY

Formal property verification provides a means of ensuring that a hardware or software system satisfies certain key properties, regardless of the input presented. Automated methods developed over the last decades have made it possible to verify properties of large, complex systems with minimal user interaction, and to find errors in these systems. These methods have become important adjuncts to simulation and testing in the hardware design verification process, enabling the design of more robust and reliable systems. Software formal verification has evolved into a mature field. Verification techniques are finding increasing adoption in the software industry and constitute an ongoing focus of academic research.

REFERENCES

1. B. Bentley, Validating the Intel Pentium 4 microprocessor, in *Proceedings of the 38th Design Automation Conference (DAC)*, ACM Press, New York, 2001, pp. 244–248.
2. G. Kamhi, O. Weissberg, L. Fix, Z. Binyamini, and Z. Shtadler, Automatic datapath extraction for efficient usage of HDD, in *Proceedings of the Ninth International Conference Computer Aided Verification (CAV)*, Lecture Notes in Computer Science, Vol. 1254, Springer, Berlin, Germany, 1997, pp. 95–106.
3. D. Geist and I. Beer, Efficient model checking by automated ordering of transition relation partitions, in *Proceedings of Computer-Aided Verification (CAV)*, Springer-Verlag, Berlin, Germany, 1994.
4. R.P. Kurshan, Formal verification on a commercial setting, in *Proceedings of Design Automation Conference (DAC)*, 1997, pp. 258–262.
5. T. Grotker, S. Liao, G. Martin, and S. Swan, *System Design with SystemC*, Kluwer Academic Publishers, Dordrecht, the Netherlands, 2002.
6. G.J. Holzmann, *Design and Validation of Computer Protocols*, Prentice-Hall, Englewood Cliffs, NJ, 1991.
7. D.L. Dill, A.J. Drexler, A.J. Hu, and C. Han Yang, Protocol verification as a hardware design aid, in *IEEE International Conference on Computer Design: VLSI in Computers and Processors*, IEEE Computer Society, Washington, DC, 1992, pp. 522–525.
8. C. Demartini, R. Iosif, and R. Sisto, A deadlock detection tool for concurrent Java programs. *Softw.—Practice Exp.*, 29, 577–603, 1999.
9. K. Havelund and T. Pressburger, Model checking Java programs using Java PathFinder, *Int. J. Software Tools Technol. Trans.*, 2, 366–381, 2000.
10. T. Ball, R. Majumdar, T. Millstein, and S.K. Rajamani, Automatic predicate abstraction of C programs, PLDI 2001, *SIGPLAN Notices*, 36, 203–213, 2001.
11. E. Clarke, D. Kroening, and F. Lerda, A tool for checking ANSI-C program, in *Proceedings of 10th International Conference, on Tools and Algorithms for Construction and Analysis of Systems (TACAS)*, Lecture Notes in Computer Science, Vol. 2988, Springer, Berlin, Germany, 2004.
12. A. Pnueli, M. Siegel, and O. Shtrichman, The code validation tool (CVT)-automatic verification of a compilation process, *Int. J. Software Tools Technol. Trans.*, 2, 192–201, 1998.
13. E. Clarke and D. Kroening, Hardware verification using ANSI-C programs as a reference, in *Proceedings of ASP-DAC 2003*, IEEE Computer Society Press, Washington, DC, 2003, pp. 308–311.
14. E. Clarke, D. Kroening, and K. Yorav, Behavioral consistency of C and Verilog programs using bounded model checking, in *Proceedings of the 40th Design Automation Conference (DAC)*, ACM Press, New York, 2003, pp. 368–371.
15. L. Fix, M. Mishaeli, E. Singerman, and A. Tiemeyer, MicroFormal: FV for Microcode. Intel Internal Report, 2004.
16. Bill Gates, Keynote address at WinHec, Seattle, WA, 2002.
17. L. Zuck, A. Pnueli, Y. Fang, and B. Goldberg, Voc: A translation validator for optimizing compilers, *J. Univ. Comput. Sci.*, 3, 223–247, 2003. Preliminary version in ENTCS, 65, 2002.
18. A. Pnueli, Specification and development of reactive systems, in *Information Processing 86*, Kugler, H.-J. (ed.), IFIP, North-Holland, Amsterdam, the Netherlands, 1986, pp. 845–858.
19. R. Alur and T.A. Henzinger, Logics and models of real time: A survey, in *Real Time: Theory in Practice*, de Bakker, J.W., Huizing, K., de Roever, W.-P., and Rozenberg, G. (eds.), Lecture Notes in Computer Science, Vol. 600, Springer, Berlin, Germany, 1992, pp. 74–106.
20. A. Pnueli, The temporal logic of programs, in *Proceedings of the 18th Symposium on Foundations of Computer Science*, 1977, pp. 46–57.
21. M. Benari, Z. Manna, and A. Pnueli, The temporal logic of branching time, *Acta Inform.*, 20, 207–226, 1983.
22. E.M. Clarke and E.A. Emerson, Design and synthesis of synchronization skeletons using branching time temporal logic, in *Proceedings of Workshop on Logic of Programs*, Lecture Notes in Computer Science, Vol. 131, Springer, Berlin, Germany, 1981, pp. 52–71.
23. M.J. Fischer and R.E. Lander, Propositional dynamic logic of regular programs, *J. Comput. Syst. Sci.*, 18, 194–211, 1979.
24. L. Lamport, Sometimes is sometimes "not never"—On the temporal logic of programs, in *Proceedings of the Seventh ACM Symposium on Principles of Programming Languages*, January 1980, pp. 174–185.
25. B. Misra and K.M. Chandy, Proofs of networks of processes, *IEEE Trans. Software Eng.*, 7, 417–426, 1981.
26. R. Milner, *A Calculus of Communicating Systems*, Lecture Notes in Computer Science, Vol. 92, Springer, Berlin, Germany, 1980.
27. A. Pnueli, The temporal semantics of concurrent programs, *Theor. Comput. Sci.*, 13, 45–60, 1981.
28. J.P. Queille and J. Sifakis, Specification and verification of concurrent systems in Cesar, in *Proceedings of Fifth International Symposium on Programming*, Lecture Notes in Computer Science, Vol. 137, Springer, Berlin, Germany, 1981, pp. 337–351.

29. P. Wolper, Synthesis of communicating processes from temporal logic specifications, PhD thesis, Stanford University, Stanford, CA, 1982.

30. R. Armoni et al., The ForSpec temporal logic: A new temporal property specification language, in *Proceedings of TACAS*, 2002, pp. 296–311.

31. I. Beer, S. Ben-David, C. Eisner, D. Fisman, A. Gringauze, and Y. Rodeh, The temporal logic sugar, in *Proceedings of Conference on Computer-Aided Verification (CAV'01)*, Lecture Notes in Computer Science, Vol. 2102, Springer, Berlin, Germany, 2001, pp. 363–367.

32. Property Specification Language Reference Manual, Accellera, 2003.

33. M.Y. Vardi and P. Wolper, An automata-theoretic approach to automatic program verification, in *Proceedings of IEEE Symposium on Logic in Computer Science*, Computer Society Press, Cambridge, 1986, pp. 322–331.

34. R.P. Kurshan, *Computer-Aided Verification of Coordinating Processes*, Princeton University Press, Princeton, NJ, 1994.

35. O. Coudert, C. Berthet, and J.C. Madre, Verification of synchronous sequential machines based on symbolic execution, in *Proceedings of International Workshop on Automatic Verification Methods for Finite State Systems*, Lecture Notes in Computer Science, Vol. 407, Springer, Berlin, Germany, 1989.

36. K.L. McMillan, Symbolic model checking: An approach to the state explosion problem, PhD thesis, Carnegie Mellon University, Pittsburgh, PA, CMU CS-929131, 1992.

37. R.E. Bryant, Graph-based algorithms for Boolean function manipulation, *Proc. IEEE Trans. Comp.*, C-35, 677–691, 1986.

38. K.L. McMillan and J. Schwalbe, Formal verification of the gigamax cache consistency protocol, in *International Symposium on Shared Memory Multiprocessing*, 1991, pp. 242–251.

39. M.W. Moskewicz, C.F. Madigan, Y. Zhao, L. Zhang, and S. Malik, Chaff: Engineering an efficient SAT solver, in *Proceedings of the 38th Design Automation Conference (DAC'01)*, June 2001.

40. J.P. Marques-Silva and K.A. Sakallah, GRASP: A new search algorithm for satisfiability, *IEEE Trans. Comput.*, 48, 506–521, 1999.

41. A. Biere, A. Comatti, E. Clarke, and Y. Zhu, Symbolic model checking without BDDs, in *Proceedings of the Workshop on Tools and Algorithms for the Construction and Analysis of Systems (TACAS'99)*, Lecture Notes in Computer Science, Springer, Berlin, Germany, 1999.

42. M. Sheeran, S. Singh, and G. Stalmarck, Checking safety properties using induction and a sat-solver, in *Proceedings of International Conference on Formal Methods in Computer Aided Design (FMCAD 2000)*, Hunt, W.A. and Johnson, S.D. (eds.), 2000.

43. P. Chauhan, E.M. Clarke, J.H. Kukula, S. Sapra, H. Veith, and D. Wang, Automated abstraction refinement for model checking large state spaces using SAT-based conflict analysis, in *Proceedings of FMCAD 2002*, Lecture Notes in Computer Science, Vol. 2517, Springer, Berlin, Germany, 2002, pp. 33–51.

44. K.L. McMillan and N. Amla, Automatic abstraction without counterexamples, in *Proceedings of Tools and Algorithms for the Construction and Analysis of Systems (TACAS'03)*, Lecture Notes in Computer Science, Vol. 2619, Springer, Berlin, Germany, 2003, pp. 2–17.

45. K.L. McMillan, Interpolation and SAT-based model checking, in *Proceedings of CAV 2003*, Lecture Notes in Computer Science, Vol. 2725, Springer, Berlin, Germany, 2003, pp. 1–13.

46. R.E. Bryant and C.-J. Seger, Formal verification of digital circuits using symbolic ternary system models, in *Workshop on Computer Aided Verification. Second International Conference, CAV'90*, Clarke, E.M. and Kurshan, R.P. (eds.), Lecture Notes in Computer Science, Vol. 531, Springer, Berlin, Germany, 1990.

47. D.L. Beatty, R.E. Bryant, and C.-J. Seger, Formal hardware verification by symbolic ternary trajectory evaluation, in *Proceedings of the 28th ACM/IEEE Design Automation Conference*, IEEE Computer Society Press, Washington, DC, 1991.

48. S.C. Kleene, *Introduction to Metamathematics*, Van Nostrand, Princeton, NJ, 1952.

49. J. McCarthy, Computer programs for checking mathematical proofs, in *Proceedings of the Symposia in Pure Mathematics*, Vol. V, Recursive Function Theory, American Mathematical Society, Providence, RI, 1962.

50. J.A. Robinson, A machine-oriented logic based on the resolution principle, *J. ACM*, 12, 23–41, 1965.

51. D. Scott, Constructive validity, in *Symposium on Automatic Demonstration*, Lecture Notes in Mathematics, Vol. 125, Springer, New York, 1970, pp. 237–275.

52. C.A.R. Hoare, Notes on data structuring, in *Structural Programming*, Dahl, O.J., Dijkstra, E.W., and Hoare, C.A.R. (eds.), Academic Press, New York, 1972, pp. 83–174.

53. R. Milner and R. Weyhrauch, Proving computer correctness in mechanized logic, in *Machine Intelligence*, Meltzer, B. and Michie, D. (eds.), Vol. 7, Edinburgh University Press, Edinburgh, Scotland, 1972, pp. 51–70.

54. G.P. Huet, A unification algorithm for types lambda-calculus, *Theor. Comput. Sci.*, 1, 27–58, 1975.

55. D.W. Loveland, *Automated Theorem Proving: A Logical Basis*, North-Holland, Amsterdam, the Netherlands, 1978.

56. R.S. Boyer and J.S. Moore, *A Computational Logic*, Academic Press, New York, 1979.

57. M. Gordon, R. Milner, and C.P. Wadsworth, *Edinburgh LCF: A Mechanised Logic of Computation*, Lecture Notes in Computer Science, Vol. 78, Springer-Verlag, Berlin, Germany, 1979.

58. M. Kaufmann and J. Moore, An industrial strength theorem prover for a logic based on common lisp, *IEEE Trans. Software Eng.*, 23, 203–213, April 1997.

59. S. Owre, J. Rushby, and N. Shankar, PVS: A prototype verification system, in *Proceedings of 11th International Conference on Automated Deduction*, Kapur, D. (ed.), Lecture Notes in Artificial Intelligence, Vol. 607, Springer, Berlin, Germany, 1992, pp. 748–752.

60. M.J.C. Gordon and T.F. Melham (eds.), *Introduction to HOL: A Theorem Proving Environment for Higher Order Logic*, Cambridge University Press, London, U.K., 1993.

61. G. Nelson and D.C. Oppen, Simplification by cooperating decision procedures, *ACM Trans. Prog. Lang. Syst.*, 1, 245–257, 1979.

62. J. Joyce and C.-J.H. Seger, The HOL-Voss system: Model-checking inside a general-purpose theorem-prover, in *Proceedings of the Sixth International Workshop on Higher Order Logic Theorem Proving and Its Applications*, 1993, pp. 185–198.

63. M. Aagaard, R.B. Jones, and C.-J.H. Seger. Lifted-FL: A pragmatic implementation of combined model checking and theorem proving, in *Theorem Proving in Higher Order Logics* (*TPHOLs*), Bertot, Y., Dowek, G., and Hirschowitz, A. (eds.), Springer, Berlin, Germany, 1999, pp. 323–340.

64. K.L. McMillan, A methodology for hardware verification using compositional model checking, *Sci. Comput. Prog.*, 37, 279–309, 2000.

65. A. Pnueli, In transition from global to modular temporal reasoning about programs, in *Logics and Models of Concurrent Systems, Sub-series F: Computer and System Science*, Apt, K.R. (ed.), Springer-Verlag, Berlin, Germany, 1985, pp. 123–144.

66. P. Cousot and R. Cousot, Abstract interpretation: A unified lattice model for static analysis of programs by construction and approximation of fixpoints, in *Proceedings of Fourth ACM Symposium on Principles of Programming Languages*, 1977, pp. 238–252.

67. S. Graf and H. Saidi, Construction of abstract state graphs with PVS, in *CAV97: Computer-Aided Verification*, Lecture Notes in Computer Science, Vol. 1254, Springer, Berlin, Germany, 1997, pp. 72–83.

68. T.A. Henzinger, R. Jhala, R. Majumdar, and G. Sutre, Lazy abstraction, in *Proceedings of the 29th Annual Symposium on Principles of Programming Languages* (*POPL*), ACM Press, 2002, pp. 58–70.

69. Y. Yu, P. Manolios, and L. Lamport, Model checking TLA1 specifications, in *Proceedings of Correct Hardware Design and Verification Methods*.

70. J. Corbett, M. Dwyer, J. Hatcliff, S. Laubach, C. Pasareanu, Robby, and H. Zheng, Bandera: Extraction finite-state models from Java source code, in *Proceedings of 22nd International Conference on Software Engineering* (*ICSE2000*), June 2000.

71. M. Dwyer, J. Hatcliff, R. Joehanes, S. Laubach, C. Pasareanu, W. Visser, and H. Zheng, Tool supported program abstraction for finite-state verification, in *ICSE 01: Software Engineering*, 2001.

72. T. Lev-Ami and M. Sagiv, TVLA: A framework for Kleene-based static analysis, in *Proceedings of the Seventh International Static Analysis Symposium*, 2000.

73. G. Holzmann, Logic verification of ANSI-C code with SPIN, in *Proceedings of Seventh International SPIN Workshop*, K. Havelund (ed.), Lecture Notes in Computer Science, Vol. 1885, Springer, Berlin, Germany, 2000, pp. 131–147.

74. D.Y.W. Park, U. Stern, J.U. Skakkebaek, and D.L. Dill, Java model checking, in *Proceedings of the First International Workshop on Automated Program Analysis, Testing and Verification*, June 2000.

75. J.R. Burch, E.M. Clarke, K.L. McMillan, D.L. Dill, and L.J. Hwang, Symbolic model checking: 10^{20} states and beyond, in *Proceedings of the Fifth Annual IEEE Symposium on Logic in Computer Science*, IEEE Computer Society Press, Washington, DC, 1990, pp. 1–33.

76. M. Davis and H. Putnam, A computing procedure for quantification theory, *J. ACM*, 7, 201–215, 1960.

77. E.M. Clarke, O. Grumberg, S. Jha, Y. Lu, and H. Veith, Counterexample-guided abstraction refinement, in *Computer Aided Verification, LNCS 1855, 12th International Conference, CAV 2000*, Emerson, E.A. and Sistla, A.P. (eds.), Springer, Chicago, IL, July 15–19, 2000, pp. 154–169.

78. O. Grumberg and H. Veith (eds.), *25 Years of Model Checking: History, Achievements, Perspectives*, Lecture Notes in Computer Science, Springer, Berlin, Germany, 2008.

79. 1850-2010—IEEE Standard for Property Specification Language (PSL), IEEE, 2010.

80. 1800-2012—IEEE Standard for SystemVerilog—Unified Hardware Design, Specification, and Verification Language, IEEE, 2012.

81. Open Verification Library (OVL) Version 2.8, Accellera Systems Initiative, December 2013.

82. 1685-2014—IEEE Standard for IP-XACT, Standard Structure for Packaging, Integrating, and Reusing IP within Tool Flows, IEEE, 2014.

83. SystemRDL 1.0, SPIRIT Consortium, 2009.

84. Si2 Common Power Format Specification, Version 2.0, Silicon Integration Initiative, 2011.

85. 1801a-2014—IEEE Standard for Design and Verification of Low-Power Integrated Circuits—Amendment 1, IEEE, 2014.

86. R.K. Ranjan, C. Coelho, and S. Skalberg, Beyond verification: Leveraging formal for debugging, in *Design Automation Conference*, 2009.

87. Unified Coverage Interoperability Standard (UCIS), Version 1.0, Accellera Systems Initiative, June 2012.

88. T. Ball et al., SLAM2: Static driver verification with under 4% false alarms, in *Proceedings of the 2010 Conference on Formal Methods in Computer-Aided Design*, FMCAD Inc., 2010.

89. K.L. McMillan, Applications of Craig interpolants in model checking, in *Tools and Algorithms for the Construction and Analysis of Systems*, Springer, Berlin, Germany, 2005, pp. 1–12.

90. A.R. Bradley, SAT-based model checking without unrolling, in *Verification, Model Checking, and Abstract Interpretation*, Springer, Berlin, Germany, 2011.

91. N. Een, A. Mishchenko, and R. Brayton, Efficient implementation of property directed reachability, in *Formal Methods in Computer-Aided Design (FMCAD)*, IEEE, 2011.

92. A. Biere, C. Artho, and V. Schuppan, Liveness checking as safety checking, *Electron. Notes Theor. Comp. Sci.*, 66(2), 160–177, 2002.

93. K. Claessen and N. Sorensson, A liveness checking algorithm that counts, in *Formal Methods in Computer-Aided Design (FMCAD)*, IEEE, 2012.

94. S. Malik and L. Zhang, Boolean satisfiability from theoretical hardness to practical success, *Commun. ACM*, 52(8), 76–82, 2009.

95. C.W. Barrett et al., Satisfiability Modulo theories, in *Handbook of Satisfiability*, A. Biere, M. Heule, H. Van Maaren, and T. Walsh (eds.), IOS Press, Vol. 185, 2009, pp. 825–885.

96. L. De Moura and N. Bjørner, Z3: An efficient SMT solver, in *Tools and Algorithms for the Construction and Analysis of Systems*, Springer, Berlin, Germany, 2008.

97. R. Bruttomesso, A. Cimatti, A. Franzén, A. Griggio, and R. Sebastiani, The MathSat 4 SMT solver, in *Computer Aided Verification*, Springer, Berlin, Germany, January 2008, pp. 299–303.

98. B. Dutrerte, Yices 2.2, in *Proceedings of Computer-Aided Verification (CAV)*, Springer-Verlag, 2014.

99. C. Barrett et al., CVC4, in *Computer Aided Verification*, Springer, Berlin, Germany, January 2011, pp. 171–177.

100. C. Barrett, A. Stump, and C. Tinelli, The SMT-LIB standard: Version 2.0, in *Proceedings of the Eighth International Workshop on Satisfiability Modulo Theories*, Vol. 13, Edinburgh, England, 2010.

101. H. Ganzinger, G. Hagen, R. Nieuwenhuis, A. Oliveras, and C. Tinelli, DPLL (T): Fast decision procedures, in *Computer Aided Verification*, Springer, Berlin, Germany, January 2004, pp. 175–188.

102. R. Nieuwenhuis, A. Oliveras, and C. Tinelli, Abstract DPLL and abstract DPLL modulo theories, in *Logic for Programming, Artificial Intelligence, and Reasoning*, Springer, Berlin, Germany, January 2005, pp. 36–50.

103. G. Nelson and D.C. Oppen, Simplification by cooperating decision procedures. *ACM Trans. Prog. Lang. Syst. (TOPLAS)*, 1(2), 245–257, 1979.

104. R.E. Shostak, Deciding combinations of theories. In *Sixth Conference on Automated Deduction*, Springer, Berlin, Germany, January 1982, pp. 209–222.

105. P. Cousot, R. Cousot, J. Feret, L. Mauborgne, A. Miné, D. Monniaux, and X. Rival, The ASTRÉE analyzer, in *Programming Languages and Systems*, Springer, Berlin, Germany, 2005, pp. 21–30.

106. D. Delmas, E. Goubault, S. Putot, J. Souyris, K. Tekkal, and F. Védrine, Towards an industrial use of FLUCTUAT on safety-critical avionics software, in *Formal Methods for Industrial Critical Systems*, Springer, Berlin, Germany, 2009, pp. 53–69.

107. K.L. McMillan, Lazy abstraction with interpolants, in *Computer Aided Verification*, Springer, Berlin, Germany, January 2006, pp. 123–136.

108. A. Cimatti and A. Griggio, Software model checking via IC3, in *Computer Aided Verification*, Springer, Berlin, Germany, January 2012, pp. 277–293.

109. K. Hoder and N. Bjørner, Generalized property directed reachability, in *Theory and Applications of Satisfiability Testing—SAT 2012*, Springer, Berlin, Germany, 2012, pp. 157–171.

110. B. Cook, A. Podelski, and A. Rybalchenko, Proving program termination, *Commun. ACM*, 54(5), 2011, 88–98.

111. J.C. Reynolds, Separation logic: A logic for shared mutable data structures, in *Proceedings of the 17th Annual IEEE Symposium on Logic in Computer Science*, IEEE, 2002, pp. 55–74.

112. V. Pratt, Anatomy of the Pentium bug, in *TAPSOFT'95: Theory and Practice of Software Development*, Springer, Berlin, Germany, 1995, pp. 97–107.

113. N. Sorensson and N. Een, Minisat v1.13—A SAT solver with conflict-clause minimization, SAT 2005 Competition, 2005.

114. J. Lions, Ariane 5 flight 501 failure, 1996.

115. US General Accounting Office, Patriot missile defense: Software problem led to system failure at Dhahran, Saudi Arabia, US GAO Reports, Report no. GAO/IMTEC-92-26, 1992.

Test

Design-for-Test

21

Bernd Koenemann and Brion Keller

CONTENTS

21.1 INTRODUCTION

Design-for-test or design-for-testability (DFT) is a name for design techniques that add certain testability features to a microelectronic hardware product design. The premise of the added features is to make it easier to develop and apply manufacturing tests for the designed hardware. The purpose of manufacturing tests is to validate that the product hardware contains no defects that could adversely affect the product's correct functioning.

Tests are applied at several steps in the hardware manufacturing flow and, for certain products, may also be used for hardware maintenance in the customer's environment. The tests generally are driven by test programs that execute in automatic test equipment (ATE) or, in the case of system maintenance, inside the assembled system itself. In addition to finding and indicating the presence of defects (i.e., the test fails), tests may be able to log diagnostic information about the nature of the encountered test failures. The diagnostic information can be used to locate the source of the failure.

DFT plays an important role in the development of test programs and as an interface for test application and diagnostics.

Historically speaking, DFT techniques have been used at least since the early days of electric/electronic data-processing equipment. Early examples from the 1940s and 1950s are the switches and instruments that allowed an engineer to "scan" (i.e., selectively) probe the voltage/current at some internal nodes in an analog computer (analog scan). DFT is often associated with design modifications that provide improved access to internal circuit elements such that the local internal state can be controlled (controllability) or observed (observability) more easily. The design modifications can be strictly physical in nature (e.g., adding a physical probe point to a net) or add active circuit elements to facilitate controllability/observability (e.g., inserting a multiplexer into a net). While controllability and observability improvements for internal circuit elements definitely are important for the test, they are not the only type of DFT. Other guidelines, for example, deal with the electromechanical characteristics of the interface between the product under test and the test equipment, for example, guidelines for the size, shape, and spacing of probe points, or the suggestion to add a high-impedance state to drivers attached to probed nets such that the risk of damage from backdriving is mitigated.

Over the years, the industry has developed and used a large variety of more or less detailed and more or less formal guidelines for desired and/or mandatory DFT circuit modifications. The common understanding of DFT in the context of electronic design automation (EDA) for modern microelectronics is shaped to a large extent by the capabilities of commercial DFT software tools as well as by the expertise and experience of a professional community of DFT engineers researching, developing, and using such tools. Much of the related body of DFT knowledge focuses on digital circuits, while DFT for analog/mixed-signal circuits takes something of a backseat. The following text follows this scheme by allocating most of the space to digital techniques.

21.2 OBJECTIVES OF DESIGN-FOR-TEST FOR MICROELECTRONICS PRODUCTS

DFT affects and depends on the methods used for test development, test application, and diagnostics. The objectives, hence, can only be formulated in the context of some understanding of these three key test-related activities.

21.2.1 TEST GENERATION

Most tool-supported DFT practiced in the industry today, at least for digital circuits, is predicated on a *structural* test paradigm. *Functional* testing attempts to validate that the circuit under test operates according to its functional specification. For example, does the adder really add for all possible operands? *Structural* testing, by contrast, makes no direct attempt to ascertain the intended functionality of the circuit under test. Instead, it tries to make sure that the circuit has been assembled correctly from some low-level building blocks as specified in a structural netlist and that all of those low-level building blocks and their wiring connections have been manufactured without defect. For example, are all logic gates there that are supposed to be there and are they connected correctly? The stipulation is that if the netlist is correct (e.g., somehow it has been fully verified against the functional specification) and structural testing has confirmed the correct assembly of the structural circuit elements, then the circuit should be functioning correctly.

One benefit of the structural paradigm is that the test generation can focus on testing a limited number of relatively simple circuit elements rather than having to deal with an exponentially exploding multiplicity of functional states and state transitions. While the task of testing a single logic gate at a time sounds simple, there is an obstacle to overcome. For today's highly complex designs, most gates are deeply embedded, whereas the test equipment is connected only to the primary I/Os and/or some physical test points. Access to the embedded gates, hence, must be obtained by sensitizing paths through intervening layers of logic. If the intervening logic contains state elements, then the issue of an exponentially exploding state space and state transition sequencing effectively poses an unsolvable problem for automatic test generation. To simplify test generation, DFT addresses the accessibility problem by removing the need for complicated state transition sequences when trying to control or observe what is happening at some internal circuit element.

Depending on the DFT choices made during circuit design/implementation, the generation of structural tests for complex logic circuits can be more or less automated. One key objective of DFT methodologies, hence, is to allow designers to make trade-offs between the amount and type of DFT and the cost/benefit (time, effort, quality) of the test generation task.

21.2.1.1 TEST APPLICATION

Complex microelectronics products typically are tested multiple times. Chips, for example, may be tested on the wafer before the wafer is diced into individual chips (wafer probe/sort) and again after being packaged (final test). More testing is due after the packaged chips have been assembled into a higher-level package such as a printed circuit board (PCB) or a multichip module (MCM). For products with special reliability needs, additional intermediate steps such as burn-in may be involved, which, depending on the flow details, may include even more testing (e.g., preburn and postburn-in test or *in situ* test during burn-in). The cost of test in many cases is dominated by the test equipment cost, which in turn depends on the number of I/Os that need to be contacted, the performance characteristics of the tester I/O (i.e., channel) electronics, and the depth/speed of pattern memory behind each tester channel. In addition to the cost of the tester frame itself, interfacing hardware (e.g., wafer probes and prober stations for wafer sort, automated pick-and-place handlers for final test, burn-in boards [BIBs] for burn-in) is needed that connects the tester channels to the circuit under test.

One challenge for the industry is keeping up with the rapid advances in chip technology (I/O count/size/placement/spacing, I/O speed, internal circuit count/speed/power, thermal control, etc.) without being forced to continually upgrade the test equipment; thus, modern DFT techniques have to offer options that allow next-generation chips and assemblies to be tested on existing test equipment and/or reduce the requirements/cost for new test equipment. At the same time, DFT has to make sure that test times stay within certain bounds dictated by the cost target for the products under test.

21.2.2 DIAGNOSTICS

Especially for advanced semiconductor technologies, it is expected that some of the chips on each manufactured wafer will contain defects that render them nonfunctional. The primary objective of testing is to find and separate those nonfunctional chips from the fully functional ones, meaning that one or more responses captured by the tester from a nonfunctional chip under test differ from the expected response. The percentage of chips that fail the test, hence, should be closely related to the expected functional yield for that chip type. In reality, however, it is not uncommon that all chips of a new chip type arriving at the test floor for the first time fail (so-called zero-yield situation). In that case, the chips have to go through a debug process that tries to identify the reason for the zero-yield situation. In other cases, the test fallout (percentage of test fails) may be higher than expected/acceptable or fluctuate suddenly. Again, the chips have to be subjected to an analysis process to identify the reason for the excessive test fallout.

In both cases, vital information about the nature of the underlying problem may be hidden in the way the chips fail during test. To facilitate better analysis, additional fail information beyond a simple pass/fail is collected into a fail log. The fail log typically contains information about when (e.g., tester cycle), where (e.g., at what tester channel), and how (e.g., at what logic value) the test failed. Diagnostics attempt to derive from the fail log the logical/physical location inside the chip at which the problem most likely started. This location provides a starting point for further detailed failure analysis (FA) to determine the actual root cause. FA, in particular physical FA, can be very time-consuming and costly, since it typically involves a variety of highly specialized equipment and an equally specialized FA engineering team. The throughput of the FA labs is very limited, especially if the initial problem localization from diagnostics is poor. That adversely affects the problem turnaround time and the number of problem cases that can be analyzed. Additional inefficiency arises if the cases handed over to the FA lab are not relevant for the tester fallout rate.

In some cases (e.g., PCBs, MCMs, embedded, or stand-alone memories), it may be possible to repair a failing circuit under test. For that purpose, diagnostics must quickly find the failing unit and create a work order for repairing/replacing the failing unit. For PCBs/MCMs, the replaceable/repairable units are the chips and/or the package wiring. Repairable memories offer spare rows/columns and some switching logic that can substitute a spare for a failing row/column. The diagnostic resolution must match the granularity of replacement/repair. Speed of diagnostics for replacement is another issue. For example, cost reasons may dictate that repairable memories must be tested, diagnosed, repaired, and retested in a single test insertion. In that scenario, the failure data collection and diagnostics must be more or less realistic as the test is applied. Even if diagnostics are to be performed offline, failure data collection on expensive production test equipment must be efficient and fast or it will be too expensive.

DFT approaches can be more or less diagnostics friendly. The related objectives of DFT are to facilitate/simplify failure data collection and diagnostics to an extent that can enable intelligent FA sample selection, as well as improve the cost, accuracy, speed, and throughput of diagnostics and FA.

21.2.3 PRODUCT LIFE-CYCLE CONSIDERATIONS

Test requirements from other stages of a chip's product life cycle (e.g., burn-in, PCB/MCM test, [sub]system test) can benefit from additional DFT features beyond what is needed for the chip manufacturing test proper. Many of these additional DFT features are best implemented at the chip level and affect the chip design. Hence, it is useful to summarize some of these additional requirements, even though the handbook primarily focuses on EDA for IC design.

21.2.3.1 BURN-IN

Burn-in exposes the chips to some period of elevated ambient temperature to accelerate and weed out early life fails prior to shipping the chips. Typically, burn-in is applied to packaged chips. Chips designated for direct chip attach assembly may have to be packaged into a temporary chip carrier for burn-in and subsequently removed from the carrier. The packaged chips are put on BIBs, and several BIBs at a time are put into a burn-in oven. In static burn-in, the chips are simply exposed to an elevated temperature, then removed from the oven and retested. Burn-in is more effective if the circuit elements on the chips are subjected to local electric fields during burn-in. Consequently, at a minimum, some chip power pads must be connected to a power-supply grid on the BIBs. The so-called dynamic burn-in further requires some switching activity, typically, meaning that some chip inputs must be connected on the BIBs and be wired out of the burn-in oven to some form of test equipment. The most effective form of burn-in, called *in situ* burn-in, further requires that some chip responses can be monitored for failures while in the oven. For both dynamic and *in situ* burn-in, the number of signals that must be wired out of the oven is of concern because it drives the complexity/cost of the BIBs and test equipment. Some of the newer technologies cannot reliably perform under the elevated temperatures of the burn-in oven, so their responses during burn-in may be ignored; these devices must be retested after burn-in is complete to identify which chips are still in operating condition.

Burn-in friendly chip DFT makes it possible to establish a chip burn-in mode with minimal I/O footprint and data bandwidth needs.

21.2.3.2 PRINTED CIRCUIT BOARD/MULTICHIP MODULE TEST

The rigors (handling, placing, heating) of assembling multiple chips into a higher-level package can create new assembly-related defects associated with the chip attach (e.g., poor solder connection) and interchip wiring (e.g., short caused by solder splash). In some cases, the chip internals may also be affected (e.g., bare chips for direct chip attach are more vulnerable than packaged chips). The basic PCB/MCM test approaches concentrate largely on assembly-related defects and at best use very simple tests to validate that the chips are still "alive."

Although functional testing of PCBs/MCMs from the edge connectors is sometimes possible and used, the approach tends to make diagnostics very difficult. In-circuit testing is a widely practiced alternative or complementary method. In-circuit testing historically has used so-called bed-of-nails interfaces to contact physical probe points connected to the interchip wiring nets. If every net connected to the chip is contacted by a nail, then the tester can essentially test the chip as if stand-alone. However, it often is difficult to prevent some other chip driving the same net that the tester needs to control for testing the currently selected chip. To overcome this problem, the in-circuit circuit tester drivers are strong enough to override (backdrive) other chip drivers. Backdriving is considered a possible danger, and reliability problem for some types of chip drivers and some manufacturers may discourage backdriving. Densely packed, double-sided PCBs or other miniaturized packages may not leave room for landing pads on enough nets, and the number and density of nets to be probed may make the bed-of-nails fixtures too unwieldy.

DFT techniques implemented at the chip level can remove the need for backdriving from a physical bed-of-nails fixture or use electronic alternatives to reduce the need for complete physical in-circuit access.

21.2.3.3 [SUB]SYSTEM SUPPORT

Early prototype bring-up and, in the case of problems, debug pose a substantial challenge in the development of complex microelectronics systems. It often is very difficult to distinguish between hardware, design, and software problems. Debug is further complicated by the fact that valuable information about the detailed circuit states that could shed light on the problem may be hidden deep inside the chips in the assembly hierarchy. Moreover, the existence of one problem (e.g., a timing problem) can prevent the system from reaching a state needed for other parts of system bring-up, verification, and debug.

System manufacturing, just like PCB/MCM assembly, can introduce new defects and possibly damage the components. The same may apply to hardware maintenance/repair events (e.g., hot plugging a new memory board).

Operating the system at the final customer's site can create additional requirements, especially if the system must meet stringent availability or safety criteria.

DFT techniques implemented at the chip level can help enable a structural hardware integrity test that quickly and easily validates the physical assembly hierarchy (e.g., chip to board to backplane) and verifies that the system's components (e.g., chips) are working. DFT can also increase the observability of internal circuit state information for debug or the controllability of internal states and certain operating conditions to continue debug in the presence of problems.

21.3 CHIP-LEVEL LOGIC DESIGN-FOR-TEST TECHNIQUES

DFT has a long history with a large supporting body of theoretical work as well as industrial application. Only a relatively small and narrow subset of the full body of DFT technology has found its way into the current EDA industry.

21.3.1 BRIEF HISTORICAL COMMENTARY

Much of the DFT technology available in today's commercial DFT tools has its roots in the electronic data-processing industry. Data-processing systems have been complex composites made up of logic, memory, I/O, analog, human interface, and mechanical components long before the semiconductor industry invented the system-on-chip (SoC) moniker. Traditional DFT as practiced by the large data-processing system companies since at least the 1960s represents highly sophisticated architectures of engineering utilities that simultaneously address the needs of manufacturing, product engineering, maintenance/service, availability, and customer support. The first commercial DFT tools for IC design fielded by the EDA industry, by contrast, were primitive scan-insertion tools that only addressed the needs of automatic test pattern generation (ATPG) for random logic. Tools have become more sophisticated and comprehensive, offering more options for logic test (e.g., built-in self-test [BIST]), support for some nonrandom logic design elements (e.g., BIST for embedded memories), and support for higher-level package testing (e.g., boundary scan for PCB/MCM testing). However, with few exceptions, there still is a lack of comprehensive DFT architectures for integrating the bits and pieces and a lack of consideration for applications besides manufacturing test (e.g., support for nondestructive memory read for debug purposes is not a common offering by the tool vendors).

21.3.2 ABOUT DESIGN-FOR-TEST TOOLS

There are essentially three types of DFT-related tools:

1. *DFT synthesis (DFTS)*. DFT involves design modification/edit steps (e.g., substituting one flip-flop type with another one) akin to simple logic transformation or synthesis. DFTS performs the circuit modification/edit task.
2. *Design rule checking (DRC)*. Chip-level DFT is mostly used to prepare the circuit for some ATPG tool or to enable the use of some type of manufacturing test equipment [29]. The ATPG tools and test equipment generally impose constraints on the design under test. DRC checks the augmented design for compliance with those constraints. Note that this DRC should not be confused with physical verification DRC of ICs, as is discussed in Chapter 20 of *Electronic Design Automation for IC Implementation, Circuit Design, and Process Technology.*
3. *DFT intellectual property creation, configuration, and assembly*. In addition to relatively simple design modifications, DFT may add test-specific function blocks to the design. Some of these DFT blocks can be quite sophisticated and may rival some third-party IP in complexity. And, like other IP blocks, the DFT blocks often must be configured for a particular design and then assembled into the design.

21.3.3 CHIP DESIGN ELEMENTS AND ELEMENT-SPECIFIC TEST METHODS

Modern chips can contain different types of circuitry with vastly different type-specific testing needs. Tests for random logic and tests for analog macros are very different. For example, tests for random logic and tests for analog macros are very different. DFT has to address the specific needs of each such circuit type and also facilitate the integration of the resulting type-specific tests into an efficient, high-quality composite test program for all pieces of the chip. SoC is an industry moniker for chips made up of logic, memory, analog/mixed-signal, and I/O components. The main categories of DFT methods needed, and to a reasonable extent commercially available, for today's IC manufacturing test purposes can be introduced in the context of a hypothetical SoC.

SoCs are multiterrain devices consisting of predesigned and custom design elements:

- Digital logic, synthesized (e.g., cell based) or customized (e.g., transistor level)
- Embedded digital cores (e.g., processors)
- Embedded memories (static RAM [SRAM], embedded DRAM [eDRAM], read-only memory [ROM], content addressable memory [CAM], and flash, with or without embedded redundancy)
- Embedded register files (large number, single port, and multiport)
- Embedded field-programmable gate array (eFPGA)
- Embedded analog/mixed signal (phase-locked loop [PLL]/DLL, DAC, ADC)
- High-speed I/Os (e.g., SerDes)
- Conventional I/Os (large number, different types, some differential)

The following overview will introduce some key type-specific DFT features for each type of component and then address chip-level DFT techniques that facilitate the integration of the components into a top-level design.

21.3.4 DIGITAL LOGIC

The most common DFT strategies for digital logic help prepare the design for ATPG tools. ATPG tools typically have difficulties with hard-to-control or hard-to-observe nets/pins, sequential depth, and loops.

21.3.4.1 CONTROL/OBSERVE POINTS

The job of an ATPG tool is to locally set up suitable input conditions that *excite* a fault (i.e., trigger an incorrect logic response according to the fault definition; e.g., to trigger a stuck-at-1 fault at a particular logic gate input, that input must receive a logic 0 from the preceding gates) and that *propagate* the incorrect value to an observable point (i.e., side inputs of gates along the way must be set to their noncontrolling values). The runtime and success rate of test generation depend not least on the search space the algorithm has to explore to establish the required excitation and propagation conditions.

Control points provide an alternative means for the ATPG tool to more easily achieve a particular logic value. In addition to providing enhanced controllability for test, it must be possible to disable the additional logic such that the original circuit function is retained for normal system operation. In other words, DFT often means the implementation of multiple distinct modes of operation, for example, a *test mode* and a *normal mode*.

The second control point type is mostly used to override unknown/unpredictable signal sources, in particular for signal types that impact the sequential behavior, for example, clocks. In addition to the two types of control points, there are other types for improved 1-controllability (e.g., using an OR gate) and for *randomization* (e.g., using an exclusive OR [XOR] gate). The latter type, for example, is useful in conjunction with pseudorandom test methods that will be introduced later. As can be seen from the examples, the implementation of control points tends to add cost due to one or more additional logic levels that affect the path delay and require additional area/power for transistors and wiring. The additional cost for implementing DFT is generally referred to as "overhead," and over the years there have been many, sometimes heated, debates juxtaposing the overhead against the benefits of DFT.

Observe points are somewhat *cheaper* in that they generally do not require additional logic in the system paths. The delay impact, hence, is reduced to the additional load posed by the fan-out and (optional) buffer used to build a path to an observation point.

21.3.4.2 SCAN DESIGN

Scan design is the most common DFT method associated with synthesized logic. The concept of scan goes back to the very early days of the electronics industry, and it refers to certain means for controlling and/or observing otherwise hidden internal circuit states. Examples are manual dials to connect measurement instruments to probe points in analog computers, the switches, and lights on the control panel of early digital computers (and futuristic computers in sci-fi flicks) and to more automated electronic mechanisms to accomplish the objective, for example, the use of machine instructions to write or read internal machine registers. Beginning with the late 1960s or so, scan has been implemented as a dedicated, hardware-based operation that is independent of, and does not rely on, specific intelligence in the intended circuit function [14,20,27,55].

Among the key characteristics of scan architectures are the choice of which circuit states to control/observe and the choice of an external data interface (I/Os and protocol) for the control/observe information. In basic scan methods, all (full scan) or most (partial scan) internal sequential state elements (latches or flip-flops) are made controllable and observable via a serial interface to minimize the I/O footprint required for the control/observe data. The most common implementation strategy is to replace the functional state elements with dual-purpose state elements (scan cells) that can operate as originally intended for functional purposes and as a serial shift register for scan. The most commonly used type of scan cell consists of an edge-triggered flip-flop with two-way multiplexer (scan mux) for the data input (mux-scan flip-flop).

The scan mux is typically controlled by a single control signal called scan_enable that selects between a scan-data and a system data input port. The transport of control/observe data from/to the test equipment is achieved by a serial shift operation. To that effect, the scan cells are connected into serial shift register strings called scan chains. The scan-in port of each cell is either connected to an external input (scan-in) for the first cell in the scan chain or to the output of a single predecessor cell in the scan chain. The output from the last scan cell in the scan chain must be connected to an external output (scan-out). The data input port of the scan mux is connected to the functional logic as needed for the intended circuit function.

There are several commercial and proprietary DFTS tools available that largely automate the scan-chain construction process. These tools operate on register-transfer-level (RTL) and/or gate-level netlists of the design. The tools typically are driven by some rules on how to substitute nonscan storage elements in the prescan design with an appropriate scan cell and how to connect the scan cells into one or more scan chains. In addition to connecting the scan and data input ports of the scan cells correctly, attention must be given to the clock input ports of the scan cells. To make the shift registers operable without interference from the functional logic, a particular circuit state (*scan state*), established by asserting designated scan state values at certain primary inputs and/or by executing a designated initialization sequence, must exist that

1. Switches all scan muxes to use the scan side (i.e., the local scan_enable signals are forced to the correct value)
2. Assures that each scan cell clock pin is controlled from one designated external clock input (i.e., any intervening clock gating or other clock manipulation logic is overridden/disabled)
3. All other scan cell control inputs like set/reset are disabled (i.e., the local control inputs at the scan cell are forced to their inactive state)
4. All scan-data inputs are sensitized to the output of the respective predecessor scan cell or the respective scan_in port for the first scan cell in the chain (i.e., side inputs of logic gates along the path are forced to a nondominating value and muxes select the scan-data path)
5. The output of the last scan_cell in the scan chain is sensitized to its corresponding scan_out port or the side inputs of logic gates along the path are forced to a nondominating value, such that pulsing the designated external clock (or clocks) once results in shifting the data in the scan chains by exactly one-bit position

This language may sound pedantic, but the DFTS and DFT DRC tools tend to use even more detailed definitions for what constitutes a valid scan chain and scan state. Only a crisp definition allows the tools to validate the design thoroughly and, if problems are detected, write error messages with enough diagnostic information to help a user find and fix the design error that caused the problem.

Very few modern chips contain just a single scan chain. In fact, it is fairly common to have several selectable scan-chain configurations, typically referred to as *test modes*. The reason is that scan can be used for a number of different purposes. Facilitating manufacturing test for synthesized logic is one purpose. In that case, the scan cells act as serially accessible control and observe points for logic test; test application essentially follows a protocol such as the following:

1. Establish the manufacturing test-scan mode and associated scan state, and serially load the scan chains with new test input conditions.
2. Switch out of the scan state into the functional path capture state (typically done by switching scan_enable to the system side of the scan mux).
3. Apply any other primary input conditions required for the test.
4. Wait until the circuit stabilizes, and measure/compare the external test responses at primary outputs.
5. Capture the internal test responses into the scan cells by pulsing one or more clocks.
6. Reestablish the manufacturing test-scan mode's associated scan state, and serially unload the test responses from the scan chains into the tester for comparison.

Steps 1 and 6 in many cases can be overlapped, meaning that while the responses from one test are unloaded through the scan-out pins, new test input data are simultaneously shifted in from the scan-in pins. The serial load/unload operation requires as many clock cycles as there are scan cells in the longest scan chain. Manufacturing test time and consequently test cost for scan-based logic tests are typically dominated by the time used for scan load/unload. Hence, to minimize test times and cost, it is preferable to implement as many short, parallel scan chains as possible. The limiting factors are the availability of chip I/Os for scan-in/scan-out or the availability of test equipment channels suitable for scan. Modern DFT tools can help optimize the number of scan chains and balance their length according to the requirements and constraints of chip-level manufacturing test. Today's scan-insertion flows also tend to include a postplacement *scan reordering* step to reduce the wiring overhead for connecting the scan cells. The currently practiced state of the art generally limits reordering to occur within a scan chain. Research projects have indicated that further improvements are possible by allowing the exchange of scan cells between scan chains. All practical tools tend to give the user some control over partial ordering, keeping subchains untouched, and placing certain scan cells at predetermined offsets in the chains.

In addition to building the scan chains proper, modern DFT tools also can insert and validate *pin-sharing* logic that makes it possible to use functional I/Os as scan-in/scan-out or scan control pins, thus avoiding the need for additional chip I/Os dedicated to the test. In many practical cases, a single dedicated pin is sufficient to select between normal mode and test mode. All other test control and interface signals are mapped onto the functional I/Os by inserting the appropriate pin-sharing logic.

Besides chip manufacturing test, scan chains often are also used for access to internal circuit states for higher-level assembly (e.g., board-level) testing. In this scenario, it generally is not possible or economically feasible to wire all scan-in/scan-out pins used for chip testing out to the board connectors. Board-level wiring and connectors are very limited and relatively *expensive*. Hence, the I/O footprint dedicated to scan must be kept at a minimum and it is customary to implement another scan configuration in the chips, wherein all scan cells can be loaded/unloaded from a single pair of scan-in/scan-out pins. This can be done by concatenating the short scan chains used for chip manufacturing test into a single, long scan chain or by providing some addressing mechanism for selectively connecting one shorter scan chain at a time to the scan-in/scan-out pair. In either case, a *scan-switching network* and associated control signals are required to facilitate the reconfiguration of the scan interface.

In many practical cases, there are more than two scan configurations to support additional engineering applications beyond chip and board-level manufacturing test, for example, debug or system configuration [13,35]. The scan architectures originally developed for large data-processing systems in the 1960s and 1970s, for example, were designed to facilitate comprehensive engineering access to all hardware elements for testing, bring-up, maintenance, and diagnostics. The value of comprehensive scan architectures is only now being rediscovered for the complex system-level chips possible with nanometer technologies.

21.3.4.3 TIMING CONSIDERATIONS AND AT-SPEED TESTING

Timing issues can affect and plague both the scan infrastructure as well as the application of scan-based logic test.

The frequently used mux-scan methodology uses edge-triggered flip-flops as storage elements in the scan cells. And the edge clock is used for both the scan operation and for capturing test responses into the scan cells, making both susceptible to hold-time errors due to clock skew. Clock skew exists not only between multiple clock domains but also within each clock domain. The latter tend to be more subtle and easier to overlook. To deal with interdomain issues, the DFT tools have to be aware of the clock domains and clock-domain boundaries. The general rule of thumb for scan-chain construction is that each chain should only contain flip-flops from the same clock domain. Also, leading-edge and falling-edge flip-flops should be kept in separate scan chains even if driven from the same clock source. These strict rules of *division* can be relaxed somewhat if the amount of clock skew is small enough to be reliably overcome by inserting a lockup latch or flip-flop between the scan cells. A lockup latch or flip-flop updates its output on the edge opposite to what the next flop captures its input value—avoiding a value change on the same clock edge as the downstream capture. The susceptibility of the scan operation to hold-time problems can further be reduced by increasing the delay between scan cells, for example, by adding buffers to the scan connection between adjacent scan cells. Be aware that scan-chain reordering to reduce wiring based on physical layout can introduce hold-time violations; such reordering should also allow for that by inserting delay or lockup latches as appropriate.

In practice, it is not at all unusual for the scan operation to fail for newly designed chips. To avoid the likelihood of running into these problems, it is vitally important to perform a very thorough timing verification on the scan mode. In the newer nanometer technologies, signal integrity issues such as static/dynamic IR drop have to be taken into account in addition to process and circuit variability. A more *radical* approach is to replace the edge-triggered scan clocking with a level-sensitive multiphase clocking approach as, for example, in level-sensitive scan design (LSSD). In this case, the master and slave latches in the scan cells are controlled from two separate clock sources that are pulsed alternately. By increasing the nonoverlap period between the clock phases, it is possible to overcome any hold-time problems during scan without the need for lockup latches or additional intercell delay. With improved clock tree synthesis tools that reduce clock skew to all flops in a clock domain, the use of LSSD has practically disappeared.

Clock skew and hold-time issues also affect the reliable data transmission across interdomain boundaries. For example, if the clocks of two interconnected domains are pulsed together for capture, then it may be impossible to predict whether old or new data are captured. If the data change and clock edge get too close together, the receiving flip-flop could even be forced into metastability. ATPG tools traditionally try to avoid these problems by using a *capture-by-domain* policy in which only one clock domain is allowed to be captured in a test. This is only possible if DFT makes sure that it is indeed possible to issue a capture clock to each clock domain separately (e.g., using a separate test clock input for each domain or by degating the clocks of other domains). The *capture-by-domain* policy can adversely affect test time and data volume for designs with many clock domains, by limiting fault detection for each test to a single domain. Some ATPG tools nowadays offer sophisticated multiclock compaction techniques that overcome the *capture-by-domain* limitation (e.g., if clock-domain analysis shows that there is no connection between certain domains, then capture clocks can be sent to all of those domains without creating potential hold-time issues; if two domains are connected, then their capture clocks can be staggered— issued sequentially with enough pulse separation to assure predictability of the interface states). Special treatment of the boundary flip-flops between domains in DFT is an alternative method.

In static, fully complementary CMOS logic there is no direct path from power to ground except when switching. If a circuit is allowed to stabilize and settle down from all transitions, a very low power should be seen in the stable (quiescent) state. That expectation is the basis of IDDq (quiescent power) testing. Certain defects, for example, shorts, can create power-ground paths and therefore be detectable by an abnormal amount of quiescent power. For many years, low-speed stuck-at testing combined with IDDq testing have been sufficient to achieve reasonable quality levels for many CMOS designs. The normal quiescent background current unfortunately increases with each new technology generation, which reduces the signal-to-noise ratio of IDDq measurements. Furthermore, modern process technologies are increasingly susceptible to interconnect opens and resistive problems that are less easily detectable with IDDq to begin with. Many of these defects cause additional circuit delays and cannot be tested with low-speed stuck-at tests. Consequently, there is an increasing demand for at-speed delay fault testing [39,40,42,52–54].

In scan-based delay testing, the circuit is first initialized by a scan operation. Then a rapid sequence of successive input events is applied at tight timings to create transitions at flip-flop outputs in the circuit, have them propagate through the logic, and capture the responses into receiving flip-flops. The responses finally are unloaded by another scan operation for comparison. Signal transitions at flip-flop outputs are obtained by loading the initial value into the flip-flop, placing the opposite final value at the flip-flop's data input, and pulsing the clock. In the case of mux-scan, the final value for the transition can come from the scan side (*release from scan*) or the system side (*release from capture*) of the scan mux, depending on the state of the scan-enable signal. The functional logic typically is connected to the system side of the scan mux. The transitions will, hence, generally arrive at the system side of the receiving flip-flop's scan mux, such that the scan-enable must select the system side to enable capture. The release from scan method, therefore, requires switching the scan-enable from the scan side to the system side early enough to meet setup time at the receiving flip-flop, but late enough to avoid hold-time issues at the releasing flip-flop. In other words, the scan-enable signal is subject to a two-sided timing constraint and accordingly must be treated as a timing-sensitive signal for synthesis, placement, and wiring. Moreover, it must be possible to synchronize the scan-enable appropriately to the clocks of each clock domain. To overcome latency and synchronization issues with high fan-out scan-enables in high-speed logic, the scan-enables sometimes are pipelined. The DFT scan-insertion tools must be able to construct viable pipelined or nonpipelined scan-enable trees and generate the appropriate timing constraints and assertions for physical synthesis and timing verification. Such scan-enable pipelines often are used only for transitioning out of the scan state; when scan_enable is asserted, the pipeline is bypassed so scan state is entered directly.

Most ATPG tools do not have access to timing information and, in order to generate predictable results, tend to assume that the offset between release and capture clocks is sufficient to avoid completely setup time violations at the receiving flip-flops. That means the minimal offset between the release and capture clock pulses are dominated by the longest signal propagation path between the releasing and receiving flip-flops. If slow maintenance paths, multicycle paths, or paths from other slower clock domains get mixed in with the *normal* paths of a particular target clock domain, then it may be impossible to test the target domain paths at their native speed. To overcome this problem, some design projects disallow multicycle paths and insist that all paths (including maintenance paths) that can be active during the test of a target domain, must fit into the single-cycle timing window of that target domain. Another approach is to add enough timing capabilities to the ATPG software to identify all potential setup and hold-time violations at a desired test timing. The ATPG tool can then avoid sending transitions through problem paths (e.g., holding the path inputs stable) or set the state of all problem flip-flops to *unknown*. Yet other approaches use DFT techniques to separate multicycle paths out into what looks like another clock domain running at a lower frequency.

Most scan-based at-speed test methods perform the scan load/unload operations at a relatively low frequency, not the least to reduce power consumption during shift. ATPG patterns tend to have close to 50% switching probability at the flip-flop outputs during scan, which in some cases can be 10× more than what is expected during normal functional operation. Such abnormally high switching activity can cause thermal problems, excessive IR drop, or exceed the tester power-supply capabilities when the scan chains are shifted at full system speed. The reality of pin electronic capabilities in affordable test equipment sets another practical limit to the data rate at which the scan interface can be operated. One advantage of lower scan speeds for design is that it is not necessary to design the scan chains and scan-chain interface logic for full system speed. That can help reduce the placement, wiring, and timing constraints for the scan logic.

Slowing down the scan rate helps reduce average power, but does little for dynamic power (di/dt). In circuits with a large number of flip-flops, in particular when combined with tightly controlled low-skew clocks, simultaneous switching of flip-flop outputs can result in unexpected power/noise spikes and dynamic IR drops, leading to possible scan-chain malfunctions. These effects may need to be considered when allocating hold-time margins during scan-chain construction and verification.

All current commercial ATPG tools have the ability to generate tests with reduced switching activity during the scan shift cycles, specifically to deal with this issue of dynamic power during scan shift; however, when scan bits are constrained to reduce switching activity, the number of test patterns required to achieve a given fault coverage may increase by 30% or more. The di/dt issue is also of concern during at-speed clocking outside of scan shift cycles. Normally, ATPG is not concerned with trying to reduce the switching activity during the *capture* clock pulses between scan operations; however, most ATPG tools have the ability to try to reduce switching here as well and this is typically achieved by using existing functional clock gating to prevent a clock pulse from reaching a significant proportion of the flops of that clock domain [70]. With clocks gated off for a significant percentage of the scan flops, this too can cause the number of test patterns to increase, by perhaps an additional 30%. For extremely low switching demands of 7% or less, expect the number of test patterns to increase by a factor of 2 or more compared with no such switching activity constraints.

The effectiveness of at-speed tests not only depends on the construction of proper test event sequences but may also critically depend on being able to deliver these sequences at higher speed with higher accuracy than supported by the test equipment. Many modern chips use on-chip clock frequency multiplication (e.g., using PLLs) and phase alignment, and there is an increasing interest in taking advantage of this on-chip clocking infrastructure for testing. To that effect, clock system designers or DFTS add programmable test waveform generation features to the root of the clock tree for each clock domain. Programmable in this context tends to mean the provision of a serially loadable control register that determines the details of which and how many clock pulses/phases are generated. The actual sequence generation is triggered by a (possibly asynchronous) start signal that can be issued from the tester (or some other internal controller source) after the scan load operation has completed. The clock generator will then produce a deterministic sequence of internal clock edges that are synchronized to the PLL output. Some high-performance designs may include additional features (e.g., programmable delay lines) for manipulating the relative edge positions over and above what is possible by simply changing the frequency of the PLL input clock. Typically each functional clock domain has its own programmable clock sequence generator, although many domains may be driven from the output of the same PLL.

The ATPG tools generally cannot deal with the complex clock generation circuitry. Therefore, the on-product clock generation (OPCG) logic is combined into an OPCG macro and separated from the rest of the chip by cut points. The cut points look like external clock pins to the ATPG tool. It is the responsibility of the user to specify a list of available OPCG programming codes and the resulting test sequences to the ATPG tool, which in turn is constrained to use only the thus specified event sequences. More recently, some commercial ATPG tools recognize the programmable nature of the OPCG logic and can specify what values to load into the control registers to obtain a certain number of pulses, delayed starting of those pulses and other parameters for sequence generation. For verification, the OPCG macro is simulated for all specified programming codes, and the simulated sequences appearing at the cut points are compared to the input sequences specified to the ATPG tool. The generated ATPG tests and sequences are normally simulated using the PLL behavioral models during [Verilog] functional verification simulation.

Some circuit elements may require finer-grained timing that requires a different approach. One example is clock jitter measurement, another memory access time measurement. Test equipment uses analog or digitally controlled, tightly calibrated delay lines (timing verniers), and it is possible to integrate similar features into the chip. A different method for measuring arbitrary delays is to switch the to-be-measured delay path into an inverting recirculating loop and measure the oscillation frequency (e.g., counting oscillations against a timing reference). Small delays that would result in extremely high frequencies are made easier to test by switching them into and out of a longer delay path and comparing the resulting frequencies. Oscillation techniques can also be used for delay calibration to counteract performance variations due to process variability.

21.3.4.4 CUSTOM LOGIC

Custom transistor-level logic, often used for the most performance-sensitive parts of a design, poses a number of unique challenges to the DFT flow and successful test generation. Cell-based designs tend to use more conservative design practices and the libraries for cell-based designs generally come with premade and preverified gate-level test generation models. For transistor-level custom designs, by contrast, the gate-level test generation models must somehow be generated and verified "after the fact" from the transistor-level schematics. Although the commercial ATPG tools may have some limited transistor-level modeling capabilities, their wholesale use generally leads to severe tool runtime problems and hence is strongly discouraged. The construction of suitable gate-level models is complicated by the fact that custom logic often uses dynamic logic or other performance/area/power-driven unique design styles, and it is not always easy to determine what should be explicitly modeled and what should be implied in the model (e.g., the precharge clocks and precharge circuits for dynamic logic). Another issue for defect coverage as well as diagnostics is to decide which circuit-level nets to explicitly keep in the logic model vs. simplifying the logic model for model size and tool performance.

The often extreme area, delay, and power sensitivity of custom-designed structures are often met with a partial scan approach, even if the synthesized logic modules have full scan. The custom design team has to make a trade-off between the negative impact on test coverage and tool runtime vs. keeping the circuit overhead small. Typical *rules of thumb* will limit the number of nonscannable levels and discourage feedback loops between nonscannable storage elements (e.g., feed-forward pipeline stages in data paths are often good candidates for partial scan). Area, delay, and power considerations also lead to a more widespread use of pass gates and other three-state logic (e.g., three-state buses) in custom-design circuitry. The control inputs of pass-gate structures, such as the select lines of multiplexers, and enables of three-state bus drivers tend to require specific decodes for control (e.g., *one hot*) to avoid three-state contention (i.e., establishing a direct path from power to ground) that could result in circuit damage due to burn-out. To avoid potential burn-out, DFT and ATPG must cooperate to assure that safe control states are maintained during scan and during test. If the control flip-flops are included in the scan chains, then DFT hardware may be needed to protect the circuits (e.g., all bus drivers are disabled during scan).

Other areas of difficulty are complex memory substructures with limited scan and possibly unusual or pipelined decodes that are not easily modeled with the built-in memory primitives available in the DFT/ATPG tools.

Finally, often we find that low-power designs have multiple power domains with the ability to power off some domains that are not required functionally, typically based on user activity or system workload. The power mode selection circuitry is often implemented using logic and power transistors that are not part of the typical logic allowed to be controlled by ATPG patterns. As such, testing of this logic and state retention logic and some of the inter-power-domain voltage-level shifting cells can require special processing by ATPG tools [59]. In order to test some of this special power control and retention logic, it may be required to apply special power mode sequencing that turns on some power domains and turns off others; the tools had better be able to know which sections of logic are not powered up—for example, to ensure scan chains are not routed through powered-down domains.

21.3.4.5 LOGIC BUILT-IN SELF-TEST

Chip manufacturing test (e.g., from ATPG) typically assumes full access to the chip I/Os plus certain test equipment features for successful test application. That makes it virtually impossible to port chip manufacturing tests to higher-level assemblies and into the field. Large data-processing systems historically stored test data specifically generated for in-system testing on disk and applied them to the main processor complex through a serial maintenance interface from a dedicated service processor that was delivered as part of the system. Over the years, it became too cumbersome to store and manage vast amounts of test data for all possible system configurations and engineering change order levels, and the serial maintenance interface became too slow for efficient data transfer. Hence, alternatives were pursued that avoid the large data volume and data transfer bottleneck associated with traditional ATPG tests.

The most widely used alternative today is logic BIST using pseudorandom patterns [2,3,8,21,24,30,31,34,49]. It is known from coding theory that pseudorandom patterns can be generated easily and efficiently in hardware, typically using a so-called pseudorandom pattern generator (PRPG) macro utilizing a linear feedback shift register (LFSR). The PRPG is initialized to a starting state called PRPG seed and in response to subsequent clock pulses, produces a state sequence that meets certain tests of randomness. However, the sequence is not truly random. In particular, it is predictable and repeatable if started from the same seed. The resulting pseudorandom logic states are loaded into scan chains for testing in lieu of ATPG data. Multiple PRPGs nowadays are generally built into chips—one in each core or partition. That makes it possible to use a large number of relatively short on-chip scan chains because the test data do not need to be brought in from the outside through a narrow maintenance interface or a limited number of chip I/Os. Simple LFSR-based PRPGs connected to multiple parallel scan chains result in undesirable, strong value correlations (structural dependencies) between scan cells in adjacent scan chains. The correlations can reduce the achievable test coverage. To overcome this potential problem, some PRPG implementations are based on cellular automata (CA) rather than LFSRs. LFSR-based implementations use a phase-shifting network constructed out of XOR to eliminate the structural dependencies. The phase-shifting network can be extended into a spreading network that makes it possible to drive a larger number of scan-chain inputs from a relatively compact LFSR.

More scan chains generally mean shorter scan chains that require fewer clock cycles for load/unload, thus speeding up test application (each scan test requires at least one scan load/unload), assuming that the DFTS tool used for scan insertion succeeds in reducing and balancing the length of the scan chains. Modern DFTS tools are capable of building scan architectures with multiple separately balanced selectable chain configurations/modes to support different test methods on the same chip. It should be noted that the achievable chain length reduction can be limited by the length of preconnected scan-chain segments in hard macros. Hence, for logic BIST it is important to assure that large hard macros are preconfigured with several shorter chain segments rather than a single long segment. As a rule of thumb, the maximum chain length for logic BIST should not exceed 500–1000 scan cells. That means a large core with 1M scan cells requires 1K scan chains or more. The overhead for the PRPG hardware is essentially proportional to the number of chains (a relatively constant number of gates per scan chain for the LFSR/CA, phase-shifting/spreading network, and scan-switching network).

The flip side of the coin for pseudorandom logic BIST is that pseudorandom patterns are less efficient for fault testing than ATPG-generated, compacted test sets. Hence, 10× as many or more pseudorandom patterns are needed for equivalent nominal fault coverage, offsetting the advantage of shorter chains. Moreover, not all faults are easily tested with pseudorandom patterns. Practical experience with large-scale data-processing systems has indicated that a stuck-at coverage of around up to 95% can be achievable with a reasonable (as dictated by test time) number of pseudorandom test patterns. Going beyond 95% requires too much test application time to be practical. Coverage of 95% can be sufficient for burn-in, higher-level assembly, system, and field testing, but may be unacceptable for chip manufacturing test. Consequently, it is not unusual to see ATPG-based patterns used for chip manufacturing tests and pseudorandom logic BIST for all subsequent tests.

Higher test coverage, approaching that of ATPG, can be achieved with pseudorandom patterns only by making the logic more testable for such patterns. The 50–50 pseudorandom signal probability at the outputs of the scan cells gets modified by the logic gates in the combinational logic between the scan cells. Some internal signal probabilities can be skewed so strongly to 0 or 1 that the effective controllability or observability of downstream logic becomes severely impaired. Moreover, certain faults require many more specific signal values for fault excitation or propagation than achievable with a limited set of pseudorandom patterns (it is like rolling dice and trying to get 100+ sixes in a row). Wide comparators and large counters are typical architectural elements afflicted with that problem. Modern DFTS and analysis tools offer the so-called pseudorandom testability analysis and automatic test point insertion features. The testability analysis tools use *testability measures* or signal local characteristics captured during good machine fault simulation to identify nets with low pseudorandom controllability or observability for pseudorandom patterns. The test point insertion tools generate a suggested list of control or observe points that should be added to the netlist to improve test coverage [6]. Users generally can control how many and what type of test points are acceptable (control points tend to be more "expensive" in circuit area and delay impact) for manual or automatic insertion. It is not unusual for test points to be "attracted" to timing-critical paths. If the timing-critical nets are known ahead of time, they optionally can be excluded from modification. As a rule of thumb, one test point is needed per 1K gates to achieve the 99%+ coverage objective often targeted for chip manufacturing test. Each test point consumes roughly 10 gates, meaning that 1% additional logic is required for the test points.

It should be noted that some hard-to-test architectural constructs such as comparators and counters can be identified at the presynthesis RTL phase of the design, creating a basis for RTL analysis and test point insertion tools.

Using pseudorandom patterns to eliminate the need for storing test input data from ATPG solves only part of the data volume problem. The expected test responses for ATPG-based tests must equally be stored in the test equipment for comparison with the actual test responses. As it turns out, LFSR-based hardware macros very similar to a PRPG macro can be used to implement error detecting code (EDC) generators. The most widely used EDC macro implementation for logic BIST is called multiple-input signature register (MISR), which can sample all scan-chain outputs in parallel. As the test responses are unloaded from the scan chains they are simultaneously clocked into the MISR where they are accumulated. The final MISR state after a specified number of scan tests have been applied is called the signature, and this signature is compared to an expected signature that has been precalculated by simulation for the corresponding set of test patterns.

The MISR uses only shifting and XOR logic for data accumulation, meaning that each signature bit is the XOR sum of some subset of the accumulated test response bit values. One property of XOR sums is that even a single unknown or unpredictable summand makes the sum itself unknown or unpredictable. If the signature of a defect-free product under test is unknown or unpredictable, then the signature is useless for testing. Hence, there is a "Golden Rule" for signature-based testing that no unknown or unpredictable circuit state can be allowed to propagate to the MISR. This rule creates additional design requirements over and above what is required for scan-based ATPG. The potential impact is further amplified by the fact that pseudorandom patterns, unlike ATPG patterns, offer little to no ability for intelligently manipulating the test stimulus data. Hence, for logic BIST, the propagation of unknown/unpredictable circuit states (also known as x states) must generally be stopped by hardware means. For example, microprocessor designs tend to contain tens of thousands of three-state nets, and modern ATPG tools have been adapted to that challenge by constructively avoiding three-state contention and floating nets in the generated test patterns. For logic BIST, either test hardware must be added to prevent contention or floating nets or the outputs of the associated logic must be degated (rendering it untestable) so that they cannot affect the signature. Some processor design projects forbid the use of three-state logic, enabling them to use logic BIST. Other design teams find that so radical an approach is entirely unacceptable.

Three-state nets are not the only source of x states. Other sources include unmodeled or incompletely modeled circuit elements, uninitialized storage elements, multiport storage element write conflicts, and setup/hold-time timing violations. DFTS and DRC tools for logic BIST must analyze the netlist for potential x-state sources and add DFT structures that remove the x-state generation potential (e.g., adding exclusive gating logic to prevent multiport conflicts, or by assuring the proper initialization of storage elements),or add degating logic that prevents x-state propagation to the MISR (e.g., at the outputs of uninitialized or insufficiently modeled circuit elements). Likewise, hardware solutions may be required to deal with potential setup/hold-time problems (e.g., clock-domain boundaries, multicycle paths, nonfunctional paths). High-coverage logic BIST, overall, is considered to be significantly more design intrusive than ATPG methods where many of the issues can be dealt within pattern generation rather than through design modifications.

21.3.4.6 AT-SPEED TESTING WITH LOGIC BUILT-IN SELF-TEST

At-speed testing with logic BIST essentially follows the same scheme as at-speed testing with ATPG patterns. (Historical note: contrary to frequent assertions by logic BIST advocates that pseudorandom pattern logic BIST is needed to enable at-speed testing, ATPG-based at-speed test has been practiced long before logic BIST became popular and is still being used very successfully today.) Most approaches use slow scan (to limit power consumption, among other things) followed by the rapid application of a short burst of at-speed edge events. Just as with ATPG methods, the scan and at-speed edge events can be controlled directly by test equipment or from an OPCG macro. The advantage of using slow scan is that the PRPG/MISR and other scan-switching and interface logic need not be designed for high speed. That simplifies timing closure and gives the placement and wiring tools more flexibility. Placement/wiring consideration may still favor using several smaller, distributed PRPG/MISR macros. Modern DFTS, DRC, fault grading, and signature-simulation tools for logic BIST generally allow for distributed macros.

Certain logic BIST approaches, however, may depend on performing all or some of the scan cycles at full system speed. With this approach, the PRPG/MISR macros and other scan interface logic must be designed to run at full speed. Moreover, scan chains in clock domains with different clock frequencies may be shifted at different frequencies. That affects scan-chain balancing, because scan chains operating at half frequency in this case should only be half as long as chains operating at full frequency. Otherwise they would require twice the time for scan load/unload.

The fact that pseudorandom logic BIST applies 10× (or more) as many tests than ATPG for the same nominal fault coverage can be advantageous for the detection of unmodeled faults and of defects that escape detection by the more compact ATPG test set. This was demonstrated empirically in early pseudorandom test experiments using industrial production chips. However, it is not easy to extrapolate these early results to the much larger chips of today. Today's chips can require 10K or more ATPG patterns. Even ATPG vectors are mostly pseudorandom, meaning that applying the ATPG vectors is essentially equivalent to applying 10K+ logic BIST patterns, which is much more than what was used in the old hardware experiments.

Only recently has the debate over accidental fault/defect coverage been refreshed. The theoretical background for the debate is the different *n-detect* profiles for ATPG and logic BIST. ATPG test sets are optimized to the extent that some faults are only detected by one test (1-detect) in the set. More faults are tested by a few tests and only the remainder of the fault population gets detected 10× (10-detect) or more. For logic BIST tests, by contrast, almost all faults are detected many times. Static bridging-fault detection, for example, requires coincidence of a stuck-at fault test for the victim net with the aggressor net being at the fault value. If the stuck-at fault at the victim net is detected only once, then the probability of detecting the bridging fault is determined by the probability of the aggressor net being at the faulty value, for example, 50% for pseudorandom values. A 2-detect test set would raise the probability to 75% and so on. Hence, multidetection of stuck-at faults increases the likelihood of detecting bridging faults. The trend can be verified by running bridging-fault simulation for stuck-at test sets with different *n*-detect profiles and comparing the results. Hardware experiments confirmed the trend for production chips.

It must be noted, however, that modern ATPG tools can and have been adapted to optionally generate test sets with improved *n*-detect profiles. The hardware experiments cited by the BIST advocates in fact were performed with ATPG-generated *n*-detect test sets, not with logic BIST tests. It should also be noted that if the probability of the aggressor net being at the faulty value is low, then even multiple detects may not do enough. ATPG experiments that try to constructively enhance the signal probability distribution have shown some success in that area. Finally, a new generation of tools is emerging that extract realistic bridging faults from the circuit design and layout. ATPG tools can and will generate explicit tests for the extracted faults, and it has been shown that low *n*-detect test sets with explicit cleanup tests for bridging faults can produce very compact test sets with equally high or higher bridging-fault coverage than high *n*-detect test sets.

Experience with logic BIST on high-performance designs reveals that test points may be helpful to improve nominal test coverage, but can have some side effects for characterization and performance screening. One reported example shows a particular defect in dynamic logic implementing a wide comparator that can only be tested with certain patterns. Modifying the counter with test points as required for stuck-at and transition fault coverage creates artificial nonfunctional short paths. The logic BIST patterns use artificial paths for coverage and never sensitize the actual critical path. Knowing that wide comparators were used in the design, some simple weighting logic had been added to the PRPG macro to create the almost-all-1s or almost-all-0s patterns suited for testing the comparators. The defect was indeed only detected by the weighted tests and escaped the normal BIST tests. Overall, the simple weighting scheme measurably increased the achievable test coverage without test points.

If logic BIST is intended to be used for characterization and performance screening, it may also be necessary to enable memory access in logic BIST. It is not uncommon that the performance-limiting paths in high-speed design traverse embedded memories. The general recommendation for logic BIST is to fence embedded memories off with boundary scan during BIST. That again creates artificial paths that may not be truly representative of the actual performance-limiting paths. Hardware experience with high-performance processors shows that enabling memory access for some portion of the logic BIST tests does indeed capture unique fails. Experiments with (ATPG-generated) scan tests for processor's performance binning have similarly shown that testing paths through embedded memories is required for better correlation with functional tests.

As a general rule of thumb, if BIST is to be used for characterization, then the BIST logic may have to be designed to operate at higher speeds than the functional logic it is trying to characterize.

Automated diagnosis of production test fails has received considerable attention recently and is considered a key technology for nanometer semiconductor technologies. In traditional stored pattern ATPG testing, each response bit collected from the device under test is immediately compared to an expected value, and most test equipment in the case of a test fail (i.e., mismatch between actual and expected response values) allows for optionally logging the detailed fail information (e.g., tester cycle, tester channel, fail value) into a so-called fail set for the device under test. The fail sets can then be postprocessed by automated logic diagnostic software tools to determine the most likely root cause locations.

In logic BIST, the test equipment normally does not see the detailed bit-level responses because these are intercepted and accumulated into a signature by the on-chip MISR. The test equipment only sees and compares highly compressed information contained in the accumulated signatures. Any difference between an actual signature and the expected signature indicates that the test response must contain some erroneous bits. It generally is impossible to reconstruct bit-level fail sets from the highly compressed information in the signatures. The automated diagnostic software tools, however, need the bit-level fail sets. The diagnosis of logic BIST fails, hence requiring an entirely different analysis approach or some means for extracting a bit-level fail set from the device under test. Despite research efforts aimed at finding alternative methods, practitioners tend to depend on the second approach. To that effect, the logic BIST tests are structured such that signatures are compared after each group of n tests, where n can be 1, 32, 256, or some other number. The tests further must be structured such that each group of n tests is independent. In that case, it can be assumed that a signature mismatch can only be caused by bit-level errors in the associated group of n tests. For fail-set extraction, the n tests in the failing group are repeated and this time the responses are directly scanned out to the test equipment without being intercepted by the MISR (*scan dump operation*). Existing production test equipment generally can only log a limited number of failing bits and not the raw responses. Hence, the test equipment must have stored expect data available for comparison. Conceptually, that could be some "fake" expect vector like all 0s, but then even correct response tests would result in failing bits that would quickly exceed the very limited fail buffers on the testers, meaning that the actual expect data must be used. However, the number of logic BIST tests tends to be so high that it is impractical to store all expect vectors in the production test equipment. Bringing the data in from some offline medium would be too slow. The issue can be overcome if it is desired to diagnose only a small sample of failing devices, for example, prior to sending them to the FA lab. In that case, the failing chips can be sent to a nonproduction tester for retesting and fail-set logging. Another approach that may be useful is to have the full dump expect values for just the first 1K or so logic BIST test cycles and be able to diagnose failures that occur within those first 1K cycles and skip diagnosis on those chips that fail after that.

Emerging, very powerful, statistical yield analysis and yield management methods require the ability to log large numbers of fail sets during production testing, which means fail-set logging must be possible with minimal impact to the production test throughput. That generally is possible and fairly straightforward for ATPG tests (as long as the test data fit into the test equipment in the first place), but may require some logistical ingenuity for logic BIST and other signature-based test methods.

21.3.4.7 TEST DATA COMPRESSION

Scan-based logic tests consume significant amounts of storage and test time on the ATE used for chip manufacturing test. The data volume in first order is roughly proportional to the number of logic gates on the chip and the same holds for the number of scan cells. Practical considerations and test equipment specifications oftentimes limit the number of pins available for scan-in/scan-out and the maximum scan frequency. Consequently, the scan chains for more complex chips tend to be longer and it takes commensurately longer to load/unload the scan chains. There is a strong desire to keep existing test equipment and minimize expensive upgrades or replacements. Existing equipment on many manufacturing test floors tends to have insufficient vector memory for the newer chip generations. Tester memory reload is very time-consuming and should be avoided if possible. The purpose of test data compression is to reduce the memory footprint of the scan-based logic tests such that they comfortably fit into the vector memory of the [existing] test equipment.

Test equipment tends to have at least two types of memory for the data contained in the test program. One type is the vector memory that essentially holds the logic levels for the test inputs and for the expected responses. The memory allocation for each input bit could include additional space for waveform formats (the actual edge timings are kept in a separate time-set memory space) over and above the logic level. On the output side, there typically are at least two bits of storage for each expected response bit. One bit is a mask value that determines if the response value should be compared or ignored (e.g., unknown/unpredictable responses are masked) and the other bit defines the logic level to compare with. The other memory type contains nonvector program information like program op-codes for the real-time processing engine in the test equipment's pin electronics.

Some memory optimization for scan-in data is possible by taking advantage of the fact that the input data for all cycles of the scan load/unload operation use the same format. The format can be defined once upfront and only a single bit is necessary to define the logic level for each scan cycle. A test with a single scan load/unload operation may thus consume three bits of vector memory in the test equipment, and possibly only 2 bits of no response masking is needed.

Scan-based logic test programs tend to be simple in structure, with one common loop for the scan/load unload operation and short bursts of other test events in between. Consequently, scan-based test programs tend to consume only very little op-code memory and the memory limitation for large complex chips is only in the vector memory.

It is worth noticing that most production test equipment offers programming features such as branching, looping, and logic operations, for functional testing. The scan-based ATPG programs, however, do not typically take advantage of these features for two reasons. First, some of the available features are equipment specific. Second, much of the data volume is for the expected responses. While the ATPG tools may have control over constructing the test input data, it is the product under test, not the ATPG software that shapes the responses. Hence, taking full advantage of programming features for really significant data reduction would first require some method for removing the dependency on storing the full amount of expected responses.

21.3.4.7.1 Input Data Compression

Input data compression, in general, works by replacing the bit-for-bit storage of each logic level for each scan cell with some means for algorithmically constructing multiple-input values on the fly from some compact source data. The algorithms can be implemented in software or hardware on the tester or in software or hardware inside the chips under test. To understand the nature of the algorithms, it is necessary to review some properties of the tests generated by ATPG tools.

ATPG proceeds by selecting one yet untested fault and generating a test for that one fault. To that effect, the ATPG algorithm will determine sufficient input conditions to excite and propagate the selected fault. That generally requires that specific logic levels must be asserted at some scan cells and primary inputs. The remaining scan cells remain unspecified at this step in the algorithm. The thus constructed, partially specified, vector is called a *test cube*. It has become customary to refer to the specified bits in the test cube as "care bits" and to the unspecified bits as "don't care bits." All ATPG tools used in practice perform vector *compaction*, which means that they try to combine the test cubes for as many faults as possible into a single test vector. The methods for performing vector compaction vary but the result generally is the same. Even after compaction, for almost all tests, there are many more don't care bits than care bits. In other words, scan-based tests generated by ATPG tools are characterized by a low *care bit density* (percentage of specified bits in the compacted multifault test cube). After compaction, the remaining unspecified bits will be filled by some arbitrary fill algorithm. Pseudorandom pattern generation is the most common type of algorithm used for fill. That is noteworthy because it means that the majority of bit values in the tests are generated by the ATPG software algorithmically from a very compact seed value. However, neither the seed nor the algorithm is generally included in the final test data. And, without that knowledge, it tends to be impossible to recompact the data after the fact, with any commonly known compression algorithm (e.g., zip, lzw). The most commonly practiced test data compression methods in use today focus on using simple software/hardware schemes to regenerate, algorithmically, fill data in the test equipment or in the chip under test. Ideally, very little memory is needed to store the seed value for the fill data and only the care bits must be stored explicitly.

The "cheapest" method for input fill data compression is to utilize common software features in existing test equipment without modifications of the chip under test. Run length encoding (RLE) is one software method used successfully in practice. In this approach, the pseudorandom fill algorithm in ATPG is replaced with a repeat option that simply repeats the last value until a specified bit with the opposite value is encountered. The test program generation software is modified to search for repeating input patterns in the test data from ATPG. If a repeating pattern of sufficient length is found, it will be translated into a single pattern in vector memory, and a repeat op-code in op-code memory, rather than storing everything explicitly in vector memory. Practical experience with RLE shows that care bit density is so low that repeats are possible and the combined op-code plus vector memory can be 10× less than storing everything in vector memory. As a side effect, tests with repeat fill create less switching activity during scan and, hence, are less power hungry than their pseudorandom brethren. One caveat for use of repeat op-codes is that you must be able to repeat both the input values and any expected response values; for this reason, repeat op-codes work best with compression that uses a MISR since there are no expects on outputs during scan loading.

RLE can be effective for reducing the memory footprint of scan-based logic tests in test equipment. However, the fully expanded test vectors comprised of care and don't care bits are created in the test equipment and these are the expanded vectors that are sent to the chip(s) under test. The chips have the same number of scan chains and the same scan-chain length as for normal scan testing without RLE. Test vector sets with repeat fill are slightly less compact than sets with pseudorandom fill. Hence, test time suffers slightly with RLE, meaning that RLE is not the right choice if test time reduction is as important as data volume reduction.

Simultaneous data volume and test time reduction is possible with on-chip decompression techniques [1,7,9,10,18,48]. In that scenario, a hardware macro for test input data decompression is inserted between the scan-in pins of the chip and the inputs of a larger number of shorter scan chains (i.e., there are many times more scan chains than scan-in pins). The decompression macro can be combinational or sequential in nature, but in most practical implementations it tends to be linear in operation. Linear in this context means that the value loaded into each scan cell is a predictable linear combination (i.e., XOR sum and optional inversion) of some of the compressed input values supplied to the scan-in pins. In the extreme case, each linear combination contains only a single term, which can, for example, be achieved by replication or shifting.

Broadcast scan, where each scan-in pin simply fans-out to several scan chains without further logic transformation, is a particularly simple decompression macro using replication of input values. In broadcast scan, all scan chains connected to the same scan-in pin receive the same value. That creates strong correlation (replication) between the values loaded into the scan cells of those chains. Most ATPG software implementations have the ability to deal directly with correlated scan cell values and will automatically imply the appropriate values in the correlated scan cells. The only new software needed is for DFTS to create automatically the scan fan-out and DFT DRC for analyzing the scan fan-out network and setting up the appropriate scan cell correlation tables for ATPG. The hard value correlations created by broadcast scan can make some faults hard to test and make test compaction more difficult because the correlated values create care bits even if the values are not required for testing the target fault. It must therefore be expected that ATPG with broadcast scan has longer runtimes, creates slightly less compact tests, and achieves slightly lower test coverage than ATPG with normal scan. With a scan fan-out ratio of 1:32 (i.e., each scan-in fans out to 32 scan chains), it is possible to achieve an effective data volume and test time reduction of 20× or so when using 4 or fewer scan-in pins; data volume reduction ratios closer to the fan-out ratio are possible when 8 or more scan-in pins are available as that tends to avoid correlation in chains near each other. It is assumed that the scan chains can be reasonably well balanced and that there are no hard macros with preconnected scan segments that are too long.

The more sophisticated decompression macros contain XOR gates and the values loaded into the scan cells are a linear combination of input values with more than one term. Industrial ATPG does understand hard correlations between scan cells but not Boolean relationships like linear combinations with more than one term. Hence, the ATPG flow has to be enhanced to deal with the more sophisticated decompression techniques. Each care bit in a test cube creates a linear equation with the care bit value on the one side and an XOR sum of some input values on the other side. These specific XOR sums for each scan cell can be determined upfront by symbolic simulation of the scan load operation. Since a test cube typically has more than one care bit, a system of linear equations is formed, and a linear equation solver is needed to find a solution for the system of equations. In some cases, the system of equations has no solution; for example, if the total number of care bits exceeds the number of input values supplied from the tester, which can happen when too many test cubes are compacted into a single test, it will result in an overly specified system of linear equations that is unlikely to have a solution. Another approach for a decompression macro uses multiplexors to select a different fan-out to each chain. By allowing, for example, three of the scan-in pins to drive the select input of a set of muxes in the scan-in data path, eight different scan fan-out decompressors can be selected on the fly from the set of remaining scan-in pins. In practice, this approach also can deal with most care bit needs in a single scan cycle by selecting a mux configuration that allows all required care bits to be correctly set. In summary, similar to broadcast scan, ATPG for the more sophisticated schemes also adds CPU time, possibly reduces test coverage, and increases the number of tests in the test set. The sophisticated techniques require more hardware per scan chain (e.g., one flip-flop plus some other gates) than the simple fan-out in broadcast scan. However, the more sophisticated methods tend to offer more flexibility and should make more optimal compression results possible.

It should also be pointed out that a sequential decompressor has the ability to use scan-in values from multiple shift cycles to solve for any specific care bit; this allows use of a smaller number of scan-in pins and still generally be able to solve the equations that will prove more difficult for the combinational logic decompressors. Note that with just a single scan-in pin, all combinational logic decompressors are the same and appear to be fan-out of that single scan-in. To support a small number of scan-in pins, decompressors can be extended by use of a deserializer that can convert a single input stream of values into multiple values to be loaded in parallel—at the expense of more shift cycles during the serial loading; the time for these additional serial load cycles can be mitigated by using a faster clock for the deserializer. Use of a sequential linear decompressor has some of the benefits of a deserializer, but possibly cannot achieve the much faster shift rates of a simple shift register; however, it is certainly possible to combine a short deserializer with a sequential decompressor to obtain the high shift-in frequency along with the benefits of the sequential decompressor.

Although the so-called weighted random pattern (WRP) test data compression approach is proprietary and not generally available, it is worth a brief description [38]. The classical WRP does not exploit the low care bit density that makes the other compression methods possible, but a totally different property of scan-based logic tests generated by ATPG. The property is sometimes called *test cube clustering*, meaning that the specified bit values in the test cubes for groups of multiple faults are mostly identical and only very few care bit values are different (i.e., the test cubes in a cluster have a small Hamming distance from each other). That makes it possible to find more compact data representations for describing all tests in a cluster (e.g., a common base vector and a compact difference vector for each cluster member). Many years of practical experience with WRP confirms that input data volume reductions in excess of 10× are possible by appropriately encoding the cluster information.

It has been suggested that two-level compression should be possible by taking advantage of both the cluster effect and the low care bit density of scan tests. Several combined schemes have been proposed, but they have not shown up in any DFT tool on the market.

21.3.4.7.2 Response Data Compression/Compaction

Since the amount of memory needed for expected responses can be more than for test input data, any efficient data compression architecture must include test response data compression (sometimes also called compaction). WRP, the oldest data compression method with heavy-duty practical production use in chip manufacturing test, for example, borrows the signature approach from logic BIST for response compaction. Both the WRP decompression logic and the signature generation logic were provided in proprietary test equipment. The chips themselves were designed for normal scan and the interface between the chips and the test equipment had to accommodate the expanded test input and test response data. The interface for a given piece of test equipment tends to have a relatively fixed width and bandwidth, meaning that the test time will grow with the gate count of the chip under test. The only way to reduce test time in this scenario is to reduce the amount of data that have to go through the interface bottleneck. That can be achieved by inserting a response compression/compaction macro between the scan chains and the scan-out pins of the chip so that only compressed/compacted response data have to cross the interface.

How that can be done with EDCs using MISR macros was already known from and proven in practice by logic BIST, and the only "new" development required was to make the on-chip MISR approach work with ATPG tests with and without on-chip test input data decompression. From a DFT tool perspective, that means adding DFTS and DFT DRC features for adding and checking the on-chip test response compression macro and adding a fast signature-simulation feature to the ATPG tool. The introduction of signatures instead of response data also necessitated the introduction of new data types in the test data from ATPG and corresponding new capabilities in the test program generation software, for example, to quickly reorder tests on the test equipment without having to resimulate to recompute the signatures.

One unique feature of the MISR-based response compression method is that it is not necessary to monitor the scan-out pins during the scan load/unload operation (the MISR accumulates the responses on-chip and the signature can be compared after one or more scan load/unload operations are completed). The test equipment channels normally used to monitor the scan-outs can be reallocated for scan-in, meaning that the number of scan-ins and scan chains can be doubled, which reduces test time by a factor of 2 if the scan chains can be rebalanced to be half as long as before. Furthermore, no expected values are needed in the test equipment vector memory for scan, thus reducing the data volume by a factor of 2 or 3 (the latter comes into play if the test equipment uses a two-bit representation for the expected data and the input data can be reformatted to a one-bit representation).

Mathematically speaking, the data manipulations in a MISR are very similar to those in a linear input data decompression macro. Each bit of the resulting signature is a linear combination (XOR sum) of a subset of test response bit values that were accumulated into the signature. Instead of using a sequential state machine like an MISR to generate the linear combinations, it is also possible to use a combinational XOR network to map the responses from a large number of scan-chain outputs to a smaller number of scan-out pins. Without memory, the combinational network cannot accumulate responses on-chip. Hence, the scan-out pins must be monitored and compared to expected responses for each scan cycle. The data reduction factor is given by the number of internal scan-chain outputs per scan-out pin.

Selective compare uses multiplexing to connect one out of several scan-chain outputs to a single scan-out pin. The selection can be controlled from the test equipment directly (i.e., the select lines are connected to chip input pins) or through some intermediate decoding scheme. Selective compare is unique in that the response value appearing at the currently selected scan-chain output is directly sent to the test equipment without being combined with other response values, meaning that any mismatches between actual and expected responses can be directly logged for analysis. The flip side of the coin is that responses in the currently deselected scan chains are ignored, which reduces the overall observability of the logic feeding into the deselected scan cells and could impair the detection of unforeseen defects. It also should be noted that in addition to monitoring the scan-out pins, some input bandwidth is consumed for controlling the selection.

Mapping a larger number of scan-chain outputs onto a smaller number of scan-out pins through a combinational network is sometimes referred to as *space compaction*. A MISR or similar sequential state machine that can accumulate responses over many clock cycles performs *time compaction*. Of course, it is possible to combine space and time compaction into a single architecture.

21.3.4.7.3 X-State Handling

Signatures are known for not being able to handle unknown/unpredictable responses (x states). In signature-based test methods, for example, logic BIST, it is therefore customary to insist that x-state sources should be avoided or disabled, or if that is not possible, then the associated x states must under no circumstances propagate into the signature generation macro.

x-state avoidance can be achieved by DFT circuit modifications (e.g., using logic gates instead of pass-gate structures) or by adjusting the test data such that the responses become predictable (e.g., asserting enable signals such that no three-state conflicts are created and no unterminated nets are left floating). The latter is in many cases possible with ATPG patterns; however, that typically increases ATPG runtime and can adversely affect test coverage. Design modifications may be considered too intrusive, especially for high-performance designs, which have hampered the widespread acceptance of logic BIST.

Disabling x-state propagation can be achieved by local circuit modifications near the x-state sources or by implementing a general-purpose *response-masking* scheme. Response masking typically consists of a small amount of logic that is added to the inputs of the signature generation macro (e.g., MISR) [66]. This logic makes it possible to selectively degate (i.e., force a known value) onto the MISR input(s) that could carry x states. A relatively simple implementation, for example, consists of a serially loadable *mask vector* register with one mask bit for each MISR input. The value loaded into the mask bit determines whether the response data from the associated scan-chain output are passed through to the MISR or are degated. The mask vector is serially preloaded prior to scan load/unload and could potentially be changed by reloading during scan load/unload. A single control signal directly controlled from the test equipment can optionally activate or deactivate the effect of the masking on a scan cycle by scan cycle basis. More sophisticated implementations could offer more than one dynamically selectable mask bit per scan chain or decoding schemes to dynamically update the mask vector. The dynamic control signals as well as the need to preload or modify mask vectors do consume input data bandwidth from the tester and add to the input data volume. This impact generally is very minimal if a single mask vector can be preloaded and used for one or more than one full-scan load/unload without modification. The flip side of the coin in this scenario is some loss of observability due to the fact that a number of predictable responses that could carry defect information may be masked in addition to the x states.

If a combinational space compaction network is used and the test equipment monitors the outputs, then it becomes possible to let x states pass through and mask them in the test equipment. Flexibility could, for example, be utilized to reduce the amount of control data needed for a selective compare approach (e.g., if a *must-see* response is followed by an x state in a scan chain, then it is possible to leave that chain selected and to mask the x state in the tester rather than expending input bandwidth to change the selection).

If an XOR network is used for space compaction, then an x state in a response bit will render all XOR sums containing that particular bit value equally unknown/unpredictable, masking any potential fault detection in the other associated response bits. The "art" of constructing x-tolerant XOR networks for space compaction is to minimize the danger of masking "must-see" response bit values. It has been shown that suitable and practical networks can indeed be constructed if the number of potential x states appearing at the scan-chain outputs in any scan cycle is limited [67].

A similar solution adds some memory to the XOR-network approach. Unlike in MISRs, there is no feedback loop in the memory. The feedback in the MISR amplifies and "perpetuates" x states (each x state, once captured, is continually fed back into the MISR and will eventually contaminate all signature bits). The x-tolerant structures with memory, by contrast, are designed such that x states are flushed out of the memory fairly quickly (e.g., using a shift register arrangement without feedback) [68].

Another solution is to use input bandwidth to select a small subset of the scan chains for observation during any shift cycle where an x may cause loss of observation for a fault effect [69]. Such schemes may greatly reduce the amount of scan output data that are observed on cycles where not all scan chains are being observed, possibly impacting the coverage of unmodeled defects—especially if the input bandwidth for output selection is shared with care bit loading for the next test such that output observation tends to be random on cycles not observing targeted faults.

It should be noted that many of the x-tolerant schemes are intended for designs with a relatively limited number of x states. However, certain test methodologies can introduce a large and variable number of x states. One example is delay test with aggressive strobe timings that target short paths but cause setup time violations on longer paths.

One way to improve the efficiency of the data masking approach is to identify which scan bits have the potential for capturing x values and place them into a small number of chains. Mask vector register bits can then be allocated to ensure these chains can be easily masked; other chains that are not expected to see an x very often can share mask vector bits so the size of the mask vector is greatly reduced (e.g., by 10×). Combining the ability to mask chain outputs with an x-tolerant XOR network can minimize the impact of x values to the compression results.

21.3.4.7.4 Logic Diagnostics

Automated logic diagnostics software generally is set up to work from a bit-level fail set collected during test [16,26]. The fail set identifies which tests failed and were data logged and, within each data-logged failing test, which scan cells or output pins encountered a mismatch between the expected value and the actual response value. Assuming binary values, the failing test can be expressed as the bitwise XOR sum of the correct response vector and an *error vector*. The diagnosis software can use a *precalculated fault dictionary or posttest simulation*. In both cases, fault simulation is used to associate individual faults from a list of model faults with error-simulated vectors. Model faults like stuck-at or transition faults are generally attached to gate-level pins in the netlist of the design under test. Hence, the fault model carries gate-level locality information with it (the pin the fault is attached to). The diagnosis algorithms try to find a match between the error vectors from the data-logged fail sets and simulated error vectors that are in the dictionary or are created on the fly by fault simulation. If a match is found, then the associated fault will be added to the so-called "call-out" list of faults that partially or completely match the observed fail behavior.

Response compression/compaction reduces the bitwise response information to a much smaller amount of data. In general, the reduction is lossy in nature, meaning that it may be difficult or impossible to unambiguously reconstruct the bit-level error vectors for diagnosis. That leaves essentially two options for diagnosing fails with compressed/compacted responses. The first option is to first detect the presence of fails in a test, then reapply the same test and extract the bit-level responses without compression/compaction. This approach has already been discussed in the section on logic BIST. Quick identification of the failing tests in signature-based methods can be facilitated by adding a reset capability to the MISR macro and comparing signatures at least once for each test. The reset is applied in between tests and returns the MISR to a fixed starting state even if the previous test failed and produced an incorrect signature. Without the reset, the errors would remain in the MISR, causing incorrect signatures in subsequent tests even if there are no further errant responses captured. With the reset and signature compared at least once per test, it is easier to determine the number of failing tests and schedule all or some of them for retest and data logging.

As already explained in the section on logic BIST, bit-level data logging from retest may make it necessary to have the expected responses in the test equipment's vector memory. To what extent that is feasible during production test is a matter of careful data management and sampling logistics. Hence, there is considerable interest in enabling meaningful *direct diagnostics* that use the compacted responses without the need for retest and bit-level data logging.

Most response compression/compaction approaches used in practice are linear in nature, meaning that the *difference vector* between a nonfailing compacted response and the failing compacted response is only a function of the bit-level error vector. In other words, the difference vector contains reduced information about the error vector. The key question for direct diagnostics is to what extent it is possible to associate this reduced information with a small enough number of *matching* faults to produce a meaningful call-out list. The answer depends on what type of and how much fault-distinguishing information is preserved in the mapping from the uncompacted error vector to the compacted difference vector and how that information can be accessed in the compacted data. Also important is what assumptions can be realistically made about the error distributions and what type of encoding is used for data reduction. For example, if it is assumed that for some grouping of scan chains, at most one error is most likely to occur per scan cycle in a group, then using an error-correcting code (ECC) for compaction would permit complete reconstruction of the bit-level error vector. Another issue is what matching criteria are used for deciding whether to include a fault in the call-out. A complete match between simulated and actual error vector generally also entails a complete match between simulated and observed compacted difference vectors. That is, the mapping preserves the matching criterion (albeit the fact that the information reduction leads to higher ambiguity). On the other hand, partial matching success based on proximity in terms of Hamming distance may not be preserved. For example, MISR-based signatures essentially are hash codes that do not preserve proximity. Nor should we forget the performance considerations. The bit-level error information is used in posttest simulation methods not only to determine matching but also to greatly reduce the search space by only including faults from the back-trace cones feeding into the failing scan cells or output pins.

Selective compare at first blush appears to be a good choice for diagnostics, because it allows for easy mapping of differences in the compacted responses back to bit-level errors. However, only a small subset of responses is visible in the compacted data and it is very conceivable that the ignored data contain important fault-distinguishing information.

To make a long story short, direct diagnostics from compacted responses are an area for potentially fruitful research and development. Some increasingly encouraging successes of direct diagnosis for certain linear compression/compaction schemes have been reported recently.

21.3.4.7.5 Scan-Chain Diagnostics

Especially for complex chips in new technologies it must be expected that defects or design issues affect the correct functioning of the scan load/unload operation. For many design projects, scan is not only important for test but also for debug. Not having fully working scan chains can be a serious problem.

Normal logic diagnostics assume fully working scan chains and are not immediately useful for localizing scan-chain problems. Having a scan-chain problem means that scan cells downstream from the problem location cannot be controlled and scan cells upstream from the problem location cannot be observed by scan. The presence of such problems can mostly be detected by running scan-chain integrity tests, but it generally is difficult or impossible to derive the problem location from the results. For example, a hold-time problem that results in race through makes the scan chain look too short and it may be possible to deduce that from the integrity test results. However, knowing that one or more scan cells were skipped does not necessarily indicate which cells were skipped.

Given that the scan load/unload cannot be reliably used for control and observation, scan-chain diagnostics tend to rely on alternative means of controlling/observing the scan cells in the broken scan chains. If mux-scan is used, for example, most scan cells have a second data input (namely the system data input) other than the scan-data input, and it may be possible to control the scan cell from that input (e.g., utilizing scan cells from working scan chains). Forcing known values into scan cells through nonscan inputs for the purpose of scan-chain diagnostics is sometimes referred to as *lateral insertion* [8]. And most scan cell outputs not only feed to the scan-data input of the next cell in the chain but also feed functional logic that in turn may be observable. It should also be noted that the scan cells downstream from the problem location are observable by scan.

DFT can help with diagnostics using lateral insertion techniques. A drastic approach is to insist that all scan cells have a directly controllable set and clear to force a known state into the cells. Other scan architectures use only the clear in combination with inversion between all scan cells to help localize scan-chain problems. Having many short scan chains can also help as long as the problem is localized and affects only a few (ideally one) chain. In that scenario the vast majority of scan chains are still working and can be used for lateral insertion and observation. In this context, it is very useful to design the scan architecture such that the short scan chains can be individually scanned out to quickly determine which chains are working and which are not.

Most ATPG tools can generate a scan-chain integrity test to verify the scan logic is all working. This is true for designs using test compression as well. The chain integrity test has an alternating sequence of zeros and ones that change at some frequency; for example, 000011110000 might be used if a toggling rate of ¼ (25%) is acceptable.

When using test compression, we are dealing with many scan chains that are fairly short. If one of these short chains is failing, it is already more localized than when a chain fails without test compression. In addition, there is a better chance that scan bits that feed to the failing chain or are fed from the failing chain are in a different chain (and that is working). Utilizing the smaller chains of a test compression mode helps isolate where the possible defect is located faster than using a full-scan (noncompression) mode.

21.4 EMBEDDED MEMORY DESIGN-FOR-TEST TECHNIQUES

The typical DFTS tools convert only flip-flops or latches into scan cells. The storage cells in embedded memories generally are not automatically converted. Instead, special memory-specific DFT is used to test the memories themselves as well as the logic surrounding the memories.

21.4.1 TYPES OF EMBEDDED MEMORIES

Embedded memories come in many different flavors, varying in functionality, usage, and in the way they are implemented physically. From a test perspective, it is useful to grossly distinguish between register files and dense custom memories.

Register files are often implemented using design rules and cells similar to logic cells. The sensitivity to defects and failure modes is similar to that of logic, making them suitable for testing with typical logic tests. They tend to be relatively shallow (small address space) but can be wide (many bits per word), and have multiple ports (to enable read or write from/to several addresses simultaneously). Complex chips and cores can contain tens or even many hundreds of embedded register files.

Dense custom memories, by contrast, tend to be hand optimized and may use special design rules to improve the density of the storage cell array. Some memory types, for example, eDRAM, use additional processing steps. Because of these special properties, dense custom memories are considered to be subject to special and unique failure modes that require special testing (e.g., retention time testing, and pattern sensitivities). On the other hand, the regular repetitive structure of the memory cell arrays, unlike *random* logic, lends itself to algorithmic testing. Overall, memory testing for stand-alone as well as embedded memories has evolved in a different direction than logic testing.

In addition to *normally* addressed random-access memories (RAMs) including register files, SRAMs, and dynamic RAMs (DRAMs), there are ROMs, CAMs, and other special memories to be considered.

21.4.2 EMBEDDED MEMORIES AND LOGIC TEST

Logic testing with scan design benefits from the internal controllability and observability that comes from converting internal storage elements into scan cells. With embedded memories, the question arises as to whether and to what extent the storage cells inside the memories should likewise be converted into scan cells. The answer depends on the type of memory and how the scan function is implemented. Turning a memory element into a scan cell typically entails adding a data port for scan data, creating master–slave latch pairs or flip-flops, and connecting the scan cells into scan chains. The master–slave latch pair and scan-chain configuration can be fixed or dynamic in nature.

In the fixed configuration, each memory storage cell is (part of) one scan cell with a fixed, dedicated scan interconnection between the scan cells. This approach requires modification of the storage cell array in the register file, meaning that it can be done only by the designer of the register file. The overhead, among other things, depends on what type of cell is used for normal operation of the register file. If the cells are already flip-flops, then the implementation is relatively straightforward. If the cells are latches, then data port and a single-port scan-only latch can be added to each cell to create a master–slave latch pair for shifting. An alternative is to only add a data port to each latch and combine cell pairs into the master–slave configuration for shifting. The latter implementation tends to consume less area and power overhead, but only half of the words can be controlled or observed simultaneously (pulsing the master or slave clock overwrites the data in the corresponding latch types). Hence, this type of scan implementation is not entirely useful for debug where a nondestructive read is preferred, and such a function may have to be added externally if desired.

In the dynamic approach, a shared set of master or slave latches is temporarily associated with the latches making up one word in the register file cell array, to establish a master–slave configuration for shifting. Thus, a serial shift register through one word of the memory at a time can be formed using the normal read/write to first read the word and then write it back shifted 1-bit position. By changing the address, all register file bits can be serially controlled and observed. Because no modification of the register cell array is needed, and normal read/write operations are used, the dynamic scan approach could be implemented by the memory user. The shared latch approach and reuse of the normal read/write access for scan keep the overhead limited and the word-level access mechanism is very suitable for debug (e.g., reading/writing one particular word). The disadvantage of the dynamic approach is that the address-driven scan operation is not supported by the currently available DFT and ATPG tools.

Regardless of the implementation details, scannable register files are modeled at the gate level and tested as part of the chip logic. However, they increase the number of scan cells as well as the size of the ATPG netlist (the address decoding and read/write logic are modeled explicitly at the gate level), and thereby can possibly increase test time. Moreover, ATPG may have to be enhanced to recognize the register files and create intelligently structured tests for better compaction. Otherwise, longer than normal test sets can result.

Modern ATPG tools tend to have some sequential test generation capabilities and can handle nonscannable embedded memories to some extent. It should be noted in this context that the tools use highly abstract built-in memory models. Neither the memory cell array nor the decoding and access logic are explicitly modeled, meaning that no faults can be assigned to them for ATPG. Hence, some other means must be provided for testing the memory proper. Sequential test generation can take substantially longer and result in lower test coverage. It is recommended to make sure that the memory inputs can be controlled from and the memory outputs can be observed at scan cells or chip pins through combinational logic only. For logic BIST and other test methods that use signatures without masking, it must also be considered that nonscannable embedded memories are potential x-state sources until they are initialized. If memory access is desired as part of such a test (e.g., it is not unusual for performance-limiting paths to include embedded memories such that memory access may be required for accurate performance screening/characterization), then it may be necessary to provide some mechanism for initializing the memories. Multiple write ports can also be a source of x states if the result cannot be predicted when trying to write different data to the same word. To avoid multiport write conflicts, it may be necessary to add some form of port priority, for example, by adding logic that detects the address coincidence and degates the write clock(s) for the port(s) without priority or by delaying the write clock for the port with enough priority to assure that its data will be written last.

For best predictability in terms of ATPG runtimes and achievable logic test coverage, it may be desirable to remove entirely the burden of having to consider the memories for ATPG. That can be accomplished, for example, by providing a memory bypass (e.g., combinational connection between data inputs and data outputs) in conjunction with observe points for the address and control inputs. Rather than a combinational bypass, boundary scan can be used. With bypass or boundary scan, it may not be necessary to model the memory behavior for ATPG and a simple black-box model can be used instead. However, the bypass or boundary scan can introduce artificial timing paths or boundaries that potentially mask performance-limiting paths.

More recently, it has become desirable to test through embedded memories to catch delay defects that affect the functional paths through these memories—often part of the critical paths of any design. ATPG is tasked with creating transitions in the surrounding logic and writing them into the RAMs and reading out of RAMs/ROMs to create transitions on the outputs. These paths include the address inputs to the various ports on the memory as well as the data path, controls, and clocks. Although it is still advisable to provide an alternate means to observe the logic that feeds a memory, that nonfunctional observation path does not match the characteristics and timing of the memory. By writing through RAMs using OPCG at-speed clocking and switching address inputs on read and write ports, we test much closer to the true functional speeds through what are typically the toughest critical paths. To do this requires an ATPG model of the memories and it requires ATPG to support creation of such tests—including the OPCG clocking macros to support creation of tests that may include a minimum of 3, but may require up to 6 pulses of the same at-speed clock. ATPG may be required to generate the tests with a single scan load or it may be allowed to perform a second scan load after a transition has been written into a RAM.

21.4.3 TESTING EMBEDDED MEMORIES

It is possible to model smaller nonscannable embedded memories at the gate level. In that case, each memory bit cell is represented by a latch or flip-flop, and all decoding and access logic is represented with logic gates. Such a gate-level model makes it possible to attach faults to the logic elements representing the memory internals and to use ATPG to generate tests for those faults. For large, dense memories this approach is not very practical because the typical logic fault models may not be sufficient to represent the memory failure modes and because the ATPG fault selection and cube compaction algorithms may not be suited for generating compact tests for regular structures like memories. As a consequence, it is customary to use special memory tests for the large, dense memories and, in many cases, also for smaller memories. These memory tests are typically specified to be applied to the memory interface pins. The complicating factor for embedded memories is that the memory interface is buried inside the chip design and some mechanism is needed to transport the memory tests to/from the embedded interface through the intervening logic.

21.4.3.1 DIRECT ACCESS TESTING

Direct access testing requires that all memory interface pins are individually accessible from chip pins through combinational access paths. For today's complex chips it is rare to have natural combinational paths to/from chip pins in the functional design. It is, hence, up to DFT to provide the access paths and also provide some chip-level controls to selectively enable the access paths for testing purposes and disable them for normal functional chip operation. With direct access testing, the embedded memory essentially can be tested as if it were a stand-alone memory. It requires that the memory test program is stored in the external test equipment. Memory tests for larger memories are many cycles long and could quickly exceed the test equipment limits if the patterns for each cycle have to be stored in vector memory. This memory problem does not arise if the test equipment contains dedicated algorithmic pattern generator (APG) hardware that can be programmed to generate a wide range of memory test sequences from a very compact code. In that case, direct access testing has the benefit of being able to take full advantage of the flexibility and high degree of programmability of the APG and other memory-specific hardware/software features of the equipment.

If the chip contains multiple embedded memories or, memories with too many pins, and there are not enough chip pins available to accommodate access for all memory interface pins, then a multiplexing scheme with appropriate selection control has to be implemented. It should be noted that the need to connect the access paths to chip pins can possibly create considerable wiring overhead if too many paths have to be routed over long distances. It should also be noted that, particularly for high-performance memories, it could be difficult to control the timing characteristics of the access paths accurately enough to meet stringent test timing requirements. Variants of direct access testing may permit the use of tightly timed pipeline flip-flops or latches in the data and nonclock access paths to overcome the effect of inaccurate access path timings. The pipelining and latency of such sequential access paths must be taken into account when translating the memory test program from the memory interface to chip interface.

Another potential method for reducing the chip pin footprint required for direct access testing is to serialize the access to some memory interface pins, for example, the data pins. To that effect, scan chains are built to shift serially wide bit patterns for the data words from/to a small number of chip pins. It may take many scan clock cycles to serially load/unload an arbitrary new bit pattern and the memory may have to "wait" until that is done. The serial access, hence, would not be compatible with test sequences that depend on back-to-back memory accesses with arbitrary data changes. Test time also is a concern with serial access for larger memories.

21.4.3.2 MEMORY BIST

Although direct access testing is a viable approach in many cases, it is not always easy to implement, can lead to long test times if not enough bandwidth is available between the chip under test and the test equipment, and may consume excessive vector memory if no APG hardware is available. Moreover, it may not be easy or possible to design the access paths with sufficient timing accuracy for a thorough performance test. *Memory BIST* has become a widely used alternative. For memory BIST, one or more simplified small APG hardware macros, also known as *memory BIST controllers*, are added to the design and connected to the embedded memories using a multiplexing interface. The interface selectively connects the BIST hardware resources or the normal functional logic to the memory interface pins [17,19,44–47].

A variety of more or less sophisticated controller types are available. Some implementations use pseudorandom data patterns and signature analysis similar to logic BIST in conjunction with simple address stepping logic. Signature analysis has the advantage that the BIST controller does not have to generate expected responses for compare after read. This simplification comes at the expense of limited diagnostic resolution in case the test fails.

The memory test sequences used with APG hardware in the test equipment are constructed from heavily looped algorithms that repeatedly traverse the address space and write/read regular bit patterns into/from the memory cell array. Most memory BIST implementations used in practice follow the same scheme. *Hardwired controllers* use customized finite state machines to generate the data-in pattern sequences, expected pattern sequences, as well address traversal sequences for some common memory test algorithms (e.g., march-type algorithms). The so-called programmable controllers generally contain several hardwired test programs that can be selected at runtime by loading programming register in the controller. *Microcoded controllers* offer additional flexibility by using a dedicated microengine with a memory test–specific instruction set and an associated code memory that must be loaded at runtime to realize a range of user customizable algorithms.

The complexity of the controller not only depends on the level of programmability but also on the types of algorithms it can support. Many controllers are designed for so-called linear algorithms in which the address is counted up/down so that the full address space is traversed a certain number of times. In terms of hardware, the linear traversal of the address space can be accomplished by one address register with increment/decrement logic. Some more complex nonlinear algorithms jump back and forth between a test address and a disturb address, meaning that two address registers with increment/decrement and more complex control logic are required. For data-in generation, a data register with some logic manipulation features (invert, shift, rotate, mask) is needed. The register may not necessarily have the full data word width as long as the data patterns are regular and the missing bits can be created by replication. If bit-level compare of the data-outs is used for the BIST algorithm, then similar logic plus compare logic is required to generate the expected data-out patterns and perform the comparison. Clock and control signals for the embedded memory are generated by timing circuitry that, for example, generates memory clock and control waveforms from a single free-running reference clock. The complexity of the timing circuitry depends on how flexible it is in terms of generating different event sequences to accommodate different memory access modes and how much programmability it offers in terms of varying the relative edge offsets.

Certain memory tests, like those for retention and pattern-sensitive problems, require that specific bit values are set up in the memory cell array according to physical adjacency. The physical column/row arrangement is not always identical to the logical bit/word arrangement. Address/bit scrambling as well as cell layout details may have to be known and taken into account to generate physically meaningful patterns.

The impact of memory BIST on design complexity and design effort depends, among other things, on methodology and flow. In some ASIC flows, for example, the memory compiler returns the memories requested by the user complete with fully configured, hardwired, and preverified memory BIST hardware already connected to the memory. In other flows, it is entirely up to the user to select, add, and connect memory BIST after the memories are instantiated in the design. Automation tools for configuring, inserting, and verifying memory BIST hardware according to memory type and configuration are available. The tools generate customized BIST RTL or gate-level controllers and memory interfaces from input information about memory size/configuration, number/type of ports, read/write timings, BIST/bypass interface specification, address/bit scrambling, and physical layout characteristics. Some flows may offer some relief through optional *shared controllers* where a single controller can drive multiple embedded memories. For shared controllers, the users have to decide and specify the sharing method (e.g., testing multiple memories in parallel or one after the other).

In any scenario, memory BIST can add quite a bit of logic to the design and this additional logic should be planned and taken into account early enough in the physical design planning stage to avoid unpleasant surprises late in the design cycle. In addition to the additional transistor count, wiring issues and BIST timing closure can cause conflicts with the demands of the normal function mode. Shared controllers make placement and wiring between the controller and the memory interface more complicated and require that the trade-off between wiring complexity and additional transistor count is well understood and the sharing strategy is planned appropriately (e.g., it may not make sense to share a controller for memories that are too far away from each other). If a memory BIST controller is sufficiently far away from a memory to be tested, pipeline stages may be required to be added to allow the timing of the memory operations to run at the intended functional speeds.

21.4.3.3 COMPLEX MEMORIES AND MEMORY SUBSTRUCTURES

In addition to embedded single-/multiport embedded SRAMs, some chips contain more complex memory architectures. eDRAMs can have more complex addressing and timing than *simple* SRAMs. eDRAMs are derived from stand-alone DRAM architectures, and, for example, they may have time multiplexed addressing (i.e., the address is supplied in two cycles and assembled in the integrated memory controller) and programmable access modes (e.g., different latency, fast page/column). The BIST controller and algorithm design also has to accommodate the need for periodic refresh, all of which can make BIST controllers for eDRAM more complex.

Other fairly widely used memory types include logic capabilities. CAMs, for example, contain logic to compare the memory contents with a supplied data word. BIST controllers for CAMs use enhanced algorithms to thoroughly test the compare logic in addition to the memory array. The algorithms depend on the compare capabilities and the details of the circuit design and layout.

High-performance processors tend to utilize tightly integrated custom memory subsystems where, for example, one dense memory is used to generate addresses for another dense memory. Performance considerations do not allow for separating the memories with scan. Likewise, compare logic and other high-performance logic may be included as well. Memory subsystems of this nature generally cannot be tested with standard memory test algorithms and hence require detailed analysis of possible failure modes and custom development of suitable algorithms.

Some microcontrollers and many chips for smart cards, for example, contain embedded non-volatile flash memory. Flash memory is fairly slow and poses no challenge for designing a BIST controller with sufficient performance, but flash memory access and operation are different from SRAM/DRAM access. Embedded flash memory is commonly accessible via an embedded microprocessor that can, in many cases, be used for testing the embedded flash memory.

21.4.3.4 PERFORMANCE CHARACTERIZATION

In addition to use in production test, a BIST approach may be desired for characterizing the embedded memory performance. The BIST controller, memory interface, and timing circuitry in this scenario have to be designed for higher speed and accuracy than what is required for production testing alone. It may also be necessary to add more algorithm variations or programmability to enable stressing particular areas of the embedded memory structure. The BIST designers in this case also may want to accommodate the use of special lab equipment used for characterization.

If BIST is to be used for characterization and speed binning, it should be designed with enough performance headroom to make sure that the BIST circuitry itself is not the performance limiter that dominates the measurements. Designing the BIST circuitry to operate at *system cycle time*, as advertised by some BIST tools, may not be good enough.

The quality of speed binning and performance characterization not only depends on the performance of BIST controller and memory interface but also on the accuracy and programmability of the timing edges used for the test. The relative offset between timing edges supplied from external test equipment generally can be programmed individually and in very small increments. Simple timing circuitry for memory BIST, by contrast, tends to be aligned to the edges of a reference clock running at or near system speed. The relative edge offset can only be changed by changing the frequency, and affects all edge offsets similarly.

Some memory timing parameters can be measured with reasonable accuracy without implementing commensurate on-chip timing circuitry. For example, data-out bit values can be fed back to the address. The memory is loaded with appropriate values such that the feedback results in oscillation and the oscillation frequency is an indicator of read access time. By triggering a clock pulse from a data-out change, it may be similarly possible to obtain an indication of write access time.

21.4.3.5 DIAGNOSIS, REDUNDANCY, AND REPAIR

Embedded memories can be yield limiters and useful sources of information for yield improvement. To that effect, the memory BIST implementation should permit the collection of memory fail-bit maps. The fail-bit maps are created by data logging the compare status or compare data and, optionally, BIST controller status register contents.

A simple data-logging interface with limited diagnostic resolution can be implemented by issuing a BIST-start edge when the controller begins with the test and a cycle-by-cycle pass/fail signal that indicates whether a mismatch occurred during compare or not. Assuming the relationship between cycle count and BIST progress is known, the memory addresses associated with cycles indicating miscompare can be derived. This simple method allows for creating an address-level fail log. However, the information is generally not sufficient for FA.

To create detailed fail-bit maps suitable for FA, the miscompares must be logged at the bit level. This can, for example, be done by unloading the bit-level compare vectors to the test equipment through a fully parallel or a serial interface. The nature of the interface determines how long it takes to log a complete fail-bit map. It also has an impact on the design of the controller. If a fully parallel interface is used that is fast enough to keep up with the test, then the test can progress uninterrupted while the data are logged. If a serial or slower interface is used, then the test algorithm must be interrupted for data logging. This approach sometimes is referred to as *stop-on-nth-error*. Depending on the timing sensitivity of the tests, the test algorithm may have to be restarted from the beginning or from some suitable checkpoint each time the compare data have been logged for one fail. In other cases, the algorithm can be paused and can resume after logging.

Stop-on-*n*th-error with restart and serial data logging is relatively simple to implement, but it can be quite time-consuming if miscompares happen at many addresses (e.g., defective column). Although that is considered tolerable for FA applications in many cases, there have been efforts to reduce the data-logging time without needing a high-speed parallel interface by using on-chip data reduction techniques. For example, if several addresses fail with the same compare vector (e.g., defective column), then it would be sufficient to log the compare data once and only log the addresses for subsequent fails. To that effect, the on-chip data reduction hardware would have to remember one or more already encountered compare vectors and for each new compare vector, check whether it is different or not.

Large and dense embedded memories, for example, those used for processor cache memory, offer *redundancy* and *repair* features, meaning that the memory blocks include spare rows/columns and address manipulation features to replace the address/bit corresponding to the failing rows/columns with those corresponding to the spares. The amount and type of redundancy depend on the size of the memory blocks, their physical design, and the expected failure modes, trying to optimize postrepair memory density and yield. Memories that have either spare rows or spare columns, but not both, are said to have *1D* redundancy, and memories that have both spare rows and columns are said to have *2D* redundancy. It should be noted that large memories may be composed of smaller blocks and it is also possible to have block-level redundancy/repair.

Hard repair means that the repair information consisting of the failing row/column address(es) is stored in nonvolatile form. Common programming mechanisms for hard repair include laser-programmable fuses, electrically programmable fuses, or flash-type memory. The advantage of hard repair is that the failing rows/columns need to be identified only once upfront. However, hard repair often requires special equipment for programming, and one-time programmable fuses are not suitable for updating the repair information later in the chip's life cycle. *Soft repair*, by contrast, uses volatile memory, for example, flip-flops, to store the repair information. This eliminates the need for special equipment and the repair information can be updated later. However, the repair information is not persistent and must be redetermined after each power on.

All embedded memory blocks with redundancy/repair must be tested prior to repair, during manufacturing production test, with sufficient data logging to determine whether they can be repaired and if so, which rows/columns must be replaced. Chips with hard repair fuses may have to be brought to a special programming station for fuse blow and then be returned to a tester for postrepair retest. Stop-on-nth-error data logging with a serial interface is much too slow for practical repair during production test. Some test equipment has special hardware/software features to determine memory repair information from raw bit-level fail data. In that scenario, no special features for repair may be required in the memory BIST engine, except for providing a data-logging interface that is fast enough for real-time logging.

If real-time data logging from memory BIST is not possible or not desired, then on-chip redundancy analysis (ORA) is required. Designing a compact ORA macro is relatively simple for 1D redundancy. In essence all that is needed is memory to hold row/column addresses corresponding to the number of spares, and logic that determines whether a new failing row/column address already is in memory; if not, the memory uses the new memory if there still is room, and, if there is no more room, sets a flag indicating that the memory cannot be repaired. Dealing with 2D redundancy is much more complicated, and ORA engines for 2D redundancy are not commonly available. eDRAM, for example, may use 2D redundancy and come with custom ORA engines from the eDRAM provider. The ORA engines for 2D redundancy are relatively large, and typically one such engine is shared by several eDRAM blocks, meaning that the blocks must be tested one after the other. To minimize test time, the eDRAM BIST controller should stagger the tests for the blocks such that ORA for one block is performed when the other blocks are idle for retention testing.

If hard repair is used, the repair data are serially logged out. Even if hard repair is used, the repair information may also be written into a repair data register. Then repair is turned on, and the memory is retested to verify its full postrepair functionality.

Large complex chips may contain many memory blocks with redundancy, creating a large amount of repair data. In that scenario, it can be useful to employ data compression techniques to reduce the amount of storage needed for repair or to keep the size of the fuse bay for repair small enough.

21.5 EMBEDDED LOGIC CORE DESIGN-FOR-TEST TECHNIQUES

The notion of embedded cores has become popular with the advent of SoC products. Embedded cores are predesigned and preverified blocks that are assembled with other cores and user-defined blocks into a complex chip. Being able to isolate and test an embedded core with access restricted to a few chip pins is crucial to efficient testing of future SoC designs [5,15,61,63].

21.5.1 TYPES OF EMBEDDED DIGITAL CORES

Embedded cores are generally classified into *hard cores*, which are already physically completed and delivered with layout data, and *soft cores*, which in many cases are delivered as synthesizable RTL code. In some cases, the cores may already be synthesized to a gate-level netlist that still needs to be placed and wired (*firm cores*).

For hard cores, it is important whether or not a "test-ready" detailed gate-level netlist is made available (*white box*), or the block internals are not disclosed (*black box*). In some cases, partial information about some block features may be made available (*gray box*). Also, the details of what DFT features have been implemented in the core can be very important.

21.5.2 MERGING EMBEDDED CORES

Soft cores and some white cores can be *merged* with other compatible cores and user-defined blocks into a single netlist for test generation.

For hard cores, the success of merging depends on whether the DFT features in the core are compatible with the test methodology chosen for the combined netlist. For example, if the chosen methodology is full-scan ATPG, then the core should be designed with full-scan and the scan architecture as well as the netlist should be compatible with the ATPG tool. If logic BIST or test data compression is used, then the scan-chain segments in the core should be short enough and balanced to enable scan-chain balancing in the combined netlist.

21.5.3 DIRECT ACCESS TEST

Direct access test generally requires the inputs of the embedded core to be individually and simultaneously controlled from, and the core outputs to be individually and simultaneously observed at, chip pins, through combinational access paths. This is only possible if enough chip pins are available and the test equipment has enough appropriately configured digital tester channels. Some direct access test guidelines may allow for multiplexing subsets of a core's outputs onto common chip pins. In that case, the test must be repeated with a different output subset selected until all outputs have been observed. That is, the required chip output pin footprint for direct access can be somewhat reduced at the expense of test time. All inputs must, however, remain controllable at all times. For multiple identical cores, it may be possible to broadcast the input signals to like core inputs from a common set of chip input pins and test the cores in parallel, as long as enough chip output pins are available for observing the core outputs and testing the cores together does not exceed the power budget. Concurrent testing of nonidentical cores seems conceptually possible if all core inputs and outputs can be directly accessed simultaneously. However, even with parallel access it may not be possible to align the test waveforms for nonidentical cores well enough to fit within the capabilities of typical production test equipment (the waveforms for most equipment must fit into a common tester cycle length and limited time-set memory; the tester channels of some equipment, on the other hand, can be partitioned into multiple groups that can be programmed independently). If the chip contains multiple cores that cannot be tested in parallel, then some chip-level control scheme must be implemented to select one core (or a small enough group of compatible cores) at a time for direct access test.

In addition to implementing access paths for direct access testing, some additional DFT may be required to make sure that a complete chip test is possible and does not cause unwanted side effects. For example, extra care may be required to avoid potential three-state burn-out conditions resulting from a core with three-state outputs and some other driver on the same net trying to drive opposite values. In general, it must be expected that currently deselected cores and other logic could be exposed to unusual input conditions during the test of the currently selected core(s). The integration of cores is easier if there is a simple control state to put each core into safe state that protects the core internals from being affected by unpredictable input conditions, asserts known values at the core outputs, and keeps power/noise for/from the core minimal. For cores that can be integrated into chips in which IDDq testing is possible, the safe state or another state should prevent static current paths in the core. The core test selection mechanism should activate the access paths for the currently selected core(s), while asserting the appropriate safe/IDDq state for nonselected cores.

The inputs of black-box cores cannot be observed and the black-box outputs cannot be fully controlled. Other observe/control means must be added to the logic feeding the core inputs and fed by the core outputs (*shadow logic*) to make that logic fully testable. Isolating and diagnosing core-internal defects can be very difficult for black-box cores or cores using proprietary test approaches that are not supported by the tools available and used for chip-level test generation.

There are some DFTS and DRC tools that help with the generation, connection, and verification of direct access paths. Also, there are some test generation tools that help translate core test program pin references from the core pins to the respective chip pins, and integrate the tests into an overall chip-level test program.

21.5.4 SERIALIZING THE EMBEDDED CORE TEST INTERFACE

There may not be enough chip pins available or the overhead for full parallel access to all core pins may be unacceptable. In that case, it may be possible and useful to serialize the access to some core pins. One common serialization method is to control core input pins and observe core output pins with scan cells. Updating the input pattern or comparing the output pattern in this scenario entails a scan load/unload operation. The core test program must be able to allow insertion of multiple tester cycles worth of wait time for the scan load/unload to complete. Moreover, the output of scan cells change state during scan. If the core input pin and core test program cannot tolerate multiple arbitrary state changes during scan load/unload, then a hold latch or flip-flop may have to be provided between the scan cell output and core input to retain the previous state required by the core test program until scan load/unload is done and the new value is available. For digital logic cores, data pins may be suitable for serialized access, assuming there is no asynchronous feedback and the internal memory cells are made immune to state changes (e.g., by turning off the clocks) during the core-external scan load/unload. Although it is conceptually possible to synthesize clock pulses and arbitrary control sequences via serialized access with hold, it generally is recommended or required to provide direct combinational control from chip pins for core clock and control pins.

Functional test programs for digital logic cores can be much shorter than memory test programs for large embedded memories. Serialized access for digital logic cores can, hence, be more practical and is more widely practiced for digital logic cores. In many cases, it is possible to find existing scan cells and to sensitize logic paths between those scan cells and the core pins with little or no need for new DFT logic. This minimizes the design impact and permits testing through functional paths rather than test-only paths.

However, it should be noted that with serialized access, it generally is not possible to create multiple arbitrary back-to-back input state changes and to observe multiple back-to-back responses. Hence, serialized access may not be compatible with all types of at-speed testing.

There are some DFTS/DRC and test program translation tools that work with serialized access. Test program translation in this case is not as simple as changing pin references (and possibly polarity) and adjusting time sets as in the case of combinational direct access testing. Input value changes and output measures on core pins with serialized access entail inserting a core-external scan load/unload procedure. Additional scan load operations may be required to configure the access paths.

The capabilities of the pattern translation tool may have a significant impact on the efficiency of the translated test program. For example, if the embedded macro is or contains an embedded memory with BIST, then at least a portion of the test may consist of a loop. If the translation tool cannot preserve the loop, then the loop must be unrolled into parallel vectors prior to translation. That can result in excessive vector memory demand from the test equipment. Black-box cores that come with internal scan and with a scan-based test program are another special case. If the core scan-in/scan-out pins are identified and the associated scan load/unload procedures are appropriately defined and referenced in the core test program, then the pattern translation software may (or may not) be able to retain the scan load/unload procedure information in the translated test program, resulting in a more (or less) efficient test program.

21.5.5 STANDARDIZED EMBEDDED CORE ACCESS AND ISOLATION ARCHITECTURES

There have been several proprietary and industry-wide attempts to create an interoperable architecture for embedded core access and isolation. These architectures generally contain at least two elements. One element is the core-level test interface, often referred to as a core test *wrapper*, and the other one is a chip-level test access mechanism (TAM) to connect the interfaces among themselves and to chip pins, as well as some control infrastructure to selectively enable/disable the core-level test interfaces.

In addition to a core's functional mode of operation, there are several test modes of operation expected to be operational when the core is embedded within a chip. One or more internal-logic testing modes (INTEST modes) are used to isolate the core from surrounding logic allowing tests to be applied to it that verify its internal logic is operating correctly. There also needs to be a mode where the core is not isolated from the surrounding logic and the scan registers within the core can be used to drive values out into the surrounding logic and observe values from the surrounding logic. This mode is used to test the logic above the core within the hierarchy of the chip design. Such a mode is considered to be externally facing and is called an EXTEST mode. In some core DFT approaches, there may also be a need for a so-called BYPASS mode that allows external scan chains to shift through the core, with the core adding perhaps 1 bit to the operating scan chain. A core that is operating in a BYPASS mode is not currently being tested, but it may help facilitate the passing of data to test another core on the chip.

The typical approach used with core-based designs is to generate tests for the core by itself (in isolation—also known as out of context). These tests are then migrated to the chip such that when these tests are applied to chip pins and observing chip pins, an instance of the core (in context) is being tested. This test migration (sometimes referred to as test retargeting) is what makes it possible to create tests for a core and put them on the shelf to be used to test any chip in which that core is instantiated.

When selecting/designing a core-level test interface and TAM, it is important to have a strategy for both, that is, for testing the core itself and for testing the rest of the chip. For testing the core itself, the core test interface should have an internal test mode in which the core input signals are controlled from the TAM and the core output signals are observed by the TAM without needing participation from other surrounding logic outside of the core. The issue for testing the surrounding logic is how to observe the signals connected to the core inputs and how to control the signals driven by the core outputs. If the core is a black-box core, for example, the core inputs cannot be observed and the core outputs cannot be controlled. Even if the core is a white box, there may be no internal DFT that would make it easy. To simplify test generation for the surrounding logic, the core test interface should have an *external test mode* that provides alternate means for observing the core input signals and for controlling the core output signals, without needing participation from the core internals. Even if the core internals do not have to participate in the external test mode, it may be important to assure that the core does not go into some illegal state (e.g., three-state burn-out) or create undesired interference (e.g., noise, power, and drain) with the rest of the chip. It therefore may be recommended or required that the core test interface should have controls for a *safe mode* that protects the core itself from arbitrary external stimulation and vice versa. Last but not the least, there has to be a *normal mode* in which the core pins can be accessed for normal system function, without interference from the TAM and core test interface. Layout considerations may make it desirable to allow for routing access paths for one core "through" another core. The core test interface may offer a dedicated *TAM bypass mode* (combinational or registered) to facilitate the daisy chaining of access connections.

Overall, the TAM, so to speak, establishes switched connections between the core pins and the chip pins for the purpose of transporting test data between test equipment and embedded cores. It should be noted that some portions of the test equipment could be integrated on the chip (e.g., a BIST macro) such that some data sources/sinks for the TAM are internal macro pins and others are chip I/Os. The transport mechanism in general can be parallel or serialized. It is up to the core provider to inform the core user about any restrictions and constraints regarding the type and characteristics of the TAM connections as well as what the required/permissible data sources/sinks are for each core pin. Different core pins, for example, clock pins and data pins, tend to have different restrictions and constraints. Depending on the restrictions and constraints, there probably will be some number of pins that must have parallel connections (with or without permitted pipelining), some number of pins that could be serialized (with some of those possibly requiring a hold latch or flip-flop). If certain core input pins need to be set to a specific state to initialize the core to some mode of operation, then it may be sufficient to assure that the respective state is asserted (e.g., decoded from some test mode control signals) without needing full blown independent access from chip pins. It further should be noted that certain cores may have different modes of operations, including different test modes (e.g., logic test and memory BIST), such that there could be different mode-specific access restrictions and constraints on any given pin. The TAM has to accommodate that and if necessary, be dynamically adjustable.

Serialization of the core pin access interface reduces the number of chip pins and the bit width (e.g., affecting the wiring overhead) of the TAM required for access, at the expense of test time. In addition to serializing some portion of the TAM for an individual core, there often is a choice to be made about testing multiple cores in parallel vs. serially. If the chip contains many cores, the trade-off becomes increasingly complex. For example, the TAM for all or subsets of cores could be arranged in a daisy-chain or star configuration, and decisions need to be made about the bit width of each TAM. More decisions need to be made about testing groups of cores in parallel or in sequence. The decisions are influenced by the availability of pin, wiring and test equipment resources for parallel access, by the power/noise implications of testing several cores in parallel, by test time, and more.

Core test wrappers can be predesigned into a core by the core designer/provider (so-called wrapped core) or not (unwrapped core). In the latter case, the core user could wrap the core prior to, during, or after assembling it into a chip. In some cases, the wrapper overhead can be kept smaller by taking advantage of already existing core-internal design elements (e.g., flip-flops on the boundary of the core to implement a serial shift register for serialized access). If performance is critical, then it could be better to integrate the multiplexing function required for test access on the core input side with the first level of core logic or take advantage of hand-optimized custom design optimization that is possible in a custom core but not in synthesized logic. In any case, building a wrapper that requires modification of the core logic proper in hard cores in general can only be done by the core designer. Prewrapping a core can have the disadvantage of not being able to optimize the wrapper for a particular usage instance and of not being able to share wrapper components between cores. The once popular expectation that each third-party core assembled into a chip would be individually wrapped and tested using core-based testing has given way to a more pragmatic approach where unwrapped cores are merged into larger partitions based on other design flow and design (team, location, schedule, etc.) management decisions, and only the larger partitions are wrapped and tested using core-based testing.

A chip-level control infrastructure is needed in addition to the wrapper and TAM components to create an architecture for core-based test. The control infrastructure is responsible for distributing and (locally) decoding instructions that configure the core/wrapper modes and the TAM according to the intended test objective (e.g., testing a particular core vs. testing the logic in between cores). The distribution mechanism of the control infrastructure is, in some architectures, combined with a mechanism for retrieving local core test status/result information. For example, local core-attached instruction and status/result registers can be configured into serial scan chains (dedicated chains for core test or part of "normal" chains) for core test instruction load and status/result unload.

The IEEE 1500 Embedded Core Test standard is in its 10th year [63]. It defines a standard for embedded core wrapping, allowing for both a serial and a parallel scan-chain access mechanism to talk to the core from chip pins; the standard is flexible and does not define specifics of the TAM.

A newly approved IEEE standard for testing and accessing internal cores (the standard refers to them as embedded instruments) is IEEE P1687 (the P should be removed by the time this handbook update is published) [64]. This standard defines a means to connect to and interact with various blocks within a chip that implement an interface defined by the standard; by using segment insertion blocks (SIBs), it becomes possible to add the internal scan of a block into the middle of an existing scan chain. This may be perfect for use with embedded memory BIST controllers or even block logic BIST interfaces.

21.5.6 HIERARCHICAL CORE TEST COMPRESSION ARCHITECTURES

Given that cores can be wrapped to allow isolation from the surrounding logic, it is reasonable to pursue a strategy of divide and conquer to test the large core-based chip designs of today and the future. The IEEE 1500 standard for embedded core test provides a reasonable approach for building wrappers to isolate a core; this standard does not dictate the type of internal scan architecture of the core or of the TAM it connects to. This flexibility allows a core DFT designer to choose to use a test compression scan architecture within the core and to connect the core scan-in and scan-out pins to chip pins via a TAM of any style appropriate for further exploiting the compression within the cores—including additional compression outside the cores.

If cores internally decompress scan data fed through their scan-in pins and compress the internal chain data before feeding it out their scan-out pins, this begins an architecture that allows a single core instance to be tested using compressed stimulus and response data. With freedom to design the TAM any way necessary to further exploit compression, it becomes possible to have the chip scan-in pins fan-out to multiple core instances identically and have the core scan-outs be compressed/merged with the outputs of other core instances prior to leaving through the chip's scan-out pins. In such a chip DFT architecture, the input TAM fans out all scan-in data identically to each core instance (to identical and nonidentical cores). The output TAM merges the outputs of multiple cores together, using, for example, an XOR space compactor, allowing any number of cores to be tested while utilizing the same chip scan-in and scan-out pin resources. This allows any number of identical cores to be tested using the scan-in stimulus for a single core instance and the scan-out responses merged from the n instances of the same core down to a data volume equivalent to that from a single core.

When testing a subset of the full set of cores on a chip in this manner, those cores not participating in the current testing should be placed into a BYPASS mode. This BYPASS mode places zeros on all core outputs feeding to the chip XOR space compactor so the bypassed cores do not disturb the output compression for those cores that are participating. This gets around the limitation mentioned earlier of needing more chip pins to test more core instances.

While it is a bit more complicated, testing multiple nonidentical cores can also be done as long as ATPG is aware of the scan-in fan-out to all the core types in play. The key is that when the patterns that were generated for the cores individually are migrated to the chip level, their scan output data streams must be compressed together to form the chip-compressed scan-out streams. All core types being tested together will scan for the same scan length (those cores with shorter chains will be overscanned).

The scan-in TAM can include pipeline stages if necessary to keep the scan rate high. Similarly, the scan-out TAM with the XOR space compactor can also include pipeline stages.

Hierarchical core-based compression allows compressed tests for the cores to be migrated (also known as retargeted) to the chip level. This tests the internals of each core instance. In addition, the top-level logic must still be tested using standard ATPG run against the chip gate-level model; however, this chip model does not have to include the full gate-level model of each core—it can use gray-box models of each core that include just the boundary logic of each core plus some control logic, enough to allow the cores' EXTEST and BYPASS mode(s) to work. These boundary models allow ATPG at the chip level to be run on a model that may be less than 5% of the full chip gate-level model (if more than 95% of chip logic is contained within cores).

Nearly all commercial test tools have some form of core-based hierarchical test support. This appears to be a valid means to achieve higher test compression and yet create the chip-level tests with fewer resources by

1. Creating tests for the out-of-context core models that are relatively small
2. Migrating the core tests to the chip and allowing multiple core instances to be tested in parallel
3. Running ATPG at the chip level on a much smaller gate-level model of the chip by utilizing the gray-box boundary models of each core

By processing the much smaller core models and chip model, the total elapsed time to create the chip tests is greatly reduced and the memory needed to create the tests is also greatly reduced compared with running ATPG on a flat chip model.

21.6 EMBEDDED FIELD-PROGRAMMABLE GATE ARRAY DESIGN-FOR-TEST TECHNIQUES

Although eFPGAs are not very prevalent yet, they can have some unique properties that affect chip-level testing.

21.6.1 EMBEDDED FIELD-PROGRAMMABLE GATE ARRAY CHARACTERISTICS

The term "field programmable" means that the eFPGA function can be programmed in the field, that is long after manufacturing test. At manufacturing test time, the final function generally is not known. FPGAs contain both functional resources and programming resources. Both types of resources tend to have their own I/O interface. The functional resources are configured by the programming resources prior to actual "use" to realize the intended function, and the rest of the chip logic communicates with the functional resources of an eFPGA through the functional I/O interface of the eFPGA. The normal chip functional logic in general does not see or interact directly with the programming resources or the programming I/O interface of the eFPGAs.

The customers of chips with eFPGAs expect that all functional and programming resources of the eFPGAs are working. Hence, it is necessary at chip manufacturing test time to fully test all functional and programming resources (even if only a few of them will eventually be used).

21.6.2 EMBEDDED FIELD-PROGRAMMABLE GATE ARRAY TEST AND TEST INTEGRATION ISSUES

The chip-level DFT for chips with eFPGAs will have to deal with the duality of the programming and functional resources. Many eFPGAs, for example, use an SRAM-like array of storage elements to hold the programming information. However, the outputs of the memory cells are not connected to data selection logic for the purpose of reading back a data word from some memory address. Instead, the individual memory cell outputs are intended to control the configuration of logic function blocks and of interconnect switches that constitute the functional resources of the eFPGA. The data-in side of the eFPGA programming memory array may also be different than in a normal SRAM that is designed for random access read/write. The design is optimized for loading a full eFPGA configuration program from some standardized programming interface. The programming interfaces tend to be relatively narrow even if the internal eFPGA programming memory is relatively wide. Hence, the address/data information may have to be brought in sequentially in several chunks.

Another idiosyncrasy of FPGAs is that the use of BIST techniques for testing does not necessarily mean that the test is (nearly) as autonomous as logic BIST, for example. Several BIST techniques have been proposed and are being used for testing the functional resources of the FPGA. To that effect, the reprogrammability of the FPGA is used to temporarily configure pattern generation and response compression logic from some functional resources and configure other functional resources as test target. Once configured, the BIST is indeed largely autonomous and needs only a little support from external test equipment. The difference is that potentially large amounts of programming data need to be downloaded from the external test equipment to create a suite of BIST configurations with sufficient test coverage. The programming data can consume large amounts of vector memory and downloading them costs test time. Non-BIST tests still need programming data and have the additional problem of needing test equipment support at the functional I/O interface during test application.

The chip-level DFT approach for chips with eFPGAs has to deal with the vector memory and test time issue as well as implement a suitable test interface for the I/O interface of the functional resources (even for eFPGAs with BIST there is a need to test the interface between the chip logic and the functional resources of the eFPGA). Also to be considered is that an unconfigured eFPGA is of little help in testing the surrounding logic, and a decision has to be made whether to exploit (and depend on) the programmability of the eFPGA's functional resources or to hardwire the DFT circuitry.

21.6.3 EMBEDDED FIELD-PROGRAMMABLE GATE ARRAY TEST ACCESS TECHNIQUES

To test fully an eFPGA, it is necessary to test both the programming resources and the functional resources, meaning that test access to the interfaces of the two different types of resources is needed. Of course, it is possible to treat the eFPGA like other embedded cores and implement an interface wrapper connected to the chip-level TAM. This approach, however, does not take advantage of the fact that the chip already provides access to the eFPGA programming interface for functional configuration. The chip-level programming interface may be a good starting point for test access, especially if signature-based BIST is used to test the functional resources such that there is limited return traffic from the eFPGAs during test.

Regardless of whether the programming interface is adapted or another TAM is used, it may be desirable to test multiple eFPGA cores in parallel to reduce the demand on vector memory and test time. For example, if there are multiple identical core instances, then it makes sense to broadcast the same programming data to all instances simultaneously. The normal programming interface may have no need for and, therefore, not offer a broadcast load option for multiple eFPGA macros in parallel. Hence, such an option may have to be added to make the normal programming interface more useable for testing. Likewise, if another TAM is used, then the TAM should be configurable for the optional broadcast of programming data.

21.7 CHIP-LEVEL DESIGN-FOR-TEST FOR HIGHER-LEVEL PACKAGE TEST

All the DFT approaches discussed to this point have been with regard to helping better test the chip itself during chip manufacturing test. Now, we consider DFT logic added to chips to help test the packages the chips will go into. There is not enough space here to go into great depth about any specific area, but we will touch on a few important items that are useful to know and are applicable in various situations.

21.7.1 BOARD TEST AND CHIP BOUNDARY SCAN

In the 1980s, board designs were getting more complex. Chips were being placed directly onto the boards using *direct chip attach* and other approaches that made it difficult for board testing to contact the pins on the chips. Without access to the chip pins or the wires (which were often embedded within layers of the board), it became impossible to provide a means to test and verify the operational correctness of the boards. One DFT approach that was generally applicable to board-level testing was to insert boundary scan *wrappers* around all of the data I/Os of each chip and then use those boundary wrappers to serially load values to drive across the connections between chips on the board; these values could then be captured into the wrapper cells on other chips and serially unloaded to verify the values received were correct.

Because board developers needed the chips to come with boundary scan built into them and these chips were often created by many different companies, it became obvious that the industry needed a standard means of implementing the boundary scan and effectively forcing all chip providers to include it within their chips so that boards had a chance of being tested with modest costs. This led to the development of the IEEE 1149.1 Standard Test Access Port and Boundary Scan Architecture [62]. It took nearly a decade, but chip manufactures did eventually place this standardized form of boundary scan into most chips and board-level testing became much easier. The IEEE 1149.1 Test Access Port (TAP) includes a finite state machine that allows the number of chip pins to be fairly minimal (four or five pins); conserving chip pins was essential to getting the standard accepted within the chip design community since chip pins are fairly expensive. The architecture is also extendable because it includes an instruction register, with only a few mandatory instructions; chip designers are free to define their own instructions to perform other test or even functional operations.

The IEEE 1149 family of standards provides for many extensions, but most of them are targeted for helping test the package the chip will be inserted into. More recently, the IEEE 1149.1-2013 update includes many new features for internal chip testing.

21.7.2 MULTICHIP MODULES AND CHIP BOUNDARY SCAN

Most MCMs are manufactured by the same company that makes the chips that go into them. Even before the advent of the IEEE 1149.1 standard, chips on MCMs had boundary scan included to make it reasonable to test the internals of the chips and then test the interconnects between the chips after the chips had been placed onto the MCM. These chips implemented their boundary scan simply as a special test mode of each chip and they used the scan-in and scan-out pins already there for chip manufacturing test; they did not require the use of a finite state machine or an instruction register. Each chip on the MCM typically has connections to the MCM I/Os and these MCM pins can then give control of each chip's scan interface to allow the boundary registers to be scan loaded and unloaded.

Because IEEE 1149.1 had become so well accepted, today MCMs are more likely to be found using chips with the IEEE 1149.1 standard boundary scan rather than a custom one, but the purpose is the same—to make it easier to test the package.

21.7.3 3D CHIP STACKS AND CHIP BOUNDARY SCAN

With the advent of through-silicon vias (TSVs), it is now possible to interconnect chips in a package vertically. Whereas an MCM may place each chip on a small board or ceramic substrate with multiple layers of wiring, a 3D chip stack uses TSVs to directly connect from one chip to the one above it or below it. Although this may look similar to an MCM from the outside, from a test perspective it is different for one reason: only the chip on the bottom of the stack has pins accessible to the test equipment; all other chips in the stack can be sent data only by arranging to have the chip below it pass the data along to it.

In spite of the difference, the basic DFT plan for aiding test of a 3D package is still to include some form of chip boundary scan on each chip in the stack. Because only the bottom chip in the stack talks to the outside world, that is the only chip in a stack that would be required to implement an IEEE 1149.1 TAP interface. As of the writing of this version of the handbook, there are not many 3D chip stacks in production at the time of the publication of this edition; some experimental ones have been created and tested using a chip-level boundary scan more closely related to IEEE 1500 with the bottom chip including an 1149.1 TAP interface that can be used to generate the control signals to operate the boundary scan interfaces for all other chips in the stack [60].

There is also a new IEEE standard being developed specifically for 3D chip stacks (or a special type of MCM that uses a silicon interposer for interchip wiring, often referred to as a 2.5D package). The proposed standard is IEEE P1838 and currently looks to require an IEEE 1149.1 TAP interface on each chip in the package [65]. Many details are yet to be worked out for this standard, but as was true for board testing, a standard DFT approach is a requirement when building a higher-level package with chips from multiple vendors.

21.8 CONCLUSION

We hope you have found the information included in this chapter of use in understanding the basics of DFT for chip and higher-level package testing. There are many other topics that could be covered in this chapter. These include the issues of embedded analog/mixed-signal DFT (which is covered in Chapter 23) and DFT and I/Os, both normal and high speed. In addition, there are additional issues of top-level DFT to consider, including the integration of DFT elements, boundary scan for high-level assembly test, and test interface considerations (including IEEE 1149.1 protocol, BSDL, HSDL, COP/ESP, interrupts, burn-in). However, considerations of space and time do not let us go any further into these many interesting details at this point; perhaps in a future edition of this handbook, we will be able to cover them in some detail.

We do hope you will look to some of the references provided for more details where you are interested. Clearly, we could have included many more references as there are many interesting papers on the topics we have described; however, due to space constraints, a small sample of references are being provided. You are encouraged to search for papers on any of these topics using the IEEE Xplore facility or even your favorite Internet search engine.

REFERENCES

1. K.J. Lee et al., Using a single input to support multiple scan chains, *Proceedings of ICCAD*, 1998, pp. 74–78 (broadcast scan).
2. I.M. Ratiu and H.B. Bakoglu, Pseudorandom built-in self-test and methodology and implementation for the IBM RISC System/6000 processor, *IBM J. R&D*, 34, 78–84, 1990.
3. B.L. Keller and D.A. Haynes, Design automation for the ES/9000 series processor, *Proceedings of ICCD*, 1991, pp. 550–553.

4. A. Samad and M. Bell, Automating ASIC design-for-testability—The VLSI test assistant, *Proceedings of ITC*, 1989, pp. 819–828 (DFT automation including direct access macro test).

5. V. Immaneni and S. Raman, Direct access test scheme—Design of block and core cells for embedded ASICs, *Proceedings of ITC*, 1990, pp. 488–492.

6. B.H. Seiss et al., Test point insertion for scan-based BIST, *Proceedings of the European Test Conference*, 1991, pp. 253–262.

7. D. Kay and S. Mourad, Controllable LFSR for BIST, *Proceedings of the IEEE Instrument and Measurement Technology Conference*, 2000, pp. 223–229 (streaming LFSR-based decompressor).

8. G.A. Sarrica and B.R. Kessler, Theory and implementation of LSSD scan ring & STUMPS channel test and diagnosis, *Proceedings of the International Electronics Manufacturing Technology Symposium*, 1992, pp. 195–201 (chain diagnosis for LBIST, lateral insertion concept).

9. R. Rajski and J. Tyszer, Parallel decompressor and related methods and apparatuses, U.S. Patent 5,991,909, 1999.

10. I. Bayraktaroglu and A. Orailoglu, Test volume and application time reduction through scan chain concealment, *Proceedings of DAC*, 2001, pp. 151–155 (combinational linear decompressor).

11. D.L. Fett, Current mode simultaneous dual-read/write memory device, U.S. Patent 4,070,657, 1978 (scannable register file).

12. N.N. Tendolkar, Diagnosis of TCM failures in the IBM 3081 processor complex, *Proceedings of DAC*, 1983, pp. 196–200 (system-level ED/FI).

13. J. Reilly et al., Processor controller for the IBM 3081, *IBM J. R&D*, 26, 22–29, January 1982 (system-level scan architecture).

14. H.W. Miller, Design for test via standardized design and display techniques, *Elect. Test*, 6, 108–116, 1983 (system-level scan architecture; scannable register arrays, control, data, and shadow chains; reset and inversion for chain diagnosis).

15. A.M. Rincon et al., Core design and system-on-a-chip integration, *IEEE Des. Test*, 14, 26–35, 1997 (test access mechanism for core testing).

16. E.K. Vida-Torku et al., Bipolar, CMOS and BiCMOS circuit technologies examined for testability, *Proceedings of the 34th Midwest Symposium on Circuits and Systems*, 1992, pp. 1015–1020 (circuit-level inductive fault analysis).

17. J. Dreibelbis et al. Processor-based built-in self-test for embedded DRAM, *IEEE J. Solid-State Circ.*, 33, 1731–1740, November 1998 (microcoded BIST engine with 1D on-chip redundancy allocation).

18. U. Diebold et al., Method and apparatus for testing a VLSI device, European Patent EP 0 481 097 B1, 1995 (LFSR re-seeding with equation solving).

19. D. Westcott, The self-assist test approach to embedded arrays, *Proceedings of ITC*, 1981, pp. 203–207 (hybrid memory BIST with external control interface).

20. E.B. Eichelberger and T.W. Williams, A logic design structure for LSI testability, *Proceedings of DAC*, 1977, pp. 462–468.

21. D.R. Resnick, Testability and maintainability with a new 6k gate array, *VLSI Des.*, IV, 34–38, March/April 1983 (BILBO implementation).

22. J.J. Zasio, Shifting away from probes for wafer test, *COMPCON S'83*, 1983, pp. 317–320 (boundary scan and RPCT).

23. P. Goel, PODEM-X: An automatic test generation system for VLSI logic structures, *Proceedings of DAC*, 1981, pp. 260–268.

24. N. Benowitz et al., An advanced fault isolation system for digital logic, *IEEE Trans. Comput.*, C-24, 489–497, May 1975 (pseudo-random logic BIST).

25. P. Goel and M.T. McMahon, Electronic chip-in-place test, *Proceedings of ITC*, 1982, pp. 83–90 (sort-of boundary scan).

26. Y. Arzoumanian and J. Waicukauski, Fault diagnosis in an LSSD environment, *Proceedings of ITC*, 1981, pp. 86–88 (post-test simulation/dynamic fault dictionary; single location paradigm with exact matching).

27. K.D. Wagner, Design for testability in the Amdahl 580, *COMPCON 83*, 1983, pp. 383–388 (random access scan).

28. H.-J. Wunderlich, PROTEST: A tool for probabilistic testability analysis, *Proceedings of DAC*, 1985, pp. 204–211 (testability analysis).

29. H.C. Godoy et al., Automatic checking of logic design structures for compliance with testability ground rules, *Proceedings of DAC*, 1977, pp. 469–478 (early DFT DRC tool).

30. B. Koenemann, J. Mucha, and G. Zwiehoff, Built-in test for complex digital integrated circuits, *IEEE J. Solid State Circ.*, 15, 315–319, June 1980 (hierarchical TM-Bus concept).

31. B. Koenemann et al., Built-in logic block observation techniques, *Proceedings of ITC*, 1979, pp. 37–41.

32. R.W. Berry et al., Method and apparatus for memory dynamic burn-in, U.S. Patent, U.S. 5,375,091, 1994 (design for burn-in).

33. J.A. Waicukauski et al., Fault simulation for structured VLSI, *VLSI Syst. Des.*, December 1985, pp. 20–32 (PPSFP).

34. P.H. Bardell and W.H. McAnney, Self-testing of multi-chip logic modules, *Proceedings of ITC*, 1982, pp. 200–204 (STUMPS).
35. K. Maling and E.L. Allen, A computer organization and programming system for automated maintenance, *IEEE Trans. Electr. Comput.*, 1963, pp. 887–895 (early system-level scan methodology).
36. R.D. Eldred, Test routines based on symbolic logic statements, *ACM J.*, 6, 33–36, January 1959.
37. M. Nagamine, An automated method for designing logic circuit diagnostics programs, *Design Automation Workshop*, 1971, pp. 236–241 (parallel pattern compiled code fault simulation).
38. P. Agrawal and V.D. Agrawal, On improving the efficiency of Monte Carlo test generation, *Proceedings of FTCS*, 1975, pp. 205–209 (weighted random patterns).
39. V.S. Iyengar et al., On computing the sizes of detected delay faults, *IEEE Trans. CAD*, 9, 299–312, March 1990 (small delay fault simulator).
40. Y. Aizenbud et al., AC test quality: Beyond transition fault grading, *Proceedings of ITC*, 1992, pp. 568–577 (small delay fault simulator).
41. R.C. Wong, An AC test structure for fast memory arrays, *IBM J. R&D*, 34, 314–324, March/May 1990 (SCAT-like memory test timing).
42. M.H. McLeod, Test circuitry for delay measurements on a LSI chip, U.S. Patent 4,392,105, 1983 (on chip delay measurement method).
43. W.S. Klara et al., Self-contained array timing, U.S. Patent 4,608,669, 1986.
44. J.A. Monzel et al., AC BIST for a compilable ASIC embedded memory library, Digital papers North Atlantic test workshop, 1996 (SCAT-like approach).
45. L. Ternullo et al., Deterministic self-test of a high-speed embedded memory and logic processor subsystem, *Proceedings of ITC*, 1995, pp. 33–44 (BIST for complex memory subsystem).
46. W.V. Huott et al., Advanced microprocessor test strategy and methodology, *IBM J. R&D*, 41, 611–627, 1997 (microcoded memory BIST).
47. H. Koike et al., A BIST scheme using microprogram ROM for large capacity memories, *Proceedings of ITC*, 1990, pp. 815–822.
48. S.B. Akers, The use of linear sums in exhaustive testing, *Comput. Math Appl.*, 13, 475–483, 1987 (combinational linear decompressor for locally exhaustive testing).
49. D. Komonytsky, LSI self-test using level sensitive design and signature analysis, *Proceedings of ITC*, 1982, pp. 414–424 (scan BIST).
50. S.K. Jain and V.D. Agrawal, STAFAN: An alternative to fault simulation, *Proceedings of DAC*, 1984, pp. 18–23.
51. B. Nadeau-Dostie et al., A serial interfacing technique for built-in and external testing, *Proceedings of CICC*, 1989, pp. 22.2.1–22.2.5.
52. T.M. Storey and J.W. Barry, Delay test simulation, *Proceedings of DAC*, 1977, pp. 492–494 (simulation with calculation of test slack).
53. E.P. Hsieh et al., Delay test generation, *Proceedings of DAC*, 1977, pp. 486–491.
54. K. Kishida et al., A delay test system for high speed logic LSI's, *Proceedings of DAC*, 1986, pp. 786–790 (Hitachi delay test generation system).
55. A. Toth and C. Holt, Automated database-driven digital testing, *IEEE Comput.*, 7, 13–19, January 1974 (scan design).
56. A. Kobayashi et al., Flip-flop circuit with FLT capability, *Proceedings of IECEO Conference*, 1968, p. 962.
57. M.J.Y. Williams and J.B. Angell, Enhancing testability of large scale circuits via test points and additional logic, *IEEE Trans. Comput.*, C-22, 46–60, 1973.
58. R.R. Ramseyer et al., Strategy for testing VHSIC chips, *Proceedings of ITC*, 1982, pp. 515–518.
59. K. Chakravadhanula et al., Why is conventional ATPG not sufficient for advanced low power designs?, *Proceedings Asian Test Symposium*, 2009, pp. 295–300.
60. S. Deutsch et al., DfT architecture and ATPG for interconnect tests of JEDEC wide-I/O memory-on-logic die stacks, *Proceedings of ITC*, 2012, paper 12.4.
61. B. Keller et al., Efficient testing of hierarchical core-based SOCs, *Proceedings of ITC*, 2014, paper 6.1.
62. IEEE 1149.1, http://grouper.ieee.org/groups/1149/1/ (Accessed on December 4, 2015).
63. IEEE 1500, http://grouper.ieee.org/groups/1500/ (Accessed on December 4, 2015).
64. IEEE P1687, http://grouper.ieee.org/groups/1687/ (Accessed on December 4, 2015).
65. IEEE P1838, http://grouper.ieee.org/groups/3Dtest/ (Accessed on December 4, 2015).
66. V. Chickermane, B. Foutz, and B. Keller, Channel masking synthesis for efficient on-chip test compression, *Proceedings of ITC*, 2004, pp. 452–461.
67. S. Mitra and K.S. Kim, X-Compact: An efficient response compaction technique for test cost reduction, *Proceedings of ITC*, 2002, pp. 311–320.
68. J. Rajski et al., Convolutional compaction of test responses, *Proceedings of ITC*, 2003, pp. 745–754.
69. P. Wohl, J.A. Waicukauski, and S. Ramnath, Fully X-tolerant combinational scan compression, *Proceedings of ITC*, 2007, paper 6.1.
70. K. Chakravadhanula et al., Capture power reduction using clock gating aware ATPG, *Proceedings of ITC*, 2009, Paper 4.3.

Automatic Test Pattern Generation

22

Kwang-Ting (Tim) Cheng, Li-C. Wang, Huawei Li, and James Chien-Mo Li

CONTENTS

22.1 INTRODUCTION

Test development for complex designs can be time consuming, sometimes stretching over several months of tedious work. In the past three decades, various test development automation tools have attempted to address this problem and eliminate bottlenecks that hinder the product's time to market. These tools that automate dozens of tasks essential for developing adequate tests generally fall into four categories: design for testability (DFT), test pattern generation, pattern grading, and test program development and debugging. The focus of this chapter is on automatic test pattern generation (ATPG).

Because ATPG is one of the most difficult problems for electronic design automation, it has been researched for more than 40 years. Researchers, both theoreticians and industrial tool developers, have focused on issues such as scalability, ability to handle various fault models, and methods for extending the algorithms beyond Boolean domains to handle various abstraction levels.

Historically, ATPG has focused on a set of *faults* derived from a gate-level fault model. For a given target fault, ATPG consists of two phases: *fault activation* and *fault propagation*. *Fault activation* establishes a signal value at the fault site opposite to that produced by the fault. *Fault propagation* propagates the fault effect forward by sensitizing a path from the fault site to a primary output (PO). The objective of ATPG is to find an input (or test) sequence that, when applied to the circuit, enables testers to distinguish between the correct circuit behavior and the faulty circuit behavior caused by a particular fault. Effectiveness of ATPG is measured by the fault coverage achieved for the fault model and the number of generated vectors, which should be directly proportional to test application time.

ATPG efficiency is another important consideration. It is influenced by the fault model under consideration, the type of circuit under test (full scan, synchronous sequential, or asynchronous sequential), the level of abstraction used to represent the circuit under test (gate, register transistor, switch), and the required test quality.

As design trends move toward nanometer technology, new ATPG problems are emerging. During design validation, engineers can no longer ignore the effects of crosstalk and power supply noise on reliability and performance. Current modeling and vector-generation techniques must give way to new techniques that consider timing information, especially small delay faults (SDFs) during test generation, that are scalable to larger designs, and that can capture extreme design conditions. For nanometer technology, many current design validation problems are becoming manufacturing test problems as well, so new fault-modeling and ATPG techniques will be needed. In addition, multi-core CPUs and graphics processing units (GPUs) have become more powerful computing platforms for exploring parallel ATPG techniques to speed up computation for large-scale industrial circuits.

This chapter is divided into seven sections. Section 22.2 introduces gate-level fault models and concepts in traditional combinational ATPG. Section 22.3 discusses ATPG on gate-level sequential circuits. Section 22.4 describes circuit-based Boolean satisfiability (SAT) techniques for solving circuit-oriented problems. Section 22.5 presents advanced ATPG research topics, such as for delay faults, crosstalk, and power supply noise. Test compression and parallel ATPG are also discussed in this section. Section 22.6 introduces ATPG for design applications including logic optimization and design verification. Section 22.7 presents sequential ATPG approaches that go beyond the traditional gate-level model.

22.2 COMBINATIONAL ATPG

A fault model is a hypothesis of how the circuit may go wrong in the manufacturing process. In the past several decades, the most popular fault model used in practice is the *single stuck-at fault* model. In this model, one of the signal lines in a circuit is assumed to be stuck at a fixed logic value, regardless of what inputs are supplied to the circuit. Hence, if a circuit has n signal lines, there are potentially $2n$ stuck-at faults defined on the circuit, of which some can be viewed as being equivalent to others [1].

The stuck-at fault model is a *logical fault* model because no delay information is associated with the fault definition. It is also called a "permanent fault" model because the faulty effect is assumed to be permanent, in contrast to *intermittent* and *transient* faults that can appear randomly through time. The fault model is *structural* because it is defined based on a structural gate-level circuit model.

A stuck-at fault is said to be detected by a test pattern if, when applying the pattern to the circuit, different logic values can be observed, in at least one of the circuit's POs between the original circuit and the faulty circuit. A pattern set with 100% stuck-at fault coverage consists of tests to detect every possible stuck-at fault in a circuit.

Stuck-at fault coverage of 100% does not necessarily guarantee high quality. Earlier studies demonstrate that not all fault coverages are created equal [2,3] with respect to the quality levels they achieve. As fault coverage approaches 100%, additional stuck-at fault tests have diminishing chances to detect nontarget defects [4]. Experimental results have shown that in order to capture all nontarget defects, generating multiple tests for a fault may be required [5]. Generating tests to observe faulty sites multiple times may help to achieve higher quality [6,7]. Correlating fault coverages to test quality is a fruitful research area beyond the scope of this chapter. We use stuck-at fault as an example to illustrate the ATPG techniques.

A test pattern that detects a stuck-at fault satisfies two criteria simultaneously: *fault activation* and *fault propagation*. Consider Figure 22.1 as an example. In this example, input line *a* of the AND gate is assumed to be stuck-at 0. In order to activate this fault, a test pattern must produce logic value 1 at line *a*. Then, under the good circuit assumption, line *a* has logic value 1 when the test pattern is applied. Under the faulty circuit assumption, line *a* has logic value 0. The symbol $D = 1/0$ is used to denote the situation. *D* needs to be *propagated* through a sensitized path to one of the POs. In order for *D* to be propagated from line *a* to line *c*, input line *b* has to be set at logic value 1. The logic value 1 is called the "noncontrolling value" for an AND gate. Once *b* is set at the noncontrolling value, line *c* will have whatever logic value that line *a* has.

The ATPG process involves simultaneous justification of the logic value 1 at lines *a* and *b* and propagation of the fault difference *D* to a PO. In a typical circuit with reconvergent fanouts, the process involves a search for the right decisions to assign logic values at primary inputs (PIs) and at internal signal lines in order to accomplish both justification and propagation. The ATPG problem is an NP-complete problem [8]. Hence, all known algorithms have an exponential worst-case runtime.

Algorithm 22.1 BRANCH-AND-BOUND ATPG (*circuit, a fault*)

```
Solve();
if (bound_by_implication()=FAILURE) then
    return (FAILURE);
if (error difference at PO) and (all lines are justified) then
    return (SUCCESS);
while (there is an untried way to solve the problem) do {
    make a decision to select an untried way to propagate or to justify;
    if (Solve()=SUCCESS) then
        return (SUCCESS);
}
return (FAILURE);
```

Algorithm 22.1 illustrates a typical branch-and-bound approach to implement ATPG. The efficiency of this algorithm is affected by two things:

- The *bound_by_implication()* procedure determines whether there is a conflict in the current value assignments. This procedure helps the search to avoid making decisions in subspaces that contain no solution.

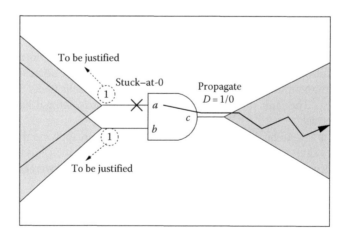

FIGURE 22.1   Fault activation and fault propagation for a stuck-at.

■ The decision-making step inside the while loop determines how branches should be ordered in the search tree. This determines how quickly a solution can be reached. For example, one can make decisions to propagate the fault difference to a PO before making decisions to justify value assignments.

22.2.1 IMPLICATION AND NECESSARY ASSIGNMENTS

The ATPG process operates on at least a five-value logic defined over $\{0, 1, D, D', X\}$; X denotes unassigned value [9]. D denotes that in the good circuit, the value should be logic 1, and in the faulty circuit the value should be logic 0. D' is the complement of D. Logical AND, OR, and NOT can be defined based on these five values [1,9]. When a signal line is assigned with a value, it can be one of the four values $\{0, 1, D, D'\}$.

After certain value assignments have been made, *necessary assignments* are those implied by the current assignments. For example, for an n-input AND gate, its output being assigned with logic 1 implies that all its inputs have to be assigned to logic 1. If its output is assigned to logic 0 and $n - 1$ inputs are assigned with logic 1, then the remaining input has to be assigned with logic 0. These necessary assignments derived from an analysis of circuit structure are called "implications." If the analysis is done individually for each gate, the implications are *direct implications*. There are other situations where implications can be *indirect*.

Figure 22.2a shows a simple example of indirect implication. Suppose that a value 0 is being justified backward through line d where line b and line c have been already assigned with logic 1. To justify $d = 0$, there are three choices to set the values of the AND's inputs. Regardless of which way is used to justify $d = 0$, the line a must be assigned with logic value 0. Hence, in this case $d = 0$, $b = 1$, $c = 1$ together implies $a = 0$. Figure 22.2b shows another example where $f = 1$ implies $e = 1$. This implication holds regardless of other assignments.

Figure 22.3 shows an example where the fault difference D is propagated through line a. Suppose that there are two possible paths to propagate D, one through line b and the other

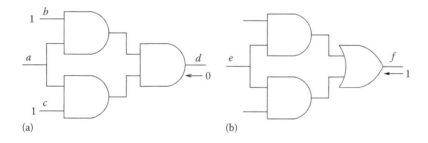

FIGURE 22.2   Implication examples [10]. (a) $d = 0$ implies $a = 0$ when $b = 1$, $c = 1$; (b) $f = 1$ implies $e = 1$.

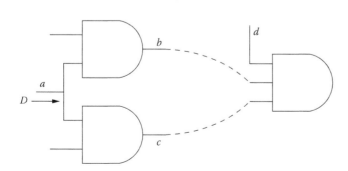

FIGURE 22.3   Implication $d = 1$ because of the unique propagation path. (From Schulz, M.H. et al., *IEEE Trans. Comput. Aided Des.*, 7, 126, 1988.)

through line c. Suppose that both paths eventually converge at a 3-input AND gate. In this case, regardless of which path is chosen as the propagation path, line d must be assigned with logic value 1. Therefore, in this case $a = D$ implies $d = 1$.

The *bound_by_implication()* procedure in Algorithm 22.1 performs implications to derive all necessary assignments. A *conflict* occurs if a line is assigned with two different values after all the necessary assignments have been derived. In this case, the procedure returns failure. It can be seen that the greater the number of necessary assignments that can be derived by implications, the more likely that a conflict can be detected. Because of this, efficiently finding necessary assignments through implications has been an important research topic for improving the performance of ATPG since the introduction of the first complete ATPG algorithm in Reference 9.

In addition to the direct implications that can easily be derived based on the definitions of logic gates, indirect implications can be obtained by analyzing circuit structure. This analysis is called "learning" where correlations among signals have been established by simple and efficient methods. Learning methods should be simple and efficient enough so that the speedup in search by exploring the resulting implications should outweigh the cost of establishing the signal correlations. There are two types of learning approaches proposed in the literature:

1. *Static learning*: Signal correlations are established before the search. For example, in Figure 22.2b, forward logic simulation with $e = 0$ obtains $f = 0$. Because $e = 0$ implies $f = 0$, we obtain that $f = 1$ implies $e = 1$. For example, in Figure 22.3, the implication $a = D \rightarrow d = 1$ is universal. This can be obtained by analyzing the *unique propagation path* from signal line a. These implications can be applied at any time during the search process.

2. *Dynamic learning*: Signal correlations are established during the search. If the learned implications are conditioned on a set of value assignments, these implications can only be used to prune the search subspace based on those assignments. For example, in Figure 22.2a, $d = 0$ implies $a = 0$ only when b and c have been assigned with 1. However, unconditional implications of the form $(x = v_1) \rightarrow (y = v_2)$ can also be learned during dynamic learning.

The concepts of static and dynamic learning were suggested in [10]. A more sophisticated learning approach called "recursive learning" was presented in [11]. Recursive learning can be applied statically or dynamically. Because of its high cost, it is more efficient to apply recursive learning dynamically. Conflict-driven recursive learning and conflict learning for ATPG were recently proposed in [12].

Knowing when to apply a particular learning technique is crucial in dynamic learning to ensure that the gain from learning outweighs the cost of learning. For example, it is more efficient to apply recursive learning on hard-to-detect faults where most of the subspaces during search contain no solution [11]. Because of this, search in these subspaces is inefficient. On the other hand, recursive learning can quickly prove that no solution exists in these subspaces. From this perspective, it appears that recursive learning implements a complementary strategy with respect to the decision tree–based search strategy, since one is more efficient for proving the absence of a solution but the other is more efficient for finding a solution. Conflict learning is another example in which learning is triggered by a conflict [12]. The conflict is analyzed and the cause of the conflict is recorded. The assumption is that during a search in a neighboring region, the same conflict might recur. By recording the cause of the conflict, the search subspace can be pruned more efficiently in the neighboring region. Conflict learning was first proposed in [13] with application in Boolean satisfiability (SAT). The authors in [12] implement the idea in their ATPG with circuit-based techniques.

22.2.2 ATPG ALGORITHMS AND DECISION ORDERING

One of the first complete ATPG algorithms is the D-algorithm [9]. Subsequently, other algorithms were proposed, including PODEM [14], FAN [15], and SOCRATES [10].

22.2.2.1 D-ALGORITHM

The D-algorithm is based on the five-value logic defined on $\{0, 1, D, D', X\}$. The search process makes decisions at PIs as well at internal signal lines. The D-algorithm is able to find a test even though a fault difference may necessitate propagation through multiple paths. Figure 22.4 illustrates such an example.

Suppose that fault difference D is propagated to line b. If the decision is made to propagate D through path d, g, j, l, we will require setting $a = 1$ and $k = 1$. Since $a = 1$ implies $i = 0$, $k = 1$ implies $h = 1$, which further implies $e = 1$. A conflict occurs when $e = 1$ and $b = D$. If the decision is made to propagate D through path e, h, k, l, we will require setting $i = 0$ and $j = 1$. $j = 1$ implies $g = 1$, which further implies $d = 1$. Again, $d = 1$ and $b = D$ cause a conflict. In this case, D has to be propagated through both paths. The required assignments are setting $a = 1, c = 1, f = 1$. This example illustrates a case when *multiple-path sensitization* is required to detect a fault.

The D-algorithm is the first ATPG algorithm that can produce a test for a fault even though it requires multiple-path sensitization. However, because the decisions are based on a five-value logic system, the search can be time consuming. In practice, most faults may require only single-path sensitization, and hence, explicit consideration of multiple-path sensitization in search may become an overhead for ATPG [16].

22.2.2.2 PODEM

The D-algorithm can be characterized as an *indirect search* approach because the goal of an ATPG is to find a test at PIs while the search decisions in D-algorithm are made on PIs and internal signal lines. PODEM implements a *direct search* approach where value assignments are made only on PIs, so that potentially the search tree is smaller. PODEM was proposed in [14]. A recent implementation of PODEM, called ATOM, is presented in [17].

22.2.2.3 FAN

FAN [15] introduces two new concepts to PODEM. First, decisions can be made on internal *head lines* that are the end points of a tree logic cone. Therefore, a value assigned to a head line is guaranteed to be justifiable because of the tree circuit structure. Second, FAN uses a *multiple-backtrace* procedure so that a set of objectives can be satisfied simultaneously. In contrast, the original PODEM tries to satisfy one objective at a time.

22.2.2.4 SOCRATES

SOCRATES [10] is a FAN-based implementation with improvements in the implication and multiple-backtrace procedures. It also offers an improved procedure to identify a *unique sensitization path* [15].

The efficiency of an ATPG implementation depends primarily on the decision ordering it takes. There can be two approaches to influence the decision ordering: one by analyzing the fault difference propagation paths and the other by measuring the testability of signal lines.

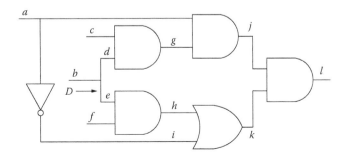

FIGURE 22.4   Fault propagation through multiple paths.

Analyzing the potential propagation paths can help to make decisions more likely to reach a solution. For example, a simple x-path check [14] can determine whether there exists a path from the point of a D line to a PO, where no signal lines on the path have been assigned any value. *Unique sensitization* [15] can identify signal lines necessary for D propagation regardless of which path it takes. The *Dominator* approach [18] can identify necessary assignments for D propagation. The authors in [16] propose a single-path-oriented ATPG where fault propagation is explicitly made to have a higher priority than value justification.

Decision ordering can be guided through *testability measures* [1,19]. There are two types of testability measures: *controllability measures* and *observability measures*. Controllability measures indicate the relative difficulty of justifying a value assignment to a signal line. Observability measures indicate the relative difficulty of propagating the fault difference D from a line to a PO. One popular strategy is to select a more difficult problem to solve before selecting the easier ones [1]. However, there can be two difficulties with a testability-guided search approach. First, the testability measures may not be sufficiently accurate. Second, always solving the hard problems first may bias the decisions too much in some cases. The authors in [12] suggest a *dynamic decision ordering* approach in which failures in the justification process will trigger changes in decision ordering.

22.2.3 BOOLEAN SATISFIABILITY–BASED ATPG

ATPG can also be viewed as solving a Boolean satisfiability (SAT) problem. SAT-based ATPG was originally proposed in [20]. This approach duplicates that part of the circuit that is influenced by the fault and constructs a satisfiability circuit instance by combining the good circuit with the faulty part. An input assignment that differentiates the faulty circuit from the good circuit is a test to detect the fault. Several other SAT-based ATPG approaches were developed later [21–23]. In [23], a SAT-based ATPG called SPIRIT was proposed, which includes almost all known ATPG techniques with improved heuristics for learning and search. ATPG and SAT will be discussed further in Section 22.4.

Practical implementation of an ATPG tool often involves a mixture of learning heuristics and search strategies. Popular commercial ATPG tools support full-scan designs where ATPG is mostly combinational. Although ATPG efficiency is important, other considerations such as test compression rate and diagnosability are also crucial for the success of an ATPG tool.

22.3 SEQUENTIAL ATPG

The first ATPG algorithm for sequential circuits was reported in 1962 by Seshu and Freeman [24]. Since then, tremendous progress has been made in the development of algorithms and tools. One of the earliest commercial tools, LASAR [25], was reported in the early 1970s.

Due to the high complexity of the sequential ATPG, it remains a challenging task for large, highly sequential circuits that do not incorporate any DFT scheme. However, these test generators, combined with low-overhead DFT techniques such as partial scan, have shown a certain degree of success in testing large designs. For designs that are sensitive to area and/or performance overhead, the solution of using sequential circuit ATPG and partial scan offers an attractive alternative to the popular full-scan solution, which is based on combinational circuit ATPG.

It requires a sequence of vectors to detect a single stuck-at fault in a sequential circuit. Also, due to the presence of memory elements, the controllability and observability of the internal signals in a sequential circuit are in general much more difficult than those in a combinational circuit. These factors make the complexity of sequential ATPG much higher than that of combinational ATPG.

Sequential circuit ATPG searches for *a sequence of vectors* to detect a particular fault through the space of all possible vector sequences. Various search strategies and heuristics have been devised to find a shorter sequence and/or to find a sequence faster. However, according to reported results, no single strategy/heuristic outperforms others for all applications/circuits. This observation implies that a test generator should include a comprehensive set of heuristics.

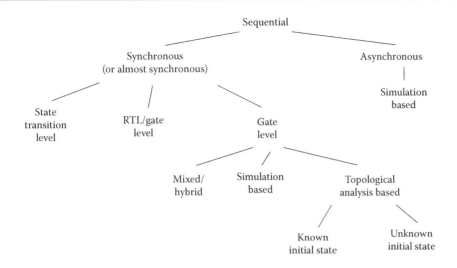

FIGURE 22.5 Sequential test generation: taxonomy.

In this section, we will discuss the basics and give a survey of methods and techniques for sequential ATPG. We focus on the methods that are based on gate-level circuit models. Examples will be given to illustrate the basics of representative methods. The problem of sequential justification, sometimes referred to as *sequential SAT*, will be discussed in more detail in Section 22.4.

Figure 22.5 shows the taxonomy for sequential test generation approaches. Few approaches can directly deal with the timing issues present in highly asynchronous circuits. Most sequential circuit test generation approaches neglect the circuit delays during test generation. Such approaches primarily target synchronous or almost synchronous (i.e., with some asynchronous reset/clear and/or few asynchronous loops) sequential circuits, but they cannot properly handle highly asynchronous circuits whose functions are strongly related to the circuit delays and are sensitive to races and hazards. One engineering solution to using such approaches for asynchronous circuits is to divide the test generation process into two phases. A potential test is first generated by ignoring the circuit delays. The potential test is then simulated using proper delay models in the second phase to check its validity. If the potential test is invalid due to race conditions, hazards, or oscillations, test generation is called again to produce a new potential test.

The approaches for (almost) synchronous circuits can be classified according to the level of abstraction at which the circuit is described. A class of approaches uses the state transition graph for test generation [26–29]. This class is suitable for pure controllers for which the state transition graphs are either readily available or easily extractable from a lower level description. For data-dominated circuits, if both register-transfer-level (RTL) and gate-level descriptions are provided, several approaches can effectively use the RTL description for state justification and fault propagation [30–32].

Most of the commercial test generators are based on the gate-level description. Some of them employ the iterative array model [33,34] and use topological analysis algorithms [35–38] or they might be enhanced from a fault simulator [24,39–42]. Some use the mixed/hybrid methods that combine the topological analysis–based methods and the simulation-based methods [43–45]. Most of these gate-level approaches assume an unknown initial state in the flip-flops, whereas some approaches assume a known initial state to avoid initialization of the state-holding elements [46–48]. The highlighted models and approaches in Figure 22.5 are those commonly adopted in most of today's sequential ATPG approaches.

22.3.1 TOPOLOGICAL ANALYSIS–BASED APPROACHES

Many sequential circuit test generators have been devised on the basis of fundamental combinational algorithms. Figure 22.6a shows the Hoffman model of a sequential circuit. Figure 22.6b shows an array of combinational logic through *time-frame expansion*. In any

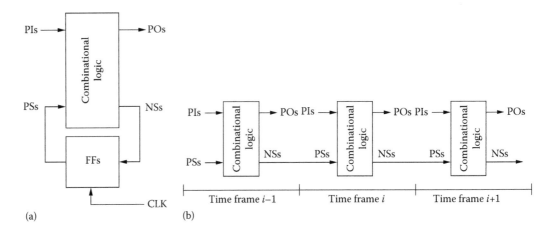

FIGURE 22.6 Synchronous sequential circuit model and time-frame expansion. (a) Huffman model and (b) iterative array model.

time frame, logic values can be assigned only to the PIs. The values on the next-state (NS) lines depend on the values of the current state lines at the end of the previous time frame. The iterative combinational model is used to approximate the timing behavior of the circuit. Topological analysis algorithms that activate faults and sensitize paths through these multiple copies of the combinational circuit are used to generate input assignments at the PIs. Note that a single stuck-at fault in a sequential circuit will correspond to a multiple stuck-at fault in the iterative array model where each time frame contains the stuck-at fault at the corresponding fault site.

The earliest algorithms extended the D-algorithm [9] based on the iterative array model [33,34]. It starts with one copy of the combinational logic and sets it to time frame 0. The D-algorithm is used for time frame 0 to generate a combinational test. When the fault effect is propagated to the NSs, a new copy of the combinational logic is created as the next time frame and the fault propagation continues. When there are values required at the present state lines, a new copy of the combinational logic is created as the previous time frame. The state justification is then performed backward in the previous time frame. The process continues until there is no value requirement at the present state lines, and a fault effect appears at a PO.

Muth [49] pointed out that the five-value logic based on $\{0, 1, D, D', X\}$ used in the D-algorithm is not sufficient for sequential ATPG. A nine-value logic is suggested to take into account the possible repeated effects of the fault in the iterative array model. Each of the nine values is defined by an ordered pair of binary values—the first value of the pair represents the ternary value (0, 1, or X) of a single line in the fault-free circuit, and the second value represents the ternary value of the signal line in the faulty circuit. Hence for a signal, there are possibly nine distinct ordered pairs (0/0, 0/1, 0/X, 1/0, 1/1, 1/X, X/0, X/1, and X/X).

The extended D-algorithm and the nine-value-based algorithm use mixed forward and reverse time processing (RTP) techniques during test generation. The requirements created during the forward process (fault propagation) have to be justified by the backward process later. The mixed time processing techniques have some disadvantages. The test generator may need to maintain a large number of time frames during test generation because all time frames are partially processed and the implementation is somewhat complicated.

The RTP technique used in the Extended Backtrace Algorithm (EBT) [35] overcomes the problems caused by the mixed time processing technique. RTP works backward in time from the last time frame to the first time frame. For a given fault, it preselects a path from the fault site to a PO. This path may involve several time frames. The selected path is then sensitized backward starting from the PO. If the path is successfully sensitized, backward justification is performed for the required value at the fault site. If the sensitization process fails, another path is selected.

RTP has two main advantages: (1) at any time during the test generation process, only two time frames need to be maintained: the current time frame and the previous one. For such a unidirectional algorithm, the backward justification process is done in a breadth-first manner.

The value requirements in time frame n are completely justified before the justification of the requirements in time frame $n - 1$. Therefore, the justified values at internal nodes of time frame n can be discarded when the justification of time frame $n - 1$ starts. As a result, the memory usage is low and the implementation is easier. Note that the decision points and their corresponding circuit status still need to be stacked for the purpose of backtracking. (2) It is easier to identify repetition of state requirements. A state requirement is defined as the state specified at the present state lines of a time frame during the backward justification process. If a state requirement has been visited earlier during the current backward justification process, the test generator has found a loop in the state transition diagram. This situation is called "state repetition." The backward justification process should not continue to circle that loop, so backtracking should take place immediately. Since justification in time frame n is completed before the justification in time frame $n - 1$, state repetition can be easily identified by simply recording the state requirement after the completion of backward justification of each time frame and then comparing each newly visited state requirement with the list of previously visited state requirements. Therefore, the search can be conducted more effectively. Similarly, the test generator can maintain a list of illegal states, that is, the states that have been previously determined as unjustifiable. Each newly visited state requirement should also be compared against this list to determine whether the state requirement is an identified illegal state, in order to avoid repetitive and unnecessary searches.

There are two major problems with the EBT algorithm: (1) only a single path is selected for sensitization. Faults that require multiple-path sensitization for detection may not be covered. (2) The number of possible paths from the fault site to the POs can be very large; trying path by path may not be efficient.

After the EBT approach, several other sequential ATPGs were proposed, including the BACK algorithm [36], HITEC [37], and FASTEST [38].

22.3.1.1 BACK

The BACK algorithm [36] is an improvement over the EBT algorithm. It also employs the RTP technique. Instead of preselecting a path, the BACK algorithm preselects a PO. It assigns a D or D' to the selected PO and justifies the value backward. A testability measure (called drivability) is used to guide the backward D-justification from the selected PO to the fault site. Drivability is a measure associated with a signal that estimates the effort of propagating a D or D' from the fault site to the signal. The drivability measurement is derived based on the SCOAP [50] controllability measurement of both fault-free and faulty circuits. For a given fault, the drivability measure of each signal is computed before test generation starts.

22.3.1.2 HITEC

HITEC [37] employs several techniques to improve the performance of test generation. Even though it uses both forward time processing and RTP, it clearly divides the test generation process into two phases. The first is the forward time processing phase in which the fault is activated and propagated to a PO. The second phase is the justification of the initial state determined in the first phase using the RTP. Due to the use of the forward time processing for fault propagation, several efficient techniques (such as the use of dominators, unique sensitization, and mandatory assignments [10,15,18,51] used in combinational ATPG) can be extended and applied in phase 1. In the RTP algorithms, such techniques are of no use. Also, no drivability is needed for the fault propagation phase, which further saves some computing time.

22.3.1.3 FASTEST

FASTEST [38] uses only forward time processing and uses PODEM [14] as the underlying test generation algorithm. For a given fault, FASTEST first attempts to estimate the total number of time frames required for detecting the fault and also to estimate at which time frame the fault is activated. The estimation is based on SCOAP [50]-like controllability and observability measures. An iterative array model with the estimated number of time frames is then constructed. The present state lines of the very first time frame have unknown values and cannot be assigned to either binary value. A PODEM-like algorithm is employed where the initial objective is to activate the target fault at the estimated time frame. After an initial objective has been determined, it backtraces starting from the line of the initial objective until it reaches an unassigned PI or a present state line in the first time frame. For the latter case, backtracking is performed immediately. This process is very similar to the PODEM algorithm except that the process now works on a circuit model with multiple time frames. If the algorithm fails to find a test within the number of time frames currently in the iterative array, the number of time frames is increased, and test generation is attempted again based on the new iterative array.

Compared to the RTP algorithms, the main advantage of the forward time processing algorithm is that it will not waste time in justifying unreachable states and will usually generate a shorter justification sequence for bringing the circuit to a hard-to-reach state. For circuits with a large number of unreachable states or hard-to-reach states, the RTP algorithms may spend too much time in proving that unreachable states are unreachable or generating an unduly long sequence to bring the circuit to a hard-to-reach state. However, the forward time processing algorithm requires a good estimate of the total number of time frames and the time frame for activating each target fault. If that estimation is not accurate, the test generator may waste much effort in the smaller-than-necessary iterative array model.

22.3.2 UNDETECTABILITY AND REDUNDANCY

For combinational circuits or full-scan sequential circuits, a fault is called "undetectable" if no input sequence can produce a fault effect at any PO. A fault is called "redundant" if the presence of the fault does not change the input/output behavior of the circuit. The detectability is associated with a test generation procedure, whereas the redundancy is associated with the functional specification of a design.

A fault is combinationally redundant if it is reported as undetectable by a complete combinational test generator [52]. The definitions of detectability and redundancy for (nonscan) sequential circuits are much more complicated [1,53,54], and these two properties (the redundancy and undetectability of stuck-at faults) are no longer equivalent [1,53,54].

It is pointed out in [54] that undetectability could be precisely defined only if a test strategy is specified, and redundancy cannot be defined unless the operational mode of the circuit is known. The authors give formal and precise definitions of undetectability with respect to four different test strategies, namely, full scan, reset, multiple observation time (MOT), and single observation time (SOT). They also explain redundancies with respect to three different circuit operational modes, namely, reset, synchronization, and nonsynchronization [54].

A fault is called undetectable under full scan if it is combinationally undetectable [55]. In the case where hardware reset is available, a fault is said to be undetectable under the reset strategy if no input sequence exists such that the output response of the fault-free circuit is different from the response of the faulty circuit, both starting from their reset states. In the case where hardware reset is not available, there are two different test strategies: the MOT strategy and the SOT strategy.

Under the SOT strategy, a sequence detects a fault only if a fault effect appears at the same PO O_i and at the same vector v_j for all power-up initial state pairs of the fault-free and faulty circuits (O_i could be any PO, and v_j could be any vector in the sequence). Most gate-level test generators and the aforementioned sequential ATPG algorithms assume the SOT test strategy.

Under the MOT strategy, a fault can be detected by multiple input sequences—each input sequence produces a fault effect at some PO for a subset of power-up initial state pairs and the union of the subsets covers all possible power-up initial state pairs (for an n-flip-flop circuit, there are 2^{2n} power-up initial state pairs). Under the MOT strategy, it is also possible to detect a fault using a single test sequence for which fault effects appear at different POs and/or different vectors for different power-up initial state pairs.

22.3.3 APPROACHES ASSUMING A KNOWN RESET STATE

To avoid generating an initialization sequence, a class of ATPG approaches assumes the existence of a known initial state. For example, this assumption is valid for circuits like controllers that usually have a hardware reset (i.e., there is an external reset signal, and the memory elements are implemented by resettable flip-flops). Approaches like STALLION [46], STEED [47], and VERITAS [48] belong to this category.

22.3.3.1 STALLION

STALLION first extracts the state transition graph (STG) for the fault-free circuit. For a given fault, it finds an activation state S and a fault propagation sequence T that will propagate the fault effect to a PO. This process is based on PODEM and the iterative array model. There is no backward state justification in this step. Using the STG, it then finds a state transfer sequence T_0 from the initial state S_0 to the activation state S. Because the derivation of the state transfer sequence is based on the state graph of the fault-free circuit, the sequence may be corrupted by the fault and hence may not bring the faulty circuit into the required state S. Therefore, fault simulation for the concatenated sequence $T_0 - T$ is required. If the concatenated sequence is not a valid test, an alternative transfer sequence or propagation sequence will be generated.

STALLION performs well for controllers for which the state transition graph can be extracted easily. However, the extraction of STG is not feasible for large circuits. To overcome this deficiency, STALLION constructs a partial STG only. If the required transfer sequence cannot be derived from the partial STG, the partial STG is then dynamically augmented.

22.3.3.2 STEED

STEED is an improvement upon STALLION. Instead of extracting the complete or partial state transition graph, it generates an ON set and an OFF set for each PO and each NS line for the fault-free circuit during the preprocessing phase. The ON set (OFF set) of a signal is the complete set of cubes (in terms of the PIs and the present state lines) that produces a logic 1 (logic 0) at a signal. The ON sets and OFF sets of the POs and NSs can be generated using a modified PODEM algorithm.

For a given fault, PODEM is used to generate one combinational test. The state transfer sequence and fault propagation sequence are constructed by intersecting the proper ON/OFF sets. In general, ON/OFF set is a more compact representation than the state transition graph. Therefore, STEED can handle larger circuits than STALLION. STEED shows good performance for circuits that have relatively small ON/OFF sets. However, generating, storing, and intersecting the ON/OFF sets can be very expensive (in terms of both CPU time and memory) for certain functions such as parity trees. Therefore, STEED may have difficulties generating tests for circuits containing such function blocks. Also, like STALLION, the transfer and fault propagation sequences derived from the ON/OFF sets of the fault-free circuit may not be valid for the faulty circuit and therefore need to be verified by a fault simulator.

22.3.3.3 VERITAS

VERITAS is a binary decision diagram (BDD)-based test generator that uses the BDD [56] to represent the state transition relations as well as sets of states. In the preprocessing phase, a state enumeration algorithm based on such BDD representations is used to find the set of states that are reachable from the reset state and the corresponding shortest transfer sequence for each of the reachable states. In the test generation phase, as with STEED, a combinational test is first generated. The state transfer sequence to drive the machine into the activation state is readily available from the data derived from reachability analysis done in the preprocessing phase. Due to the advances in BDD representation, construction, and manipulation, VERITAS in general achieves better performance than STEED.

In addition to the assumption of a known reset state, another common principle used by the three aforementioned approaches is to incorporate a preprocessing phase to (explicitly or implicitly) compute the state transition information. Such information could be used during test generation to save some repeated and unnecessary state justification effort. However, for large designs with huge state space, such preprocessing could be excessive. For example, complete reachability analysis used in the preprocessing phase of VERITAS typically fails (due to memory explosion) for designs with several hundreds of flip-flops. Either using partial reachability analysis or simply performing state justification on demand during test generation is a necessary modification to these approaches for large designs.

22.3.4 SUMMARY

The presence of flip-flops and feedback loops substantially increases the complexity of the ATPG. Due to the inherent intractability of the problem, it remains infeasible to automatically derive high-quality tests for large, nonscan sequential designs. However, because considerable progress has been made during the past few decades, and since robust commercial ATPG tools are now available, the partial-scan design methodology that relies on such tools for test generation might become a reasonable alternative to the full-scan design methodology [57].

As is the case with most other computer-aided design (CAD) tools, there are many engineering issues involved in building a test generator to handle large industrial designs. Industrial designs may contain tristate logic, bidirectional elements, gated clocks, I/O terminals, etc. Proper modeling is required for such elements, and the test generation process would also benefit from some modifications. Many of these issues are similar to those present in the combinational ATPG problem that have been addressed (e.g., [58–60]).

Developing special versions of ATPG algorithms/tools for circuits with special circuit structures and/or properties could be a good way to further improve the ATPG performance. Many partial-scan circuits have unique circuit structures. A more detailed comparison of various sequential ATPG algorithms, practical implementation issues, and applications with partial-scan designs can be found in the survey [57].

22.4 ATPG AND SAT

SAT has attracted tremendous research effort in recent years, resulting in the development of various efficient SAT solver packages. Popular SAT solvers [13,61–64] are designed based upon the conjunctive normal form (CNF).

Given a finite set of variables, V, over the set of Boolean values $B \in \{0, 1\}$, a *literal, l* or $\bar{l}$ is an instance of a variable v or its complement $\neg v$, where $v \in V$. A *clause* c_i, is a disjunction of literals $(l_1 \vee l_2 \vee \cdots \vee l_n)$. A formula f, is a conjunction of clauses $c_1 \wedge c_2 \wedge \cdots \wedge c_n$. Hence, a clause is considered as a set of literals, and a formula as a set of clauses. An assignment A *satisfies* a formula f if $f(A) = 1$. In a SAT problem, a formula f is given and the problem is to find an assignment A to satisfy f or prove that no such assignment exists.

22.4.1 SEARCH IN SAT

Modern SAT solvers are based on the search paradigm proposed in GRASP [13], which is an extension from the original DPLL [65] search algorithm. Algorithm 22.2 [13] describes the basic GRASP search procedure.

Algorithm 22.2 SAT()

```
// B is the backtracking decision level
// d is the current decision level
Search(d,B);
```

```
 if (decided(d) = SUCCESS) then
     return (SUCCESS);
 while (true) do {
 if deduce(d) ≠ (CONFLICT) then {
     if (Search(d + 1, B) = SUCCESS) then
      return (SUCCESS);
 if (B ≠ d) then {
     erase();
     return (CONFLICT);
 }
}
if (diagnose(d,B) = CONFLICT)  then {
     erase();
     return (CONFLICT);
 }
 erase();
}
```

In the algorithm, function *decide*() selects an unassigned variable and assigns it with a logic value. This variable assignment is referred to as a "decision." If no unassigned variable exists, *decide*() will return *SUCCESS*, which means that a solution has been found. Otherwise, *decide*() will return *CONFLICT* to invoke the *deduce()* procedure to check for conflict. A decision level d is associated with each decision. The first decision has decision level 1, and the decision level increases by one for each new decision.

The purpose of *deduce*() is to check for conflict by finding all necessary assignments induced by the current decisions. This step is similar to perform *implications* in ATPG. For example, in order to satisfy f, every clause of it must be satisfied. Therefore, if a clause has only one unassigned literal and all the other literals are assigned with 0, then the unassigned literal must be assigned with value 1. A conflict occurs when a variable is assigned with both 1 and 0 or a clause becomes unsatisfiable.

The purpose of *diagnose*() is to analyze the reason that causes the conflict. The reason can be recorded as a *conflict clause*. The procedure can also determine a backtracking level other than backtracking to the previous decision level, a feature that can be used to implement *nonchronological backtracking* [66]. The *erase*() procedure deletes the value assignments at the current decision level.

In a modern SAT solver, one of the key concepts is *conflict-driven learning*. *Conflict-driven learning* is a method to analyze the causes of a conflict and then record the reason as a conflict clause to prevent the search from reentering the same search subspace. Since the introduction of conflict-driven learning, a hot research topic has been to find ways to derive conflict clauses that could efficiently prune the search space.

22.4.2 COMPARISON OF ATPG AND CIRCUIT SAT

From all appearances, the problem formulation of ATPG is more complicated. ATPG involves fault activation and fault propagation, whereas circuit SAT concerns only justifying the value 1 at the single PO of a circuit. However, as mentioned in Section 2.3, the ATPG problem can also be converted into a SAT problem [20].

Conflict-driven learning was originally proposed for SAT. One nice property of conflict-driven learning is that the reason for a conflict can be recorded as a *conflict clause* whose representation is consistent with that of the original problem. This simplifies the SAT solver implementation, in contrast to ATPGs where various learning heuristics are used, each of which may require a different data structure for efficient implementation. The authors in [22] argued that this simplification in implementation might provide benefits for runtime efficiency.

For circuit SAT, the conflict clauses can also be stored as gates. Figure 22.7 illustrates this. During the application of SAT search, constraints on the signal lines can be accumulated and added onto the circuit. These constraints are represented as OR gates where the outputs are set with logic value 1. These constraints encode the signal correlations that have to hold due to the given circuit structure.

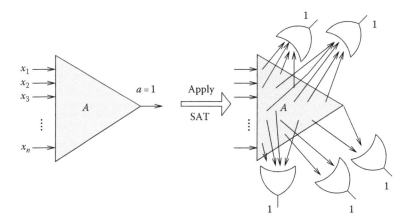

FIGURE 22.7 Learned gates accumulated by solving $a = 1$ in a circuit SAT problem.

The idea of conflict-driven learning was implemented in the ATPG in [12]. For many other applications in CAD automation of integrated circuits, applying SAT to solve a circuit-oriented problem often requires transformation of the circuit gate-level netlist into its corresponding CNF format [67]. In a typical circuit-to-CNF transformation, the topological ordering among the internal signals is obscured in the CNF formula. In CNF format, all signals become (input) *variables*.

For solving circuit-oriented problems, circuit structural information has proved to be very useful. The authors in [68] developed a structural graph model called an "implication graph" for efficient implication and learning in SAT. Methods were also provided in [69,70] to utilize structural information in SAT algorithms, which required minor modifications to the existing SAT algorithms. The authors in [71] implemented a circuit-based SAT solver that used structural information to identify unobservable gates and to remove the clauses for those gates. The work in [72] represented Boolean circuits in terms of 2-input AND gates and inverters. Based on this circuit model, a circuit SAT solver could be integrated with BDD sweeping [73].

The authors in [74] developed a circuit-based SAT solver that adopted the techniques used in CNF-based SAT solver zChaff [62], for example, the watched literal technique for efficient implication. The authors in [75] tried to recover the structural information from CNF formulas, utilizing the structural information to eliminate clauses and variables. Theoretical results regarding circuit-based SAT algorithms were presented in [76]. The authors in [77] developed a SAT solver employing circuit-based implicit and explicit learning, applying the solver on industrial hard cases [78]. The authors in [79] developed a sequential circuit SAT solver for the sequential justification problem. In the following, we summarize the ideas in [77,79] to illustrate how circuit information can be used in circuit SAT.

22.4.3 COMBINATIONAL CIRCUIT SAT

Consider the circuit in Figure 22.8, where shaded area B contains shaded area A and shaded area C contains shaded area B. Suppose we want to solve a circuit SAT problem with the output objective $c = 1$. When we apply a circuit SAT solver to prove that $c = 0$ or to find an input assignment to make $c = 1$, potentially the search space for the solver is the entire circuit. Now suppose we identify, in advance, two internal signals a and b, such that $a = 1$ and $b = 0$ will be very unlikely outcomes when random inputs are supplied to the circuit. Then, we can divide the original problem into three subproblems: (1) solving $a = 1$, (2) solving $b = 0$, and then (3) solving $c = 1$.

Since $a = 1$ is unlikely to happen, when a circuit SAT solver makes decisions trying to satisfy $a = 1$, it is likely to encounter conflicts. As a result, much conflict-driven information can be learned and stored as *conflict gates* (as illustrated in Figure 22.7). If we assume that solving $a = 1$ is done only based upon the *cone of influence* headed by the signal a (the shaded area A in Figure 22.8), then the conflict gates will be based upon the signals contained in the area A only.

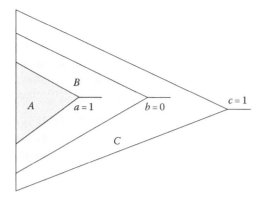

FIGURE 22.8   An example for incremental SAT solving.

As the solver finishes solving $a = 1$ and starts solving $b = 0$, all the learned information regarding the circuit area A can be used to help in solving $b = 0$. In addition, if $a = 1$ is indeed unsatisfiable, then signal a can be assigned with 0 when the solver is solving $b = 0$. Similarly, learned information from solving $a = 1$ and $b = 0$ can be reused to help in solving $c = 1$.

Intuitively, we believe that solving the three subproblems following their topological order could be accomplished much faster than directly solving the original problem. This is because when solving $b = 0$, hopefully fewer (or no) decisions are required to go into area A. Hence, the search space is more restricted within the portion of area B that is not part of area A. Similarly, solving $c = 1$ requires most decisions to be made only within the portion of area C, which is not part of area B. Moreover, the conflict gates accumulated by solving $a = 1$ could be smaller because they are based upon the signals in area A only. Similarly, the conflict gates accumulated during the solving of $b = 0$ could be smaller. Conceptually, this strategy allows solving a complex problem incrementally.

The authors in [77] implement a circuit SAT solver based on the idea just described. The decision ordering in the SAT solver is guided through *signal correlations* identified before the search process. A group of signals s_1, s_2, ..., s_i (where $i > 1$) are said to be correlated if their values satisfy a certain Boolean function $f(s_1, s_2, ..., s_i)$ during random simulation, for example, the values of s_1 and s_2 satisfy $s_1 = s_2$ during random simulation. Examples of signal correlations are *equivalence correlation*, *inverted equivalence correlation*, and *constant correlation*.

22.4.4 SEQUENTIAL CIRCUIT SAT

Recently, a sequential SAT solver utilized combined ATPG and SAT techniques to implement a sequential SAT solver by retaining the efficiency of Boolean SAT and being complete in the search [79]. Given a circuit following the Huffman synchronous sequential circuit model, *sequential SAT* (or *sequential justification*) is the problem of finding an *ordered input assignment sequence* such that a desired objective is satisfied or proving that no such sequence exists. Under this model, a sequential SAT problem can fit into one of the following two categories:

- In a weak SAT problem, an initial state value assignment is given. The problem is to find an ordered sequence of input assignments such that together with the initial state, the desired objective is satisfied or proved to be unsatisfiable.
- In a strong SAT problem, no initial state is given. Hence, it is necessary to identify an input sequence to satisfy the objective starting from the unknown state. To prove unsatisfiability, a SAT solver needs to prove that no input sequence can satisfy the given objective for *all* reachable initial states.

A strong SAT problem can be translated to a weak SAT problem by encoding technique [68]. In sequential SAT, a sequential circuit is conceptually unfolded into multiple copies of the combinational circuit through time-frame expansion. In each time frame, the circuit becomes combinational, and hence, a combinational SAT solver can be applied. In each time frame, a state element such as a flip-flop is translated into two corresponding signals: a pseudo-PI (PPI) and a pseudo-PO (PPO). The initial state is specified with the PPIs in time frame 0. The objective is specified with the signals in time frame n (the last time frame, where n is unknown before solving the problem). During search in the intermediate time frames, intermediate solutions are produced at the PPIs, and they become the intermediate PPO objectives to be justified in the previous time frames.

Given a sequential circuit, a *state clause* is a clause consisting only of state variables. A state clause encodes a state combination where no solution can be found. Due to the usage of state clauses, the time-frame expansion can be implemented implicitly by keeping only one copy of the combinational circuit.

To illustrate the usage of state clauses in sequential SAT, Figure 22.9 depicts a simple example circuit with three PIs a, b, c, one PO f, and three state-holding elements (i.e., flip-flops) x, y, z. The initial state is $x = 1, y = 0, z = 1$. Suppose the SAT objective is to satisfy $f = 1$.

Starting from time frame n where n is unknown, the circuit is treated as a combinational circuit with state variables duplicated as PPOs and PPIs. This is illustrated as (1) in the figure. Since this represents a combinational SAT problem, a combinational circuit SAT solver can be applied.

Suppose after the combinational SAT solving, we can identify a solution $a = 1, b = 0, c = 0$, $PPI_x = 0, PPI_y = 1, PPI_z = 0$ to satisfy $f = 1$ (step (2)). The PPI assignment implies a state assignment $x = 0, y = 1, z = 0$. Since it is not the initial state, at this point, we may choose to continue the search by expanding into time frame $n - 1$ (this follows a depth-first search strategy).

Before solving in time frame $n - 1$, we need to examine the solution state $x = 0, y = 1, z = 0$ more closely. This is because this solution may not represent the minimal assignment to satisfy the objective $f = 1$. Suppose after the analysis, we can determine that $z = 0$ is unnecessary. In other words, by keeping the assignment "$x = 0, y = 1$," we may discover that $f = 1$ can still be satisfied. This step is called "state minimization."

After state minimization, backward time-frame expansion is achieved by adding a *state objective* $PPO_x = 0, PPO_y = 1$ to the combinational copy of the circuit. Also, a state clause "$(x + y')$" is generated to prevent reaching those same state solutions defined by the state assignment "$x = 0, y = 1$." The new combinational SAT instance is then passed to the combinational circuit SAT solver for solving.

Suppose in time frame $n - 1$ no solution can be found. Then, we need to backtrack to time frame n to find another solution other than the state assignment "$x = 0, y = 1$." In a way, we have proved that from state "$x = 0, y = 1$," there exists no solution. This implies that there is no need to remove the state clause "$(x + y')$." However, at this point we may want to perform further analysis to determine whether both $PPO_x = 0$ and $PPO_y = 1$ are necessary to cause the unsatisfiability.

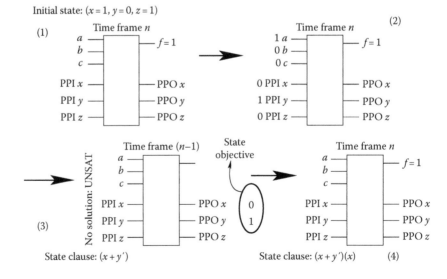

FIGURE 22.9   Sequential circuit SAT with state clauses.

Suppose that after *conflict analysis* we discover that $PPO_x = 0$ alone is sufficient to cause the conflict. Then, for the sake of efficiency, we want to add another state clause "(x)." This is illustrated in (4) of the figure. This step is called "state conflict analysis."

The new combinational SAT instance now has the state clauses "$(x + y')(x)$" that record the nonsolution state subspaces previously identified. The solving continues until either one of the following two conditions is reached:

1. After state minimization, a solution is found with a state assignment *containing* the initial state. For example, a solution with the state assignment "$x = 1, z = 1$" contains the initial state "$x = 1, y = 0, z = 1$." In this case, a solution for the sequential SAT problem is found. We note that a solution without any state assignment and with only PI assignments is considered as one containing any initial state.
2. If in time frame n, the initial objective $f = 1$ and state clauses together cannot be satisfied, then the original problem is unsatisfiable. This is equivalent to saying that, without adding any PPO objective, if the initial objective $f = 1$ and state clauses together cannot be satisfied, then the original problem is unsatisfiable. Here, with the state clauses, the time-frame expansion is conducted implicitly rather than explicitly, that is, only one copy of the combinational part of the circuit is required to be kept in the search process.

The aforementioned example illustrates several important concepts in the design of a sequential circuit SAT solver.

- The *state minimization* involves finding the minimal state assignments for an intermediate PPI solution. The search can be more efficient with an intermediate PPI solution containing a smaller number of assigned states.
- The use of state clauses serves two purposes: (1) to record those state subspaces that have been explored and (2) to record those state subspaces containing no solution. The first is to prevent the search from entering a *state justification loop*. The second follows the same spirit as that in the combinational SAT: when enough state clauses are accumulated, combinational SAT can determine that there is no solution to satisfy the initial objective and the state clauses. In this case, the problem is unsatisfiable. Note that while in sequential SAT, unsatisfiability is determined through combinational SAT, in combinational SAT, unsatisfiability is determined by implications based on the *conflict clauses*.
- Although conceptually the search follows backward time-frame expansion, the aforementioned example demonstrates that for implementation, explicit time-frame expansion is not necessary. In other words, a sequential SAT solver needs only one copy of the combinational circuit. Moreover, there is no need to memorize the number of time frames being expanded.
- The aforementioned example demonstrates the use of a depth-first search strategy. As mentioned before, decision ordering can significantly affect the efficiency of a search process. In sequential circuit SAT, the issue is when to proceed with time-frame expansion. In depth-first fashion, time-frame expansion is triggered by trying to solve the most recent state objective generated. In breadth-first fashion, the newly generated state objectives are resolved after solving all state objectives in the current time frame. A hybrid search strategy can be implemented by managing the state objective queue in various orders.

A *frame objective* is an objective to be satisfied, which is passed to the combinational SAT solver. A frame objective can be either the initial objective or a state objective. A *frame solution* is an assignment on the PIs and PPIs, which satisfies a given frame objective without conflicting with the current state clauses.

When a frame objective is proved by the combinational SAT to be unsatisfiable, conflict analysis is applied to derive additional state clauses based on only the PPOs. In other words, conflict analysis traces back to PPOs to determine which of them actually contribute to the conflict. Here, the conflict analysis can be similar to that in the combinational SAT [80], but the goal is to analyze the conflict sources up to the PPOs only.

As in the combinational SAT, where conflict clauses are accumulated through the solving process, in sequential SAT design, state clauses are accumulated through the solving process. The sequential solving process consists of a sequence of combinational solving tasks based on given frame objectives. At the beginning, the frame objective is the initial objective. As the solving proceeds, many state objectives become frame objectives. Hence, frame objectives also accumulate through the solving process.

A frame objective can be removed from the *objective array* only if it is proved to be unsatisfiable by the combinational SAT. If it is satisfiable, the frame objective stays in the *objective array*. The sequential SAT solver stops only when the *objective array* becomes empty. This means that it has exhausted all the state objectives and has also proved that the initial objective is unsatisfiable based on the accumulated state clauses.

During each step of combinational SAT, conflict clauses also accumulate through the combinational SAT solving process. When the sequential solving switches from one frame objective to another, these conflict clauses stay. Hence, in the sequential solving process, the conflict clauses generated by the combinational SAT are also accumulated. Although these conflict clauses can help to speed up the combinational SAT solving, experiences show that for sequential SAT, managing the state clauses dominates the overall sequential search efficiency [79].

Algorithm 22.3 SEQUENTIAL SAT(C, obj, s_0)

```
// C is the circuit with PPIs and PPOs expanded
// obj is the initial objective
// s_0 is the initial state
// FO is the objective array
FO <-- {obj};
while (FO ≠ Φ ) do {
    f_obj <-- select_a_frame_objective(FO);
    f_sol <-- combinational_solve_a_frame_objective(C, f_obj);
    if (f_sol = NULL) then {
        clause <-- PPO_state_conflict_analysis(C, f_obj);
        add_state_clause(C, clause);
        FO <-- FO - {f_obj};
    }
    else {
        stateassignment <-- state_minimization(C, f_obj, f_sol);
        if (s_0 ∈ stateassignment ) then
                return (SAT);
        else {
            clause <-- convert_to_clause(stateassignment);
            add_state_clause(C, clause);
            FO <-- FO + stateassignment;
        }
    }
}
return (UNSAT)
```

The overall algorithm of the sequential circuit SAT solver is described in Algorithm 22.3. Note that in this algorithm, solving each frame objective produces only one solution. However, it is easy to extend this algorithm so that solving each frame objective produces many solutions at once. The search strategy is based on the selection of one frame objective at a time from the objective array *FO*, where different heuristics can be implemented. The efficiency of the sequential circuit SAT solver highly depends on the selection heuristic [79].

22.5 ADVANCED ATPG RESEARCH

The move toward nanometer technology has introduced new failure modes and a new set of design and test problems [81]. Device features continue to shrink as the number of interconnect layers and gate density increases. The result is increased current density and a higher voltage drop along the power nets, as well as increased signal interference from coupling capacitance. The performance of devices becomes more and more vulnerable to various small delay variations caused by process variation, resistive opens and shorts, crosstalk, and power supply noise. All these give rise to timing failures and cause excessive propagation delays that degrade circuit performance. We introduce advanced research in delay faults, crosstalk faults, and power-aware ATPG in Sections 22.5.1, 22.5.2, and 22.5.3, respectively. With ever-increasing test data volume, Section 22.5.4 describes important test compression techniques, which have been widely implemented in modern chips. Section 22.5.5 shows parallel algorithms for multicore and many-core computers. Finally, Section 22.5.6 shows a novel application: ATPG for hardware Trojan detection.

22.5.1 ATPG FOR DELAY FAULTS AND NOISE

Demands for higher circuit operating frequencies, lower cost, and higher quality mean that testing must ascertain that the circuit's timing is correct. Timing defects can stay undetected after logic–fault testing such as testing of stuck-at faults, but they can be detected using delay tests. In particular, detection of SDFs requires test generation to sensitize paths that are long enough to make the accumulated delay exceed the test clock cycle.

22.5.1.1 TEST APPLICATION STRATEGIES FOR DELAY TESTING

Unlike ATPG for stuck-at faults, ATPG for delay faults is closely tied to the test application strategy [82]. Before tests for delay faults are derived, the test application strategy has to be decided. The strategy depends on the circuit type as well as on the test equipment's speed.

In structural delay testing, detecting delay faults requires applying 2-vector patterns (V1, V2) to the combinational part of the circuit at the circuit's intended operating speed. However, because high-speed testers require huge investments, most testers could be slower than the designs being tested. Testing high-speed designs on slower testers requires special test application and test generation strategies—a topic that has been investigated for many years [85]. Because an arbitrary vector pair cannot be applied to the combinational part of a sequential circuit, ATPG for delay faults may be significantly more difficult for these than for full-scan circuits. Various testing strategies for sequential circuits have been proposed. According to the way of obtaining the second vector V2, they can be generally classified into three categories: enhanced scan (ES), launch on shift (LOS), and launch on capture (LOC) [83].

In the ES strategy, standard scan cells are replaced by ES cells, which can hold two scanned-in bits separately. Hence, the functional dependency for obtaining V2 from V1 is removed. Thereby high delay fault coverage can be achieved with a more compacted test data volume using the ES strategy. However, due to the unacceptable hardware overhead, ES is rarely supported in modern VLSI chips. In the LOS strategy, the second vector V2 is generated by shifting one bit from the first vector V1 [84]. The implementation of LOS generally requires a timing critical scan enable signal. In the LOC strategy, which is widely practiced by industry because of the low implementation cost, the second vector V2 is obtained by capturing the circuit response to the first vector V1 [85]. However, due to the stringent functional dependency for generating V2 from V1, either the generated test data volume is very large or the delay fault coverage is limited using the LOC strategy, especially for circuits with high complexity.

Some efforts on partial enhanced-scan strategies were explored to increase delay fault coverage [86,87] or reduce the test data volume [88,89]. In these methods, several effective metrics for fault coverage improvement or test data volume reduction were developed from controllability measures and observability measures. They select a small number of standard scan cells (typical 1%) to be replaced with ES cells [89]. The ES cell may require a specially designed structure to avoid the need for an expensive global timing critical scan enable signal as required by the LOS strategy [87].

22.5.1.2 PATH SELECTION FOR DETECTING SDFs

In delay testing, a path is a combinational path that starts from a PI or the output of a flip-flop (called PPI) and ends at a PO or the input of a flip-flop (called PPO). The object of delay testing is to apply test patterns to detect delay fault via at least one path of which delay exceeds the period of system clock. There is no doubt that this object will be continuously important for higher and higher clock frequencies. Unfortunately, the number of paths is exponential to the number of gates in the circuit, so it is impractical to exercise each path during ATPG. Such problems as which paths are critical and how many paths are enough to be tested in delay testing, though having been targeted for many years, have yet no satisfactory answers. Things become even worse when considering delay variations or SDFs caused by process variations, power supply noise, crosstalk noise, and resistive shorts and opens. Note that resistive shorts and opens are often referred as small delay defects (SDDs) in the literature.

There are basically two types of delay fault models adopted in ATPG: transition delay fault (TDF) and path delay fault (PDF). The advantage of ATPG for PDFs is its ability to generate tests that detect cumulative or distributed delay defects. It is natural to detect SDFs if long paths are targeted for test generation. Considering the explosive growth of path numbers, it is recommended that only critical paths, or the longer paths, are considered in ATPG and delay testing as well. Static timing analysis (STA) is usually used to find long paths in industry.

However, long path selection by STA is not reliable. Some of the challenges are listed here:

1. Selection before test generation is always misleading. On the one hand, many paths are functionally redundant. On the other hand, path delay varies among different test patterns due to crosstalk, multiple input switching noise, and power supply noise.
2. With tight timing constraint during synthesis, path delay tends to be balanced. In other words, more and more paths are long paths.
3. Due to process variation, a designated medium path can become a long path after fabrication.

Considering the obstacles in path delay testing, more researchers have been focused on enhancing ATPG for TDFs in the recent decade. A TDF on a line refers to a slow signal change on the line. Traditional test generation of TDFs does not consider the path delay of the sensitized path for fault propagation. It is thus implicitly assumed that any path propagating the corresponding signal change on the line will fail the generated test if there is a TDF on the line. This requires the size of the delay fault larger than the slack of the tested path (often a shorter path). Certain large-size delay faults are regarded as gross delay faults (GDFs) [90]. A TDF test set has the ability to cover the whole GDF space and part of SDF space depending on the sensitized long paths. For instance, a method to select the K longest path through each gate was proposed to improve the possibility of detecting delay faults caused by small delay variations [91]. Several metrics, such as statistical delay quality level [92] and delay test coverage [93], have been proposed to measure the SDF coverage.

Either timing-aware test generation or test generation for faster-than-at-speed testing can enhance the detection of SDFs. Timing-aware test generation tries to activate the fault and propagate the fault effect through the longest path [93–95]. The test generation time increases considerably when these tools are run in timing-aware mode. Faster-than-at-speed testing, on the other hand, reduces the slack of the sensitized paths for each TDF test by increasing the frequency of the test clock [96,97]. However, most of these methods are based on inaccurate timing information provided by STA.

Resulting from the development of statistical STA techniques [98], many new methods for path selection have been proposed. One method proposes to take path correlations into consideration based on Monte Carlo simulation results [99]. First, a path whose delay is longer than the clock cycle by a certain probability is considered as a candidate critical path and all candidate critical paths are enumerated. Second, path refinement is processed on all candidate critical paths using Monte Carlo simulation, and a limited number of paths are selected to guarantee that when these paths satisfy the delay constraint the circuit will meet the delay constraint with a high probability. The method suggested in [100] adopts two heuristics for path reduction. First, those paths that cannot be the longest path under any process condition are deleted in the first heuristic. After that, a path is identified as an insignificant path if there exists another path such that their maximum delay difference is small, and all insignificant paths are deleted in the second heuristic. The method suggested in [101] adopts a branch-and-bound search strategy and the criticalities of subpaths are calculated during the path selection process. One path of the test path set is substituted by another path each time if the test quality of the test path set can be improved. The method suggested in [102] converts the path selection problem to a minimal space intersection problem based on a statistical timing model and calculates the probability that all the paths in a path set meet the circuit's delay constraint. Given a path number threshold, this method can efficiently select paths maximizing the probability of capturing potential delay failures caused by the accumulated distributed small delay variations.

Detailed introductions of testing SDFs can be found in [103]. Rather than sensitizing long paths during ATPG to detect SDFs, there are also many methods proposed for pattern grading and selection for screening SDFs from a large delay test set (such as an *n*-detection TDF test set) [104–106].

22.5.2 ATPG FOR CROSSTALK FAULTS

Noise faults must be detected during both design verification and manufacturing testing. When coupled with process variations, noise effects can exercise worst-case design corners that exceed operating conditions. These corners must be identified and checked as part of design validation. This task is extremely difficult, however, because noise effects are highly sensitive to the input pattern and to timing. Timing analysis that cannot consider how noise effects influence propagation delays will not provide an accurate estimation of performance, nor will it reliably identify problem areas in the design.

An efficient ATPG method must be able to generate validation vectors that can exercise worst-case design corners. To do this, it must integrate accurate timing information when the test vectors are derived. For manufacturing testing, ATPG techniques must be augmented and adapted to new failure conditions introduced by nanometer technology. Tests for conventional fault models, such as stuck-at and transition faults, obviously cannot detect these conditions. Thus, to check worst-case design corners, test vectors must sensitize the faults and propagate their effects to the POs, as well as activate the conditions of worst-case noise effects. They must also scale to increasingly larger designs.

The increased design density in deep-submicron designs leads to more significant interference between the signals because of capacitive coupling, or crosstalk. Crosstalk can induce both Boolean errors and delay faults. Therefore, ATPG for worst-case crosstalk effects must produce vectors that can create and propagate crosstalk pulses as well as crosstalk-induced delays [107–111].

22.5.2.1 CROSSTALK-INDUCED PULSES

Hazard-sensitive lines such as inputs to dynamic gates, clock, set/reset, and data inputs to flip-flops are likely to suffer from errors caused by crosstalk-induced pulses. Crosstalk pulses can result in logic errors or degraded voltage levels, which increase propagation delays. ATPG for worst-case crosstalk pulse aims to generate a pulse of maximum amplitude and width at the fault site and propagate its effects to POs with minimal attenuation [112].

A mixed-signal test generation process was proposed in [113] where characteristics of crosstalk-induced pulses are accurately modeled for a pair of coupling lines. Modeling in the presence of multiple aggressors and victims was discussed in [114], which can be effectively used in testing crosstalk faults. Modeling without timing information tends to get pessimistic results of fault analysis.

The timing of transitions in aggressors is usually captured by timing windows obtained by STA. In [115], a crosstalk target identification framework was proposed, which is composed of a set of extractors and filters to identify the target faults. In [116], a more accurately signal-switching method was introduced, which is characterized by the set of discontinuous timing window to reduce the number of violations. In [117], a timing analysis technique was proposed in which circuit functionality, delay, and crosstalk-induced pulses are simultaneously considered using time-sliced Boolean logic.

The maximum crosstalk-induced effect appears under the condition that all the aggressors switch in same direction and time, which is called maximum aggressive time (MAT) in [118].

In order to find the test patterns satisfying the maximum glitch condition, the multiple crosstalk-induced glitch fault (MCGF) model was introduced in [118], which identifies subpaths to sensitize and generate the necessary aggressor transitions coupled to a victim line. In order to find the MAT and appropriate subpaths causing the maximum effect on the victim line, an improved STA method was presented using two transition maps to record all the likely rise and fall transition arrival time of a line and the effective coupling capacitance of each aggressor lines to estimate the noise on victim lines at certain time. The test for an MCGF is a two-vector pattern that sensitizes the transition signals along the subpath to each aggressor line at the MAT and propagates the signal on a victim line to an output as well. Therefore, given an accurate timing model, the crosstalk-induced pulses can be effectively identified and exactly activated using the generated test patterns.

22.5.2.2 CROSSTALK-INDUCED DELAY

Studies show that increased coupling effects between signals can cause signal delay to increase (slowdown) or decrease (speedup) significantly. Both conditions can cause errors. Signal slowdown can cause delay faults if a transition is propagated along paths with small slacks. Signal speedup can cause race conditions if transitions are propagated along short paths. To guarantee design performance, ATPG techniques must consider how worst-case crosstalk affects propagation delays [108,109].

To generate deterministic test patterns for crosstalk-induced delay faults, timing information cannot be ignored. However, including timing information into an ATPG engine will significantly increase the complexity of the ATPG algorithm. Considering the timing of the aggressors is the main obstacle for efficient test generation.

The authors in [109] proposed a test pattern generation algorithm with a timing-oriented backtrace procedure targeting coupled transition faults. The authors in [119] presented a timing-independent approach to generate tests for crosstalk-induced slowdown effects. Focused on all aggressor lines of a victim line, the authors in [120] proposed a solution that combines an integer linear program with the traditional stuck-at fault ATPG. These three methods could not activate the worst-case crosstalk-induced delay since they consider testing of the crosstalk effect on a single victim line, similar to the TDF testing, without considering accumulative delay defects or effects on a path.

The authors in [110] presented a constrained PDF (CPDF) model as a combination of a critical path and a set of crosstalk noise sources interacting with the path. A genetic algorithm with a dynamic timing simulator was used to deal with timing information and determine unjustified PIs. This test generation method is a nondeterministic approach in terms of crosstalk activation. The authors in [121] proposed a timed ATPG method to generate critical paths and corresponding input vectors to sensitize these paths under crosstalk effects. It incorporated special timing processing techniques into ATPG algorithms and employed circuit-level timing simulation, which is computationally expensive.

The authors in [122] used the timed-Boolean logic to characterize signal transitions in a time interval and used Boolean SAT to check the correlations between aggressor and victim transitions. The authors in [123] proposed a structural test pattern generation procedure to magnify parasitic crosstalk effects on delay-sensitive paths by inducing switching on nearby nets. These two methods can find the patterns efficiently by ignoring the timing of aggressors.

The authors in [124] proposed a test generation method for critical paths considering multiple-aggressor crosstalk effects to maximize the noise of the victim lines. Physical and timing information are used to prune false aggressors, which is helpful for reducing the ATPG time cost. Specifically, timing false crosstalk effects are reduced based on static timing window analysis and recalculated timing windows using a delay test pattern of a victim path.

The authors in [125] introduced a precise crosstalk-induced PDF (PCPDF) model, denoted as $(p, \{sp\text{-}a_i, <v_i, a_i>\})$, which is similar to CPDF, but consists of a critical path p (with a number of victim lines v_i on p) and subpaths $sp\text{-}a_i$ propagating transitions to the aggressor lines a_i at certain times. Since the exact timing of signal switching is determined by the subpath reaching the line, sensitizing specific subpaths assures the aggressors' switching time [126]. Based on the PCPDF model, the method proposed in [127] adopts a path delay test generation flow toward activation of

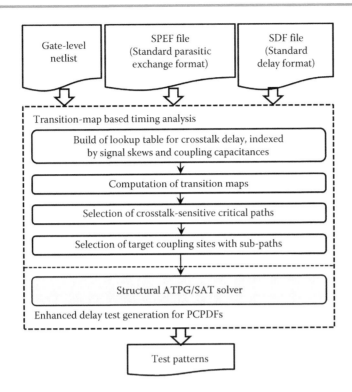

FIGURE 22.10   Flow diagram for test generation toward activation of worst-case coupling effects on paths.

worst-case crosstalk effects, as shown in Figure 22.10. It performs a transition map–based timing analysis to identify crosstalk-sensitive critical paths, followed by a deterministic test generation process for PCPDFs. Timing and logic constraints are unified during test generation to deterministically activate crosstalk effects. Consequently, a structural ATPG tool, or a Boolean SAT solver, can be efficiently applied for test generation.

Algorithm 22.4 describes the structural ATPG procedure, which can be easily extended from a conventional path delay ATPG engine [127].

Algorithm 22.4 ATPG _ for _ PCPDF(f)

```
// the circuit is stored as a levelized netlist
// f is a PCPDF fault (p, {sp-a_i, <v_i, a_i>})
// T is the test pattern to be generated
Sort coupling sites a_i of p from PI to PO;
S = first coupling site; S_last = last coupling site;
T = the pattern robustly sensitizing p;
while(true) do {
  Select an unconsidered sub-path to S;
  while(sub-path is found)  do {
    Generate a pattern T_sub which sensitizes the sub-path, and is
      consistent with T;
    if(T_sub is found) then {
      Update the accumulative crosstalk-induced delay on path p;
      T = T_sub ∩ T;
    }
    else  Select an unconsidered sub-path to S;
  }
  if S == S_last then
    return T;
  else  S = next coupling site;
}
```

The worst crosstalk-induced delay on a path can be estimated based on transition maps and crosstalk-sensitive critical paths can be found efficiently. This method can trade accuracy for efficiency by increasing the size of timescale used in the transition map, which exhibits good scalability to large circuits.

The aforementioned methods can be used as helpful extensions of current ATPG systems, but they still have their limitations to be applied by industry. Methods ignoring timing cannot guarantee meeting of the timing requirements for activation of the crosstalk effects, while methods depended on an accurate delay model require extensions or evaluations under process variations.

22.5.3 POWER-AWARE ATPG

It has been long observed that test power is much higher than the functional power because *circuits under test* are exercised intensively during test [128]. Besides overheating, excessive power supply noise during test application may lead to extra delay and even cause yield loss [129]. During scan test, there are two major sources of test power consumption: *shift power* and *capture power*. The former is caused by scan chain shifting, while the latter is caused by the capture cycles. In terms of heat dissipation issue, shift power is a bigger concern since the number of shift cycles is much more than that of capture cycles. For at-speed testing, however, capture power should be considered because excessive power supply noise may lead to extra path delay and even yield loss. Many power-aware ATPG techniques have been summarized in the survey papers and books [130–133].

Power reduction can be performed in different phases of ATPG: test cube generation [134,135], test compression [136,137], test pattern ordering [138,139], and X-filling. After test generation for specific target faults, there are many don't care bits (X) in the test cubes. Traditionally, don't care bits are randomly filled to shorten the test length. Since there are many don't care bits (typically more than 90%) in the test cubes, X-filling is the most effective way to reduce test power.

In order to reduce shift power, *adjacent fill* [140] and *minimum transition fill* [141] repeat the same bits as its scan chain neighbors to reduce shift power. For example, the following test cube can be easily filled as

$$0XXX00XX1XXX1 \rightarrow 0000000011111$$

In order to reduce capture power, *FF-silencing* tries to equalize the FF values before and after the capture cycle. There are four possible scenarios. For type A, both values before and after capture are determined. Apparently, there is no X-filling possible for type A. For type B, the value before capture is X (unfilled) but the value after capture is determined (0 or 1). We can fill in the same value as the captured value for type B. For type C, the value before capture is determined but the value after capture is X. For type D, both the values before and after capture are X. Different approaches have been proposed to X-fill type C and type D flip-flops: *justification based* and *probability based*. Justification-based X-filling approaches can be more effective than the latter but requires more computation time.

The first justification approach, low capture power X-filling (*LCP-fill*), was proposed in [142]. Type B flip-flops are first assigned and simulated iteratively until only type C and type D flip-flops remain. Then one or more type C flip-flops are chosen for justification. After each assignment, logic simulation is performed and more type B and type C pairs are generated. This process is repeated until only type D flip-flop remains. Eventually, assignments and justification are conducted to set the same values to each type D flip-flop.

The probability-based approach does not require justification. *Preferred fill* [143] is a one-pass approach (without iteration) based on *signal probability* [144]. Type B flip-flops are first directly assigned. Then signal probability is calculated for each X in the PPO. If the 0-probability of a PPO is larger than the 1-probability, its corresponding PPI is assigned to be 0 and vice versa.

There are many hybrid X-filling techniques that tried to (1) apply both probability and justification approaches: *JP-fill* [145], or (2) reduce both shift power and capture power: *i-fill* [146]. Some other techniques work with clock gating in the design to turn off necessary clocks, such as in [136] and *CTX-fill* in [147]. For those test patterns where don't care bits have already been filled, *bit-stripping* [141] or *X-identification* [115] techniques can be used to identify some X bits without fault coverage loss or test set inflation.

IR-drop analysis requires extensive matrix computation that is very time consuming. Many metrics have been proposed as an alternative to measure IR drop during ATPG. *Weighted switching activity* (WSA) is one of the most widely used metrics [130]. WSA is simply the weighted summation of switching activities of all nodes in the circuit. *Flip-flop toggle count* (FFTC) is another simple and popular metric for ATPG [142]. WSA and FFTC do not consider hazards and the physical information so they are not good indications of IR drop. Other metrics have been proposed, such as *critical capture transition* [145], *switching cycle average power* [148], and *FAIR* [149].

On the opposite side, some research tried to maximize the test power to test the robustness of power grid network [150,151].

22.5.4 ATPG AND TEST COMPRESSION

Test compression is an important ATPG option to minimize the test length. *Dynamic test compression* tries to merge as many faults as possible in a test cube during test generation. *Static test compression* tries to shorten the number of test patterns after test generation. With ever-increasing test data volume, many test compression techniques are now necessary to save test time and ATE memory [152,153].

Code-based test compression partitions the test input data into symbols, each of which is replaced by a new symbol. Typical examples are Colomb code [154] and Huffman code [155]. Code-based techniques suffer from the difference in data rate so they are not widely used in practice.

Linear decompressors use a small finite-state machine (FSM) (such as *linear feedback shift register*) with a small number of inputs to generate a large volume of test data. *Embedded deterministic test* is a commercial tool in this category [156].

Broadcast-based techniques use common inputs to deliver same test data. Typical examples are *broadcast scan* [157], *Illinois scan* [158], and *virtual scan* [159]. There are techniques that reduce test power during test compression, such as *CJP-fill* [160]. This is an ATPG optimization trade-off between test power reduction and test compression.

22.5.5 PARALLEL ATPG

22.5.5.1 PARALLEL FAULT GENERATION

Fault simulation, a process of determining if a given set of target faults is detected by a given pattern set, is an important while time-consuming task in the VLSI testing flow. Significant efforts have been devoted to accelerate this process in the past. Exploring the parallelism by assigning independent subtasks to different processors/threads in the multiple-processor or multiple-thread platform have been studied extensively. These parallel logic/fault simulation approaches can be classified into three categories: algorithm parallel, model parallel, and data parallel.

The algorithm-parallel approaches distribute the workloads by partitioning the simulation process into multiple pipeline stages, each of which is mapped to one processor and the communications between the processors are minimized [161,162]. The model-parallel approaches partition the circuit into multiple subcircuits, each of which is assigned to a processor [163,164]. While successful implementations have been demonstrated in various customized multiprocessor and supercomputer platforms, approaches of these two categories are too complex for the general computing platforms.

The data-parallel approaches compute multiple copies of identical or near-identical circuits concurrently [165–167]. Either multiple patterns and/or multiple faults are simulated in parallel. Among the three categories, approaches in this category have the least data dependence between processors and, thus, can often be implemented in a low-cost vector processor.

GPUs have recently been explored as a new general-purpose computing platform. Modern GPUs contain several multiprocessors, each of which executes a block of threads (or a block for short). They follow a single-instruction multiple-data architecture and often can have thousands of threads running concurrently. This feature makes GPUs suitable for acceleration of computation-intensive EDA applications, such as fault simulation, logic simulation, test generation, power grid analysis, and statistical timing analysis.

The first attempt to use GPUs for fault simulation was reported in [168], which performs a large number of tiny threads to implement a data-parallel simulation using a conventional forward fault simulation flow for all faulty circuits. This approach, implemented in an NVIDIA GeForce GTX 8800 GPU card, evaluates logic gates in the same logic level in parallel, for both fault-free and faulty circuits. Three types of kernels are implemented: (1) the logic simulation kernel for evaluation of each fault-free gate, (2) the fault simulation kernel for evaluation of each faulty gate and each gate in the transitive fanout of the faulty gate, and (3) the fault detection kernel for comparing the results at each PO in the transitive fanout of the faulty gate. All the gate evaluations are performed using lookup table–based computations. This implementation demonstrated a 35X speedup, in comparison with a CPU-based commercial fault simulation engine running on a 1.5 GHz UltraSPARC-IV+ processor with 1.6 GB of RAM.

The authors in [169] proposed nGFSIM, a GPU-based fault simulator that can report the fault coverages of 1 to n-detection for any specified integer n using only a single run of fault simulation. nGFSIM, which explores the massive parallelism in the GPU architecture and optimizes the memory access and usage, enables accelerated fault simulation without the need of fault dropping. The pseudocode of the host program, called GFSIM, which runs on the CPU and uses the GPU as a coprocessor, is shown in Algorithm 22.5, while P is the number of patterns to be simulated [169]:

Algorithm 22.5 *GFSIM()*

```
    // P is number of patterns to be simulated
    // LogicSim (l, g₁) does the logic simulation on GPU
    // ForwardDetect(s, c, l_s) calculates the detectability of stems on GPU
    // BackwardDetect(l, g₁) calculates the detectability of other
       gates on GPU
1.  (NumLevel, LevelList, NumStem, StemList) = CircuitPrepocessing();
2.  PatternGeneration;
3.  Transfer the patterns and levelized netlist from CPU to GPU;
4.  v=0;
5.  while (v<P) do {
6.      v = v + 512*LNumBlock;
7.      Invoke LogicSim (l, g₁) on GPU;
8.      ns = 0;
9.      while (ns<NumStem) do {
10.         ns = ns + FNumBlock;
11.         Invoke ForwardDetect(s, c, l_s) on GPU;
12.     }
13.     Invoke BackwardDetect(l, g₁) on GPU;
14.     Transfer detectability arrays from GPU to CPU;
15. }
```

A preprocessing step is done on the CPU to levelize the circuit and identify reconvergent stems (in line 1). The patterns to be simulated can be either randomly generated or read from a pattern file (in line 2). After the levelized netlist and the patterns are transferred to the GPU, logic simulation (in line 7), forward simulation for detectability calculation on fanout reconvergent stems (in line 9–12), and backward simulation for detectability calculation on other faults (in line 13) are processed serially. Three kernels, LogicSim, ForwardDetect, and BackwardDetect, are designed for the three tasks, respectively.

GFSIM does not experience much difference in runtime if fault dropping is implemented or not. With the computational power provided by a GPU, critical path tracing used in the backward fault simulation can derive the detectability of all nonreconvergent-fanout-stem faults for a large number of patterns in a single pass. The nature of no fault dropping enables a single run of n-detection fault simulation for multiple ns of interest. Using an additional 32-bit int per fault for counting detection times, GFSIM is extended to nGFSIM, producing a complete fault detection table that records the per-pattern detectability of faults, and the number of detections of each fault.

22.5.5.2 PARALLEL TEST GENERATION

Parallel programming is a popular technique to speed up ATPG [170]. Parallel ATPG algorithms on CPU can be divided into three categories according to different partitioning schemes: fault partitioning [163,171,172], search space partitioning [173,174], and circuit partitioning [175]. The fault partitioning approach dynamically or statically partitions the fault list into multiple partitions and each of which is handled by different cores. In search space partitioning, each core searches a portion of solution space. The circuit partitioning approach partitions the circuit into multiple subcircuits while trying to minimize communication and maximize concurrency between subcircuits. Each subcircuit is taken care of by different cores. Fault partitioning and circuit partitioning can be applied to both test generation and fault simulation, whereas search space partitioning can only be applied to test generation. In [172], two to five times speedup of TDF test generation is reported using eight CPU cores to parallelize a fault partition-based ATPG system.

Many researches on GPU-based logic simulation and fault simulation have been done in recent years. Based on the partitioning scheme, simulation techniques can be classified into three categories: fault partitioning [168,176], pattern partitioning [168], and circuit partitioning [169,177,178]. In fault and pattern partitioning, fault list or pattern set is divided into small partitions. Each partition is simulated by different blocks. The circuit partitioning approach partitions the circuit into subcircuits. There are several methods to partition the circuit: logic level, fanout-free region, macrogates, etc. Each subcircuit is then simulated by different blocks in parallel.

The first GPU-based ATPG was proposed in [179]. This is a fast GPU-based N-detect transition fault ATPG, which is an extension to a CPU-based parallel ATPG algorithm, *SWK* [180]. This GPU-based ATPG implemented three levels of parallelism: device-level, block-level, and word-level partitioning. The partitioning techniques in each level are (1) device level, fault partitioning; (2) block level, fault and circuit partitioning; and (3) word level, fault and search space partitioning.

Device-level fault partitioning statically partitions faults into different fault lists, each of which is assigned to a different device. Block-level fault partitioning assigns different target faults to different blocks, whereas block-level circuit partitioning assigns different logic levels to different blocks. Word-level fault partitioning assigns target faults to different bits in a word, whereas word-level search space partitioning assigns different branches of the decision tree to different bits. That means, the proposed algorithm converts decision making into bitwise logic operation so different branches of the decision tree can be explored at the same time. Suppose the number of devices is v, the number of blocks is t, and the size of a word is w. This GPU-based ATPG generates $v*t*w$ patterns concurrently. For example, if $v = 2$, $t = 64$, and $w = 32$, then 4096 patterns can be generated simultaneously. This is a massively parallel ATPG that cannot be achieved on traditional CPU architecture. Overall, the GPU-based ATPG is 5.9 and 1.6 times faster than a single-core and 8-core CPU-based commercial ATPG in generating 8-detect transition fault patterns, respectively.

It adopts a PODEM-based algorithm that generates test patterns by repeatedly backtracing objectives to inputs and propagating fault effects to outputs. This algorithm converts backtrace and propagation into bitwise logic operation so different branches of the decision tree can be explored at the same time.

In this algorithm, a signal Y is represented by seven words of w bits: $Y^0, Y^1, Y^d, Y^{\bar{d}}, Y^{b_0}, Y^{b_1}$, and Y^p. Each bit in a word represents an individual *clone*, which performs an independent search. The meaning of the first four words is as follows:

$Y^0 = 1$: signal Y is zero (good 0/faulty 0).
$Y^1 = 1$: signal Y is one (good 1/faulty 1).
$Y^d = 1$: signal Y is d (good 1/faulty 0).
$Y^{\bar{d}} = 1$: signal Y is $\bar{d}$ (good 0/faulty 1).

For a given clone, these four values are mutually exclusive so at most one of them equals one at a time. If all of them are zero, Y is unknown. When Y is unknown, the other three words indicate whether Y is on the propagation path or the *objective path*. The former is a path that faulty effect (d or $\bar{d}$) will potentially propagate to reach an output. The latter is a path that the objective backtrace follows to reach an input. For each clone,

$Y^p = 1$: signal Y is on a propagation path.
$Y^{b_0} = 1$: signal Y is on an objective 0 backtrace path.
$Y^{b_1} = 1$: signal Y is on an objective 1 backtrace path.

Again, these three values are mutually exclusive so at most one of them equals one at a time.

Figure 22.11 shows the test generation flow of each block in the GPU-based test generation for N-detect TDFs. Given a set of faults F, word size w, and the target number of detection N, each fault in F is handled by N clones. For example, if $w = 32$, $F = \{f_1, f_2, f_3, f_4\}$, and $N = 8$, then the first eight clones generate patterns for f_1, and the following eight clones generate patterns for f_2 and so on. For transition faults test generation, the two initial objectives are fault-free values at the fault site in two time frames. A backtrace is then performed from the fault site to the inputs. After inputs are assigned, fault effect propagation is performed from inputs to the outputs. The test generation of a clone is *successful* if a d or d' has reached any output. This process ends if test generations for all w clones are successful or time limit has reached.

Figures 22.12 through 22.15 illustrate an example to generate a test pattern for a single transition fault. This is a two-time-frame circuit, each of which consists of three AND gates, three inputs (E, F, G), and one flip-flop (H). LV stands for logic level. Suppose that the target fault is G_2 slow to fall. We use a two-bit word ($w = 2$) to represent two clones: clone #1 and clone #2. Since the fault is slow to fall, the initial objectives in time frame one and two are one and zero, respectively. They are denoted as $\{b_1, b_1\}$ and $\{b_0, b_0\}$ in Figure 22.12.

Figure 22.13 shows the first backtrace after Figure 22.12. Backtracing G_2 in time frame one is an *implication backtrace* because both gate inputs must be one to justify the output objective b_1. Backtracing G_2 in time frame two requires a decision because either $H = 0$ or $G = 0$ justifies the output objective b_0. An *objective split* is performed to assign two clones with different objectives. In this example, clone #1 backtraces b_0 on input H and clone #2 backtraces b_0 on input G.

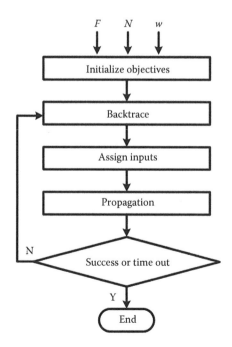

FIGURE 22.11   Test generation kernel flow (for a single block).

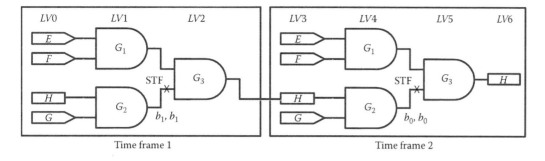

FIGURE 22.12   Test generation example (initialize objectives).

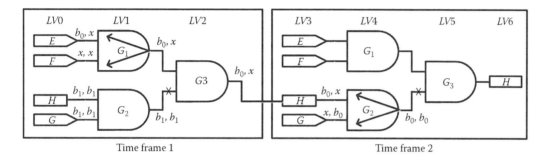

FIGURE 22.13   Test generation example (first backtrace).

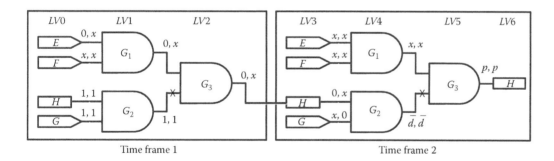

FIGURE 22.14   Test generation example (first propagation).

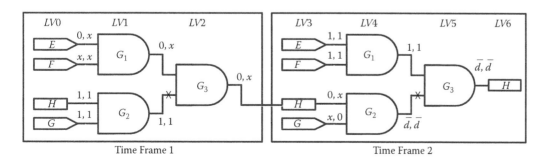

FIGURE 22.15   Test generation example (second backtrace and then propagation).

Figure 22.14 shows the propagation after Figure 22.13. The fault is excited for both clones so G_2 in time frame two is $\{\bar{d}, \bar{d}\}$ in the figure. Output H is $\{p, p\}$ in the figure. The symbol 'p' indicates that output H is on the propagation path.

Figure 22.15 shows the second backtrace followed by another propagation. To propagate the faulty effect to output, both clones of E and F in time frame two are set to one. Two test patterns $EFG = \{(0x1, 11x), (xx1, 110)\}$ and $H = \{1, 1\}$ are successfully generated. More details of this algorithm can be found in [179].

It has been shown that GPU N-detect ATPG technique achieved 1.6 times speedup compared with a 8-core commercial CPU N-detect ATPG, while test length and quality are about the same after test selection. However, GPU ATPG is currently limited by the size of memory and memory access bandwidth.

22.5.6 ATPG FOR HARDWARE TROJAN DETECTION

Economic factors mandate that design, manufacturing, testing, and deployment of silicon chips constitute a global effort involving multiple companies and countries. If a single contributor in this process decides it is advantageous to insert malicious functionality into the chip, referred to as hardware Trojans, the consequences can be disastrous. Detection is challenging because hardware Trojans can be inserted during any phase of the chip design lifecycle: as malicious third-party IP, modifications to the netlist or layout, or mask alterations during fabrication to name a few [181].

Using traditional ATPG methods to target hardware Trojans presents several unique challenges. In manufacturing test, stimulus vectors target stuck-at or delay fault models that are mainly based on circuit structure. Trojans inserted postsilicon are not present in the gate-level circuit description so they cannot be targeted by test pattern generation tools or candidates for observation points.

Many methods for postsilicon Trojan detection using functional test vectors assume that the adversary will incorporate design signals that have very low 0 or 1-controllability into the Trojan activation mechanism, making the combination of these rare values very unlikely to occur during testing. These strategies first identify random-pattern resistant nodes in the circuit, and their corresponding rare values, then derive an optimal test set to trigger low probability node values multiple times [182,183].

Another challenge to detecting Trojans using predictable test patterns is that the adversary can examine the test sequence or scan/BIST circuitry and craft the Trojan to avoid detection. An example of this is the Trojan proposed in [184], which creates permanent stuck-at faults in select state bits in the Intel Ivy Bridge RNG by altering transistor dopant levels. The choice of bits and values to tie them to is carefully chosen to avoid detection during BIST by taking advantage of aliasing inherent in test response compaction.

22.6 DESIGN APPLICATIONS

ATPG technology has been applied successfully in several areas of IC design automation, including logic optimization, logic equivalence checking, design property checking, and timing analysis.

22.6.1 LOGIC OPTIMIZATION

To optimize logic, design aids can either remove redundancy or restructure the logic by adding and removing redundancy.

22.6.1.1 REDUNDANCY REMOVAL

Redundancy is the main link between test and logic optimization. If there are untestable stuck-at faults, there is likely to be redundant logic. The reasoning is that, if a stuck-at fault does not have any test (the fault is untestable), the output responses of the faulty circuit (with this untestable fault) will be identical to the responses of the fault-free circuit for all possible input patterns applied to these two circuits. Thus, the faulty circuit (with an untestable stuck-at fault) is indeed a valid implementation of the fault-free circuit. Therefore, when ATPG identifies a stuck-at-1 (stuck-at-0) fault as untestable, one can simplify the circuit by setting the faulty net to logic 1(0) and thus effectively removing the faulty net from the circuit. This operation, called redundancy removal, also removes all the logic driving the faulty net.

Figure 22.16 illustrates an example. However, note that output Z in Figure 22.16a is hazard-free, but output Z in Figure 22.16b may have glitches. Testers must ensure that redundancy is removed only if having glitches is not a concern (e.g., as in synchronous design).

Because this method only removes logic from the circuits, the circuit is smaller when the process ends; the topological delay of the longest paths will be shorter than or at most equal to that of the original circuit. The power dissipation of the optimized circuit will also be lower.

22.6.1.2 LOGIC RESTRUCTURING

Removing a redundant fault can change the status of other faults. Those that were redundant might no longer be redundant, and vice versa. Although these changes complicate redundancy removal, they also pave the way for more rigorous optimization methods. Even for a circuit with no redundancies, designers can add redundancies to create new redundancies elsewhere in the circuit. By removing the created new redundancies, they may obtain an optimized circuit. This technique is called logic restructuring. For example, Figure 22.17 shows a circuit example that has no redundant logic. In Figure 22.18a, a signal line is artificially added that does not change the function of the circuit but does create redundant logic. Figure 22.18b shows the resulting circuit after redundancy removal. This circuit is simpler than the one in Figure 22.17.

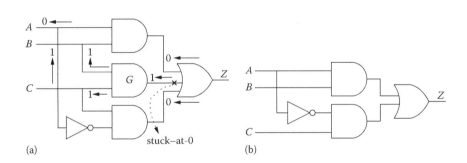

FIGURE 22.16   How ATPG works for redundancy removal: (a) the stuck-at 0 fault is untestable; (b) remove gate G and simply the logic.

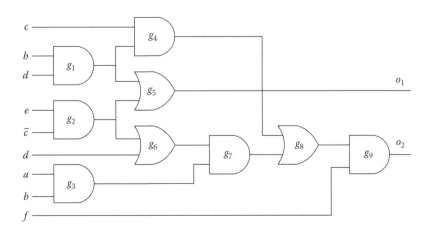

FIGURE 22.17   A circuit that is not redundant.

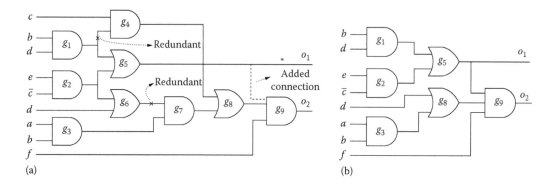

(a)   (b)

FIGURE 22.18   An example of logic restructuring by adding and removing redundancy: (a) adding a redundant connection and (b) optimized circuit after redundancy removal.

Efficient algorithms for finding effective logic restructuring [185] have been proposed in the past few years. By properly orienting the search for redundancy, these techniques can be adapted to target several optimization goals.

22.6.2 DESIGN VERIFICATION

Techniques used to verify designs include checking logic equivalence and determining that a circuit does or does not violate certain properties.

22.6.2.1 LOGIC EQUIVALENCE CHECKING

It is important to check the equivalence of two designs described at the same or different levels of abstraction. Checking the functional equivalence of the optimized implementation against the RTL specification, for example, guarantees that no error is introduced during logic synthesis and optimization, especially if part of the process is manual. Checking the equivalence of the gate-level implementation and the gate-level model extracted from the layout assures that no error is made during physical design.

Traditionally, designers check the functional equivalence of two Boolean functions by constructing their canonical representations, as truth tables or BDDs, for example. Two circuits are equivalent if and only if their canonical representations are isomorphic.

Consider the comparison of two Boolean networks in Figure 22.19. A joint network can be formed by connecting the corresponding PI pairs of the two networks and by connecting the corresponding PO pairs to XOR gates. The outputs of these XOR gates become the new POs of the joint network. The two networks are functionally equivalent if the PO response of the joint network is 0 for any input vector. Therefore, to prove that two circuits are equivalent, designers must merely prove that no input vector produces 1 at this model's output signal g.

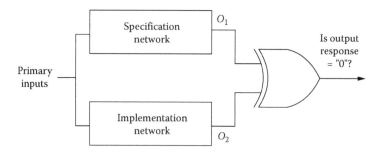

FIGURE 22.19   Circuit model for equivalence checking of two networks.

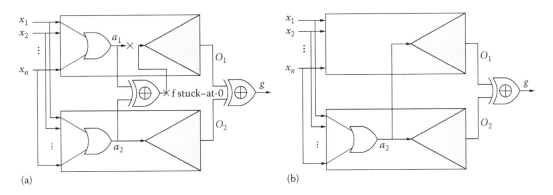

FIGURE 22.20 Pruning a joint network by finding internal equivalent pair. (a) Model for checking if signals a_1 and a_2 are equivalent. (b) Reducing complexity by replacing a_1 with a_2.

Another way to do equivalence checking is to formulate it as a problem that searches for a distinguishing vector, for which the two circuits under verification produce different output responses. If no distinguishing vector can be found after the entire space is searched, the two circuits are equivalent. Otherwise, a counterexample is generated to disprove equivalence. Because a distinguishing vector is also a test vector for the stuck-at-0 fault on the joint network's output g, equivalence checking becomes a test generation process for g's stuck-at-0 fault. However, directly applying ATPG to check the output equivalence (finding a test for stuck-at-0 fault) could be CPU intensive for large designs.

Figure 22.20 shows how complexity can be reduced substantially by finding an internal functional similarity between the two circuits being compared [186]. Designers first use naming information or structure analysis to identify a set of potentially equivalent internal signal pairs. They then build a model, as in Figure 22.20a, where signals a_1 and a_2 are candidate internal equivalent signals. To check the equivalence between these signals, we run ATPG for a stuck-at-0 fault at signal line f. If ATPG concludes that no test exists for that fault, the joint network can be simplified to the one in Figure 22.20b, where signal a_1 has been replaced with signal a_2.

With the simplified model, the complexity of ATPG for the output g stuck-at 0 fault will be reduced. The process identifies internal equivalent pairs sequentially from PIs to POs. By the time it gets to the output of the joint network, the joint network could be substantially smaller, and ATPG for the g stuck-at 0 fault will be quite trivial. Various heuristics for enhancing this idea and combining it with BDD techniques have been developed in the past few years [187]. Commercial equivalence checking tools can now handle circuit modules of more than a million gates within tens of CPU minutes.

22.6.2.2 PROPERTY CHECKING

An ATPG engine can find an example for proving that the circuit violates certain properties or, after exhausting the search space, can prove that no such example exists and thus that the circuit meets certain properties [188,189]. One example of this is checking for tristate bus contention, which occurs when multiple tristate bus drivers are enabled and their data are not consistent. Figure 22.21 shows a sample application. If the ATPG engine finds a test for the output stuck-at-0 fault, the test found will be the vector that causes bus contention. If no test exists, the bus can never have contention. Similarly, ATPG can check to see if a bus is floating—all tristate bus drivers are disabled—simply by checking for a vector that sets all enable lines to an inactive state.

ATPG can also identify races, which occur when data travels through two levels of latches in one clock cycle. Finally, an ATPG engine can check for effects (memory effect or an oscillation) from asynchronous feedback loops that might be in a pure synchronous circuit [188]. For each asynchronous loop starting and ending at a signal S, the ATPG engine simply checks to see whether there is a test to sensitize this loop. If such a test exists, the loop will cause either a memory effect (the parity from S to S is even) or an oscillation (the parity is odd).

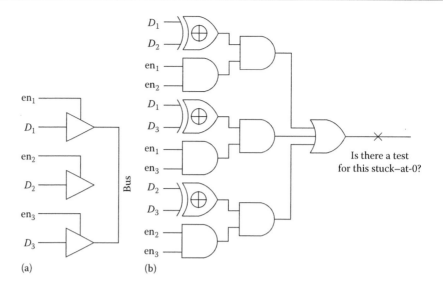

FIGURE 22.21   Application to bus contention checking. (a) A bus example. (b) An ATPG model for checking bus contention.

22.6.2.3 TIMING VERIFICATION AND ANALYSIS

Test vectors that sensitize selected long paths are often used in simulations to verify circuit timing. In determining the circuit's clock period, designers look for the slowest true path. Various sensitization criteria have been developed for determining a true path. Such criteria set requirements (in terms of logic values and the arrival times of the logic values) at side inputs of the gates along the path. These requirements are somewhat similar to those for deriving tests for PDFs. Thus, an ATPG engine can be used directly for this application.

22.6.3 SUMMARY

ATPG remains an active research area in both the CAD and the test communities, but the new emphasis is on moving ATPG operations toward higher levels of abstraction and on targeting new types of faults in deep-submicron devices. Test tools have evolved beyond merely gate-level test generation and fault simulation. Most design work now takes place at RTL and above, and test tools must support RTL handoff. New noise faults, including those from power supply noise, and crosstalk-induced noise, as well as substrate and thermal noise, will need models for manufacturing testing. The behaviors of these noise faults need to be modeled at levels of abstraction higher than the electrical, circuit, and transistor levels. Finding test vectors that can cover these faults is a challenge for ATPG.

22.7 HIGH-LEVEL ATPG

Test generation could be significantly sped up if a circuit model at a higher level of abstraction is used. In this section, we discuss briefly the principles of approaches using RTL models and state transitions graphs. Here, we do not intend to give a detailed survey for such approaches, only a brief description of representative methods.

Approaches using RTL models have the potential to handle larger circuits because the number of primitives in an RTL description is much smaller than the gate count. Some methods in this class of approaches use only RTL description of the circuit [190–195], while others assume that both gate-level and RTL models are available [30–32]. Note that automatic extraction of the RTL description from a lower level of description is still not possible, and therefore, the RTL descriptions must be given by the designers. It is also generally assumed that data and control can be separated in the RTL model.

For approaches using both RTL and gate-level models [30–32], typically a combinational test is first generated using the gate-level model. The fault-free justification sequence and the fault propagation sequence are generated using the (fault-free) RTL description. Justification and fault propagation sequences generated in such a manner may not be valid and therefore need to be verified by a fault simulator. These approaches, in general, are suitable for data-dominated circuits but are not appropriate for control-dominated circuits.

For approaches using only RTL models [190–195], functional fault models at RTL are targeted, instead of the single stuck-at fault model at the gate level. The approaches in [190,191] target microprocessors and functional fault models are defined for various functions at the control-sequencing level. Because tests are generated for the functional fault models, a high coverage for gate-level stuck-at faults cannot be guaranteed. The methods suggested in [192,193] focus on minimizing the value conflicts during the value justification and fault propagation processes, using the high-level information. The technique in [195] intends to guarantee that the functional tests for their proposed functional faults achieve a complete gate-level stuck-at fault coverage. To do so, mappings from gate-level faults to functional operations of modules need to be established. This approach also uses an efficient method for resolving the value conflicts during propagation/justification at the RTL level. A method of characterizing a design's functional information using a model extended from the traditional FSM model, with the capability of modeling both the datapath operations and the control state transitions, is suggested in [194]. However, this method does not target any fault model and only generates functional vectors for design verification.

For FSMs for which the state transition graphs are available, test sequences can be derived using the state transition information. In general, this class of approaches can handle only relatively small circuits due to the known state-explosion problem in representing a sequential circuit using its state table. However, successful applications of such approaches to protocol performance testing [196] and to testing the boundary-scan test access port controller [197] have been reported. The earliest method is the checking experiment [26] that is based on distinguishing sequences. The distinguishing sequence is defined as an input sequence that produces different output responses for each initial state of the FSM. This approach is concerned with the problem of determining whether or not a given state machine is distinguishable from all other possible machines with the same number of states or fewer. No explicit fault model is used. The distinguishing sequence may not exist, and the bound on length, if it exists, is proportional to the factorial of the number of states. This method is impractical because of the long test sequence. Improved checking experiments, based on either the simple input/output (I/O) sequence [27] or the unique input/output (UIO) sequence [196] of the FSM, significantly reduce the test length.

In [29], a functional fault model in the state transition level is used in a test generator FTG for FSMs. In the single-state-transition (SST) fault model, a fault causes the destination state of a SST to be faulty. It has been shown [29] that the test sequence generated for the SST faults in the given state transition graph achieves high fault coverages for the single stuck-at faults as well as the transistor faults in its multilevel logic implementation. As an approximation, FTG uses the fault-free state transition graph to generate the fault propagation sequence. AccuraTest [28] further improves the technique by using both fault-free and faulty circuits' state transition graphs to generate accurate test sequences for the SST faults as well as for some multiple-state-transition faults.

Due to the increasing complexity and size of modern designs, high-level ATPG has received more attention after 2000 [198–203]. The work in [198] utilizes *program slicing*, which was originally proposed as a static program analysis technique, for hierarchical test generation. Program slicing extracts environmental constraints for a given module, after which the module with the constraints can be synthesized into a gate-level model for test generation using a commercial sequential ATPG tool. Since program slicing extracts only relevant constraints for a module and ignores the rest of the design, it can significantly reduce the complexity of ATPG for each individual module. The authors in [199] propose a *word-level* ATPG combined with an arithmetic constraint solver in an ATPG application for property checking. The word-level ATPG involves a *word-level* implication engine. The arithmetic constraint solver is used as a powerful implication engine on the arithmetic datapath. The authors in [200] developed an RTL behavioral benchmark, namely, ITC'99 benchmark, and proposed ARTIST, which uses genetic algorithms to generate tests for RTL behavioral descriptions. The authors in [201] introduced RTL transfer faults from the *data dependency graph* of an RTL behavioral description and proposed X-Pulling, a simulation-based backtrack-free ATPG method for testing transfer faults. The authors in [203] propose a sequential ATPG guided by RTL information represented as an *assignment decision diagram* (ADD). State transition graphs extracted from ADDs are used to guide the ATPG process. The author in [202] developed a constraint solver for application in functional verification based on *random test program generation* methodology. The constraint solver utilizes word-level ATPG techniques to derive functional test sequences under user-supplied functional constraints.

Recently, abstraction-guided simulation has been explored as an effective approach to RTL functional test generation for hard-to-reach states [204–208]. The work in [204] is the pioneering work of abstraction-guided simulation using a random test generator, which builds an abstract model using the modules that closely interact with the target state. Preimages of the target state are calculated on this abstract model and the abstract distances are acquired to guide the simulation toward the target state. In order to avoid being stuck at dead-end states because of the abstraction bias, the work in [205] iteratively extracts and purifies the abstract model to get more design information until the simulator can reach the target state. The work in [206] introduces cultural algorithms to replace the random test generator and shows a better performance. The information of abstract preimages is used in the fitness calculation to evaluate each individual. The work in [207] uses path constraint solving that operates in an abstraction-guided simulation framework to generate tests to cover hard-to-reach states. The work in [208] uses Markov analysis of the abstract model as the guidance of simulation on the concrete design. The results of the Markov analysis provide a mixture of information including both the abstract distance and the difficulty to generate a concrete trace from each state to the target state. Consequently, the simulation using the Markov analysis as guidance shows better efficiency in target states exercising than the abstract distance–guided simulation.

REFERENCES

1. Abramovici, M., Breuer, M.A., and Friedman, A.D., *Digital Systems Testing and Testable Design*, Vol. 2. Computer Science Press, New York, 1990.
2. Maxwell, P.C., Aitken, R.C., Johansen, V., and Chiang, I., The effect of different test sets on quality level prediction: When is 80% better than 90%?, *Proceedings of the IEEE International Test Conference*, Nashville, TN, 1991, p. 358.
3. Maxwell, P.C. and Aitken, R.C., Test sets and reject rates: All fault coverages are not created equal, *IEEE Des. Test. Comput.*, 10, 42–51, 1993.
4. Wang, L.-C., Mercer, M.R., Kao, S.W., and Williams, T.W., On the decline of testing efficiency as fault coverage approaches 100%, *Proceedings of the IEEE VLSI Test Symposium*, Princeton, NJ, 1995, pp. 74–83.
5. Ma, S.C., Franco, P., and McCluskey, E.J., An experimental chip to evaluate test techniques: Experiment results, *Proceedings of the International Test Conference*, Washington, DC, 1995, pp. 663–672.
6. Grimaila, M.R., Lee, S., Dworak, J., Butler, K.M., Stewart, B., Balachandran, H., Houchins, B., Mathur, V., Park, J., and Wang, L.-C., REDO-random excitation and deterministic observation-first commercial experiment, *Proceedings of the IEEE VLSI Test Symposium*, San Diego, CA, 1999, pp. 268–274.
7. Dworak, J., Wicker, J.D., Lee, S., Grimaila, M.R., Mercer, M.R., Butler, K.M., Stewart, B., and Wang, L.-C., Defect-oriented testing and defective-part-level prediction, *IEEE Des. Test. Comput.*, 18, 31–41, 2001.
8. Fujiwara, H. and Toida, S., The complexity of fault detection problems for combinational logic circuits, *IEEE Trans. Comput.*, 100, 555–560, 1982.
9. Roth, J.P., Diagnosis of automata failures: A calculus and a method, *IBM J. Res. Dev.*, 10, 278–291, 1966.
10. Schulz, M.H., Trischler, E., and Sarfert, T.M., SOCRATES: A highly efficient automatic test pattern generation system, *IEEE Trans. Comput. Aided Des.*, 7, 126–137, 1988.
11. Kunz, W. and Pradhan, D.K., Recursive Learning: An attractive alternative to the decision tree for test generation in digital ci, *Proceedings of the IEEE International Test Conference*, Piscataway, NJ, 1992, p. 816.
12. Wang, C., Reddy, S.M., Pomeranz, I., Lin, X., and Rajski, J., Conflict driven techniques for improving deterministic test pattern generation, *Proceedings of the ACM International Conference on Computer-Aided Design*, San Jose, CA, 2002, pp. 87–93.

13. Marques-Silva, J.P. and Sakallah, K.A., GRASP: A search algorithm for propositional satisfiability, *IEEE Trans. Comput.*, 48, 506–521, 1999.
14. Goel, P., An implicit enumeration algorithm to generate tests for combinational logic circuits, *IEEE Trans. Comput.*, 100, 215–222, 1981.
15. Fujiwara, H. and Shimono, T., On the acceleration of test generation algorithms, *IEEE Trans. Comput.*, 100, 1137–1144, 1983.
16. Henftling, M., Wittmann, H.C., and Antreich, K.J., A single-path-oriented fault-effect propagation in digital circuits considering multiple-path sensitization, *Proceedings of the IEEE International Conference on Computer-Aided Design*, San Jose, CA, 1995, pp. 304–309.
17. Hamzaoglu, I. and Patel, J.H., New techniques for deterministic test pattern generation, *J. Electron. Test. (JETTA)*, 15, 63–73, 1999.
18. Kirkland, T. and Mercer, M.R., A topological search algorithm for ATPG, *Proceedings of the ACM Design Automation Conference*, Miami Beach, FL, 1987, pp. 502–508.
19. Ivanov, A. and Agarwal, V.K., Dynamic testability measures for ATPG, *IEEE Trans. Comput. Aided Des.*, 7, 598–608, 1988.
20. Larrabee, T., Test pattern generation using Boolean satisfiability, *IEEE Trans. Comput. Aided Des.*, 11, 4–15, 1992.
21. Stephan, P., Brayton, R.K., and Sangiovanni-Vincentelli, A.L., Combinational test generation using satisfiability, *IEEE Trans. Comput. Aided Des.*, 15, 1167–1176, 1996.
22. Silva, J. and Sakallah, K.A., Robust search algorithms for test pattern generation, *Proceedings of the IEEE International Symposium on Fault-Tolerant Computing*, Seattle, WA, 1997, pp. 152–161.
23. Gizdarski, E. and Fujiwara, H., SPIRIT: A highly robust combinational test generation algorithm, *IEEE Trans. Comput. Aided Des.*, 21, 1446–1458, 2002.
24. Seshu, S. and Freeman, D., The diagnosis of asynchronous sequential switching systems, *IEEE Trans. Electron. Comput.*, EC-11, 459–465, 1962.
25. Thomas, J.J., Automated diagnostic test program for digital networks, *Comput. Des.*, 10, 63–67, 1971.
26. Hennine, F., Fault detecting experiments for sequential circuits, *Proceedings of the IEEE Symposium on Switching Circuit Theory and Logical Design*, Princeton, NJ, 1964, pp. 95–110.
27. Hsieh, E., Checking experiments for sequential machines, *IEEE Trans. Comput.*, 100, 1152–1166, 1971.
28. Pomeranz, I. and Reddy, S.M., On achieving a complete fault coverage for sequential machines using the transition fault model, *Proceedings of the IEEE/ACM Design Automation Conference*, Orlando, FL, 1991, pp. 341–346.
29. Cheng, K.-T. and Jou, J.-Y., A functional fault model for sequential machines, *IEEE Trans. Comput. Aided Des.*, 11, 1065–1073, 1992.
30. Hill, F.J. and Huey, B., SCIRTSS: A search system for sequential circuit test sequences, *IEEE Trans. Comput.*, 100, 490–502, 1977.
31. Breuer, M.A. and Friedman, A.D., Functional level primitives in test generation, *IEEE Trans. Comput.*, 100, 223–235, 1980.
32. Ghosh, A., Devadas, S., and Newton, A.R., Sequential test generation at the register-transfer and logic levels, *Proceedings of the ACM/IEEE Design Automation Conference*, Orlando, FL, 1990, pp. 580–586.
33. Kubo, H., A procedure for generating test sequences to detect sequential circuit failures, *NEC Res. Dev.*, 12, 69, 1968.
34. Putzolu, G.R. and Roth, J.P., A heuristic algorithm for the testing of asynchronous circuits, *IEEE Trans. Comput.*, 20, 639–647, 1971.
35. Marlett, R.A., EBT: A comprehensive test generation technique for highly sequential circuits, *Proceedings of the IEEE/ACM Design Automation Conference*, Las Vegas, NV, 1978, pp. 335–339.
36. Cheng, W.-T., The BACK algorithm for sequential test generation, *Proceedings of the IEEE International Conference on Computer Design*, Port Chester, NY, 1988, pp. 66–69.
37. Niermann, T. and Patel, J.H., HITEC: A test generation package for sequential circuits, *Proceedings of the IEEE Computer Society Press Design Automation and Test in Europe*, Paris, France, 1991, pp. 214–218.
38. Kelsey, T.P., Saluja, K.K., and Lee, S.Y., An efficient algorithm for sequential circuit test generation, *IEEE Trans. Comput.*, 42, 1361–1371, 1993.
39. Cheng, K.-T. and Agrawal, V.D., *Unified Methods for VLSI Simulation and Test Generation*. Kluwer Academic Publishers, Boston, 1989.
40. Saab, D.G., Saab, Y.G., and Abraham, J.A., CRIS: A test cultivation program for sequential VLSI circuits, *Proceedings of the IEEE International Conference on Computer-Aided Design*, Santa Clara, CA, 1992, pp. 216–219.
41. Rudnick, E.M., Patel, J.H., Greenstein, G.S., and Niermann, T.M., Sequential circuit test generation in a genetic algorithm framework, *Proceedings of the IEEE/ACM Design Automation Conference*, San Diego, CA, 1994, pp. 698–704.

42. Prinetto, P., Rebaudengo, M., and Reorda, M.S., An automatic test pattern generator for large sequential circuits based on genetic algorithms, *Proceedings of the IEEE International Test Conference*, Washington, DC, 1994, pp. 240–249.

43. Saab, D.G., Saab, Y.G., and Abraham, J.A., Iterative [simulation-based genetics+ deterministic techniques]= complete ATPG0, *Proceedings of the IEEE/ACM International Conference on Computer-Aided Design*, San Jose, CA, 1994, pp. 40–43.

44. Rudnick, E.M. and Patel, J.H., Combining deterministic and genetic approaches for sequential circuit test generation, *Proceedings of the ACM/IEEE Design Automation Conference*, San Francisco, CA, 1995, pp. 183–188.

45. Hsiao, M.S., Rudnick, E.M., and Patel, J.H., Alternating strategies for sequential circuit ATPG, *Proceedings of the IEEE Computer Society Design Automation and Test in Europe*, Paris, France, 1996, p. 368.

46. Ma, H.-K.T., Devadas, S., Newton, A.R., Sangiovanni-Vincentelli, A., Test generation for sequential circuits, *IEEE Trans. Comput. Aided Des.*, 7, 1081–1093, 1988.

47. Ghosh, A., Devadas, S., and Newton, A.R., Test generation and verification for highly sequential circuits, *IEEE Trans. Comput. Aided Des.*, 10, 652–667, 1991.

48. Cho, H., Hachtel, G.D., and Somenzi, F., Redundancy identification/removal and test generation for sequential circuits using implicit state enumeration, *IEEE Trans. Comput. Aided Des.*, 12, 935–945, 1993.

49. Muth, P., A nine-valued circuit model for test generation, *IEEE Trans. Comput.*, 100, 630–636, 1976.

50. Goldstein, L.H., Controllability/observability analysis of digital circuits, *IEEE Trans. Circ. Syst.*, 26, 685–693, 1979.

51. Schulz, M.H. and Auth, E., Advanced automatic test pattern generation and redundancy identification techniques, *Proceedings of the IEEE International Symposium on Fault-Tolerant Computing*, Munich, Germany, 1988, pp. 30–35.

52. Agrawal, V.D. and Chakradhar, S.T., Combinational ATPG theorems for identifying untestable faults in sequential circuits, *IEEE Trans. Comput. Aided Des.*, 14, 1155–1160, 1995.

53. Cheng, K.-T., Redundancy removal for sequential circuits without reset states, *IEEE Trans. Comput. Aided Des.*, 12, 13–24, 1993.

54. Pomeranz, I. and Reddy, S.M., Classification of faults in synchronous sequential circuits, *IEEE Trans. Comput.*, 42, 1066–1077, 1993.

55. Devadas, S., Ma, H.K.T., and Newton, A.R., Redundancies and don't cares in sequential logic synthesis, *J. Electron. Test. (JETTA)*, 1, 15–30, 1990.

56. Bryant, R.E., Graph-based algorithms for Boolean function manipulation, *IEEE Trans. Comput.*, 100, 677–691, 1986.

57. Cheng, K.-T., Gate-level test generation for sequential circuits, *ACM T Des. Automat. El.*, 1, 405–442, 1996.

58. Breuer, M., Test generation models for busses and tri-state drivers, *Proceedings of the IEEE ATPG Workshop*, San Francisco, CA, 1983, pp. 53–58.

59. Ogihara, T., Murai, S., Takamatsu, Y., Kinoshita, K., and Fujiwara, H., Test generation for scan design circuits with tri-state modules and bidirectional terminals, *Proceedings of the IEEE/ACM Design Automation Conference*, Miami Beach, FL, 1983, pp. 71–78.

60. Chakradhar, S.T., Rotherweiler, S., and Agrawal, V.D., Redundancy removal and test generation for circuits with non-Boolean primitives, *IEEE Trans. Comput. Aided Des.*, 16, 1370–1377, 1997.

61. Zhang, H., SATO: An efficient prepositional prover, *Automated Deduction—CADE-14*. Springer, New York, 1997, pp. 272–275.

62. Moskewicz, M.W., Madigan, C.F., Zhao, Y., Zhang, L., and Malik, S., Chaff: Engineering an efficient SAT solver, *Proceedings of the IEEE/ACM Design Automation Conference*, Las Vegas, NV, 2001, pp. 530–535.

63. Goldberg, E. and Novikov, Y., BerkMin: A fast and robust SAT-solver, *Discrete Appl. Math.*, 155, 1549–1561, 2007.

64. Ryan, L. The siege satisfiability solver. Available from: http://www.cs.sfu.ca/research/groups/CL/software/siege/ (Accessed on November 20, 2015).

65. Davis, M., Logemann, G., and Loveland, D., A machine program for theorem-proving, *Commun. ACM*, 5, 394–397, 1962.

66. McAllester, D.A., An outlook on truth maintenance, AIMemo 551, MIT AI Laboratory, Cambridge, MA, 1980.

67. Tseitin, G.S., On the complexity of derivation in propositional calculus, in *Automation of Reasoning*, Siekmann J., Wrightson G. (eds.), Springer, Heidelberg, Germany, 1983, pp. 466–483.

68. Tafertshofer, P., Ganz, A., and Henftling, M., A SAT-based implication engine for efficient ATPG, equivalence checking, and optimization of netlists, *Proceedings of the IEEE Computer Society International Conference on Computer-Aided Design*, San Jose, CA, 1997, pp. 648–655.

69. Guerra e Silva, L., Silveira, L.M., and Marques-Silva, J., Algorithms for solving Boolean satisfiability in combinational circuits, *Proceedings of the ACM Design Automation and Test in Europe*, Munich, Germany, 1999, p. 107.

70. Marques-Silva, J.P. and Silva, L.G., Solving satisfiability in combinational circuits, *IEEE Des. Test. Comput.*, 20, 16–21, 2003.

71. Gupta, A., Gupta, A., Yang, Z., and Ashar, P., Dynamic detection and removal of inactive clauses in SAT with application in image computation, *Proceedings of the IEEE/ACM Design Automation Conference*, Las Vegas, NV, 2001, pp. 536–541.

72. Kuehlmann, A., Ganai, M.K., and Paruthi, V., Circuit-based Boolean reasoning, *Proceedings of the ACM/IEEE Design Automation Conference*, Las Vegas, NV, 2001, pp. 232–237.

73. Kuehlmann, A. and Krohm, F., Equivalence checking using cuts and heaps, *Proceedings of the IEEE/ACM Design Automation Conference*, Austin, TX, 1997, pp. 263–268.

74. Ganai, M.K., Ashar, P., Gupta, A., Zhang, L., and Malik, S., Combining strengths of circuit-based and CNF-based algorithms for a high-performance SAT solver, *Proceedings of the IEEE/ACM Design Automation Conference*, New Orleans, LA, 2002, pp. 747–750.

75. Ostrowski, R., Grégoire, É., Mazure, B., and Sais, L., Recovering and exploiting structural knowledge from CNF formulas, *Proceedings of the Springer Principles and Practice of Constraint Programming-CP 2002*, Ithaca, NY, 2002, pp. 185–199.

76. Broering, E. and Lokam, S.V., Width-based algorithms for SAT and CIRCUIT-SAT, *Proceedings of the Springer Theory and Applications of Satisfiability Testing*, Vancouver, British Columbia, Canada, 2004, pp. 162–171.

77. Lu, F., Wang, L.-C., Cheng, K.-T., and Huang, R.-Y., A circuit SAT solver with signal correlation guided learning, *Proceedings of the IEEE Design Automation and Test in Europe*, Munich, Germany, 2003, pp. 892–897.

78. Lu, F., Wang, L.-C., Cheng, K.-T.T., Moondanos, J., and Hanna, Z., A signal correlation guided ATPG solver and its applications for solving difficult industrial cases, *Proceedings of the IEEE/ACM Design Automation Conference*, Anaheim, CA, 2003, pp. 436–441.

79. Lu, F., Iyer, M.K., Parthasarathy, G., Wang, L.-C., Cheng, K.-T., and Chen, K.-C., An efficient sequential SAT solver with improved search strategies, *Proceedings of the IEEE Design Automation and Test in Europe*, Munich, Germany, 2005, pp. 1102–1107.

80. Zhang, L., Madigan, C.F., Moskewicz, M.H., and Malik, S., Efficient conflict driven learning in a Boolean satisfiability solver, *Proceedings of the IEEE International Conference on Computer-Aided Design*, San Jose, CA, 2001, pp. 279–285.

81. Roy, K., Cheng, K.-T., Rodgers, M., and Dey, S., Test challenges for deep sub-micron technologies, *Proceedings of the ACM/IEEE Design Automation Conference*, Los Angeles, CA, 2000, pp. 142–149.

82. Butler, K.M., Cheng, K.-T., and Wang, L.-C., Guest editors' introduction: Speed test and speed binning for complex ICs, *IEEE Des. Test. Comput.*, 20, 6–7, 2003.

83. Krstic, A. and Cheng, K.-T., *Delay Fault Testing for VLSI Circuits*, Vol. 14, Springer, New York, 1998.

84. Savir, J., Skewed-load transition test: Part I, calculus, *Proceedings of the IEEE International Test Conference*, Baltimore, MD, 1992, p. 705.

85. Savir, J. and Patil, S., Broad-side delay test, *IEEE Trans. Comput. Aided Des.*, 13, 1057–1064, 1994.

86. Cheng, K.-T., Devadas, S., and Keutzer, K., A partial enhanced-scan approach to robust delay-fault test generation for sequential circuits, *Proceedings of the IEEE International Test Conference*, Nashville, TN, 1991, p. 403.

87. Xu, G., Flip-flop selection to maximize TDF coverage with partial enhanced scan, *Proceedings of the IEEE Asian Test Symposium*, Sapporo, Japan, 2007, pp. 335–340.

88. Wang, S. and Wei, W., Low overhead partial enhanced scan technique for compact and high fault coverage transition delay test patterns, *Proceedings of the IEEE European Test Symposium*, Avignon, France, 2008, pp. 125–130.

89. Pei, S., Li, H., and Li, X., Flip-flop selection for partial enhanced scan to reduce transition test data volume, *IEEE Trans. VLSI Syst.*, 20, 2157–2169, 2012.

90. Park, E.S. and Mercer, M.R., An efficient delay test generation system for combinational logic circuits, *IEEE Trans. Comput. Aided Des.*, 11, 926–938, 1992.

91. Qiu, W. and Walker, D., An efficient algorithm for finding the k longest testable paths through each gate in a combinational circuit, *Proceedings of the IEEE International Test Conference*, Charlotte, NC, 2003, pp. 592–592.

92. Sato, Y., Hamada, S., Maeda, T., Takatori, A., Nozuyama, Y., and Kajihara, S., Invisible delay quality-SDQM model lights up what could not be seen, *Proceedings of the IEEE International Test Conference*, Austin, TX, 2005, p. 1210 (9pp.).

93. Lin, X., Tsai, K.-H., Wang, C., Kassab, M., Rajski, J., Kobayashi, T., Klingenberg, R., Sato, Y., Hamada, S., and Aikyo, T., Timing-aware ATPG for high quality at-speed testing of small delay defects, *Proceedings of the IEEE Asian Test Symposium*, Fukuoka, Japan, 2006, pp. 139–146.

94. Ahmed, N., Tehranipoor, M., and Jayaram, V., Timing-based delay test for screening small delay defects, *Proceedings of the IEEE/ACM Design Automation Conference*, San Francisco, CA, 2006, pp. 320–325.

95. Kajihara, S., Morishima, S., Takuma, A., Wen, X., Maeda, T., Hamada, S., and Sato, Y., A framework of high-quality transition fault ATPG for scan circuits, *Proceedings of the IEEE International Test Conference*, Brussels, Belgium, 2006, pp. 1–6.

96. Ahmed, N. and Tehranipoor, M., A novel faster-than-at-speed transition-delay test method considering IR-drop effects, *IEEE Trans. Comput. Aided Des.*, 28, 1573–1582, 2009.

97. Fu, X., Li, H., and Li, X., Testable path selection and grouping for faster than at-speed testing, *IEEE Trans. VLSI Syst.*, 20, 236–247, 2012.

98. Blaauw, D., Chopra, K., Srivastava, A., and Scheffer, L., Statistical timing analysis: From basic principles to state of the art, *IEEE Trans. Comput. Aided Des.*, 27, 589–607, 2008.

99. Wang, L., Liou, J.-J., and Cheng, K.-T., Critical path selection for delay fault testing based upon a statistical timing model, *IEEE Trans. Comput. Aided Des.*, 23, 1550–1565, 2004.

100. Lu, X., Li, Z., Qiu, W., Walker, D., and Shi, W., Longest-path selection for delay test under process variation, *IEEE Trans. Comput. Aided Des.*, 24, 1924–1929, 2005.

101. Zolotov, V., Xiong, J., Fatemi, H., and Visweswariah, C., Statistical path selection for at-speed test, *IEEE Trans. Comput. Aided Des.*, 29, 749–759, 2010.

102. He, Z., Lv, T., Li, H., and Li, X., Test path selection for capturing delay failures under statistical timing model, *IEEE Trans. VLSI Syst.*, 21, 1210–1219, 2013.

103. Tehranipoor, M., Peng, K., and Chakrabarty, K., *Test and Diagnosis for Small-Delay Defects*, Springer, New York, 2011.

104. Peng, K., Yilmaz, M., Tehranipoor, M., and Chakrabarty, K., High-quality pattern selection for screening small-delay defects considering process variations and crosstalk, *Proceedings of the European Design and Automation Association Design Automation and Test in Europe*, Dresden, Germany, 2010, pp. 1426–1431.

105. Yilmaz, M., Chakrabarty, K., and Tehranipoor, M., Test-pattern selection for screening small-delay defects in very-deep submicrometer integrated circuits, *IEEE Trans. Comput. Aided Des.*, 29, 760–773, 2010.

106. Xu, D., Li, H., Ghofrani, A., Cheng, K.-T., Han, Y., and Li, X., Test-quality optimization for variable n-detections of transition faults, *IEEE Trans. VLSI Syst.*, 22, 1738-1749, 2014.

107. Lee, K.T., Nordquist, C., and Abraham, J.A., Automatic test pattern generation for crosstalk glitches in digital circuits, *Proceedings of the IEEE VLSI Test Symposium*, Princeton, NJ, 1998, pp. 34–39.

108. Chen, L.H. and Marek-Sadowska, M., Aggressor alignment for worst-case coupling noise, *Proceedings of the ACM International Symposium Physical Design*, San Diego, CA, 2000, pp. 48–54.

109. Chen, W.-Y., Gupta, S.K., and Breuer, M.A., Test generation for crosstalk-induced delay in integrated circuits, *Proceedings of the IEEE International Test Conference*, Atlantic, NJ, 1999, pp. 191–200.

110. Krstic, A., Liou, J.-J., Jiang, Y.-M., and Cheng, K.-T., Delay testing considering crosstalk-induced effects, *Proceedings of the IEEE International Test Conference*, Baltimore, MD, 2001, pp. 558–567.

111. Chen, L.-C., Gupta, S.K., Mak, T., and Breuer, M.A., Crosstalk test generation on pseudo industrial circuits: A case study, *Proceedings of the IEEE International Test Conference*, Baltimore, MD, 2001, pp. 548–548.

112. Chen, W., Gupta, S.K., and Breuer, M.A., Analytic models for crosstalk delay and pulse analysis under non-ideal inputs, *Proceedings of the IEEE International Test Conference*, Washington, DC, 1997, pp. 809–818.

113. Chen, W., Gupta, S.K., and Breuer, M.A., Test generation in VLSI circuits for crosstalk noise, *Proceedings of the IEEE International Test Conference*, Washington, DC, 1998, pp. 641–650.

114. Kundu, S., Zachariah, S.T., Chang, Y.-S., and Tirumurti, C., On modeling crosstalk faults, *IEEE Trans. Comput. Aided Des.*, 24, 1909–1915, 2005.

115. Nazarian, S., Huang, H., Natarajan, S., Gupta, S.K., and Breuer, M.A., XIDEN: Crosstalk target identification framework, *Proceedings of the IEEE International Test Conference*, Baltimore, MD, 2002, pp. 365–374.

116. Chen, P., Kukimoto, Y., and Keutzer, K., Refining switching window by time slots for crosstalk noise calculation, *Proceedings of the ACM International Conference on Computer-Aided Design*, San Jose, CA, 2002, pp. 583–586.

117. Tseng, K. and Horowitz, M., False coupling exploration in timing analysis, *IEEE Trans. Comput. Aided Des.*, 24, 1795–1805, 2005.

118. Zhang, M. and Li, X., Test generation for crosstalk glitches considering multiple coupling effects, *Proceedings of the IEEE Asian Test Symposium*, Sapporo, Japan, 2007, pp. 259–264.

119. Irajpour, S., Gupta, S.K., and Breuer, M.A., Timing-independent testing of crosstalk in the presence of delay producing defects using surrogate fault models, *Proceedings of the IEEE International Test Conference*, Charlotte, NC, 2004, pp. 1024–1033.

120. Ganeshpure, K.P. and Kundu, S., On ATPG for multiple aggressor crosstalk faults in presence of gate delays, *Proceedings of the IEEE International Test Conference*, Santa Clara, CA, 2007, pp. 1–7.

121. Paul, B.C. and Roy, K., Testing cross-talk induced delay faults in static CMOS circuit through dynamic timing analysis, *Proceedings of the IEEE International Test Conference*, Baltimore, MD, 2002, pp. 384–390.

122. Ran, Y., Kondratyev, A., Tseng, K.H., Watanabe, Y., and Marek-Sadowska, M., Eliminating false positives in crosstalk noise analysis, *IEEE Trans. Comput. Aided Des.*, 24, 1406–1419, 2005.
123. Lee, J. and Tehranipoor, M., A novel pattern generation framework for inducing maximum crosstalk effects on delay-sensitive paths, *Proceedings of the IEEE International Test Conference*, Santa Clara, CA, 2008, pp. 1–10.
124. Chun, S., Kim, T., and Kang, S., ATPG-XP: Test generation for maximal crosstalk-induced faults, *IEEE Trans. Comput. Aided Des.*, 28, 1401–1413, 2009.
125. Li, H. and Li, X., Selection of crosstalk-induced faults in enhanced delay test, *J. Electron. Test. (JETTA)*, 21, 181–195, 2005.
126. Li, H., Shen, P., and Li, X., Robust test generation for precise crosstalk-induced path delay faults, *Proceedings of the IEEE VLSI Test Symposium*, Berkeley, CA, 2006, p. 305 (6pp.).
127. Zhang, M., Li, H., and Li, X., Path delay test generation toward activation of worst case coupling effects, *IEEE Trans. VLSI Syst.*, 19, 1969–1982, 2011.
128. Zorian, Y., A distributed BIST control scheme for complex VLSI devices, *Proceedings of the IEEE VLSI Test Symposium*, Atlantic, NJ, 1993, pp. 4–9.
129. Saxena, J., Kundu, S., Arvind, N., Butler, K.M., Jayaram, V.B., Sreeprakash, P., and Hachinger, M., A case study of IR-drop in structured at-speed testing, *Proceedings of the IEEE International Test Conference*, Charlotte, NC, 2003, pp. 1098–1098.
130. Girard, P., Survey of low-power testing of VLSI circuits, *IEEE Des. Test. Comput.*, 19, 82–92, 2002.
131. Girard, P., Wen, X., and Touba, N., Low-power testing, in *System-On-Chip Test Architectures— Nanometer Design for Testability*, L.-T. Wang, C.E. Stroud, N.A. Touba, (eds.), Morgan Kaufmann, San Francisco, CA, Chapter 7, 2007.
132. Girard, P., Nicolici, N., and Wen, X., *Power-Aware Testing and Test Strategies for Low Power Devices*, Springer, New York, 2010.
133. Tehranipoor, M. and Butler, K.M., Power supply noise: A survey on effects and research, *IEEE Des. Test. Comput.*, 27, 51–67, 2010.
134. Wen, X., Kajihara, S., Miyase, K., Suzuki, T., Saluja, K.K., Wang, L.-T., Abdel-Hafez, K.S., and Kinoshita, K., A new ATPG method for efficient capture power reduction during scan testing, *Proceedings of the IEEE VLSI Test Symposium*, Berkeley, CA, 2006, p. 65 (6pp.).
135. Devanathan, V., Ravikumar, C., and Kamakoti, V., Glitch-aware pattern generation and optimization framework for power-safe scan test, *Proceedings of the IEEE VLSI Test Symposium*, Berkeley, CA, 2007, pp. 167–172.
136. Wang, J., Lu, X., Qiu, W., Yue, Z., Fancler, S., Shi, W., and Walker, D., Static compaction of delay tests considering power supply noise, *Proceedings of the IEEE VLSI Test Symposium*, Palm Springs, CA, 2005, pp. 235–240.
137. Wu, M.-F., Huang, J.-L., Wen, X., and Miyase, K., Power supply noise reduction for at-speed scan testing in linear-decompression environment, *IEEE Trans. Comput. Aided Des.*, 28, 1767–1776, 2009.
138. Chakravarty, S. and Dabholkar, V., Minimizing power dissipation in scan circuits during test application, Citeseer, 1994.
139. Girard, P., Landrault, C., Pravossoudovitch, S., and Severac, D., Reducing power consumption during test application by test vector ordering, *Proceedings of the International Symposium on Circuits and Systems*, Vol. 2, Monterey, CA, 1998, pp. 296–299.
140. Butler, K.M., Saxena, J., Fryars, T., Hetherington, G., Jain, A., and Lewis, J., Minimizing power consumption in scan testing: Pattern generation and DFT techniques, *Proceedings of the IEEE International Test Conference*, Charlotte, NC, 2004, pp. 355–364.
141. Sankaralingam, K., Oruganti, R.R., and Touba, N.A., Static compaction techniques to control scan vector power dissipation, *Proceedings of the IEEE VLSI Test Symposium*, Berkeley, CA, 2000, pp. 35–40.
142. Wen, X., Yamashita, Y., Kajihara, S., Wang, L.-T., Saluja, K.K., and Kinoshita, K., On low-capture-power test generation for scan testing, *Proceedings of the IEEE VLSI Test Symposium*, Palm Springs, CA, 2005, pp. 265–270.
143. Remersaro, S., Lin, X., Zhang, Z., Reddy, S.M., Pomeranz, I., and Rajski, J., Preferred fill: A scalable method to reduce capture power for scan based designs, *Proceedings of the IEEE International Test Conference*, Brussels, Belgium, 2006, pp. 1–10.
144. Parker, K.P. and McCluskey, E.J., Probabilistic treatment of general combinational networks, *IEEE Trans. Comput.*, 100, 668–670, 1975.
145. Wen, X., Miyase, K., Suzuki, T., Kajihara, S., Ohsumi, Y., and Saluja, K.K., Critical-path-aware X-filling for effective IR-drop reduction in at-speed scan testing, *Proceedings of the IEEE/ACM Design Automation Conference*, San Diego, CA, 2007, pp. 527–532.
146. Li, J., Xu, Q., Hu, Y., and Li, X., iFill: An impact-oriented X-filling method for shift-and capture-power reduction in at-speed scan-based testing, *Proceedings of the ACM Design Automation and Test in Europe*, Munich, Germany, 2008, pp. 1184–1189.

147. Furukawa, H., Wen, X., Miyase, K., Yamato, Y., Kajihara, S., Girard, P., Wang, L.-T., and Tehranipoor, M., CTX: A clock-gating-based test relaxation and X-filling scheme for reducing yield loss risk in at-speed scan testing, *Proceedings of the IEEE Asian Test Symposium*, Hokkaido, Japan, 2008, pp. 397–402.

148. Ahmed, N., Tehranipoor, M., and Jayaram, V., Supply voltage noise aware ATPG for transition delay faults, *Proceeding of the IEEE VLSI Test Symposium*, Berkeley, CA, 2007, pp. 179–186.

149. Ding, W.S., Test pattern modification for average IR-drop reduction, MS, National Taiwan University, Taipei, Taiwan, 2012.

150. Jiang, Y.-M., Cheng, K.-T., and Deng, A.-C., Estimation of maximum power supply noise for deep submicron designs, *Proceedings of the ACM International Symposium on Low Power Electronics and Design*, Monterey, CA, 1998, pp. 233–238.

151. Krstic, A., Jiang, Y.-M., and Cheng, K.-T., Pattern generation for delay testing and dynamic timing analysis considering power-supply noise effects, *IEEE Trans. Comput. Aided Des.*, 20, 416–425, 2001.

152. Wang, L.-T., Wu, C.-W., and Wen, X., *VLSI Test Principles and Architectures: Design for Testability*, Morgan Kaufmann, San Francisco, CA, 2006.

153. Touba, N.A., Survey of test vector compression techniques, *IEEE Des. Test. Comput.*, 23, 294–303, 2006.

154. Chandra, A. and Chakrabarty, K., Efficient test data compression and decompression for system-on-a-chip using internal scan chains and Golomb coding, *Proceedings of the IEEE Design Automation and Test in Europe*, Munich, Germany, 2001, pp. 145–149.

155. Wang, Z. and Chakrabarty, K., Test data compression for IP embedded cores using selective encoding of scan slices, *Proceedings of the IEEE International Test Conference*, Austin, TX, 2005, Paper 24.3.

156. Rajski, J., Tyszer, J., Kassab, M., and Mukherjee, N., Embedded deterministic test, *IEEE Trans. Comput. Aided Des.*, 23, 776–792, 2004.

157. Lee, K.-J., Chen, J.-J., and Huang, C.-H., Using a single input to support multiple scan chains, *Proceedings of the ACM International Conference on Computer-Aided Design*, San Jose, CA, 1998, pp. 74–78.

158. Hamzaoglu, I. and Patel, J.H., Reducing test application time for full scan embedded cores, *Proceedings of the International Symposium on Fault-Tolerant Computing*, Madison, WI, 1999, pp. 260–267.

159. Wang, L.T., Wen X., Furukawa, H., Hsu, F.S., Lin, S.H., Tsai, S.W., Abdel-Hafez, K.S., and Wu, S.L., VirtualScan: A new compressed scan technology for test cost reduction, *Proceedings of the IEEE International Test Conference*, Charlotte, NC, 2004, pp. 916–925.

160. Wu, M.F., Huang, J.L., Wen, X., and Miyase, K., Reducing power supply noise in linear-decompressor-based test data compression environment for at-speed scan testing, *Proceedings of the IEEE International Test Conference*, Santa Clara, CA, 2008, pp. 1–10.

161. Mueller-Thuns, R.B., Saab, D.G., Damiano, R.F., and Abraham, J.A., VLSI logic and fault simulation on general-purpose parallel computers, *IEEE Trans. Comput. Aided Des.*, 12, 446–460, 1993.

162. Amin, M.B. and Vinnakota, B., Workload distribution in fault simulation, *J. Electron. Test. (JETTA)*, 10, 277–282, 1997.

163. Patil, S. and Banerjee, P., Performance trade-offs in a parallel test generation/fault simulation environment, *IEEE Trans. Comput. Aided Des.*, 10, 1542–1558, 1991.

164. Narayanan, V. and Pitchumani, V., Fault simulation on massively parallel SIMD machines algorithms, implementations and results, *J. Electron. Test. (JETTA)*, 3, 79–92, 1992.

165. Ozguner, F. and Daoud, R., Vectorized fault simulation on the cray x-mp supercomputer, *Proceedings of the IEEE International Conference on Computer-Aided Design*, Santa Clara, CA, 1988, pp. 198–201.

166. Ishiura, N., Ito, M., and Yajima, S., Dynamic two-dimensional parallel simulation technique for high-speed fault simulation on a vector processor, *IEEE Trans. Comput. Aided Des.*, 9, 868–875, 1990.

167. Amin, M.B. and Vinnakota, B., Data parallel fault simulation, *IEEE Trans. VLSI Syst.*, 7, 183–190, 1999.

168. Gulati, K. and Khatri, S.P., Towards acceleration of fault simulation using graphics processing units, *Proceedings of the IEEE/ACM Design Automation Conference*, Anaheim, CA, 2008, pp. 822–827.

169. Li, H., Xu, D., Han, Y., Cheng, K.-T., and Li, X., nGFSIM: A GPU-based fault simulator for 1-to-n detection and its applications, *Proceedings of the IEEE International Test Conference*, Austin, TX, 2010, pp. 1–10.

170. Klenke, R.H., Williams, R.D., and Aylor, J.H., Parallel-processing techniques for automatic test pattern generation, *IEEE Comput.*, 25, 71–84, 1992.

171. Wolf, J.M., Kaufman, L.M., Klenke, R.H., Aylor, J.H., and Waxman, R., An analysis of fault partitioned parallel test generation, *IEEE Trans. Comput. Aided Des.*, 15, 517–534, 1996.

172. Cai, X., Wohl, P., Waicukauski, J.A., and Notiyath, P., Highly efficient parallel ATPG based on shared memory, *Proceedings of the International Test Conference*, Austin, TX, 2010, pp. 1–7.

173. Balboni, G., Cabodi, G., Gai, S., and Sonza Reorda, M., A parallel system for test pattern generation, *Trans. Parallel Distrib. Syst.*, 19, 177–185, 1993.

174. Patil, S. and Banerjee, P., A parallel branch and bound algorithm for test generation, *IEEE Trans. Comput. Aided Des.*, 9, 313–322, 1990.

175. Smith, S.P., Underwood, B., and Mercer, M.R., An analysis of several approaches to circuit partitioning for parallel logic simulation, *Proceedings of the International Conference on Computer Design*, Chester, NY, 1987, pp. 664–667.

176. Li, M. and Hsiao, M.S., FSimGP^ 2: An efficient fault simulator with GPGPU, *Proceeding of the IEEE Asian Test Symposium*, Shanghai, China, 2010, pp. 15–20.

177. Chatterjee, D., DeOrio, A., and Bertacco, V., Event-driven gate-level simulation with GP-GPUs, *Proceedings of the IEEE/ACM Design Automation Conference*, San Francisco, CA, 2009, pp. 557–562.

178. Kochte, M.A., Schaal, M., Wunderlich, H.-J., and Zoellin, C.G., Efficient fault simulation on many-core processors, *Proceedings of the ACM/IEEE Design Automation Conference*, Austin, TX, 2010, pp. 380–385.

179. Liao, K.-Y., Hsu, S.-C., and Li, J.C.-M., GPU-based n-detect transition fault ATPG, *Proceedings of the ACM Design Automation Conference*, Austin, TX, 2013, p. 28.

180. Liao, K.-Y., Chang, C.-Y., and Li, J.-M., A parallel test pattern generation algorithm to meet multiple quality objectives, *IEEE Trans. Comput. Aided Des.*, 30, 1767–1772, 2011.

181. Tehranipoor, M. and Koushanfar, F., A survey of hardware Trojan taxonomy and detection, *IEEE Des. Test. Comput.*, 27, 10–25, 2010.

182. Wolff, F., Papachristou, C., Bhunia, S., and Chakraborty, R., Towards Trojan-free trusted ICS: Problem analysis and detection scheme, *Proceedings of the IEEE Design Automation and Test in Europe*, Munich, Germany, 2008, pp. 1362–1365.

183. Chakraborty, R., Wolff, F., Paul, S., Papachristou C., and Bhunia, S., Mero: A statistical approach for hardware Trojan detection, *Proceedings of Cryptographic Hardware and Embedded Systems*, 2009, Lecture Notes in Computer Science, Vol. 5747, Springer, Berlin, Germany, 2009, pp. 396–410.

184. Becker, G.T., Regazzoni, F., Paar, C., and Burleson, W.P., Stealthy dopant-level hardware Trojans, *Proceedings of Cryptographic Hardware and Embedded Systems*, 2013, Lecture Notes in Computer Science, Vol. 8086, Springer, Berlin, Germany, 2013, pp. 197–214.

185. Entrena, L.A. and Cheng, K.-T., Combinational and sequential logic optimization by redundancy addition and removal, *IEEE Trans. Comput. Aided Des.*, 14, 909–916, 1995.

186. Brand, D., Verification of large synthesized designs, *Proceedings of the International Conference on Computer-Aided Design*, Santa Clara, CA, 1993, pp. 534–537.

187. Huang, S.-Y. and Cheng, K.-T., *Formal Equivalence Checking and Design Debugging*, Springer, New York, 1998, p. 229.

188. Keller, B., McCauley, K., Swenton, J., and Youngs, J., ATPG in practical and non-traditional applications, *Proceedings of the IEEE International Test Conference*, Washington, DC, 1998, pp. 632–640.

189. Wohl, P. and Waicukauski, J., Using ATPG for clock rules checking in complex scan designs, *Proceedings of the IEEE VLSI Test Symposium*, Monterey, CA, 1997, pp. 130–136.

190. Thatte, S.M. and Abraham, J.A., Test generation for microprocessors, *IEEE Trans. Comput.*, 100, 429–441, 1980.

191. Brahme, D. and Abraham, J.A., Functional testing of microprocessors, *IEEE Trans. Comput.*, 100, 475–485, 1984.

192. Lee, J. and Patel, J.H., A signal-driven discrete relaxation technique for architectural level test generation, *Proceedings of the International Conference on Computer-Aided Design*, Santa Clara, CA, 1991, pp. 458–461.

193. Lee, J. and Patel, J.H., Architectural level test generation for microprocessors, *IEEE Trans. Comput. Aided Des.*, 13, 1288–1300, 1994.

194. Cheng, K.T. and Krishnakumar, A., Automatic generation of functional vectors using the extended finite state machine model, *ACM Trans. Des. Automat. Electr. Syst.*, 1, 57–79, 1996.

195. Hansen, M.C. and Hayes, J.P., High-level test generation using symbolic scheduling, *Proceedings of the IEEE International Test Conference*, Washington, DC, 1995, pp. 586–595.

196. Sabnani, K. and Dahbura, A., A protocol test generation procedure, *Comput. Networks ISDN*, 15, 285–297, 1988.

197. Dahbura, A.T., Uyar, M.U., and Yau, C.W., An optimal test sequence for the JTAG/IEEE P1149. 1 test access port controller, *Proceedings of the IEEE International Test Conference*, Washington, DC, 1989, pp. 55–62.

198. Vedula, V.M., Abraham, J.A., and Bhadra, J., Program slicing for hierarchical test generation, *Proceedings of the IEEE VLSI Test Symposium*, Monterey, CA, 2002, pp. 237–243.

199. Huan, C.-Y. and Cheng, K.-T., Using word-level ATPG and modular arithmetic constraint-solving techniques for assertion property checking, *IEEE Trans. Comput. Aided Des.*, 20, 381–391, 2001.

200. Corno, F., Reorda, M.S., and Squillero, G., RT-level ITC'99 benchmarks and first ATPG results, *IEEE Des. Test. Comput.*, 17, 44–53, 2000.

201. Yin, Z., Min, Y., Li, X., and Li, H., A novel RTL behavioral description based ATPG method, *J. Comput. Sci. Technol.*, 18, 308–317, 2003.

202. Iyer, M.A., RACE: A word-level ATPG-based constraints solver system for smart random simulation, *Proceedings of the IEEE International Test Conference*, Charlotte, NC, 2003, pp. 299–308.

203. Zhang, L., Ghosh, I., and Hsiao, M., Efficient sequential ATPG for functional RTL circuits, *Proceedings of the IEEE International Test Conference*, Charlotte, NC, 2003, pp. 290–298.

204. Shyam, S. and Bertacco, V., Distance-guided hybrid verification with GUIDO, *Proceedings of the IEEE Design Automation and Test in Europe*, Munich, Germany, 2006, pp. 1211–1216.

205. Nanshi, K. and Somenzi, F., Guiding simulation with increasingly refined abstract traces, *Proceedings of the IEEE/ACM Design Automation Conference*, San Francisco, CA, 2006, pp. 737–742.

206. Wu, W. and Hsiao, M.S., Efficient design validation based on cultural algorithms, *Proceedings of the IEEE Design Automation and Test in Europe*, Munich, Germany, 2008, pp. 402–407.

207. Zhou, Y., Wang, T., Lv, T., Li, H., and Li, X., Path constraint solving based test generation for hard-to-reach states, *Proceedings of the IEEE Asian Test Symposium*, Yilan County, Taiwan, 2013, pp. 239–244.

208. Wang, J., Li, H., Lv, T., Wang, T., and Li, X., Functional test generation guided by steady-state probabilities of abstract design, *Proceedings of the IEEE Design Automation and Test in Europe*, Dresden, Germany, 2014, Paper 11.4.

Analog and Mixed-Signal Test

Haralampos-G. Stratigopoulos and Bozena Kaminska

CONTENTS

23.1 INTRODUCTION

Integrated circuits (ICs) are fabricated using a series of photolithographic, printing, etching, implanting, and chemical vapor deposition steps. This process is subject to imperfections that may cause complete failure in the operation of individual ICs or variations in performance among ICs on the same wafer or across different lots. The performance of an IC could also shift in the postsilicon production flow during the packaging process. For all these reasons, each fabricated IC must be tested in order to ensure that it meets its design specifications.

The current practice for testing the analog and mixed-signal (AMS) functions of ICs is specification-based testing [1–4]. Specification-based testing involves direct measurement of the performances that are promised in the specification data sheet one by one. However, despite the ease of interpreting the test result, specification-based testing incurs a very high cost since it relies on specialized automatic test equipment (ATE) with advanced capabilities and running the tests takes a long time. In fact, testing the AMS functions of modern systems-on-chip (SoCs) is responsible for the largest fraction of the test cost despite the fact that AMS circuits occupy a much smaller area on the die compared to their digital counterparts [5]. With the ever-increasing levels of integration of SoC designs, more and more of which include AMS circuits, the ATE cost, test development time, and test execution time are being increasingly impacted and will keep increasing as we move toward more advanced technology nodes. AMS testing is nowadays an area of focus and innovation for the microelectronics industry.

It is thus necessary for design and test engineers to work together, early in the SoC architecture design phase, in order to keep the testing costs under control. Alternative low-cost test techniques need to be developed that can effectively replace standard specification-based tests. These techniques should target reducing test times and/or alleviating the need to rely on specialized ATE. Introducing alternative test techniques, however, should not sacrifice the high accuracy of specification-based testing, which is measured by test metrics such as test escape (e.g., faulty circuits passing the test) and yield loss (e.g., functional circuits failing the test). Therefore, any alternative test technique should be assessed by estimating the resultant test metrics that will be met in production.

This chapter will focus on generic alternative test techniques, such as (1) inferring the outcome of specification-based testing from low-cost measurements (broadly known as "alternate test"), (2) eliminating redundant specification-based tests and applying a reduced set, (3) structural test techniques where the test problem reduces to checking for the presence of faults contained in a predefined list, and (4) practical integrated test techniques for different types of AMS circuits, such as analog-to-digital converters (ADCs), digital-to-analog converters (DACs), phase-locked loops (PLLs), and radio-frequency (RF) circuits. This chapter will also review tools to assess the feasibility of alternative test techniques. All these techniques and tools are seen as an enabling technology to break the cost trend.

23.2 INTEGRATED TEST: AN OVERVIEW

Integrated test techniques are grouped into design-for-test (DFT) and built-in self-test (BIST) techniques.

DFT techniques can be broadly grouped into two approaches. The first DFT approach is to facilitate test access into the design by implementing test signal buses according to the IEEE Standards 1149.1 [6] and 1149.4 [7], as illustrated in Figure 23.1. The IEEE 1149.4 architecture in Figure 23.1 comprises a test bus interface circuit with analog test stimulus (AT1) and analog test output (AT2) pins, analog boundary modules on each analog I/O, digital boundary modules on each digital I/O, and a test access port controller with test data in, test data out, test mode select, test clock, and test reset pins. This test bus architecture provides the means for bypassing functional blocks in the circuit under test (CUT) in order to apply test stimuli directly to internal blocks and reading out the test responses. Therefore, the test bus can be used to enhance the overall testability, as well as to enable system diagnostics and silicon debugging in postmanufacturing. In addition, the test bus can be used for testing for open and short circuits among the interconnections of circuits in a printed circuit assembly. The second DFT approach is based on reconfiguring the CUT to enhance its testability. A first well-known example is the generic

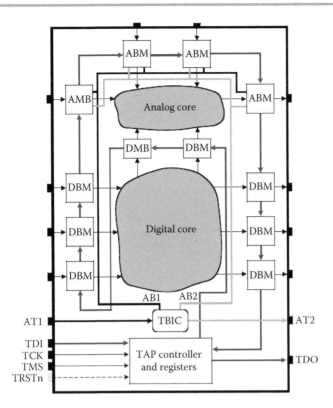

FIGURE 23.1   IEEE 1149.4 architecture.

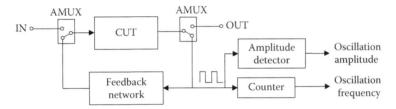

FIGURE 23.2   Oscillation-based test architecture.

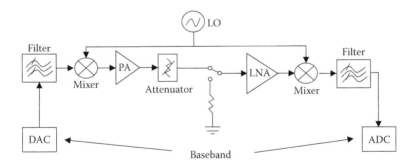

FIGURE 23.3   Loop-back test for RF transceivers.

oscillation test where the CUT is reconfigured to oscillate by connecting it into a positive feed-back loop, as shown in Figure 23.2. The oscillation frequency and magnitude are information-rich signatures that can be used to gain insight about the functionality of the CUT and to detect abnormal behavior [8,9]. A second example is the loopback test for RF transceivers where the test signals are generated in the baseband and the transmitter's output is switched to the receiver's input through an attenuator to analyze the test response also in the baseband [10–16], as shown in Figure 23.3. We will revisit these two DFT techniques in more detail in Sections 23.7 and 23.10.

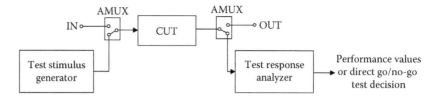

FIGURE 23.4 BIST employing an on-chip test stimulus generator and an on-chip test response analyzer.

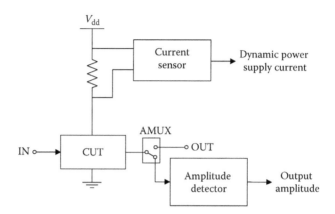

FIGURE 23.5 BIST employing on-chip sensors.

BIST techniques can also be broadly grouped into two approaches. The first approach consists of embedding a signal generator and a test response analyzer into the chip [17–23], as shown in Figure 23.4, whereas the second approach consists of embedding sensors into the chip to extract off-chip information-rich, low-cost test signatures from which the status of the performances can be implicitly inferred [24–28], as shown in Figure 23.5. The impetus for BIST techniques is to facilitate the use of low-performance ATE or perhaps to eliminate any need whatsoever by adding self-test capabilities, strategic control, and observation points within the IC.

DFT and BIST are very often ad hoc and largely a matter of early engagement with the design community to specify the test architecture. Great strides have been made to make DFT and BIST techniques successful for AMS ICs, but robust, production deployment of these techniques is not yet widespread. This is due in part to the challenge of evaluating their efficiency with respect to the standard specification-based test, which requires accurate simulation models and speeding up circuit simulation. In addition, DFT and BIST techniques should not consume a disproportionate amount of silicon die area and should neither be intrusive to sensitive circuits and design methodologies nor impede the postsilicon debugging process. Trade-offs between DFT and BIST techniques and traditional specification-based testing need to be considered, and the test resources need to be intelligently partitioned between integrated and external test methods. Finally, given the rather high development time of DFT and BIST techniques, it is important to focus on their portability, such that they can be reused in different intellectual property blocks or cores. Despite these challenges, the pressing demand to reduce test cost has sparked an immense effort to materialize DFT and BIST techniques since they arguably constitute very attractive alternatives. This rationale stems from the fact that much of the ATE will be on-chip or in the form of partitioned test that can be executed much faster.

DFT and BIST techniques become of vital importance in the case of ICs that are part of a larger safety-critical, mission-critical, or remote-controlled system. During its lifetime, an IC may fail due to aging, wear and tear, harsh environments, and overuse or due to defects that are not detected by the production tests and manifest themselves later in the field of operation. In such cases, DFT and BIST can be used to support online test during normal operation by detecting early reliability hazards and gaining insight about environmental conditions that can jeopardize the system's health [29–41]. They can also provide valuable feedback for achieving fault tolerance through calibration, tuning, or even reconfigurability [42–46].

Finally, DFT and BIST techniques can provide valuable feedback for diagnosis purposes [47–59]. Diagnosing the root causes of failures in the first prototypes helps to reduce design iterations and to meet the time-to-market goal. In a high-volume production environment, diagnosing the root causes of failures can assist the designers in gathering valuable information for enhancing yield in future IC generations. Diagnosis is also of vital importance in the case of failures in the field for safety-critical applications. Here, it is important to identify the root causes of failures so as to repair the system if possible and apply corrective actions that will prevent failure reoccurrence and, thereby, will expand the safety features.

DFT and BIST techniques vary depending the type of the AMS block and very often even for a particular design style or architecture. In Sections 23.8 through 23.10, we will review popular DFT and BIST solutions for different AMS blocks in detail.

23.3 ALTERNATE TEST

As mentioned in the introduction, the standard approach for testing AMS circuits is to measure directly the performances that are promised in the data sheet. The IC is declared faulty or functional by simply comparing the measured performance values to the design specifications. In this context, the necessary ATE resources are employed and overall the test approach is easy to interpret and implement since the same test benches are used as during the design and prototype characterization phases.

Alternate test aims to circumvent specialized ATE resources and speed up the test execution time by relying solely on measurements that can be rapidly extracted using a low-cost assortment of test equipment [60–62]. The grounds for achieving the objective of inferring the performances implicitly from low-cost alternate measurements are that both the performances and the alternate measurements are subject to the same process variations. Thus, in the presence of process variations, both performances and alternate measurements vary and the objective boils down to identifying alternate measurements that correlate well with the performances, such that any performance shift can be predicted from the corresponding shift in the alternate measurement pattern, as illustrated in Figure 23.6. For the method to succeed, it is needed first to (a) identify

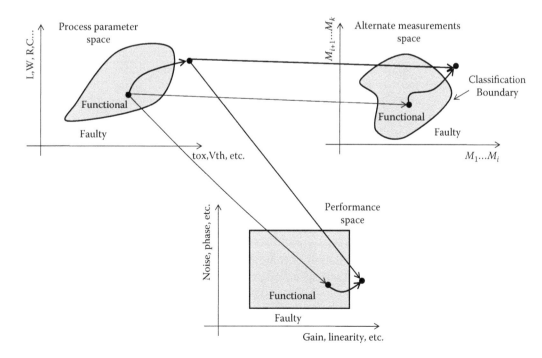

FIGURE 23.6 Principle of alternate test.

such information-rich alternate measurements and second to (b) build the mapping between alternate measurements and performances.

The identification of appropriate alternate measurements is a circuit-specific problem since the input, output, frequency band, transfer function, etc., depend on the type of the IC, as well as its architecture. In the recent years, the alternate test paradigm has been proven for different types of ICs, including baseband analog [60,63], RF [64–67], data converters [68,69], PLLs [70], and microelectromechanical systems (MEMS) [71]. Very often, the alternate measurements are extracted ad hoc without a clear rationale. Through simulations, it is demonstrated that they can be used indeed to predict the performance values and in the next step the concept is demonstrated experimentally in silicon. The reason is the large number of process parameters and their intricate interactions that makes it impossible to justify that an alternate measurement captures all variation scenarios that can occur in practice. A typical approach is to identify as many alternate measurements as possible and then compact this set using feature selection algorithms [60,72–75], such as genetic algorithms [76] and floating search algorithms [77]. Another approach is to craft the test stimulus such that the output response becomes appropriate for alternate test [60,78].

Examples of alternate measurements for baseband analog circuits include sampling the output response when applying at the input a piecewise linear test stimulus [60,63], a multitone sinusoidal [72], or a pseudo-random bit sequence [73]. Popular approaches to extract alternate measurements from RF circuits include (1) applying a baseband multitone sinusoidal, up-converting it using a mixer that exists on the test load board or on-chip, down-converting the RF output using again a mixer, and sampling the demodulated baseband test response [64,65,73] and (2) applying sensors that tap into the RF signal path, for example, amplitude detectors [24,25,79–83] and current sensors [26,82,84].

The intricate relationship among performances and alternate measurements makes it impossible to build a mapping in the form of a closed-form mathematical equation. For this reason, the mapping is built through statistical learning. In particular, a set of N circuit instances that is representative of the fabrication process is collected. The d performances $P = [P_1,...,P_d]$ and alternate measurements M are obtained on each circuit instance. Part of the circuit instances is used to train a regression function $f_i: M \rightarrow P_i$ for each performance P_i. The circuit instances that are left out are used as an independent validation set. Target performances for the circuit instances in the validation set are assumed to be unknown and they are only used to estimate the test error. In particular, the alternate measurements are given as arguments to the regression functions to obtain performance predictions $\hat{P}$. If the test error $P - \hat{P}$, averaged over all circuit instances in the validation set, is deemed to be small, then the alternate measurements are satisfactory.

It should be noticed that outliers should be excluded from the training phase since they are inconsistent with the statistical nature of the bulk of the training data stemming from circuits with process variations and will adversely affect the regression fit results. In fact, outliers are non-statistical in nature since their real cause is physical defects that are induced or enhanced during the IC manufacturing in a random fashion. Likewise, the learned regression functions are not designed to predict the performances of outliers as the test outcome will be somewhat random. Thus, in the testing phase, all ICs should be checked to verify that they are not outliers before the learned regression functions are applied to reach a test decision. This indispensable step in the flow of alternate test makes use of a defect filter [85].

Instead of predicting the actual values of the performances, it is also possible to predict directly whether the performances satisfy their specifications, that is, a form of go/no-go test. In this case, a classifier is used that implements a function $g: M \rightarrow [pass, fail]$ [72,73,86–89]. The classifier should be able to allocate a nonlinear decision boundary in the space of alternate measurements such that the population of functional circuits is separated from the population of circuits that violate at least one specification, as illustrated in Figure 23.6. Various classifiers can be used in this context, including support vector machines (SVMs) [90], decision trees [91], ontogenic neural networks [92], and feed-forward neural networks [93]. Similar to the regression functions, the better the correlation among performances and alternate measurements is, the smaller will be the overlap between the two populations and the better the classification rate will be. Furthermore, the classifier can be implemented on-chip toward a stand-alone BIST [94].

The advantage of using regression functions is that it offers the possibility of predicting the performance values, which allows binning of functional circuits and a better insight into the performance distributions. The classifier has the advantage that it can screen out circuits with defects on top of circuits with excessive process variations. However, this is at the expense of requiring to include circuits with defects in the training set, which may be difficult to collect in the production environment in a short period of time. To this end, a one-class classifier can be used that avoids this requirement [95].

Finally, as any other indirect test method, alternate test is prone to error. To improve confidence in the test decision, it is possible to identify the small fraction of circuits that will be likely erroneously predicted and forward them to a second test tier where more thorough testing is performed. Several techniques exist for this purpose, including the use of guard bands in the case of classifiers [73], multiple regression functions [96], and a pair of defect filters [95].

23.4 SPECIFICATION-BASED TEST COMPACTION

A plausible direction toward decreasing test cost is to identify and eliminate information redundancy in the set of specification-based tests, thereby relying only on a subset of them in order to reach a pass/fail decision. Such redundancy is likely to exist since groups of performances refer to the same portion of the IC and are subject to similar process imperfections. However, it is highly unlikely that it will manifest itself in a coarse and easily observable form of superfluous tests that can be summarily discarded. Hence, more advanced statistical analysis methods are likely to be required.

In the linear error-mechanism model algorithm [97], availability of a linear model $y = Ax$ is assumed [98], where y is the $m \times 1$ measurement error vector, x is a $n \times 1$ process parameter error vector, A is a $m \times n$ sensitivity matrix, and m corresponds to the number of measurements required for complete specification testing. The method aims to predict the complete vector y by carrying out only a subset $\tilde{y}$ of y. The cardinality p of $\tilde{y}$ ($p \geq n$) is a compromise between the permitted measurement cost and the maximum tolerable prediction error. The selection process is performed through QR factorization [99] and minimizes the normalized prediction variance. This is equivalent to maximizing the determinant $|\tilde{A}^T \tilde{A}|$, where $\tilde{A}$ designates the $p \times n$ row-reduced matrix A. In [100], an iterative selection approach is followed, which considers subsets rather than individual measurements. Next, the complete measurement vector is predicted by $y = A(\tilde{A}^T \tilde{A})^{-1} \tilde{A}^T \tilde{y}$. A leisurely look at this approach and some refinements are provided in [101].

In [102], a fault-driven test selection approach is proposed. The set of needed tests is cumulatively built by adding to the current set the test for which the yield, as computed by considering only the specifications of the remaining tests, is maximized. The algorithm terminates when the desired fault coverage is reached. In [103], in addition to fault coverage, the test selection is driven by the degree to which faults are exposed. After the redundant tests have been eliminated, a test ordering algorithm can be run, aiming to reduce the average test time. The tests that have higher priority of being placed at the beginning of the sequence are those that have a high probability of detecting failures, are low cost, and are independent of the previous tests in the sequence. Test ordering algorithms that appear in the literature include trial of various test permutations based on a heuristic approach to estimate whether a permutation is likely to improve average test time [102], dynamic programming of a directed flow graph [104], which results in the optimal order, and a variation of the latter, called A^* algorithm [102], which, in the worst case, requires the same computational cost while maintaining optimality.

In [105], specification testing is simplified using a technique called predictive subset testing. It requires measurement of all performances for a set of circuit instances, which is assumed to be representative, reflecting accurately the statistical mechanisms of the manufacturing process. The technique is based upon establishing regression mappings between correlated pairs of performances. In this situation, we seek to predict one (untested) performance using another performance that is explicitly tested. Then, new test limits are assigned to the tested performance, such that they guarantee the compliance of the untested performance to its specifications with the desired confidence levels.

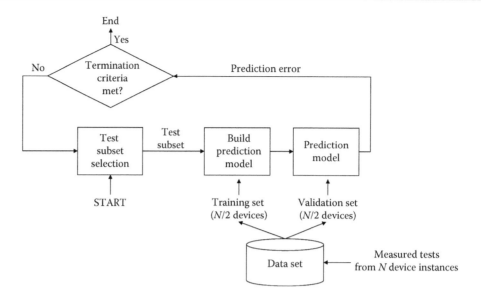

FIGURE 23.7   Feature selection algorithm for specification-based test compaction.

The problem of specification-based test compaction can also be viewed as a binary pass/fail classification problem [106–108]. This approach entails two components, namely, a feature selection algorithm for searching in the power set of specification-based tests and a prediction model for predicting based solely on a select subset the outcome of the remaining specification tests that are excluded from this subset, as shown in Figure 23.7. Different feature selection algorithms can be employed, for example, genetic algorithms [76] and floating search algorithms [77], and the prediction model can be built using binary classifiers, such as SVMs [90], decision trees [91], ontogenic neural networks [92], and feed-forward neural networks [93]. The search progresses toward a low-cost, low-dimensional, specification-based test subset based on which the classifier predicts correctly the pass/fail outcome of the complete specification-based test suite.

23.5 PROBABILISTIC TEST METRICS ESTIMATION

As it has been made clear so far, the general objective of the research conducted in the AMS test domain is to develop alternative tests that can replace effectively the costly specification-based tests. Despite the high number of proposals, few have materialized to date. One of the primary reasons is the lack of automated tools to evaluate alternative tests in terms of the resultant test error. Indeed, although it is straightforward to provide claims about the test cost reduction and overhead in the case of DFT and BIST techniques, it is not as straightforward to provide estimates of test escape and yield loss and project such test metrics in parts per million (PPM) values, as shown in Figure 23.8. Test metrics estimation has to take place during the test development phase through simulation before moving to production test. Otherwise, significant test resources and time must be dedicated without any guarantee that the alternative test will be proven indeed effective. Preferably, we would like to estimate test metrics early in the process so as to refine alternative tests or even abandon them indefinitely in time.

Broadly speaking, the faulty behavior of a circuit can be due to two reasons: (1) defects in manufacturing that translate into topological changes in the form of short- and open circuits, in which case we refer to as "catastrophic faults," and (2) variations in the process parameters, in which case we refer to as "parametric faults." While catastrophic fault coverage can be evaluated given a list of probable catastrophic faults, evaluating parametric fault coverage (or, equivalently, parametric test escape) and yield loss is a far more complex problem. The reason is twofold. First, the set of circuits that give rise to yield loss and the set of parametric faults are infinite. Second, we would like to be able to estimate parametric test escape and yield loss metrics as low as a few tens of hundreds of PPM, whereas a standard Monte Carlo approach by default samples the most likely statistical

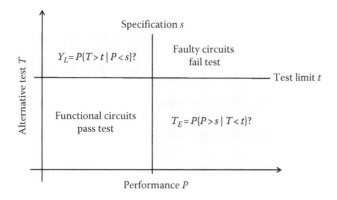

FIGURE 23.8   Areas of test escape T_E and yield loss Y_L assuming that the specification-based test targeting performance P has been replaced by an alternative test T.

events and, thereby, it would require an intractable number of runs that we cannot afford. The problem essentially boils down to speeding up Monte Carlo simulation.

The general idea is to substitute the circuit with an equivalent model that can be simulated much faster, as shown in Figure 23.9. Using N initial simulations that we can afford to run, we build a model of the circuit and thereafter we simulate it to generate $N' \gg N$ circuit instances. Thereafter, we can approximate test metrics using relative frequencies.

For example, the model could consist of a set of regression functions that map any point in the design space (e.g., process and design parameters) to the output parameters of the circuit (e.g., performances and test measurements) [102,109]. As an alternative for the regression functions, one can use symbolic models that are fitted using genetic programming [110]. If the accuracy of these models is deemed to be insufficient at the tails of the design space, then we can choose to perform circuit-level simulations for the points that are identified to lie at the tails of the design space [111].

An alternative approach is to use the initial N simulations to estimate the joint probability density function (PDF) of output parameters [112–114]. The joint PDF can be sampled very fast to generate the N' new circuit instances. Other solutions related to PDF estimation are based on the theory of Copulas [115] or on linear error-mechanism models [116].

The most recent approach to test metrics estimation is based on the statistical blockade technique and extreme value theory [117,118]. The statistical blockade technique [119] is used to bias the Monte Carlo simulation so as to quickly simulate a set of most probable "extreme" circuits that lie close to the tails of the distribution of performances and alternative tests and give rise to test escape and yield loss. Thereafter, this set of extreme circuits is used to fit a probability model for the parametric test escape and yield loss using the extreme value theory [120], as shown in Figure 23.10. This technique has the advantage that it focuses directly at the tails of the distributions where the test escape and yield loss events occur.

For circuits that are hard to simulate, such as data converters and PLLs, to estimate test metrics simulations are performed at the behavioral level [121,122].

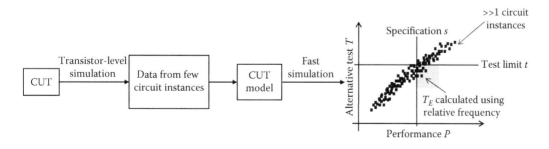

FIGURE 23.9   Principle of test metrics estimation using fast statistical simulation.

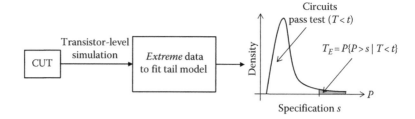

FIGURE 23.10   Principle of test metrics estimation using statistical blockade and extreme value theory.

23.6 FAULT MODELING AND STRUCTURAL TEST

The introduction of the *stuck-at* fault model for digital circuits enabled digital testing to cope with the exponential growth in the digital circuit size and complexity. Indeed, the stuck-at fault model enabled functional testing to be replaced by structural testing, acted as a measure to quantify the quality of the test plan and paved the way for the development of efficient DFT and BIST strategies.

Intense efforts have been made to borrow from the success of the *stuck-at* fault model and develop an appropriate and comprehensive fault model for AMS circuits that accounts for all fault mechanisms in the manufacturing process. In general, a fault model is defined as a set of circuit instances, each representing a scenario in the manufacturing process that results in faulty circuit behavior, such that all possible catastrophic and parametric fault scenarios are accounted for.

Fault models can be used to evaluate the test escape as a result of replacing specification-based tests with alternative tests. The approach consists of injecting one fault at a time into the netlist of the circuit and checking whether the alternative test is capable of detecting the fault, as shown in Figure 23.11. Detection implies that the alternative test is capable of clearly distinguishing the response of the faulty circuits from the response of functional circuits. In addition, fault models can be used for structural test. For a given fault model, the structural test generation problem boils down to generating a set of tests that detect all faults in the model. To this end, many test generation algorithms have been written to craft a test stimulus and select the output test signature such that the distance between the functional circuits and the faulty circuits is maximized [123–131].

A fault model that accounts for catastrophic faults can be developed based on inductive fault analysis on physical layouts [132–137]. This requires a description of defect statistics, which, when

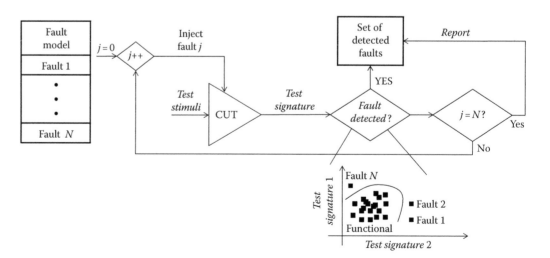

FIGURE 23.11   Estimating test escape of an alternative test approach based on a fault model.

mapped onto the layout, provide a list of probable faults. At the circuit level, the fault list consists of short and open circuits. This approach has led to defect-oriented test techniques [58,59,138], which can be applied for wafer-level testing to detect dies with gross defects or for final testing of robust designs that are highly unlikely to fail due to process variations.

The definition of a parametric fault model poses a greater challenge [139]. The reason is that there are an infinite number of combinations of process parameters that result in specification violation, that is, the size of the parametric fault model is limitless. Thus, we need to consider parametric faults according to their probability of occurrence and define a parametric fault model that contains the most probable parametric faults. Previous proposals for parametric fault modeling made certain assumptions to be able to deal with this challenge [137,140–145]. The widespread approach has been to build a parametric fault model at a higher level of abstraction, for example, by modeling faults as variations in passive components and in transistor parameters, that is, transconductance, geometry, oxide thickness, and threshold voltage, or by considering behavioral simulation instead of transistor-level simulation and modeling faults as variations in the parameters of the behavioral model. Furthermore, a common assumption is that parameters vary independently, which is known as single-fault assumption, and that a circuit fails a specification when one parameter exceeds a specific tolerance. These simplified fault models make simulation more traceable, yet their ability to capture correctly faulty behavior due to process variations has never been proven.

A fault model that consists of the most probable parametric faults generated naturally by a Monte Carlo analysis that makes use of the actual process design kit remains an open challenge [146]. A technique was proposed recently in [147] that employs the statistical blockade algorithm to speed up Monte Carlo analysis. The underlying idea is to bias the Monte Carlo analysis such as to avoid simulating circuits that are functional and, thereby, are of no use for generating a fault model.

23.7 OSCILLATION-BASED TEST

Oscillation-based test is a generic DFT/BIST technique [8,9]. It relies on converting the circuit to an oscillator and subsequently measuring the oscillation frequency and magnitude, as shown in Figure 23.2. The reconfiguration of the circuit is accomplished by using digital circuits to control analog switches and multiplexers, which are designed to have a minimal impact on the measurement. The oscillation frequency and magnitude are information-rich alternate measurements that are a function of the process parameters and also the elements that are added in the feedback path to enable oscillation. As a result, by repeating these measurements under varying voltage conditions, for different frequency selection, and for different feedback topologies with different attenuation levels, it is possible to infer implicitly the status of the performances. In particular, test limits can be placed on the oscillation frequency and magnitude to guarantee with some level of certainty that the performance values lie within the specification range. The test limits can be found through simulation and can be selected such that a desired trade-off between test metrics is achieved [148].

Oscillation-based test has several advantages: (1) it is applicable virtually to all AMS circuits including baseband analog [8,9,149–156], data converters [157–159], RF [160], and MEMS [161]; (2) it is vectorless since the circuit is self-excited without requiring signal generators; (3) the oscillation frequency can be processed to obtain digital test signatures that can be easily extracted off-chip for analysis; (4) it is immune to noise since the frequency is averaged over many periods; and (5) it delivers excellent fault coverage for catastrophic and parametric faults [162].

23.8 DFT AND BIST FOR DATA CONVERTERS

ADCs and DACs are characterized by dynamic specifications, such as total harmonic distortion, signal-to-noise ratio, and signal-to-noise-and-distortion ratio, and static specifications, such as differential nonlinearity, integral nonlinearity, offset, and gain error.

The standard approach to measure the dynamic specifications of an ADC is to apply a high-resolution sinusoidal at the input and compute the fast Fourier transform (FFT) at the output. The resolution of the sinusoidal typically needs to be at least two bits higher than the effective resolution of the ADC, which poses a great design challenge for BIST implementation. A classical approach for generating on-chip an analog sinusoidal is to employ a closed-loop oscillator that involves a highly selective bandpass filter and a comparator [163,164], as shown in Figure 23.12. To improve the resolution of the sinusoidal generators, they can be combined with harmonic cancellation techniques [165–168]. Another classical approach is to employ an open-loop oscillator, as shown in Figure 23.13. The starting point is to use an ideal $\Sigma\Delta$ modulator in software that converts a high-resolution sinusoidal to a bit stream, which, thereafter, is loaded and periodically reproduced in an on-chip circular shift register [19]. Bit streams can also be generated by on-chip digital oscillators [169,170]. The bit stream can be converted on-chip to a high-resolution sinusoidal by passing it through an 1-bit DAC followed by a low-pass filter to remove the quantization noise [171,172]. Digital techniques can be used to cancel harmonic components at the output of the DAC [173]. Interestingly, in the case of switched-capacitor (SC) $\Sigma\Delta$ ADCs, the bit stream can be fed directly into the modulator by adding simple circuitry at its input [174–177]. Another possibility is to replace the conventional sinusoidal test stimulus with a stepwise exponential [178] or, in the case of $\Sigma\Delta$ ADCs, with a pseudo-random pattern sequence [179]. For analyzing the test response, performing a FFT on-chip incurs a high area overhead [180]. If the FFT cannot be performed in a digital signal processor (DSP), then for a full BIST implementation, it is required to replace the FFT algorithm with an alternative less computationally intensive algorithm. Well-known algorithms are the sine-wave fitting [172,181,182] and the Goertzel algorithm [183]. A variant of the sine-wave fitting algorithm with reduced complexity and, thereby, more efficient digital implementation is proposed in [175] in the case of stereo SC $\Sigma\Delta$ ADC. Another variant is proposed in [184] and makes use of the COordinate Rotation DIgital Computer (CORDIC) algorithm.

The standard approach to measure the static specifications of an ADC is to apply a ramp at the input and obtain the histogram of the number of occurrences of each code at the output. On-chip adaptive ramp generators are proposed in [185,186]. A ramp generator that relies on the oscillation-based test principle is described in [158]. Alternatively, an exponential waveform can be employed in the analysis [187,188]. Another interesting approach is based upon first identifying and computationally removing the source nonlinearity and then accurately estimating the ADC static performances [166,189,190]. For analyzing the test response, it is required to store both the experimental and reference histograms and use the DSP to perform the comparison. Efficient BIST implementations of the histogram analysis are proposed in [191,192]. For ADCs

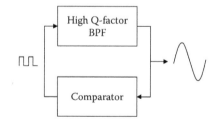

FIGURE 23.12 Closed-loop sinusoidal signal generator.

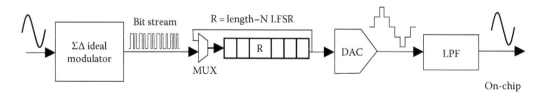

FIGURE 23.13 Open-loop sinusoidal signal generator based on $\Sigma\Delta$ bit streams.

having a repetitive structure, such as pipeline, cyclic, and successive approximation ADCs, we can also apply what is broadly known as reduced code testing [193–196]. Only a few codes need to be judiciously targeted, and from this information we can extrapolate the complete histogram.

Instead of targeting a functional BIST approach aiming at measuring directly the dynamic and static specifications, it is also possible to consider a structural DFT approach where the aim is to obtain measurements that reveal important design parameters, such as the poles and settling errors of the integrators in the case of $\Sigma\Delta$ ADCs [197]. A DFT approach proposed also specifically for pipeline ADCs is to reconfigure consecutive pipeline stages to form $\Sigma\Delta$ modulators and then test instead the $\Sigma\Delta$ modulators through digital means [198]. A generic approach virtually applicable to any ADC is to obtain information from process control monitors (PCMs) that are scattered across the die and use this information to make judgments about the performances of the ADC [199].

Regarding DACs, the standard approach to measure the dynamic specifications is to apply a digital signal at the input that encodes a high-resolution sinusoidal and compute the FFT at the output. For a BIST implementation, the digital signal can be stored on-chip in a memory. However, for analyzing the test response, performing the FFT on-chip [22,23,200–202] seems to be unavoidable, unlike in the case of ADCs where there exist alternatives, as discussed earlier. The standard approach to measure the static specifications is to apply a digital ramp at the input that contains all possible digital codes and measure the output with a digital voltmeter. A BIST implementation is proposed in [203].

Finally, in the case of SoCs that comprise a set of ADCs and DACs, a technique to test them altogether in a fully digital setup is proposed in [204].

23.9 DFT AND BIST FOR PLLS

PLLs are fundamental building blocks for processors and communication systems since they are used to synthesize clocks for data synchronization and to provide the frequency sources for up-conversion and down-conversion in RF transceivers. The standard functional tests for PLLs include measuring the loop gain, the lock time, defined as the time it takes for a PLL to acquire phase lock after an abrupt change in the phase of its reference signal, and the lock range, defined by the minimum and maximum frequencies that a PLL can lock to within its lock time.

However, perhaps the most important performance of a PLL is the jitter, which defines, in turn, the bit error rate in communication systems. The most common types of jitter are the timing jitter, defined as the edge timing variation relative to the ideal edge timing, the period jitter, defined as the variation of each period relative to the average period, and the cycle-to-cycle jitter, defined as the variation in each period with respect to the preceding period. Within these types, jitter can be random or deterministic.

The jitter, despite being the most important performance of a PLL, happens to be the most challenging one to measure on an ATE. The reason is twofold. First, the standard method that uses high-speed sampling oscilloscope requires tens of seconds, which is prohibitive for production testing. Second, as the PLL output is transferred off-chip, significant jitter is added due to capacitive coupling to other signals on-chip and transmission line effects off-chip. In addition, sampling the PLL output to extract the PLL jitter introduces even more jitter and, thus, the measurement ends up being imprecise. For these reasons, measuring jitter exactly at the PLL output using BIST is the recommended method, especially for multi-GHz PLLs that may have RMS jitter as low 0.1–5 ps.

An on-chip jitter measurement technique based on an adjustable delay line is shown in Figure 23.14 [205]. The PLL output edge timing is compared to a delayed version of the input reference clock edge that has low jitter. For digitally controlled incrementing delays, we can deduce a cumulative histogram of the output jitter relative to the reference clock edges. However, the fact that the delay line adds jitter and the fact that the PLL and the BIST share the same power supply that increases the jitter in the PLL limit the resolution of the measurement to 10 ps. Similar resolution can be achieved by other on-chip jitter measurement techniques based on Vernier delay lines and Vernier oscillators that also incur similar area overhead [206–208]. Better resolution down to the resolution required by multi-GHz PLLs can be achieved with the scheme in Figure 23.15 that employs a phase comparator [209,210]. The idea is to undersample the PLL output that has

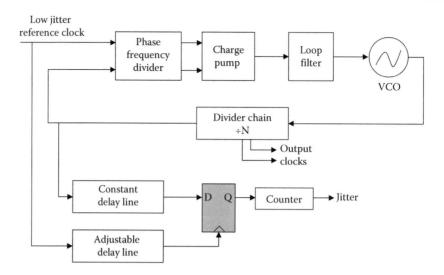

FIGURE 23.14   On-chip jitter measurement circuit that uses an adjustable delay line.

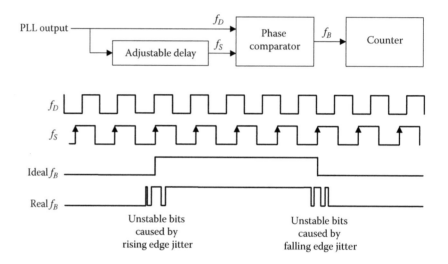

FIGURE 23.15   On-chip jitter measurement circuit that uses a phase comparator.

frequency f_D by a slightly lower clock frequency $f_S = f_D \times N/(N + 1)$, where N is a positive integer. The resulting low-frequency digital waveform has a beat frequency $f_B = f_D - f_S$. At the time where we sample the rising and fall edges of the PLL output, the output of the phase comparator will be unstable due to the presence of jitter at the PLL output. The level of instability of the bits at times where the rising and falling edges occur, which is directly related to the PLL jitter, and, thus, by using a counter we can extract a signature that can be mapped to the PLL jitter.

Finally, BIST techniques to measure the jitter transfer function of PLLs are described in [211,212]. BIST techniques to measure various PLL parameters, including VCO gain, lock time, lock range, and phase offset, are described in [70,205,213,214]. Structural test techniques targeting catastrophic faults that employ DFT are described in [215,216].

23.10 DFT AND BIST FOR RF CIRCUITS

Perhaps the earliest system-level DFT strategy for RF transceivers is the loopback architecture shown in Figure 23.3 [10–16]. Digitally modulated baseband test signals are transmitted through the RF transceiver chain and the baseband response signals are sampled to evaluate the performance of the RF transceiver. In this case, the DSP can serve as both a test stimulus generator and a test response analyzer. The transmitter is connected to the receiver through an attenuator such that the power amplifier's output is suitable for the dynamic range of the low-noise amplifier. This approach presents many challenges related to the area overhead of the loopback connection and the signal-path mismatch, cross-talk, and signal leakage, which can all obscure the faulty behavior. Analytical techniques to decouple the transmitter and the receiver in loopback mode and measure I/Q imbalances and nonlinearity are presented in [217–220]. Alternate test of RF transceivers in loopback mode is discussed in [221,222].

Instead of performing a system-level test based on a loopback connection, another possibility is to target BIST of the blocks inside the RF transceiver. A popular BIST strategy is to employ on-chip sensors, as shown in Figure 23.5. Two types of sensors are most commonly used. The first type is an amplitude (or envelope) detector that provides a DC signature that carries information about the RF amplitude [24,25,79–83]. The second type is a current sensor that takes advantage of the small parasitic resistor (it can reach several ohms) of the line that connects the core of the CUT to the power supply pad [26,82,84]. The current sensor offers dynamic power supply current monitoring and its output can be switched to an amplitude detector in order to obtain a DC signature that carries information about the RF amplitude of the power supply current. Using two amplitude detectors at the input and output of the CUT, we can measure the gain and the 1 dB compression point. Alternatively, the DC signatures of the sensors can be used to perform an alternate test.

A common disadvantage of the loopback test and sensor-based test is that they require to add on-chip circuits that tap into sensitive RF paths and, therefore, they unavoidably degrade the RF performances. For this reason, these DFT and BIST approaches must be considered early in the design process, and the degradation must be compensated during design in order to meet the intent design specifications. This finds designers rather reluctant since it increases design iterations and it does not allow aggressive design to exploit the features that the technology has to offer.

To this end, nonintrusive BIST techniques have been proposed recently that leave the design intact. The idea is to employ sensors to monitor the CUT while being totally transparent to the CUT, as shown in Figure 23.16. Two types of sensors can be considered. The first type consists of variation-aware sensors that include process control monitors (PCMs), such as single-layout components (i.e., transistors, capacitors) and dummy analog stages (i.e., level shifters, bias stages, gain stages) [28,82]. If PCMs are placed in close physical proximity to the CUT, then they can offer an image of the process variations affecting the CUT. In this way, measurements on PCMs can track the RF performances since they are both subject to the same interdie and correlated intradie process variations. In turn, the RF performances can be implicitly predicted using the alternate test paradigm. This approach can be virtually applied to any RF circuit since any RF circuit can be decomposed into PCMs. The objective is to layout the PCMs in close physical proximity to the corresponding structures in the RF circuit such that they are well matched in order to minimize uncorrelated intradie process variations.

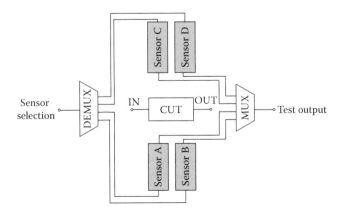

FIGURE 23.16 BIST based on non-intrusive sensors.

However, variation-aware sensors cannot detect the presence of defects within the CUT since they are not electrically connected to it. For the purpose of detecting defects, a second type of nonintrusive sensor can be used that is based on temperature monitoring [223–227]. The underlying observation is that when the CUT operates, part of its electric power is dissipated, that is, it is converted to heat due to the electrothermal Joule effect. The heat is mostly conducted through the silicon substrate and the temperature in a sensing point near a dissipating device of the CUT varies due to the power dissipated. The idea is that a defect will necessarily shift the dissipated power that, in turn, will change the temperature profile close to the CUT. Thus, a temperature sensor can capture this change to indicate the presence of the defect. The heat transfer has a low-pass filter behavior with a time constant defined by the thermal path of a few tens of kilohertz. It appears that only DC spectral component of the dissipated power induces a temperature variation near the CUT. Nevertheless, since the dissipated power is the product of current flowing through the dissipating device and the voltage across its terminals, this DC component carries information about both the DC biasing and the RF amplitude of the signals. By extension, by measuring temperature near a dissipating device of the CUT, we can monitor both the DC biasing point and RF operation of the CUT.

23.11 SUMMARY

We presented a concise overview of AMS testing and practical DFT and BIST techniques. The aim of this chapter was to explain basic concepts and provide references where the interested reader can find a more detailed presentation of the material as well as experiments that demonstrate these concepts. Finally, we should note that, in general, we did not include traditional test approaches that have been for many years the practice in industry. These approaches are presented in great detail in [2,4]. The aim was instead to collect in a comprehensive manner the latest research results in the field and describe techniques that have a high potential and are close to industrialization.

REFERENCES

1. B. Vinnakota (ed.), *Analog and Mixed-Signal Test*. Prentice Hall, Englewood Cliffs, NJ, 1998.
2. K. B. Schaub and J. Kelly, *Production Testing of RF and System-on-a-Chip Devices for Wireless Communications*. Artech House, Inc., Norwood, MA, 2004.
3. S. Sunter, Mixed-signal testing and DfT, in *Advances in Electronic Testing: Challenges and Methodologies*, D. Gizopoulos (ed.). Springer, Dordrecht, the Netherlands, 2006, pp. 301–336.
4. M. Burns and G. W. Roberts, *An Introduction to Mixed-Signal IC Test and Measurement*, 2nd edn. Oxford University Press, New York, 2011.
5. F. Poehl, F. Demmerle, J. Alt, and J. Obermeir, Production test challenges for highly integrated mobile phone SoCs, in *Proceedings of IEEE European Test Symposium*, Prague, Czech Republic, 2010, pp. 17–22.
6. IEEE Standard for Test Access Port and Boundary-Scan Architecture, IEEE Std. 1149.1-2013.
7. IEEE Standard for Mixed-Signal Test Bus, IEEE Std. 1149.4-2010.
8. K. Arabi and B. Kaminska, Oscillation-test strategy for analog and mixed-signal integrated circuits, in *Proceedings of IEEE VLSI Test Symposium*, Princeton, NJ, 1996, pp. 476–482.
9. G. Huertas, D. Vázquez, A. Rueda, and J. L. Huertas, *Oscillation-Based Test in Mixed-Signal Circuits*. Springer, Dordrecht, the Netherlands, 2006.
10. M. Jarwala, L. Duy, and M. S. Heutmaker, End-to-end test strategy for wireless systems, in *Proceedings of IEEE International Test Conference*, Washington, DC, 1995, pp. 940–946.
11. M. S. Heutmaker and D. K. Le, An architecture for self-test of a wireless communication system, *IEEE Communications Magazine*, 37(6), 98–102, 1999.
12. D. Lupea, U. Pursche, and H.-J. Jentschel, RF-BIST: Loopback spectral signature analysis, in *Proceedings of Design, Automation, & Test in Europe Conference*, Munich, Germany, 2003, pp. 478–483.
13. S. Ozev and C. Olgaard, Wafer-level RF test and DfT for VCO modulating transceiver architectures, in *Proceedings of IEEE VLSI Test Symposium*, Napa Valley, CA, 2004, pp. 217–222.
14. J.-S. Yoon and W. R. Eisenstadt, Embedded loopback test for RF ICs, *IEEE Transactions on Instrumentation and Measurement*, 54(5), 1715–1720, 2005.
15. A. Valdes-Garcia, J. Silva-Martinez, and E. Sanchez-Sinencio, On-chip testing techniques for RF wireless transceivers, *IEEE Design & Test of Computers*, 23(4), 268–277, 2006.

16. J. J. Dabrowski and R. M. Ramzan, Built-in loopback test for IC RF transceivers, *IEEE Transactions on Very Large Scale Integration (VLSI) Systems*, 18(6), 933–946, 2010.

17. B. R. Veillette and G. W. Roberts, A built-in self-test strategy for wireless communication systems, in *Proceedings of IEEE International Test Conference*, Washington, DC, 1995, pp. 930–939.

18. X. Haurie and G. W. Roberts, Arbitrary-precision signal generation for mixed-signal built-in-self-test, *IEEE Transactions on Circuits and Systems—II: Analog and Digital Signal Processing*, 45(11), 1425–1432, 1998.

19. B. Dufort and G. W. Roberts, On-chip analog signal generation for mixed-signal built-in self-test, *IEEE Journal of Solid-State Circuits*, 34(3), 318–330, 1999.

20. M. M. Hafed, N. Abaskharoun, and G. W. Roberts, A 4-GHz effective sample rate integrated test core for analog and mixed-signal circuits, *IEEE Journal of Solid-State Circuits*, 37(4), 499–514, 2002.

21. M. M. Hafed and G. W. Roberts, Techniques for high-frequency integrated test and measurement, *IEEE Transactions on Instrumentation and Measurement*, 52(6), 1780–1786, 2003.

22. M. G. Méndez-Rivera, A. Valdes-Garcia, J. Silva-Martinez, and E. Sánchez-Sinencio, An on-chip spectrum analyzer for analog built-in testing, *Journal of Electronic Testing: Theory and Applications*, 21(3), 205–219, 2005.

23. A. Valdes-Garcia, F. A.-L. Hussien, J. Silva-Martínez, and E. Sánchez-Sinencio, An integrated frequency response characterization system with a digital interface for analog testing, *IEEE Journal of Solid-State Circuits*, 41(10), 2301–2313, 2006.

24. A. Valdes-Garcia, R. Venkatasubramanian, J. Silva-Martinez, and E. Sanchez-Sinencio, A broadband CMOS amplitude detector for on-chip RF measurements, *IEEE Transactions on Instrumentation and Measurement*, 57(7), 1470–1477, 2008.

25. Y.-C. Huang, H.-H. Hsieh, and L.-H. Lu, A built-in self-test technique for RF low-noise amplifiers, *IEEE Transactions on Microwave Theory and Techniques*, 56(2), 1035–1042, 2008.

26. M. Cimino, H. Lapuyade, Y. Deval, T. Taris, and J.-B. Bégueret, Design of a 0.9 V 2.45 GHz self-testable and reliability-enhanced CMOS LNA, *IEEE Journal of Solid-State Circuits*, 43(5), 1187–1194, 2008.

27. D. Mannath, D. Webster, V. Montano-Martinez, D. Cohen, S. Kush, T. Ganesan, and A. Sontakke, Structural approach for built-in tests in RF devices, in *Proceedings of IEEE International Test Conference*, Austin, TX, 2010, paper 14.1.

28. L. Abdallah, H.-G. Stratigopoulos, S. Mir, and C. Kelma, Experiences with non-intrusive sensors for RF built-in test, in *Proceedings of IEEE International Test Conference*, Anaheim, CA, 2012, paper 17.1.

29. J. L. Huertas, A. Rueda, and D. Vasquez, Testable switched-capacitor filters, *IEEE Journal of Solid-State Circuits*, 28(7), 719–724, 1993.

30. A. Chatterjee, Concurrent error detection and fault-tolerance in linear analog circuits using continuous checksums, *IEEE Transactions on Very Large Scale Integration (VLSI) Systems*, 1(2), 138–150, 1993.

31. B. Vinnakota and R. Harjani, The design of analog self-checking circuits, in *Proceedings of IEEE International Conference on VLSI Design*, Calcutta, India, 1994, pp. 67–70.

32. V. Kolarik, S. Mir, M. Lubaszewski, and B. Courtois, Analog checkers with absolute and relative tolerances, *IEEE Transactions on Computer-Aided Design of Integrated Circuits and Systems*, 14(5), 607–612, 1995.

33. C.-L. Wey, S. Krishnan, and S. Sahli, Test generation and concurrent error detection in current-mode A/D converters, *IEEE Transactions on Computer-Aided Design of Integrated Circuits and Systems*, 14(10), 1291–1298, 1995.

34. A. Laknaur and H. Wang, A methodology to perform online self-testing for field-programmable analog array circuits, *IEEE Transactions on Instrumentation and Measurement*, 54(5), 1751–1760, 2005.

35. E. Simeu, A. Peters, and I. Rayane, Automatic design of optimal concurrent fault detector for linear analog systems, in *Proceedings of IEEE International Symposium on Fault-Tolerant Computing*, Madison, WI, 1999, pp. 184–191.

36. K.-J. Lee, W.-C. Wang, and K.-S. Huang, A current-mode testable design of operational transconductance amplifier-capacitor filters, *IEEE Transactions on Circuits and Systems—II: Analog and Digital Signal Processing*, 46(4), 401–413, 1999.

37. M. Lubaszewski, S. Mir, V. Kolarik, C. Nielsen, and B. Courtois, Design of self-checking fully differential circuits and boards, *IEEE Transactions on Very Large Scale Integration (VLSI) Systems*, 8(2), 113–128, 2000.

38. S. Ozev and A. Orailoglu, Design of concurrent test hardware for linear analog circuits with constrained hardware overhead, *IEEE Transactions on Very Large Scale Integration (VLSI) Systems*, 12(7), 756–765, 2004.

39. H.-G. D. Stratigopoulos and Y. Makris, Concurrent detection of erroneous responses in linear analog circuits, *IEEE Transactions on Computer-Aided Design of Integrated Circuits and Systems*, 25(5), 878–891, 2006.

40. H.-G. D. Stratigopoulos and Y. Makris, An adaptive checker for the fully differential analog code, *IEEE Journal of Solid-State Circuits*, 41(6), 1421–1429, 2006.

41. V. Natarajan, G. Srinivasan, and A. Chatterjee, On-line error detection in wireless RF transmitters using real-time streaming data, in *Proceedings of IEEE International On-Line Testing Symposium*, Lake Como, Italy, 2006, pp. 159–164.

42. T. Das, A. Gopalan, C. Washburn, and P. R. Mukund, Self-calibration of input-match in RF front-end circuitry, *IEEE Transactions on Circuits and Systems—II: Express Briefs*, 52(12), 821–825, 2005.

43. S. Bou-Sleiman and M. Ismail, Enabling efficient built-in-self-calibration for RFICs, in *Proceedings of IEEE International Conference on Electronics, Circuits and Systems*, Beirut, Lebanon, 2011, pp. 492–495.

44. A. Goyal, M. Swaminathan, A. Chatterjee, D. Howard, and J. Cressler, A new self-healing methodology for RF amplifier circuits based on oscillation principles, *IEEE Transactions on Very Large Scale Integration (VLSI) Systems*, 20(10), 1835–1848, 2012.

45. C. Maxey, G. Creech, S. Raman, J. Rockway, K. Groves, T. Quach, L. Orlando, and A. Mattamana, Mixed-signal SoCs with in situ self-healing circuitry, *IEEE Design & Test of Computers*, 29(6), 27–39, 2012.

46. S. Bowers, K. Sengupta, B. Parker, and A. Hajimiri, Integrated self-healing for mm-wave power amplifiers, *IEEE Transactions on Microwave Theory and Techniques*, 61(3), 352–363, 2013.

47. N. Sen and R. Saeks, Fault diagnosis for linear systems via multifrequency measurements, *IEEE Transactions on Circuits and Systems*, 26(7), 457–465, 1979.

48. H. Dai and M. Souders, Time-domain testing strategies and fault diagnosis for analog systems, *IEEE Transactions on Instrumentation and Measurement*, 39(1), 157–162, 1990.

49. M. Slamani and B. Kaminska, Analog circuit fault diagnosis based on sensitivity computation and functional testing, *IEEE Design & Test of Computers*, 9(1), 30–39, 1992.

50. S. S. Somayajula, E. Sanchez-Sinencio, and J. P. de Gyvez, Analog fault diagnosis based on ramping power supply current signature clusters, *IEEE Transactions on Circuits and Systems—II: Analog and Digital Signal Processing*, 43(10), 703–712, 1996.

51. R. Spina and S. Upadhyaya, Linear circuit fault diagnosis using neuromorphic analyzers, *IEEE Transactions on Circuits and Systems—II: Analog and Digital Signal Processing*, 44(3), 188–196, 1997.

52. S. Chakrabarti, S. Cherubal, and A. Chatterjee, Fault diagnosis for mixed-signal electronic systems, in *Proceedings of IEEE Aerospace Conference*, Snowmass at Aspen, CO, 1999, pp. 169–179.

53. E. F. Cota, M. Negreiros, L. Carro, and M. Lubaszewski, A new adaptive analog test and diagnosis system, *IEEE Transactions on Instrumentation and Measurement*, 49(2), 223–227, 2000.

54. M. Aminian and F. Aminian, A modular fault-diagnosis system for analog electronic circuits using neural networks with wavelet transform as a preprocessor, *IEEE Transactions on Instrumentation and Measurement*, 56(5), 1546–1554, 2007.

55. E. S. Erdogan, S. Ozev, and P. Cauvet, Diagnosis of assembly failures for system-in-package RF tuners, in *IEEE International Symposium on Circuits and Systems*, Seattle, WA, 2008, pp. 2286–2289.

56. K. Huang, H.-G. Stratigopoulos, and S. Mir, Fault diagnosis of analog circuits based on machine learning, in *Proceedings of Design, Automation & Test in Europe Conference*, Dresden, Germany, 2010, pp. 1761–1766.

57. S. Krishnan, K. D. Doornbos, R. Brand, and H. G. Kerkhoff, Block level Bayesian diagnosis of analogue electronic circuits, in *Proceedings of Design, Automation & Test in Europe Conference*, Dresden, Germany, 2010, pp. 1767–1772.

58. H. Hashempour, J. Dohmen, B. Tasic, B. Kruseman, C. Hora, M. Beurden, and Y. Xing, Test time reduction in Analogue/Mixed-signal devices by defect oriented testing: An industrial example, in *Proceedings of Design, Automation & Test in Europe Conference*, Grenoble, France, 2011.

59. K. Huang, H.-G. Stratigopoulos, S. Mir, C. Hora, Y. Xing, and B. Kruseman, Diagnosis of local spot defects in analog circuits, *IEEE Transactions on Instrumentation and Measurement*, 61(10), 2701–2712, 2012.

60. P. N. Variyam, S. Cherubal, and A. Chatterjee, Prediction of analog performance parameters using fast transient testing, *IEEE Transactions on Computer-Aided Design of Integrated Circuits and Systems*, 21(3), 349–361, 2002.

61. S. S. Akbay, A. Halder, A. Chatterjee, and D. Keezer, Low-cost test of embedded RF/Analog/Mixed-signal circuits in SOPs, *IEEE Transactions on Advanced Packaging*, 27(2), 352–363, 2004.

62. R. Voorakaranam, S. S. Akbay, S. Bhattacharya, S. Cherubal, and A. Chatterjee, Signature testing of analog and RF circuits: Algorithms and methodology, *IEEE Transactions on Circuits and Systems—I*, 54(5), 1018–1031, 2007.

63. R. Voorakaranam, R. Newby, S. Cherubal, B. Cometta, T. Kuehl, D. Majernik, and A. Chatterjee, Production deployment of a fast transient testing methodology for analog circuits: Case study and results, in *IEEE International Test Conference*, Charlotte, NC, 2003, pp. 1174–1181.

64. R. Voorakaranam, S. Cherubal, and A. Chatterjee, A signature test framework for rapid production testing of RF circuits, in *Proceedings of Design, Automation and Test in Europe Conference*, Paris, France, 2002, pp. 186–191.

65. S. Cherubal, R. Voorakaranam, A. Chatterjee, J. Mclaughlin, J. L. Smith, and D. M. Majernik, Concurrent RF test using optimized modulated RF stimuli, in *IEEE International Conference on VLSI Design*, Mumbai, India, 2004, pp. 1017–1022.

66. A. Halder and A. Chatterjee, Low-cost alternate EVM test for wireless receiver systems, in *Proceedings of IEEE VLSI Test Symposium*, Palm Springs, CA, 2005, pp. 255–260.

67. S. Ellouz, P. Gamand, C. Kelma, B. Vandewiele, and B. Allard, Combining internal probing with artificial neural networks for optimal RFIC testing, in *Proceedings of IEEE International Test Conference*, Santa Clara, CA, 2006, pp. 4.3.1–4.3.9.

68. S. Goyal, A. Chatterjee, and M. Purtell, A low-cost test methodology for dynamic specification testing of high-speed data converters, *Journal of Electronic Testing: Theory and Applications*, 23(1), 95–106, 2006.

69. S. Kook, A. Banerjee, and A. Chatterjee, Dynamic specification testing and diagnosis of high-precision sigma-delta ADCs, *IEEE Design & Test of Computers*, 30(4), 36–48, 2013.

70. S.-W. Hsiao, X. Wang, and A. Chatterjee, Analog sensor based testing of phase-locked loop dynamic performance parameters, in *Proceedings of IEEE Asian Test Symposium*, Jiaosi Township, Taiwan, 2013, pp. 50–55.

71. V. Natarajan, S. Bhattacharya, and A. Chatterjee, Alternate electrical tests for extracting mechanical parameters of MEMS accelerometer sensors, in *Proceedings of IEEE VLSI Test Symposium*, Berkeley, CA, 2006, pp. 192–199.

72. H.-G. D. Stratigopoulos and Y. Makris, Non-linear decision boundaries for testing analog circuits, *IEEE Transactions on Computer-Aided Design of Integrated Circuits and Systems*, 24(11), 1760–1773, 2005.

73. H.-G. Stratigopoulos and Y. Makris, Error moderation in low-cost machine-learning-based Analog/RF testing, *IEEE Transactions on Computer-Aided Design of Integrated Circuits and Systems*, 27(2), 339–351, 2008.

74. H. Ayari, F. Azais, S. Bernard, M. Compte, M. Renovell, V. Kerzerho, O. Potin, and C. Kelma, Smart selection of indirect parameters for dc-based alternate RF IC testing, in *Proceedings of IEEE VLSI Test Symposium*, Maui, HI, 2012, pp. 19–24.

75. M. J. Barragan and G. Leger, Efficient selection of signatures for analog/RF alternate test, in *Proceedings of IEEE European Test Symposium*, Avignon, France, 2013.

76. D. E. Goldberg, *Genetic Algorithms in Search, Optimization, and Machine Learning*. Addison-Wesley, Boston, MA, 1989.

77. P. Pudil, J. Novovicova, and J. Kittler, Floating search methods in feature selection, *Pattern Recognition Letters*, 15, 1119–1125, 1994.

78. S. S. Akbay, J. L. Torres, J. M. Rumer, A. Chatterjee, and J. Amtsfield, Alternate test of RF front ends with IP constraints: Frequency domain test generation and validation, in *Proceedings of IEEE International Test Conference*, Santa Clara, CA, 2006, pp. 4.4.1–4.4.10.

79. S. S. Akbay and A. Chatterjee, Built-in test of RF components using mapped feature extraction sensors, in *IEEE VLSI Test Symposium*, Palm Springs, CA, 2005, pp. 243–248.

80. S. Bhattacharya and A. Chatterjee, A DFT approach for testing embedded systems using DC sensors, *IEEE Design & Test of Computers*, 23(6), 464–475, 2006.

81. P. F. D. Mota and J. M. D. Silva, A true power detector for RF PA built-in calibration and testing, in *Proceedings of Design, Automation, & Test in Europe Conference*, Grenoble, France, 2011, pp. 1–6.

82. L. Abdallah, H.-G. Stratigopoulos, S. Mir, and C. Kelma, RF front-end test using built-in sensors, *IEEE Design & Test of Computers*, 28(6), 76–84, 2011.

83. C. Zhang, R. Gharpurey, and J. A. Abraham, Built-in self test of RF subsystems with integrated sensors, *Journal of Electronic Testing: Theory and Applications*, 28(5), 557–569, 2012.

84. A. Gopalan, M. Margala, and P. R. Mukund, A current based self-test methodology for RF front-end circuits, *Microelectronics Journal*, 36(12), 1091–1102, 2005.

85. H.-G. Stratigopoulos, S. Mir, E. Acar, and S. Ozev, Defect filter for alternate RF test, in *Proceedings of IEEE European Test Symposium*, Sevilla, Spain, 2009, pp. 101–106.

86. C. Y. Pan and K. T. Cheng, Test generation for linear time-invariant analog circuits, *IEEE Transactions on Circuits and Systems—II: Analog and Digital Signal Processing*, 46(5), 554–564, 1999.

87. W. M. Lindermeir, H. E. Graeb, and K. J. Antreich, Analog testing by characteristic observation inference, *IEEE Transactions on Computer-Aided Design of Integrated Circuits and Systems*, 18(9), 1353–1368, 1999.

88. P. N. Variyam and A. Chatterjee, Specification-driven test generation for analog circuits, *IEEE Transactions on Computer-Aided Design of Integrated Circuits and Systems*, 19(10), 1189–1201, 2000.

89. V. Stopjakova, P. Malosek, D. Micusik, M. Matej, and M. Margala, Classification of defective analog integrated circuits using artificial neural networks, *Journal of Electronic Testing: Theory and Applications*, 20, 25–37, 2004.

90. N. Cristianini and J. Shawe-Taylor, *Support Vector Machines and Other Kernel-Based Learning Methods*. Cambridge University Press, New York, 2000.

91. L. Rokach and O. Maimon, Top-down induction of decision trees classifiers—A survey, *IEEE Transactions on Systems, Man, and Cybernetics-Part C: Applications and Reviews*, 35(46), 476–487, 2005.

92. R. Parekh, J. Yang, and V. Honavar, Constructive neural-network learning algorithms for pattern classification, *IEEE Transactions on Neural Networks*, 11(2), 436–451, 2000.

93. C.M. Bishop, *Neural Networks for Pattern Recognition*. Oxford University Press Inc, New York, NY, 1995.

94. D. Maliuk, H.-G. Stratigopoulos, H. Huang, and Y. Makris, Analog neural network design for RF built-in self-test, in *Proceedings of IEEE International Test Conference*, Austin, TX, 2010, paper 23.2.

95. H.-G. Stratigopoulos and S. Mir, Adaptive alternate analog test, *IEEE Design & Test of Computers*, 29(4), 71–79, 2012.

96. H. Ayari, F. Azais, S. Bernard, M. Compte, V. Kerzerho, O. Potin, and M. Renovell, Making predictive analog/RF alternate test strategy independent of training set size, in *Proceedings of IEEE International Test Conference*, Anaheim, CA, 2012, paper 10.1.

97. T. M. Souders and G. N. Stenbakken, A comprehensive approach for modeling and testing analog and mixed-signal devices, in *Proceedings of IEEE International Test Conference*, Washington, DC, 1990, pp. 169–176.

98. G. N. Stenbakken and T. M. Souders, Developing linear error models for analog devices, *IEEE Transactions on Instrumentation and Measurement*, 43(2), 157–163, 1994.

99. G. N. Stenbakken and T. M. Souders, Test-point selection and testability measures via QR factorization of linear models, *IEEE Transactions on Instrumentation and Measurement*, IM-36(2), 406–410, 1987.

100. J. V. Spaandonk and T. A. M. Kevenaar, Iterative test-point selection for analog circuits, in *Proceedings of IEEE VLSI Test Symposium*, Princeton, NJ, 1996, pp. 66–71.

101. A. Wrixon and M. P. Kennedy, A rigorous exposition of the LEMMA method for analog and mixed-signal testing, *IEEE Transactions on Instrumentation and Measurement*, 48(5), 978–985, 1999.

102. L. Milor and A. L. Sangiovanni-Vincentelli, Minimizing production test time to detect faults in analog circuits, *IEEE Transactions on Computer-Aided Design of Integrated Circuits and Systems*, 13(6), 796–813, 1994.

103. G. Devarayanadurg, M. Soma, P. Goteti, and S. D. Huynh, Test set selection for structural faults in analog IC's, *IEEE Transactions on Computer-Aided Design of Integrated Circuits and Systems*, 18(7), 1026–1039, 1999.

104. S. D. Huss and R. S. Gyurcsik, Optimal ordering of analog integrated circuit tests to minimize test time, in *ACM/IEEE Design Automation Conference*, San Francisco, CA, 1991, pp. 494–499.

105. J. B. Brockman and S. W. Director, Predictive subset testing: Optimizing IC parametric performance testing for quality, cost, and yield, *IEEE Transactions on Semiconductor Manufacturing*, 2(3), 104–113, 1989.

106. S. Biswas, P. Li, R. D. Blanton, and L. Pileggi, Specification test compaction for analog circuits and MEMS, in *Proceedings of Design, Automation & Test in Europe Conference*, Munich, Germany, 2005, pp. 164–169.

107. S. Biswas and R. D. Blanton, Statistical test compaction using binary decision trees, *IEEE Design & Test of Computers*, 23(6), 452–462, 2006.

108. H.-G. Stratigopoulos, P. Drineas, M. Slamani, and Y. Makris, RF specification test compaction using learning machines, *IEEE Transactions on Very Large Scale Integration (VLSI) Systems*, 18(6), 998–1002, 2010.

109. A. A. Mutlu and M. Rahman, Statistical methods for the estimation of process variation effects on circuit operation, *IEEE Transactions on Electronics Packaging Manufacturing*, 28(4), 364–375, 2005.

110. T. McConaghy and G. G. E. Gielen, Template-free symbolic performance modeling of analog circuits via canonical-form functions and genetic programming, *IEEE Transactions on Computer-Aided Design of Integrated Circuits and Systems*, 28(8), 1162–1175, 2009.

111. E. Yilmaz and S. Ozev, Fast and accurate DPPM computation using model based filtering, in *Proceedings of IEEE European Test Symposium*, Trondheim, Norway, 2011, pp. 165–170.

112. A. Bounceur, S. Mir, E. Simeu, and L. Rolindez, Estimation of test metrics for the optimisation of analogue circuit testing, *Journal of Electronic Testing: Theory and Applications*, 23(6), 471–484, 2007.

113. H.-G. Stratigopoulos, S. Mir, and A. Bounceur, Evaluation of analog/RF test measurements at the design stage, *IEEE Transactions on Computer-Aided Design of Integrated Circuits and Systems*, 28(4), 582–590, 2009.

114. S. Mukhopadhyay, A generic data-driven nonparametric framework for variability analysis of integrated circuits in nanometer technologies, *IEEE Transactions on Computer-Aided Design of Integrated Circuits and Systems*, 28(7), 1038–1046, 2009.

115. A. Bounceur, S. Mir, and H.-G. Stratigopoulos, Estimation of analog parametric test metrics using copulas, *IEEE Transactions on Computer-Aided Design of Integrated Circuits and Systems*, 30(9), 1400–1410, 2011.

116. C. Wegener and M. P. Kennedy, Test development through defect and test escape level estimation for data converters, *Journal of Electronic Testing: Theory and Applications*, 22(4–6), 313–324, 2006.

117. H.-G. Stratigopoulos, Test metrics model for analog test development, *IEEE Transactions on Computer-Aided Design of Integrated Circuits and Systems*, 31(7), 1116–1128, 2012.

118. H.-G. Stratigopoulos, P. Faubet, Y. Courant, and M. Firas, Multidimensional analog test metrics estimation using extreme value theory and statistical blockade, in *ACM/IEEE Design Automation Conference*, Austin, TX, 2013.

119. A. Singhee and R. A. Rutenbar, Statistical blockade: Very fast statistical simulation and modeling of rare circuit events and its application to memory design, *IEEE Transactions on Computer-Aided Design of Integrated Circuits and Systems*, 28(8), 1176–1189, 2009.

120. S. Coles, *An Introduction to Statistical Modeling of Extreme Values*, ser. Springer Series in Statistics. Springer, London, U.K., 2001.

121. C. M. Kurker, J. J. Paulos, R. S. Gyurcsik, and J.-C. Lu, Hierarchical yield estimation of large analog integrated circuits, *IEEE Journal of Solid-State Circuits*, 28(3), 203–209, 1993.

122. M. Dubois, H.-G. Stratigopoulos, and S. Mir, Hierarchical parametric test metrics estimation: A $\Sigma\Delta$ converter BIST case study, in *Proceedings of IEEE International Conference on Computer Design*, Lake Tahoe, CA, 2009, pp. 78–83.

123. L. Milor and V. Visvanathan, Detection of catastrophic faults in analog integrated circuits, *IEEE Transactions on Computer-Aided Design*, 8(2), 114–130, 1989.

124. S. J. Tsai, Test vector generation for linear analog devices, in *Proceedings of IEEE International Test Conference*, Nashville, TN, 1991, pp. 592–597.

125. M. Slamani and B. Kaminska, Multifrequency analysis of faults in analog circuits, *IEEE Design & Test of Computers*, 12(2), 70–80, 1995.

126. S. Mir, M. Lubaszewski, and B. Courtois, Fault-based ATPG for linear analog circuits with minimal size multifrequency test sets, *Journal of Electronic Testing: Theory and Applications*, 9(1–2), 43–57, 1996.

127. Z. Wang, G. Gielen, and W. Sansen, Probabilistic fault detection and the selection of measurements for analog integrated circuits, *IEEE Transactions on Computer-Aided Design of Integrated Circuits and Systems*, 17(9), 862–872, 1998.

128. S. D. Huynh, S. Kim, M. Soma, and J. Zhang, Automatic analog test signal generation using multifrequency analysis, *IEEE Transactions on Circuits and Systems—II: Analog and Digital Signal Processing*, 46(5), 565–576, 1999.

129. K. Saab, N. B. Hamida, and B. Kaminska, Closing the gap between analog and digital testing, *IEEE Transactions on Computer-Aided Design of Integrated Circuits and Systems*, 20(2), 307–314, 2001.

130. S. Bhunia and K. Roy, Dynamic supply current testing of analog circuits using wavelet transform, in *Proceedings of IEEE VLSI Test Symposium*, Monterey, CA, 2002, pp. 302–307.

131. Y. Joannon, V. Beroulle, C. Robach, S. Tedjini, and J.-L. Carbonero, Decreasing test qualification time in AMS and RF systems, *IEEE Design & Test of Computers*, 25(1), 29–37, 2008.

132. A. Meixner and W. Maly, Fault modeling for the testing of mixed integrated circuits, in *Proceedings of IEEE International Test Conference*, Nashville, TN, 1991, pp. 564–572.

133. M. Soma, An experimental approach to analog fault models, in *Proceedings of IEEE Custom Integrated Circuits Conference*, San Diego, CA, 1991, paper 13.6.

134. M. Sachdev and B. Atzema, Industrial relevance of analog IFA: A fact or a fiction, in *Proceedings of IEEE International Test Conference*, Washington, DC, 1995, pp. 61–70.

135. C. Sebeke, J. Teixeira, and M. Ohletz, Automatic fault extraction and simulation of layout realistic faults for integrated analogue circuits, in *Proceedings of IEEE European Design & Test Conference*, Paris, France, 1995, pp. 464–468.

136. A. Milne, D. Taylor, J. Saunders, and A. Talbot, Generation of optimised fault lists for simulation of analogue circuits and test programs, *IEE Proceedings*, 146(6), 355–360, 1999.

137. E. Acar and S. Ozev, Defect-oriented testing of RF circuits, *IEEE Transactions on Computer-Aided Design of Integrated Circuits and Systems*, 27(5), 920–931, 2008.

138. E. Yilmaz, G. Shofner, L. Winemberg, and S. Ozev, Fault analysis and simulation of large scale industrial mixed-signal circuits, in *Proceedings of Design, Automation & Test in Europe Conference*, Grenoble, France, 2013, pp. 565–570.

139. M. Soma, Challenges in analog and mixed-signal fault models, *IEEE Circuits & Devices Magazine*, 12(1), 16–19, 1996.

140. N. Nagi and J. A. Abraham, Hierarchical fault modeling for analog and mixed-signal circuits, in *Proceedings of IEEE VLSI Test Symposium*, Atlantic City, NJ, 1992, pp. 96–101.

141. R. Voorakaranam, S. Chakrabarti, J. Hou, A. Gomes, S. Cherubal, and A. Chatterjee, Hierarchical specification-driven analog fault modeling for efficient fault simulation and diagnosis, in *Proceedings of IEEE International Test Conference*, Washington, DC, 1997, paper 36.1.

142. S. Sunter and N. Nagi, Test metrics for analog parametric faults, in *Proceedings of IEEE VLSI Test Symposium*, Dana Point, CA, 1999, pp. 226–234.

143. K. Saab, N. Ben-Hamida, and B. Kaminska, Parametric fault simulation and test vector generation, in *Proceedings of Design, Automation & Test in Europe Conference*, Paris, France, 2000, pp. 650–656.

144. J. Savir and Z. Guo, Test limitations of parametric faults in analog test, *IEEE Transactions on Instrumentation and Measurement*, 52(5), 1444–1454, 2003.

145. H.-C. Hong, A static linear behavior analog fault model for switched-capacitor circuits, *IEEE Transactions on Computer-Aided Design of Integrated Circuits and Systems*, 31(4), 597–609, 2012.

146. M. B. Yelten, S. Natarajan, B. Xue, and P. Goteti, Scalable and efficient analog parametric fault identification, in *Proceedings of IEEE/ACM International Conference on Computer-Aided Design*, San Jose, CA, 2013, pp. 387–392.

147. H.-G. Stratigopoulos and S. Sunter, Efficient Monte Carlo-based analog parametric fault modelling, in *Proceedings of IEEE VLSI Test Symposium*, Napa, CA, 2014.

148. H.-G. Stratigopoulos and S. Mir, Analog test metrics estimates with PPM accuracy, in *Proceedings of IEEE/ACM International Conference on Computer-Aided Design*, San Jose, CA, 2010, pp. 241–247.

149. K. Arabi and B. Kaminska, Testing analog and mixed-signal integrated circuits using oscillation-test method, *IEEE Transactions on Computer-Aided Design of Integrated Circuits and Systems*, 16(7), 745–753, 1997.

150. K. Arabi and B. Kaminska, Design for testability of embedded integrated operational amplifiers, *IEEE Journal of Solid-State Circuits*, 33(4), 573–581, 1998.

151. K. Arabi and B. Kaminska, Oscillation-test methodology for low-cost testing of active analog filters, *IEEE Transactions on Instrumentation and Measurement*, 48(4), 798–806, 1999.

152. M. S. Zarnik, F. Novak, and S. Macek, Design of oscillation-based test structures for active RC filters, *IEE Proceedings-Circuits, Devices, and Systems*, 147(5), 297–302, 2000.

153. G. Huertas, D. Vázquez, E. J. Peralías, A. Rueda, and J. L. Huertas, Practical oscillation-based test of integrated filters, *IEEE Design & Test of Computers*, 19(6), 64–72, 2002.

154. J. Roy and J. A. Abraham, A comprehensive signature analysis scheme for oscillation-test, *IEEE Transactions on Computer-Aided Design of Integrated Circuits and Systems*, 22(10), 1409–1423, 2003.

155. K. Suenaga, E. Isern, R. Picos, S. Bota, M. Roca, and E. García-Moreno, Application of predictive oscillation-based test to a CMOS opAmp, *IEEE Transactions on Instrumentation and Measurement*, 59(8), 2076–2082, 2010.

156. S. Callegari, F. Pareschi, G. Setti, and M. Soma, Complex oscillation-based test and its application to analog filters, *IEEE Transactions on Circuits and systems I: Regular Papers*, 57(5), 956–969, 2010.

157. K. Arabi and B. Kaminska, Efficient and accurate testing of analog-to-digital converters using oscillation-test method, in *Proceedings of European Design and Test Conference*, Paris, France, 1997, pp. 348–352.

158. K. Arabi and B. Kaminska, Oscillation built-in self test (OBIST) scheme for functional and structural testing of analog and mixed-signal integrated circuits, in *Proceedings of IEEE International Test Conference*, Washington, DC, 1997, pp. 786–795.

159. G. Huertas, D. Vazquez, E. J. Peralias, A. Rueda, and J. L. Huertas, Oscillation-based test in oversampling ΣΔ modulators, *Microelectronics Journal*, 33, 799–806, 2002.

160. A. Goyal, M. Swaminathan, and A. Chatterjee, Low-cost specification based testing of RF amplifier circuits using oscillation principles, *Journal of Electronic Testing: Theory and Applications*, 26(1), 13–24, 2010.

161. V. Beroulle, Y. Bertrand, L. Latorre, and P. Nouet, Evaluation of the oscillation-based test methodology for micro-electro-mechanical systems, in *Proceedings of IEEE VLSI Test Symposium*, Monterey, CA, 2002, pp. 439–444.

162. K. Arabi and B. Kaminska, Parametric and catastrophic fault coverage of analog circuits in oscillation-test methodology, in *Proceedings of IEEE VLSI Test Symposium*, Monterey, CA, 1997, pp. 166–171.

163. A. K. Lu, G. W. Roberts, and D. A. Johns, A high quality analog oscillator using oversampling D/A converters technique, *IEEE Transactions on Circuits and Systems II: Analog and Digital Signal Processing*, 41(7), 437–444, 1994.

164. S. W. Park, J. L. Ausín, F. Bahmani, and E. Sánchez-Sinencio, Nonlinear shaping SC oscillator with enhanced linearity, *IEEE Journal of Solid-State Circuits*, 42(11), 2421–2431, 2007.

165. M. M. Elsayed and E. Sánchez-Sinencio, A low THD, low power, high output-swing time-mode-based tunable oscillator via digital harmonic-cancellation technique, *IEEE Journal of Solid-State Circuits*, 45(5), 1061–1071, 2010.

166. M. J. Barragán, D. Vázquez, and A. Rueda, Analog sinewave signal generators for mixed-signal built-in test applications, *Journal of Electronic Testing: Theory and Applications*, 27(3), 305–320, 2011.

167. M. J. Barragán, G. Leger, D. Vázquez, and A. Rueda, Sinusoidal signal generation for mixed-signal BIST using a harmonic-cancellation technique, in *Proceedings of IEEE Latin American Symposium on Circuits and Systems*, Cusco, Peru, 2013, pp. 1–4.

168. B. K. Vasan, S. K. Sudani, D. J. Chen, and R. L. Geiger, Low-distortion sine wave generation using a novel harmonic cancellation technique, *IEEE Transactions on Circuits and Systems I: Regular Papers*, 60(5), 1122–1134, 2013.

169. A. Lu, G. Roberts, and D. Johns, A high-quality analog oscillator using oversampling D A conversion techniques, in *Proceedings of IEEE International Symposium on Circuits and Systems*, Chicago, IL, 1993, pp. 1298–1301.

170. H.-C. Hong, S.-C. Liang, and H.-C. Song, A cost effective BIST second-order ΣΔ modulator, in *Proceedings of IEEE Workshop on Design and Diagnostics of Electronic Circuits and Systems*, Bratislava, Slovakia, 2008, pp. 1–6.

171. S. Mir, L. Rolindez, C. Domigues, and L. Rufer, An implementation of memory-based on-chip analogue test signal generation, in *Proceedings of IEEE Asia and South Pacific Design Automation Conference*, Kitakyushu, Japan, 2003, pp. 663–668.

172. H. Mattes, S. Sattler, and C. Dworski, Controlled sine wave fitting for ADC test, in *Proceedings of IEEE International Test Conference*, Charlotte, NC, 2004, pp. 963–971.

173. D. A. Lampasi, A. Moschitta, and P. Carbone, Accurate digital synthesis of sinewaves, *IEEE Transactions on Instrumentation and Measurement*, 57(3), 522–529, 2008.

174. C.-K. Ong, K.-T. Cheng, and L.-C. Wang, A new sigma-delta modulator architecture for testing using digital stimulus, *IEEE Transactions on Circuits and Systems I: Regular Papers*, 51(1), 206–213, 2004.

175. L. Rolindez, S. Mir, J.-L. Carbonero, D. Goguet, and N. Chouba, A stereo ΣΔ ADC architecture with embedded SNDR self-test, in *Proceedings of IEEE International Test Conference*, Santa Clara, CA, 2007, paper 32.1.

176. H.-C. Hong and S.-C. Liang, A decorrelating design-for-digital-testability scheme for ΣΔ modulators, *IEEE Transactions on Circuits and Systems I: Regular Papers*, 56(1), 60–73, 2009.

177. M. Dubois, H.-G. Stratigopoulos, and S. Mir, Ternary stimulus for fully digital dynamic testing of SC ΣΔ ADCs, in *IEEE International Mixed-Signals, Sensors, and Systems Test Workshop*, Taipei, Taiwan, 2012.

178. A. Roy, S. Sunter, A. Fudoli, and D. Appello, High accuracy stimulus generation for A/D converter BIST, in *Proceedings of IEEE International Test Conference*, Baltimore, MD, 2002, pp. 1031–1039.

179. C. K. Ong, J. L. Huang, and K. T. Cheng, Testing second-order delta-sigma modulators using pseudo-random patterns, *Microelectronics Journal*, 33(10), 807–814, 2002.

180. H. Chauhan, Y. Choi, M. Onabajo, I.-S. Jung, and Y.-B. Kim, Accurate and efficient on-chip spectral analysis for built-in testing and calibration approaches, *IEEE Transactions on Very Large Scale Integration (VLSI) Systems*, 22(3), 49–506, 2014.

181. IEEE Standard for Digitizing Waveform Recorders, IEEE Std. 1057-2007 (revision of IEEE 1057-2004).

182. H.-C. Hong, F.-Y. Su, and S.-F. Hung, A fully integrated built-in self-test ΣΔ ADC based on the modified controlled sine-wave fitting procedure, *IEEE Transactions on Instrumentation and Measurement*, 59(9), 2334–2344, 1996.

183. A. Tchegho, H. Mattes, and S. Sattler, Optimal high-resolution spectral analyzer, in *Proceedings of Design, Automation and Test in Europe Conference*, Munich, Germany, 2008, pp. 62–67.

184. N. Chouba and L. Bouzaida, A BIST architecture for sigma delta ADC testing based on embedded NOEB self-test and CORDIC algorithm, in *Proceedings of International Conference on Design and Technology of Interated Systems in Nanoscale Era*, Hammamet, Tunisia, 2010, pp. 1–7.

185. F. Azais, S. Bernard, Y. Bertrand, X. Michel, and M. Renovell, A low-cost adaptive ramp generator for analog BIST applications, in *Proceedings of IEEE VLSI Test Symposium*, Marina Del Ray, CA, 2001, pp. 266–271.

186. B. Provost and E. Sánchez-Sinencio, On-chip ramp generators for mixed-signal BIST and ADC self-test, *IEEE Journal of Solid-State Circuits*, 38(2), 263–273, 2003.

187. H.-K. Chen, C.-H. Wang, and C.-C. Su, A self calibrated ADC BIST methodology, in *Proceedings of IEEE VLSI Test Symposium*, Monterey, CA, 2002, pp. 117–122.

188. R. Holcer, L. Michaeli, and J. Saliga, DNL ADC testing by the exponential shaped voltage, *IEEE Transactions on Instrumentation and Measurement*, 52(3), 946–949, 2003.

189. L. Jin, K. Parthasarathy, T. Kuyel, D. Chen, and R. L. Geiger, Accurate testing of analog-to-digital converters using low linearity signals with stimulus error identification and removal, *IEEE Transactions on Instrumentation and Measurement*, 54(3), 1188–1199, 2005.

190. M. A. Jalon, A. Rueda, and E. Peralias, Enhanced double-histogram test, *Electronics Letters*, 45(7), 349–351, 2009.

191. F. Azais, S. Bernard, Y. Bertrand, and M. Renovell, Optimizing sinusoidal histogram test for low cost ADC BIST, *Journal of Electronic Testing: Theory and Applications*, 17(3–4), 255–266, 2001.

192. Y. Wang, J. Wang, F. Lai, and Y. Ye, Optimal schemes for ADC BIST based on histogram, in *Proceedings of IEEE Asian Test Symposium*, Calcutta, India, 2005, pp. 52–57.

193. S. Goyal and A. Chatterjee, Linearity testing of A/D converters using selective code measurement, *Journal of Electronic Testing: Theory and Applications*, 24(6), 567–576, 2008.

194. H. Xing, H. Jiang, D. Chen, and R. L. Geiger, High-resolution ADC linearity testing using a fully digital-compatible BIST strategy, *IEEE Transactions on Instrumentation and Measurement*, 58(8), 2697–2705, 2009.

195. J. Lin, S. Chang, T. Kung, H. Ting, and C. Huang, Transition-code based linearity test method for pipelined ADCs with digital error correction, *IEEE Transactions on Very Large Scale Integration (VLSI) Systems*, 19(12), 2158–2169, 2010.

196. A. Laraba, H.-G. Stratigopoulos, S. Mir, H. Naudet, and G. Bret, Reduced-code linearity testing of pipeline ADCs, *IEEE Design & Test*, 30(6), 80–88, 2013.

197. G. Leger and A. Rueda, Low-cost digital detection of parametric faults in cascaded sigma-delta modulators, *IEEE Transactions on Circuits and Systems I: Regular Papers*, 56(7), 1326–1338, 2009.

198. A. Gines and G. Leger, Sigma-delta testability for pipeline A/D converters, in *Proceedings of Design, Automation and Test in Europe Conference*, Dresden, Germany, 2014.

199. A. Zjajo, M. B. Asian, and J. P. de Gyvez, BIST method for die-level process parameter variation monitoring in analog/mixed-signal integrated circuits, in *Proceedings of Design, Automation & Test in Europe Conference*, Nice, France, 2007, pp. 1301–1306.

200. M. A. Domínguez, J. L. Ausísin, J. F. Duque-Carrilo, and G. Torelli, A 1-MHz area-efficient on-chip spectrum analyzer for analog testing, *Journal of Electronic Testing: Theory and Applications*, 22(4), 437–448, 2006.

201. J. Qin, C. E. Stroud, and F. F. Dai, FPGA-based analog functional measurements for adaptive control in mixed-signal systems, *IEEE Transactions on Industrial Electronics*, 54(4), 1885–1897, 2010.

202. M. J. Barragán, D. Vázquez, and A. Rueda, A BIST solution for frequency domain characterization of analog circuits, *Journal of Electronic Testing: Theory and Applications*, 26(4), 429–441, 2010.

203. K. Arabi, B. Kaminska, and J. Rzeszut, BIST for d/a and a/d converters, *IEEE Design & Test of Computers*, 13(4), 40–49, 1996.

204. V. Kerzérho, P. Cauvet, S. Bernard, F. Azais, M. Comte, and M. Renovell, "Analogue network of converters": A DFT technique to test a complete set of ADCs and DACs embedded in a complex sip or soc, in *Proceedings of IEEE European Test Symposium*, Freiburg, Germany, 2007, pp. 211–216.

205. S. Sunter and A. Roy, BIST for phase-locked loops in digital applications, in *Proceedings of IEEE International Test Conference*, Atlantic City, NJ, 1999, pp. 532–540.

206. P. Dudek, S. Szczepanski, and J. Hatfield, A high resolution CMOS time-to-digital converter utilizing a Vernier delay line, *IEEE Journal of Solid-State Circuits*, 35(2), 240–247, 2000.

207. A. Chan and G. Roberts, A synthesizable, fast and high-resolution timing measurement device using a component-invariant Vernier delay line, in *Proceedings of IEEE International Test Conference*, Baltimore, MD, 2001, pp. 858–867.

208. S. Tabatabaei and A. Ivanov, Embedded timing analysis: A SoC infrastructure, *IEEE Design & Test of Computers*, 19(3), 22–34, 2002.

209. S. Sunter and A. Roy, On-chip digital jitter measurement, from megahertz to gigahertz, *IEEE Design & Test of Computers*, 21(4), 314–321, 2004.

210. J.-L. Huang, J.-J. Huang, and Y.-S. Liu, A low-cost jitter measurement technique for BIST applications, *Journal of Electronic Testing: Theory and Applications*, 22(3), 219–228, 2006.

211. B. R. Veillette and G. Roberts, On-chip measurement of the jitter transfer function of charge-pump phase-locked loops, *IEEE Journal of Solid-State Circuits*, 33(3), 483–491, 1998.

212. J. Kim, On-chip measurement of jitter transfer and supply sensitivity of PLL/DLLs, *IEEE Transactions on Circuits and Systems II: Express Briefs*, 56(6), 449–453, 2009.

213. A. Asquini, F. Badets, S. Mir, J.-L. Carbonero, and L. Bouzaida, PFD output monitoring for RF PLL BIST, in *Proceedings of IEEE International Mixed-Signals, Sensors, and Systems Test Workshop*, Vancouver, Canada, 2008.

214. S.-W. Hsiao, N. Tzou, and A. Chatterjee, A programmable BIST design for PLL static phase offset estimation and clock duty cycle detection, in *Proceedings of IEEE VLSI Test Symposium*, Berkeley, CA, 2013.

215. S. Kim and M. Soma, An all-digital built-in self-test for high-speed phase-locked loops, *IEEE Transactions on Circuits and Systems—II: Analog and Digital Signal Processing*, 48(2), 141–150, 2001.

216. F. Azais, Y. Bertrand, M. Renovell, A. Ivanov, and S. Tabatabaei, On-chip digital jitter measurement, from megahertz to gigahertz, *IEEE Design & Test of Computers*, 20(1), 60–67, 2003.

217. E. S. Erdogan and S. Ozev, Detailed characterization of transceiver parameters through loop-back-based BiST, *IEEE Transactions on Very Large Scale Integration (VLSI) Systems*, 18(6), 901–911, 2010.

218. J. W. Jeong, S. Ozev, S. Sen, and T. M. Mak, Measurement of envelope/phase path delay skew and envelope path bandwidth in polar transmitters, in *Proceedings of IEEE VLSI Test Symposium*, Berkeley, CA, 2013.

219. A. Nassery, J. W. Jeong, and S. Ozev, Zero-overhead self test and calibration of RF transceivers, in *Proceedings of IEEE International Test Conference*, Anaheim, CA, 2013.

220. J. W. Jeong, S. Ozev, S. Sen, V. Natarajan, and M. Slamani, Built-in self-test and characterization of polar transmitter parameters in the loop-back mode, in *Proceedings of Design, Automation, & Test in Europe Conference*, Dresden, Germany, 2014.

221. A. Haider, S. Bhattacharya, G. Srinivasan, and A. Chatterjee, A system-level alternate test approach for specification test of RF transceivers in loopback mode, in *Proceedings of IEEE International Conference on VLSI Design*, Kolkata, India, 2005, pp. 289–294.

222. G. Srinivasan, A. Chatterjee, and F. Taenzler, Alternate loop-back diagnostic tests for wafer-level diagnosis of modern wireless transceivers using spectral signatures, in *Proceedings of IEEE VLSI Test Symposium*, Berkeley, CA, 2006, pp. 222–227.

223. J. Altet, A. Rubio, E. Schaub, S. Dilhaire, and W. Claeys, Thermal coupling in integrated circuits: Application to thermal testing, *IEEE Journal of Solid-State Circuits*, 36(1), 81–91, 2001.

224. J. Altet, A. Rubio, J. L. Rosselló, and J. Segura, Structural RFIC device testing through built-in thermal monitoring, *IEEE Communications Magazine*, 41(9), 98–104, 2003.
225. M. Onabajo, J. Altet, E. Aldrete-Vidrio, D. Mateo, and J. Silva-Martinez, Electrothermal design procedure to observe RF circuit power and linearity characteristics with a homodyne differential temperature sensor, *IEEE Transactions on Circuits and Systems I: Regular Papers*, 99, 1, 2011.
226. J. Altet, D. Mateo, D. Gómez, X. Perpinyà, M. Vallvehi, and X. Jordà, DC temperature measurements for power gain monitoring in RF power amplifiers, in *Proceedings of IEEE International Test Conference*, Anaheim, CA, 2012, paper 17.3.
227. L. Abdallah, H.-G. Stratigopoulos, S. Mir, and J. Altet, Defect-oriented non-intrusive RF test using on-chip temperature sensors, in *Proceedings of IEEE VLSI Test Symposium*, Berkeley, CA, 2013.

Index

T - #0929 - 101024 - C664 - 279/216/29 - PB - 9781138586000 - Gloss Lamination